ECOLOGY OF LIANAS

ABOUT THE COMPANION WEBSITE

This book is accompanied by a companion website:

www.wiley.com/go/schnitzer/lianas

The website includes:

- PowerPoints of all figures from the book for downloading;
- PDFs of tables from the book.

ECOLOGY OF LIANAS

Edited by

Stefan A. Schnitzer
University of Wisconsin–Milwaukee, WI, USA

Frans Bongers
Wageningen University and Research Centre, The Netherlands

Robyn J. Burnham
University of Michigan, MI, USA

Francis E. Putz
University of Florida, FL, USA

WILEY Blackwell

Registered office: John Wiley & Sons, Ltd, The Atrium, Southern Gate, Chichester, West Sussex, PO19 8SQ, UK

Editorial offices: 9600 Garsington Road, Oxford, OX4 2DQ, UK
The Atrium, Southern Gate, Chichester, West Sussex, PO19 8SQ, UK
111 River Street, Hoboken, NJ 07030-5774, USA

For details of our global editorial offices, for customer services and for information about how to apply for permission to reuse the copyright material in this book please see our website at www.wiley.com/wiley-blackwell.

Library of Congress Cataloging-in-Publication Data has been applied for.

A catalogue record for this book is available from the British Library.

Wiley also publishes its books in a variety of electronic formats. Some content that appears in print may not be available in electronic books.

Cover image: A large Bauhina reflexa (Fabaceae) liana branching into numerous canopy trees in a Panamanian secondary forest © S.A. Schnitzer

Set in 9/11 Photina by Laserwords Pvt Ltd, Chennai, India
Printed and bound in Singapore by Markono Print Media Pte Ltd

1 2015

CONTENTS

CONTRIBUTORS

LUCIANA F. ALVES, *Instituto Ciências agrárias (IAC), Departamento de Botânica, Campinas, Brasil.*

ANA ANDRADE, *Biological Dynamics of Forest Fragments Project – BDFFP, Instituto Nacional de Pesquisas da Amazônia – INPA, Manaus-AM 69080–971, Brazil.*

VERONICA ANGYALOSSY, *Universidade de São Paulo, Instituto de Biociências, Departamento de Botânica. Rua do Matão 277, Cidade Universitária, 05508–090, São Paulo (SP), Brazil.* [vangyalossy@usp.br]

VÍCTOR ARROYO-RODRÍGUEZ, *Centro de Investigaciones en Ecosistemas, Universidad Nacional Autónoma de México, Morelia, Michoacán, Mexico.* [victorarroyo_rodriguez@hotmail.com]

NORBERTO ASENSIO, *Faculty of Environment and Resource Studies, Mahidol University, Nakorn Pathom, Thailand.*

GREGORY P. ASNER, *Department of Global Ecology, Carnegie Institution for Science, 260 Panama Street, Stanford, CA 94305, USA.* [gpa@stanford.edu]

N. AYYAPPAN, *French Institute of Pondicherry, Puducherry 605 001, India.*

FRANS BONGERS, *Wageningen University and Research Centre, Centre of Ecosystem Studies, P.O. Box. 47, 6700 AA Wageningen, The Netherlands.* [Frans.Bongers@wur.nl]

ROBYN J. BURNHAM, *Department of Ecology & Evolutionary Biology, University of Michigan, Ann Arbor, MI 48109–1048, USA.* [rburnham@umich.edu]

ZHIQUAN CAI, *Xishuangbanna Tropical Botanical Garden, Chinese Academy of Sciences, Mengla 666303, China.*

JOSÉ LUÍS C. CAMARGO, *Biological Dynamics of Forest Fragments Project – BDFFP, Instituto Nacional de Pesquisas da Amazônia – INPA, Manaus-AM 69011–970, Brazil.*

MASON CAMPBELL, *Centre for Tropical Environmental and Sustainability Science (TESS) and School of Marine and Tropical Biology, James Cook University, Cairns, Queensland 4878, Australia.* [mason.campbell@my.jcu.edu.au]

PABLO CARRILLO-REYES, *Centro de Investigaciones en Ecosistemas, Universidad Nacional Autónoma de México. Antigua Carretera a Pátzcuaro 8701. Col. Ex Hacienda de San José de La Huerta, C. P. 58190, Morelia, Michoacán, México.*

WALTER P. CARSON, *Department of Biological Sciences, University of Pittsburgh, Pittsburgh, PA, USA.*

CAROLINA V. CASTILHO, *Empresa Brasileira de Pesquisa Agropecuária – EMBRAPA, Centro de Pesquisa Agroflorestal de Roraima, Boa Vista-RR 69301–970, Brazil.*

JÉRÔME CHAVE, *Laboratoire Evolution et Diversité Biologique, CNRS and Université Paul Sabatier, Toulouse, France.*

GEORGE CHUYONG, *Department of Botany and Plant Physiology, University of Buea, Buea, Cameroon.*

GUADALUPE CORNEJO-TENORIO, *Centro de Investigaciones en Ecosistemas, Universidad Nacional Autónoma de México. Antigua Carretera a Pátzcuaro 8701. Col. Ex Hacienda de San José de La Huerta, C. P. 58190, Morelia, Michoacán, México.*

FLAVIA R. C. COSTA, *Coordenação de Pesquisas em Ecologia, Instituto Nacional de Pesquisas da Amazônia – INPA, Manaus-AM 69011–970, OG Brazil.* [flaviacosta001@gmail.com]

JURGI CRISTÓBAL-AZKARATE, *Primate Immunogenetics and Molecular Ecology (PrIME) Research Group, Department of Biological Anthropology, University of Cambridge, Cambridge, UK.*

MARK E. DE GUZMAN, *Department of Botany and Plant Sciences, University of California, 2150 Batchelor Hall, Riverside, CA 92521, USA.*

SAARA J. DEWALT, *Department of Biological Sciences, Clemson University, Clemson, SC 29634, USA.* [saarad@CLEMSON.EDU]

JACOB C. DUNN, *Primate Immunogenetics and Molecular Ecology (PrIME) Research Group, Department of Biological Anthropology, University of Cambridge, Cambridge, UK; Centro de Investigaciones Tropicales, Universidad Veracruzana, Xalapa, Veracruz, México.*

CORNEILLE E.N. EWANGO, *Wildlife Conservation Society (WCS)-DR Congo and Centre de Formation et de Recherche en Conservation Forestiére (CEFRECOF); Reserve de Faune à Okapis, Ituri-Epulu-Aru Landscape, DR Congo.*

TAYLOR S. FEILD, *School of Marine and Tropical Biology, James Cook University, Townsville, Queensland, Australia.*

RACHAEL V. GALLAGHER, *Department of Biological Sciences, Macquarie University, North Ryde, Sydney 2109, Australia.* [rachael.gallagher@mq.edu.au]

JEFFREY J. GERWING, *University Studies, Portland State University, Portland, Oregon 97207, USA.*

ERNESTO GIANOLI, *Departamento de Biología, Universidad de La Serena, La Serena, Chile; Departamento de Botánica, Universidad de Concepción, Concepción, Chile.* [egianoli@gmail.com]

ARTURO GONZALEZ-ZAMORA, *Instituto de Ecología A.C., Xalapa, Veracruz, Mexico; Instituto de Investigaciones Biológicas, Universidad Veracruzana, Xalapa, Veracruz, Mexico.*

ESTEBAN GORTAIRE AMEZCUA, *Botanist and Consultant, Quito, Ecuador.* [egortaire@yahoo.com]

MICHAELA K. HALSEY, *Department of Biology, Frostburg State University, Frostburg MD 21532, USA.*

TERESE HART, *Wildlife Conservation Society, 16 Av. Basoko, Kinshasa Gombe, DR Congo.*

STEPHEN P. HUBBELL, *Smithsonian Tropical Research Institute, Apartado 2072, Balboa, Republic of Panama; Department of Ecology and Evolutionary Biology, University of California Los Angeles, USA.*

GUILLERMO IBARRA-MANRÍQUEZ, *Centro de Investigaciones en Ecosistemas, Universidad Nacional Autónoma de México. Antigua Carretera a Pátzcuaro 8701. Col. Ex Hacienda de San José de La Huerta, C. P. 58190, Morelia, Michoacán, México.*

KALAN ICKES, *Department of Biological Sciences, Clemson University, Clemson, SC 29634, USA.*

SANDRINE ISNARD, *IRD, UMR AMAP, Laboratoire de Botanique et d'Écologie Végétale Appliquées, BPA5, 98800, Nouméa, Nouvelle-Calédonie.* [sandrine.isnard@ird.fr]

MARIAN KAZDA, *Institute of Systematic Botany and Ecology, Ulm University, Germany.* [marian.kazda@uni-ulm.de]

DAVID KENFACK, *Center for Tropical Forest Science-Smithsonian Institution Global Earth Observatory, Smithsonian National Museum of Natural History, Washington, DC. USA.*

KRISTINA KROENING, *University of Wisconsin-Milwaukee, PO Box 413, Milwaukee, WI 53201 USA.*

LAURA M. LADWIG, *Department of Biology, University of New Mexico, Albuquerque, New Mexico, USA.* [lladwig@unm.edu]

THOMAS D. LAMBERT, *Department of Biology, Frostburg State University, Frostburg MD 21532*, USA. [tdlambert@frostburg.edu]

WILLIAM F. LAURANCE, *Centre for Tropical Environmental and Sustainability Science & School of Marine and Tropical Biology, James Cook University, Cairns, Queensland 4878, Austrália; Biological Dynamics of Forest Fragments Project – BDFFP, Instituto Nacional de Pesquisas da Amazônia – INPA, Manaus-AM, 69011–970, Brazil.*

STACEY A. LEICHT-YOUNG, *The Arnold Arboretum of Harvard University, 1300 Centre St. Boston, MA, USA.* [stacey.leicht@gmail.com]

SUSAN G. LETCHER, *Department of Environmental Studies, Purchase College (SUNY), 735 Anderson Hill Road, Purchase, NY 10577*, USA. [susan.letcher@purchase.edu]

ANDRÉ C. LIMA, *Universidade de São Paulo, Instituto de Biociências, Departamento de Botânica. Rua do Matão 277, Cidade Universitária, 05508–090, São Paulo (SP), Brazil.*

MANUEL J. MACÍA, *Universidad Autónoma de Madrid, Madrid, Spain.*

AINHOA MAGRACH, *Centre for Tropical Environmental and Sustainability Science (TESS) and School of Marine and Tropical Biology, James Cook University, Cairns, Queensland 4878, Australia.*

JEAN-REMY MAKANA, *Wildlife Conservation Society, 16 Av. Basoko, Kinshasa Gombe, DR Congo.*

AGUSTINA MALIZIA, *National University of Tucuman, Instituto de Ecología Regional (IER), Tucumán, Argentina.*

SCOTT A. MANGAN, *Department of Biological Sciences, Washington University, St. Louis, USA; Smithsonian Tropical Research Institute, Apartado 2072, Balboa, Republic of Panama.*

ROBERTA E. MARTIN, *Department of Global Ecology, Carnegie Institution for Science, 260 Panama Street, Stanford, CA 94305, USA.* [remartin@stanford.edu]

MIGUEL MARTÍNEZ-RAMOS, *Centro de Investigaciones en Ecosistemas, Universidad Nacional Autónoma de México, CP 58190 Morelia, Michoacán, México.*

JOSEPH MASCARO, *American Association for the Advancement of Science, USA.* [S1@uwm.edu]

SCOTT J. MEINERS, *Department of Biological Sciences, Eastern Illinois University, Carleston, IL, USA.*

NICOLE L. MICHEL, *Department of Ecology, Evolutionary Biology, Tulane University, New Orleans, LA, USA.* [nicole.l.michel1@gmail.com]

CHELLAM MUTHUMPERUMAL, *Department of Plant Sciences, School of Biological Sciences, Madurai Kamaraj University, Madurai 625 021, Tamil Nadu, India.*

S. MUTHURAMKUMAR, *Department of Botany, V.H.N.S.N. College, Virudhunagar- 626 001, Tamil Nadu, India.*

ANSELMO NOGUEIRA, *Universidade de São Paulo, Instituto de Biociências, Departamento de Botânica. Rua do Matão 277, Cidade Universitária, 05508–090, São Paulo (SP), Brazil.* [anselmoeco@yahoo.com.br]

MARCELO R. PACE, *Universidade de São Paulo, Instituto de Biociências, Departamento de Botânica. Rua do Matão 277, Cidade Universitária, 05508–090, São Paulo (SP), Brazil.*

MARC P. E. PARREN, *Tropenbos International Congo-Basin Programme, Yaoundé, Cameroon.*

NARAYANASWAMY PARTHASARATHY, *Department of Ecology and Environmental Sciences, Pondicherry University, Puducherry 605 014, India.* [parthapu@yahoo.com]

SARAH C. PASQUINI, *Department of Botany and Plant Sciences, University of California, 2150 Batchelor Hall, Riverside, CA 92521, USA.*

NOEL B. PAVLOVIC, *US Geological Survey, Lake Michigan Ecological Research Station, Porter, Indiana, USA.*

DIEGO R. PÉREZ-SALICRUP, *Centro de Investigaciones en Ecosistemas, Universidad Nacional Autónoma de México, CP 58190 Morelia, Michoacán, México.*

OLIVER L. PHILLIPS, *School of Geography, University of Leeds, Leeds LS2 9JT, UK.*

JENNIFER S. POWERS, *Departments of Ecology, Evolution & Behavior and Plant Biology, University of Minnesota, St Paul, MN 55108, USA.* [powers@umn.edu]

FRANCIS E. PUTZ, *Department of Biology, P.O. 118526, University of Florida, Gainesville, FL 32611, USA.* [fep@ufl.edu]

FRANCISCO JAVIER RENDÓN-SANDOVAL, *Centro de Investigaciones en Ecosistemas, Universidad Nacional Autónoma de México. Antigua Carretera a Pátzcuaro 8701. Col. Ex Hacienda de San José de La Huerta, C. P. 58190, Morelia, Michoacán, México.*

W. DOUGLAS ROBINSON, *Department of Fisheries and Wildlife, Oregon State University, Corvallis, OR, USA.*

HUGO G. ROMERO-SALTOS, *Yachay Tech University, Liudad del conocimiento, Urcuquí, Ecuador.* [hromero@yachaytech.edu.ec]

NICK P. ROWE, *Botanique et bioinformatique de l'architecture des plantes – AMAP, UMR 5120 CNRS, TA A-51/PS2, Boulevard de la Lironde, 34398 Montpellier, France.* [nrowe@cirad.fr]

MOSES NSANYI SAINGE, *Tropical Plant Exploration Group, Mundemba, Southwest Region, Cameroon.*

LOUIS S. SANTIAGO, *Department of Botany and Plant Sciences, University of California, 2150 Batchelor Hall, Riverside, CA 92521, USA. Smithsonian Tropical Research Institute, P.O. Box 0843–03092, Balboa, Ancon, Panama, Republic of Panama.* [santiago@ucr.edu]

STEFAN A. SCHNITZER, *University of Wisconsin-Milwaukee, PO Box 413, Milwaukee, WI 53201 USA; Smithsonian Tropical Research Institute, Apartado 2072, Balboa, Republic of Panama.* [S1@uwm.edu]

THOMAS W. SHERRY, *Department of Ecology, Evolutionary Biology, Tulane University, New Orleans, LA, USA.*

THOMAS SPECK, *Plant Biomechanics Group, Botanic Garden, Faculty of Biology, University of Freiburg, Schänzlestraße 1, D-79104 Freiburg, Germany; Freiburg Materials Research Center (FMF) and Freiburg Center for Interactive Materials and Bio-Inspired Technologies (FIT).* [thomas.speck@biologie.uni-freiburg.de]

M. SRIDHAR REDDY, *Department of Environmental Sciences, Yogi Vemana University, Kadapa 516 003, Andhra Pradesh, India.*

DUNCAN THOMAS, *School of Biological Sciences, Washington State University, Vancouver, Washington, USA.* [duncanwt@gmail.com]

TARIN TOLEDO-ACEVES, *Instituto de Ecología, A.C., Departamento de Ecología Funcional, Veracruz, México.* [tarintoledo@gmail.com]

GEERTJE M.F. VAN DER HEIJDEN, *University of Wisconsin-Milwaukee, P.O Box 413, Milwaukee WI 53201, USA; Smithsonian Tropical Research Institute, Apartado Postal 0843–03092, Panama.* [g.m.f.vanderheijden@googlemail.com]

JULIANO VAN MELIS, *Departamento de Biología Vegetal, University of Campinas, São Paulo, Brazil.*

MÁRCIA C. VILELA-SANTOS, *Núcleo de Estudo da Biodiversidade da Amazônia Mato- Grossense, Instituto de Ciências Naturais Humanas e Sociais, Universidade Federal de Mato Grosso – UFMT, Sinop-MT 78557–267, Brazil.*

P. VIVEK, *Department of Ecology and Environmental Sciences, School of Life Sciences, Pondicherry University, Puducherry 605 014, India.*

STEPHEN P. YANOVIAK, *Department of Biology, University of Louisville, Louisville, Kentucky, KY 40292 USA.* [steve.yanoviak@louisville.edu]

FOREWORD

In a world without trees, would there still be lianas? Years ago Robin Foster casually observed in conversation that many of the oldest forests he knew in South America were impenetrable tangles dominated by lianas. He said that it seemed to him that there were almost no stopping rules to prevent a complete takeover by lianas, to the point of having what he called a "vine forest" at the climax of succession. I recalled this conversation on reading this wonderful book reviewing the current state of knowledge about the ecology and evolution of the climbing habit in angiosperms. One of the underlying if not always overtly stated themes of the book is why lianas are increasing in abundance in tropical forests, particularly in the New World, and the potential ramifications of this dramatic change.

There are so many differences and trade-offs in the functional ecology of lianas and trees that, on the face of it, one might imagine simple rules for stabilizing their competitive interactions. Lianas in general are more light-demanding than trees, and so in closed-canopy forests, trees might be expected to have the competitive advantage, or at least to coexist in competitive equipoise under natural disturbance regimes with lianas – except perhaps under current anthropogenic stress. Climate change, with greater frequency and intensity of droughts that kill canopy trees and open up forests to more light, and increasing forest fragmentation, which also increases openness and light, may change everything. Maybe these stresses have tipped the balance in favor of lianas, so that tropical forests are headed for alternate stable states – less desirable states – that provide reduced ecosystem services such as carbon sequestration and maintenance of biodiversity. Perhaps, like Foster's speculation, many forests have passed a tipping point where lianas are in a positive niche-construction feedback loop, creating ever more forest disturbance to their own advantage.

One thing I am struck with is how much more we know about lianas now than when Robin Foster and I launched the 50-ha Forest Dynamics Project on Barro Colorado Island (BCI), 34 years ago. Lianas were barely on our radar screen then, though, to our credit, we tried three times to get an NSF grant to census the lianas, before Stefan Schnitzer and I finally succeeded more than 20 years after the BCI FDP was launched. It is a pity that we do not have a longer record of the abundance and diversity of lianas on the BCI plot, and of their direct impact on tree growth and mortality extending right back to the first census in 1982. We have had to rely on indirect methods, such as litterfall data, to confirm the long-term increase in lianas in the BCI plot, not on direct, long-term census data. However, the 2007 census of the lianas of BCI has corrected this deficiency and future censuses will allow us to discern the causal connections between changes in the liana and tree communities on BCI. We can also inquire whether the mechanisms of liana–tree interaction uncovered on BCI apply to other tropical forests, such as the plots in Cameroon and Congo, where liana censuses also exist.

Although this volume confirms that we know much more about lianas than we did three decades ago, it also makes clear how far we have to go to understand the roles of lianas in forest functioning and dynamics. One clear way forward is to develop a formal theoretical framework to explore the interactions of trees and lianas and that can be challenged by the increasing volume of census and functional trait data on lianas being collected around the world. We need to ask, what is the minimal set of essential functional trait differences and trade-offs among trees and lianas that must be included in any such theoretical framework?

If the balance of competition among lianas and trees is indeed shifting, as several of the authors of this book suggest, this shifting balance could have profound implications for the future and conservation of tropical forests. This wide-ranging and thought-provoking book will inspire even more research on lianas, their impact on tree communities, and greater understanding of the mechanisms driving the shifting balance. It is my hope that it will also inspire the development of a theoretical framework to guide strong-inference testing of the many hypotheses to explain forest diversity and dynamics including lianas, not excluding them.

Steve Hubbell
Topanga, CA

PREFACE

Lianas were once largely overlooked in tropical and temperate forest research; instead, trees were the focus of most ecological studies. However, during the past two decades, scientific interest in lianas has increased considerably, largely catalyzed by the publication of the first major book on lianas (*The Biology of Vines*, Putz & Mooney 1991). Indeed, lianas are now recognized as key components of tropical forests, and research on liana ecology, evolution, systematics, and physiology has become increasingly common.

The amount of information on liana biology has increased so rapidly in recent years that we recognized a need to compile a major overview of the field. This edited volume, *The Ecology of Lianas*, is our attempt to provide this overview. Because no single book can adequately cover a field as broad and fast moving as the biology of lianas, we limited the volume to topics we felt were representative of the most important research now being conducted in tropical and temperate forests.

In Part 1 of the book, we provide a review of publishing trends, revealing that the rate of publications in lianas has increased far more rapidly that that of the broader field of ecology (Chapter 1). Since, detailed descriptions of plant communities are the basis for ecological studies, we included in Part 2 a broad representation of major liana datasets from around the world: Cameroon, Democratic Republic of the Congo, India, Ecuador, Brazil, Panama, Mexico, and Australia (Chapters 2–9). We also examine the role of lianas in forest succession (Chapter 10) and provide a synthesis of worldwide patterns of liana abundance and diversity (Chapter 11).

Competition between lianas and trees is important in tropical forest ecology and management; therefore, in Part 3, we explore some of the aspects of liana–tree competition and how the effects of lianas cascade to the community and ecosystem levels. In particular, we explore the latest research on competition between lianas and trees, and the relative importance of aboveground versus belowground interactions between these two important plant groups (Chapter 12). We also review recent advances in our knowledge of ecosystem-level effect of lianas in terms of carbon storage (Chapter 13), and the reciprocal effects between lianas and forest soils (Chapter 14). Although lianas are more abundant and diverse in tropical forests, they also play a potentially important role in temperate forests, which is reviewed in Chapter 15.

The focus of Part 4 is the evolution of diversity of lianas, which includes the paleobotanical history of lianas (Chapter 16), the early evolution of the liana growth form with a focus on the unique liana vascular system (Chapter 17), and the importance of the climbing habit for the diversification of flower plant taxa (Chapter 18). In Part 5, we examine the latest developments in liana anatomy, physiology, and biomechanics. Specifically, the chapters in this section provide a comprehensive review of liana anatomy (Chapter 19), ecophysiology with an emphasis on water relations (Chapter 20), the use of liana leaf chemistry to ecological traits and life-history strategy

(Chapter 21), the nutrient relations of lianas (Chapter 22), and the architecture and biomechanics of lianas (Chapter 23).

Lianas depend on animals for many services, including pollination, seed dispersal, and sometimes protection. In turn, animals use lianas as a source of food, shelter, protection from predation, and bridges through the forest canopy. Part 6 describes the unique interactions between lianas and invertebrates (Chapter 24), bird populations (Chapter 25), arboreal mammals (Chapter 26), and primate populations (Chapter 27).

The final section covers a selection of issues on the role of lianas in forest management and conservation. Chapter 28 describes the role of lianas as invasive species in North America. Chapter 29 discusses the role of lianas in forest conservation. Chapter 30, the final chapter of the book, discusses the recent phenomenon of increasing liana abundance in neotropical forests and the potential causes and consequences of greater liana density and biomass for forest functioning.

In the 1991 volume on climbing plants (*The Biology of Vines*), Putz and Mooney stated: "Much remains to be learned about vines; hopefully this volume will provide a solid foundation upon which future studies will be based." We believe that their goal was accomplished and that their book was probably more influential than the authors had ever imagined it would be. We hope that the current volume, *The Ecology of Lianas*, illustrates the impressive amount of new information acquired on liana ecology, thus documenting the increasing awareness of the importance of lianas as key contributors to the dynamics, diversity, evolution, conservation, and management of forest ecosystems. Our ultimate goal is that this new volume will be successful in further stimulating the study of lianas in tropical and temperate forests.

Stefan Schnitzer
Frans Bongers
Robyn Burnham
Francis Putz

ACKNOWLEDGEMENTS

We owe our gratitude to the many people who have contributed to this edited volume. We thank the many reviewers who provided critical feedback for each chapter of the book. We do not list them by name, thus allowing them to remain anonymous. We thank our editors at Wiley-Blackwell for their support throughout the publication process. Finally, and foremost, we thank the contributing authors, who made this book possible. Without their tireless study of tropical forests our knowledge of lianas and their roles in forests around the world would still be in the dark ages.

Part I

Introduction

Chapter 1

THE PAST, PRESENT, AND POTENTIAL FUTURE OF LIANA ECOLOGY

Stefan A. Schnitzer,[1,2] *Francis E. Putz,*[3] *Frans Bongers,*[4] *and Kristina Kroening*[1]

[1]University of Wisconsin-Milwaukee, Milwaukee, WI, USA
[2]Smithsonian Tropical Research Institute, Balboa, Republic of Panama
[3]University of Florida, Gainesville, FL, USA
[4]Wageningen University and Research Centre, Wageningen, The Netherlands

Ecology of Lianas, First Edition. Edited by Stefan A. Schnitzer, Frans Bongers, Robyn J. Burnham, and Francis E. Putz.

Companion Website: www.wiley.com/go/schnitzer/lianas

OVERVIEW

Over the past decade there has been increasing recognition of the substantial contribution of lianas to tropical and temperate forest dynamics, diversity, management, and ecosystem function. The increased interest in the ecology of lianas has apparently led to a significant increase in the number of publications focused on liana ecology. We quantified the change in this publication rate using the comprehensive and searchable database of liana publications found at: www.LianaEcologyProject.com. We compared liana publication trends to the background rate of general ecology publications from 1900–2005 using the 72 journals in JSTOR. We found a 22-fold jump in liana ecology publications over the past three decades, whereas the increase in all general ecology publications was approximately a twofold increase over a similar period.

The rapid rise in the liana publication rate reflects the increasing recognition of the importance of lianas in tropical forest dynamics, diversity, and regeneration. One of the major findings to emerge from these new studies is that lianas reduce tree recruitment, growth, reproduction, and survival, effects that, in turn, reduce tree diversity. Furthermore, the effects of lianas can cascade across trophic levels and alter fundamental ecosystem processes. A second major finding is that liana abundance, biomass, and productivity have increased substantially in neotropical forests. There are now 12 studies documenting striking increases in lianas relative to trees over the past 30 years – a phenomenon that has serious implications for the future of tropical forests.

The increasing importance of lianas as a key component of tropical forests calls for a deeper understanding of liana ecology and life-history, as well as the myriad roles lianas play in tropical forest dynamics and functioning. Increases in liana density and biomass may impose important changes on tropical forests, particularly on tree growth and survival, as well as on ecosystem properties. Increasing liana abundance also raises new challenges for forest managers, resulting in the need for thoughtful liana management plans for tropical forest timber tree production. In terms of basic research, lianas provide an excellent alternative to trees for the study of ecological theory, and thus the entire field of ecology (as well as the study of lianas) will benefit when the study of lianas moves beyond the realm of applied ecology and a tropical curiosity, and becomes part of the body of mainstream ecological research.

INTRODUCTION

Lianas are a conspicuous component of many forests, particularly in the tropics, and their importance has been noted in the literature since Darwin's (1865) publication, "On the movements and habits of climbing plants." Darwin and others recognized the unique ability of lianas to use tree stems as trellises to climb to the forest canopy. Once in the canopy, lianas produce large numbers of leaves that cover those of their host-trees, thus competing for light (Putz 1983; Gerwing & Farias 2000). Lianas also compete intensely with trees for belowground resources – primarily water and nutrients (e.g., Dillenburg et al. 1993, Schnitzer et al. 2005; Toledo-Aceves, Chapter 12 in this volume). Thus, lianas limit tree growth, fecundity, and survival (Putz 1984; Wright et al. 2005; Ingwell et al. 2010; Schnitzer & Carson 2010; Ladwig & Meiners, Chapter 15 in this volume; van der Heijden et al., Chapter 13 in this volume). Lianas also can be aggressive invaders, and both native and exotic lianas are among the world's most noxious weeds (e.g., Horvitz & Koop 2004; Forseth & Innis 2004; Leicht-Young & Pavlovic, Chapter 28 in this volume). Additionally, lianas have long been exploited by humans as food (e.g., passion fruits, gourds, legumes), construction material (ropes), stimulants (e.g., guarana from *Paullinia cupana*), medicines (e.g., curare from *Strychnos toxifera* and *Chondrodendron tomentosum*), and for the production of alcoholic beverages such as wine and cognac from grapes (*Vitis spp.*) and beer from hops (*Humulus lupulus*).

In the ecological literature, lianas were largely ignored historically while studies of trees abounded (Schnitzer & Bongers 2002). More recently, lianas have been recognized as important and integral components of forests, particularly in the tropics (e.g., Isnard & Silk 2009; Paul & Yavitt 2010; Schnitzer & Bongers 2011; van der Heijden et al. 2013; Wyka et al. 2013), and the renewed scientific interest in lianas is reflected in a rapidly rising publication rate. However, since the number of publications in all areas of ecology is increasing, it is difficult to evaluate whether the rate of liana-related publications is actually rising faster than the background rate of the entire spectrum of ecology publications.

We evaluated whether the spate of new publications on liana ecology reflects growing interest in this particular growth-form, or whether the increase is an artifact of the general escalation in ecological publishing, by comparing the rate of peer-reviewed, liana-specific publications over the past century to that of general ecology publications. We compiled a comprehensive list of over 650 liana publications from 1865 until 2009 using extensive literature searches, which is available in the comprehensive and searchable liana database at the *Liana Ecology Project* website: www.LianaEcologyProject.com. We classified liana publications as being either focused primarily on liana ecology or publications that included lianas but in which lianas were not the main emphasis of the study. Since both metrics yielded similar results, we used only the former in our analyses. We excluded herbaceous vines, rattans and other climbing palms, as well as non-ecology/environmental publications in our analyses, even though these publications, as well as publications after 2009, are included in the www.LianaEcologyProject.com database. For a metric of all ecology publications, we examined the 72 journals in JSTOR from 1900 until 2005 and searched for all publications with the term "Ecology" in the title, abstract, or keywords.

EXPONENTIALLY INCREASING LIANA PUBLICATION RATE

Liana publication rate was extremely low until the 1980s, after which it began a rapid rise (Fig. 1.1A). The highest number of publication was in 2009, the last complete year of our literature survey, during which 44 liana-specific articles were published, along with 12 additional publications that included lianas, but in which lianas were not the focus of the study (56 total publications in 2009). The rate of increase in liana publications was much greater than for all ecological publications, whose rate also increased over the past century, with the steepest annual increase beginning in the early 1960s, followed by a steady linear increase thereafter (Fig. 1.1B). For example, the number of liana publications increased 22-fold between 1980 and 2009 (Fig. 1.1C), compared to the twofold increase in the number of all ecology publications over a similar period (Fig. 1.1D).

IMPORTANCE OF LIANAS IN TROPICAL FORESTS

The rapid relative increase in the liana publication rate indicates a growing recognition of the importance of lianas in forests, particularly in the tropics. Lianas add considerably to the diversity of tropical forests, often more than would be predicted based on their stem density. For example, in a survey of all lianas $\geq$1 cm diameter on Barro Colorado Island, Panama, lianas composed 35% of the woody species and 25% of the woody stem density (e.g., Schnitzer et al. 2012; Schnitzer et al., Chapter 7 in this volume). Lianas can compete intensely with other growth forms, such as trees (Fig. 1.2), and both large-scale observational and experimental studies have shown that lianas reduce tree regeneration, diversity, growth, and biomass in numerous tropical forests (Grauel & Putz 2004; Peña-Claros et al. 2008; van der Heijden et al. 2008; Ingwell et al. 2010; Schnitzer & Carson 2010; Tobin et al. 2012; Schnitzer et al. 2014). Lianas reduce tree fecundity (e.g., Wright et al. 2005; Kainer et al. 2006, 2014) and the density and richness of trees recruiting into treefall gaps (Schnitzer et al. 2000; Schnitzer & Carson 2010). Lianas increase tree mortality; trees heavily infested by lianas (e.g., Fig. 1.2) have twice the probability of death than trees with fewer lianas in their crowns (Ingwell et al. 2010).

Lianas also appear to have a much greater competitive effect on trees than would be predicted by their biomass. Tobin et al. (2012) removed similar amounts of liana or tree biomass from target trees and found that the target trees consistently responded immediately to liana cutting, but the target trees did not respond to tree cutting. Furthermore, lianas may have the capacity to alter tree community composition by competing particularly intensely with some tree species (Schnitzer & Bongers 2002). For example, lianas reduce recruitment and growth of shade-tolerant trees, while some pioneer trees are largely unaffected (e.g., van der Heijden et al. 2008, van der Heijden et al., Chapter 13 in this volume; Schnitzer & Carson 2010).

At the ecosystem level, lianas reduce rates of carbon uptake by trees, which typically store more than 90% of aboveground forest carbon (van der Heijden & Phillips 2009; Durán & Gianoli 2013; Schnitzer et al. 2014). These losses are not offset by increases in liana biomass; lianas store only a fraction of the carbon that they displace due to their thin stems and low wood volumes relative to the trees with which they compete. For

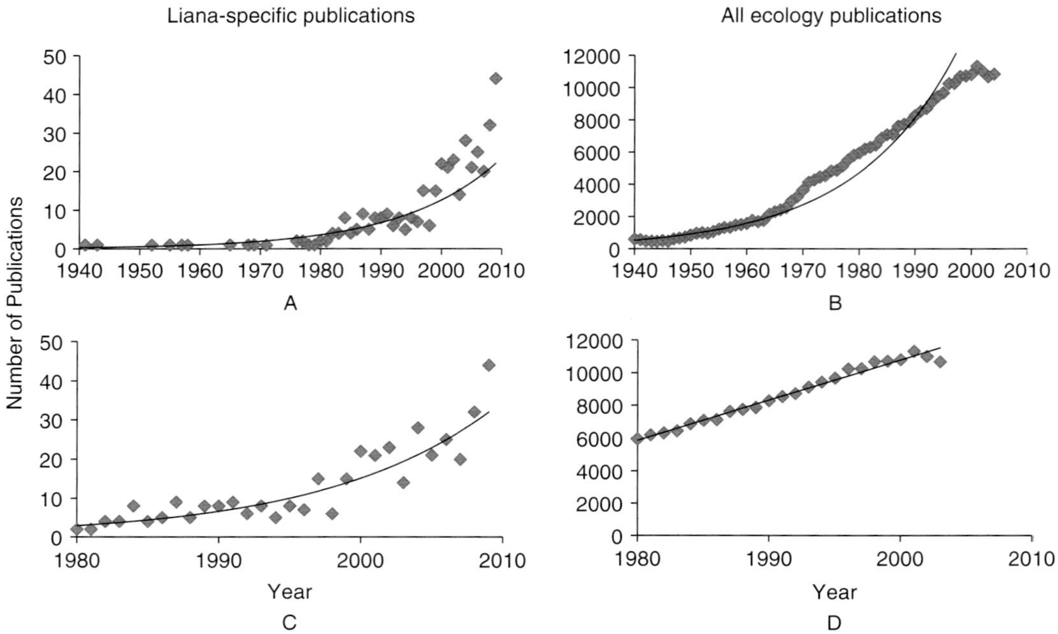

Fig. 1.1 The annual number of publications focused on liana ecology from 1940 until 2009 (A) versus the number of general ecological topics from 1940 until 2005 (B). The number of liana publications was estimated using extensive literature searches. The number of general ecological publications was calculated by searching the 72 journals in JSTOR for all publications with the term "Ecology" in which the title, abstract, or keywords were included. Data were collected from 1865–2009 for lianas and from 1900–2005 for general ecological publications, but graphing data starting in 1940 was sufficient to accurately portray the patterns. The different panels per column use the same data, but focus on different time periods. Exponential and linear curves, where appropriate, were fitted to the data. Liana publication rate increased exponentially since 1980 (C), whereas the rate of all ecology publications has increased linearly over this time period (D).

example, in a liana-removal study in central Panama, Schnitzer et al. (2014) reported that lianas reduced tree biomass accumulation in gaps by nearly 300% over an 8-year period, but lianas themselves compensated for only 24% of the biomass uptake that they displaced in trees (see also van der Heijden et al. 2013; van der Heijden et al., Chapter 13 in this volume). In the Peruvian Amazon, lianas appeared to reduce tree annual biomass increment by around 10%; however, that liana annual increment compensated only around 29% for what was displaced in trees (van der Heijden & Phillips 2009). In addition, the increase in tree mortality due to lianas (Ingwell et al. 2010) reduces overall forest carbon storage even further, and forest areas with high liana densities can have far lower biomass than forest areas with low liana densities (e.g., Chave et al. 2001; Durán & Gianoli 2013). Furthermore, lianas are increasing in abundance in many neotropical forests (Schnitzer & Bongers 2011; Schnitzer, Chapter 30 in this volume), and thus, the effects of lianas on trees at all life-history stages (recruitment, growth, reproduction, and mortality), as well as the influence of lianas on tropical forest ecosystem processes, are likely to increase (Schnitzer et al. 2011; van der Heijden et al., Chapter 13 in this volume).

LIANA MANAGEMENT: FORESTRY AND WILDLIFE IMPACTS

The increase in liana abundance will motivate forest managers to more actively manage lianas in timber-producing forests. The deleterious impacts of lianas on trees, plus the observation that when a liana-infested tree falls or is felled, it often pulls down its neighbors, have led to the often-made but

Fig. 1.2 Lianas blanketing trees on Barro Colorado Island, Panama. (Source: C. Ziegler. Reproduced with permission.) *See plate section for color representation of this figure.*

seldom-followed recommendation that lianas be cut prior to selective logging. Pre-felling liana cutting, if done at least 6 months in advance of the harvest to allow cut liana stems time to decompose and weaken, can considerably reduce felling damage to neighboring trees (e.g., Appanah & Putz 1984; Garrido-Perez et al. 2008). Furthermore, cutting lianas that entangle the target trees' canopy prior to logging will reduce post-logging liana infestations of logging gaps (Alvira et al. 2004; Schnitzer et al. 2004), where lianas can arrest tree regeneration and density, as well as reduce tree survival and ultimately tree diversity (Schnitzer et al. 2000; Schnitzer & Carson 2010; Schnitzer et al. 2014). Selective liana cutting can also benefit future timber trees by reducing their susceptibility to logging damage as well as acclimatizing them to post-logging increases in light intensities and temperatures, which will stimulate their growth and increase their survival.

In terms of overall forest biodiversity, retaining lianas should be a priority because lianas substantially increase plant diversity (e.g., Gianoli 2004; Schnitzer et al. 2012) as well as providing an important resource for many forest animals (Putz et al. 2001). Lianas produce leaves, flowers, and fruits that may be critical for the survival of many animal species, especially during the dry season when lianas grow more and produce more leaves and fruits than trees (e.g., Putz & Windsor 1987; Schnitzer 2005; Dunn et al. 2012; Arroyo-Rodríguez, Chapter 27 in this volume). By connecting tree canopies, liana stems also provide critical inter-crown pathways for non-volant arboreal animals (e.g., ants, sloths, rodents, monkeys; Montgomery & Sunquist 1978; Dunn et al. 2012; Yanoviak, Chapter 24 in this volume; Fig. 1.3). Liana tangles in the understory provide structure, forage, and predator escape for birds and small mammals (e.g., Putz et al. 2001; Kilgore et al. 2010; Michel et al., Chapter 25 in this volume; Lambert & Halsey, Chapter 26 in this volume), and bind the forest canopy together, thus potentially reducing treefalls when liana connectivity is high (Garrido-Pérez et al. 2008). Consequently, selectively cutting lianas from timber trees, rather than cutting all lianas, is an appropriate compromise between sustaining timber production and maintaining biodiversity in forests managed for timber.

Fig. 1.3 Three-toed sloth in central Panama using a liana to ascend to the forest canopy. (Source: Photo credit: S.A. Schnitzer. Reproduced with permission.) *See plate section for color representation of this figure.*

NEW FRONTIERS IN LIANA ECOLOGY

The accelerating pace of liana research reflects increasing recognition that lianas are an ecologically important component of tropical forests. Lianas can be used to test basic ecological theory on the factors controlling plant species distribution, the maintenance of species diversity, plant competition, and large-scale structural change in tropical forests. Until recently, however, trees were used almost exclusively to test major ecological theories that pertain to tropical forests, whereas lianas were largely ignored. Lianas differ from trees in their anatomy, physiology, and leaf characteristics (Angylossy et al., Chapter 19 in this volume; Asner & Martin, Chapter 21 in this volume; Santiago et al., Chapter 20 in this volume), and thus if the theories to explain the distribution and diversity maintenance of plant species are to be accepted as general, these theories must be tested beyond solely using trees.

Recent evidence suggests that the mechanisms that control tree and liana distribution and diversity maintenance differ substantially. For example, among forests, liana density and biomass increase with decreasing mean annual precipitation and increasing seasonality – a pattern that differs strikingly from trees (Schnitzer 2005; DeWalt et al. 2010; DeWalt et al., Chapter 11 in this volume). Within a given forest, there is now strong evidence that disturbance and treefall gap formation maintain liana diversity (Schnitzer & Carson 2001; Ledo & Schnitzer 2014; Schnitzer et al., Chapter 7 in this volume), with little evidence for edaphic specialization and negative density dependence as causative mechanisms (Dalling et al. 2012; Ledo & Schnitzer 2014). In contrast, the putative mechanisms responsible for the maintenance of tree species diversity and distribution are edaphic specialization and negative density dependence (e.g., John et al. 2007; Mangan et al. 2010); whereas disturbance appears to have little influence on tree distribution or diversity (Hubbell et al. 1999; Schnitzer & Carson 2001; Bongers et al. 2009). Therefore, the mechanisms responsible for distribution of species and the maintenance of diversity appear to differ between lianas and trees (Ledo & Schnitzer 2014). Determining the mechanisms that control liana distribution and diversity maintenance and investigating how these mechanisms differ from other growth forms, such as trees, will lead to insights into how and when these varied mechanisms operate, thus contributing to a more comprehensive understanding of the maintenance of plant species diversity and the factors that control plant species distribution.

The causes and consequences of increasing liana abundance and biomass will likely become an active area of future study. Determining the factors responsible for liana increases, as well as the effects of increased lianas on forest dynamics, diversity, and ecosystem function are critical to our understanding of tropical forest ecology. Increasing levels of drought, temperature, disturbance, nitrogen deposition, and CO_2 have all been proposed as potential causes of increasing liana abundance (Schnitzer & Bongers 2011; Schnitzer et al. 2011; Schnitzer, Chapter 30 in this volume). To fully understand the drivers of increasing liana abundance, however, these factors need to be tested both separately and in combination along major environmental gradients (Schnitzer et al., Chapter 7 in this volume).

Another important line of research is determining the relative strengths of aboveground and belowground competition between lianas and trees, and how these effects vary with changes in resource availability (Toledo-Aceves, Chapter 12 in this volume). As with other plant growth forms, there is less research on liana roots than on their shoots. Given that large structural roots are not required to support liana stems, a larger proportion of their investments belowground should be available for root elongation, leading to aggressive water and nutrient foraging. To the extent that this so-far-untested prediction is true, it might explain why lianas appear to be such effective competitors with trees for belowground resources (e.g., Dillenburg et al. 1993; Schnitzer et al. 2005; Toledo-Aceves & Swaine 2008; Toledo-Aceves, Chapter 12 in this volume).

A deeper understanding of liana life-history strategies and how these strategies enable lianas to compete successfully with trees may reveal the causes of increasing liana abundance and biomass, and the ways in which these increases will alter future tropical forest dynamics. Furthermore, a balanced assessment of both the positive and negative ecological effects of lianas on community and ecosystem processes, as well as their contribution to biodiversity within and across trophic levels, is necessary to assess how increasing lianas will affect future forests. Considering the importance of these issues to tropical forest ecology, therefore, we predict that the steep climb in liana ecology publications will continue.

ACKNOWLEDGMENTS

Financial support for this chapter and the *Liana Ecology Project* website was provided by National Science Foundation grants DEB-0613666, DEB-0845071, and DEB-1019436, UWM Research Growth Initiative, and UWM Office for Undergraduate Research.

REFERENCES

Alvira, D., Putz, F.E. & Fredericksen, T.S. (2004) Liana loads and post-logging liana densities after liana cutting in a lowland forest in Bolivia. *Forest Ecology and Management*, **190**, 73–86.

Appanah, S. & Putz, F.E. (1984) Climber abundance in virgin dipterocarp forest and the effect of pre-felling climber cutting on logging damage. *Malaysian Forester*, **47**, 335–342.

Benítez-Malvido, J. & Martínez-Ramos, M. (2003) Impact of forest fragmentation on understory plant species richness in Amazonia. *Conservation Biology*, **17**, 389–400.

Bongers, F., Poorter, L., Hawthorne, W.D. & Sheil, D. (2009) The intermediate disturbance hypothesis applies to tropical forests, but disturbance contributes little to tree diversity. *Ecology Letters*, **12**, 1–8.

Chave, J., Riéra, B. & Dubois, M.A. (2001) Estimation of biomass in a neotropical forest of French Guiana: spatial and temporal variability. *Journal of Tropical Ecology*, **17**, 79–96.

Dalling, J.W., Schnitzer, S.A., Baldeck, C. et al. (2012) Resource-based habitat associations in a neotropical liana community. *Journal of Ecology*, **100**, 1174–1182.

Darwin, C.R. (1865) On the movements and habits of climbing plants. *Journal of the Linnean Society of London (Botany)*, **9**, 1–118.

DeWalt, S.J., Schnitzer, S.A., Chave, J. et al. (2010) Annual rainfall and seasonality predict pan-tropical patterns of liana density and basal area. *Biotropica*, **42**, 309–317.

Dillenburg, L.R., Whigham, D.F., Teramura, A.H. & Forseth, I.N. (1993) Effects of below- and aboveground competition from the vines *Lonicera japonica* and *Parthenocissus quinquefolia* on the growth of the tree host *Liquidambar styraciflua*. *Oecologia*, **93**, 48–54.

Dunn, J.C., Asensio, N., Arroyo-Rodríguez, V. et al. (2012) The ranging costs of a fallback food: liana consumption supplements diet but increases foraging effort in howler monkeys. *Biotropica*, **44**, 704–714.

Durán, S.M. & Gianoli, E. (2013) Carbon stocks in tropical forests decrease with liana density. *Biology Letters*, **9**, 20130301.

Forseth, I.N. & Innis, A.F. (2004) Kudzu (*Pueraria montana*): history, physiology, and ecology combine to make a major ecosystem threat. *Critical Reviews in Plant Science*, **23**, 401–413.

Garrido-Pérez, E.I., Dupuy, J.M., Durán-García, R. et al. (2008) Structural effects of lianas and Hurricane Wilma on trees in Yucatan Peninsula, Mexico. *Journal of Tropical Ecology*, **24**, 559–562.

Gerwing, J.J. & Farias, D.L. (2000) Integrating liana abundance and forest stature into an estimate of total aboveground biomass for an eastern Amazonian forest. *Journal of Tropical Ecology*, **16**, 327–335.

Gianoli, E. (2004) Evolution of a climbing habit promotes diversification in flowering plants. *Proceedings of the Royal Society B*, **271**, 2011–2015.

Grauel, W.T. & Putz, F.E. (2004) Effects of lianas on growth and regeneration of *Prioria copaifera* in Darien, Panama. *Forest Ecology and Management*, **190**, 99–108.

Horvitz, C.C. & Koop, A. (2004) Removal of nonnative vines and post-hurricane recruitment in tropical hardwood forests of Florida. *Biotropica*, **33**, 268–281.

Hubbell, S.P., Foster, R.B., O'Brien, S.T. et al. (1999) Light-gap disturbances, recruitment limitation, and tree diversity in a neotropical forest. *Science*, **22**, 554–557.

Ingwell, L.L., Wright, S.J., Becklund, K.K. et al. (2010) The impact of lianas on 10 years of tree growth and mortality on Barro Colorado Island, Panama. *Journal of Ecology*, **98**, 879–887.

Isnard, S. & Silk, W.K. (2009) Moving with climbing plants from Charles Darwin's time into the 21st Century. *American Journal of Botany*, **96**, 1205–1221.

John, R., Dalling, J.W., Harms, K.E. et al. (2007) Soil nutrients influence spatial distributions of tropical tree species. *Proceedings of the National Academy of Sciences of the United States of America*, **104**, 864–869.

Kainer, K.A., Wadt, L.H.O., Gomes-Sliva, D.A.P. & Capanu, M. (2006) Liana loads and their association with *Bertholletia excelsa* fruit and nut production, diameter growth and crown attributes. *Journal of Tropical Ecology*, **22**, 147–154.

Kainer, K.A., Wadt, L.H.O. & Staudhammer, C.L. (2014) Testing a silvicultural recommendation: Brazil nut responses 10 years after liana cutting. *Journal of Applied Ecology*. doi: 10.1111/1365-2664.12231

Kilgore, A., Lambers, T.D. & Adler, G.H. (2010) Lianas influence fruit and seed use by rodents in a tropical forest. *Tropical Ecology*, **51**, 265–271.

Ledo, A. & Schnitzer, S.A. (2014) Disturbance, not negative density dependence or habitat specialization maintains liana diversity in a tropical forest. *Ecology*, in press.

Mangan, S.A., Schnitzer, S.A., Herre, E.A. et al. (2010) Negative plant-soil feedback predicts relative species abundance in a tropical forest. *Nature*, **466**, 752–756.

Montgomery, G.G. & Sunquist, M.E. (1978) Habitat selection and use by two-toed and three-toed sloths, in *The Ecology of Arboreal Foliovores* (ed G.G. Montgomery), Smithsonian Institution Press, Washington, DC, pp. 329–359.

Paul, G.S. & Yavitt, J.B. (2010) Tropical vine growth and the effects on forest succession: a review of the ecology and management of tropical climbing plants. *The Botanical Review*, **77**, 11–30.

Peña-Claros, M., Fredericksen, T.S., Alarco, A. et al. (2008) Beyond reduced-impact logging: silvicultural treatments to increase growth rates of tropical trees. *Forest Ecology and Management*, **256**, 1458–1467.

Putz, F.E. (1983) Liana biomass and leaf area of a tierra firme forest in the Rio Negro Basin, Venezuela. *Biotropica*, **15**, 185–189.

Putz, F.E. (1984) The natural history of lianas on Barro Colorado Island, Panama. *Ecology*, **65**, 1713–1724.

Putz, F.E., Sirot, L.K. & Pinard, M.A. (2001) Tropical forest management and wildlife: silvicultural effects on forest structure, fruit production, and locomotion of non-volant arboreal animals, in *The Cutting Edge: Conserving Wildlife in Managed Tropical Forests* (eds R. Fimbel, A. Grajal & J. Robinson), Columbia University Press, New York.

Putz, F.E. & Windsor, D.M. (1987) Liana phenology on Barro Colorado Island, Panama. *Biotropica*, **19**, 334–341.

Schnitzer, S.A. (2005) A mechanistic explanation for global pattern of liana abundance and distribution. *American Naturalist*, **166**, 262–276.

Schnitzer, S.A. & Bongers, F. (2002) The ecology of lianas and their role in forests. *Trends in Ecology and Evolution*, **17**, 223–230.

Schnitzer, S.A. & Bongers, F. (2011) Increasing liana abundance and biomass in tropical forests: emerging patterns and putative mechanisms. *Ecology Letters*, **14**, 397–406.

Schnitzer, S.A., Bongers, F. & Wright, S.J. (2011) Community and ecosystem ramifications of increasing lianas in neotropical forests. *Plant Signaling & Behavior*, **6**, 598–600.

Schnitzer, S.A. & Carson, W.P. (2001) Treefall gaps and the maintenance of species diversity in a tropical forest. *Ecology*, **82**, 913–919.

Schnitzer, S.A. & Carson, W.P. (2010) Lianas suppress tree regeneration and diversity in treefall gaps. *Ecology Letters*, **13**, 849–857.

Schnitzer, S.A., Dalling, J.W. & Carson, W.P. (2000) The impact of lianas on tree regeneration in tropical forest canopy gaps: Evidence for an alternative pathway of gap-phase regeneration. *Journal of Ecology*, **88**, 655–666.

Schnitzer, S.A., Kuzee, M.E. & Bongers, F. (2005) Disentangling above-and below-ground competition between lianas and trees in a tropical forest. *Journal of Ecology*, **93**, 1115–1125.

Schnitzer, S.A., Mangan, S.A., Dalling, J.W., et al. (2012) Liana abundance, diversity, and distribution on Barro Colorado Island, Panama. *PLoS One* 7(12): e52114. doi:10.1371/journal.pone.0052114.

Schnitzer, S.A., Parren, M.P.E. & Bongers, F. (2004) Recruitment of lianas into logging gaps and the effects of pre-harvest liana cutting. *Forest Ecology and Management*, **190**, 87–98.

Schnitzer, S.A., van der Heijden, G.M.F., Mascaro, J., Carson, W.P. (2014) Lianas reduce biomass accumulation in a tropical forest. *Ecology*, in press.

Tobin, M.F., Wright, A.J., Mangan, S.A., & Schnitzer, S.A. (2012) Lianas have a greater competitive effect than trees of similar biomass on tropical canopy trees. *Ecosphere*, **3**, Article 20: 1-11. 10.1890/ES11-00322.1

Toledo-Aceves, T. & Swaine, M.D. (2008) Above- and below-ground competition between the liana *Acacia kamerunensis* and tree seedlings in contrasting light environments. *Plant Ecology*, **196**, 233–244.

van der Heijden, G.M.F., Healey, J.R. & Phillips, O.L. (2008) Infestation of trees by lianas in a tropical forest in Amazonian Peru. *Journal of Vegetation Science*, **19**, 747–756.

van der Heijden, G.M.F. & Phillips, O.L. (2009) Liana infestation impacts tree growth in a lowland tropical moist forest. *Biogeosciences*, **6**, 2217–2226.

van der Heijden, G.M.F., Schnitzer, S.A., Powers, J.S. & Phillips, O.L. (2013) Liana impacts on carbon cycling, storage and sequestration in tropical forests. *Biotropica*, **45**, 682–692.

Wright, S.J., Jaramillo, M.A., Pavon, J. et al. (2005) Reproductive size thresholds in tropical trees: variation among individuals, species and forests. *Journal of Tropical Ecology*, **21**, 307–315.

Wyka, T.P., Oleksyn, J., Karolewsk, P., & Schnitzer, S.A. (2013) Phenotypic correlates of the lianescent growth form: a review. *Annals of Botany*, **112**, 1667–1681.

Part II

Patterns of Liana Demography and Distribution: From local to Global

Chapter 2

LIANA ABUNDANCE AND DIVERSITY IN CAMEROON'S KORUP NATIONAL PARK

Duncan Thomas,[1] Robyn J. Burnham,[2] George Chuyong,[3] David Kenfack,[4] and Moses Nsanyi Sainge[5]

[1]School of Biological Sciences, Washington State University, Vancouver, Washington, USA
[2]University of Michigan, Ann Arbor, MI, USA
[3]University of Buea, Buea, Cameroon
[4]Center for Tropical Forest Science, Smithsonian Institution Global Earth Observatory, Smithsonian National Museum of Natural History, Washington, DC, USA
[5]Tropical Plant Exploration Group, Mundemba, Southwest Region, Cameroon

Ecology of Lianas, First Edition. Edited by Stefan A. Schnitzer, Frans Bongers, Robyn J. Burnham, and Francis E. Putz.

Companion Website: www.wiley.com/go/schnitzer/lianas

OVERVIEW

The lianas (≥1 cm stem diameter) in an 18-ha plot of lowland rainforest in southern Cameroon were surveyed to document the structure and species composition of the liana flora, and to compare the abundance and diversity of lianas to that of trees in the same area. The liana crowns were concentrated in the middle and lower canopy of the forest. We found a total of 256 species in 77 genera and 31 families. When compared to trees of the same diameter, lianas are both less abundant than trees (9023 versus 119,027) and less speciose (256 versus 409). We also found that the tree community in the mid/lower canopy is fairly strongly dominated by a single species (*Oubanguia alata*), while several liana species share dominance, each comprising a smaller portion of the dominance structure. As a result, liana diversity measured by several commonly used indices equals or exceeds that of the trees. This creates a forest canopy where trees and lianas both make large but different contributions to fruit and pollen/nectar resources. The liana community adds at least ten angiosperm families that are not represented in the Korup tree flora, broadening the phylogenetic diversity of the forest. About 80% of the 77 liana genera in the plot are unknown as trees, adding further to diversity above the species level. Lianas in Korup are overwhelmingly dispersed by animals or by ballistic means, which contrasts with the high incidence of wind dispersal among neotropical liana species.

INTRODUCTION

Data on liana density have become more common from forests worldwide. However, it still remains a challenge to find accurate or extensive data on the species comprising the liana community in tropical forests. While comparisons among tree communities are slowly emerging from large-plot consortia (DeCáceres et al. 2012), liana data are still largely lacking and this is especially true for African forests. Here, we describe a study of lianas conducted in the Korup National Park, located in Cameroon's Southwest Region, adjacent to the Nigerian border (Fig. 2.1). We focus on liana abundance, species richness, dominance, and diversity, with comparisons to these attributes for the tree community in the same area. For our study, lianas are narrowly defined as climbers that develop secondary wood and are rooted in the soil. Rattan palms, hemiepiphytes/stranglers, and other categories of climbing plants that do not meet the narrow definition were included in the census in Korup but excluded from this account.

The study site is near the town of Mundemba at 160 m a.s.l, 5.074° N, 8.855° E, 30 km inland from estuarine mangrove swamps, and 70 km from the Atlantic Ocean in the Bight of Bonny. Following botanical and primatological studies in Korup National Park in the 1970s and 1980s (Gartlan & Struhsaker 1972; Gartlan et al. 1986; Newbery & Gartlan 1996; Usongo & Amubode 2000), a long-term forest monitoring plot was established in 1996. Each census of the 50-hectare plot (1000 × 500 m) includes all trees and saplings at least 1-cm diameter at breast height (dbh): all tagged, measured, mapped, and identified (Thomas et al. 2003; Kenfack et al. 2007; http://www.ctfs.si.edu/site/Korup/). Two complete tree censuses have been conducted, in 1997–1999 and in 2008–2009. In 2000–2002 and 2011–2012, two liana censuses for stems ≥1 cm were conducted in 18 ha of the 50-ha plot. The 18-ha liana census area is in lower elevation forest on gentle topography at the southern end of the 50-ha plot, in a mosaic of swampy creeks and uplands (Figs. 2.1B, 2.1C).

Proximity to the ocean, the presence of surrounding hills, and the onshore wet winds during much of the year create a very wet climate, with a mean annual rainfall of 5272 mm (1973–1994 data), as measured 20 km away (Chuyong et al. 2004). Korup experiences a short dry season with 3 months (December–February) averaging less than 100 mm precipitation. The remaining 9 months all average well over 100 mm precipitation, with average monthly rainfall peaking at over 900 mm in July and August. Because of the 3-month dry season, Korup vegetation is classified as moist tropical evergreen forest. However, the length and intensity of the wet season are unusual among tropical moist forests, and "wet seasonal evergreen forest" might be a better description. Mean daily maximum temperature is 32.6 °C, with a diurnal range of about 10 °C. Month-to-month and season-to-season variation is less than the diurnal range. Mean daily maximum temperatures are highest in February (32.8 °C), a dry season month with low cloud cover, and lowest in August in the middle of the wet season (27.8 °C).

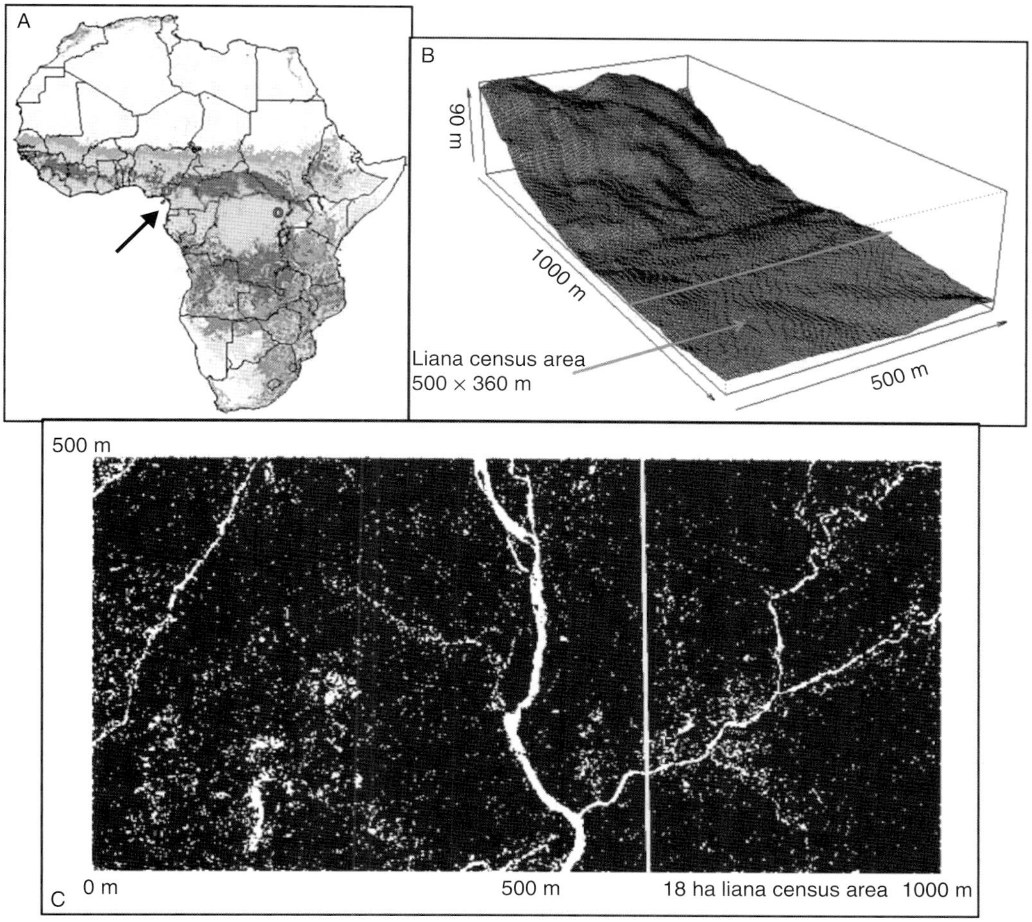

Fig. 2.1 (A) Africa, showing study area in western Cameroon (arrow); additional dot is Ituri Plot in DR Congo. (B) 3-D map of the Korup Forest Dynamics Plot showing the location of the liana census area at the flatter south end. (C) 50-ha Korup Forest Dynamics Plot showing the location of the 18-ha liana census area (right of vertical line) and pale areas of sparse tree cover, including rock outcrops (lower left) and creeks/swamps elsewhere. (Source: Fig. 2.1 A, adapted from United States Geological Survey vegetation map. Reproduced with permission.)

Distribution patterns of the Korup tree species have been studied relative to the major African phytochoria (Kenfack et al. 2007), based on the three main blocks of African moist tropical forest described by White (1979, 1983). The large Congolian forest block falls largely in the basin of the Congo River, the Lower Guinean forest covers the coastal belt from southeastern Nigeria through Gabon, and the Upper Guinea forest is distributed mostly in Liberia, the Ivory Coast, and Ghana. The 50-ha plot is located at the western end of the Lower Guinea forest and the tree flora shows strong floristic affinities with this forest block. Approximately 33% of the tree species in the plot are known only from the Lower Guinea forest (Kenfack et al. 2007), and though we have not yet completed this analysis for lianas, we expect to find similar affinities for the liana flora.

Trees are less dense in the wetlands (Fig. 2.1C) and in treefall gaps. Korup experiences thunderstorms, especially during the onset of the wet season, resulting in blow-downs or lightning strikes. These disturbances tend to be small, limited to a few trees or large branches, and the large tropical storms that cause widespread forest damage in other parts of the tropics are unknown in Korup (Thomas et al. 2003; Egbe et al. 2012). In the general area of Korup, large gaps result from shifting cultivation, rather than from storms. The Korup plot shows no signs of former cultivation, and is currently protected within a national park, so large canopy gaps favoring the establishment and growth of lianas and pioneer tree species are absent. Consequently, conditions in the plot are less conducive to the establishment of lianas than in most other tropical forests. Many liana species in Korup produce shade-tolerant tree-like saplings, which survive in the dark, forest understory, where they compete with tree saplings and with understory trees (unpublished data).

Liana field census methods followed those of Gerwing et al. (2006). Liana individuals were defined as plants with visible (even if shallowly buried) living connections between stems. We mapped the rooting position, measured the stem diameter at 1.3 m from the rooting position, and identified all individual lianas in the field as far as possible. Field botanists were responsible for allocating individual lianas to morphospecies, although 10% of the individuals remain unassigned to a morphospecies because of problems locating the foliage. Identification numbers for the tree(s) that hosted the crown of each liana were noted, where the host could be determined. We made herbarium collections for each morphospecies and the taxonomic identification of the multiple voucher specimens per morphotype is ongoing. For analyses of liana floristics and diversity we use data for individuals that have been assigned at least to morphospecies, while for calculations of overall density and basal area we use all individuals.

Soils in the plot are mostly well-drained ultisols, with the exception of wetlands. Seasonally torrential rainfall has moved much of the clay fraction from the upper horizons downward or washed it into the creeks, resulting in sandy clay with very low levels of nutrients because of low ion exchange capacity (Newbery et al. 1997).

The 18-ha liana census area has a fairly continuous canopy at 15–25 m, except in the wetlands, and is dominated by the tree *Oubanguia alata* Bak. f. (Lecythidaceae), which contributes 15.4% of the trees ≥20 cm dbh. Scattered larger trees emerge through this canopy, principally *Lecomptedoxa klaineana* (Pierre ex Engl.) Pierre ex Dubard (Sapotaceae), with 10.3% of the trees ≥50 cm dbh. The largest tree is an individual of *Desbordesia glaucescens* (Engl.) Tiegh. (Irvingiaceae) at 197 cm diameter. However, five of the ten largest trees are *Lecomtedoxa klaineana*, so this species is one of the major dominants in terms of biomass.

The liana data presented here are based on the second (2011–2012) census in Korup, which includes the most accurate identifications of the largest number of individuals. Comparisons with data from the first 2001–2002 liana census are made where appropriate. The tree data reported here are drawn from the 2008/2009 second tree census. Analyses are presented as summary statistics for the complete set of 18 hectares that were censused, with all calculations performed with software available in Microsoft Excel or in open source R packages.

RESULTS

In the 18-ha survey area we found a liana density of 547 ha^{-1} for individuals ≥1.0 cm dbh (Table 2.1, s.d. = 105.2, range 388–721 ha^{-1}). For individual lianas ≥5 and ≥10 cm diameter respectively, hectare-level densities are 76 and 8.5 individuals. The largest liana encountered had a diameter of 37 cm in 2012 (not yet identified to morphospecies). Liana densities at Korup are low compared to other African sites, especially for smaller lianas (see Table 2.1).

Between the two censuses (2001–2002 and 2011–2012) the liana density in the 18-ha plot declined by about 16%. Ewango et al. (2010b) found an even more dramatic decline in liana density (33%) in the Ituri forest (D.R. Congo) over three censuses in 13 years (see also Bongers & Ewango, Chapter 3 in this volume). Our results on liana community decline will be detailed in a separate publication.

Liana crowns are concentrated in the middle/lower canopy of the study area, in the "*Oubanguia alata* stratum," so called because of the dominance by this mid-canopy tree species. Almost 50% of the liana individuals are hosted by trees between 10 and 30 cm dbh, most of which are less than 30 m tall. About 90% of the liana stems are hosted by trees smaller than 40 cm dbh. Trees larger than 20 cm dbh consistently show a 50% occupancy rate by liana crowns. For trees 10–20 cm dbh, the crown occupancy rate is only 26%, for trees 5–10 cm dbh occupancy is 7%, and occupancy is less than 1% for trees 1–5 cm in dbh. However, the number

Table 2.1 Liana density and basal area (m^2 per hectare) by stem minimum diameters for Korup and two other central African forests. Korup values include means and standard deviation for 18 ha.

Liana density and basal area per ha	Diameter minimum (cm)			
	1	2	5	10
Korup density	517+/−240	312+/−160	91	10
Ituri density		677		13.5
Ebom density		408	113	10
Korup basal area	0.60+/−0.28	0.56+/−0.27		
Ituri basal area		0.71		
Ebom basal area		0.3 − 1.6		

Source: Ituri (Democratic Republic of the Congo): Ewango et al. (2010a); Ebom (Cameroon): Parren (2003)

of trees in each diameter class decreases inversely with the log of the tree size, so larger trees host far fewer liana individuals in total, even when the occupancy rate is similar.

Liana density, richness, and diversity at the family, genus, and species level are shown in Tables 2.2–2.5. A total of 31 families, 77 genera and 256 species were recorded. These numbers are probably underestimates, at least for the number of species, since the 10% of liana individuals that are as yet unidentified likely include additional taxa. At the 1-ha scale, fifteen 100×100 m plots supported an average of 104.9 liana species, with a standard deviation of 13.4. At the family level, the Loganiaceae has the greatest density of individuals, with 1736 lianas in 15 species, all in the single genus *Strychnos*. The Rubiaceae includes the highest diversity of both genera and species, with 1170 individuals in 51 species and 15 genera. At the genus level, *Strychnos* also includes the highest density, with 1736 lianas from 15 species. *Salacia* (Celastraceae) and *Dichapetalum* (Dichapetalaceae) are the most speciose genera, with 17 species each. Dominance in abundance of individuals at the species level is shared between *Raphiostylis beninesis* (Icacinaceae) with 669 lianas, and *Strychnos camptoneura*, with 622 lianas, each representing over 7% of the individuals censused. The next most abundant species, *Strychnos urceolata*, is represented by 281 individuals. At the family level, about 50% of the liana families are unrepresented or only very poorly represented in the tree flora. This percentage increases to 94% unrepresented in the tree flora at the genus level and of course 100% for species, showing that lianas add significant floristic diversity to the Korup forest canopy at all taxonomic levels.

We have characterized liana diversity and dominance using several indices and by rank order of abundance (Table 2.5, Fig. 2.2). We compared liana to tree diversity within Korup using three minimum diameter limits for the trees in the 18 ha of the liana census. When we include all trees and saplings ≥1 cm diameter, trees/saplings are about 13 times more abundant than lianas (119,027 vs. 9023 respectively). Including just trees ≥5 cm diameter results in trees being about 2.5 times more abundant than lianas (22281 vs. 9023 respectively). And including trees ≥9 cm diameter only, the trees and all lianas ≥1 cm diameter are about equal in abundance (9066, 9023, respectively). These ratios of tree-to-liana abundance suggest a lower abundance of lianas at Korup than censused at BCI in Panama where trees >5 cm are similar in number to lianas >1 cm (see values in Schnitzer et al. 2012; Schnitzer et al., Chapter 7 in this volume).

Species richness in the 18-ha plot is related, of course, to the number of individuals censused. Trees ≥1 cm include 409 species, more species than lianas with the same minimum diameter (268 species). At 5 cm diameter and larger, trees include roughly equal species richness (309 species) to all lianas ≥1 cm diameter. The 5-cm diameter tree to 1-cm diameter liana density equivalence was also suggested by Gerwing and Farias (2000), based on estimated similarity of crown sizes in Brazil. In Korup, trees ≥9 cm dbh include fewer species (244 species) than lianas ≥1 cm, although they include similar numbers of individuals (9066 vs. 9023, respectively). Lianas add diversity through the presence of many species, none of which is overwhelmingly dominant in Korup (Fig. 2.2). Although we have not estimated crown volumes for lianas and trees, our results suggest that the species richness of lianas in

Table 2.2 Summary of the Korup liana data by family for all 31 families. Families in italics are well represented among the canopy tree flora (defined here as trees ≥5 cm diameter). The 17 families in bold are absent – or almost absent – in the tree canopy.

Rank order	Family	Individuals	Species	Genera
1	**Loganiaceae**	1736	15	1
2	*Rubiaceae*	1170	51	15
3	*Fabaceae*	914	23	7
4	*Icacinaceae*	867	6	5
5	*Apocynaceae*	827	27	9
6	**Dichapetalaceae**	727	21	1
7	*Annonaceae*	479	16	6
8	**Connaraceae**	438	24	8
9	**Convolvulaceae**	350	4	1
10	**Celastraceae**	348	19	1
11	**Dioscoreaceae**	140	5	1
12	**Combretaceae**	130	10	1
13	**Linaceae**	130	4	1
14	**Dilleniaceae**	124	2	1
15	**Vitaceae**	98	5	1
16	*Euphorbiaceae*	69	2	2
17	**Aristolochiaceae**	46	3	1
18	**Malpighiaceae**	46	2	1
19	*Rutaceae*	25	2	1
20	**Menispermaceae**	23	2	2
21	**Verbenaceae**	22	1	1
22	*Gentianaceae*	17	1	1
23	*Melastomataceae*	15	1	1
24	**Piperaceae**	11	1	1
25	*Anacardiaceae*	9	1	1
26	**Curcubitaceae**	7	1	1
27	**Ancistrocladaceae**	5	1	1
28	*Malvaceae*	5	2	2
29	*Passifloraceae*	4	2	1
30	*Lauraceae*	3	1	1
31	*Polygalaceae*	2	1	1
	TOTALS	**8787**	**256**	**78**

the Korup 18-ha plot is equivalent to or exceeds that of the trees, when a species richness number is used that takes into account the size of the canopy of lianas versus trees, regardless of their supporting tissues.

Evenness is greater in lianas and small trees than in larger trees, because of the dominance of the tree *Oubanguia alata*. The effective diversity (the numbers equivalent of a diversity index), calculated as the exponent of the Shannon-Weiner index (see Table 2.5), is 87 for lianas and 70 for trees, in spite of the higher total species richness values of the trees. The use of effective diversity for comparison is strongly recommended when comparisons of different areas or life forms is made (Jost 2007). It is clear that, in Korup, the lianas make a major contribution to the species diversity of the forest canopy, even though their stem density is relatively low.

DISCUSSION

As in many tropical forests studied to date (Bongers & Ewango, Chapter 3 in this volume; Parthasarathy et al., Chapter 4 in this volume; Burnham & Romero-Saltos, Chapter 5 in this volume; Nogueira et al., Chapter 6 in this volume; Schnitzer et al., Chapter 7 in this volume;

Table 2.3 The 20 most abundant liana genera in the Korup plot, out of the 77 total. With the exception of *Strychnos* and possibly *Millettia*, these genera are not represented among the tree species in the plot, nor in the immediate vicinity where tree observations are easily made during field surveys.

Rank order	Genus	Individuals	Species
1	*Strychnos* (Loganiaceae)	1736	15
2	*Rhaphiostylis* (Icacinaceae)	695	1
3	*Landolphia* (Apocynaceae)	647	14
4	*Dichapetalum* (Dichapetalaceae)	548	17
5	*Millettia* (Fabaceae)	403	5
6	*Neuropeltis* (Convolvulaceae)	350	4
7	*Salacia* (Celastraceae)	310	17
8	*Leptactina* (Rubiaceae)	277	2
9	*Agelaea* (Connaraceae)	176	7
10	*Leptoderris* (Fabaceae)	174	4
11	*Monanthotaxis* (Annonaceae)	174	3
12	*Atractogyne* (Rubiaceae)	171	2
13	*Friesodielsia* (Annonaceae)	171	3
14	*Dioscorea* (Dioscoreaceae)	140	1
15	*Iodes* (Icacinaceae)	135	1
16	*Combretum* (Combretaceae)	130	10
17	*Hugonia* (Linaceae)	130	4
18	*Sherbournia* (Rubiaceae)	129	6
19	*Tetracera* (Dilleniaceae)	124	2
20	*Dioclea* (Fabaceae)	119	2

Ibarra-Manríquez et al., Chapter 8 in this volume), the liana community adds substantial species richness to the Korup forest. We documented the presence of 268 species of lianas in addition to the tree flora of 409 species known from a forested area of only 18 hectares. The additional diversity added by lianas is contributed to a large degree by species from families scarcely, if at all, represented among the tree community. Phylogenetic diversity, then, is increased by the liana community through the addition of families such as Aristolochiaceae, Convolvulaceae, Celastraceae, Dichapetalaceae, Dioscoreaceae, Loganiaceae, and Menispermaceae, which are absent or almost absent from the tree flora (Celastraceae and Loganiaceae are both also present as scarce small trees).

Dominant species of lianas in neotropical forests have been reported as comprising 7–17% of the individuals, similar to the 14.7% combined value for the two dominant liana species in Korup (Perez-Salicrup et al. 2001; Burnham 2002; Mascaro et al. 2004). The two dominant liana species in Korup are animal dispersed, adding support to the supposition of an intact faunal community in the area (Ndenecho 2011). In contrast with the wind-dispersed dominant lianas in Yasuní, Ecuador (e.g., *Machaerium cuspidatum*, *Combretum laxum*), the dominants in Korup are largely animal dispersed. For example, Korup lianas bear seeds embedded in a tangy, fleshy pulp and then enclosed in a durable fruit (*Strychnos spp.*, *Landolphia*, *Atractogyne*), or they bear fruits with a fleshy external appendage (*Raphiostylis*), or each carpel is a succulent apocarp (*Friesodielsia*). Indeed, the nine most abundant species in Korup appear to be animal dispersed, a phenomenon also reported from the Ituri forest in DR Congo (Ewango 2010a), whereas six of the top ten dominant lianas in Ecuador are wind dispersed (Burnham & Romero-Saltos, Chapter 5 in this volume). Among the largest diameter trees in Korup (upper canopy or emergents), there are only two wind-dispersed species among the 20 most abundant species, the remainder being animal, ballistic, or unknown, but not wind dispersed. Among mid-canopy Korup trees, none of the 20 most abundant species are wind dispersed (data from Thomas et al. 2003). The predominance of animal dispersal among trees and lianas in Korup may be at odds with a general impression of lianas, as more wind dispersed than trees in tropical forests (Gentry 1983, 1991). The high annual precipitation in Korup may contribute to favoring this difference, in spite of a three-month dry season. Similar patterns in Nigerian forests may have been found as early as 1957 by Keay, who found old secondary growth forest to have 56% wind-dispersed trees and 48% wind-dispersed lianas (see van der Pijl 1982). Reports from two tropical rainforest sites in DR Congo (Ewango 2010a; Beaune et al. 2013) show 74% and 79.2% of the liana species (Ituri and Salonga, respectively) are animal dispersed. Tree dispersal by

Table 2.4 The 20 most abundant liana species in the Korup plot, of the 256 species total, along with their inferred dispersal agent based on fruit morphology.

Rank order	Individuals	Species	Inferred dispersal agent
1	669	*Rhaphiostylis beninensis* (Hook. f ex Planch.) Planch. ex Benth.	*Animal*
2	622	*Strychnos camptoneura* Gilg & Busse	*Animal*
3	281	*Strychnos urceolata* Leeuwenb.	*Animal*
4	276	*Leptactina latifolia* K.Schum.	*Animal*
5	256	*Strychnos johnsonii* Hutch. & M.D.Moss	*Animal*
6	231	*Dichapetalum affine* (Planch. Ex Benth.) Breteler	*Animal*
7	229	*Strychnos tricalisoides* Hutch. & M.D.Moss	*Animal*
8	180	*Landolphia* sp. "LANDPR"	*Animal*
9	167	*Atractogyne gabonii Pierre*	*Animal*
10	152	*Fabaceae* sp. "MILL"	*Unknown*
11	148	*Friesodielsia enghiana* (Diels) Verdc.	*Animal*
12	145	*Neuropeltis velutina* Hallier f.	*Wind*
13	141	*Millettia* sp. "MILLLE"	*Ballistic*
14	139	*Neuropeltis* sp. "NEUR"	*Wind*
15	135	*Iodes africana* Welw. Ex Oliv.	*Animal*
16	131	*Salacia longipes* (Oliv.) N.Halle	*Animal*
17	129	*Dioscorea smilacifolia* (De Wild.)	*Wind*
18	127	*Landolphia dulcis* (Sabine ex G.Don) Pichon	*Animal*
19	120	*Leptoderris ledermannii* Harms	*Ballistic*
20	118	*Dioclea* sp. "DIOC"	*Ballistic*

Table 2.5 Diversity indices for lianas and three size classes of trees (minimum dbh in cm) from 18 hectares of forest in Korup National Park. Comparison to values from Ituri (DR Congo) provided, when known, in column 6 (Ewango 2010a). Ituri data comes from two non-contiguous 10-ha plots (500 m apart). The Shannon-Wiener exponential eH' is the effective species number calculated considering frequency values (see Jost 2006, 2007).

	Trees ≥1	Trees ≥5	Trees ≥9	Lianas ≥1	Ituri lianas ≥2
Area in hectares	18	18	18	18	20
Number of individuals	119027	22281	9066	9023	15008
Minimum diameter cm	1	5	9	1	2
Number of species	409	309	244	268	195
Shannon evenness index	0.71	0.67	0.65	0.8	
Shannon-Wiener index H'	4.26	3.86	3.59	4.47	3.1
Shannon-Wiener exponential eH'	70.7	47.38	36.27	87.36	
Inverse Simpson's index D	30.23	17.14	11.06	48.38	11.4
Fisher's alpha	53	50.77	46.17	51.9	17.9

animals in the Salonga National Park, DR Congo is even higher (84%, Beaune et al. 2013).

Lianas add substantial diversity to the Korup forest, adding a community of dominant taxa that are relatively homogeneously distributed, contrasting with the tree community in which one species is strongly dominant. The effect of this difference is to increase species richness but also to increase true diversity, as reflected in the frequency with which a new species is found in the forest. How does this affect the fauna

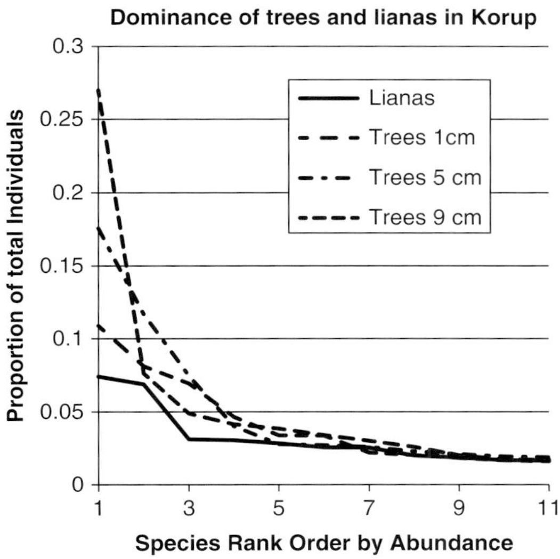

Fig. 2.2 Proportional abundance by species from the data used to compute the diversity indices in Table 2.5 (11 most abundant species). The highest level of species dominance is found among the largest diameter trees and lowest dominance is in the liana flora.

relying on resources in the area? The patterns of liana dispersal contrast with those in neotropical forests, but are in line with other African sites, suggesting that dispersal may be phylogenetically controlled to a larger degree than previously recognized. Liana management worldwide has become a prominent issue over recent years, and the clear contribution of lianas to intact forests suggests that the specific differences among lianas may help guide management. Species-specific attributes and interactions should be a high priority among liana biologists in the future.

ACKNOWLEDGMENTS

Funding for our fieldwork from the Center for Tropical Forest Science, Smithsonian Institution Global Earth Observatory is gratefully acknowledged. Permission to conduct the research in the Korup National Park was given by the Ministry for Forestry and Wildlife and the Ministry of Scientific Research and Innovation. We also thank project botanist Peter Mambo Ekole and numerous field staff for their tireless efforts to document the liana diversity of Korup.

REFERENCES

Beaune, D., Bretagnolle, F., Bollache, L. et al. (2013) Seed dispersal strategies and the threat of defaunation in a Congo forest. *Biodiversity and Conservation*, **22**, 225–238.

Burnham, R.J. (2002) Dominance, diversity and distribution of lianas in Yasuní, Ecuador: who is on top? *Journal of Tropical Ecology*, **18**, 845–864.

DeCáceres, M., Legendre, P., Valencia, R. et al. (2012) The variation of tree beta diversity across a global network of forest plots. *Global Ecology and Biogeography*, **21**, 1191–1202.

Condit, R. (1998) *Tropical Forest Census Plots*, Springer Verlag, New York.

Condit, R., Ashton, P., Baslev, H. et al. (2005) Tropical tree alpha-diversity: Results from a worldwide network of large plots. *Biologiske Skrifter*, **55**, 565–582.

Chuyong, G. B., Condit, R., Kenfack, D., et al. (2004) Korup forest dynamics plot, Cameroon. In *Forest Diversity and Dynamism: Findings from a Network of Large-scale Tropical Forest Plots* (eds. U E.C. Losos & E.G., Leigh, Jr.). University of Chicago Press, Chicago.

Egbe, E.A., Chuyong, G.B., Fonge, B.A. & Namuene, K.S. (2012) Forest disturbance and natural regeneration in an African rainforest at Korup National Park, Cameroon. *International Journal of Biodiversity and Conservation*, **4**, 377–384.

Ewango, C.E.N., Bongers, F., Poorter, L., et al. (2010a) Structure and composition of the liana assemblage of a mixed rain forest in the Congo Basin. In *The Liana Assemblage of a Congolian Rainforest: Diversity, Structure and Dynamics* (C.E.N. Ewango). PhD thesis, University of Wageningen.

Ewango, C.E.N., Poorter, L., Sosef, M.S.M., et al. (2010b) Thirteen years of dynamics of the liana assemblage in a Congo Basin rain forest. In *The Liana Assemblage of a Congolian Rainforest: Diversity, Structure and Dynamics*. (C.E.N. Ewango). PhD thesis, University of Wageningen.

Gartlan, J.S., Newbery, D.M.C., Thomas, D.W. & Waterman, P.G. (1986) The influence of topography and soil phosphorus on the vegetation of Korup Forest Reserve, Cameroon. *Vegetatio*, **65**, 131–148.

Gartlan, J.S. & Struhsaker, T.T. (1972) Polyspecific associations and niche separation of rain-forest anthropoids in Cameroon, West Africa. *Journal of Zoology (London)*, **168**, 221–266.

Gentry, A.H. (1983) Disperal ecology and diversity in Neotropical forest communities. *Sonderbände des naturwissenschaftlichen Vereins in Hamburg*, **7**, 303–314.

Gentry, A.H. (1991) Breeding and dispersal systems of lianas, in *The Biology of Vines* (eds F.E. Putz & H. Mooney), Cambridge University Press, Cambridge, pp. 393–423.

Gerwing, J.J. & Farias, D.L. (2000) Integrating liana abundance and forest stature into an estimate of total aboveground biomass for an eastern Amazonian forest. *Journal of Ecology*, **16**, 327–356.

Gerwing, J.J., Schnitzer, S.A., Burnham, R.J. et al. (2006) Censusing lianas in tropical forests. *Biotropica*, **38**, 256–261.

Hawthorne, W.D. & Jongkind, C. (2006) *Guide to the Woody Plants of Western Africa*, Royal Botanic Gardens, Kew, London.

Jost, L. (2006) Entropy and diversity. *Oikos*, **113**, 363–375.

Jost, L. (2007) Partitioning diversity into independent alpha and beta components. *Ecology*, **88**, 2427–2439.

Kenfack, D., Thomas, D.W., Chuyong, G. & Condit, R. (2007) Rarity and abundance in a diverse African forest. *Biodiversity and Conservation*, **16**, 2045–2074.

Mascaro, J., Schnitzer, S.A. & Carson, W.P. (2004) Liana diversity, abundance and mortality in a tropical wet forest in Costa Rica. *Forest Ecology and Management*, **190**, 3–14.

Ndenecho, E.N. (2011) *Local Livelihoods and Protected Area Management: Biodiversity Conservation Problems in Cameroon*, African Books Collective, Langaa Research & Publishing Common Initiative Group, Bamenda, Cameroon.

Newbery, D.M.C., Alexander, I.J. & Rother, J.A. (1997) Phosphorus dynamics in a lowland African rain forest: the influence of ecto-mycorrhizal trees. *Ecological Monographs*, **67**, 367–409.

Newbery, D.M.C. & Gartlan, J.S. (1996) A structural analysis of rain forest at Korup and Douala-Edea, Cameroon. *Proceedings of the Royal Society of Edinburgh*, **104B**, 177–224.

Parren, M.P.E. (2003) Lianas and logging in West Africa. PhD dissertation. Wageningen University, Tropenbos-Cameroon Series 6. Tropenbos International, Wageningen, the Netherlands.

Perez-Salicrup, D.R., Sork, V.L. & Putz, F.E. (2001) Lianas and trees in a liana forest of Amazonian Bolivia. *Biotropica*, **33**, 34–47.

Schnitzer, S.A., Mangan, S.A., Dalling, J.W. et al. (2012) Liana abundance, diversity, and distribution on Barro Colorado Island, Panama. *PLoS One*, **7**(12), e52114. doi: 10.1371/journal.pone.0052114

Thomas, D.W., Kenfack, D., Chuyong, G.B. et al. (2003) *Tree Species of Southwestern Cameroon: Tree Distribution Maps, Diameter Tables, and Species Documentation of the 50-Hectare Korup Forest Dynamics Plot*, Center for Tropical Forest Science of the Smithsonian Tropical Research Institute and Bioresources Development and Conservation Programme-Cameroon, Washington, DC.

Usongo, L.I. & Amubode, F.O. (2000) Nutritional ecology of Preuss' red colobus monkey (*Colobus badius* preussi Rahm 1970) in Korup National Park, Cameroon. *African Journal of Ecology*, **39**, 121–125.

Van der Pijl, L. (1982) *Principles of Dispersal in Higher Plants*, Springer-Verlag, Berlin.

White, F. (1979) The Guineo-Congolian Region and its relationships to other phytochoria. *Bullétin du Jardin Botanique National de Belgique*, **49**, 11–55.

White, F. (1983) *The Vegetation of Africa*, UNESCO, Paris.

Chapter 3

DYNAMICS OF LIANAS IN DR CONGO

Frans Bongers[1] *and Corneille E.N. Ewango*[2]

[1]Wageningen University and Research Centre, Wageningen, The Netherlands
[2]Wildlife Conservation Society (WCS)-DR Congo and Centre de Formation et de Recherche en Conservation Forestière (CEFRECOF), Reserve de Faune à Okapis, Ituri-Epulu-Aru Landscape, DR Congo

Ecology of Lianas, First Edition. Edited by Stefan A. Schnitzer, Frans Bongers, Robyn J. Burnham, and Francis E. Putz.

Companion Website: www.wiley.com/go/schnitzer/lianas

OVERVIEW

The African continent is vastly underrepresented in the number and geographic range of detailed studies of lianas. Most existing studies are based on herbarium specimens, and there are few liana inventories based on plot sampling. Previous studies have been carried out largely in West Africa, specifically the Upper Guinea region, and few inventories include Central and East Africa. Here we present the first results of the longest-running, large-plot (20 ha) study on lianas, where all lianas ≥2 cm diameter were monitored in DR Congo for 13 years (1994–2007). We found a total of 197 liana species, 81genera, and 34 families. Liana density averaged 402.2 stems per ha with basal area averaging 0.338 m^2 per ha. Fisher's alpha diversity averaged 21.7 per ha. Richness and diversity values changed little over 13 years, but stem density and basal area decreased 33% over this period. This decrease resulted from high mortality (31.4 individuals per year and per ha) and low recruitment (9.6 individuals). At the species level, 28 of the 78 analysed species decreased in abundance, 9 species increased, and 41 species remained constant over the 13-year period. The net decrease in liana abundance in this forest contrasts with results from the neotropics, where lianas have increased dramatically. However, the global changes proposed to be driving liana increase may be overwhelmed by local factors. We suggest that the Ituri changes may be governed by recovery from past disturbance. Additionally, we describe the dynamics of one widespread liana species (*Sericostachys scandens*) that is invasive in DR Congo and neighboring countries. This species is currently increasing. Our study provides one of the few large-scale, long-term datasets from Africa. To understand the diversity and dynamics of African liana communities, we urgently require more research focused on the region, including detailed descriptions of the species present and monitoring their changes in density and biomass over time.

INTRODUCTION

A century and a half ago, Charles Darwin wrote a book on climbers because he was intrigued by this spectacular and unique growth form (Darwin 1865). Like other plants, lianas harvest light energy and, as with most species, the more light, the better. Unlike many woody plant species, however, lianas do not usually build strong support structures; instead, they use the stems of other woody plants, mostly trees, to grow toward the light. Although some liana species are abundant in temperate regions (e.g., *Vitis spp.*), far more liana species are found in tropical forests, where this growth-form is abundant and plays an important ecological role (Schnitzer & Bongers 2002). Of the studies conducted in tropical forests, the vast majority has been carried out in the neotropics (e.g., Schnitzer et al., Chapter 7 in this volume; DeWalt et al., Chapter 11 in this volume).

In African forests, most studies of lianas have concentrated on West Africa (reviewed in Bongers et al. 2005). For example, recent studies have been published on lianas in Ghana (Swaine & Grace 2007; Toledo-Aceves & Swaine 2007a, 2007b, 2008a, 2008b; Addo-Fordjour & Anning 2008; Addo-Fordjour et al. 2009a, 2009b, 2009c, 2013), the Côte d'Ivoire (Schnitzer et al. 2005), and Cameroon (Tchouto et al. 2006). Liana studies from Central and Eastern Africa, however, are rare. The few recent studies that we found on Central and Eastern Africa liana composition include: montane forest in Ethiopia (Senbeta et al. 2005), liana abundance in three lowland forests in Uganda (Eilu 2000), and liana response to logging gaps in Uganda (Babweteera et al. 2000). Additional research has evaluated the importance of lianas for gorillas and chimpanzees (Liengola 2008; Moscovice et al. 2010), or their importance as plants used by humans for food and binding materials (Eilu & Bukenya-Ziraba 2004; McGeoch et al. 2008; Muhwezi et al. 2009). Even Gentry (Gentry 1991), in his worldwide liana inventory, included only eight mainland African sites: three in Cameroon, one in Central African Republic, two in Makokou, Gabon, one in Nigeria, and two in Tanzania. Thus, data on lianas from Africa, particularly Central and Eastern Africa, are severely lacking.

This chapter focuses on the lianas from DR Congo where few recent studies have been conducted. In the 1950 and 1960s, several vegetation studies included lianas. For example, in seasonally flooded, hydromorphic soil forests along rivers in Tshuapa, the central DR Congo, Evrard (1968) reported 226 climber species, 181 of which were twiners, 20 tendrillers, 19 with hooks, 6 with adhesive roots, and 14 stranglers. Lianas comprised 34.1% of all woody species in these forests. He also reviewed the earlier literature for DR Congo and reported that 28.6% of species found in a Yangambi forest were climbers, 10.0% were climbers in a forest at the transition between the high forest and the savannah (Kaniama), 8.7% were climbers in the

Ruwenzori transition forest, 9.1% were climbers in the Ruwenzori montane forest, and 7.4% were climbers in the dry savannah of Bugesera (now Rwanda; Evrard 1968). Most of these studies were focused on trees, and were based on species lists.

Studies on liana dynamics (e.g., recruitment, growth, and mortality) in African forests are also scarce and we know only three published works. In Gabon, Caballé and Martin (2001) evaluated changes in liana abundance in 1.6 ha and found that richness and density of lianas decreased over a 13-year period. In Cameroon, Thomas and co-workers (Chapter 2 in this volume) monitored liana dynamics in an 18-ha plot in Korup National Park, Cameroon, and their first results show an extremely high liana diversity (see also DeWalt et al., Chapter 11 in this volume). In the eastern DR Congo forest of Ituri, 40 ha of forest are being monitored and trees and lianas are being marked, identified and their diameter measured (Makana et al. 2004a, 2004b, 2011), but the dynamics of the lianas there have not been reported yet. The study of liana dynamics has become particularly important because neotropical forests have shown drastic increases in abundance (Phillips et al. 2002; Wright et al. 2004; Schnitzer & Bongers 2011; Schnitzer, Chapter 30 in this volume), which may have an increasing role in these forests at the expense of trees (Schnitzer & Bongers 2002; Schnitzer et al. 2011). Additional studies on the dynamics of African lianas are urgently needed to evaluate whether lianas are increasing in African forests as well (Schnitzer & Bongers 2011).

This chapter reports on two developments in liana dynamics in the Congo Basin. First, we present results from the liana dynamics plots of Ituri, in the eastern DR Congo. The Ituri plots, which are the largest of the Central/Eastern African liana communities and their dynamics, were first sampled in the early 1990s with the establishment of four 10-ha plots – two in mixed forest (Edoro) and two in forest dominated by *Gilbertiodendron dewevrei* (Lenda) (Makana et al. 2004a, 2004b). In these plots, all trees ≥1 cm diameter at breast height (dbh) and all lianas ≥2 cm dbh were tagged in 1994, identified, and their diameter recorded and mapped. All individuals were measured again in 2000 and 2007, and growth, recruitment, and mortality were evaluated. Here, we report community-wide dynamics of lianas in the mixed forest plots. Second, we address the dynamics of the invading climber species *Sericostachys scandens* (Amaranthaceae), which is now prevalent in large areas of the Congo Basin. In recent years, this species and its dynamics have received special attention as it was seen as an invasive and an important food species for gorillas, not only in DR Congo (Liengola 2008; Masumbuko et al. 2012b) but also in Uganda and Rwanda (Scholte et al. 2010).

LIANAS IN THE ITURI MIXED FOREST

We monitored lianas in the Ituri lowland moist forest (Okapi Faunal Reserve, 1°25'N, 28°35'E) in the eastern DR Congo (Fig. 3.1). The forest is located at 800 m a.s.l. and receives an average of 1785 mm of rainfall per year (data from Epulu weather station, 1986–2007; Ewango 2010). Rainfall is distributed fairly evenly throughout the year and only the period December through February receives <100 mm of rain. Soils are oxisols derived from granitic pre-Cambrian shield rock, have a loamy sandy to sandy-clay texture, are acidic (pH ~ 4), and are low in available phosphorus and nitrogen (Makana et al. 2004a). Trees and lianas have been inventoried in two 10-ha plots (200 x 500) of mixed canopy semi-evergreen forest (mixed forest) and two 10-ha plots of mono-dominant Mbau (*Gilbertiodendron dewevrei*, Leguminosae) evergreen forest since 1994 (Makana et al. 2004a), as part of the worldwide tropical forest network of the Center for Tropical Forest Science (Condit 1998; Losos & Leigh 2004). In this forest *Cynometra alexandri* C.H. Wright and *Julbernardia seretii* (De Wild.) Troupin (both Leguminosae) are the most abundant tree species in the canopy and *Scaphopetalum dewevrei* (Sterculiaceae) is the most abundant tree species in the understory.

Here we present data for lianas in the two 10-ha plots located in the mixed Edoro forest (cf. Ewango 2010). All liana stems (including rooted clonal stems) were identified, tagged, and measured (diameter), following a standard protocol (Gerwing et al. 2006), and inventories were repeated in 2000 and 2007. We present liana results for the inventory of 2007 and for changes since 1994. The Edoro plots are separated by 500 m and all results are given either for the combined 20 ha or as 1 hectare averages.

In the two plots combined, we found 197 species of lianas in 20 ha (2007 census, Table 3.1), and 29 unique morphospecies that were identified only to genus or to family. Of one morphospecies not even the family is known. Species belonged to 82 genera and 34 families. On average, each hectare harboured 66.7 species (sd 9.03). Fisher's α diversity was, on average, 21.70 (sd 3.72) and Shannon diversity was 3.31 (sd

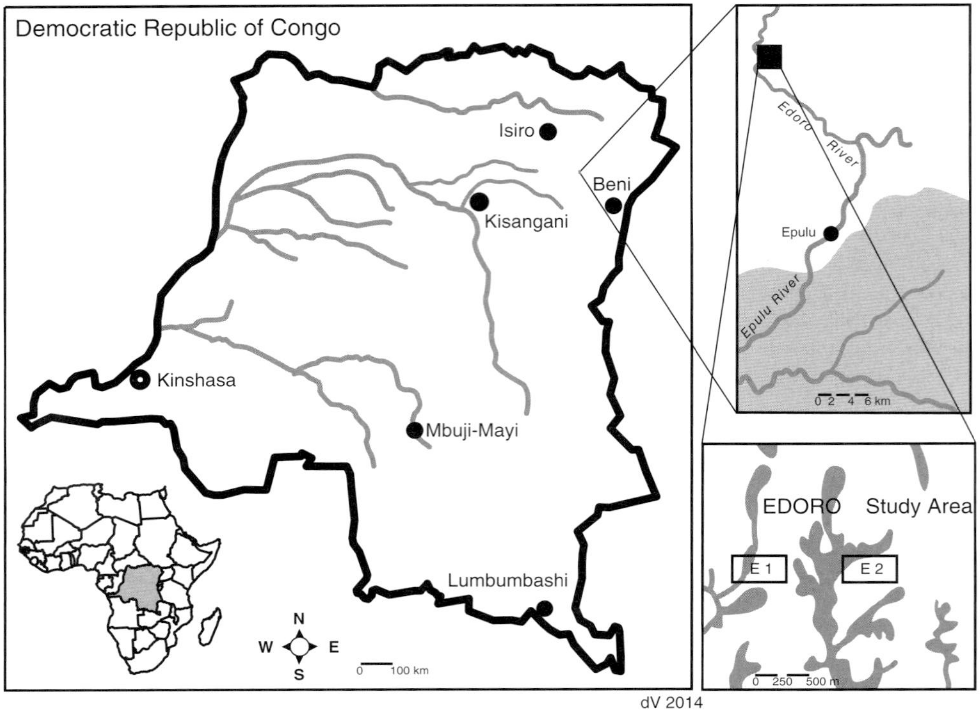

Fig. 3.1 Location of the two Edoro mixed forest dynamics plots in the Ituri Forest, eastern DR Congo. The shaded areas in the region inset (upper right) indicate the area with monodominant forest, while the white area shows the mixed forest. The shaded areas in and around the plots in the lower right inset indicate swampy areas.

0.18). In 2007, 1 hectare harboured 402.2 (sd 90.8) liana stems ≥2 cm dbh, 88.5 (sd 28.1) stems ≥5 cm, and 8.3 (sd 3.5) stems ≥10 cm (Table 3.1). Basal area values for these diameter limits were 0.338 m^2 per ha (sd 0.077 m^2), 0.297 m^2 per ha (sd 0.101), and 0.100 m^2 per ha (sd 0.047).

In 2007, the most abundant species was *Rourea thomsonii* (Baker) Jongkind, Connaraceae with 816 individuals over the 20 ha. The species with the most basal area was *Combretum marginatum* Engl.and Diels, Combretaceae with 1.097 m^2 in 20 ha (Table 3.2). The top five species included 36.7% of all individuals and 32.7 % of the total basal area. The most important genera in abundance and basal area are *Agelaeae*, *Dichapetalum*, and *Combretum*, while Connaraceae, Dichapetalaceae, and Combretaceae are the dominants at the family level, with Connaraceae accounting for almost 30% of both stem abundance and biomass.

The number of families, genera, and species in the 20 ha remained constant over the 13-year period (Table 3.1). Hectare-based averages in species richness and diversity values increased slightly between 1994 and 2007: Shannon's Index increased from 2.94 ± 0.24 to 3.31 ± 0.18; Fisher's alpha increased from 16.53 ± 4.30 to 21.70 ± 3.72 (Table 3.1). Liana abundance decreased over this time period, with an increase in larger lianas (≥5 cm or ≥10 cm dbh) but a strong decrease in the number of smaller individuals (<5 cm). These changes indicate both low recruitment and high mortality of smaller individuals during the study period. During the 13-year monitoring period, only 9.6 ± 3.71 individuals were recruited per year per ha, while 31.4 ± 10.84 individuals died per year per ha, leading to a net decrease of all liana individuals in the plots. Total basal area decreased over the monitoring period, particularly the basal area of lianas <5 cm dbh,

Table 3.1 Liana density, basal area, and species richness per hectare for all stems from 20 ha of Ituri mixed forest, DR Congo, censused in 2007 and in 1994. Density and basal area data are presented as the mean per hectare for three commonly used size classes. Richness and diversity values are based only on unique (morpho)species.

	2007	1994
Families	34	34
Genera	81	84
Species richness total	197	194
per ha	66.7 ± 9.03	61.0 ± 13.38
Alpha diversity per ha >2 cm	21.70 ± 3.72	16.53 ± 4.30
Shannon per ha >2 cm	3.31 ± 0.18	2.94 ± 0.24
Stem density (#/ha) >2 cm	402.2 ± 90.85	669.3 ± 197.14
>5 cm	88.45 ± 28.12	74.1 ± 21.56
>10 cm	8.30 ± 3.45	6.95 ± 3.38
Basal area (m^2/ha) >2 cm	0.3381 ± 0.0774	0.4859 ± 0.1314
>5 cm	0.2966 ± 0.1010	0.2427 ± 0.0718
>10 cm	0.1001 ± 0.0474	0.0806 ± 0.0401

while the basal area of larger individuals increased (Table 3.1).

The liana community changed from a relatively mono-dominated community in 1994 with *Manniophyton fulvum* Mull.Arg. (Euphorbiaceae) comprising 23% of all liana stems ≥2 cm dbh, to a more even community in 2007, with the most abundant species *Rourea thomsonii* representing 8.2% of all stems. Despite the dramatic change in the dominant species over this 13-year period (from rank 1 to rank 24, from 3299 to 89 individuals over 20 ha), dominance-diversity curves showed little change (Fig. 3.2a). Aside from *Manniophyton*, species that were abundant in 1994 maintained their rank in 2007 (both abundance and basal area, Table 3.2). The loss of *Manniophyton*, however, reduced the contribution of the top five species in 1994 from 44.1% to 35.0% (abundance) and increased basal area from 32.8% to 36.2 % in 2007.

At the species level, many observed changes suggest a dynamic community over the 13-year period (Fig. 3.2b). Twelve species were lost and 18 new species appeared (species on the x-axis and the y-axis, respectively, Fig. 3.2b). Many species deviated from the x–y line, indicating changes in rank of these species. For 78 species that had ≥10 individuals in 1994, we evaluated the changes in abundance between 1994 and 2007 using plot-based randomization tests. A total of 37 species changed significantly, with 28 decreasing and 9 increasing. A total of 41 species remained constant over the census period. The most dramatic increases were in *Urera trinervis* (Hochst.) Friis and Immelman, Urticaceae (2007/1994 ratio of 2.5), *Salacia cerasifera* Welw. Ex Oliv., Celastraceae (2.21) and *Campylostemon bequartii* DeWild., Celastraceae (1.61). The most dramatic decreasing species were *Manniophyton fulvum* Mull. Arg., Euphorbiaceae (0.03), *Piper guineense* Schuman. and Thonn., Piperaceae (0.09) and *Adenia cynanchifolia* (Benth.) Harms, Passifloraceae (0.24).

Species composition changed over the 13 years, shown by NMDS ordination of species composition of each of the 20 one-ha plots for both 1994 and 2007 (Fig. 3.3). Although most plots changed over time, the shifts were not dramatic. The spatial variation in the composition of lianas, however, is larger (both in 2007 and in 1994) and the two 10-ha blocks are distinguishable in Fig. 3.3.

The ranks of the dominant genera and families changed little over the 13-year period (Table 3.2). *Agelaea* increased in abundance and basal area from number two (after *Manniophyton*) in 1994 to number one in 2007, and Connaraceae remained number one in abundance and basal area in over the 13-year period. The other species within the dominant five showed only minor changes.

Dynamics of an Invading Liana in the Congo Basin

The widely distributed liana *Sericostachys scandens* (Amaranthaceae) receives ample attention in the Congo Basin because of its invasive nature (Scholte et al. 2010; Masumbuko et al. 2012a, 2012b) and because it is widely used by humans (Jansen 2004). For example, in DR Congo the leaves of *S. scandens* collected from the wild are eaten as a vegetable. The leaves are applied as a poultice on wounds and from the bark a medicine for venereal diseases is prepared. In Rwanda and Uganda *S. scandens* is a key species for honey production from wild flowers in the forest. In Kenya, initiation ceremonies are related to the occurrence of flowering, which there is thought to happen only once every 7–8 years.

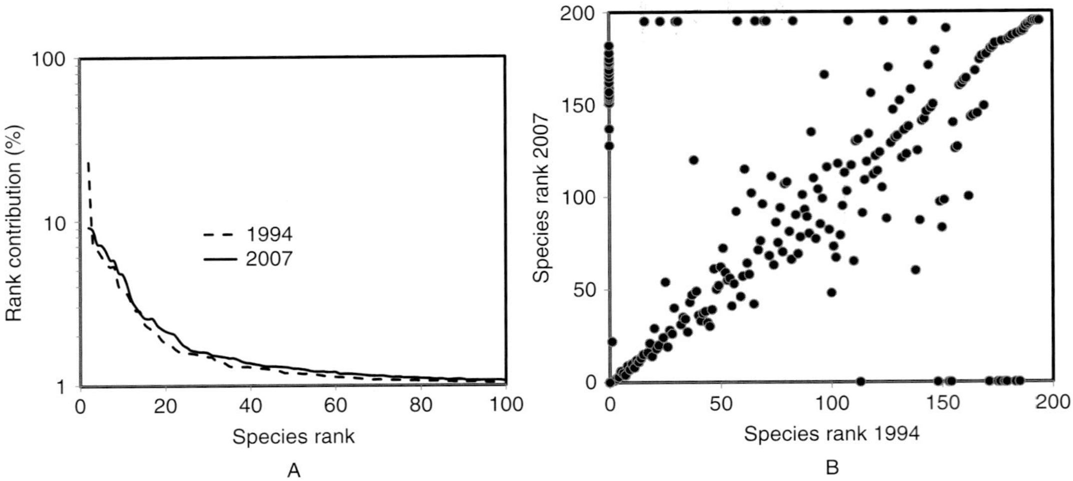

Fig. 3.2 Liana species ranks in 20 ha of Ituri mixed forest, DR Congo. (A) Rank dominance curves for 2007 and 1994. The graph is truncated at species rank 100. (B) Rank dominance changes between 1994 and 2007. In both figures only species with unique names are included, no combined genera or families. A total of 197 species were recorded in 2007 and 194 species in 1994.

Table 3.2 Dominant species, genera, and families of lianas in 20 ha of Ituri mixed forest, DR Congo, censused in 2007 and in 1994. The dominant five taxa from 2007 and 1994 censuses are shown. Only fully determined species, genera, families are ranked.

	Stem abundance (in 20 ha)		**Basal area (in 20 ha)**		
	2007	**1994**		**2007**	**1994**
Totals	9982 ind in 20 ha	15008 ind in 20 ha		14.67 m^2 in 20 ha	16.18 m^2 in 20 ha
Species					
1 Rourea thomsonii	816 (8.2%)	922 (6.4%) 2nd	*1 Combretum marginatum*	7.5%	5.0% 6th
2 Dichapetalum staudtii	779 (7.8)	854 (5.7) 3th	*2 Agelaea pentagyna*	7.1	6.0 2nd
3 Agelaea pentagyna	625 (6.3)	767 (5.1) 4th	*3 Dichapetalum staudtii*	6.5	5.4 5th
4 Agelaea paradoxa	612 (6.1)	636 (4.2) 7th	*4 Rourea thomsonii*	6.4	5.6 4th
5 Dichapetalum heudelotii	558 (5.6)	638 (4.3) 6th	*5 Milletia psylopetala*	5.3	5.6 3rd
24 Manniophyton fulvum	89 (0.9)	3299 (22.0) 1st	*42 Manniophyton fulvum*	0.4	13.6 1st
6 Combretum racemosum	480 (4.8)	685 (4.6)5th			
Genera					
1 Agelaea	17.5%	14.1% 2nd	1 Agelaea	17.8%	15.3% 1st
2 Dichapetalum	16.4	12.1 3th	2 Combretum	13.9	11.2 4th
3 Combretum	12.8	10.0 4th	3 Dichapetalum	13.1	11.6 3rd
4 Rourea	8.8	6.5 5th	4 Milletia	9.0	8.5 5th
5 Salacia	6.2	3.26 7th	5 Rourea	7.3	6.1 6th
16 Manniophyton	0.9	22.0 1st	36 Manniophyton	0.4	13.6 2nd
Families					
1 Connaraceae	29.3%	23.2% 1st	1 Connaraceae	27.2%	23.5% 1st
2 Dichapetalaceae	16.4	12.1 3rd	2 Combretaceae	13.9	11.2 4th
3 Combretaceae	12.9	10.0 4th	3 Dichapetalaceae	13.1	11.6 3rd
4 Celastraceae	9.7	6.7 5th	4 Fabaceae	12.3	11.1 5th
5 Apocynaceae	6.7	4.4 7th	5 Celastraceae	10.4	9.0 6th
12 Euphorbiabeae	1.4	22.8 2nd	13 Euphorbiaceae	1.1	14.6 2nd

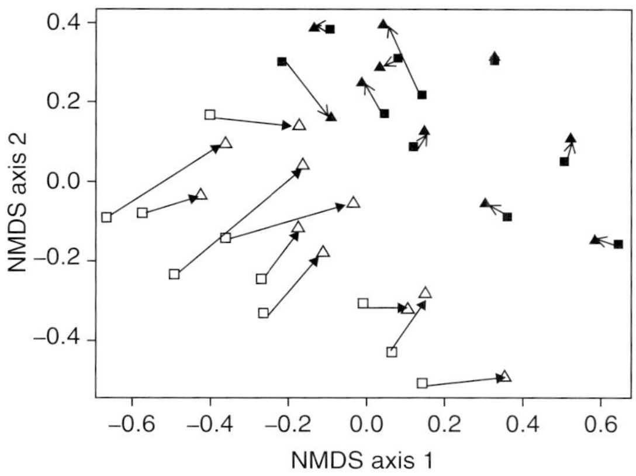

Fig. 3.3 NMDS (Nonmetric Multi-Dimensional Scaling) graph of liana species composition for each 1-ha plots censused in 1994 and 2007 from Ituri mixed forest, DR Congo. The square symbols indicate plots in 1994, the triangles are plots in 2007, open symbols are plots from Edoro1, closed symbols are plots from Edoro 2. Edoro 1 and Edoro 2 are each 10-ha plots, 500 m apart. The NMDS uses stem abundances (square root transformation) of liana species per one-ha plot (only fully determined species, no combined genera or families, 197 species in 2007, 194 species in 1994).

Sericostachys scandens is a wind-dispersed monocarpic semi-woody climber, growing 20–30 m in length and living for 10–20 years, after which massive flowering and fruiting occur and the plants die. It grows rapidly in high light conditions and quickly suppresses other vegetation. This species has been shown to colonize recent forest gaps, overtopping the adjacent canopy, killing trees, and creating large mono-dominant gaps in Kahuzi-Biega National Park, DR Congo (Liengola 2008; Masumbuko et al. 2012a, 2012b), and does the same in Rwanda (Scholte et al. 2010) and southern Uganda (D. Sheil, pers. comm). In all these areas *Sericostachys* also attracts the attention of primate researchers because of the potential negative impact on gorilla habitat (Scholte et al. 2010). With its vigorous growth, this species is expected to suppress gorilla food (a wide variety of leaves, roots, stems, pith, and some fruits of herbaceous plants, lianas, and bamboos), thereby threatening gorilla survival.

Although the species is native to the area, in the Ituri forest plots this species was not found, probably because Ituri is at the lower end of the altitudinal range of the species. *S. scandens* is found in riverine or lakeside forest, at 700–2600 m altitude (Jansen 2004). In 1999, *Sericostachys* covered 21.8 km^2 of a total of 1145.1 km^2 in the Kahuzi-Biega National Park, in DR Congo (Liengola 2008). Between 1986 and 1999, *Sericostachys* had encroached mostly in secondary and in cleared forests, especially after elephants and buffalos had been killed during the early 1990s war. Masumbuko and coworkers (Masumbuko et al. 2012a, 2012b) analysed the effects of *Sericostachys* on trees and found that tree density was very low in plots dominated by *Sericostachys*, yet species richness was not affected. This species has functional traits that differ from three co-occurring important non-invasive liana species (*Gouania longispicata*, *Tacazzea apiculata*, and *Adenia bequaertii*): leaf traits SLA, N content and dry matter content are no different but biomass allocation to sexual reproduction is much higher (Masumbuko et al. 2012b). *Sericostachys* combines vegetative and sexual reproduction and is wind dispersed, has larger leaves and more side branches per unit stem length. The authors expressed doubts as to whether these characteristics were sufficient to explain the invasive nature of the species.

Elephants may have a role in the increase of *Sericostachys*. Elephants are thought to enlarge and maintain canopy gaps in the forest, but also to open up the *Sericostachys* tangles in those gaps and create space for other species (Liengola 2008; Scholte et al. 2010). The recent decline in elephants and buffalo populations (Maisels et al. 2013) has reduced trampling and grazing, and may thereby have benefitted the regeneration of *Sericostachys*, thus possibly explaining its increase. However, in Bwindi National Park in southern Uganda, *Sericostachys* is also common but not perceived as a problem (D. Sheil, pers. comm.; Scholte et al. 2010). As long ago as 1898 and 1907, explorers reported *S. scandens* to be abundant in Nyungwe National Park (Rwanda), despite the presence of elephants and buffaloes, refuting assumptions that the recent extirpation of these species has triggered the proliferation of *S. scandens* (Scholte et al. 2010). In 1994, Yumoto and coworkers (Yumoto et al. 1994) published a list of plant species in Kahuzi-Biega National Park, DR Congo, for the targeted area. *S. scandens* was not included, suggesting that the species has since spread throughout the area.

Apart from the mountainous areas mentioned, *S. scandens* is found extensively in the central region of DR Congo (Evrard 1968) and over a wide area in central, west, and east Africa (based on herbarium specimens, J. Wieringa, Herbarium Vadense, pers. comm.). Information on the invasive potential of this species in other areas is unknown.

DISCUSSION

On the African continent, we know of only three sites at which the liana communities have been monitored for long periods. One is a relatively small study in Gabon (Caballé & Martin 2001) where over 13 years the lianas ≥5 cm diameter were surveyed in a 1.6-ha plot. In an ongoing study in Korup National Park, Cameroon, all lianas ≥1 cm diameter in a 18 ha plot have been tagged and marked and have been censused twice (Thomas et al., Chapter 2 in this volume). Our study in Ituri, DR Congo, has been monitored for the longest time, with 40 ha of forest monitored since 1994, of which the 20 ha of mixed semi-evergreen Edoro forest is reported here. The general characteristics of the liana communities in these forests are compared in DeWalt et al. (Chapter 11 in this volume) and in Thomas et al. (Chapter 2 in this volume).

In general terms the liana community has changed very little in species composition, species richness, or diversity over the 13-year period. In contrast, we found a general decrease in liana abundance and basal area over this period, particularly for the smaller diameter lianas. Larger lianas (≥5 cm dbh) increased. Species rank orders have changed, with 12 species lost from the plot and 18 new species recruited over the 13-year observation period. Almost half of the species that we studied have changed in density (i.e., a statistically significant occurrence compared to random changes) over the monitoring period, and the majority of the species densities has decreased. What was previously the most dominant species, *Manniophyton fulvum*, has decreased dramatically. This species has a wide distribution in Africa (from Sierra Leone to the eastern DR Congo and northern Angola), and is locally abundant in many areas (in Cameroon – Parren (2003), in the Côte d'Ivoire – F.N. Kouamé, pers. comm., in Ghana – M.D. Swaine, pers. comm.). The species occurs in the forest but has strong affinities with forest gaps and edges, and responds positively to disturbance. For instance, one month after cutting a *Manniophyton fulvum* in Cameroon, the individual had resprouted vigorously and already grown 6 m in length (Parren 2003). Despite decreases in abundance and in basal area, and pervasive changes at the species level, the overall community composition surprisingly did not change much over the 13-year period (Fig. 3.3). Perhaps longer time scales are needed to determine directional changes in liana species composition.

The general decrease in abundance and basal area of liana stems in Ituri contrasts with the general increase in liana stems found in neotropical sites (Schnitzer & Bongers 2011; Schnitzer, Chapter 30 in this volume). While we cannot generalize for African forests based on the data currently available, we question why the Ituri forest results would be so different from the neotropical data. Environmental hypotheses for liana change include increased light levels due to disturbance, the disturbance hypothesis (Putz 1984; Schnitzer & Bongers 2002, 2011; Ledo & Schnitzer 2014) which would lead to higher recruitment and faster growth of light-demanding liana species. Alternatively, the fertilization hypothesis supposes that increased nutrient deposition and atmospheric carbon dioxide levels (Phillips et al. 2002; Schnitzer et al. 2011) lead to liana increase and increased growth. Increased drought, according to the dry season growth advantage hypothesis (Schnitzer 2005), would lead to an advantage for lianas compared to trees. A combination of these factors could explain the disparate results that we found in Ituri (Schnitzer & Bongers 2011).

Taking a trait-phylogenetic approach to the problem, it is possible that African climbers have different species traits than neotropical species, but data to test this hypothesis in detail are not yet available. Any of these causes of liana change could be responsible to different degrees; however, land use change and disturbance rates may be quite different between the continents. If local factors are more important than global conditions (such as increasing CO_2), the result could be different between continents – or even for different forests in the same region. Globally changing factors are predicted to create higher nutrient deposition levels (Hietz et al. 2011), for instance, resulting from increased fire (cf. Asner et al. 1997) or increased CO_2 levels, both leading to fertilization. This is expected to provoke rapid growth and reproduction, resulting in increased liana abundance. This growth fertilization hypothesis, however, is not adequate to explain the observed changes in the Ituri forest, because growth does not differ between species that increased or decreased in abundance, and because the most dramatically changing liana populations are decreasing, not increasing (data not shown). Global increased temperature increases, changing rainfall patterns, and increased number and severity of El Niño drought events (Condit et al. 1995; Holmgren et al. 2001; Condit et al. 2004; Engelbrecht et al. 2007; Condit et al. 2013) may be responsible for changes in liana abundance (DeWalt et al. 2010; Toledo et al. 2011). These factors vary across regions and within continents, with predictions of future rainfall patterns from increasing to decreasing. The Ituri forest area did not experience strong drought effects in the last two decades, despite decreasing rainfall (Asefi-Najafabady & Saatchi 2013). Climate modellers found no change in drought risk (Otto et al. 2013), and future climate model projections indicate an increase in rainfall in this area (James et al. 2013), though concordance among models is weak and more ground-based data are needed (Washington et al. 2013). Global dimming as a result of increased cloudiness (Stanhill & Cohen 2001; Graham et al. 2003; Feeley et al. 2007) may impact liana abundance. More dimming in Africa could lead to a decrease in lianas, especially the light-demanding species. The assumption that lianas respond faster and more positively to such changes than do trees is of crucial importance and more comparative experiments are necessary.

Another possibility for the patterns observed in Ituri is that the changes might reflect local forest recovery from past disturbances, as has been shown for the La Selva forest in Costa Rica (Yorke et al. 2013). There, a decrease in liana abundance was found in areas recovering from logging. This recovery is in contrast with the Ituri forest being perceived as a largely undisturbed old growth forest (Hart 1985; Makana et al. 2004a), even though large-gap disturbances were not uncommon (Hart et al. 1989). This would fit the notion that this particular area of the Ituri forest is recovering from past disturbance. Many African tropical forests are recovering from past disturbances (Richards 1952; Whitmore & Burslem 1998), as has been shown for several other African forest sites (van Gemerden et al. 2003; Willis et al. 2004; Brncic et al. 2007; Lewis et al. 2009). Recent studies demonstrate that recovery from disturbance is evident among trees in tropical forests worldwide (Laurance et al. 2004; Chave et al. 2008). Forest recovery over time after disturbance is generally accompanied by shifts in species composition. The Ituri forest indeed shows species turnover (Fig. 3.2b), changes in species composition (Fig. 3.3), and a higher abundance of more persistent liana taxa characteristic of undisturbed forest and decreasing abundance of some light-responsive species (e.g., *Manniophyton fulvum*). The Ituri forest has witnessed increased poaching (Effiom et al. 2013). A strong reduction in the numbers of forest elephants (Beyers et al. 2011) may be generating a closed forest, with a reduction of lianas, especially the more light-responsive species. This notion is supported by the drastic increase in aboveground biomass over the study period (Makana et al. 2011). In contrast, the increase in *Sericostachys* indicates that some liana species are increasing in other parts of this region, and may be a result of increasing disturbance and opening of forest areas, but evidence for this is weak.

The changes in the liana community in the Ituri forest are in contrast to global trends, but can be explained if either the global drivers are less important than previously thought, or if local stand dynamics override the more global causes of liana change. Whether our findings represent a general phenomenon for African forests needs to be studied in detail by extending the species-level changes in liana abundance to a large number of African sites across a wide array of local conditions.

CONCLUSION AND FUTURE DIRECTIONS

The last two decades have seen an increase in liana-focused research (Schnitzer et al., Chapter 1 in this volume), especially in West Africa. Liana densities in the few African forests that have been studied are similar to other continents, but species diversity appears to be higher (see also DeWalt et al., Chapter 11 in this volume). Although the lianas in several African sites now are being inventoried along with trees, more sites across the African continent need long-term monitoring to gain a better understanding of the ecology of African lianas. Currently, only Korup and Ituri (both part of the CTFS network) are being repeatedly monitored.

Additional sources of information on liana distributions are herbarium collections, which are available in several major herbaria in the world. Providing easy access to these collections for use in large-scale databases is underway, but still needs considerable work. For some plant families that contain lianas, this work is on-going (J. Wieringa, pers. comm.), but in many cases species are not labelled as lianas, and thus liana-focused information cannot easily be retrieved, hampering large scale comparative analyses. These high-quality data can be used to analyse liana distributions and their changes, as well as ecological preferences for vegetation type, climate, and soil. In West Africa, herbarium-based analyses have been completed for lianas and other woody plants (Holmgren & Poorter 2007).

To understand the response of African liana species to climate and/or land use change, autecological studies are urgently needed. One of the few African liana species with focused attention is *Sericostachys scandens*, the invasive and economically useful species addressed in this chapter. Functional leaf, stem, and root traits of African lianas need to be measured to be able to evaluate their responses to the environment and changes therein, and to interactions with other species, preferably following specific guidelines (Cornelissen et al. 2003; Pérez-Harguindeguy et al. 2013). The tolerance of different liana species in African forests to shade and their responses to disturbance need additional study. Liana response to drought, important for understanding their high abundance in drier forests (Schnitzer 2005; DeWalt et al. 2010; De Walt et al., Chapter 11 in this volume) is receiving increased attention (Cai et al. 2009; van der Sande et al. 2013), but such research still must be extended to African forests. Only with more detailed ecological studies of the lianas of Africa, will we be able to solve the enigma of decreasing liana abundances in the Ituri forest compared to increases in neotropical forests.

ACKNOWLEDGMENTS

We thank T. Hart, J. Hart, J-R Makana and other Wildlife Conservation Society staff for having set up the plots and for all their collaboration during field work, R. Condit and CTFS for fully supporting the continuation of the plot re-census, M. Sosef and L. Poorter for discussions, S. Schnitzer and R. Burnham for comments on the manuscript, Herbarium Vadense for support during identifications, and Wageningen University for a PhD grant to Corneille Ewango.

REFERENCES

Addo-Fordjour, P. & Anning, A.K. (2008) Diversity and distribution of climbing plants in a semi-deciduous rain forest, KNUST Botanic Garden, Ghana. *International Journal of Botany*, **4**, 186–195.

Addo-Fordjour, P., Anning, A.K., Larbi, J.A. & Akyeampong, S. (2009a) Liana species richness, abundance and relationship with trees in the Bobiri forest reserve, Ghana: Impact of management systems. *Forest Ecology and Management*, **257**, 1822–1828.

Addo-Fordjour, P., Obeng, S., Addo, M.G. & Akyeampong, S. (2009b) Effects of human disturbances and plant invasion on liana community structure and relationship with trees in the Tinte Bepo forest reserve, Ghana. *Forest Ecology and Management*, **258**, 728–734.

Addo-Fordjour, P., Obeng, S., Anning, A.K. & Addo, M.G. (2009c) Floristic composition, structure and natural regeneration in a moist semi-deciduous forest following anthropogenic disturbances and plant invasion. *International Journal of Biodiversity and Conservation*, **1**, 21–37.

Addo-Fordjour, P., Rahmad, Z.B., Amui, J. et al. (2013) Patterns of liana community diversity and structure in a tropical rainforest reserve, Ghana: effects of human disturbance. *African Journal of Ecology*, **51**, 217–227.

Asefi-Najafabady, S. & Saatchi, S. (2013) Response of African humid tropical forests to recent rainfall anomalies. *Philosophical Transactions of the Royal Society of London. Series B, Biological Sciences*, **368**, 20120306.

Asner, G.P., Seastedt, T.R. & Townsend, A.R. (1997) The decoupling of terrestrial carbon and nitrogen cycles. *BioScience*, **47**, 226–234.

Babweteera, F., Plumptre, A. & Obua, J. (2000) Effect of gap size and age on climber abundance and diversity in Budongo Forest Reserve, Uganda. *African Journal of Ecology*, **38**, 230–237.

Beyers, R.L., Hart, J.A., Sinclair, A.R.E. et al. (2011) Resource wars and conflict ivory: the impact of civil conflict on elephants in the Democratic Republic of Congo. The case of the Okapi Reserve. *PLoS ONE*, **6**, e27129.

Bongers, F., Parren, M.P.E. & Traoré, D. (2005) *Forest Climbing Plants of West Africa: Diversity, Ecology and Management*, CAB International, Wallingford.

Brncic, T.M., Willis, K.J., Harris, D.J. & Washington, R. (2007) Culture or climate? The relative influences of past processes on the composition of the lowland Congo rainforest. *Philosophical Transactions of the Royal Society B: Biological Sciences*, **362**, 229–242.

Caballé, G. & Martin, A. (2001) Thirteen years of change in trees and lianas in a Gabonese rainforest. *Plant Ecology*, **152**, 167–173.

Cai, Z.-Q., Schnitzer, S.A. & Bongers, F. (2009) Seasonal differences in leaf-level physiology give lianas a competitive advantage over trees in a tropical seasonal forest. *Oecologia*, **161**, 25–33.

Chave, J., Olivier, J., Bongers, F. et al. (2008) Above-ground biomass and productivity in a rain forest of eastern South America. *Journal of Tropical Ecology*, **24**, 355–366.

Condit, R. (1998) *Tropical Forest Census Plots: Methods and Results from Barro Colorado Island, Panama and Comparison with Other Forests*, Springer Verlag, Berlin.

Condit, R., Aguilar, S., Hernandez, A. et al. (2004) Tropical forest dynamics across a rainfall gradient and the impact of an El Niño dry season. *Journal of Tropical Ecology*, **20**, 51–72.

Condit, R., Engelbrecht, B.M.J., Pino, D. et al. (2013) Species distributions in response to individual soil nutrients and seasonal drought across a community of tropical trees. *PNAS*, **110**, 5064–5068.

Condit, R., Hubbell, S.P. & Foster, R.B. (1995) Mortality rates of 205 neotropical tree and shrub species and the impact of a severe drought. *Ecological Monographs*, **65**, 419–439.

Cornelissen, J.H.C., Lavorel, S., Garnier, E. et al. (2003) A handbook of protocols for standardised and easy measurement of plant functional traits worldwide. *Australian Journal of Botany*, **51**, 335–380.

Darwin, C. (1865) On the movements and habits of climbing plants. *Journal of the Linnean Society of London*, **9**, 1–118.

DeWalt, S.J., Schnitzer, S.A., Chave, J. et al. (2010) Annual rainfall and seasonality predict pan-tropical patterns of liana density and basal area. *Biotropica*, **42**, 309–317.

Effiom, E.O., Nunez-Iturri, G., Smith, H.G. et al. (2013) Bushmeat hunting changes regeneration of African rainforests. *Proceedings of the Royal Society: Biological Sciences*, **280**, 20130246.

Eilu, G. (2000) Liana abundance in three tropical rain forests of Western Uganda. *Selbyana*, **21**, 30–37.

Eilu, G. & Bukenya-Ziraba, R. (2004) Local use of climbing plants of Budongo Forest Reserve, Western Uganda. *Journal of Ethnobiology*, **24**, 307–327.

Engelbrecht, B.M.J., Comita, L.S., Condit, R. et al. (2007) Drought sensitivity shapes species distribution patterns in tropical forests. *Nature*, **447**, 80–82.

Evrard, C. (1968) *Recherches écologiques sur le peuplement forestier des sols hydromorphes de la Cuvette centrale congolaise*, Publications INEAC, Série scientifique, p. 110.

Ewango, C.E.N. (2010) The liana assemblage of a a Congolian rainforest. Diversity, structure and dynamics. PhD thesis, Wageningen University, Wageningen, The Netherlands.

Feeley, K.J., Joseph Wright, S., Nur Supardi, M.N. et al. (2007) Decelerating growth in tropical forest trees. *Ecology Letters*, **10**, 461–469.

Gentry, A.H. (1991) The distribution and evolution of climbing plants, in *The Biology of Vines* (eds F.E. Putz & H.A. Mooney), Cambridge University Press, Cambridge, pp. 3–49.

Gerwing, J.J., Schnitzer, S.A., Burnham, R.J. et al. (2006) A standard protocol for liana censuses. *Biotropica*, **38**, 256–261.

Graham, E.A., Mulkey, S.S., Kitajima, K. et al. (2003) Cloud cover limits net CO_2 uptake and growth of a rainforest tree during tropical rainy seasons. *Proceedings of the National Academy of Sciences*, **100**, 572–576.

Hart, T.B. (1985) The ecology of a single-species-dominant forest and of a mixed forest in Zaire, Africa. PhD thesis Michigan State University.

Hart, T.B., Hart, J.A. & Murphy, P.G. (1989) Monodominant and species-rich forests of the humid tropics: causes for their co-occurrence. *The American Naturalist*, **133**, 613–633.

Hietz, P., Turner, B.L., Wanek, W. et al. (2011) Long-term change in the nitrogen cycle of tropical forests. *Science*, **334**, 664–666.

Holmgren, M. & Poorter, L. (2007) Does a ruderal strategy dominate the endemic flora of the West African forests? *Journal of Biogeography*, **34**, 1100–1111.

Holmgren, M., Scheffer, M., Ezcurra, E. et al. (2001) El Niño effects on the dynamics of terrestrial ecosystems. *Trends in Ecology and Evolution*, **16**, 89–94.

James, R., Washington, R. & Rowell, D.P. (2013) Implications of global warming for the climate of African rainforests. *Philosophical Transactions of the Royal Society of London. Series B, Biological Sciences*, **368**, 20120298.

Jansen, P.C.M. (2004) *Sericostachys scandens* Gilg and Lopr, in *PROTA 2: Vegetables/Légumes* (eds G.J.H. Grubben & O.A. Denton), [CD-Rom]. PROTA, Wageningen, the Netherlands.

Laurance, W.F., Oliveira, A.A., Laurance, S.G. et al. (2004) Pervasive alteration of tree communities in undisturbed Amazonian forests. *Nature*, **428**, 171–175.

Ledo, A. & Schnitzer, S.A. (2014) Disturbance and clonal reproduction determine liana distribution and maintain liana diversity in a tropical forest. *Ecology*, in press. 10.1890/13-1775.1

Lewis, S.L., Lopez-Gonzalez, G., Sonke, B. et al. (2009) Increasing carbon storage in intact African tropical forests. *Nature*, **457**, 1003–1006.

Liengola, I. (2008) Impact of the invasive liana *Sericostachys scandens* on forest composition: implications for the recovery of Grauer's gorilla in the Kahuzi-Biega National Park, Democratic Republic of Congo. *Tropical Resources Bulletin*, **27**, 43–50.

Losos, E. & Leigh, E.G. (eds) (2004) *Tropical Forest Diversity and Dynamism: Findings from a Large Scale Plot Network*, University of Chicago Press, Chicago.

Maisels, F., Strindberg, S., Blake, S. et al. (2013) Devastating decline of forest elephants in Central Africa. *PLoS ONE*, **8**, e59469.

Makana, J.-R., Ewango, C.N., McMahon, S.M. et al. (2011) Demography and biomass change in monodominant and mixed old-growth forest of the Congo. *Journal of Tropical Ecology*, **27**, 447–461.

Makana, J.R., Hart, T.B., Ewango, C.E. et al. (2004a) Ituri forest dynamics plots, DRC, in *Tropical Forest Diversity and Dynamism: Findings from a Large Scale Plot Network* (eds E.C. Losos & E.G. Leigh), University of Chicago Press, Chicago, pp. 492–505.

Makana, J.R., Hart, T.B., Hibbs, D. & Condit, R. (2004b) Forest structure and diversity in the Ituri forest dynamics plots, in *Tropical Forest Diversity and Dynamism: Findings from a Large Scale Plot Network* (eds E.C. Losos & E.G. Leigh), University of Chicago Press, Chicago, pp. 159–174.

Masumbuko, N.C., Habiyaremye, M.F. & Lejoly, J. (2012a) Woody climbing plants influence the structure of the mountain forest in the Kahuzi-Biega National Park, DR Congo. *Regional Environmental Change*, **12**, 951–959.

Masumbuko, N.C., Herpigny, B., Barbier, N. et al. (2012b) Life strategy traits of the liana *Sericostachys scandens* spreading in the montane forests in the Kahuzi-Biega National Park (DR Congo). *Journal of Mountain Science*, **9**, 665–675.

McGeoch, L., Gordon, I. & Schmitt, J. (2008) Impacts of land use, anthropogenic disturbance, and harvesting on an African medicinal liana. *Biological Conservation*, **141**, 2218–2229.

Moscovice, L.R., Mbago, F., Snowdon, C.T. & Huffman, M.A. (2010) Ecological features and ranging patterns at a chimpanzee release site on Rubondo Island, Tanzania. *Biological Conservation*, **143**, 2711–2721.

Muhwezi, O., Cunningham, A.B. & Bukenya-Ziraba, R. (2009) Lianas and livelihoods: the role of fibrous forest plants in food security and society around Bwindi Impenetrable National Park, Uganda. *Economic Botany*, **63**, 340–352.

Otto, F.E., Jones, R.G., Halladay, K. & Allen, M.R. (2013) Attribution of changes in precipitation patterns in African rainforests. *Philosophical Transactions of the Royal Society of London. Series B, Biological Sciences*, **368**, 20120299.

Parren, M.P.E. (2003) Lianas and logging in West Africa. PhD thesis, Wageningen University.

Pérez-Harguindeguy, N., Díaz, S., Garnier, E. et al. (2013) New handbook for standardised measurement of plant functional traits worldwide. *Australian Journal of Botany*, **61**, 167–235.

Phillips, O.L., Martinez, R.V., Arroyo, L. et al. (2002) Increasing dominance of large lianas in Amazonian forests. *Nature*, **418**, 770–774.

Putz, F.E. (1984) The natural history of lianas on Barro Colorado Island, Panama. *Ecology*, **65**, 1713–1724.

Richards, P.W. (1952) *The Tropical Rain Forest: An Ecological Study*, Cambridge University Press, Cambridge.

Schnitzer, S.A. (2005) A mechanistic explanation for global patterns of liana abundance and distribution. *The American Naturalist*, **166**, 262–276.

Schnitzer, S.A. & Bongers, F. (2002) The ecology of lianas and their role in forests. *Trends in Ecology and Evolution*, **17**, 223–230.

Schnitzer, S.A. & Bongers, F. (2011) Increasing liana abundance and biomass in tropical forests: emerging patterns and putative mechanisms. *Ecology Letters*, **14**, 397–406.

Schnitzer, S.A., Bongers, F. & Wright, S.J. (2011) Community and ecosystem ramifications of increasing lianas in neotropical forests. *Plant Signalling and Behavior*, **6**, 598–600.

Schnitzer, S.A., Kuzee, M.E. & Bongers, F. (2005) Disentangling above- and below-ground competition between lianas and trees in a tropical forest. *Journal of Ecology*, **93**, 1115–1125.

Scholte, P., Rugyerinyange, L., Bizimungu, F. et al. (2010) Reaching consensus: impact of the liana *Sericostachys scandens* on forest dynamics in Nyungwe National Park, Rwanda. *Oryx*, **44**, 325–326.

Senbeta, F., Schmitt, C., Denich, M. et al. (2005) The diversity and distribution of lianas in the Afromontane rain forests of Ethiopia. *Diversity and Distributions*, **11**, 443–452.

Stanhill, G. & Cohen, S. (2001) Global dimming: a review of the evidence for a widespread and significant reduction in global radiation with discussion of its probable causes and possible agricultural consequences. *Agricultural and Forest Meteorology*, **107**, 255–278.

Swaine, M.D. & Grace, J. (2007) Lianas may be favoured by low rainfall: evidence from Ghana. *Plant Ecology*, **192**, 271–276.

Tchouto, M.G.P., Boer, W.F., Wilde, J.J.F.E. & Maesen, L.J.G. (2006) Diversity patterns in the flora of the Campo-Ma'an rain forest, cameroon: do tree species tell it all? *Biodiversity and Conservation*, **15**, 1353–1374.

Toledo, M., Poorter, L., Peña-Claros, M. et al. (2011) Climate and soil drive forest structure in Bolivian lowland forests. *Journal of Tropical Ecology*, **27**, 333–345.

Toledo-Aceves, T. & Swaine, M.D. (2007a) Above- and below-ground competition between the liana *Acacia kamerunensis* and tree seedlings in contrasting light environments. *Plant Ecology*, **196**, 233–244.

Toledo-Aceves, T. & Swaine, M.D. (2007b) Effect of three species of climber on the performance of Ceiba pentandra seedlings in gaps in a tropical forest in Ghana. *Journal of Tropical Ecology*, **23**, 45.

Toledo-Aceves, T. & Swaine, M.D. (2008a) Biomass allocation and photosynthetic responses of lianas and pioneer tree seedlings to light. *Acta Oecologica*, **34**, 38–49.

Toledo-Aceves, T. & Swaine, M.D. (2008b) Effect of lianas on tree regeneration in gaps and forest understorey in a tropical forest in Ghana. *Journal of Vegetation Science*, **19**, 717–728.

van der Sande, M.T., Poorter, L., Schnitzer, S.A. & Markesteijn, L. (2013) Are lianas more drought-tolerant than trees? A test for the role of hydraulic architecture and other stem and leaf traits. *Oecologia*, **172**, 961–972.

van Gemerden, B.S., Olff, H., Parren, M.P.E. & Bongers, F. (2003) The pristine rain forest? Remnants of historical human impacts on current tree species composition and diversity. *Journal of Biogeography*, **30**, 1381–1390.

Washington, R., James, R., Pearce, H. et al. (2013) Congo Basin rainfall climatology: can we believe the climate models? *Philosophical Transactions of the Royal Society of London. Series B, Biological Sciences*, **368**, 20120296.

Whitmore, T.C. & Burslem, D.F.R.P. (1998) Major disturbances in tropical rainforests, in *Dynamics of Tropical Communities : The 37th Symposium of the British Ecological Society* (eds D.M. Newbery, H.H.T. Prins & N.D. Brown), Cambridge University Press, Cambridge, pp. 549–565.

Willis, K.J., Gillson, L. & Brncic, T.M. (2004) How 'virgin' is virgin rainforest? *Science*, **304**, 402–403.

Wright, S.J., Calderón, O., Hernandéz, A. & Paton, S. (2004) Are lianas increasing in importance in tropical forests? A 17-year record from Panama. *Ecology*, **85**, 484–489.

Yorke, S.R., Schnitzer, S.A., Mascaro, J.L. et al. (2013) Increasing liana abundance and basal area in a tropical forest: the contribution of long-distance clonal colonization. *Biotropica*. doi: 10.1111/btp.12015

Yumoto, T., Yamagiwa, J., Manza, N. & Maruhashi, T. (1994) List of plant species identified in Kahuzi-Biega National Park, Zaïre. *Tropics*, **3**, 295–308.

Chapter 4

LIANA COMPOSITION AND DIVERSITY AMONG TROPICAL FOREST TYPES OF PENINSULAR INDIA

Narayanaswamy Parthasarathy,[1]
S. Muthuramkumar,[2] *Chellam Muthumperumal,*[3]
P. Vivek,[1] *N. Ayyappan,*[4] *and M. Sridhar Reddy*[5]

[1]Pondicherry University, Puducherry, India
[2]V.H.N.S.N. College, Virudhunagar, Tamil Nadu, India
[3]Madurai Kamaraj University, Madurai, Tamil Nadu, India
[4]French Institute of Pondicherry, Puducherry, India
[5]Yogi Vemana University, Kadapa, Andhra Pradesh, India

Ecology of Lianas, First Edition. Edited by Stefan A. Schnitzer, Frans Bongers, Robyn J. Burnham, and Francis E. Putz.

Companion Website: www.wiley.com/go/schnitzer/lianas

OVERVIEW

Recent investigations have shown that lianas are a prominent component of tropical forests, where they play an important role in ecosystem processes, including foliage and fruit production and carbon sequestration. While lianas were once ignored in many forest inventories, they have generated considerable interest recently, in recognition of their importance in forest functional ecology. Yet, research on lianas in an Indian context remains rudimentary. Considering this situation, research was undertaken on the diversity and ecology of lianas in four major tropical forest types of peninsular India: wet evergreen forest, semi-evergreen forest, seasonal dry forest and dry evergreen forest located in the Western Ghats, the Eastern Ghats, and the Coromandel Coast of peninsular India.

This chapter is the culmination of our efforts. In total, we sampled 109.4 hectares spread among 40 different study sites over the four forest types. This large-scale inventory yielded 61,789 liana individuals representing 237 species in 135 genera and 56 families. Among the four forest types, liana density was highest in the dry forests, and decreased significantly with increasing annual rainfall and increasing dry season length, consistent with earlier reports. Our analyses for species richness were more difficult to decipher because of slightly different plot sizes and sampling methods; however, we found that liana species richness was greater in wet forests than in dry forests.

The highest percentage of unique species (58%) was found in wet evergreen forests, whereas unique species were uncommon in semi-evergreen forests (1.95%). A non-metric multidimensional scaling ordination based on similarity indices among sites produced distinct clusters for wet and dry evergreen forests and another cluster with the least differentiation between the sites of seasonally dry and semi-evergreen forests. The contribution of the top ten abundant liana species from each forest type revealed that species dominance increased from wet to dry forests. The Apocynaceae, Asclepiadaceae, Fabaceae, and Oleaceae constituted the most speciose liana families in peninsular India, which differs from other Asian tropical sites, where the Annonaceae, Combretaceae, Loganiaceae, and Rutaceae are well represented. By contrast, in the neotropics, families such as the Bignoniaceae, Fabaceae, Sapindaceae, Malpighiaceae, Euphorbiaceae, Celastraceae, and Vitaceae are most diverse.

Among the five climbing mechanisms we defined, stem twiners were most common, followed by scramblers, while tendril climbers were abundant in dry forests. Similar to other moist tropical forests, liana fruit resources are utilized by various arboreal mammals and birds in the wet Indian forests and largely by birds in dry evergreen forests. Overall, biotic dispersal is more common in terms of both species richness and abundance across forest types, which implies faunal dependence and emphasizes the need for a holistic approach in biological conservation.

INTRODUCTION

There are 16 major forest types in India, ranging from tropical scrub to montane, wet temperate forests. There are also 103 sub-types of forests that are distinguished based on their dominant species (Champion & Seth 1968). In the Indian sub-continent, ten biogeographic regions are recognized (Rodgers & Panwar 1988), and among them the Western Ghats, Deccan, and Coasts fall under peninsular India. In the past few decades, many studies on plant diversity have been conducted in peninsular Indian forests, but most have focused on tree diversity, and a few on shrubs and herbs. Lianas remained largely neglected. At the same time, lianas were reported to comprise about 25% of species diversity and 10–45% of woody stem density in tropical forests (Schnitzer & Bongers 2002). This scientific gap and paucity of information on lianas stimulated us to undertake research on the diversity and ecology of lianas in the Western Ghats, Eastern Ghats, and Coromandel Coast forests of peninsular India. This chapter consolidates that research.

In peninsular Indian forests, the species richness of lianas is high in the Western Ghats, moderate in the Eastern Ghats, and low in Coromandel Coast, whereas liana abundance shows an opposite trend in these sites (Parthasarathy et al. 2004; Muthumperumal & Parthasarathy 2010). Stem twining is the most common climbing mechanism, forming the largest proportion of species richness, abundance, and basal area in many tropical forests (Putz & Chai 1987; Nabe-Nielsen 2001; Parthasarathy et al. 2004; Senbeta et al. 2005). The majority (92%) of stems of twining plants found in a range of vegetation types and geographical areas across the world grow by twining in a right-handed (anti-clockwise) direction (Edwards et al. 2007). Lianas have a wide range of pollination systems, seed size, and seed dispersal modes. According to Gentry (1982, 1991a), liana species are mostly wind dispersed, though he noted that species numbers decrease as rainfall increases, changing the prevalence to animal-dispersed species in wetter localities. The seed dispersal strategies of liana species have implications for the type of dispersal and overall introduction success of the lianas.

This chapter describes variation in species composition, density, climbing mechanisms and dispersal syndromes of lianas across four principal forest types of peninsular India, ranging from tropical dry evergreen to wet evergreen forests. We compare the characteristics of Indian lianas to those reported for other forests around the world.

MATERIAL AND METHODS

Study sites and forest types

We utilized a dataset of liana inventories conducted in 40 sites in four major tropical forest types: wet evergreen forest (henceforth referred as WEF), semi-evergreen forest (SEF), seasonal dry forest (SDF), and dry evergreen forest (DEF), distributed in 40 different study sites that include the Western Ghats, Eastern Ghats, and the Coromandel Coast of peninsular India (Fig. 4.1). Table 4.1 summarizes the details of the 40 study sites, including the forest type, altitude, the forest stature, the mean annual rainfall, the length of the dry season and the sample design. Tropical wet evergreen forest (WEF) sites include Varagalaiar, Valparai, Agumbe (in the Western Ghats), and the hills of Kalrayans, Pachamalais and Shervarayans (in the southernmost Eastern Ghats). The Varagalaiar and Valparai plateaus are located within the Indira Gandhi National Park and Wildlife Sanctuary, Anamalais, Western Ghats, south India. Of the five sites in Valparai, two sites, Tata Finlay and Injipara, are surrounded by coffee and tea plantations and are subject to greater human interference than the other sites. These sites benefit from both southwest (June–September) and northeast monsoon (October–December) rains.

Among the Eastern Ghats sites, liana inventories were made in the hill ranges of Bodamalai, Chitteri, Kalrayan, the Kolli hills, Pachamalai, and Shervarayan, which are situated in the Salem district of Tamil Nadu, south India. The principal vegetation of all sites is the SDF (seasonal dry forest, in which we include tropical mixed and dry deciduous and thorn forests) at lower altitudes and the SEF to WEF (particularly in the high ranges of Kalrayan, Pachamalai and Shervarayan). These sites receive rainfall between August and December, with a mean annual value of 1057.9 mm. Most sites are subject to various forest disturbances, such as land-use change for cultivation (e.g., cereals, pineapple, tapioca, etc.), illegal timber extraction, collection of non-timber forest produce, soil removal (for ore mining and road construction), cattle grazing, tourism, and plant invasions.

On the Coromandel Coast of peninsular India, 18 sites were selected, ten in coastal districts (Cuddalore and Villupuram districts, Tamil Nadu) and eight in inland areas (Pudukottai district, Tamil Nadu). These sites are sacred groves or temple forests protected by the religious belief of the local people and belong to tropical dry evergreen forest type. Photographs of the selected liana species from four forest types are provided in Fig. 4.2.

Of the 40 study sites, DEF sites on the Coromandel Coast and SDF sites of the Eastern Ghats experience longer dry seasons (6–8 months), followed by the SEF sites of Eastern Ghats; the dry season is shortest (3–4 months) in the WEF sites. Among the four forest types, the forest stature is short (7 m) in DEF sites, moderately tall (12–16 m) in SDF and SEF sites, and tallest in WEF sites of the Western Ghats (35 and 25 m respectively at Varagalaiar and Agumbe), and we find a negative relation ($r^2 = 0.77$; $p < 0.05$) between forest stature and length of dry season.

Sampling and data analysis

We measured all lianas that were rooted within the plot 1.3 m from the rooting point. This standard method allowed us to validly compare liana diversity across four forest types. In this census, all climbers rooted on the forest floor and having woody/fibrous stems (Vitaceae and Cucurbitaceae) and a few thin climbers, but that had attained ≥1 cm diameter (*Pothos*) were considered as lianas. Data on liana density were converted to individuals per hectare by interpolation. We converted liana density for Injipara, Lower Manamboli, and Tata Finlay from 0.8 to 1 hectare. To facilitate comparison of the results of the liana species richness and density at the unified scale of 1 ha among the four forest types that we inventoried and with those of other

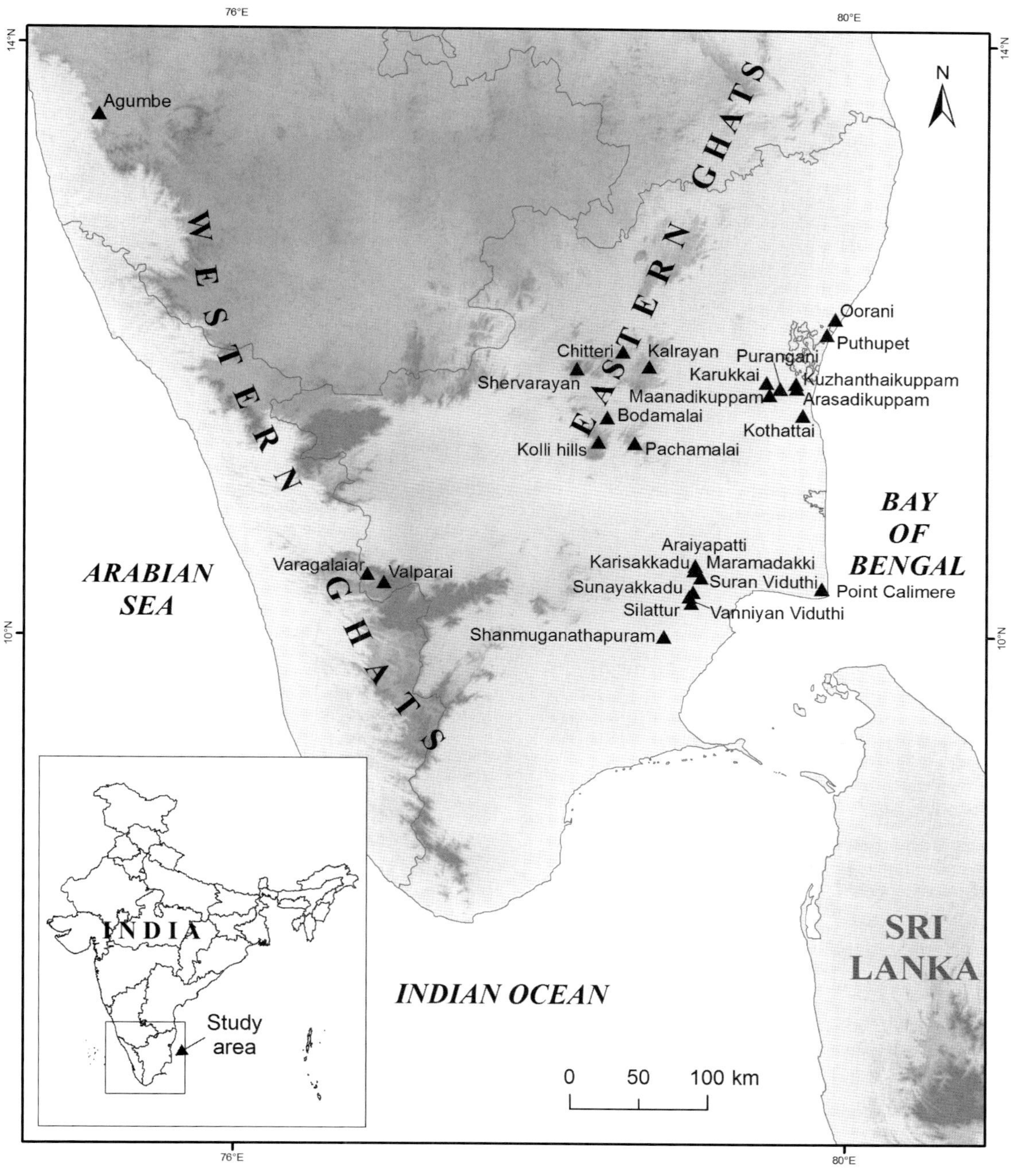

Fig. 4.1 Map showing the location of liana inventory sites distributed in peninsular India.

Table 4.1 Details of four tropical forest types in peninsular India, sites of liana inventories with site name, forest stature, rainfall, length of dry season and sample design

Forest type, study sites & elevation (m)	Forest stature (m)	Mean annual rainfall (mm) & Length of dry season (months)	Number of plots and dimensions (m)
Wet evergreen forest (WEF), Western Ghats			
Agumbe Plot I (650)	25	7000–7502 (5)	1, 100 × 100
Agumbe Plot II (200)	27	7000–7501 (5)	1, 100 × 100
Akkamalai-Iyerpadi (1260–1360)	27	3497 (3–4)	25, 20 × 20
Injipara (1000–1100)	27	3497 (3–4)	20, 20 × 20
Lower Manamboli (760–800)	27	3497 (3–4)	20, 20 × 20
Tata Finlay (980–1200)	27	3497 (3–4)	20, 20 × 20
Upper Manamboli (920–1120)	27	3497 (3–4)	25, 20 × 20
Varagalaiar (600–660)	35	1660 (3–4)	30, 100 × 100 (i.e., contiguous 500 × 600 block)
Wet evergreen forest (WEF), Eastern Ghats			
Kalrayan (701.5–984.5)	13	1058 (4)	2, 5 × 1000
Pachamalai (720.5–789.5)	12	1058 (4)	1, 5 × 1000
Shervarayan (1030.5–1105.5)	14	1058 (4)	1, 5 × 1000
Semi-evergreen forest (SEF), Eastern Ghats			
Chitteri (328–1038)	12	1058 (5)	11, 5 × 1000
Kalrayan (379–926)	13	1058 (5)	7, 5 × 1000
Kolli hills (324.5–1316.5)	16	1058 (5)	11, 5 × 1000
Pachamalai (639.5–871.5)	12	1058 (5)	6, 5 × 1000
Shervarayan (476–1415.5)	14	1058 (5)	8, 5 × 1000
Seasonal dry forest (SDF), Eastern Ghats			
Bodamalai (306.4–743.4)	6	1058 (6)	10, 5 × 1000
Chitteri (411.5- 785.6)	12	1058 (6)	9, 5 × 1000
Kalrayan (208–826)	13	1058 (6)	15, 5 × 1000
Kolli hills (272.5–660)	16	1058 (6)	7, 5 × 1000
Pachamalai (176.5–481.7)	12	1058 (6)	17, 5 × 1000
Shervarayan (433.5–650)	14	1058 (6)	5, 5 × 1000
Dry evergreen forest (DEF), Coromandel Coast			
Arasadikuppam (34)	7	1151 (6)	1, 100 × 100
Kuzhanthaikuppam (34)	7	1165 (6)	1, 100 × 100
Oorani (13)	7	1132 (6)	1, 100 × 100
Puthupet (15)	7	1141 (6)	1, 100 × 100
Karukkai (51)	7	1091 (6)	1, 100 × 100
Maanadikuppam (91)	7	1125 (6)	1, 100 × 100
Purangani (71)	7	1137 (6)	1, 100 × 100
Kothattai (8)	7	1227 (6)	1, 100 × 100
Point Calimere 1 (6)	7	1433 (6)	1, 100 × 100
Point Calimere 2 (6)	7	1433 (6)	1, 100 × 100
Dry evergreen forest, Inland			
Araiyapatti (64)	7	905 (6)	1, 100 × 100
Karisakkadu (62)	7	903 (6)	1, 100 × 100
Maramadakki (81)	7	887 (6)	1, 100 × 100
Shanmuganatha puram (61)	7	884 (6)	1, 100 × 100
Silattur (62)	7	907 (6)	1, 100 × 100
Sunayakkadu (72)	7	899 (6)	1, 100 × 100
Suran Viduthi (67)	7	927 (6)	1, 100 × 100
Vanniyan Viduthi (75)	7	915 (6)	1, 100 × 100

Fig. 4.2 Photographs of selected liana species from four forest types (WEF, SEF, SDF and DEF) of peninsular India

WEF: A) *Gnetum ula* (Gnetaceae – Gymnosperm), exhibiting cauliflory and drupes (mammal dispersed). B) *Elaeagnus indica* with silvery/coppery shining leaves and drupes for vertebrate dispersal. C) Stout thorns of *Zanthoxylum tetraspermum* (Rutaceae)

SEF: D) Armed straggler, *Pterolobium hexapetalum* (Caesalpiniaceae) with curved thorn from bullate protuberance. E) *Jasminum angustifolium* (Oleaceae), a stem twiner with scented white flower. F) *Lantana camara* (Verbenaceae), an invasive species

SDF: G) *Grewia rhamnifolia* (Tiliaceae). H) *Abrus precatorius* (Papilionaceae), dehising legumes exposing reddish black seeds. I) *Capparis sepiaria* (Capparaceae)

DEF: J) *Ventilago madraspatana* (Rhamnaceae) that occur in all four forest types of peninsular India. K) Thorny straggler *Acacia caesia* occurs in forest gaps and edges (Mimosaceae). L) Drupes of *Hugonia mystax* (Linaceae) are vertebrate dispersed.

See plate section for color representation of this figure.

Table 4.2 Liana density, basal area and species richness per hectare (& on plot) in four tropical forest types of peninsular India.

Variable	≥1 cm (mean ± SD)	≥1.6 cm (mean ± SD)	≥2 cm (mean ± SD)	≥5 cm (mean ± SD)	≥10 cm (mean ± SD)
Wet evergreen forest					
Density (ha^{-1})	345.6 (156.1)	261.13 (86.70)	190.31 (76.30)	52.18 (28.11)	11.13 (7.65)
Basal area (m^2 ha^{-1})	0.401 (0.20)	0.427 (0.19)	0.409 (0.18)	0.318 (0.17)	0.146 (0.10)
Species richness (ha^{-1})	32.13 (10.82)	29.97 (6.24)	26.71 (5.25)	14.47 (4.93)	5.15 (3.08)
Species richness (38.4 ha)	126	116	110	73	38
Fisher's α (ha^{-1})	7.39	7.41	7.47	7.40	2.33
Fisher's α (38.4 ha)	19.07	18.51	18.45	14.9	10.2
Semi-evergreen forest					
Density (ha^{-1})	–	647.7 (152.7)	486.6 (173.8)	94.7 (62.7)	12.9 (6.7)
Basal area (m^2 ha^{-1})	–	0.69 (0.3)	0.66 (0.34)	0.43 (0.25)	0.16 (0.09)
Species richness (ha^{-1})	–	31.4 (5.3)	27.3 (5.7)	15.95 (5.9)	5.75 (3.7)
Species richness (21.5 ha)	–	126	112	74	40
Fisher's α (ha^{-1})	–	7.11	6.50	6.17	5.22
Fisher's α (21.5 ha)	–	19.27	17.69	14.76	12.37
Seasonal dry forest					
Density (ha^{-1})	–	594.8 (138.8)	322.2 (128)	9.6 (6.9)	0.24 (0.24)
Basal area (m^2 ha^{-1})	–	0.21 (0.06)	0.16 (0.06)	0.02 (0.02)	0.003 (0.003)
Species richness (ha^{-1})	–	25.2 (4.8)	18.14 (5.7)	4.88 (3)	1.16 (0.40)
Species richness (31.5 ha)	–	70	60	26	6
Fisher's α (ha^{-1})	–	5.61	4.65	3.34	0
Fisher's α (31.5 ha)	–	9.31	8.66	7.07	14.11
Dry evergreen forest					
Density (ha^{-1})	845.7 (336.7)	522.5 (242)	404.7 (194.9)	77.7 (59.5)	11.6 (12.4)
Basal area (m^2 ha^{-1})	0.768 (0.475)	0.759 (0.44)	0.729 (0.44)	0.444(0.37)	0.195 (0.20)
Species richness (ha^{-1})	24.4 (4.7)	20.9 (3.9)	18.9 (3.6)	10 (4.1)	3.55 (2.5)
Species richness (18 ha)	56	50	44	35	20
Fisher's α (ha^{-1})	4.84	4.51	4.30	3.53	2.511667
Fisher's α (18 ha)	7.51	6.94	6.39	6.75	5.73

studies (e.g., Schnitzer et al., Chapter 7 in this volume), we present the data for five different liana diameter thresholds (Table 4.2).

We assessed liana diversity using Fisher's α index (Magurran 2004). The composition and abundance of the liana community are described in terms of species, genera, and families. We used non-metric multidimensional scaling (NMDS) based on the Morisita-Horn similarity index to characterize liana community structure (using "Vegan" – version 2.0-9; Oksanen et al. 2009). We conducted a partial Mantel test to examine how species composition varied spatially between the study plots, following the methods of Chua et al. (2013). The climbing mechanism (the stem and branch twiners, the scramblers, the tendril climbers and root climbers) and the dispersal modes (biotic and abiotic) are described based on field observations and referencing floristic manuals. Data analyses were carried out using R.2.15.0 (R development core team, 2012).

RESULTS

We sampled a total area of 109.4 ha and found a total of 61,789 liana individuals representing 237 species (including three unidentified morphospecies and two taxa recognized only to family) in 135 genera and 56 families (Table 4.1). Liana density varied significantly across the four forest types (F 3, 100 = 27.98, $p < 0.01$). Liana density was greatest in DEF sites with 845.7 individuals ha^{-1} followed by SEF sites with 647.7 individuals ha^{-1}, SDF (594.8 individuals ha^{-1}), and least in WEF sites (309.3 individuals ha^{-1}). A post-hoc

Tukey HSD pair-wise test revealed that liana density for WEF sites was significantly lower than in other forest types ($p < 0.05$), whereas the other forest types did not vary significantly among each other ($p > 0.05$ in all cases). Nonetheless, the pattern of increasing liana density with decreasing mean annual precipitation and increasing dry season length is consistent with larger, global datasets (Schnitzer 2005; DeWalt et al. 2010; De Walt et al., Chapter 11 in this volume).

Liana diversity in tropical wet and dry forests on peninsular India

Because our sampling scheme differed in the size and number of subplots and total area sampled among forest types, we cannot make reliable direct comparisons of species richness among forest types. Nonetheless, our dataset generated some patterns that suggest that liana diversity increases with increasing rainfall. In the most species-poor DEF sites, there were 56 species among the 18 sites (11–31 species ha^{-1}per site). The SDF sites had 70 species (12–35 species ha^{-1}per site) in 27 families among the six sites. The SEG sites had 126 species representing 36 families (23–43 species ha^{-1}per site). Species richness in the WEF sites was also 126 species, but in 51 families, with high intra-site variation (13–43 species ha^{-1}). An ANOVA of mean liana species richness per hectare of forest types revealed that the values differed significantly (F 3, 100 = 16.68, $p < 0.01$). A post-hoc Tukey HSD pair-wise test revealed that WEF-SEF did not differ significantly ($p > 0.05$), while the pairs SEF-SDF, SEF-DEF, SDF-WEF and WEF-DEF were all significantly different ($p < 0.05$).

When compared at 1-ha scale for all lianas ≥1.6 cm diameter, liana species richness increased from the dry forest to the wet forest sites, with mean species richness lowest in the DEF sites (20.94 species ha^{-1}), followed by the SDF sites (25 species ha^{-1}) and the most species in the SEF and WEF sites (mean 31.3 and 31.4 species ha^{-1}, respectively). Diversity measured as Fisher's α index ha^{-1} revealed a similar pattern: the DEF sites had the lowest value (4.84), the SDF sites had intermediate values (5.61), and the WEF and SEF sites had the highest values (7.39 and 7.11 respectively; Table 4.2). A partial Mantel test showed that spatial distance between the plots was strongly correlated with liana species composition ($r = 0.439$, $p < 0.001$).

Variation in species composition among forests of peninsular lianas

The floristic composition of lianas across the four forest types varied considerably. Of the 237 species, just two species, *Ipomoea staphylina* and *Ventilago madraspatana*, were common to all the forest types and together they formed 0.76, 1.18, 2.20 and 6.7% of liana abundance respectively in the WEF, DEF, SDF and SEF sites. Species unique to each forest type revealed that the WEF had the highest number of unique species (58% total species richness and 52.7% of abundance), followed by the SEF (with 25% of species richness and 1.95% of abundance), the DEF (with 10.2% of species richness and 42.5% of abundance) and the SDF (with just 6.6% of species richness and 2.74% of abundance). Of the 237 species, 57 were common to SDF and SEF sites (these species respectively contributing 81.4 and 45.2% of species richness and 95.3 and 81.7% of total liana abundance of the respective forest types). Among these 57 species, 12 species (21.4% of species richness) were recorded from the DEF sites and they contributed 49.8% of total liana abundance there; 13 of these 57 species (10.7% of species richness) occurred in the WEF sites (which contributed to 10.8% of the total liana species abundance in the WEF).

NMDS ordination

Non-metric multidimensional scaling (NMDS) ordination of all sites based on a similarity index between sites (the Morisita function), produced a two-axis dimension solution, with a final stress of 0.12 after 30 tries. The four forest types diverged (Fig. 4.3), with WEF sites clustering on the right side of the ordination, (except for Kalrayans, Eastern Ghats, placed closer to the SEF sites); on the left side of the ordination the DEF sites form a group, adjacent to SDF and SEF sites which forms another single group. Thus, greater similarity was found among the composition of the dry forest species.

Species and family dominance

The distribution of lianas in the four forest types showed a pronounced dominance by a limited subset of species. Of the 237 species recorded from the 40

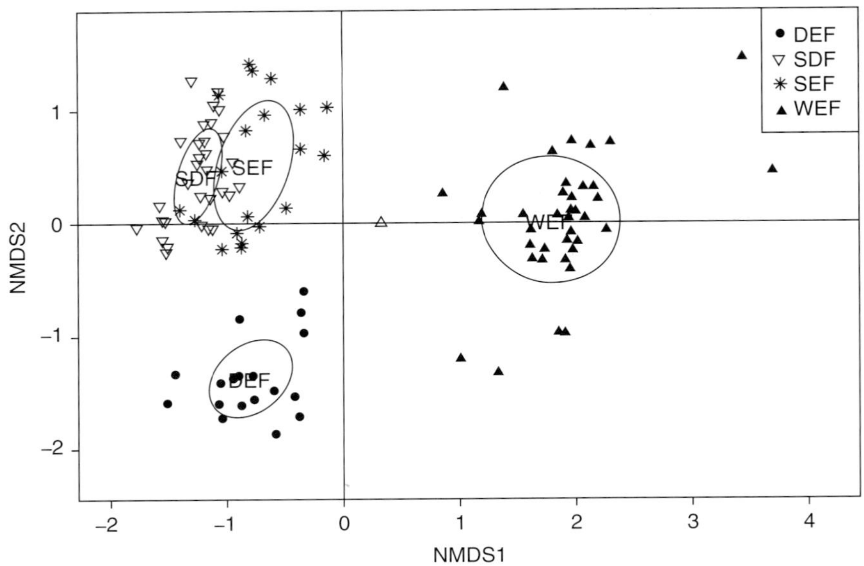

Fig. 4.3 NMDS ordination of four forest types using Morisita-Horn similarity index, presented in two dimensions (stress, 0.12). Note: Name of forest types: Dry evergreen forest (DEF), Seasonal dry forest (SDF), Semi-evergreen forest (SEF) and Wet evergreen forest (WEF).

sites, 25 species were the first or second most abundant species. Among them, *Piper nigrum*, *Olax scandens*, and *Connarus sclerocarpus* were well represented in the WEF sites of the Western Ghats; *Lantana camara* dominated the WEF of the Eastern Ghats. Other predominant species were unique to particular sites (Table 4.3). *Lantana camara*, *Pterolobium hexapetalum* and *Acacia torta* were the predominant species in the SEF and SDF sites of the Eastern Ghats, whereas *Strychnos lenticellata*, *Combretum albidum*, and *Reissantia indica* were predominant in the DEF sites. Thirty-two of the total 237 liana species comprised the top ten abundant species in all sites. These species dominated the liana communities among the respective forest types. The contribution of the top ten abundant liana species from each forest type revealed that species dominance increased from wet to dry forests. Collectively the top ten species contributed 48.6%, 66.2%, 68.9% and 73.4% individuals to the WEF, SEF, DEF and SDF respectively, demonstrating that species dominance increased from wet to dry forest types.

The top ten species of WEF sites occurred only in the WEF sites: *Olax scandens* (9.53 abundance), *Piper nigrum* (7.63%), *Chilocarpus atrovirens* (5.74%), *Calamus gamblei* (4.73%), *Connarus sclerocarpus* (4.69%), *Kunstleria keralense* (3.73%), *Artabotrys zeylanicus* (3.68%), *Calamus thwaitesii* (3.45%), *Ancistrocladus heyneanus* (2.81%), and *Gnetum ula* (2.63%). Similarly, the DEF sites had three unique species, including *Strychnos lenticellata* (18.1%), *Reissantia indica* (7.47%), and *Derris scandens* (5.81%). The top ten abundant species, pooled for all sites, contributed 63.94% of total liana abundance. Sites SEF and SDF shared six species of their top ten species. They included *Lantana camara*, *Pterolobium hexapetalum*, *Scutia myrtina*, *Ziziphus oenoplia*, *Acacia torta* and *Acacia caesia* and collectively form 45.9% of abundance in the SEF and 59.1% of abundance in the SDF. The SDF shared one species with DEF (*Hugonia mystax*) among the top ten species and it contributed 4.18% in the SDF and 5.6% in the DEF. Species unique among the top ten of the SDF sites included *Capparis sepiaria* var. *sepiaria* (2.34% of abundance) and *Grewia oppositifolia* (3.63%). Three species were unique among the top ten of SEF, including *Maytenus heyneana* (3.75%), *Secamone emetica* (2.17%), and *Ventilago madraspatana* (3.91%). The SEF sites shared only one species, *Jasminum angustifolium* with the DEF sites among the top ten abundant species and its contribution to liana abundance was 10.44% in the SEF and 4.20% in the DEF.

Table 4.3 Liana diameter threshold, species richness, top three most speciose families and most abundant liana species in four tropical forest types of peninsular India.

Forest type	Diameter threshold (cm)	Total area sampled (ha)	Species richness	Three most speciose families	Three most abundant species
Wet evergreen forest (WEF)	≥1	38.4	126	Vitaceae Papilionaceae Apocynaceae	*Piper nigrum* *Olax scandens* *Connarus sclerocarpus*
Semi-evergreen forest (SEF)	≥1.6	21.5	126	Asclepiadaceae Papilionaceae Apocynaceae	*Lantana camara* *Pterolobium hexapetalum* *Jasminum angustifolium*
Seasonal dry forest (SDF)	≥1.6	31.5	70	Apocynaceae Oleaceae Vitaceae	*Lantana camara* *Pterolobium hexapetalum* *Ziziphus oenoplia*
Dry evergreen forest (DEF)	≥1	18	56	Capparaceae Papilionaceae Asclepiadaceae	*Strychnos lenticellata* *Combretum albidum* *Reissantia indica*

At the family level, of the total 56 families, the Apocynaceae, Asclepiadaceae, Fabaceae and Oleaceae constituted the most speciose families in our sites (Table 4.3). The Annonaceae and Connaraceae were also well represented in WEF sites. The Capparaceae and Papilionaceae formed the most speciose families, along with Asclepiadaceae, in the DEF sites.

Climbing mechanisms

Five climbing mechanism were employed by lianas in the four forest types (Fig. 4.4A). All five climbing mechanisms were represented in WEF sites, with the exception of one site (Shervarayan, Eastern Ghats) at which only two modes were noted. The lianas of SDF and DEF sites displayed four different climbing mechanisms. In SEF sites, lianas used four or five different climbing mechanisms (Fig. 4.4A). Stem twiners and scramblers were most common in all forest types, as compared to other climbing mechanisms (Fig. 4.4B). Root climbers were absent in the DEF and SDF sites. Twining was also the most abundant mechanism by the measure of stems bearing this mechanism, particularly in the DEF and WEF sites (Fig. 4.4B). In contrast, in the SDF and SEF sites, scramblers were as abundant as twiners.

Liana dispersal ecology

Overall, biotic dispersal was uniformly common in the four forest types in terms of species richness and abundance (Figs. 4.5A, 4.5B). While the percentage of species with biotic dispersal varied only marginally (65–68%) across the forest types, liana stem density was almost twice as high for abiotic dispersal in dry evergreen forest compared to other forest types, indicating that drier forests have more individuals that disperse by wind.

DISCUSSION

We found that in peninsular India, liana density decreased significantly with increasing annual rainfall and increased with increasing dry season length across the tropics, and our results are consistent with global datasets (DeWalt et al. 2010; De Walt et al., Chapter 11 in this volume). However, the dry forests in our study were more disturbed and, since local liana abundance and distribution are linked tightly to disturbance (e.g., Schnitzer et al., Chapter 7 in this volume), our results may be driven by a combination of rainfall, dry season length, and disturbance. Likewise, the positive pattern

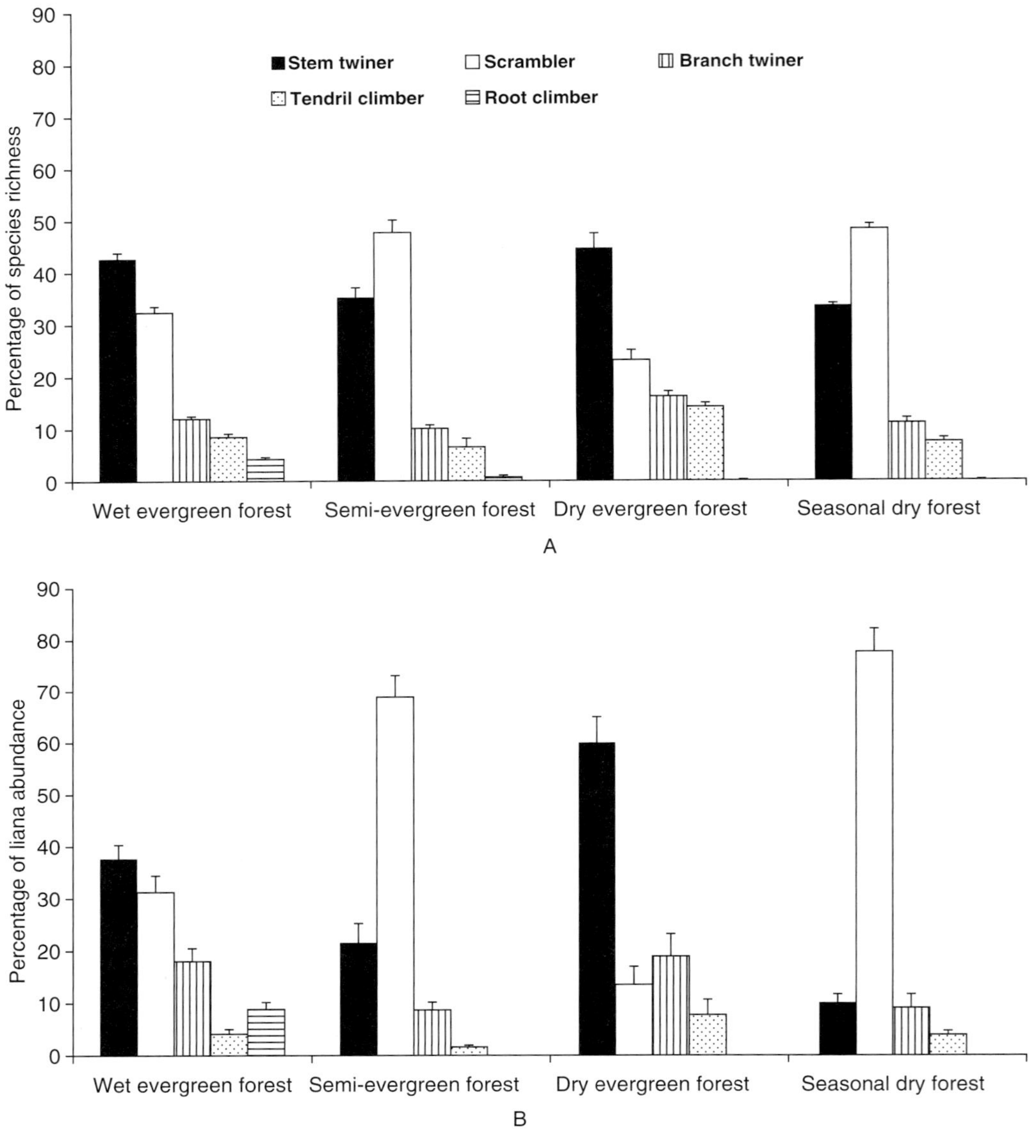

Fig. 4.4 Percentage of liana species richness and density under five climbing modes in four forest types of peninsular India.

of liana species richness with forest type may also be linked, to some degree, by disturbance. While our evidence and that of other authors (Phillips et al. 2002; Schnitzer 2005; DeWalt et al. 2010; DeWalt et al., Chapter 11 in this volume) suggest that liana abundance and diversity can be predicted by seasonality of rainfall and disturbance, other factors such as soil fertility (Putz & Chai 1987; Balfour & Bond 1993; Powers, Chapter 14 in this volume) may also play a minor role in liana distribution.

The occurrence of shade-loving root climbers (*Piper* sp., *Pothos scandens and Rhaphidophora laciniata*) in the

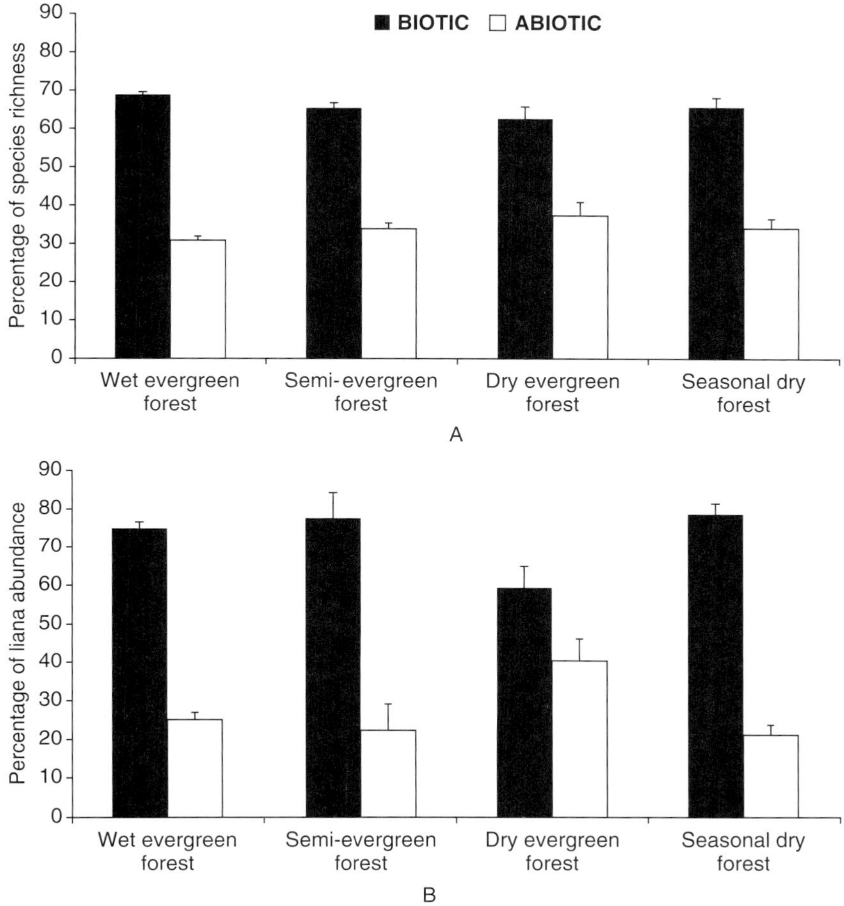

Fig. 4.5 Percentage of liana species richness and density under biotic and abiotic dispersal modes in four forest types of peninsular India.

WEF and SEF sites is attributable to forest stratification, greater rainfall, and the shorter dry season. Their absence in the drier forests supports this notion. Species unique to each forest type revealed that the number of unique species (DEF-14 species; SDF-9 species; SEF-34 species; WEF-79 species) increases from dry to wet forests. In this respect, the similarity of the SEF and the WEF sites, with almost equal species richness, suggests the abiotic similarity of the two forest types.

The two most dominant and ten most abundant species of inventoried sites and forests, respectively, displayed differences within and among the forest types, suggesting that dominants in the liana community may change locally or regionally for Indian peninsular forests. This phenomenon is contrary to the results of liana communities obtained from the western Amazonian forests, which may be dominated by fewer species (Macia & Svenning 2005). Exclusive distribution of species in one forest type or forest site can be explained by the dispersal ability and ecological adaptation to environmental conditions of different species by the so-called mass effect hypothesis, which maintains that species disperse their seeds over long distances and, therefore, there could be more species growing in habitats if they could maintain viable populations over the long term (Shmida & Wilson 1985). Our study also shows the proportion of dominance by a limited set of species, which was larger in the WEF and DEF sites. Ten species occurred exclusively in the WEF sites, while the SEF and the SDF had the smallest number of unique taxa. The SDF and the SEF showed 60% shared dominant species (defined by the basis of number of stems), which is due to the floristic similarity of these forest types in the Eastern Ghats. The NMDS ordination exhibits distinctions among the four forest types, which vary in species density as well as in site characteristics such as altitude and length of dry season (Fig. 4.3).

Stem twiners and scramblers were the predominant climbing mechanisms, both in terms of species diversity and density in four forest types (Figs. 4.4A, 4.4B). These climbing mechanisms were three to five times more common than tendril, branch twiners, and root climbers. These results are consistent with other studies reporting that stem twining is the predominant climbing mechanism (Jongkind & Hawthorne 2005; Kuzee & Bongers 2005; Gianoli, Chapter 18 in this volume), especially in older forests (DeWalt et al. 2000). Among the forest types studied, tendril climber density is greater in the DEF as compared to other forest types. This trend may be due to the low stature of the DEF and the putatively high abundance of small trellises, consistent with other research (DeWalt et al. 2000; Parthasarathy et al. 2004; Campanello et al. 2007).

Biotic dispersal mode was prevalent and equal in proportion among species in all four forest types (Fig. 4.5A). There was a higher density of individuals with abiotic dispersal syndromes in the DEF compared to the WEF (lower panel, Fig. 4.5B). Our results are consistent with other studies reporting that seed dispersal syndromes are often correlated with seasonality and precipitation, with wind-dispersed seeds common in highly seasonal forests and far less common in aseasonal forests (Gentry 1982, 1991b; Parthasarathy et al. 2004) and in seasonally dry tropical forest in Mexico (Solorzano et al. 2002). The fact that the majority of lianas in peninsular India forests showed biotic dispersal implies a faunal dependence and emphasizes the need for a holistic approach in conservation to protect forest sites with their fauna and flora (Schnitzer et al., Chapter 1, in this volume), specifically the tree communities which provide physical support for lianas, the liana community which has a positive effect on many forest animals (e.g., Arroyo-Rodríguez et al., Chapter 27 in this volume; Yanoviak, Chapter 24 in this volume), and the fauna that are vital for pollination and seed dispersal.

REFERENCES

Balfour, D.A. & Bond, W.J. (1993) Factors limiting climber distribution and abundance in a southern African forest. *Journal of Ecology*, **81**, 93–99.

Campanello, P.I., Garibaldi, J.F., Gatti, M.G. & Goldstein, G. (2007) Lianas in a subtropical Atlantic forest: host preference and tree growth. *Forest Ecology and Management*, **242**, 250–259.

Champion, H.G. & Seth, S.K. (1968) *A Revised Survey of Forest Types in India*, Manager of Publications, Delhi.

Chua, S.C., Ramage, B.S., Ngo, K.M. et al. (2013) Slow recovery of a secondary tropical forest in Southeast Asia. *Forest Ecology and Management*, **308**, 153–160.

DeWalt, S.J., Schnitzer, S.A., Chave, J. et al. (2010) Annual rainfall and seasonality predict pan-tropical patterns of liana density and basal area. *Biotropic*, **42**, 309–317.

DeWalt, S.J., Schnitzer, S.A. & Denslow, J.S. (2000) Density and diversity of lianas along a chronosequence in a central Panamanian lowland forest. *Journal of Tropical Ecology*, **16**, 1–19.

Edwards, W., Moles, A.T. & Franks, P. (2007) The global trend in plant twining direction. *Global Ecology and Biogeography*, **16**, 795–800.

Gentry, A.H. (1982) Patterns of neotropical plant species diversity, in *Evolutionary Biology* (eds M.K. Hecht, B. Wallace & G.T. Prance), Plenum Press, New York, pp. 1–84.

Gentry, A.H. (1991a) The distribution and evolution of climbing plants, in *The Biology of Vines* (eds F.E. Putz & H.A. Mooney), Cambridge University Press, Cambridge, pp. 3–49.

Gentry, A.H. (1991b) Breeding and dispersal systems of lianas, in *The Biology of Vines* (eds F.E. Putz & H.A. Mooney), Cambridge University Press, Cambridge, pp. 393–426.

Jongkind, C.C.H. & Hawthorne, W.D. (2005) A botanical synopsis of the lianes and other forest climbers. In *Forest Climbing Plants of West Africa: Diversity, Ecology and Management* (eds. F. Bongers, M.P.E. Parren, & D. Traore), pp. 19–39. CABI Publishing, Wallingford.

Kuzee, M.E. & Bongers, F. (2005) Climber abundance, diversity and colonization in degraded forests of different ages in Côte d'Ivoire. In *Forest Climbing Plants of West Africa: Diversity, Ecology and Management* (eds. F. Bongers, M.P.E. Parren, & D. Traore), pp. 73–91. CABI Publishing, Wallingford.

Macia, M.J. & Svenning, J.C. (2005) Oligarchic dominance in western Amazonian plant communities. *Journal of Tropical Ecology*, **21**, 613–626.

Magurran, A. (2004) *Measuring Biological Diversity*, Blackwell, Oxford.

Muthumperumal, C. & Parthasarathy, N. (2010) A large-scale inventory of liana diversity in tropical forests of South Eastern Ghats, India. *Systematics and Biodiversity*, **8**, 289–300.

Nabe-Nielsen, J. (2001) Diversity and distribution of lianas in a neotropical rain forest, Yasuni National Park, Ecuador. *Journal of Tropical Ecology*, **17**, 1–19.

Oksanen, J. et al. (2009.) *Vegan: Community Ecology Package*. R package version **1**.15-4.

Parthasarathy, N., Muthuramkumar, S. & Reddy, M.S. (2004) Patterns of liana diversity in tropical evergreen forests of peninsular India. *Forest Ecology and Management*, **190**, 15–31.

Phillips, O.L., Martinez, R.V., Arroyo, L. et al. (2002) Increasing dominance of large lianas in Amazonian forests. *Nature*, **418**, 770–774.

Putz, F.E. & Chai, P. (1987) Ecological studies of lianas in Lambir National Park, Sarawak, Malaysia. *Journal of Ecology*, **75**, 523–531.

R Core Team (2013) R: a language and environment for statistical computing. R Foundation for Statistical Computing, Vienna, Austria. URL http://www.Rproject.org/.

Rodgers, W.A. & Panwar, H.S. (1988) *Planning a Wildlife Protected Area Network in India*. Vols 1, 2, Wildlife Institute of India, Dehradun.

Schnitzer, S.A. (2005) A mechanistic explanation for global patterns of liana abundance and distribution. *American Naturalist*, **166**, 262–276.

Schnitzer, S.A. & Bongers, F. (2002) The ecology of lianas and their role in forests. *Trends in Ecology and Evolution*, **17**, 223–230.

Senbeta, F., Schmitt, C., Denich, M. et al. (2005) The diversity and distribution of lianas in the Afromontane rain forests of Ethiopia. *Diversity and Distributions*, **11**, 443–452.

Shmida, A. & Wilson, M.V. (1985) Biological determinants of species diversity. *Journal of Biogeography*, **12**, 1–20.

Solorzano, S., Ibarra-Manriquez, G. & Oyama, K. (2002) Liana diversity and reproductive attributes in two tropical forests in Mexico. *Biodiversity and Conservation*, **11**, 197–212.

Chapter 5

DIVERSITY AND DISTRIBUTION OF LIANAS IN YASUNÍ, ECUADOR

Robyn J. Burnham[1] *and Hugo G. Romero-Saltos*[2,3]

[1]University of Michigan, Ann Arbor, MI, USA
[2]Pontificia Universidad Católica del Ecuador, Quito, Ecuador
[3]Universidad Internacional del Ecuador, Quito, Ecuador

Ecology of Lianas, First Edition. Edited by Stefan A. Schnitzer, Frans Bongers, Robyn J. Burnham, and Francis E. Putz.

Companion Website: www.wiley.com/go/schnitzer/lianas

OVERVIEW

Based on intensive inventories of lianas in Yasuní National Park and the Woarani Ethnic Reserve of lowland eastern Ecuador, this chapter describes the taxonomic diversity and species abundance of climbing plants in one of the mega-diverse regions of the neotropics. This combined survey by two independent researchers is nearly unique in the detail paid to species-level occurrences of lianas in areas previously surveyed for tree composition and abundance. Our comparative analysis of these surveys shows that rank order among the 10 dominant species is not consistent across even small areas (10 km) of high diversity forest in western Amazonia. However, a relatively small handful of abundant species can dominate over large areas (thousands of hectares), in which those few species are almost always one of the ten most frequent. While it is possible for a liana species to be locally hyper-abundant, only a few kilometers away, it may not be among the 50 most abundant species. When many small-sized samples are collected over a large area, such pseudo-dominance can be detected. Dispersed sampling highlights species with broader dominance, however dispersed sampling can create greater taxonomic uncertainty because of singleton individuals, an important trade-off for species-level censuses. While minimizing singletons is not the goal *per se*, substantial doubt in diversity values is validly raised when a single sterile specimen is used to define a distinct morphotype. We suspect that there will be little resolution of this problem, without extensive comparative DNA sequencing both among and within regions. We also found that several of the dominant species in the Yasuní region are dominants elsewhere in the neotropics, suggesting a potential role for wide ecological tolerances among some of the dominant species. Conversely, other dominants have a restricted range and their local dominance may be due to abundant resprouting combined with depressed long-distance dispersal. Family-level diversity of lianas is not correlated with whether a species ascends to dominance in Yasuní or not. Dominant species in this hyperdiverse area of western Amazonia are derived from families whose lianas are represented by very few species (e.g., *Petrea*: Lamiaceae; *Combretum*: Combretaceae) as well as in those families with several diverse lianoid genera (e.g., Fabaceae, Bignoniaceae). Whether this pattern is typical of Amazonia or of other tropical areas remains to be determined as larger systematically resolved censuses are completed.

INTRODUCTION

Yasuní National Park, on Ecuador's eastern border, is a site of exceptionally high species richness and moderate endemism of trees, mammals, amphibians, birds, ants, beetles, fish, and insects (Bass et al. 2010). There are also hundreds of liana species at Yasuní, though few that have been identified are endemic (Nabe-Nielsen 2001; Romero-Saltos et al. 2001; Burnham 2002, 2004; Bass et al. 2010; Macía 2011). Few other neotropical rainforest sites have been fully censused for liana diversity, so any claim for the highest species richness awaits the results of both broad and detailed taxonomic surveys elsewhere. Several such surveys are ongoing at other sites in Amazonia and in other tropical regions. Our research on the diversity of climbing species (variously called lianas, vines, and/or scramblers) at two very different spatial scales, serves as a well-documented, north-western anchor-point in Amazonia for the diversity of lianas. Our interests are in the taxonomic composition of the climbing community of Yasuní, the consistency of species dominance from site to site, and the identity of species whose distributions are substantially broader than most, as well as those that are more narrowly restricted. The latter interest grows from a shared concern for conservation of Amazonian biodiversity, the extent of which cannot be understood and protected if taxonomic problems muddy the water. Species under threat of extinction due to small population sizes, or species whose spread may harm the viability of other populations, are of particular concern.

Here we focus on what we consider the two best inventories of lianas in Yasuní, both in terms of sample size and in quality of taxonomic identification: the 60 0.04 ha plots (4348 stems, of Burnham (2002, 2004) and the 30 0.04 ha plots (1919 stems) of Romero-Saltos (2011). The sample size in these two studies is larger than other comparable sampling of lianas in Yasuní (cf. Nabe-Nielsen 2001; Romero-Saltos et al. 2001; Macía 2011) either because of a lower diameter cutoff limit or because a larger area was sampled. Burnham's work (2002, 2004) provided preliminary results on the diversity, composition, and distribution of climbing plants throughout the Yasuní National Park (YNP) and the contiguous Waorani Ethnic Reserve (WER) of eastern Ecuador. Subsequently, Romero-Saltos (2011) performed an intensive census in the Yasuní 50-ha Forest Dynamics Plot, associated with the worldwide network of CTFS plots. These two independent censuses, focused at two different scales of observation, give a quantitative picture of the measurable and potential biodiversity among climbing plants in YNP, as well as a population-level perspective on dominant and uncommon species. Fortuitously, the area of plots in our studies is the same (400 m^2), creating a solid comparison of methods and results, previously a problem for liana researchers (Gerwing et al. 2006; DeWalt et al. 2010). Intensive surveys of lianas also are available in the Yasuní area from studies by Nabe-Nielsen (2001) and by Macía (2011). These surveys are not included here because Nabe-Nielsen's was a spatially limited survey (606 stems in 0.4 ha from a limited area), which measured stem diameters at 1.3 m height, instead of 20–40 cm, and lacks verified taxonomic identifications. Macía's study was focused on two sites within the Yasuní area, but included only 750 stems of lianas, all of which were larger than 2.5 cm diameter.

Given the limitations of identification of sterile census samples, we acknowledge three additional major lacunae in data on liana biogeography, evolution, and systematics in Amazonia. First, there is no clear protocol for defining the geographic range of a species of climbing plant (or any plant for that matter) in the Western Hemisphere. Before real headway can be made, the relationship between the non-random vouchered collections of systematists/ecologists and the full geographic range of a species must be determined. Second, the population sizes, shapes, densities, and variability of even the most common liana species of the Amazon Basin are unknown. We can make educated predictions based on current collections, but sampling is uneven and these are simply predictions. Third, modern phylogenetic frameworks are available for very few clades of lianas at the species level, which thwarts efforts to correlate species traits with evolutionary diversification. Exceptions are the recent treatments of the Bignoniaceae, Menispermaceae, Salacioideae, *Cayaponia*, and *Vitis* (Lohmann 2006; Olmstead et al. 2009; Ortiz et al. 2007; Hoot et al. 2009; Coughenour et al. 2010; Duchen & Renner 2010; Tröndle et al. 2010). All three lacunae represent basic biodiversity dimensions that currently are unresolved and could substantially revise the results presented here.

We addressed three main hypotheses for the dominant species in Yasuní. Dominant species here are defined as those with a high rank order relative to other species. Because the single most dominant liana species in Yasuní contributes about 10% of all liana individuals censused, the dominance structure is clearly that of a high diversity community, in which many species are represented by few individuals.

Hypothesis 1. The rank order of dominant liana species in the diverse tropical forest in Yasuní is not consistent across relatively small areas, even with large sample sizes. We reasoned that because lianas reproduce both clonally and sexually, they can be represented by isolated "frontier" individuals as well as by dense long-lived aggregations. This type of population structure should create a relatively unpredictable rank of species density over regions larger than 100 hectares, even if the top ten species in rank are often the same.

Hypothesis 2. Families represented by a large number of liana species are not necessarily well represented among liana dominants. Because species within families are often similar in their dispersal characteristics, resprouting capacities, climbing mechanisms, and ecological tolerances, we anticipated dominance by closely related species, more so than would be drawn from the flora at random. Different selective filters will be present during the evolution of new species which promotes high family diversity, (filters such as reproductive isolation, and success in novel ecological settings) compared with ecological filters determining local and regional success of a single species (i.e., liana dominance, filters such as rampant resprouting, long-distance dispersal, and effective regeneration from seed).

Hypothesis 3. Species attaining local dominance will be the most likely to have widespread geographic ranges, and further, locally dominant species are predicted to have a higher proportion of wind-dispersed fruits or seeds than the complete liana flora of Yasuní. Consistent presence over large areas suggests that widespread species would be both dispersed over long distances and ecologically tolerant. Both characteristics should aid species in achieving dominance locally. The combination of local dominance and broad geographic distribution makes these liana species good candidates for ecological disruptors if they hyper-dominate over large areas of forest, as may be happening in the "liana forests" of the southern Amazon (Nelson 1996; Nelson & Irmão 1998; Saatchi et al. 2000; Perez-Salicrup et al. 2001). Currently, what we know about the dominant species in relatively undisturbed forests is that they are likely to be different species from those that dominate in disturbed areas, where rapid growth and high light tolerance are prerequisites. Such dichotomies are seen between forests of primary versus secondary status north of Manaus, Brazil (see Roeder et al. 2010).

SITES AND METHODS

Yasuní National Park and the Waorani (Huaorani) Ethnic Reserve cover approximately 1.6 million hectares of lowland tropical forest in eastern Ecuador (Finer et al. 2009). The combined area, often collectively referred to as Yasuní, begins just 30 minutes south of the equator and extends southward 1.5 degrees of latitude and east–west 1.5 degrees of longitude (Fig. 5.1). The elevation ranges between 200 and 300 m with undulating hills throughout the area (UNESCO 2011). The National Park (962,000 ha) is designated as an IUCN Level II Protected Area, while the Reserve and Park are united as a "[Hu]Man and the Biosphere Reserve." Two biological research stations, Yasuní Scientific Research Station (YSRS) and Tiputini Biodiversity Station (TBS), administered by the Pontificia Universidad Católica del Ecuador and the Universidad San Francisco de Quito, respectively, were established in the 1990 s, and have

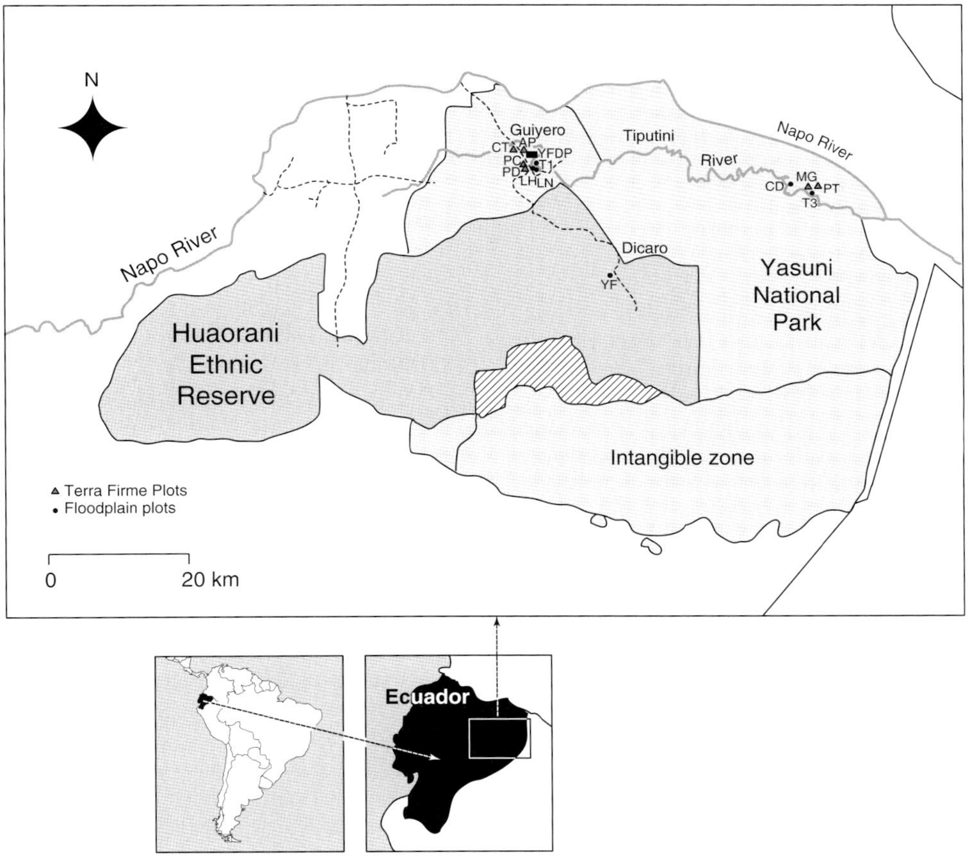

Fig. 5.1 Localities mentioned in the text. Location of the Yasuní Forest Dynamics 50-ha plot (Romero Plots) is noted by filled rectangle, just south of the name of the Huaorani village, Guiyero. Sample sites for Burnham Plots are noted by two-letter abbreviation of plot names. *Terra Firme* plots are denoted by filled triangles, Floodplain plots by filled circles.

been focal sites for evaluating the biodiversity of the region. The southern half of the WER is fully protected as an Untouchable Zone, preserved for the indigenous Waorani people of the area (Finer et al. 2009) and has not been sampled often. The most useful division of terrestrial habitats in the area is into three major habitat types (Romero-Saltos et al. 2001; Macía 2011): (1) *Terra Firme*; (2) well-drained Floodplain; and (3) *Mauritia*-dominated swamp forest. The work reported here sampled *Terra Firme* and well-drained Floodplain. To our knowledge, the only systematic sampling of lianas in the *Mauritia flexuosa*-dominated swamp was carried out using a 2.5-cm stem diameter minimum limit (Romero-Saltos et al. 2001; Macía 2011).

Two census designs were used for the data compared here, and because the sites studied were entirely distinct, we employ the terms Burnham plots and Romero plots to designate each series of sites and sampling design. Fig. 5.1 shows the location of all plots and plot codes. The census methods (outlined below) were similar enough to one another that comparison is warranted.

Burnham plots

We sampled 12 sites distributed in a broad arc through YNP and the WER (Fig. 5.1), from near the community of Guiyero in the north to just south of the community of Dicaro, about 65 km apart by road (Burnham 2002, 2004). Half of the Burnham sites were located in areas also censused for tree diversity and density (Pitman et al. 2001, 2002) and are neither evenly nor randomly distributed. Aerial inter-site distances range from 500 m to 45 km. Each site consists of five evenly spaced, parallel, 4 × 100 m transect plots within a single hectare (each transect area = 400 m^2, each site area = 2000 m^2, combined area fully sampled for all 12 sites and 60 plots = 2.4 ha). We sampled six sites in each habitat type, *Terra Firme* and Floodplain, with no distinction among *Terra Firme* habitat (ridge versus valley). Our minimum stem diameter for enumeration, measurement, and identification was 1.0 cm, measured at 20–40 cm above the base of the stem, avoiding any irregular or unusually thickened stem portions. We counted all individuals rooted within the 4 × 100 m transect plots, identified the species, and measured for stem diameter, as well as mapped stems to the 4 × 10 m segment of the transect in which they occurred. In addition, for stems whose primary root fell outside a transect, but that was secondarily rooted within the transect, with a stem size ≥1 cm diameter, we traced the stem to its primary rooting position beyond the transect and the diameter was measured there at 20–40 cm above the base of the primary stem. This sampling method overestimates liana biomass of the 4 × 100 m areas sampled and should not be used for biomass estimations. Excluded from samples were all epiphytes and hemiepiphytes. At least one vouchered specimen of every species censused was archived. If a specimen could not be identified in the field as belonging to a previously collected or identified species, a voucher was taken, and is archived at the University of Michigan herbarium (MICH) in Ann Arbor, MI, with duplicates at the Museo Ecuatoriano de Ciencias Naturales Herbarium (QCNE) in Quito, Ecuador, the Pontificia Universidad Católica del Ecuador Herbarium (QCA), and at the Field Museum (F) in Chicago, IL. Sites were established, measured, and sampled from 1998 to 2000 (Burnham 2002, 2004).

Romero plots

Romero-Saltos and colleagues established 30 regularly-spaced 20 × 20 m plots in a rectangular grid in 2006 and 2008 in the 50-ha Yasuní Forest Dynamics Plot (YFDP) at 00°41′14″ S, 76°23′72″ W (plots are 400 m^2, total area sampled = 1.2 ha). Each 20 × 20 m plot lies at the center of one of 30 hectares of a 5 × 6 hectare grid at the YFDP. Each plot is thus within 100 m of another plot with the greatest distance between any two plots being 640 m. The Romero plots were classified (Romero-Saltos 2011) as valley or ridge sub-habitats, all within *Terra Firme* habitat and we designate them similarly here. We enumerated, mapped, identified, and collected all liana individuals as above for Burnham plots, except that of following secondarily rooted stems to their primary rooted position: only stems with a primary root within plots were included. Hence, density of stems is anticipated to be lower for Romero's plots, and insofar as species richness is dependent on stem number, we anticipated species richness to be lower. Vouchered reference specimens are archived at the Pontificia Universidad Católica del Ecuador herbarium (QCA) and at the reference herbarium at the Yasuní Scientific Research Station (YSRS) in Yasuní (Romero-Saltos 2011).

SUMMARY CHARACTERIZATION OF THE LIANA COMMUNITY IN YASUNÍ

The liana community of Yasuní, Ecuador, is a highly diverse and spatially heterogeneous assemblage including at least 47 families, 130 genera, and a minimum of 350 species of woody climbing plants (classification of APG III, Haston et al. 2009; APG III 2009). This value represents the minimum species richness for Yasuní because we do not include species found beyond our spatially limited census sample. Collectively, we have vouchered ≥150 additional species within Yasuní that were not recorded from our limited censusing areas (Burnham and Romero, unpublished data). Yasuní has been systematically censused for lianas in various configurations and at various scales by at least four teams of researchers (Burnham 2002, 2004; Macía & Svenning 2005; Nabe-Nielsen 2001; Romero 2011). Our estimate therefore, is that at least 550 species of lianas currently inhabit the forests of the YNP and WER. Here we present a community characterization combining the two census designs, allowing comparison with other diverse liana communities.

A small suite of abundant species, all of which are found in most areas, dominates the Yasuní community of lianas. No dominant species comprises more than 11% of all stems, and 28 species collectively contribute to 50% of all stems. Locally, dominants may cede the top position(s) to other climbing species, but when viewed at a larger spatial scale, the abundance patterns consistently show a small handful of species as dominants. Climbing species that show up time and again with the highest basal area and abundance are *Machaerium cuspidatum*, *Combretum laxum*, *Petrea maynensis*, *Clitoria pozuzoensis*, *Paullina bracteosa*, *Bauhinia rutilans*, *B. guianensis*, and *Cuervea kappleriana*.

Estimated species richness varies from 160 to more than 240 species per hectare, the upper value an astoundingly high value, especially compared to other well-studied sites (e.g., Thomas et al., Chapter 2 in this volume; Bongers & Ewango, Chapter 3 in this volume; Schnitzer et al., Chapter 7 in this volume), but it mirrors the high tree diversity in Yasuní (Valencia et al. 2004; Condit et al. 2005). Liana species richness (excluding all unidentified stems) is slightly higher in Burnham's *Terra Firme* plots than in Floodplain plots, 35.8 versus 32 species, respectively (Table 5.1), but Romero's *Terra Firme* plots include an average of fewer species (27.3). When plots are arrayed along a moisture gradient, Floodplains have the lowest values of diversity as measured by Fisher's alpha, the moist valley habitats of *Terra Firme* (segregated by Romero only) have the highest diversity, though the difference between the two *Terra Firme* sub-habitats sampled by Romero was not significant (Romero-Saltos 2011). In other words, while Romero's plots show a lower average species richness, the number of species per individuals sampled is not substantially different (Table 5.1).

Lianas density is as high or higher in Yasuní forests as has been observed in the majority of moist, low-disturbance neotropical forests (Rice et al. 2004; Campanello et al. 2012; Schnitzer et al. 2012; Schnitzer et al., Chapter 7 in this volume). We found 60–76 individuals ≥1 cm in every 400 m^2 plot, which extrapolates to 1500–1900 individuals per hectare. Roughly half as many liana stems were found in the forests of La Selva, Costa Rica (Mascaro et al. 2004; Yorke et al. 2013) and in Puerto Rico (Rice et al. 2004), in the latter case even compared to their samples of older forests that had accumulated both large and abundant lianas. However, Yasuní forests include slightly less than half as many lianas as the hyper-abundant liana forests from lowland Bolivia (Pérez-Salicrup et al. 2001). Interestingly, the two 0.2 ha plots censused by Nabe-Nielsen (2001) in Yasuní included 606 liana stems ≥1 cm dbh. This extrapolates to 1515 lianas per hectare, squarely within the per hectare envelope of our independent censuses. An important benchmark, because of the magnitude of the liana census, is the 50-ha plot on Barro Colorado Island (BCI), Panama. There, Schnitzer and colleagues (2012; Chapter 7 in this volume) found slightly fewer liana stems ≥1 cm (1350 stems ha^{-1}), even including rooted clones as stems. There, the second species in dominance is the same Bignoniaceae species (*Paragonia pyramidata* = *Tanaecium pyramidatum*) that is second in dominance in Burnham's plots in Yasuní (Burnham 2002; Schnitzer et al. 2012).

The basal area of lianas in Amazonian forests is an important currency for comparison among sites because carbon sequestration by forests is directly related to the standing carbon stock and the potential photosynthetic area (Phillips et al. 2005; van der Heijden & Phillips 2008; Schnitzer & Bongers 2011; van der Heijden et al., Chapter 13 in this volume). When lianas increase sufficiently to suppress tree growth, carbon stocks can diminish because a high photosynthetic area of lianas results in lower carbon accumulation. Lianas contribute less carbon in proportion to their photosynthetic area than trees do

Table 5.1 Richness, density, and dominance of lianas in Yasuní, Ecuador. Species richness, stem density, basal area, dominance, and singleton/unidentified species in Yasuní liana plots for the two census designs. Burnham plots are shown both grouped and subdivided into habitats. Mean values are followed by the range in parenthesis or the standard deviation reported as ±. Species richness of 400 m^2 plots or transects was measured directly and is shown as averages across all plots within habitat and census method. Hectare species richness and density values for both census designs were estimated by 10,000 randomizations without replacement in EstimateS (Mao-Tau) up to a total of 1 hectare. Dominant species for each census design are listed by stem number and by basal area.

	Romero: n = 30 20 × 20 m plots (*Terra Firme*)	**Burnham: 60 4 × 100 m plots**	**Burnham: 30 *Terra Firme* plots**	**Burnham: 30 Floodplain plots**
Number individuals	1919	4348	2053	2295
Individuals per 400 m^2	64 (34–100)	72 (33–142)	68 (33–115)	76 (41–142)
Estimated individuals ha^{-1}	1605 (1525–1700)	1695 (1598–1792)	1574	1811
Fisher's Alpha estimate 400 m^2	46.4 & 55.3 (ridge & valley)	39.1	48.5	29.6
Species richness 400 m^2	27.3 (12–41)	34.0 (21–59)	35.8 (21–59)	32.0 (24–44)
Est. species richness ha^{-1}	183 ± 15	220 ± 15	238 ± 8	172 ± 7
Stem basal area per 400 m^2	434 ± 238 cm^2	893 ± 322 cm^2	885 ± 338 cm^2	896 ± 310 cm^2
Species Dominant (stem number)	*Combretum laxum*	*Machaerium cuspidatum*	*Machaerium cuspidatum*	*Machaerium cuspidatum*
Species Dominant (basal area)	*Combretum laxum*	*Machaerium cuspidatum*	*Machaerium cuspidatum*	*Machaerium cuspidatum*
Singleton species	2.5% (49 of 1919)	2.0% (88 of 4348)	3% (63 of 2053)	1% (25 of 2295)
Unidentified individuals	180	304	173	131

and they also suppress tree growth when they create large photosynthetic areas (Schnitzer et al. 2014; van der Heijden et al., Chapter 13 in this volume). In non-aseasonal forests like Yasuní, lianas should show high stem basal area in two areas: (1) where forests are temporally stable, allowing for the growth of large-diameter liana individuals; and (2) where rapid dynamics via tree falls have led to resprouting of fallen liana stems and exuberant sexual regeneration in sunny gaps (Schnitzer 2005; Ledo & Schnitzer 2014). Liana basal area estimates in Yasuní range from 0.95 m^2 ha^{-1} to 2.8 m^2 ha^{-1}, a range reflecting differences between habitats, between dense tree falls versus tall, shady forests, and differences between transect versus square census plots (e.g., Schnitzer et al. 2006).

DENSITY, DIVERSITY, AND DOMINANCE OF YASUNÍ LIANAS

Although sampling areas differed in shape, the basic unit of area sampled was equivalent (400 m^2) between the two censuses, allowing for comparison and a synthetic summary. The average number of species per sample unit was not statistically different (Table 5.1) between the two studies (27 and 34), nor was the number of individuals per sample unit (64 and 72). Below, we examine the two census results in detail.

The Burnham 400 m^2 transect plots were, on average, more species-rich and denser in liana stems than the Romero square plots, but the differences were not significant (Table 5.1). Between census styles, plots that share the most similarities are the 30 *Terra Firme* plots of Burnham compared to all 30 plots of Romero (both ridge and valley in *Terra Firme*). Long thin plots (Burnham) had been anticipated to show higher liana species richness because clonal regeneration of many species should create single-species patches that can dominate square or nearly-square plots (such as those of Romero). Similarly, long thin plots could have included more individuals because of the longer boundary edge for potential genet intersection. More dramatic density results were found by Schnitzer et al. (2006) in 0.01 ha census of lianas in the *Terra Firme* forests of French Guiana, where comparison of long rectangular versus square plots showed 50% more liana stems in the rectangular samples. In contrast, our results suggest that plot shape makes less difference to density measures in Yasuní than it did in French Guiana. However, our unit of measurement was larger (0.04 ha vs. their 0.01 ha) and our minimum stem diameter was 1 cm, (versus their 0.5 cm). Likewise, on BCI, liana density did not differ with plot shape (long, thin transects versus square plots; Schnitzer, unpublished data). Both factors may provide some explanation of the differences in the results.

Dominance among Yasuní lianas

The abundance-based dominance rank of liana species differed between the censuses, suggesting an important limitation to small or geographically limited censuses for extrapolation to larger areas. Romero plots identified *Combretum laxum* (Combretaceae), a wind-dispersed liana with apical twining and opposite leaves, as the most abundant liana species in YFDP. *Combretum laxum* accounted for 8.1% of all stems, 18.6% of liana stem basal area, and included the individual with the largest stem from his census. The species is especially common in Romero valley sub-habitats, and was eighth in abundance in the Burnham censuses (1.9% of all stems and 2.8% of stem basal area). *Combretum laxum* is widespread, occurring in every country from Mexico to Argentina and has been identified as dominant from censuses in the transitional forests of central Amazonia (Oliveira 2011), in the forests of southern Colombia neighboring Yasuní to the north (Duque et al. 2004), and in the Lacandon region of southern Mexico (Ibarra-Manríquez & Martínez-Ramos 2002; Ibarra-Manríquez et al., Chapter 8 in this volume). The species was recorded as the third most abundant in samples from Yasuní (Macía 2011) and also third most abundant in a study of large-stemmed lianas 90 km north of Manaus, Brazil (Oliveira et al. 2008). Unpublished censuses at Los Amigos, Madre de Dios, Peru (Burnham and Revilla-Minaya, unpublished data) show basal area dominance by *Combretum laxum* in two 0.5 ha plots.

In contrast, Burnham (2002, 2004) found that the liana dominant among more dispersed plots in Yasuní was *Machaerium cuspidatum* (Fabaceae). The species was dominant at 8 of the 12 Burnham sites, and was among the top three species on two other Burnham plots. It comprised 10.9% of all stems from the 60 400 m^2 transect plots and contributed 11.1% of the stem biomass. It was dominant in the Macía (2011) study of the Yasuní area and among the top three dominants in the Duque et al. (2004) study of southern Colombian forests. Romero plots included 7% *M. cuspidatum* stems, second ranked for stem density in his censuses. Beyond the area of the CTFS 50-ha dynamics plot, *Machaerium cuspidatum* is the dominant liana in Yasuní.

The top ten dominants (based on stem density) from the Burnham and from the Romero plots are listed in rank order in the first two columns of Table 5.2, demonstrating the uneasy crown worn by any of the liana species of the oligarchy. Only four of the top ten species from our censuses are shared. Romero's YFDP plots are derived from *Terra Firme* habitat only, subdivided into valley and ridge. Comparing only *Terra Firme* plots (columns 3 and 5, Table 5.2), five of ten species are found in common. Also listed in Table 5.2 is the number of 400 m^2 plots on which each of the dominant species is found. Although dominance is generally well represented across all plots, there are still a few species among the top ten dominants of each census for which < 40% of the 400 m^2 plots have any individuals. These species show exceptionally high abundance on just a few plots (*Clitoria javitensis*, *Callichlamys latifolia*, *Tontelea fuliginea*, *Maripa peruviana*, and *Adenocalymma impressum*).

We proposed that biological differences and similarities among taxa could be at the heart of the dominance ranking (Hypothesis 2). In contrast, it is possible that species dominance is random (a potential interpretation under the neutral theory, Hubbell 2001): that dominance will vary from place to place by chance, if most liana species are interchangeable. This scenario would produce a dominance structure that reflects the abundance of species within families: a family is more likely to be represented among a random dominance structure if that family is species-rich. Our data do not permit a direct test of these alternatives, but can shed light on the question using the presence of characteristics relevant to dominance hierarchy. Using density data from Burnham sites, we plotted the rank order of species dominance against the species richness of the species' family among the liana community of Yasuní (Fig. 5.2). While we found that the most species-rich families are indeed represented among the dominants, several dominant species are placed in families with low species richness in Yasuní. For example, Dilleniaceae (11 liana species in Yasuní), Lamiaceae (five liana species in Yasuní), Ulmaceae

Table 5.2 Dominant species ordered by decreasing stem number in two Yasuní censuses. Species with * in columns 1 and 5 are among the ten dominants in Romero's samples (column 3). Number in parenthesis after each name is the fraction of the 400 m^2 plots in that column in which the species is found. Dispersal mechanism shown in columns 2, 4, and 6.

Burnham: *Terra Firme* and Floodplain (60 plots)	Disp	Romero: All Samples (30 plots)	Disp	Burnham: *Terra Firme* only (30 plots)	Disp
Machaerium cuspidatum * (27/30 F; 29/30 TF)	W	*Combretum laxum* (20/30)	W	*Machaerium cuspidatum* * (29/30)	W
Tanaecium pyramidatum (27/30 F; 13/30 TF)	W	*Machaerium cuspidatum* (25/30)	W	*Petrea maynensis* * (14/30)	W
Tetracera volubilis (20/30 F; 6/30 TF)	V	*Petrea maynensis* (19/30)	W	*Clitoria pozozuensis* * (16/30)	B
Celtis iguanea (24/30 F; 8/30 TF)	V	*Cuervea kappleriana* (22/30)	W	*Salacia multiflora* (19/30)	V
Clitoria pozozuensis * (17/30 F; 16/30 TF)	B	*Clitoria pozozuensis* (24/30)	B	*Bauhinia guianensis* * (20/30)	B
Sciadotenia toxifera (19/30 F; 13/30 TF)	V	*Callichlamys latifolia* (13/30)	W	*Leretia cordata* (17/30)	V
Petrea maynensis * (0/30 F; 14/30 TF)	W	*Bauhinia rutilans* (16/30)	B	*Clitoria javitensis* (13/30)	B
Combretum laxum * (21/30 F; 9/30 TF)	W	*Bauhinia guianensis* (15/30)	B	*Maripa peruviana* (12/30)	V
Clitoria javitensis (12/30 F; 13/30 TF)	B	*Paullinia bracteosa* (15/30)	V	*Combretum laxum* * (9/30)	W
Doliocarpus major (18/30 F; 0/30 TF)	V	*Tontelea fuliginea* (12/30)	V	*Adenocalymma impressum* (6/30)	W

Note: F = Floodplain, TF = Terra Firme, W = wind dispersal, V = vertebrate dispersal, B = ballistic dispersal.

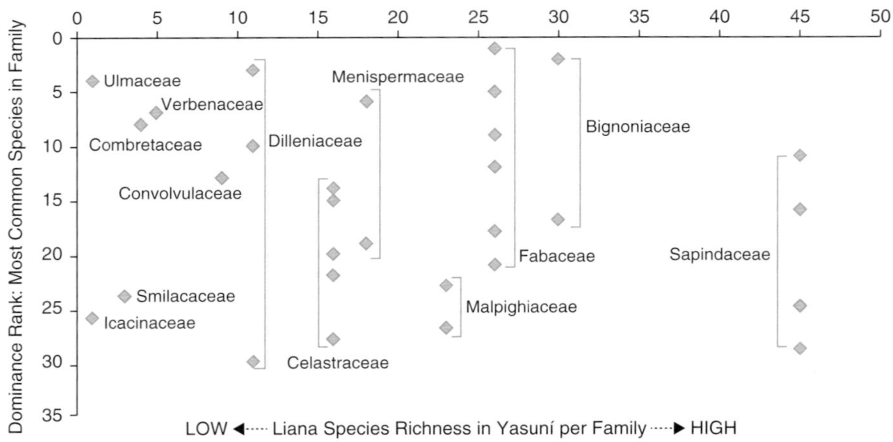

Fig. 5.2 Species-rich families are not the source of all dominant liana species in Yasuní. Liana species in Yasuní were ranked by abundance and coded by family, each point represents one of the 30 most common species of lianas. Families are plotted from left to right, by the number of liana species in Burnham's Yasuní census. If dominant species were drawn at random from all available species, we would find that dominants were represented in proportion to the number of species in the family. Instead, dominant species are found not only in species-rich families like Fabaceae and Bignoniaceae, but also in species-poor families like Ulmaceae and Verbenaceae. In addition, families such as Bignoniaceae, with many species in Yasuní, are represented by only two species among the top 30 dominants.

(one liana species in Yasuní), and Combretaceae (four liana species in Yasuní). We evaluated the correlation between the rank of the most common Yasuní species in a family and the liana species diversity of the family in Yasuní. There is no significant correlation between the two ($r^{(28)} = 0.11$, $p = 0.57$), suggesting that dominant species are not drawn at random from among the species pool in Yasuní. There may be multiple explanations for these patterns, among those explanations may be that species traits (resprouting capabilities, dispersal, probability of death on breaking, etc.) play a role in structuring the dominance hierarchy among lianas in Yasuní.

Singleton occurrences

The 60 dispersed 0.04-ha plots (2.4 ha cumulatively) censused by Burnham included 88 singleton species: represented by just a single individual across all plots (Table 5.1). Thirty *Terra Firme* plots included 63 singletons while 30 Floodplain plots included 25 singletons. Romero's 30 *Terra Firme* plots included 49 species singletons. Fisher's alpha of *Terra Firme* plots overlap between the two census methods (Table 5.1), and species richness is similar between methods (see above). Leigh (1999) demonstrated that square sample plots estimate tree diversity less well than long thin plots and this may be particularly true for lianas, because of frequent clonal growth, but may also allow for accumulation of more singletons. Romero's plots were densely sampled over a single large (30 hectares) and may partially overcome this limitation to small plot size. Higher numbers of singletons were anticipated in the more widely dispersed transect plots of Burnham, which would consequently increase Fisher's α. Taxonomic uncertainty based on the lack of the full range of diagnostic species characters in our largely sterile samples may also contribute to the presence of singleton species. There is no *a priori* reason to suspect different taxonomic resolution for the two censuses: both Romero and Burnham censuses included about 9% individuals that could not be determined to species or placed to morphospecies (Table 5.1).

Habitat diversity, stem density, and basal area

Both Romero and Burnham classified their census plots into two easily recognized habitat types. While Romero worked exclusively in *Terra Firme* forest, divided into valley and ridge sub-habitats, Burnham censused *Terra Firme* habitats (not distinguishing between ridges and valleys within plots) and Floodplain habitats. Floodplains are distinguished by their position in river flood zones, which can be inundated for up to several weeks during high water periods. In Romero's censuses within *Terra Firme* habitat, the ridge and valley were not significantly different in species richness, however, Burnham's *Terra Firme* habitats were significantly richer than the Floodplain habitats (t-test, $p < 0.002$).

Stem density per 400 m^2 plot was not significantly different between the two censuses, and yet because of the slight difference in the census methods, it might be expected that Burnham's plots showed higher density, as seen in Table 5.1. However, a wider range of density was found among Burnham plots, possibly reflecting the larger range of habitats sampled.

Basal area was significantly higher for Burnham's 400 m^2 plots than for Romero's (t-test, $p < 0.001$, Table 5.1). This is attributed directly to two important differences in sampling methods, only one of which has biological significance. First, though stem numbers were not significantly different between the census methods, Burnham measured stems from the point of original rooting by following stems to their origin, even if outside the plot boundary. Burnham thus included more large stems (largest stem 32 cm dbh, 1.8% of stems >10 cm dbh, 14.9% of stems > 5 cm dbh, 39% of stems ≤2 cm dbh) than did Romero (largest stem 31.5 cm dbh, 0.6% of stems > 10 cm dbh, 6% of stems >5 cm dbh, 59% of stems ≤2 cm dbh). In addition, while not significantly different, stem density was as high as 142 individuals in the densest 400 m^2 plot. Although including Floodplains, Burnham's wider geographic sampling does not include any exceptionally low density areas, such as *Mauritia* swamps. If our samples can be assumed to be reflective of the actual range of density, the CTFS plot sampled by Romero includes the lower density values for *Terra Firme* lianas in Yasuní.

Dispersal modes of dominant species

Based on combined data from Burnham's 12 sites (60 plots), six of the ten most common species (60%) bear fruits or seeds dispersed by wind or ballistically (including many members of Fabaceae, Bignoniaceae, and Malpighiaceae) and four are vertebrate dispersed (e.g., Celastraceae, Dilleniaceae, Ulmaceae). However, among the 25 most abundant species, 42% are winged or ballistic, and 52% are animal dispersed. In *Terra Firme* sites only (both Burnham and Romero plots), ballistically dispersed seeds are represented by about 20% of the top 25% of species, wind dispersed by 40–45% and the remaining 35–40% are vertebrate dispersed. In comparison, the 163 species identifiable at the species level, and thus classifiable for fruit and seed type, from Burnham plots include 38.5% wind-dispersed species and 2.5% ballistically-dispersed species, while vertebrate dispersal is inferred for 54% of the species (R.J. Burnham, C.V. Santanna, & C. Smith, unpublished data). The high abundance of wind- or self-dispersed species among the ten dominant species suggests that stem abundance dominance rank is positively influenced by wind-mediated seed or fruit dispersal. We acknowledge that the top ten species may not be sufficient for an adequate characterization of dispersal type, but it does represent more than 25% of all stems censused.

Finally, the deviation of dispersal agents for lianas from the pattern shown for trees in moist tropical forest demands further evaluation. Tree species are reported as being more than 70% animal dispersed in both wet and moist tropical forests (Howe & Smallwood 1982; Janzen & Vázquez-Yanes 1991). Lianas are more often wind dispersed in most tropical floras than are trees (Gentry 1982; Solórzano et al. 2002; Muller-Landau & Hardesty 2005), with values reported from 30–65% wind-dispersed liana species per flora (Gentry 1991; Ibarra-Manríquez et al. 1991; Killeen et al. 1998; Wright et al. 2007). In Central and South American tropical forests, wind dispersal predominates particularly in areas with a distinct or prolonged dry season. The liana growth-form in general is also more commonly represented among individuals (but not species) in areas with distinct dry seasons or low rainfall (DeWalt et al. 2010; De Walt et al., Chapter 11 in this volume).

GEOGRAPHIC RANGE OF DOMINANT SPECIES

We hypothesized that the geographic ranges of the dominants in Yasuní would be very large, presuming that the vigor and aggressiveness contributing to dominance on a local scale would also contribute to widespread ranges. To test this hypothesis, we selected the top five species from each dominance hierarchy, and selected five non-dominant species that we were able to consistently identify (these each with <1% of total stem number, "rare" hereafter) for comparison from each of the two census hierarchies (see Table 5.3). We did not select rare species at random because the range of the morphospecies cannot be known until certain identification is made. Rare species were chosen from species represented by one or two individuals from each census design. Once the criteria of definite identification and low frequency were met, rare species for the test were selected at random from the liana species. Our estimate of geographic range was defined as the greatest distance between two specimens available from the online databases of New York Botanical Garden, the Field Museum of Natural History, and the Missouri Botanical Garden.

Our data (Table 5.3) suggest that the average geographic range of a dominant species in Yasuní is larger than for a rare species. However, the known geographic ranges of liana species among those analyzed are too variable to assign significance to the difference. The area over which plants are censused is of critical importance when defining a dominance hierarchy for correlation with geographic ranges. The 50-ha plot might be viewed as a single sample from a small area, however intensely sampled it is, while the 12 sites sampled by Burnham may give a more complete picture of the dominance structure in Yasuní, but less detail at the local level. More dispersed sample sites can give a broader view of the regional dominance hierarchy, in that rare species are indeed found as rare over the region, instead of rare at a single site, as may be the case for some of the species that appear to be rare in Romero's samples (e.g., *Dicranostyles ampla*, *Telitoxicum minutiflorum*).

Our data indicate that, on average, locally dominant taxa have wider geographic ranges than less common species. We had anticipated this distinction to be more prominent than it was, with most rare species being limited in distribution compared to common species. Instead several of the less common species in Yasuní are geographically widespread (see Table 5.3). Three of the five less common species from Burnham censuses have geographic distributions that are as broad as the average among the five dominant species. Further, *Celtis iguanaea*, number four in dominance for Burnham plots, has one of the broadest possible neotropical distributions, but was found only three times among Romero's plots, though there each occurrence was a

Table 5.3 Geographic extent of the five most abundant species, and five low abundance species for each census of Yasuní lianas. The second and fifth columns indicate species rank under the alternate census design. All dominant species are present in both census designs, whereas NA = no abundance in the alternate census for low-abundance species. Burnham plots encompassed a substantially larger geographic area, almost twice as many 400 m^2 plots, and twice as many stems, therefore the probability of low abundance species from Burnham plots appearing in Romero's hierarchy is low. Dispersal of the rare species from Romero plots was 4 vertebrate:1 wind, and dispersal of rare species from Burnham plots was 2 vertebrate: 3 wind

		Romero plots			
Dominant species (all wind dispersed)	**Burnham rank**	**Geographic range**	**Rare species**	**Burnham rank**	**Geographic range**
Combretum laxum	8	7197.2 km	*Randia altiscandens*	219	3155 km
Machaerium cuspidatum	1	2322.7 km	*Telitoxicum minutiflorum*	117	1181 km
Petrea maynensis	7	4695.2 km	*Machaerium quinata*	NA	2569 km
Cuervea kappleriana	20	7084 km	*Neodillenia coussapoana*	NA	543.5 km
Clitoria pozuzoensis	5	1122 km	*Dicranostyles ampla*	147	3155 km
Average		4484 km			2121 km

		Burnham plots			
Dominant species (3 of 5 are wind dispersed)	**Romero rank**	**Geographic range**	**Rare species**	**Romero rank**	**Geographic range**
Machaerium cuspidatum	2	2322.7 km	*Diplopterys cabrerana*	111	3171 km
Tanaecium pyramidata	87	7763 km	*Odontocarya dielsiana*	NA	1903 km
Tetracera volubilis	54	7510 km	*Machaerium leiophyllum*	NA	>4200 km
Celtis iguanea	119	7265 km	*Tournefortia maculata*	NA	>6935 km
Clitoria pozuzoensis	5	1122 km	*Prestonia quinquangularis*	NA	>6200 km
Average		5192 km			~4482 km

liana larger than 4 cm in diameter. We conclude from these data that local dominance (as defined by small plot samples) is not closely correlated to geographic distribution.

Also interesting is the distribution of the acknowledged dominant in Yasuní, *Machaerium cuspidatum* (Nabe-Nielsen 2001; Burnham 2002, 2004; Macía 2011; Romero-Saltos 2011). This species is so abundant in Yasuní that it can easily be found within 10 minutes in almost any *Terra Firme* site that we have investigated (R.J. Burnham and Oxender, unpublished data). However, the species' geographic extent is quite limited: occurring only on the east side of the Andes from southern Colombia to northern Bolivia, and perhaps not recognized at all in Brazil (listed for the state of Amazonas, but not present in collections at INPA, Manaus). Locally uncommon species of the same genus (*M. quinata* and *M. leiophyllum*) both have broader geographic distributions (Table 5.3). A similar, possibly sister-species to *M. cuspidatum*, *M. caudatum*, is abundant and widespread in the Brazilian states of Mato Grosso, Amazonas, and Para. To our knowledge, this species does not co-exist with *M. cuspidatum* anywhere that it has been recorded or collected.

Significantly, *Combretum laxum*, the dominant on the Yasuní 50 ha plot, is reported as a dominant species in censuses from Mexico (Ibarra-Manríquez & Martínez-Ramos 2002) and in Manu, Peru (R.J. Burnham, unpublished data). Similarly, *Bauhinia guianensis* is a dominant species in censuses from Para, Brazil (Gerwing & Farias 2000) and is among the dominant ten species in *Terra Firme* samples in Yasuní.

Our evidence lends support to the hypothesis that widespread species can become dominants throughout their range. However, we currently do not support the idea that every dominant in one large area of the Amazon is common somewhere else within the Amazon. Liana species that rise to local dominance can hold that position for a variety of reasons, including being ecological generalists, frequent clonal regeneration from meristematic tissue of buds or vascular cambia, frequent flowering, successful fruit set, and/or consistent seedling establishment. Whether the rise to dominance is more often the case for widespread species than for narrow-ranged species cannot be examined until more intense local censuses are completed for lianas. Evaluation of the local dominants in light of the various routes to dominance would be extremely useful.

CONCLUSION AND FUTURE DIRECTIONS

This volume gives clear voice to the advances made in the ecology of lianas over the past 25 years. Our contribution compared different sampling methods for censusing the species of liana communities in a preserved lowland tropical forest of the Neotropics. We found more than 350 species in our combined plot sampling from a small number of sites throughout Yasuní National Park. Our two studies agree on the high local and regional diversity of the liana community, suggesting that more than 500 species of woody climbers exist within the National Park (Burnham 2004; Romero-Saltos 2011).

Our data also suggest that liana communities are extremely species-rich in the same areas that have shown high species richness among trees as well as amphibians, birds, and mammals (Bass et al. 2010). Across this area as well as across the Amazon Basin, we still have little information on the dominant species of lianas, in spite of the many stem counts completed. These censuses would be ideal for species identification, and knowing the identities of the dominant few species alone would be a major contribution. Dominance among species in Amazonia requires the completion of basic field censuses, and could be logically accomplished by local biologists.

We have found that dispersed-plot sampling designs may be difficult to compare directly to adjacent-plot sampling within a region because the accumulation of species across widely dispersed plots is generally higher than in adjacent plots. The issue might be resolved by standardizing sampling among regions (as suggested by Gerwing et al. 2006; Schnitzer et al. 2008; DeWalt et al. 2010), and by explicitly taking into account the differences in sampling effort (e.g., number of stems sampled) and subsampling (DeWalt et al., Chapter 11 in this volume). Singleton species and their uncertain determinations may continue to be a problem when limited resources and thus low sampling effort are brought to bear on species diversity sampling, but their identification may be possible using molecular approaches that are just now becoming available.

We have shown here that the top dominant species in the Yasuní forests more often bear wind-dispersed fruits and seeds than the community average, which may reflect their ability to disperse broadly, but may also reflect the loss of forest animals (see Wright et al. 2007). As the rare species are added to the compilation of dispersal syndromes, we find decreasing proportions of wind-dispersed lianas, though not less than 35%. Whether dispersal traits are causal or linked to other traits that confer dominance is unknown for climbing plants. Considerable additional research is needed on the routes to local dominance for liana species, not only in the relatively intact forests we studied here, but also in forests where disturbance is more common, and potentially increasing (e.g., Phillips et al. 2005; Schnitzer & Bongers 2011; Schnitzer, Chapter 30 in this volume).

REFERENCES

Angiosperm Phylogeny Group (2009) An update of the Angiosperm Phylogeny Group classification for the orders and families of flowering plants: APG III. *Botanical Journal of the Linnean Society*, **161**, 105–121.

Bass, M.S., Finer, M., Jenkins, C.N. et al. (2010) Global conservation significance of Ecuador's Yasuní National Park. *PLoS ONE*, **5**(e8767), 1–22.

Burnham, R.J. (2002) Dominance, diversity and distribution of lianas in Yasuní, Ecuador: who is on top? *Journal of Tropical Ecology*, **18**, 845–864.

Burnham, R.J. (2004) Alpha and beta diversity of lianas in Yasuní National Park, Ecuador. *Forest Ecology and Management*, **190**, 43–55.

Campanello, P.I., Garibaldi, J.F., Gatti, M.G. & Goldstein, G. (2007) Lianas in a subtropical Atlantic forest: host preference and tree growth. *Forest Ecology and Management*, **242**, 250–259.

Campanello, P.I., Villagra, M., Garibaldi, J.F. et al. (2012) Liana abundance, tree crown infestation, and tree regeneration ten years after liana cutting in a subtropical forest. *Forest Ecology and Management*, **284**, 213–221.

Condit, R.G., Ashton, M.S., Balslev, H. et al. (2005) Tropical tree α-diversity: results from a worldwide network of large plots. *Biologiske Skrifter*, **55**, 565–582.

Coughenour, J.M., Simmons, M.P., Lombardi, J.A. & Cappa, J.J. (2010) Phylogeny of Celastraceae subfamily Salacioideae and tribe Lophopetaleae inferred from morphological characters and nuclear and plastid genes. *Systematic Botany*, **35**, 358–367.

DeWalt, S.J., Schnitzer, S.A., Chave, J. et al. (2010) Annual rainfall and seasonality predict pan-tropical patterns of liana density and basal area. *Biotropica*, **42**, 309–317.

Duchen, P. & Renner, S.S. (2010) The evolution of *Cayaponia* (Cucurbitaceae): repeated shifts from bat to bee pollination and long-distance dispersal to Africa 2–5 million years ago. *American Journal of Botany*, **97**, 1129–1141.

Duque, A.J., Duivenvoorden, J.F., Sánchez, M., et al. (2004) Diversity and composition of woody lianas in NW Amazonia. In *Plant Diversity Scaled by Growth Forms along Spatial and Environmental Gradients: A Study in the Rain Forests of NW Amazonia* (Ph.D. dissertation, Chapter 6), pp. 69–84. University of Amsterdam and TROPENBOS International, Amsterdam.

Finer, M.A., Vijay, V., Ponce, F. et al. (2009) Ecuador's Yasuní Biosphere Reserve: a brief modern history and conservation challenges. *Environmental Research Letters*, **4**.

Gentry, A.H. (1982) Dispersal ecology and diversity in neotropical forest communities. *Sonderbild. Naturwissenschlaftlichen Vereins in Hamburg*, **7**, 303–314.

Gentry, A.H. (1991) Breeding and dispersal systems of lianas, in *The Biology of Vines* (eds F.E. Putz & H.A. Mooney), Cambridge University Press, Cambridge, pp. 393–423.

Gerwing, J.J. & Farias, D.L. (2000) Integrating liana abundance and forest stature into an estimate of total aboveground biomass for an eastern Amazonian forest. *Journal of Tropical Ecology*, **16**, 327–335.

Gerwing, J.J., Schnitzer, S.A., Burnham, R.J. et al. (2006) A standard protocol for lianas censuses. *Biotropica*, **38**, 256–261.

Haston, E., Richardson, J.E., Stevens, P.F. et al. (2009) LAPG III: a linear sequence of the families in APG III. *Botanical Journal of the Linnean Society*, **161**, 128–131.

Hoot, S., Zautke, H., Harris, D.J. et al. (2009) Phylogenetic patterns in Menispermaceae based on multiple chloroplast sequence data. *Systematic Botany*, **34**, 44–56.

Howe, H.F. & Smallwood, J. (1982) Ecology of seed dispersal. *Annual Review of Ecology and Systematics*, **13**, 201–228.

Hubbell, S.P. (2001) *The Unified Neutral Theory of Biodiversity and Biogeography*, Princeton University Press, Princeton, NJ.

Ibarra-Manríquez, G. & Martínez-Ramos, M. (2002) Landscape variation of liana communities in a neotropical rain forest. *Plant Ecology*, **160**, 91–112.

Ibarra-Manríquez, G., Sánchez-Garfias, B. & González-Garcia, L. (1991) Fenologia de lianas y arboles anemocoros en una selva calido-humeda de México. *Biotropica*, **23**, 242–254.

Janzen, D.H. & Vázquez-Yanes, C. (1991) Aspects of tropical seed ecology of relevance to tropical forested wildlands, in *Rain Forest Regeneration and Management*, Man and the Biosphere Series, vol. **6** (eds A. Gómez-Pompa, T.C. Whitmore & M. Hadley), UNESCO, Paris, pp. 137–157.

Killeen, T.J., Jardim, A., Mamani, F. & Rojas, N. (1998) Diversity, composition and structure of a tropical semi-deciduous forest in the Chiquitanía region of Santa Cruz, Bolivia. *Journal of Tropical Ecology*, **14**, 803–827.

Ledo, A. & Schnitzer, S.A. (2014) Disturbance, not negative density dependence or habitat specialization maintains liana diversity in a tropical forest. *Ecology*, **95**, 2169–2178.

Leigh, E.G. (1999) *Tropical Forest Ecology*, Oxford University Press, New York.

Lohmann, L.G. (2006) Untangling the phylogeny of neotropical lianas Bignonieae (Bignoniaceae). *American Journal of Botany*, **93**, 304–318.

Macía, M.J. (2004) Multiplicity in palm uses by the Huaorani of Amazonian Ecuador. *Botanical Journal of the Linnean Society*, **144**, 149–159.

Macía, M.J. (2011) Spatial distribution and floristic composition of trees and lianas in different forest types of an Amazonian rainforest. *Plant Ecology*, **212**, 1159–1177.

Macía, M.J. & Svenning, J.-C. (2005) Oligarchic dominance in western Amazonian plant communities. *Journal of Tropical Ecology*, **21**, 613–626.

Mascaro, J., Schnitzer, S.A. & Carson, W.P. (2004) Liana diversity, abundance, and mortality in a tropical wet forest in Costa Rica. *Forest Ecology and Management*, **190**, 3–14.

Muller-Landau, H. & Hardesty, B.D. (2005) Seed dispersal of woody plants in tropical forests: concepts, examples and future directions, in *Biotic Interactions in the Tropics: Their Role in the Maintenance of Species Diversity* (eds D.F.R.P. Burslem, M.A. Pinard & S.E. Hartley), Cambridge University Press, Cambridge, pp. 267–309.

Nabe-Nielsen, J. (2001) Diversity and distribution of lianas in a neotropical rain forest, Yasuní National Park, Ecuador. *Journal of Tropical Ecology*, **17**, 1–19.

Nelson, B.W. (1996) Natural disturbance and change in the Brazilian Amazon. In *Changes in Global Vegetative Patterns and Their Relationship to Human Activity* (eds. S.J. Hassol & J. Katzenberger), Proceedings of an Aspen Global Change Institute Workshop, 9–22 July 1995, Elements of Change Series, AGCI, Aspen, CO.

Nelson, B.W., & Irmão, M.N. (1998) Fire penetration in standing Amazon forests. In: *Anais do IX Simpósio Brasileiro de Sensoriamento Remoto* (13–18 Sept. 1998), Santos, São Paulo, Brazil (Published in CD-ROM).

Oliveira, A.N.d., doAmaral, I.L., Ramos, M.B.P. & Formiga, K.M. (2008) Aspectos florísticos e ecológicos de grandes lianas em três ambientes florestais de terra firme na Amazônia Central. *Acta Amazonica*, **38**, 421–430.

Oliveira, E.A. de. (2011) Ecologia de lianas em comunidades florestais da transição cerrado-floresta amazonica, Região Leste de Mato Grosso, Brasil. Master's dissertation, UFMT, Brazil.

Olmstead, R.G., Zjhra, M.L., Lohmann, L.G. et al. (2009) A molecular phylogeny and classification of the Bignoniaceae. *American Journal of Botany*, **96**, 1731–1743.

Ortiz, R.d.C., Kellogg, E.A. & van der Werff, H. (2007) Molecular phylogeny of the Moonseed family (Menispermaceae): implications for morphological diversification. *American Journal of Botany*, **94**, 1425–1438.

Pérez-Salicrup, D.R., Sork, V.L. & Putz, F.E. (2001) Lianas and trees in a liana forest of Amazonian Bolivia. *Biotropica*, **33**, 34–47.

Phillips, O.L., Vásquez, M.R., Monteagudo, M.A. et al. (2005) Large lianas as hyperdynamic elements of the tropical forest canopy. *Ecology*, **86**, 1250–1258.

Pitman, N.C.A., Terborgh, J., Silman, M.R. et al. (2001) Dominance and distribution of tree species in upper Amazonian terra firme forests. *Ecology*, **82**, 2101–2117.

Pitman, N.C.A., Terborgh, J., Silman, M.R. et al. (2002) A comparison of tree species diversity in two upper Amazonian forests. *Ecology*, **83**, 3210–3224.

Rice, K., Brokaw, N. & Thompson, J. (2004) Liana abundance in a Puerto Rican forest. *Forest Ecology and Management*, **190**, 33–41.

Roeder, M., Hölscher, D. & Ferraz, I.D.K. (2010) Liana regeneration in secondary and primary forests of central Amazonia. *Plant Ecology & Diversity*, **3**, 165–174.

Romero-Saltos, H.G. (2011) Community and functional ecology of lianas in the Yasuní Forest Dynamics Plots, Amazonian Ecuador. Ph.D. dissertation, University of Miami, Florida.

Romero-Saltos, H.G., Valencia, R. & Macía, M.J. (2001) Patrones de diversidad, distribución y rareza de plantas leñosas en el Parque Nacional Yasuní y Reserva Étnica Huaorani, Amazonía ecuatoriana, in *Evaluación de recursos vegetales no maderables en la Amazonía noroccidental* (eds J.F. Duivenvoorden, H. Balslev, J. Cavelier et al.), University of Amsterdam Press, Amsterdam, pp. 131–162.

Saatchi, S.S., Nelson, B., Podest, E. & Holt, J. (2000) Mapping land cover types in the Amazon Basin using 1 km JERS-1 mosaic. *International Journal of Remote Sensing*, **21**, 1201–1234.

Schnitzer, S.A. (2005) A mechanistic explanation for global patterns of liana abundance and distribution. *The American Naturalist*, **166**, 262–276.

Schnitzer, S.A. & Bongers, F. (2011) Increasing liana abundance and biomass in tropical forests: emerging patterns and putative mechanisms. *Ecology Letters*, **14**, 397–406.

Schnitzer, S.A., DeWalt, S.J. & Chave, J. (2006) Censusing and measuring lianas: a quantitative comparison of the common methods. *Biotropica*, **38**, 581–591.

Schnitzer, S.A., Mangan, S.A., Dalling, J.W. et al. (2012) Liana abundance, diversity, and distribution on Barro Colorado Island, Panama. *PLoS ONE*, **7**, e52114, 1–16.

Schnitzer, S.A., Rutishauser, S. & Aguilar, S. (2008) Supplemental protocol for liana censuses. *Forest Ecology and Management*, **255**, 1044–1049.

Schnitzer, S.A., van der Heijden, G.M.F., Mascaro, J., & Carson, W.P. (2014) Lianas reduce biomass accumulation in a tropical forest. *Ecology*, in press.

Solórzano, S., Ibarra-Manríquez, G. & Oyama, K. (2002) Liana diversity and reproductive attributes in two tropical forests in Mexico. *Biodiversity and Conservation*, **11**, 197–212.

Tröndle, D., Schröder, S., Kassemeyer, H.-H. et al. (2010) Molecular phylogeny of the genus *Vitis* (Vitaceae) based on plastid markers. *American Journal of Botany*, **97**, 1168–1178.

UNESCO (United Nations Educational, Scientific and Cultural Organization) (2011) *Biosphere Reserve Information: Ecuador: Yasuní*, UNESCO, Paris.

Valencia, R., Condit, R., Foster, R.B. et al. (eds) (2004) *Tropical Forest Diversity and Dynamism: Findings from a Large-Scale Plot Network*, University of Chicago Press, Chicago, pp. 609–620.

van der Heijden, G.M.F. & Phillips, O.L. (2008) What controls liana success in neotropical forests? *Global Ecology and Biogeography*, **17**, 372–383.

Wright, S.J., Hernandéz, A. & Condit, R. (2007) The bushmeat harvest alters seedling banks by favoring lianas, large seeds, and seeds dispersed by bats, birds, and wind. *Biotropica*, **39**, 363–371.

Yorke, S.R., Schnitzer, S.A., Mascaro, J. et al. (2013) Increasing liana abundance and basal area in a tropical forests: the contribution of long-distance clonal colonization. *Biotropica*, **45**, 317–324.

Chapter 6

LIANA ASSEMBLAGE STRUCTURE IN FOUR SITES ACROSS THE BRAZILIAN AMAZON

Anselmo Nogueira,[1] Flavia R.C. Costa,[2] Márcia C. Vilela-Santos,[3] Carolina V. Castilho,[4] Ana Andrade,[2] José Luís C. Camargo,[2] William F. Laurance,[2,5] and Robyn J. Burnham[6]

[1]Universidade de São Paulo, São Paulo – SP, Brazil
[2]Instituto Nacional de Pesquisas da Amazônia – INPA, Manaus-AM, Brazil
[3]Instituto de Ciências Naturais Humanas e Sociais, Universidade Federal de Mato Grosso, Sinop-MT, Brazil
[4]Empresa Brasileira de Pesquisa Agropecuária – EMBRAPA, Centro de Pesquisa Agroflorestal de Roraima, Boa Vista-RR, Brazil
[5]James Cook University, Cairns, Queensland, Australia
[6]University of Michigan, Ann Arbor, MI, USA

Ecology of Lianas, First Edition. Edited by Stefan A. Schnitzer, Frans Bongers, Robyn J. Burnham, and Francis E. Putz.

Companion Website: www.wiley.com/go/schnitzer/lianas

OVERVIEW

Seasonality and precipitation gradients explain much of the variation of liana density and basal area on large scales (DeWalt et al. 2010). At the smaller scale of the Amazon, liana distribution is also expected to be correlated with climatic factors. We hypothesize that a higher density and larger basal area of lianas are present at the extremes of Amazonia (i.e., more seasonal areas) compared to more central areas (i.e., less seasonal and more humid climates). In addition, we expect a higher ratio of lianas to trees (both for density and basal area) in more seasonal sites. To test these hypotheses, lianas and trees were sampled in four Amazon sites spanning 1400 km. The northernmost site was located in Viruá National Park, followed by BDFFP and Ducke Reserve both in the Central Amazon, and at the southern extreme, one site in Cotriguaçu, Mato Grosso. In most plots, all lianas with $D \geq 1$ cm at 1.3 m above the rooting point were sampled. A total of 24,984 stems were sampled over all sites. Sites near the limits of the Amazon basin – Viruá and Cotriguaçu – had higher density and basal area of lianas than the central sites, and the ratio of lianas to trees was also greater at the extreme and seasonal Amazonian sites. Lianas in two diameter size classes (2–4.9 cm and 5–9.9 cm) were responsible for the greatest differences among sites. These results agree with our hypotheses and with the global climate-related pattern of liana distribution. The two more seasonal sites had distinct patterns of liana and tree distributions: Viruá had more thin trees and lianas and few large trees and lianas, Cotriguaçu had many large trees and lianas, but fewer thin trees and lianas. The variation in the distribution of trees among size classes may be the cause of the variation in liana size distributions locally, especially in more seasonal areas.

INTRODUCTION

The natural distribution of liana abundance and biomass on a large scale has been associated with climatic gradients such as precipitation and seasonality in tropical forests (e.g., Schnitzer 2005; DeWalt et al. 2010). However, at intermediate scales liana structure can be associated with edaphic (Laurance et al. 2001; Ibarra-Manríquez & Martínez-Ramos 2002; Malizia et al. 2010; Nogueira et al. 2011) and hydrological conditions (Schietti et al. 2014), or to gradients that reflect changes in forest structure (Nogueira et al. 2011) and different types of forest disturbance (Laurance et al. 2001; Ibarra-Manríquez & Martínez-Ramos 2002; Rice et al. 2004; Malizia et al. 2010). At the local scale, liana individual stems may be limited by inadequate support in the understory before climbers reach the forest canopy (e.g., palm trees, Vleut & Pérez-Salicrup 2005).

Recently, it has been shown that the vascular cambium, which allows lianas to grow in thickness, is climatically controlled, since in all studied species the plants responded to the alternation of a dry and rainy season (Lima et al. 2010; Angyalossy, Chapter 19 in this volume). Lianas are known to grow longitudinally faster than trees in tropical forests (Putz 1984; Schnitzer et al. 2000); and competition between lianas and trees (e.g., above ground for light and CO_2 and below ground for water and nutrients) may be intensified in tropical seasonal areas due to the higher efficiency of lianas in open and disturbed seasonal environments (Cai et al. 2009; Zhu & Cao 2010). These results suggest that liana abundances may respond differently to that of trees under climatic variation or under the effects of environmental changes across large areas, such as the Amazon basin.

Little is known about the expected relationship between the number of lianas and of trees among sites across Amazonia. Due to the high efficiency of lianas in more seasonal conditions in comparison to trees (Schnitzer 2005), we hypothesize that the ratio of lianas to trees – for both abundance and basal area in a given sampling area – should be larger in seasonal sites toward the Amazonian borders relative to Central Amazonian sites. In this case, the ratio of lianas to trees could express not only the effects of climatic factors on lianas, but also on trees. Thus, it would be expected that the difference between sites should be greater when analyzing the ratio of lianas to trees rather than only lianas across this large climatic range (cf., DeWalt et al. 2010; DeWalt et al., Chapter 11 in this volume). Additionally, on the Amazon borders the forest grades into savannas or is intermingled with it, generating a more variable environment mainly due to the variation of tree size distributions. The change in the number of trees among different size classes could be directly related to the variation of liana abundance, both in the understory (thin lianas, e.g., Nogueira et al. 2011) and the canopy (large lianas, e.g., Phillips et al. 2005). Therefore, the within site variation of lianas and trees in different size classes may highlight the role of trees in the pattern of local liana assemblages.

Studies that evaluate vegetation structure (considering both lianas and trees) among sites of varying climates in the Amazon basin are still scarce. Indeed, the structure and composition of plant assemblages in some regions of the Amazon forest are largely unknown (Hopkins 2007) due to difficult access and lack of infrastructure in these sites. In the past 10 years, new areas of the Amazon forest have become available to researchers due to the installation of field infrastructure and standardized sampling methods (used for long-term ecological projects) through the federal government biodiversity research program, PPBio (http://ppbio.inpa.gov.br/en/home). This Brazilian program has opened the door to more detailed and standardized studies in several areas than were rarely sampled previously (Costa & Magnusson 2010; Magnusson et al. 2013). For instance, more seasonal areas at the limits of the Brazilian Amazon Forest (e.g., sites at the northern limits of the Amazon forest, in Roraima State, and at the southern limits in Mato Grosso State) are now included in inventories of trees, lianas, and other biological groups.

Liana and tree data from four undisturbed sites in the Brazilian Amazon were sampled and are used here to answer three questions: (1) Are there differences in the liana density and basal area (BA) across the Brazilian Amazon? Since at broader scales it is expected that the change in the liana density and BA would be correlated with climatic variables (Schnitzer 2005; DeWalt et al. 2010; DeWalt et al., Chapter 11 in this volume), we predict that sites at the northern and southern extremes of the Amazon, in regions with lower annual precipitation and more seasonal climate, would have a greater density and BA of lianas than more central Amazonian sites. (2) Is there any correlative evidence of an advantage of lianas over trees in the more seasonal sites in the Amazon basin? We examined this proposition using the ratio of lianas to trees, calculated for both density and basal area data. (3) What diameter classes include the highest concentration of lianas? Could the distribution of trees among different size classes help us to explain the liana variation in each Brazilian Amazon site?

METHODS

Sampled sites and data collection

We used four sites in the Brazilian Amazon, without evidence of recent anthropic impact, to sample liana and tree density and basal area (BA). The most distant sites were ca. 1400 km apart, representing two sites in more seasonal climate regimes with lower total annual precipitation (National Park of Viruá, hereafter *Viruá*, in the North, and *Cotriguaçu*, in the South), and two aseasonal higher precipitation sites (Biological Dynamics of Forest Fragments Project, *BDFFP*, and Reserva Florestal Adolpho Ducke, *Ducke*, both in the Central Amazon) (Fig. 6.1).

Lianas were sampled at all four sites following the protocols suggested by Gerwing et al. (2006), in which only lianas rooted within each plot were sampled. We marked lianas permanently with numbered tags and created plots that followed the altimetric contour line in the three PPBio sites (*Viruá*, *Ducke*, and *Cotriguaçu*). The detailed experimental design carried out in the PPBio sites is described in Magnusson et al. (2005) and Costa and Magnusson (2010). Lianas in the *BDFFP* site were not marked and the plots were square (100 × 100 m). In all four sampled sites, lianas were measured at 1.3 m from the rooting point with differentiation between ramets (individual stems) and the potential genets of each liana.

The number of plots and the extension of sampling area in each site were: *Viruá* with 13 plots distributed on 15 km^2, *BDFFP* with 18 plots on 340 km^2, *Ducke* with 56 plots on 64 km^2, and *Cotriguaçu* with 12 plots on 12 km^2. For more details on the sites, see http://ppbio.inpa.gov.br/en/node/451 and http://pdbff.inpa.gov.br. The plots of the *Viruá*, *Cotriguaçu*, and *Ducke* sites were systematically distributed over the landscape, with plots at least 1 km distant from each other. Since some plots of the *BDFFP* site were spatially contiguous or very close, we selected only plots with a minimum distance of 350 meters from each other (18 of 37 plots).

Lianas ≥5 cm diameter were sampled in 40 × 250 m (1 ha) plots, and lianas ≥1 cm diameter in 10 × 250 m

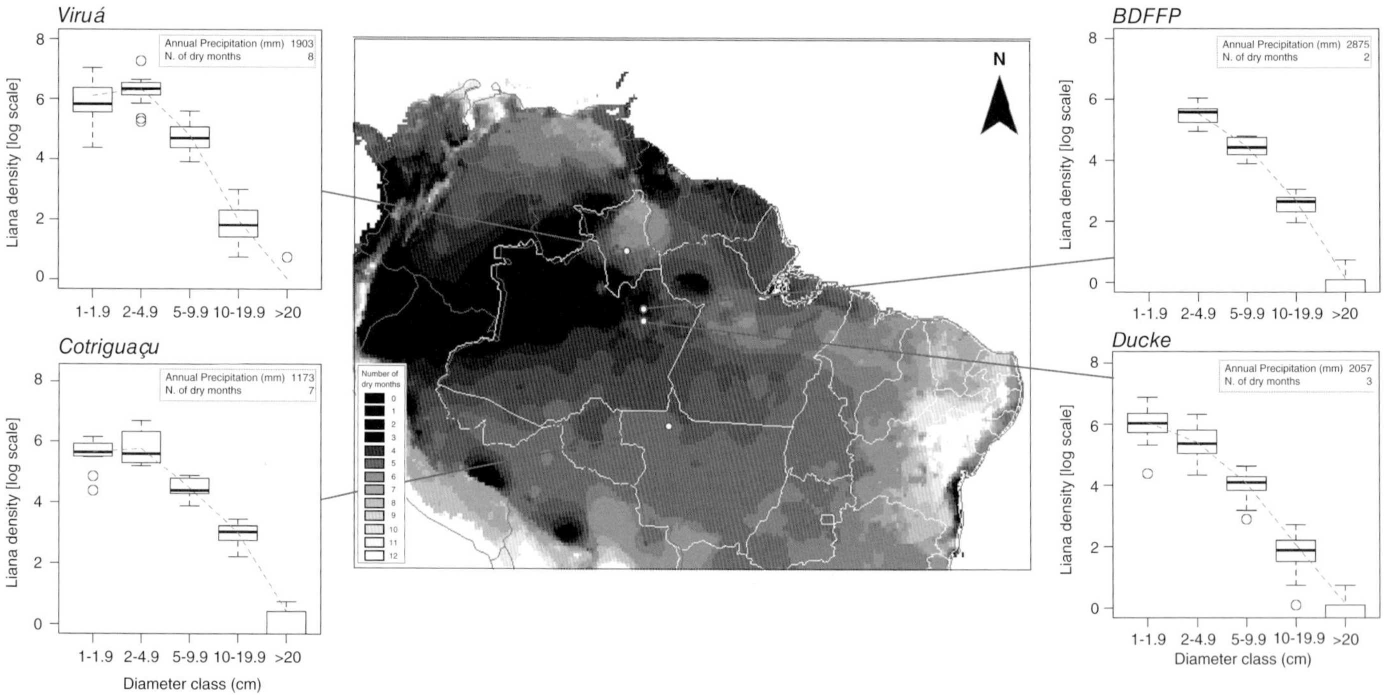

Fig. 6.1 Map of the seasonality in Amazonia based on the number of months with less than 100 mm of precipitation (original climate data available in http://www.worldclim.org). Indicated are two sites from more seasonal areas (sites 1 in the North, and site 2 in the South) and two sites from more humid areas (sites 3 and 4) in which the liana and tree dataset were sampled. Graphs represent the distribution of liana abundance per site (genets + ramets) in different diameter classes. Lianas with diameters between 1–1.9 cm (first size class) were not sampled in the BDFFP site.

(0.25 ha) sub-plots in *Ducke* and *Cotriguaçu* (Nogueira et al. 2011; Vilela-Santos, unpublished data). In *Viruá*, lianas ≥10 cm diameter were sampled in 20 × 250 m (0.5 ha) plots, and lianas ≥1 cm diameter in 4 × 250 m (0.1 ha) sub-plots (Castilho, unpublished data). In the *BDFFP*, all lianas ≥2 cm diameter were sampled in 100 × 100 m (1 ha) plots (Laurance et al. 2001). For all sites, liana density was expressed as the number of individuals or stems per hectare for each size class.

The width of the plot sampled for trees varied according to size class in the three PPBio sites, following the same logic used for lianas – larger samples for larger plants and smaller sub-samples for smaller plants. In *Viruá*, *Ducke*, and *Cotriguaçu*, trees ≥30 cm DBH were sampled over 40 × 250 m (1 ha). Trees ≥10 cm DBH were sub-sampled in 20 × 250 m plots (0.5 ha) and trees ≥1 cm DBH in 4 × 250 m plots (0.1 ha) (de Castilho et al. 2006). At the *BDFFP*, only trees ≥10 cm DBH were censused in each 100 × 100 m plot. Tree density was expressed as the number of individuals per hectare.

Analyses

We used ANOVA to test whether liana density, liana basal area, and the ratio of lianas to trees differed among sites. We used four minimum diameter (D) thresholds (D ≥1 cm; D ≥2 cm; D ≥5 cm; and D ≥10 cm) for the tests of difference of liana abundance and basal area. Residuals of all analyses were tested for homogeneity of variances, and the statistics reported here are based on log-transformed data. All statistical analyses were performed in R v. 3.0.1 (R Development Core Team 2009).

RESULTS

We surveyed a total of 24,984 liana stems in the four sites. For comparisons among sites at which the lianas ≥1 cm diameter were measured, only the data from the 0.25 ha plots from *Viruá* (N = 13), *Ducke* (N = 56), and *Cotriguaçu* (N = 12) were used. On average, *Viruá* plots had 1,203 liana stems ha^{-1}, *Cotriguaçu* 781 stems ha^{-1}, and *Ducke* 763 stems ha^{-1} (Table 6.1 and Fig. 6.1). Lianas ≥2 cm diameter (including all sample sites) were less abundant at *Ducke* (312 stems ha^{-1}) and *BDFFP* (365 stems ha^{-1}; Table 6.1).

The structure of the liana assemblage, described by density or basal area per hectare, differed among sites for all the size classes considered (Table 6.2). In general, liana density and basal area are higher at both extreme sites of the Amazonian basin (Fig. 6.2). Both seasonal sites – *Viruá* and *Cotriguaçu* – on average had higher numbers of lianas ≥2 cm and ≥5 cm diameter than the two central Amazonian sites. These data suggest that a larger number of lianas survive to enter the second diameter class (≥2 cm diameter) in more seasonal sites, such as the borders of the Amazon, than in more central humid areas. However, the density of lianas ≥1 cm diameter was higher at *Viruá* (in the North), but at the other southern seasonal site (*Cotriguaçu*) it did not differ from the *Ducke* site. This pattern suggests a higher variation in the number of lianas in small diameter size classes among sites.

The ratio of the density and basal area of lianas to that of trees per ha was higher in *Viruá* and *Cotriguaçu* than in *Ducke* (Fig. 6.3). We omitted the data from the *BDFFP* site where only trees ≥10 cm DBH were sampled. On average, 19% (SE = 0.087) of the stems ≥1 cm diameter in Viruá and 24% (SE = 0.023) in *Cotriguaçu* were lianas, while only 13% (SE = 0.005) of stems ≥1 cm diameter were lianas in *Ducke*. In addition, an average of 4.5% (SE = 0.006) and 3.4% (SE = 0.003) of the basal area of stems ≥1 cm diameter in *Viruá* and *Cotriguaçu* were lianas, respectively, while only 1.6% (SE = 0.001) of the basal area of stems were lianas in *Ducke*. The percentage of lianas ≥2 cm in diameter was similar to results for ≥1 cm stems, with 19%, 22%, and 9% of liana density ($F_{(2,78)} = 50.8$, $R^2 = 55.5$ %, and $p < 0.001$), and 4.3 %, 4.3 %, and 1.4 % of liana basal area ($F_{(2,78)} = 65.2$, $R^2 = 61.6$ %, and $p < 0.001$) (*Viruá*, *Cotriguaçu*, and *Ducke*, respectively).

Cotriguaçu and *Viruá*, though both at the extreme geographical limits of the Amazonian forests, differ in the distribution of both lianas (Table 6.1) and trees among size classes (Table 6.3). *Cotriguaçu* had more large trees and lianas and fewer thin trees and lianas; *Viruá* had fewer large trees and lianas and more thin trees and lianas. A similar pattern is observed for liana basal area based on stems over 10 cm diameter: *Viruá* had the lowest values while *Cotriguaçu* had the highest. *Viruá* has also the highest number of lianas and trees in the small diameter class (1–1.9 cm), similar to values at *Ducke*, while *Cotriguaçu* had the lowest number of small lianas and trees.

Table 6.1 Liana density and basal area for all apparent genets and for all rooted stems (apparent genets + ramets) in four sites in the Brazilian Amazon. Viruá and Cotriguaçu are more seasonal sites with less rainfall, whereas BDFFP and Ducke are aseasonal sites with more rainfall. Data are presented as the mean (±SE) per hectare for three commonly used size-classes. Lianas with diameters 1–1.9 cm (first size class) were not sampled in the BDFFP site.

	Viruá				***BDFFP***			***Ducke***				***Cotriguaçu***			
	≥1 cm	**≥2 cm**	**≥5 cm**	**≥10 cm**	**≥2 cm**	**≥5 cm**	**≥10 cm**	**≥1 cm**	**≥2 cm**	**≥5 cm**	**≥10 cm**	**≥1 cm**	**≥2 cm**	**≥5 cm**	**≥10 cm**
Apparent genets															
Density	1055 (71.1)	667 (42.4)	126 (10.2)	7 (0.8)	357.6 (23.7)	99.3 (6.7)	13.9 (0.9)	672 (32.8)	289 (15.2)	63 (2.6)	7 (0.4)	701.7 (97.4)	459.7 (66.5)	107.7 (9.0)	20.4 (2.2)
All rooted stems															
Density	1203 (86.1)	738 (51.6)	131 (10.7)	8 (0.9)	365.7 (24.3)	101.3 (6.5)	14.0 (0.4)	763 (39.1)	312 (17.1)	66 (2.7)	7 (0.5)	781.2 (96.1)	488.5 (66.4)	110.8 (9.2)	20.6 (2.1)
Total basal area (cm^2)	10578 (738.4)	8504 (591.5)	2778 (224.4)	320 (39.7)	7099 (398.3)	4917 (284.2)	1876 (169.1)	5612 (243.3)	4930 (220.3)	3090 (135.3)	1032 (76.0)	9626 (795.2)	9132 (778.5)	6087 (515.9)	2593 (307.1)

Table 6.2 Analysis of variance tests of site effects on liana density and basal area considering different diameters of inclusion. Final column is the result of post-hoc analyses among sites (Tukey Honest Significant Differences).

Liana descriptors	Diameter of inclusion (cm)	$F_{(d.f.)}$	p	Average differences among sites ($p < 0.05$)
Density (individuals ha^{-1})	$^*D \geq 1$	$4.1_{(2,78)}$	0.019	(Cotriguaçu = Ducke) ≠ < Viruá
	$D \geq 2$	$18.6_{(3, 95)}$	0.01×10^{-7}	Figure 6.2a
	$D \geq 5$	$4.5_{(3, 95)}$	0.006	Figure 6.2b
	$D \geq 10$	$19.0_{(3, 95)}$	0.01×10^{-7}	(Cotriguaçu = BDFFP) > ≠ (Ducke = Viruá)
Basal area (cm^2 ha^{-1})	$^*D \geq 1$	$15.1_{(2, 78)}$	0.03×10^{-4}	(Cotriguaçu = Viruá) > ≠ (Ducke)
	$D \geq 2$	$14.8_{(3, 95)}$	0.05×10^{-6}	Figure 6.2c
	$D \geq 5$	$17.3_{(3, 95)}$	0.04×10^{-7}	Figure 6.2d
	$D \geq 10$	$8.1_{(3, 95)}$	0.07×10^{-3}	Cotriguaçu > ≠ (BDFFP) > ≠ (Ducke = Viruá)

Note: *Analyses with diameter of inclusion greater than 1 cm were performed without *BDFFP* site.

DISCUSSION

Forests at the limits of the Amazon basin (*Viruá* and *Cotriguaçu*) had, on average, a higher density and BA of lianas than sites in central Amazonia (*BDFFP* and *Ducke*), in accordance with our hypotheses. Although the distribution pattern of lianas on a global scale has been related to climatic variables, such as seasonality and precipitation (Schnitzer 2005; DeWalt et al., Chapter 11 in this volume), contrasting results (e.g., van der Heijden & Phillips 2008) suggest the need for new data for a more detailed description of vegetation structure throughout tropical forests. Our results, including new data from two Amazonian sites (*Viruá* and *Cotriguaçu*), provide additional evidence of the association between climatic variables and the variation in liana density and BA in Amazonian forests and are consistent with Schnitzer (2005), DeWalt et al. (2010; Chapter 11 in this volume). We describe a higher ratio of lianas to trees in more seasonal sites, in terms of density and BA of both plant groups. These patterns may reflect the anatomical and physiological advantage of lianas over trees under more seasonal climatic conditions (Zhu & Cao 2009), such as that found at the limits of the Amazon basin. More specifically, lianas are known to have a higher amount of storage parenchyma (Carlquist 1985), longer water-conductive vessels (Ewers et al. 1990; Jacobsen et al. 2012), and deeper roots (Restom & Nepstad 2004) than many trees belonging to the same taxonomic group.

The distribution of lianas among size classes suggests that the transition of smaller individuals to larger size classes may generate some of the differences in the abundance and basal area among sites. Although the smaller lianas (1–1.9 cm) are very abundant in the sampled sites, they do not follow the expectation of variation from seasonal to less seasonal sites – their abundance is essentially independent of seasonality. Therefore, the greater differences in liana abundances between sites appear when lianas are over 2 cm in diameter. Interestingly, this is the exact developmental phase at which some lianas start the process of climbing in the forest (Kurzel et al. 2006), passing through various ecological filters such as the availability of adequate supports given by tree stems and branches (Nogueira et al. 2011) until they reach the canopy.

It is also at this point in their life cycle that lianas generally modify their growth system and start creating cambial variations. This altered cambium activity changes the stem structure (e.g. Pace et al. 2009), possibly providing better hydraulic efficiency and higher flexibility (Rowe et al. 2004) while allowing twisting without damaging the vessels (Carlquist 1985, 1988). This new stem anatomy is associated with the higher efficiency of lianas in the transport of water and nutrients, especially in the dry seasons, as compared to other plant groups (i.e., trees – Cernusak et al. 2008).

The changes associated with this step in the life cycle of lianas might be responsible for the variation in abundance and basal area of lianas among Amazonian forest sites encountered here. We suggest that lianas reaching this phase acquire advantages for survival and become well established. Over time, this will lead to a higher density of larger lianas (≥2 cm diameter) in sites that provide the best conditions for plants to

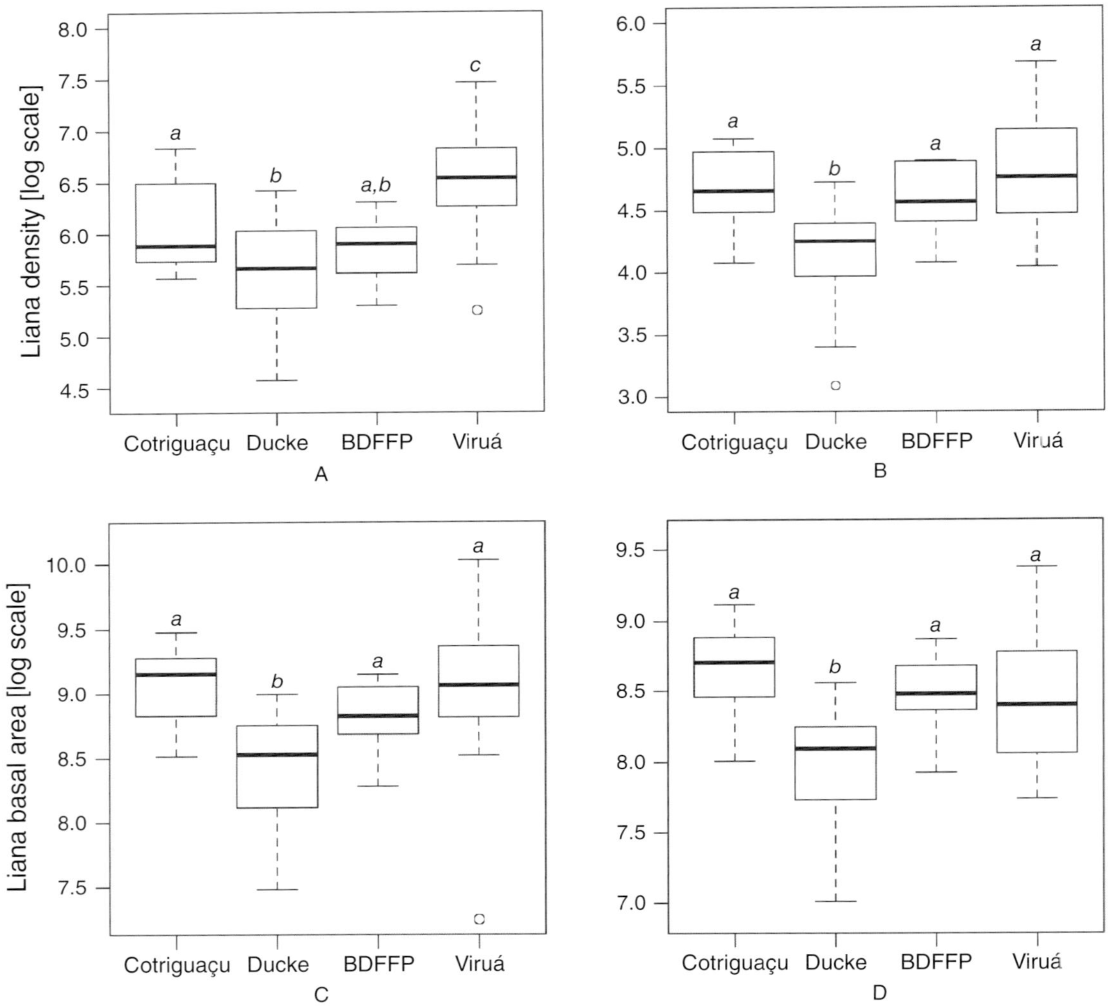

Fig. 6.2 Variation in liana density (ramets+genets) in four sites across the Brazilian Amazonia, accounting for different limits of inclusion of lianas. (A) Liana density for stems with D ≥ 2 cm; (B) Liana density for stems with D ≥ 5 cm; (C) Liana basal area for stems with D ≥ 2 cm; and (D) Liana basal area for stems with D ≥ 5 cm. Additional statistics are presented in Table 6.2; italic letters indicate the statistical differences among sites ($p < 0.05$) in the post-hoc tests (Tukey Honest Significant Differences).

overcome the first transition. These conditions may be associated with the availability of adequate supports, and thus vegetation structure created largely by trees, palms, and bamboos are critical to this transition.

On a large scale, vegetation structure (not just that of lianas) tends to co-vary with seasonality. For example, less seasonal areas tend to have a higher number of palm species (Salm et al. 2007), a plant group thought to be negatively correlated to lianas (Nogueira et al. 2011). Therefore, it is difficult to establish a causal relationship between seasonality, liana abundance and basal area in the Amazon without evidence from controlled experiments. If drier tropical forests have a greater density of small and thin trees, which are more suitable hosts for lianas of all twining types (Putz 1984; Putz & Holbrook 1991), then drier forests may

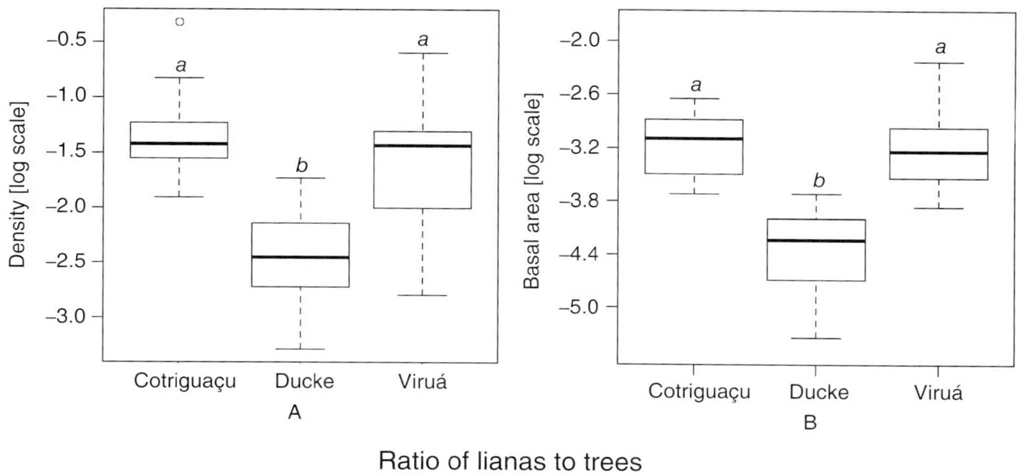

Fig. 6.3 Variation in the ratio liana:tree individuals and basal area in three sites across the Brazilian Amazonian ($D \geq 2$ cm). (A) Data from liana and tree abundance with D ≥ 2 cm [$F_{(2,78)} = 50.8$, $R^2 = 55.5$ %, and $p < 0.001$]; (B) Data from liana and tree basal area with D ≥ 2 cm [$F_{(2,78)} = 65.2$, $R^2 = 61.6$ %, and $p < 0.001$]. Italic letters indicate the statistical differences among sites ($p < 0.05$) in the post-hoc tests (Tukey Honest Significant Differences).

Table 6.3 Tree density in different diameter size classes per hectare in the four Brazilian Amazon forest sites (mean ± SE per hectare). Viruá and Cotriguaçu are more seasonal sites with less rainfall, whereas BDFFP and Ducke are aseasonal sites with more rainfall. Tree data are represented as principal rooted stems.

Diameter size classes (cm)	Viruá (North)	BDFFP (Central)	Ducke (Central)	Cotriguaçu (South)
1–1.9	2125 (415.9)	NA	1762 (80.1)	802 (96.8)
2–4.9	1948 (291.2)	NA	1936 (75.4)	842 (95.0)
5–9.9	746 (99.1)	NA	772 (30.0)	463 (58.0)
10–19.9	399 (37.9)	391 (6.5)	381 (6.2)	247 (16.2)
20–29.9	100 (8.1)	125 (3.2)	130 (3.1)	96.8 (4.7)
30–39.9	26 (3.2)	56 (1.8)	51 (1.0)	43.7 (2.7)
40–49.9	10 (1.7)	27 (1.3)	24 (0.8)	22 (1.8)
≥50	6 (1.1)	24 (0.9)	19 (0.9)	23 (1.6)

Note: NA = data not sampled.

support more lianas than wetter forests, solely on the basis of tree stem size and abundance. This may be the case in the *Viruá* site. However, large lianas generally are supported by large trees in the canopy (Phillips et al. 2005), and this could be the explanation for the greater abundance of large lianas in *Cotriguaçu*.

There are no other plant groups with the unique functional properties of liana stems and branches (i.e., conduction and flexibility). These make lianas very efficient competitors compared to trees, especially in drier conditions. These unique characteristics of lianas seem to confer an advantage on lianas in space occupation. These potential anatomical/physiological advantages agree with our liana patterns among Amazonian sites, especially when we consider the variation of the ratio of lianas to trees. The ratio is higher in the more seasonal Amazonian sites. The connection between seasonality, liana abundance distribution in different size classes, vegetation structure (with favorable and unfavorable supports), and critical liana traits (e.g., cambium variation) is quite complex, but could clarify the ecology of lianas at different spatial scales. Here, we underscore some of these interacting components and hope to stimulate other researchers to evaluate lianas in terms of their morphological-functional traits, ecological strategies, and their high species diversity in the tropics.

ACKNOWLEDGMENTS

We thank the PPBio (Brazilian Biodiversity Research Program), the Brazilian LTER (PELD, Pesquisas Ecológicas de Longa Duração, sítio 1) and BDFFP – INPA/STRI (Biological Dynamics of Forest Fragment Project) for general support of this project. We also thank Arildo Souza Dias, Frans Bongers, and two anonymous reviewers for fruitful discussions on previous versions of this chapter. This is publication number 40 and 641 in the NEBAM and BDFFP technical series, respectively.

REFERENCES

Cai, Z., Schnitzer, S.A. & Bongers, F. (2009) Seasonal differences in leaf-level physiology give lianas a competitive advantage over trees in a tropical seasonal forest. *Oecologia*, **161**, 25–33.

Carlquist, S. (1985) Observations on functional wood histology of vines and lianas: vessel dimorphism, tracheids, vasicentric tracheids, narrow vessels, and parenchyma. *Aliso*, **11**, 139–157.

Carlquist, S. (1988) *Comparative Wood Anatomy: Systematic, Ecological, and Evolutionary Aspects of Dicotyledon Wood*, Springer, Berlin.

Cernusak, L.A., Winter, K., Aranda, J. & Turner, B.L. (2008) Conifers, angiosperm trees, and lianas: growth, whole-plant water and nitrogen use efficiency, and stable isotope composition (δ^{13}C and δ^{18}O) of seedling grown in a tropical environment. *Plant Physiology*, **148**, 642–659.

Costa, F.R.C. & Magnusson, W.E. (2010) The need for large-scale, integrated studies of biodiversity: the experience of the Program for Biodiversity Research in Brazilian Amazonia. *Natureza & Conservação*, **8**, 3–12.

Castilho, C.V., Magnusson, W.E., Araújo, R.N.O. et al. (2006) Variation in aboveground tree live biomass in a central Amazonian Forest: effects of soil and topography. *Forest Ecology and Management*, **234**, 85–96.

DeWalt, S.J., Schnitzer, S.A., Chave, J. et al. (2010) Annual rainfall and seasonality predict pan-tropical patterns of liana density and basal area. *Biotropica*, **42**, 309–317.

Ewers, F.W., Fisher, J.B. & Chiu, S.T. (1990) A survey of vessel dimensions in stems of tropical lianas and other growth forms. *Oecologia*, **84**, 544–552.

Gerwing, J.J., Schnitzer, S.A., Burnham, R.J. et al. (2006) A standard protocol for lianas censuses. *Biotropica*, **38**, 256–261.

Hopkins, M.J.G. (2007) Modeling the known and unknown plant biodiversity of the Amazon Basin. *Journal of Biogeography*, **34**, 1400–1411.

Ibarra-Manríquez, G. & Martinez-Ramos, M. (2002) Landscape variation of lianas communities in a neotropical rainforest. *Plant Ecology*, **160**, 91–112.

Jacobsen, A.L., Pratt, R.B., Tobin, M.F. et al. (2012) A global analysis of xylem vessel length in woody plants. *American Journal of Botany*, **99**, 1583–1591.

Kurzel, B.P., Schnitzer, S.A. & Carson, W.P. (2006) Predicting liana crown location from stem diameter in three Panamanian lowland forests. *Biotropica*, **38**, 262–266.

Laurance, W.F., Perez-Salicrup, D., Delamonica, P. et al. (2001) Rain forest fragmentation and the structure of Amazonian liana communities. *Ecology*, **82**, 105–116.

Lima, A.C., Pace, M.R. & Angyalossy, V. (2010) Seasonality and growth rings in lianas of Bignoniaceae. *Trees*, **24**, 1045–1060.

Magnusson, W.E., Lima, A.P., Luizão, R. et al. (2005) RAPELD: a modification of the Gentry method for biodiversity surveys in long-term ecological research sites. *Biota Neotropica*, **5**, 1–5.

Magnusson, W.E., Braga-Neto, R., Pezzini, F. et al. (2013) Biodiversity and Integrated Environmental Monitoring, Áttema Editorial.

Malizia, A., Grau, H.R. & Lichstein, J.W. (2010) Soil phosphorus and disturbance influence liana communities in a subtropical montane forest. *Journal of Vegetation Science*, **21**, 551–560.

Nogueira, A., Costa, F.R.C. & Castilho, C.V. (2011) Liana abundance patterns: the role of ecological filters during development. *Biotropica*, **43**, 442–449.

Pace, M.R., Lohmann, L.G. & Angyalossy, V. (2009) The rise and evolution of the cambial variant in Bignonieae (Bignoniaceae). *Evolution & Development*, **11**, 465–479.

Phillips, O.L., Vásquez Martínez, R., Monteagudo Mendoza, A. et al. (2005) Large lianas as hyperdynamic elements of the tropical forest canopy. *Ecology*, **86**, 1250–1258.

Putz, F.E. (1984) The natural history of lianas on Barro Colorado Island, Panama. *Ecology*, **65**, 1713–1724.

Putz, F.E. & Holbrook, N.M. (1991) Biomechanical studies of vines, in *The Biology of Vines* (eds F.E. Putz & H.A. Mooney), Cambridge University Press, Cambridge, pp. 73–97.

R Development Core Team (2009) R: A language and environment for statistical computing. R Foundation for Statistical Computing, Vienna, Austria. Available at URL: http://www.R-project.org.

Restom, T.G. & Nepstad, D.C. (2004) Seedling growth dynamics of a deeply rooting liana in a secondary forest in eastern Amazonia. *Forest Ecology and Management*, **190**, 109–118.
Rice, K., Brokav, N. & Thompson, J. (2004) Liana abundance in a Puerto Rican forest. *Forest Ecology and Management*, **190**, 33–41.
Rowe, N., Isnard, S. & Speck, T. (2004) Diversity of mechanical architectures in climbing plants: an evolutionary perspective. *Journal of Plant Growth Regulation*, **23**, 108–128.
Salm, R., Salles, N.V., Alonso, W.J. & Schuck-Paim, C. (2007) Cross-scale determinants of palm species distribution. *Acta Amazonica*, **37**, 17–26.
Schietti, J., Emilio, T., Rennó, C.D. et al. (2014) Vertical distance from drainage drives floristic composition changes in an Amazonian rainforest. *Plant Ecology & Diversity*, **7**, 241–253.
Schnitzer, S.A. (2005) A mechanistic explanation for global patterns of liana abundance and distribution. *The American Naturalist*, **166**, 262–276.
Schnitzer, S.A., Dalling, J.W. & Carson, W. (2000) The importance of lianas on tree regeneration in tropical forest canopy gaps: evidence for an alternative pathway of gap-phase regeneration. *Journal of Ecology*, **88**, 655–666.
Schnitzer, S.A., Kuzee, M.E. & Bongers, F. (2005) Disentangling above- and below-ground competition between lianas and trees in a tropical forest. *Journal of Ecology*, **93**, 1115–1125.
van der Heijden, G.M.F. & Phillips, O.L. (2008) What controls liana success in Neotropical forests? *Global Ecology and Biogeography*, **17**, 372–383.
Vleut, I. & Pérez-Salicrup, D.R. (2005) Lianas and their supporting plants in the understorey at Los Tuxtlas, México. *Journal of Tropical Ecology*, **21**, 577–580.
Zhu, S. & Cao, K. (2009) Hydraulic properties and photosynthetic rates in co-occurring lianas and trees in a seasonal tropical rainforest in southwestern China. *Plant Ecology*, **204**, 295–304.
Zhu, S. & Cao, K. (2010) Contrasting cost-benefit strategy between lianas and trees in a tropical seasonal rain forest in southwestern China. *Oecologia*, **163**, 591–599.

Chapter 7

THE LIANAS OF BARRO COLORADO ISLAND, PANAMA

Stefan A. Schnitzer,[1,2] *Scott A. Mangan,*[2,3] *and Stephen P. Hubbell*[2,4]

[1]University of Wisconsin-Milwaukee, WI, USA
[2]Smithsonian Tropical Research Institute, Balboa, Republic of Panama
[3]Washington University, St. Louis, MO, USA
[4]University of California, Los Angeles, CA, USA

Ecology of Lianas, First Edition. Edited by Stefan A. Schnitzer, Frans Bongers, Robyn J. Burnham, and Francis E. Putz.

Companion Website: www.wiley.com/go/schnitzer/lianas

OVERVIEW

Lianas are currently recognized as an important component of tropical forests, where they contribute to many aspects of forest dynamics. However, up to now, in most forests, the ecology of lianas has been poorly studied compared to that of trees. In addition, most field studies related to lianas have been limited on the spatial and temporal scale. An exception is Barro Colorado Island, Panama (BCI), in that a handful of early studies there revealed important findings on liana ecology, natural history, successional patterns, diversity maintenance, and the effects of lianas on tree regeneration. The liana census of the BCI 50-ha plot has subsequently broadened and deepened our knowledge of the ecology and natural history of lianas, and the large spatial aspect of this census has allowed us to test a number of broad conceptual ecological theories, such as the mechanisms responsible for local liana distribution and the maintenance of species diversity. In this chapter, we summarize the state of knowledge about the lianas of BCI by first summarizing the early groundbreaking studies of lianas at that location and then describing, in detail, the findings from the liana census of the BCI 50-ha plot.

In the BCI 50-ha plot, there are a total of 67,447 rooted liana stems (including clones) ≥1 cm diameter (1350 rooted stems ha^{-1}) comprising 162 species and 37 families. Total rooted stems can be divided into 47,183 rooted individuals (944 rooted liana individuals ha^{-1}) and 20,264 rooted clonal stems that were still attached to another rooted stem in the census of 2007 (405 clones ha^{-1}). Rooted lianas (including clones) composed nearly 25% of the woody stems ≥1 cm diameter (trees, shrubs, and lianas), 35% of the species richness, and nearly 3% of the basal area. Thus, lianas on BCI contribute a significant proportion to forest diversity and stem number, but relatively little to forest basal area and biomass.

We found clear evidence that liana diversity is maintained mainly by disturbance, but that edaphic specialization and negative density dependence, two mechanisms thought to be responsible for the maintenance of the majority of tree diversity on the BCI 50-ha plot, appear to have relatively little influence on lianas. Disturbance appears to provide a regeneration niche for lianas; more than 50% of the liana species were more likely to be found in disturbed areas than in intact, undisturbed forest – a pattern not found for trees. Furthermore, liana stem density and species richness (controlling for stem density) were both significantly higher in disturbed areas than in undisturbed forest. Clonal reproduction enhanced the ability of lianas to colonize disturbed areas, and the majority of liana species (54%) produced significantly more clones in highly disturbed areas than in less disturbed areas. In contrast, lianas were only weakly associated with habitat type or soil variables, and few liana species had distribution patterns consistent with negative density dependence.

The liana census of the BCI 50-ha plot has revealed abundant information about liana spatial structure, relative abundance, diversity, and basal area (BA), as well as the ability of most liana species to produce copious numbers of clonal stems. Lianas respond to disturbance with high stem production, an effect that appears to maintain their diversity, and the local distribution of the vast majority of liana species can be predicted by disturbance. Furthermore, the liana census has provided comparative data for earlier liana censuses on BCI, revealing strong increases in liana density with concomitant levels of tree infestation over the past 40 years. These data are consistent with the increase in liana abundance in other tropical forests, supporting the hypothesis that the role of lianas in tropical forests is becoming more important.

INTRODUCTION

Lianas contribute substantially to tropical forest diversity, dynamics, and function, and their influence on tropical forests is becoming increasingly recognized (Schnitzer et al. Chapter 1 in this volume). Lianas can affect community-level forest dynamics and influence tree species coexistence by competing intensely with most tree species, but not all (Pérez-Salicrup et al. 2001; Ingwell et al. 2010; Schnitzer & Carson 2010). Lianas influence animal communities by providing food and shelter, which can increase animal population densities (Arroyo-Rodríguez et al., Chapter 27 in this volume; Lambers & Halsey, Chapter 26 in this volume). They also act as bridges connecting tree canopies, thereby promoting species coexistence by providing a structurally complex environment that allows subordinate animal species to escape from predators and dominant competitors (*sensu* Huffaker 1958; Yanoviak, Chapter 24 in this volume). Lianas can affect ecosystem-level dynamics by altering forest carbon, nutrient, and water dynamics, decreasing whole-forest carbon sequestration and storage, redistributing nutrients horizontally across the forest landscape, and competing intensely for soil water (Schnitzer et al. 2000, 2011, 2012, 2014; Andrade et al. 2005; van der Heijden et al. 2008; Durán & Gianoli 2013; Powers, Chapter 14 in this volume; van der Heijden et al., Chapter 13 in this volume). Moreover, lianas appear to be increasing in abundance and biomass throughout the neotropics, which could have serious consequences for community- and ecosystem-level processes in neotropical forests (reviewed by Schnitzer & Bongers 2011; Schnitzer, Chapter 30 in this volume).

Even with the recognition of the importance of lianas (e.g., Schnitzer & Bongers 2002, 2011; Wyka et al. 2013), basic data, such as liana density, diversity, spatial structure, and the processes that maintain liana diversity and explain their distribution are generally lacking. Determining the number of liana stems and species in a forest, as well as liana spatial structure gives insight into the mechanisms that maintain liana diversity and that control liana distributions within and among forests (e.g., Schnitzer 2005; Swaine & Grace 2007; Dalling et al. 2012; Schnitzer et al. 2012; Ledo & Schnitzer 2014; DeWalt et al., Chapter 11 in this volume). Currently, however, most studies of liana ecology are conducted at small spatial scales (e.g., Mascaro et al. 2004; Yorke et al. 2013) or include only very large lianas (e.g., Phillips et al. 2002; Chave et al. 2008), and thus they are unable to provide sample sizes that are large enough to examine how lianas are distributed across the landscape, whether liana distributions are consistent with mechanistic explanations for liana distribution and diversity maintenance, and how estimates of liana abundance, basal area (BA), and diversity change with spatial scale.

In this chapter, we summarize the knowledge of liana ecology that has been gained from studies on Barro Colorado Island (BCI) and the adjacent forests of the Barro Colorado Nature Monument (BCNM), located in central Panama. We first review and synthesize some of the early groundbreaking studies of lianas of BCI and the BCNM, and the recently published findings of the 2007 liana census of the BCI 50-ha forest dynamics plot. We then summarize other studies that have applied the BCI 50-ha plot liana data to test whether liana distribution patterns are consistent with one or more mechanisms thought to maintain liana species diversity, and how the patterns for lianas differ from those of trees. Finally, we review four recent studies that document and confirm that, over the past 40 years, lianas on and near the BCI 50-ha plot have increased in density and productivity, thereby increasing the percentage of trees that they infest and the intensity of tree infestation.

LIANA ECOLOGY ON BARRO COLORADO ISLAND

The forest of BCI

Barro Colorado Island (BCI) is a 1600-ha area covered by a seasonally moist lowland tropical forest. Once a former hilltop and part of a continuous forest, BCI was isolated from the mainland between 1910 and 1914 following the damming of the Rio Chagres and the creation of Lake Gatun (Leigh 1999). The forest of BCI receives an annual rainfall of 2600 mm, 90% of which falls between the months of May and December. The western side of BCI is covered by old-growth forest, with no record of large-scale anthropogenic disturbance (Hubbell & Foster 1983). The eastern side of BCI is late secondary forest (>100 years old), having been disturbed by the construction of the Panama Canal. A complete description of the geology, climate, flora and fauna of BCI may be found in Croat (1978) and Leigh (1999).

The ecology of lianas on BCI

Various aspects of the BCI forest have been studied for nearly a century (Leigh 1999), however, the lianas of BCI, similar to those in other forests, were largely ignored (Schnitzer & Bongers 2002). The publication of *Flora of Barro Colorado Island* (Croat 1978) revealed the identification and natural history of the known liana species, thus setting the stage for more detailed ecological studies. Putz (1984a) highlighted the importance of lianas in the BCI forest, indicating that lianas were abundant in the old-growth forest of BCI, that they infested nearly 50% of the trees on BCI, they reduced tree growth and increased the probability of tree mortality. Because lianas had such great potential for detrimental effects on trees, Putz (1984b) also investigated the characteristics of trees on BCI that enabled them to avoid liana infestation, concluding that fast-growing trees and palms with flexible, monopodial stems and large leaves had the fewest lianas (see also Schnitzer et al. 2000; Ingwell et al. 2010; Schnitzer & Carson 2010). DeWalt et al. (2000) examined liana density and basal area over a gradient of forest ages on BCNM, reporting that liana densities varied with forest age, peaking in secondary forests between 40 and 70 years old. They also found that the taxonomic composition of liana species and their climbing strategy changed with forest age and presumably with forest structure.

In a series of studies on BCI and on Gigante Peninsula (a mainland secondary forest that is part of the BCNM), Schnitzer and colleagues studied the effects of lianas on trees in treefall gaps and in closed-canopy forest on BCNM. They reported that liana density and diversity were particularly high in treefall gaps (Schnitzer & Carson 2000, 2001; Ledo & Schnitzer 2014; see also Putz 1984a), and that lianas in gaps greatly reduced tree growth and arrested tree regeneration, which, in turn, reduced tree species diversity (Schnitzer et al. 2000; Schnitzer & Carson 2010). Lianas may also alter tree species composition in gaps by competing intensely with shade-tolerant trees but not with some pioneer trees (Schnitzer et al. 2000; Schnitzer & Carson 2010). By reducing tree growth and survival in gaps, lianas may dramatically reduce whole forest carbon uptake and storage (Schnitzer et al. 2014). Lianas themselves, however, have low wood density and wood volume, and thus they do not compensate for the carbon that they displace in trees; therefore, lianas reduce the capacity of tropical forests to accumulate and store carbon (van der Heijden et al. 2013; van Heijden et al., Chapter 13 in this volume; Schnitzer et al. 2014).

Lianas in the intact, undisturbed forest of the BCNM also had a negative effect on trees, reducing tree growth and survival (Ingwell et al. 2010; Tobin et al. 2012). The competitive effect of lianas on trees differs from tree–tree competition, and a liana removal experiment on BCNM that controlled for the amount of biomass removed revealed that lianas have a much larger competitive effect on canopy trees than does tree–tree competition, and thus liana competition with trees is much stronger than would be predicted by their biomass (Tobin et al. 2012). Furthermore, liana competition with canopy trees was particularly strong during the dry season (Tobin et al. 2012; Alvarez-Cansino et al. in review), when lianas appear to grow vigorously even though water availability is reduced (Schnitzer 2005). The relatively strong competitive effect of lianas on canopy trees (compared to trees on canopy trees) is likely explained by the ability of even small lianas to reach the forest canopy, where they deploy their leaves over those of their tree host. Using logistic regression, Kurzel et al. (2006) found that lianas >3 cm diameter had more than a 90% chance of being in the forest canopy, whereas trees typically did not reach the canopy until they were 10-times larger. These early studies on BCI demonstrated that lianas are abundant and diverse, and compete intensely with trees.

THE LIANA CENSUS OF THE BCI 50-HA PLOT

The BCI 50-ha plot

The BCI 50-ha plot is a 1000 m by 500 m rectangle located on the relatively flat central plateau of the island. The plot is divided into 1250 20 × 20 quadrants, and each quadrant is further divided into 16 5 m × 5 m sub-quadrants (20,000 5 m x5 m sub-quadrants in total). The forest habitats of the 50-ha plot can be divided coarsely into the relatively dry high- and low-plateau areas, the moist slope area, a small seasonally inundated swamp area that drains to the southwest, and a small (~1 ha) portion of late secondary forest (>100 years old) on the northern edge (Fig. 7.1; Hubbell & Foster 1983; Dalling et al. 2012). All trees ≥1 cm diameter were measured, mapped, identified, and tagged in the 50-ha plot from 1980–1982 and have been censused every five years thereafter (described in Condit 1998). Tree data cited in this chapter are from the 2005 census (Condit et al. 2005).

Liana census of the BCI 50-ha plot

Over a 10-month period from February until December 2007, we tagged, mapped, measured, and identified all climbing and freestanding lianas ≥1 cm diameter that were rooted in the BCI 50-ha plot (the complete census methods are described in Schnitzer et al. 2008, 2012). We marked each liana stem with a uniquely numbered aluminum tag and measured liana stem diameter 1.3 m from the rooting point. We mapped the rooting point of each liana using existing 5 m × 5 m grid markers spread throughout the plot, which aided in recording the precise location of the stem. We identified lianas to species in the field using a combination of stem, leaf, and, less commonly, flower or fruit characteristics. Lianas that could not be identified

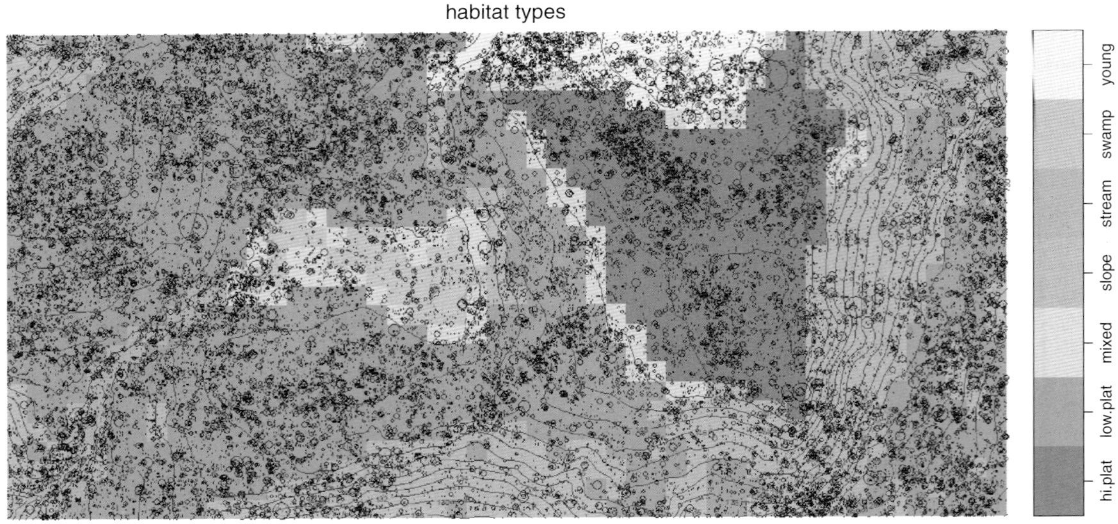

Fig. 7.1 Habitat map of the BCI 50-ha forest dynamics plot. Represented are the dry high plateau, the dry low plateau, the edge between habitats, the moist slopes, the seasonally inundated swampy area, and the white represents the late secondary forest. The plot dimensions measure 1000 m on the x-axis and 500 m on the y-axis. Large darker circles denote rooted liana individuals and small lighter colored circles indicate clonal stems that are still attached to a principal stem (all stems ≥1 cm diameter). Stem basal area is indicated by the relative size of the circle. (Source: Dalling et al. 2012. Reprinted with permission of John Wiley & Sons, Ltd.)

initially were re-examined in the following dry and wet seasons, when many of them had new leaves or flowers that enabled us to determine their identity. In total, we were able to identify 98.5% (46,495) of the individuals to species (Schnitzer et al. 2012).

We considered each separately rooted liana that was not connected to any other liana in the study to be an "apparent genet" because it appeared to be a genetically distinct individual. Lianas reproduce copiously via vegetative reproduction and connections among clonal stems can become buried and disintegrate; therefore, true genets can be confirmed only by genetic analyses. When a single liana had multiple rooted stems ≥1 cm in diameter, we considered the largest diameter stem to be the "principal stem" (and apparent genet), and each of the smaller stems to be the clonally produced ramets (following Gerwing et al. 2006; Schnitzer et al. 2006, 2008, 2012). We quantified clonal reproduction by the percentage of rooted stems ≥1 cm diameter that were still physically attached to another rooted stem in the census (ramets). For more details of our methods of sampling, quality control, and analyses, see Schnitzer et al. (2008) and Schnitzer et al. (2012).

Liana density and basal area of the BCI 50-ha plot

We found a total of 67,447 rooted liana stems ≥1 cm diameter in the BCI 50-ha plot, which averaged 1350 lianas ha^{-1} (Schnitzer et al. 2012). More than 20,000 of the rooted stems were clonally derived and were still attached to another stem in the census, and thus the total number of rooted liana apparent genets ≥1 cm diameter was 47,183 (943.7 lianas ha^{-1}; Fig. 7.2). Liana density decreased sharply with increasing diameter size classes, and the majority of liana stems were in the smallest diameter size class (1 to 2 cm diameter; Table 7.1). Liana basal area (BA) for all rooted stems ≥1 cm diameter was 49.21 m^2 (0.98 m^2 per ha). Roughly half (50.6%) of the total liana BA (and thus biomass) was contained in stems ≥5 cm diameter, even though this size class comprised less than 8% of the total number of stems in the BCI 50-ha plot (Table 7.1; Schnitzer et al. 2012).

Large lianas were exceedingly uncommon in the BCI 50-ha plot. There were 10 large (≥10 cm diameter) rooted lianas (apparent genets plus clones) per

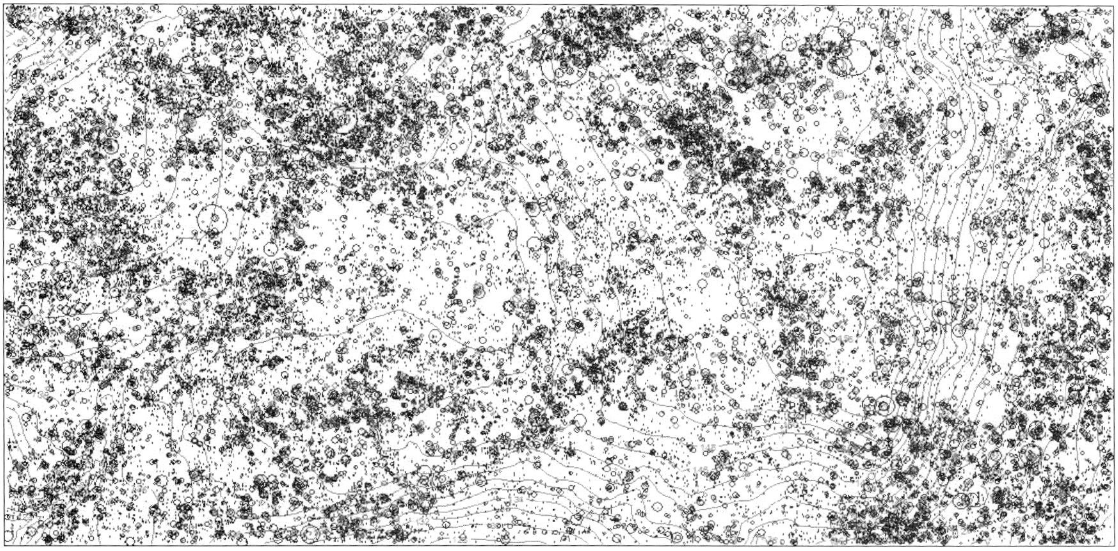

Fig. 7.2 Liana stems (≥1 cm diameter) rooted in the 50-ha forest dynamics plot on Barro Colorado Island, Panama. The plot dimensions measure 1000 m on the x-axis and 500 m on the y-axis. Large darker circles denote liana principal stems (i.e., on the largest clonal stem of a single individual) and small lighter colored circles indicate rooted clonal stems that are still attached to a principal stem. Stem basal area is indicated by the relative size of the circle; the diameter of the largest liana is 55.1 cm diameter and the smallest lianas are 1 cm. (Source: Schnitzer et al. 2012. Reproduced with permission.)

Table 7.1 Liana density, basal area, and species richness per hectare for all apparent genets and for all rooted stems (apparent genets plus ramets) in the 50-ha plot on Barro Colorado Island, Panama. Data are presented for the entire 50-ha plot and also as the mean per hectare for four commonly used size-classes.

	≥1 cm (mean ± SD)	**≥2 cm (mean ± SD)**	**≥5 cm (mean ± SD)**	**≥10 cm (mean ± SD)**
Apparent genets				
Density (ha^{-1})	928.6 (318.3)	462.6 (170.9)	78.5 (23.4)	8.3 (4.2)
Basal area (m^2 ha^{-1})	0.72 (0.23)	0.65 (0.21)	0.36 (0.13)	0.12 (0.09)
Species richness (ha^{-1})	74.6 (8.5)	60.8 (7.6)	26.8 (4.3)	5.5 (2.1)
Species richness (50 ha)	163	150	120	54
Fisher's α (ha^{-1})	19.11	18.72	14.35	7.11
Fisher's α (50 ha)	21.19	21.49	23.41	16.53
All rooted stems (genets + ramets)				
Density (ha^{-1})	1348.5 (505.9)	640.3 (256.2)	101.7 (34.2)	10.0 (5.9)
Basal area (m^2 ha^{-1})	0.96 (0.34)	0.85 (0.31)	0.46 (0.18)	0.15 (0.11)
Species richness (ha^{-1})	74.7 (8.4)	61.0 (7.4)	27.1 (4.3)	5.6 (2.1)
Species richness (50 ha)	163	150	120	54
Fisher's α (ha^{-1})	17.04	16.61	12.08	5.25
Fisher's α (50 ha)	20.11	20.38	22.03	15.40

ha, accounting for less than 1% of all rooted stems (Table 7.1). Nonetheless, lianas ≥10 cm diameter contributed 17.3% of the total liana BA. In the entire 50-ha plot, there were 514 rooted lianas (apparent genets plus clones); 487 in the 10–20 cm diameter size class, 22 in the 20–30 cm diameter size class, 3 in the 30–40 cm size class, only one in the 40–50 cm size class, and only one rooted liana ≥50 cm diameter (55.1 cm). Consequently, there are very few large lianas (≥10 cm diameter) present on BCI and in other tropical forests (e.g., Mascaro et al. 2004) compared to small stems, and the dynamics of large stems likely differ from those of more common smaller stems.

Clonal reproduction

We found a total of 20,314 rooted clonally produced ramets ≥1 cm diameter that were still connected to the principal stem. Clonal stems comprised 30% of the 67,447 lianas on the BCI 50-ha plot (small lighter colored circles in Fig. 7.2). Thus, including rooted clonal stems increased our estimate of liana density by 43%. Nearly all liana species exhibited clonal reproduction, and only 3 of the 129 species with more than 10 individuals had no rooted clones ≥1 cm diameter (Appendix 7.1; reproduced from Schnitzer et al. 2012). Both common and rare liana species had varying levels of clonal reproduction, with some common species exhibiting high levels of clonal reproduction (e.g., >50% of all stems of that species were clones) and other common species exhibiting very low levels of clonal reproduction (<5% of the stems of that species were clones). However, connections among liana ramets may be underground or be buried by leaf litter and soil, where they can decay rapidly (e.g., Peñalosa 1984; Putz 1984a); thus, our estimate of clonal reproduction is conservative, and we likely counted many lianas as apparent genets when they were actually clones that had lost their attachment to the parent stem.

The ability to produce clonal stems may be an important but underappreciated life-history characteristic of many liana species. Clonal reproduction may enable many liana species to colonize disturbed areas throughout the forest (Ledo & Schnitzer 2014) and may be a principal form of reproduction. When a tree falls, the lianas in the crown are often pulled into the gap where they resprout vigorously and with multiple stems that readily create new root systems (Putz 1984a; Schnitzer et al. 2000, 2004). The new clonal stems can eventually climb back up to the forest canopy far from the principal rooting point (Schnitzer et al. 2006, 2008; Yorke et al. 2013). Liana proliferation on large spatial and temporal scales may be the result of repeated cycles of lianas growing laterally, often through multiple tree crowns and extending long distances from their root systems, falling to the forest floor, resprouting, and then climbing back to the forest canopy.

Liana density as a proportion of all woody stems

Relative liana density (the percentage of all woody individuals excluding clones) represented 18.5% of the rooted woody stems ≥1 cm diameter in the BCI 50-ha plot. However, lianas tended to have a much higher proportion of small individuals than did trees due to the high relative density of lianas in the smallest size classes. The relative density of lianas for the 1–2 cm, 2–5 cm, 5–10 cm, ≥10 cm diameter size classes (based on a per ha average) was: 24%, 19%, 10%, and 2%, respectively (based on per ha average data from Table 7.1 and Schnitzer et al. 2012). When clonal stems are included, the relative density of lianas increased to 24.5% of all woody stems. Relative liana density per size class (based on a per ha average) including clonal stems was: 36%, 26%, 13%, and 2.5% for the 1–2 cm, 2–5 cm, 5–10 cm, ≥10 cm diameter size classes (Schnitzer et al. 2012). Consequently, lianas (including clonal stems) averaged across all size-classes constitute one-quarter of the woody stems on the BCI 50-ha plot; however, small liana stems (1–2 cm diameter) represent more than one-third of the woody stems in that size class.

Liana diversity, species accumulation, and community composition

The liana community of the BCI 50-ha plot includes 162 liana species from 37 families. The five most species-rich families contained more than half of the species: Bignoniaceae (22 species), Sapindaceae (22), Fabaceae (17), Malpighiaceae (13), and Apocynaceae

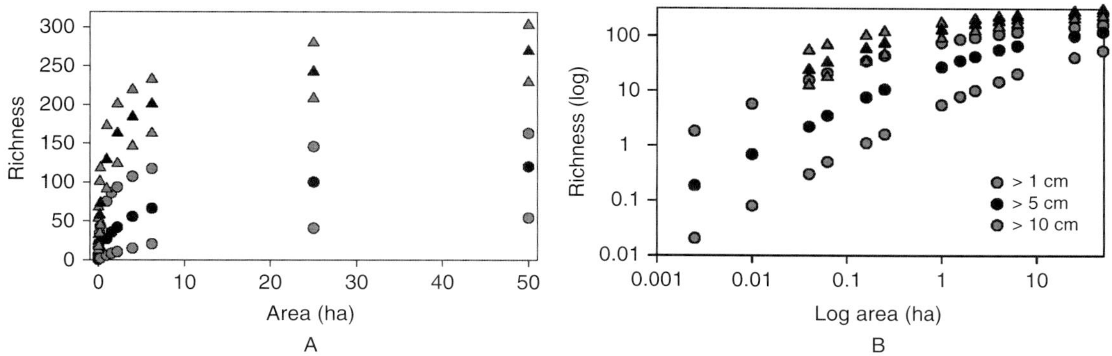

Fig. 7.3 Liana and tree species-area curves for three stem-diameter size classes over the 50-ha plot on Barro Colorado Island, Panama. Circles represent lianas; triangles represent trees. Panel A is based on untransformed data and panel B is based on transformed data. (Source: Schnitzer et al. 2012. Reproduced with permission.)

(9). The 10 most abundant species contributed nearly half (48.0%) of the total apparent genet density and 40.6% of the total BA (Appendix 7.1). *Coccoloba excelsa* (Polygonaceae) was the most common species, contributing 9.1% of all apparent genets ≥1 cm diameter, whereas *Prionostemma asperum* (Celastraceae) was the most dominant species in terms of basal area. The latter species was also the most common in the ≥5 cm and ≥10 cm diameter size classes (see Schnitzer et al. 2012 for more information on liana dominance hierarchies). In contrast, 19 families had only one or two species. Thirty-four species (21%) were represented by <10 individuals, and 10 species (6.1%) were represented by only one individual over the entire 50-ha plot (Schnitzer et al. 2012).

The census of the BCI 50-ha plot was a good representation of the entire liana community of BCI. Our sample of 162 species captured nearly 93% of the 175 liana species reported for the entire 1600-ha island (Croat 1978). Mean liana richness was 75 species per hectare, based on non-overlapping 100 × 100 m plots throughout the plot (Schnitzer et al. 2012). Species-area curves revealed that liana species richness accumulated rapidly, and we found approximately 125 species (77% of the species in the 50-ha plot) in 6 hectares (Fig. 7.3). The pattern of liana species accumulation was similar among size classes (≥1 cm, ≥5 cm, and ≥10 cm diameter); however, only approximately one-third of the species present in the 50-ha plot reached the large liana size class (≥10 cm diameter; Table 7.1). Fisher's α of large lianas at the plot-level was also much lower than the other diameter size-classes (Table 7.1). The accumulation of liana and tree species was similar, with trees always reaching higher diversity (Fig. 7.3). Nevertheless, lianas represented 35.3% of all woody species in the 50-ha plot.

The maintenance of liana diversity in the BCI 50-ha plot

The number of liana species can be exceptionally high in lowland tropical forests, as illustrated by this chapter and others in this book (e.g., Bongers & Ewango, Burnham & Romero-Saltos, Nogueira et al., Thomas et al.). However, few studies have examined the processes that maintain liana species diversity and that explain local liana distributions in tropical forests. In fact, most tests of general ecological theory in tropical forests have focused entirely on trees (Schnitzer et al., Chapter 1 in this volume). The BCI 50-ha plot provides an ideal case study for testing the maintenance of liana species diversity and the processes controlling within-forest liana species distributions because the identity and natural history of the species are well known, the soils and forest canopy have been comprehensively surveyed, and the maintenance of tree species diversity has been studied extensively at this site (see Ledo & Schnitzer 2014 and references therein).

We used the liana census of the BCI 50-ha plot to test among three hypotheses for the maintenance of plant species diversity (disturbance, negative density dependence (NDD), and niche partitioning) and compared the processes that control liana diversity and distribution with those of trees. We found that lianas of the BCI 50-ha plot were positively associated with canopy disturbance (measured as areas of low canopy height), but that lianas responded relatively little to edaphic characteristics, with only 21% of the species showing an association with soil chemistry (Dalling et al. 2012). In contrast, more than 50% of the trees on the BCI plot responded to soil chemistry (John et al. 2007) – a significantly greater percentage than lianas – suggesting that trees and not lianas were likely partitioning soil resources (Dalling et al. 2012).

In separate but complementary analyses, we tested whether per-species liana distributions responded more to habitat type, negative density dependence, or disturbance (Ledo & Schnitzer 2014). In this study, we defined habitat type as the linear combination of local topographical variables: elevation, slope, curvature and aspect, NDD as the over-dispersal of liana saplings relative to conspecific adults, and disturbance as areas where large trees had recently died (using long-term tree mortality data; Ledo & Schnitzer 2014). More than 50% of the liana species were correlated positively with disturbance, but relatively few liana species responded to habitat type or exhibited spatial patterns consistent with negative density dependence. Furthermore, liana species richness was higher in disturbed areas of the forest than in undisturbed forest even after rarefying the samples to remove the potentially confounding effects of higher liana density (see also Schnitzer & Carson 2001). Lianas appear to respond to disturbance by a combination of rampant clonal reproduction, which enhances the ability of lianas to colonize disturbed areas (Schnitzer et al. 2012), combined with sexual reproduction and successful dispersal of seeds into gaps. The majority of liana species (54%) produced significantly more clones in highly disturbed areas than in less disturbed areas (Schnitzer et al. 2012), and liana seed input into gaps is significantly overrepresented compared to trees (Puerta-Piñero et al. 2013).

In comparison, tree species on the BCI 50-ha plot demonstrated opposite patterns. Tree saplings were located consistently farther than expected from conspecific adults, demonstrating strong negative density dependence (Ledo & Schnitzer 2014; see also Comita et al. 2010; Mangan et al. 2010), trees were far more responsive to soil conditions (Harms et al. 2001; John et al. 2007), but tree species richness was not higher in gaps than non-gap sites (Hubbell et al. 1999; Schnitzer & Carson 2001). Therefore, there is a clear difference in the mechanisms that maintain liana and tree species diversity and control their local distributions. These new studies from BCI illustrate exciting possibilities for using lianas to test general theory in ecology, most of which have been tested solely with trees.

Increasing liana density, productivity, and tree colonization on BCI

Four separate studies have reported that lianas have increased on BCI over the past 40 years (Wright et al. 2004; Wright & Calderon 2006; Ingwell et al. 2010; Schnitzer et al. 2012). Wright and colleagues conducted weekly censuses of the leaves, flowers, and fruits captured in 200 $0.5\,m^2$ leaf and seed traps deployed in a portion of the BCI 50-ha plot. Over a 17-year period (1986–2002), liana productivity (measured as leaf litter production) increased 40% more than that of trees (Wright et al. 2004). Flower production increased faster for liana species than for tree species during this same period; however, there was no significant increase in fruit production for trees or lianas (Wright & Calderon 2006). Schnitzer et al. (2012) compared the liana density of the BCI 50-ha plot in 2007 to a study conducted 30 years earlier in the same general location (Putz 1984a). Liana density on BCI had increased by 75% for stems ≥ 1 cm diameter, and 140% for lianas ≥ 5 cm diameter. During this same approximate period (from 1982 until 2007), tree density (stems ≥ 1 cm diameter) on BCI decreased by 11.5%. In 2007, Ingwell et al. (2010) surveyed 3231 trees (≥ 20 cm diameter) to determine the percentage of canopy trees that carried lianas and to compare this finding to similar data collected 40 years earlier (Knight 1975). The percentage of trees infested with lianas increased from 32% in 1967–1968 to 75% in 2007. Ingwell and colleagues (2010) also surveyed the liana infestation levels in the crowns of nearly 3000 trees of 30 species in 1996 and 2007 on the BCI 50-ha plot and found that the number of trees with severe liana infestation in their crowns (>75% crown coverage by lianas) increased by 65% over this period. Trees with severe liana infestation had twice the probability of mortality as trees with no liana infestation (Ingwell et al. 2010). Collectively these studies confirm that lianas are increasing on the BCI, representing a large structural change in the BCI forest.

The trend of increasing liana abundance is consistent with other neotropical forests (Schnitzer & Bongers

2011; Schnitzer, Chapter 30 in this volume), indicating that lianas are rapidly becoming pervasive on BCI and throughout the neotropics. To date, however, there are very few studies of lianas at large spatial scales and relatively little is known about the role of lianas in most tropical forests. Most of the studies that do exist include only very large lianas, which are present at very low density and diversity. The study of the lianas of BCI and the resulting ecological advances illustrate the need for more long-term and large-scale studies of lianas, particularly studies that include small stem diameters (e.g., 1 cm), where lianas are the most dominant. A network of large forest dynamics plots that include periodic liana censuses around the world would greatly add to our understanding of liana ecology and their increasingly important role in tropical forests.

ACKNOWLEDGMENTS

We thank the many dedicated field technicians who were essential to liana census of the BCI 50-ha plot. We thank R. Burnham, S. DeWalt and A. Ercoli for helpful comments on this chapter. This work was made possible by financial support from the National Science Foundation (DEB-0613666, NSF-DEB 0845071, and NSF-DEB 1019436) and the University of Wisconsin-Milwaukee. Logistic support was provided by the Smithsonian Tropical Research Institute.

Appendix 7.1 Summary data for the lianas of the Barro Colorado Island 50-ha plot located in central Panama. Density of all rooted stems (principal stems plus clones), the total liana basal area (principal stems plus clones plus branches), the percentage of observed clonal stems for each of the 162 species.

Family	Genus	Species	Stem density	Total BA (cm^2 50 ha^{-1})	Percent clonality
Acanthaceae	*Justicia*	*graciliflora*	9	9.81	0.00
Acanthaceae	*Mendoncia*	*gracilis*	815	4038.07	10.80
Acanthaceae	*Mendoncia*	*retusa*	167	488.76	5.39
Apocynaceae	*Forsteronia*	*acouci*	827	3005.21	19.11
Apocynaceae	*Forsteronia*	*myriantha*	264	725.90	13.26
Apocynaceae	*Liedea*	*filisepala*	7	10.15	0.00
Apocynaceae	*Odontadenia*	*macrantha*	30	65.39	30.00
Apocynaceae	*Odontadenia*	*puncticulosa*	64	383.48	20.31
Apocynaceae	*Prestonia*	*lenticellata*	20	33.20	10.00
Apocynaceae	*Prestonia*	*mexicana*	1	1.77	0.00
Apocynaceae	*Prestonia*	*portobellensis*	41	70.88	0.00
Apocynaceae	*Prestonia*	*quinquangularis*	6	6.79	0.00
Aristolochiaceae	*Aristolochia*	*cordiflora*	379	714.20	7.12
Aristolochiaceae	*Aristolochia*	*tonduzii*	14	35.80	0.00
Asteraceae	*Heterocondylus*	*vitalbae*	1	2.01	0.00
Asteraceae	*Mikania*	*hookeriana*	8	28.82	0.00
Asteraceae	*Mikania*	*leiostachya*	410	1038.99	16.34
Asteraceae	*Tilesia*	*baccata*	14	39.32	7.14
Bignoniaceae	*Amphilophium*	*paniculatum*	472	7433.52	32.20
Bignoniaceae	*Arrabidaea*	*candicans*	550	5948.63	48.18
Bignoniaceae	*Arrabidaea*	*chica*	174	1560.26	36.21
Bignoniaceae	*Arrabidaea*	*corallina*	453	3827.63	41.50
Bignoniaceae	*Arrabidaea*	*florida*	514	4015.92	33.66
Bignoniaceae	*Arrabidaea*	*patellifera*	622	5411.92	54.98
Bignoniaceae	*Arrabidaea*	*verrucosa*	1797	20241.99	55.76
Bignoniaceae	*Callichlamys*	*latifolia*	634	7281.50	29.50
Bignoniaceae	*Ceratophytum*	*tetragonolobum*	557	4175.31	31.24
Bignoniaceae	*Cydista*	*aequinoctialis*	588	7878.90	38.27
Bignoniaceae	*Cydista*	*heterophylla*	3	13.05	33.33

(*continued overleaf*)

Appendix 7.1 *(continued).*

Family	Genus	Species	Stem density	Total BA (cm^2 50 ha^{-1})	Percent clonality
Bignoniaceae	*Macfadyena*	*unguis-cati*	122	1839.27	29.51
Bignoniaceae	*Mansoa*	*kerere*	43	218.56	13.95
Bignoniaceae	*Mansoa*	*verrucifera*	240	1081.12	42.08
Bignoniaceae	*Martinella*	*obovata*	36	286.65	25.00
Bignoniaceae	*Paragonia*	*pyramidata*	3571	26667.75	32.32
Bignoniaceae	*Phryganocydia*	*corymbosa*	671	3158.30	13.11
Bignoniaceae	*Pithecoctenium*	*crucigerum*	63	1195.48	14.29
Bignoniaceae	*Pleonotoma*	*variabilis*	1185	3219.52	26.16
Bignoniaceae	*Stizophyllum*	*inaquilaterum*	44	351.42	38.64
Bignoniaceae	*Stizophyllum*	*riparium*	303	1259.49	34.98
Bignoniaceae	*Tynnanthus*	*croatianus*	20	317.50	5.00
Boraginaceae	*Tournefortia*	*cuspidata*	4	19.68	25.00
Boraginaceae	*Tournefortia*	*hirsutissima*	6	50.93	16.67
Cannabaceae	*Celtis*	*iguanaea*	326	5376.42	62.88
Celastraceae	*Anthodon*	*panamense*	386	3764.83	34.20
Celastraceae	*Hippocratea*	*volubilis*	911	12312.43	30.95
Celastraceae	*Hylenaea*	*praecelsa*	149	990.65	51.01
Celastraceae	*Peritasa*	*pruinosa*	7	95.52	57.14
Celastraceae	*Prionostemma*	*asperum*	3213	42277.70	48.02
Combretaceae	*Combretum*	*decandrum*	586	5451.43	50.34
Combretaceae	*Combretum*	*fruticosum*	26	293.38	50.00
Combretaceae	*Combretum*	*laxum*	882	4580.05	40.59
Connaraceae	*Cnestidium*	*rufescens*	38	237.70	2.63
Connaraceae	*Connarus*	*panamensis*	51	260.37	5.88
Connaraceae	*Connarus*	*turczaninowii*	749	2562.89	4.01
Connaraceae	*Rourea*	*glabra*	53	313.00	5.66
Convolvulaceae	*Bonamia*	*trichanta*	27	42.90	11.11
Convolvulaceae	*Ipomea*	*phillomega*	2	4.31	0.00
Convolvulaceae	*Maripa*	*panamensis*	2961	25171.71	10.37
Cucurbitaceae	*Cayaponia*	*granatensis*	177	782.04	9.60
Cucurbitaceae	*Fevillea*	*cordifolia*	49	393.43	22.45
Cucurbitaceae	*Gurania*	*makoyana*	4	12.32	25.00
Dilleniaceae	*Davilla*	*nitida*	659	3104.73	34.75
Dilleniaceae	*Doliocarpus*	*dentatus*	216	1686.09	29.63
Dilleniaceae	*Doliocarpus*	*major*	2065	7208.50	22.03
Dilleniaceae	*Doliocarpus*	*multiflorus*	415	4990.93	28.92
Dilleniaceae	*Doliocarpus*	*olivaceus*	2936	9145.38	33.86
Dilleniaceae	*Tetracera*	*hydrophila*	272	1654.03	35.66
Dilleniaceae	*Tetracera*	*portobellensis*	74	472.42	29.73
Dilleniaceae	*Tetracera*	*volubilis*	11	36.23	45.45
Dioscoreaceae	*Dioscorea*	*hondurensis*	3	3.60	0.00
Euphorbiaceae	*Delechampia*	*dioscoreifolia*	2	10.56	0.00
Euphorbiaceae	*Omphalea*	*diandra*	725	11376.52	48.97
Fabaceae	*Acacia*	*acanthopylla*	789	15734.28	56.27
Fabaceae	*Acacia*	*hayesii*	476	8612.64	53.15
Fabaceae	*Bauhinia*	*guianensis*	1	1075.21	0.00
Fabaceae	*Clitoria*	*javitensis*	454	2213.73	19.82
Fabaceae	*Dioclea*	*wilsonii*	11	108.88	0.00
Fabaceae	*Entada*	*gigas*	180	20107.56	42.78

(continued overleaf)

Appendix 7.1 *(continued)*.

Family	Genus	Species	Stem density	Total BA (cm^2 50 ha^{-1})	Percent clonality
Fabaceae	*Entada*	*polystachya*	8	696.93	16.67
Fabaceae	*Machaerium*	*floribundum*	133	737.35	22.56
Fabaceae	*Machaerium*	*isadelphum*	9	364.64	33.33
Fabaceae	*Machaerium*	*kegelii*	67	100.52	16.42
Fabaceae	*Machaerium*	*microphyllum*	22	289.65	4.55
Fabaceae	*Machaerium*	*milleflorum*	125	1948.48	43.20
Fabaceae	*Machaerium*	*pittieri*	438	3480.93	27.85
Fabaceae	*Machaerium*	*riparium*	89	1937.80	30.34
Fabaceae	*Machaerium*	*seemanii*	203	1677.15	16.75
Fabaceae	*Rhynchosia*	*erythrinoides*	282	823.58	4.61
Gnetaceae	*Gnetum*	*leyboldii*	52	330.44	3.85
Lamiaceae	*Aegiphila*	*cephalophora*	1464	11164.83	39.55
Lamiaceae	*Aegiphila*	*elata*	727	6224.36	23.93
Loganiaceae	*Strychnos*	*brachistantha*	127	1363.30	24.41
Loganiaceae	*Strychnos*	*darienensis*	10	29.12	0.00
Loganiaceae	*Strychnos*	*panamensis*	262	886.75	5.73
Loganiaceae	*Strychnos*	*toxifera*	55	216.54	18.18
Malpighiaceae	*Adelphia*	*hiraea*	1422	7683.91	15.89
Malpighiaceae	*Bronwenia*	*wurdackii*	28	396.09	60.71
Malpighiaceae	*Heteropteris*	*laurifolia*	227	1910.26	30.84
Malpighiaceae	*Hiraea*	*fagifolia*	100	779.95	32.00
Malpighiaceae	*Hiraea*	*faginea*	120	714.28	26.67
Malpighiaceae	*Hiraea*	*grandifolia*	876	4433.83	20.78
Malpighiaceae	*Hiraea*	*reclinata*	354	2957.93	28.81
Malpighiaceae	*Hiraea*	*smilacina*	3473	16871.07	18.37
Malpighiaceae	*Mascagnia*	*divaricata*	175	3204.84	12.00
Malpighiaceae	*Mascagnia*	*ovatifolia*	115	1871.38	7.83
Malpighiaceae	*Stigmaphyllon*	*lindenianum*	389	1814.05	13.62
Malpighiaceae	*Tetrapteris*	*discolor*	215	1237.16	12.56
Malpighiaceae	*Tetrapteris*	*goudotiana*	36	236.88	8.33
Malpighiaceae	*Tetrapteris*	*seemannii*	24	163.77	12.50
Malpighiaceae	*Tetrapteris*	*tinifolia*	19	128.48	15.79
Malvaceae	*Byttneria*	*aculeata*	59	305.62	30.51
Melastomataceae	*Adelobotrys*	*adscendens*	4	11.66	0.00
Menispermaceae	*Abuta*	*racemosa*	600	4889.36	15.50
Menispermaceae	*Chondodendron*	*tomentosum*	81	608.94	8.64
Nyctaginaceae	*Pisonia*	*aculeata*	116	2416.09	32.76
Palmae	*Desmoncus*	*orthacanthos*	4	6.50	25.00
Passifloriaceae	*Passiflora*	*ambigua*	18	69.30	16.67
Passifloriaceae	*Passiflora*	*auriculata*	128	309.41	7.81
Passifloriaceae	*Passiflora*	*nitida*	5	30.39	0.00
Passifloriaceae	*Passiflora*	*vitifolia*	44	80.74	4.55
Phytolaccaceae	*Trichostigma*	*octandrum*	5	105.83	40.00
Piperaceae	*Piper*	*multiplinervium*	226	1093.12	44.69
Polygalaceae	*Securidaca*	*diversifolia*	9	77.43	33.33
Polygonaceae	*Coccoloba*	*excelsa*	7976	27840.21	46.33
Rhamnaceae	*Gouania*	*colombiana*	266	1117.36	32.71
Rubiaceae	*Chiococca*	*alba*	13	182.96	30.77
Rubiaceae	*Chomelia*	*psilocarpa*	75	668.74	54.67
Rubiaceae	*Uncaria*	*tomentosa*	206	8983.56	53.88

(continued overleaf)

Appendix 7.1 *(continued)*.

Family	Genus	Species	Stem density	Total BA (cm^2 50 ha^{-1})	Percent clonality
Sapindaceae	*Paullinia*	*baileyi*	530	825.76	4.15
Sapindaceae	*Paullinia*	*bracteosa*	504	1028.05	9.52
Sapindaceae	*Paullinia*	*fibrigera*	1216	7443.32	26.73
Sapindaceae	*Paullinia*	*fuscescens*	61	383.81	32.79
Sapindaceae	*Paullinia*	*glabrata*	57	481.28	19.30
Sapindaceae	*Paullinia*	*glomerulosa*	10	9.94	0.00
Sapindaceae	*Paullinia*	*mallophylla*	115	345.21	6.96
Sapindaceae	*Paullinia*	*pinnata*	2	3.55	0.00
Sapindaceae	*Paullinia*	*pterocarpa*	454	908.63	4.19
Sapindaceae	*Paullinia*	*rugosa*	400	1352.07	16.75
Sapindaceae	*Paullinia*	*turbacensis*	365	623.99	1.92
Sapindaceae	*Serjania*	*atrolineata*	7	56.06	85.71
Sapindaceae	*Serjania*	*circumvallata*	264	1680.83	11.74
Sapindaceae	*Serjania*	*cornigera*	1	1.77	0.00
Sapindaceae	*Serjania*	*membranacea*	15	119.82	40.00
Sapindaceae	*Serjania*	*mexicana*	847	2661.30	43.80
Sapindaceae	*Serjania*	*paucidentata*	1	24.63	0.00
Sapindaceae	*Serjania*	*pluvialiflorens*	41	249.71	21.95
Sapindaceae	*Serjania*	*pyramidata*	55	289.95	29.09
Sapindaceae	*Serjania*	*rhombea*	12	21.65	0.00
Sapindaceae	*Serjania*	*trachygona*	77	113.78	11.69
Sapindaceae	*Thinouia*	*myriantha*	1398	18046.93	36.55
Smilacaceae	*Smilax*	*domingensis*	10	15.74	50.00
Smilacaceae	*Smilax*	*febrifuga*	71	116.29	26.76
Smilacaceae	*Smilax*	*spinosa*	17	40.97	17.65
Smilacaceae	*Smilax*	*spissa*	1	0.79	0.00
Smilacaceae	*Smilax*	*spp.*	27	252.47	3.70
Solanaceae	*Solanum*	*aturense*	53	247.73	15.09
Solanaceae	*Solanum*	*lanceifolium*	46	185.33	10.87
Trigoniaceae	*Trigonia*	*rugosa*	6	28.73	16.67
Urticaceae	*Urera*	*lianoides*	95	263.10	23.16
Verbenaceae	*Petrea*	*volubilis*	2614	9424.56	11.74
Vitaceae	*Cissus*	*erosa*	26	589.21	3.85
Vitaceae	*Cissus*	*microcarpa*	67	802.99	13.43
Vitaceae	*Vitis*	*tiliifolia*	49	618.07	38.78
	Unidentified		817	6706.20	15.69

Synonymous species names: *Adelphia hiraea* = *Mascagnia hiraea*, *Aristolochia cordiflora* = *Aristolochia gigantea*, *Aristolochia tonduzii* = *Aristolochia chapmaniana*, *Bauhinia guianensis* = *Bauhinia excisa*, *Bronwenia wurdackii* = *Banisteriopsis cornifolia*, *Clitoria javitensis* = *Clitoria portobellensis*, *Coccoloba excelsa* = *Coccoloba parimensis*, *Davilla nitida* = *Davilla multiflora*, *Desmoncus orthacanthos* = *Desmoncus isthmius*, *Dioscorea hondurensis* = *Dioscorea Haenkeana*, *Doliocarpus major* = *Doliocarpus punctatus*, *Doliocarpus multiflorus* = *Doliocarpus guianensis*, *Entada polystachia* = *Entadopsis polystachya*, *Forsteronia viridescens* = *Forsteronia acouci*, *Gurania makoyana* = *Gurania seemanniana*, *Hiraea smilacina* = *Hiraea quapara*, *Hippocratea volubilis* = *Hippocratea versicolor*, *Justicia graciliflora* = *Beloperone graciliflora*, *Machaerium kegelii* = *Machaerium pachyphyllum*, *Machaerium seemannii* = *Machaerium campylocarpum*, *Mansoa kerere* = *Pachyptera kerere*, *Mascagnia ovatifolia* = *Mascagnia nervosa*, *Mendoncia litoralis* = *Mendoncia retusa*, *Odontadenia macrantha* = *Odontadenia grandiflora*, *Paullinia turbacensis* = *Paullinia wetmorei*, *Petrea volubilis* = *Petrea aspera*, *Pithecoctenium crucigerum* = *Pithecoctenium echinatum*, *Prestonia mexicana* = *Prestonia allenii*, *Prestonia quiquangularis* = *Prestonia acutifolia*, *Rhynchosia erythrinoides* = *Rhynchosia pyramidalis*, *Serjania pyramidata* = *Serjania decapleuria*, *Smilax febrigua* = *Smilax panamensis*, *Smilax domingensis* = *Smilax lanceolata*, *Solanum aturense* = *Solanum siparunoides*, *Solanum lanceifolium* = *Solanum lanciifolium*, *Tetrapterys goudontiana* = *Tetrapterys macrocarpa*, *Tilesia baccata* = *Wulffia baccata*.

Source: Schnitzer et al. (2012). Reproduced with permission.

REFERENCES

Alvarez-Cansino, L., Schnitzer, S.A., Reid, J., & Powers, J.S. In review. Liana competition with tropical trees varies with seasonally, but not with tree species identity.

Chave, J., Olivier, J., Bongers, F. et al. (2008) Aboveground biomass and productivity in a rain forest of eastern South America. *Journal of Tropical Ecology*, **24**, 355–366.

Comita, L.S., Muller-Landau, H.C., Aguilar, S. & Hubbell, S.P. (2010) Asymmetric density dependence shapes species abundances in a tropical tree community. *Science*, **329**, 330–332.

Condit, R. (1998) *Tropical Forest Census Plots*, Springer Verlag and R. G. Landes Company, Berlin, Germany, and Georgetown, Texas.

Condit, R., Hubbell, S.P., & Foster, R.B. (2005) Barro Colorado Forest Census Plot Data. URL: http://ctfs.arnarb.harvard.edu/webatlas/datasets/bci/

Croat, T.B. (1978) *Flora of Barro Colorado Island*, Stanford University Press, Stanford, CA.

Dalling, J.W., Schnitzer, S.A., Baldeck, C. et al. (2012) Resource-based habitat associations in a neotropical liana community. *Journal of Ecology*, **100**, 1174–1182.

DeWalt, S.J., Schnitzer, S.A. & Denslow, J.S. (2000) Density and diversity of lianas along a chronosequence in a central Panamanian tropical forest. *Journal of Tropical Ecology*, **16**, 1–19.

Durán, S.M. & Gianoli, E. (2013) Carbon stocks in tropical forests decrease with liana density. *Biology Letters*, **9**, 20130301.

Gerwing, J.J., Schnitzer, S.A., Burnham, R.J. et al. (2006) Censusing lianas. *Biotropica*, **38**, 256–261.

Harms, K.E., Condit, R., Hubbell, S.P. & Foster, R.B. (2001) Habitat associations of trees and shrubs in a 50-ha neotropical forest plot. *Journal of Ecology*, **89**, 947–959.

Hubbell, S.P. & Foster, R.B. (1983) Diversity of canopy trees in a neotropical forest and implications for conservation, in *Tropical Rain Forest: Ecology and Management* (eds S. Sutton, T. Whitmore and A. Chadwick), Blackwell Science, Oxford, pp. 25–41.

Hubbell, S.P., Foster, R.B., O'Brien, S.T. et al. (1999) Light-gap disturbances, recruitment limitation, and tree diversity in a neotropical forest. *Science*, **283**, 554–557.

Huffaker, C.B. (1958) Experimental studies on predation: dispersion factors and predator- prey oscillations. *Hilgardia*, **27**, 343–383.

Ingwell, L.L., Wright, S.J., Becklund, K.K. et al. (2010) The impact of lianas on 10 years of tree growth and mortality on Barro Colorado Island, Panama. *Journal of Ecology*, **98**, 879–887.

John, R., Dalling, J.W., Harms, K.E. et al. (2007) Soil nutrients influence spatial distributions of tropical tree species. *Proceedings of the National Academy of Sciences USA*, **104**, 864–869.

Knight, D.H. (1975) A phytosociological analysis of species-rich forest on Barro Colorado Island, Panama. *Ecological Monographs*, **45**, 259–284.

Kurzel, B.P., Schnitzer, S.A. & Carson, W.P. (2006) Predicting liana crown location from stem diameter in three Panamanian lowland forests. *Biotropica*, **38**, 262–266.

Ledo, A. & Schnitzer, S.A. (2014) Disturbance, not negative density dependence or habitat specialization maintains liana diversity in a tropical forest. *Ecology* in press.

Leigh, E.G. Jr. (1999) *Tropical Forest Ecology: A View from Barro Colorado Island*, Oxford University Press, Oxford.

Mangan, S.A., Schnitzer, S.A., Herre, E.A. et al. (2010) Negative plant-soil feedback predicts relative species abundance in a tropical forest. *Nature*, **466**, 752–756.

Mascaro, J., Schnitzer, S.A. & Carson, W.P. (2004) Liana diversity, abundance, and mortality in a wet tropical forest. *Forest Ecology and Management*, **190**, 314.

Peñalosa, J. (1984) Basal branching and vegetative spread in two tropical rain forest lianas. *Biotropica*, **16**, 1–9.

Pérez-Salicrup, D.R., Sork, V.L. & Putz, F.E. (2001) Lianas and trees in a liana forest of Amazonian Bolivia. *Biotropica*, **33**, 34–47.

Phillips, O.L., Vasquez Martinez, R., Arroyo, L. et al. (2002) Increasing dominance of large lianas in Amazonian forests. *Nature*, **418**, 770–774.

Puerto-Piñero, C., Mueller-Landau, H.C., Calderon, O. & Wright, S.J. (2013) Seed arrival in tropical forest tree fall gaps. *Ecology*, **94**, 1552–1562.

Putz, F.E. (1984a) The natural history of lianas on Barro Colorado Island, Panama. *Ecology*, **65**, 1713–1724.

Putz, F.E. (1984b) How trees avoid and shed lianas. *Biotropica*, **16**, 19–23.

Schnitzer, S.A. (2005) A mechanistic explanation for global patterns of liana abundance and distribution. *The American Naturalist*, **166**, 262–276.

Schnitzer, S.A. & Bongers, F. (2002) The ecology of lianas and their role in forests. *Trends in Ecology and Evolution*, **17**, 223–230.

Schnitzer, S.A. & Bongers, F. (2011) Increasing liana abundance and biomass in tropical forests: emerging patterns and putative mechanisms. *Ecology Letters*, **14**, 397–406.

Schnitzer, S.A., Bongers, F. & Wright, S.J. (2011) Community and ecosystem ramifications of increasing lianas in neotropical forests. *Plant Signaling & Behavior*, **6**, 598–600.

Schnitzer, S.A. & Carson, W.P. (2000) Have we missed the forest because of the trees? *Trends in Ecology and Evolution*, **15**, 367–377.

Schnitzer, S.A. & Carson, W.P. (2001) Treefall gaps and the maintenance of species diversity in a tropical forest. *Ecology*, **82**, 913–919.

Schnitzer, S.A. & Carson, W.P. (2010) Lianas suppress tree regeneration and diversity in treefall gaps. *Ecology Letters*, **13**, 849–857.

Schnitzer, S.A., Dalling, J.W. & Carson, W.P. (2000) The impact of lianas on tree regeneration in tropical forest canopy gaps: evidence for an alternative pathway of gap-phase regeneration. *Journal of Ecology*, **88**, 655–666.

Schnitzer, S.A., DeWalt, S.J. & Chave, J. (2006) Censusing and measuring lianas: a quantitative comparison of the common methods. *Biotropica*, **38**, 581–591.

Schnitzer, S.A., Kuzee, M. & Bongers, F. (2005) Disentangling above- and below-ground competition between lianas and trees in a tropical forest. *Journal of Ecology*, **93**, 1115–1125.

Schnitzer, S.A., Mangan, S.A., Dalling, J.W. et al. (2012) Liana abundance, diversity, and distribution on Barro Colorado Island, Panama. *PLoS ONE*, **7**, e52114.

Schnitzer, S.A., Parren, M. & Bongers, F. (2004) Recruitment of lianas into logging gaps and the effects of pre-harvest liana cutting in a Cameroon lowland forest. *Forest Ecology and Management*, **190**, 87–98.

Schnitzer, S.A., Rutishauser, S. & Aguilar, S. (2008) Supplemental protocol for liana censuses. *Forest Ecology and Management*, **255**, 1044–1049.

Schnitzer, S.A., van der Heijden, G.M.F., Mascaro, J., & Carson, W.P. (2014) Lianas reduce biomass accumulation in a tropical forest. *Ecology*, in press.

Swaine, M.D. & Grace, J. (2007) Lianas may be favoured by low rainfall: evidence from Ghana. *Plant Ecology*, **192**, 271–276.

Tobin, M.F., Wright, A.J., Mangan, S.A., & Schnitzer, S.A. (2012) Lianas have a greater competitive effect than trees of similar biomass on tropical canopy trees. *Ecosphere*, **3**, Article 20: 1–11. 10.1890/ES11-00322.1

van der Heijden, G.M.F., Healey, J.R. & Phillips, O.L. (2008) Infestation of trees by lianas in a tropical forest in Amazonian Peru. *Journal of Vegetation Science*, **19**, 747–756.

van der Heijden, G.M.F., Schnitzer, S.A., Powers, J.S. & Phillips, O.L. (2013) Liana impacts on carbon cycling, storage and sequestration in tropical forests. *Biotropica*, **45**, 682–692.

Wright, S.J. & Calderon, O. (2006) Seasonal, El Niño and longer-term changes in flower and seed production in a moist tropical forest. *Ecology Letters*, **9**, 35–44.

Wright, S.J., Calderón, O., Hernandéz, A. & Paton, S. (2004) Are lianas increasing in importance in tropical forests? A 17-year record from Panamá. *Ecology*, **85**, 484–489.

Wyka, T.P., Oleksyn, J., Karolewski, P. & Schnitzer, S.A. (2013) Phenotypic correlates of the lianescent growth form: a review. *Annals of Botany*, **112**, 1667–1681.

Yorke, S.R., Schnitzer, S.A., Mascaro, J. et al. (2013) Increasing liana abundance and biomass in a tropical forest: the contribution of long distance clonal colonization. *Biotropica*, **45**, 317–324.

Chapter 8

DIVERSITY AND DISTRIBUTION OF LIANAS IN MEXICO

Guillermo Ibarra-Manríquez, Pablo Carrillo-Reyes, Francisco Javier Rendón-Sandoval, and Guadalupe Cornejo-Tenorio

Universidad Nacional Autónoma de México, Morelia, Michoacán, Mexico

Ecology of Lianas, First Edition. Edited by Stefan A. Schnitzer, Frans Bongers, Robyn J. Burnham, and Francis E. Putz.

Companion Website: www.wiley.com/go/schnitzer/lianas

OVERVIEW

Mexico holds global prominence for its contribution to biodiversity, mainly in terms of the vascular flora. This chapter analyzes the diversity and biogeography of lianas in Mexico. Using an exhaustive bibliographic review, as well as herbarium and field studies, we recorded a total of 754 liana species, belonging to 214 genera and 55 families within Mexico. The diversity of the lianas is clearly concentrated within few families; 89.5% are included in just 20 families. The three families with the most liana species are the Apocynaceae (20.3%), Fabaceae (11.9%) and Convolvulaceae (8.9%), and each of these three families also has a high number of endemic lianas. The genus *Ipomoea* (Convolvulaceae) includes the most species (6.2%) of lianas, yet almost half of the genera are comprised of only one liana species. A total of 185 species (24.5%) and three genera (1.4%) are endemic to Mexico. *Ipomoea* and *Matelea* (Apocynaceae) stand out in their endemic component (13 species each).

Liana species affinities are more aligned with Central America (70%) than with the United States of America, the Antilles, or South America. The Mexican states of Chiapas, Guerrero, Jalisco, Michoacán, Oaxaca, and Veracruz are the most important in terms of liana richness and endemic species. The composition of lianas among Mexican states shows high beta diversity because many species are found in only one state (19.2% of the total and 23.2% of the endemics).

We compared 10 reserves whose flora is acceptable known and found that a total of 437 species (58%) were present in one or more of these Protected Areas, of which 75 were endemic to a specific reserve. A total of 233 species (61 of which are endemic) were found just in one of these Protected Areas. Although the diversity recorded in these preserves is remarkable, the disappearance of the tropical forests where the majority of lianas thrive produces a dim outlook for their future conservation. The significant role of lianas as important and distinctive elements within the diverse flora of Mexico is beyond doubt; however, lacking is detailed knowledge of liana distribution, biological traits, and their conservation status, which is of grave concern.

INTRODUCTION

Mexico is a significant contributor to global biodiversity, especially for vascular plants and vertebrates (Mittermeier 1988; WCMC 1992). The great diversity of plants and animals in Mexico is the result of various factors, such as a complex geological history (Ferrusquía-Villafranca 1993), resulting in many mountain ranges and the potential for population isolation and the evolution of new species. The heterogeneous landscape of Mexico and its latitudinal position have produced a diverse mosaic of climates and vegetation types that include almost all those that are recognized on a global scale (Rzedowski 1978, 1993; Palacio-Prieto et al. 2000). The affinities of the Mexican flora are more meridional than boreal and, according to the fossil evidence, they date back at least to the Cenozoic era, 65 million years ago (Rzedowski 1991a, 1991b, 1993).

Unfortunately, knowledge of the floristic richness of Mexico is incomplete. Floristic richness is estimated at 220–248 families, 2410–2663 genera and around 22,000–22,351 species of Angiosperms (Rzedowski 1991b; Villaseñor 2003, 2004). Despite the lack of precise knowledge of its flora, Mexico is one of the countries with the highest species richness, surpassed only in this regard by Brazil, Colombia, China, and South Africa (Mittermeier 1988; WCMC 1992; Villaseñor 2003). In addition, Mexico is likely to be the country with the greatest level of plant endemism in America; 7.8–10% of the genera, 52% of the species, and the monotypic families Lacandoniaceae and Setchellanthaceae are endemic to Mexico (Rzedowski 1991a; Villaseñor 2003, 2004).

Knowledge of the Mexican flora, in terms of its systematics and growth forms, is variable at the geographic level. For example, only a few families, such as Malvaceae (Fryxell 1988) or Orchidaceae (Hágsater et al. 2005), have been inventoried for the whole country. Similarly, while many studies have quantified tree diversity (e.g., Ibarra-Manríquez et al. 1995; Padilla-Velarde et al. 2006; Ricker & Hernández 2010; González-Espinosa et al. 2011), few studies are available for lianas. Further, the scant information for lianas that does exist is dispersed among a vast floristic and taxonomic literature (e.g., Rzedowski & Calderón de Rzedowsi 1993; Pérez-Cueto 1995; Ibarra-Manríquez & Sinaca 1995; Lott 2002; Alvarado-Cárdenas 2004; Pérez-García et al. 2010). The only existing estimate of liana taxonomic richness in Mexico so far suggests around 60 genera and 300 species (Rzedowski 1991b).

From the perspective of their ecological role, diverse studies worldwide have highlighted the importance of lianas, especially in tropical forests, where they make an important contribution through their diversity, regeneration, and succession, as well as in interactions with diverse groups of animals, especially pollinators and frugivores (Gentry 1982, 1991a, 1991b; Bongers et al. 2002, 2005; Schnitzer & Bongers 2002, 2011; Jongkind & Hawthorne 2005; Arroyo-Rodríguez et al., Chapter 27 in this volume, Toledo-Aceves, Chapter 12 in this volume). In Mexico, there are few studies of liana ecology, and those that do exist focus mainly on dispersal strategies, phenology, and landscape spatial patterns (Ibarra-Manríquez et al. 1991, 2001; Ibarra-Manríquez & Martínez-Ramos 2002; Solórzano et al. 2002; Pérez-Salicrup & de Meijere 2005; Arroyo-Rodríguez & Toledo-Aceves 2009; DeWalt et al. 2010; DeWalt et al., Chapter 11 in this volume). Furthermore, many of these studies are concentrated in areas without comprehensive floristic information. An analysis of different reserves in Mexico that have published floristic inventories, such as Calakmul, Campeche (Martínez et al. 2001), Chamela, Jalisco (Lott 2002), Los Tuxtlas, Veracruz (Ibarra-Manríquez & Sinaca 1995, 1996a, 1996b), Sian Ka'an, Quintana Roo (Durán & Olmsted 1987), Ría Celestún, Yucatán (Durán et al. 1999a), Ría Lagartos, Yucatán (Durán et al. 1999b) and Yaxchilán, Chiapas (Meave et al. 2008), indicates that the species number of lianas encompass from 8–15% of the entire flora. At a finer scale, Bongers et al. (1988) found that lianas accounted for 6.8% and 8.7%, respectively, of the species and individuals surveyed in a hectare of lowland tropical rain forest, in Veracruz. In seasonally dry tropical forests, lianas averaged 11.2% (±7.1 s.d.) of the woody species (trees, shrubs, and lianas) ≥2.5 cm dbh based on 26 0.1 ha samples; however, lianas can constitute as much as 25% of the woody species in Mexico (Lott et al. 1987; Trejo & Dirzo 2002; Pineda-García et al. 2007). These data suggest that liana species contribute substantially to the diversity of the Mexican flora.

In view of this paucity of information on the lianas of Mexico, the main objective of this chapter is to document the taxonomic richness, degree of endemism, and geographical distribution patterns, and to assess the role of protection performed by the Protected Areas for Mexican lianas. Specifically, we ask the following questions: (1) how many liana species exist in Mexico and what percentage of Mexican liana species are endemic?; (2) which families and genera are the most species-rich?; (3) are the Mexican states of equal importance in terms of the total number of liana species and of endemics?; and (4) what proportion of the woody flora do lianas represent in 10 Protected Areas of Mexico?

DEFINITION OF LIANA

Climbers are defined in a broad sense as plants that germinate in the soil but are mechanically dependent, at some part of their lives, on the support of other plants for their development (Gentry 1991a; Acevedo-Rodríguez 2005; Bongers et al. 2002, 2005; Jongkind & Hawthorne 2005). Growth forms among Angiosperms, however, can be complicated and certain species can deviate from this rather simple definition of "liana." In this chapter, we used three sources of information for the inclusion of a species as a liana: (1) classification as a liana in the floristic and taxonomic literature; (2) consultation of specimens deposited in the herbaria IBUG, IEB, MEXU, MO, and XAL; and (3) our observations from fieldwork conducted in different regions and vegetation types of Mexico. We observed woody stems in almost 40% of the species in the field. Included in the list are the hemiepiphytes, since their physiognomy and various aspects of their ecology make them similar to lianas that germinate in the soil. Monocot species are also included which, despite not possessing stems with woody secondary growth, have persistent fibrous stems (e.g., *Desmoncus orthacanthos* Mart. and *Smilax* spp.). Climbers with exclusively herbaceous stems (e.g., species of Poaceae) were omitted from this classification, along with shrubs with climbing branches of less than 5 m in length, as well as primary epiphytes (e.g., species of Bromeliaceae or Orchidaceae) and secondary epiphytes (e.g., *Monstera* spp. and *Philodendron* spp., Araceae).

TAXONOMIC DIVERSITY

Estimates of Mexican liana diversity includes 754 species, 214 genera, and 55 families. Approximately 20% of the species (153) belong to the dominant family Apocynaceae (including Asclepiadaceae). Other important families include: Fabaceae (90 species, 11.9%), Convolvulaceae (67, 8.9%), Malpighiaceae (50, 6.8%), Asteraceae (44, 5.8%), Bignoniaceae (42, 5.6%), and Sapindaceae (40, 5.3%; Table 8.1). Eighteen families only contain one species:

Table 8.1 Total liana species richness (including endemics) of the 20 most liana species-rich families and genera in Mexico.

Families	Total species (endemics)	Genus (family)	Total species (endemics)
Apocynaceae	153 (71)	*Ipomoea* (Convolvulaceae)	48 (13)
Fabaceae	88 (20)	*Passiflora* (Passifloraceae)	27 (3)
Convolvulaceae	67 (17)	*Serjania* (Sapindaceae)	27 (3)
Malpighiaceae	50 (13)	*Matelea* (Apocynaceae)	25 (13)
Asteraceae	44 (12)	*Marsdenia* (Apocynaceae)	24 (11)
Bignoniaceae	42 (1)	*Solanum* (Solanaceae)	19 (5)
Sapindaceae	40 (4)	*Aristolochia* (Aristolochiaceae)	18 (6)
Solanaceae	31 (8)	*Gonolobus* (Apocynaceae)	18 (6)
Passifloraceae	27 (3)	*Metastelma* (Apocynaceae)	17 (11)
Vitaceae	24 (4)	*Arrabidaea* (Bignoniaceae)	13 (0)
Aristolochiaceae	18 (6)	*Otopappus* (Asteraceae)	12 (7)
Rubiaceae	17 (3)	*Cissus* (Vitaceae)	12 (1)
Menispermaceae	13 (2)	*Smilax* (Smilacaceae)	12 (0)
Smilacaceae	12 (0)	*Dictyanthus* (Apocynaceae)	11 (9)
Cucurbitaceae	11 (4)	*Mandevilla* (Apocynaceae)	11 (5)
Celastraceae	9 (1)	*Rhynchosia* (Fabaceae)	10 (2)
Polemoniaceae	9 (4)	*Cobaea* (Polemoniaceae)	9 (4)
Boraginaceae	7 (0)	*Paullinia* (Sapindaceae)	9 (0)
Combretaceae	7 (1)	*Machaerium* (Fabaceae)	8 (0)
Connaraceae	6 (2)	*Heteropterys* (Malpighiaceae)	7 (2)

Arecaceae, Cannabaceae, Capparaceae, Caprifoliaceae, Dichapetalaceae, Dioscoreaceae, Gelsemiaceae, Hernandiaceae, Melastomataceae, Plantaginaceae, Rutaceae, Schizandraceae, Schlegeliaceae, Trigoniaceae, Urticaceae, Valerianaceae, Verbenaceae, and Violaceae. The families shown in Table 8.1 match those highlighted by Gentry (1991a) as being of scandent habit in the Neotropics. The majority of the climbers of the New World are concentrated within a small number of families: 10 families account for 60.3% of the species (Gentry 1991a). The present study supports this trend, because the 10 most diverse families in Mexico represent 75.1% of the total richness. This pattern of richness concentration within few families is less strong in studies of Mexican trees (e.g., Ibarra-Manríquez et al. 1995; Padilla-Velarde et al. 2006), as well as in the complete Mexican flora (Villaseñor 2003), in which trees account for ca. 50% and 48.4%, respectively.

The genus *Ipomoea* (Convolvulaceae) includes the most species (48, 6.2%), followed by *Passiflora*, *Serjania*, *Matelea*, and *Marsdenia* (Table 8.1). The 20 genera listed in Table 8.1 together contain 44.7% (337) of all Mexican liana species. By comparison, the most diverse genera in Mexico, counting all growth forms, comprise around 15.1% of the species richness (Villaseñor 2004). For lianas of Mexico, nearly half of the recorded genera (109) had only one species.

Our estimated 214 genera and 754 species of lianas greatly exceed those of Rzedowski (1991b), who suggested 60 and 300 taxa, respectively. The liana species-to-genus ratio (3.5) is less than the 8.4 for the complete Mexican flora (Villaseñor 2003). Nonetheless, this ratio is high compared to 1.5 in Ivory Coast (Den Outer 1972), 2.7 in Upper Guinea (Jongkind 2005), 2.6 in The Ducke Reserve, Brazil (Ribeiro et al. 1999), and 2.1 for vines and climbing plants of Puerto Rico and the Virgin Islands (Acevedo-Rodríguez 2005). The diversity of Mexican lianas appears to be higher than in many other regions, but this claim should be made with caution, since various factors can introduce bias, for example, the size of the regions under comparison (Gotelli & Colwell 2001) or the criteria for definition of species as a liana. One sign of the influence of these factors is that certain Protected Areas in Mexico that have different sizes and vegetation types (Durán & Olmsted 1987; Martínez et al. 1994, 2001; Ibarra-Manríquez & Sinaca 1995; 1996a, 1996b;

Vázquez et al. 1995; Lott 2002; Table 8.2), have species-to-genus ratios lower than that obtained for the whole Mexican liana flora: Chamela-Cuixmala (1.4), Boca de Chajul (1.4), Sian Ka'an (1.4), Los Tuxtlas (1.5), Calakmul (1.7), and Sierra de Manantlán (2.0).

ENDEMISM

Nearly one-quarter (185) of all liana species are endemic to Mexico. Endemic species are found in 69 genera and 25 families. The two families with the most endemic species are the Apocynaceae and Fabaceae, which have 71 and 20 species, respectively. The 20 families cited in Table 8.1 include 95.7% of all endemic liana species in Mexico. At the genus level, *Ipomoea* and *Matelea* (13 species each) have the most endemics, followed by *Marsdenia*, which has 11 species. A total of 36 genera have only one species in Mexico. The most species-rich family, the Apocynaceae, accounts for six of the 20 most diverse genera in terms of endemic species (Table 8.1). The endemic genera recorded to date are: *Balsas* J. Jiménez Ram. & K. Vega (Sapindaceae, Jiménez et al. 2011), *Peponopsis* Naudin (Cucurbitaceae) and *Thoreauea* J.K. Williams (Apocynaceae, Williams 2002; Morales 2005; Diego & Lozada 2006).

Rzedowski (1991b) calculated a liana endemism of approximately 1% of the genera and 15% of the species. Our data are similar for the genus level (1.4%), lower than the 7.8% endemism percentage for the entire Mexican flora (Villaseñor 2004). At the species level, Rzedowski's 15% is much lower than ours; we found 24.5%, still lower than the 52% estimate for the total Mexican flora by Villaseñor (2003). Families with high liana diversity also have an elevated number of endemic species (Table 8.1). Part of this similarity is due to the fact that some of the 15 most diverse families in Mexico (Villaseñor 2003; Juárez-Jaimes et al. 2007) have endemism values between 45.7 and 65.9% (Asteraceae 65.9%, Fabaceae 59.5%, Rubiaceae 48.7%, and Apocynaceae 45.7%).

DISTRIBUTION

Approximately 8.4% (63) of the lianas in Mexico are shared with the United States of America, 70% (528) with Central America, 40% (301) with South America, and 18.6% (140) with the Antilles. Within Mexico, the states have an unequal presence of species, with the richest states being those of the southeast, particularly Chiapas (453 species, 60.1%), Veracruz, and Oaxaca – the latter two with approximately 50% of the total number of species (Fig. 8.1). A total of 724 species (96%) are present in one or more of the 10 most diverse states.

Importance among the Mexican states changes when analyzing only the endemic species, since Oaxaca is the most important (74 species, 40%), followed by Jalisco, Guerrero, and Michoacán, with percentages that range from 34.6 to 38.9%. As much as 90.8% of the endemic contingent (168 species) has been recorded in one or more of the 10 states with more species. These results reinforce the trend found in previous studies regarding the floristic importance of the states of Chiapas, Guerrero, Jalisco, Michoacán, Oaxaca and Veracruz (Villaseñor 2003; Juárez-Jaimes et al. 2007; Espejo 2012).

We examined the presence or absence of species at the state level (using a cluster analysis) and found that there are five groups, at a linkage distance of 20, that show high geographical congruence (Figs. 8.1, 8.2A). The first group includes the states of the Yucatán peninsula and Tabasco, one of the biogeographical provinces of Mexico (Rzedowski 1978; Ibarra-Manríquez et al. 2002). The second group is formed by states with the highest species richness (Chiapas, Oaxaca, and Veracruz); the third is composed of Guerrero, Jalisco, and Michoacán, located on the Pacific Ocean watershed. The fourth group includes neighboring states in different parts of Mexico, while the fifth group comprises 12 states with very low species richness of lianas, generally located toward the Mexican altiplano and the Peninsula of Baja California. The three first groups include states with important areas of tropical forests, both evergreen lowland and seasonal tropical forest (Rzedowski 1978; Palacio-Prieto et al. 2000).

Eight groups are present among the clustering of endemic species, and resemble those described previously for all species (Figs. 8.1, 8.2B). At a linkage distance of number 7, the first three groups basically encompass states of the biogeographical province of the Depresión del Balsas (Rzedowski 1978): (1) Oaxaca and Guerrero; (2) Morelos, Puebla, and México; and (3) Colima, Jalisco, and Michoacán. Chiapas and Veracruz, both remarkable for their liana richness, comprise the fourth group. In the cluster diagram (Fig. 8.2B), the fifth group is comprised of states in the central and northeast part of Mexico (Hidalgo, Querétaro, San Luis Potosí, and Tamaulipas) and the

Table 8.2 Total liana species (including endemic species) for Protected Areas in Mexico.

Reserves (State)	Number of species		Surface (km²)	Dominant vegetation	Biogeographical Province
	Total	Endemics			
[1]Boca de Chajul** (Chiapas)	88 (19)	0	17779.0	Tropical evergreen forest	Costa del Golfo de México
[2]Calakmul*(Campeche)	134 (25)	3 (2)	7231.9	Tropical deciduous forest and Tropical subdeciduous forest	Península de Yucatán
[3]Chamela-Cuixmala* (Jalisco)	107 (32)	26 (14)	131.4	Tropical deciduous forest and Tropical subdeciduous forest	Serranías Meridionales and Costa Pacífica
[4]El Triunfo* (Chiapas)	53 (15)	2 (2)	1191.8	Coniferous forests, montane cloud forest, tropical deciduous forest and tropical evergreen forest	Serranías Transístmicas and Soconusco
[5]Los Tuxtlas** (Veracruz)	131 (52)	5 (5)	6.4	Tropical evergreen forest	Costa del Golfo de México
[6]Ría Celestún* (Campeche and Yucatán)	50 (2)	1 (1)	591.3	Evergreen lowland floodplain forest, mangroves and tropical deciduous forest	Península de Yucatán
[7]Ría Lagartos* (Yucatán)	49 (0)	0	569.9	Evergreen lowland floodplain forest, mangroves and tropical deciduous forest	Península de Yucatán
[8]Sierra de Manantlán* (Colima and Jalisco)	138 (50)	39 (26)	1395.8	Coniferous forests, montane cloud forest and tropical deciduous forest	Serranías Meridionales, Sierra Madre Oriental and Costa Pacífica
[9]Sian Ka'an* (Quintana Roo)	81 (9)	1 (0)	5281.5	Evergreen lowland floodplain forest, mangroves and tropical deciduous forest	Península de Yucatán
[10]Sierra Gorda* (Guanajuato and Querétaro)	69 (28)	13 (11)	3835.7	Coniferous forests, montane cloud forest, tropical deciduous forest and tropical subdeciduous forest	Altiplanicie and Sierra Madre Oriental

Notes: * Biosphere Reserves, ** Biological Station. The number of species registered solely in each reserve is showed in parenthesis.
Sources: [1]Martínez et al. (1994);[2]Martínez et al. (2001); [3]Lott (2002); [4]Martínez-Meléndez et al. (2008); Pérez-Farrera et al. (2012); [5]Ibarra-Manríquez & Sinaca (1995, 1996a, 1996b); [6]Durán et al. (1999a);[7]Durán et al. (1999b); [8]Vázquez et al. (1995); [9]Durán & Olmsted (1987); [10]Several fascicles of Flora del Bajío y Regiones Adyacentes. Biogeographical Province sensu Rzedowski (1978).

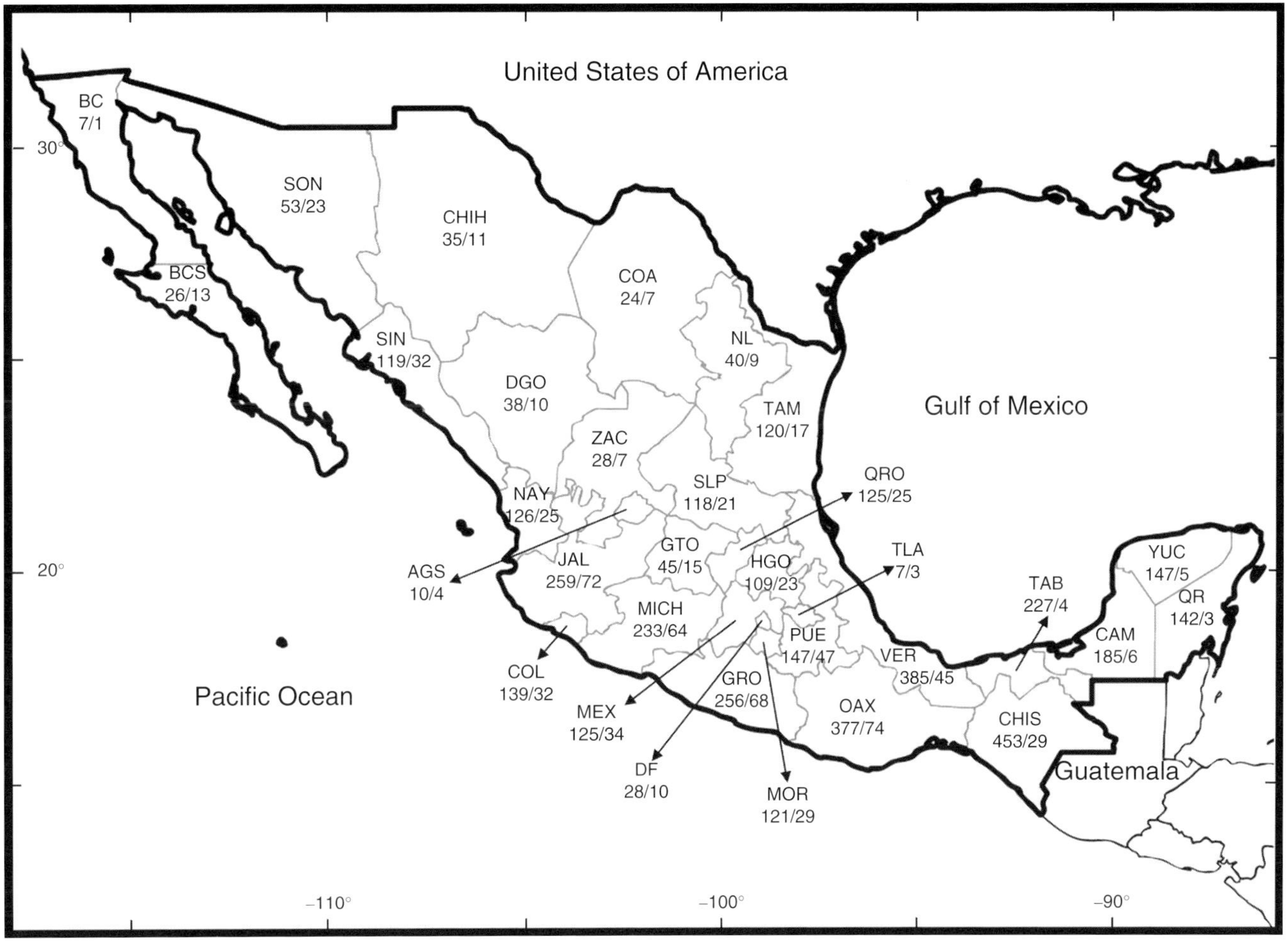

Fig. 8.1 States of Mexico, indicating for each one the number of lianas species located within their boundaries (total species/endemic species). Notes: AGS (Aguascalientes), BC (Baja California), BCS (Baja California Sur), CAM (Campeche), CHIS (Chiapas), CHIH (Chihuahua), COAH (Coahuila), COL (Colima), DF (Distrito Federal), DGO (Durango), GTO (Guanajuato), GRO (Guerrero), HGO (Hidalgo), JAL (Jalisco), MEX (México), MICH (Michoacán), MOR (Morelos), NAY (Nayarit), NL (Nuevo León), OAX (Oaxaca), PUE (Puebla), QRO (Querétaro), QR (Quintana Roo), SLP (San Luis Potosí), SIN (Sinaloa), SON (Sonora), TAB (Tabasco), TAM (Tamaulipas), TLA (Tlaxcala), VER (Veracruz), YUC (Yucatán), ZAC (Zacatecas).

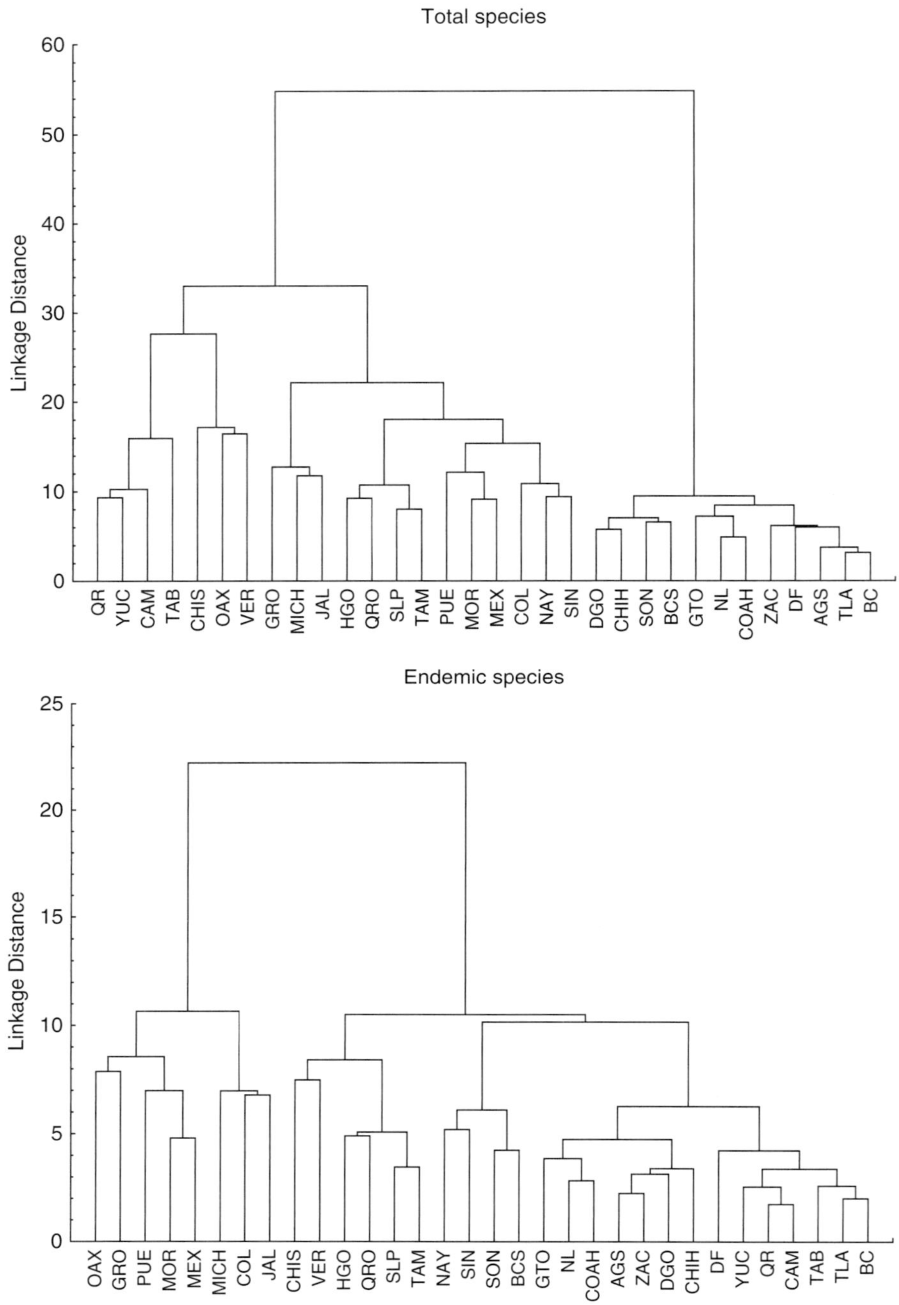

Fig. 8.2 Cluster analysis for Mexican lianas using presence–absence data. Resemblance was measured by Euclidean distance, and decision rules for classification were based on Ward's method. Analyses were conducted using STATISTICA 6.0 (StatSoft 1998).

rest of the states are grouped, in general, according to geographical proximity or low species richness.

The frequency of species per state coupled with the high total species richness signals the high geographical restriction. Approximately 19.2% (145 species) are found only in one state and 16.4% (124) in two states. In fact, only 91 species (12.1%) have been recorded in over 11 states. The species with the greatest range of distribution are *Cardiospermum halicacabum* L. (Sapindaceae), which is present in 31 Mexican states, followed by *Toxicodendron radicans* (L.) Kuntze (Anacardiaceae, 27) and *Canavalia villosa* Benth. (Fabaceae, 26).

The 185 endemic species have similar geographic restriction at the state level; 23.2% (43 species) are found in only one Mexican state, 17.3% (32) in two states and 3.2% (6) species are present in more than 11 states. *Lonicera pilosa* (Kunth) Willd. ex Kunth (Caprifoliaceae), which has the highest frequency of the endemics at state level, is present in 20 states, followed by *Passiflora exsudans* Zucc. (Passifloraceae), found in 18 states. Similarly, Villaseñor (2003) reported high percentages of species in the Mexican flora restricted to a single state, showing 33.4% of the complete flora and 43.2% of the endemic species. Thus far, species of the endemic genera *Balsas* and *Thoreauea* have been collected only in a single Mexican state (Williams 2002; Morales 2005; Diego & Lozada 2006; Jiménez et al. 2011): *Balsas guerrerensis* Cruz Durán & K. Vega and *Thoreauea guerrerensis* Diego & Lozada-Pérez in Guerrero state, *T. aberrans* J. F. Morales in Veracruz and *T. paneroi* J. K. Williams in Oaxaca.

We calculated the geographic distribution of species at the state level using Jaccard's similarity index (EstimateS; Colwell 2006). This index assesses the compositional similarity among states based on presence and absence of species. It varies between 0, when the samples do not share any species, and 1, if they have the same species composition. In general, the level of resemblance among states was lower when considering endemic species than for all species. Similarity was lower than 0.1 for 32.9% of the 496 paired comparisons of the complete flora among the 32 Mexican states, whereas 60.5% of the comparisons were lower than 0.1 for endemic species. Slightly more than 20% of the comparisons had zero values. Furthermore, after applying a Mantel test to determine whether the similarity values among states were correlated with the distance between states (ln-transformed), we detected a significant and negative spatial autocorrelation, which was higher for the endemic species ($r = -0.603$, <0.0001) than for all species ($r = -0.458$, <0.0001). Our data showed inequality in the richness of both groups of lianas (total and endemic) among states, with a declining trend with increasing latitude (Fig. 8.1). This pattern appears to represent the replacement of tropical forests by vegetation types with temperate or xeric climatic conditions (see maps in Rzedowski 1978; Palacio-Prieto et al. 2000).

CONSERVATION

A total of 462 species (61.3%) are present in one or more of the 10 Mexican reserves included in Table 8.2. A total of 250 species are exclusive to only one reserve. The species with highest frequency among the reserves was *Cissampelos pareira* L. (Menispermaceae), present in nine reserves, followed by *Byttneria aculeata* Jacq. (Malvaceae), *Mandevilla subsagittata* (Ruiz & Pav.) Woodson. (Apocynaceae) and *Cissus verticilata* (L.) Nicolson & C.E. Jarvis (Vitaceae), each of which are found in eight reserves. The 10 reserves host a total of 75 species endemic to Mexico, and 61 (81.3%) of these species are found only in one reserve. More frequent species are *Matelea velutina* (Schltdl.) Woodson (Apocynaceae), found in three reserves, followed by 13 species found in two reserves (e.g., *Aristolochia taliscana* Hook. & Arn., Aristolochiaceae; *Gaudichaudia mcvaughii* W.R. Anderson, Malpighiaceae). As far as is known, the four species in two of the three endemic genera of Mexico (*Balsas* and *Thoreauea*) have not been recorded as floristic components in any Mexican Protected Area.

Sierra de Manantlán and Calakmul include the most liana species (Table 8.2); however, Los Tuxtlas has the largest number and proportion of species found only in these reserves (52 of 131, 39.7%). For endemic species, the Sierra de Manantlán again stands out (39 of 138, 28.3%), 26 of which are registered only in this reserve. The Chamela-Cuixmala Reserve has the second highest number of endemic and exclusive species (Table 8.2). Because floristic inventories of some of the examined reserves are not complete (e.g., El Triunfo, Sierra Gorda), the values may change with ongoing data collection.

The restriction of many species to Protected Areas is reflected in values obtained from Jaccard similarity index, which do not surpass 0.41 (Table 8.3). Reserves in close geographical proximity have higher Jaccard indices (e.g., Ría Lagartos and Sian Ka'an – all

Table 8.3 Jaccard Index values (based on species presence-absence) for 10 reserves in Mexico.

Reserves	1	2	3	4	5	6	7	8	9	10
(1) Calakmul*	*	0.21	0.11	0.19	0.24	0.26	0.27	0.37	0.16	0.10
(2) Chamela-Cuixmala*	0.00	*	0.06	0.08	0.11	0.08	0.11	0.11	0.26	0.08
(3) El Triunfo*	0.00	0.00	*	0.10	0.09	0.06	0.09	0.07	0.10	0.12
(4) Los Tuxtlas**	0.00	0.00	0.00	*	0.24	0.08	0.10	0.15	0.08	0.10
(5) Boca de Chajul**	0	0	0	0	*	0.08	0.16	0.20	0.11	0.05
(6) Ría Celestún*	0.00	0.00	0.00	0.00	0.00	*	0.38	0.37	0.09	0.04
(7) Ría Lagartos*	0	0	0	0	0	0	*	0.41	0.10	0.08
(8) Sian Ka'an*	0.33	0.00	0.00	0.00	0	0.00	0	*	0.10	0.09
(9) Sierra de Manantlán*	0.00	0.23	0.00	0.00	0	0.00	0	0.00	*	0.13
(10) Sierra Gorda*	0.07	0.00	0.00	0.00	0	0.00	0	0.08	0.02	*

Notes: * Biosphere Reserves, ** Biological Station. Above diagonal shows values for total species and below diagonal for endemic species.

species Jaccard index = 0.41; Calakmul and Sian Ka'an – endemic Jaccard index = 0.33). Of the 45 possible comparisons using endemic species, there is a clear tendency toward low values (i.e., high beta diversity). In fact, 16 of the 45 possible comparisons do not share any species. Our results of the effect of conservation areas on liana richness and similarity may be biased because many Mexican Protected Areas still lack adequate biological inventories, however, our work illustrates the important role played by the 10 reserves in protection of the liana growth form. Together they host 61.3% of the all liana species registered in Mexico and 40.5% of the liana endemics. In general, liana diversity differences between the reserves reflect, among other factors, their different size, the types of dominant vegetation they contain, and the biogeographical provinces to which they belong (Table 8.2).

Significantly, lianas are not considered a threatened group; not one species is listed in the Norma Oficial Mexicana NOM-059 (Anónimo 2010). However, the absence of lianas from this document may be a direct result of the paucity of biology knowledge, particularly in terms of biogeography and ecology, both of which are fundamental to evaluating conservation concern.

Uncertainty surrounding the conservation of Mexican lianas is real and constitutes a complex problem. In addition to high species richness and the narrow geographical range of many liana species, rates of disappearance of the primary vegetation of Mexico are involved, especially in tropical forests where the majority of liana species thrive (Schnitzer & Bongers 2002). Koleff et al. (2012) indicate that due to diverse social, political, and economic factors only 17.5% of the area originally covered by evergreen lowland tropical forest in Mexico remains in 2012. The vegetation type was originally estimated at 11% of Mexico's land area (Rzedowski 1978). Although more difficult to estimate, a similar scenario exists for deciduous tropical forest, with a remaining conserved area estimated at 27% (72,850 km^2) of the original (Trejo & Dirzo 2000). Koleff et al. (2012) also indicate that sustainable forest management in Mexico has been more successful in temperate forests. Unfortunately, liana diversity is relatively low in these temperate ecosystems.

Fragmentation of vegetation during deforestation, at least for evergreen lowland tropical forest, might be a positive factor for lianas, particularly with respect to edge effects that deforestation provokes, but it must coincide with remnants of vegetation at landscape level (Arroyo-Rodríguez & Toledo-Aceves 2009). However, in general, the liana community has also collapsed in the majority of deforested landscapes. Traditional management practices in tropical forests of Mexico often entail the elimination of liana hosts, especially those of large size, as well as systematic cutting of lianas to facilitate logging, livestock, and agricultural activities. Thus, it is increasingly rare to find lianas greater than 10 cm dbh in managed forests in Mexico.

We conclude that there are several socio-economic factors that seriously threaten Mexican lianas (e.g., deforestation, or forest management strategies). Therefore, it is necessary to evaluate these factors in more detail if we hope to achieve the long-term maintenance of lianas (see also Schnitzer et al., Chapter 7 in this volume). Such an evaluation could be conducted specifically in priority regions of tropical forest conservation, which are distributed in various regions of Mexico (see Figure 1 in Koleff et al. 2012). In addition, determination of the actual role performed by all Protected Areas in Mexico in terms of the liana protection is urgently required.

ACKNOWLEDGMENTS

The authors thank two anonymous reviewers and Frans Bongers for fruitful comments to improve our manuscript. This study was supported by the Postdoctoral Scholarship Program to Pablo Carrillo-Reyes, of the Universidad Nacional Autónoma de México (UNAM) and by the Consejo Nacional de Ciencia y Tecnología (CONACyT) with a scholarship to F. J. Rendón-Sandoval. Juan Martínez Cruz kindly conducted diversity analysis and Marco Antonio Romero Romero helped edit the figures.

REFERENCES

Acevedo-Rodríguez, P. (2005) Vines and climbing plants of Puerto Rico and the Virgin Islands. *Contributions from the United States National Herbarium*, **51**, 1–483.

Alvarado-Cárdenas, L. (2004) Apocynaceae. *Flora del Valle de Tehuacán-Cuicatlán*, **38**, 1–57.

Anónimo. (2010) Norma Oficial Mexicana NOM-059-SEMARNAT-2010. Protección ambiental-especies nativas de México de flora y fauna silvestres-categorías de riesgo y especificaciones para su inclusión, exclusión o cambio. Lista de especies en riesgo. *Diario Oficial de la Federación, 2ª. Sección*. 30 de diciembre de 2010. México, D.F., pp. 1–78.

Arroyo-Rodríguez, V. and Toledo-Aceves, T. (2009) Impact of landscape spatial pattern on liana communities in tropical rainforests at Los Tuxtlas, Mexico. *Applied Vegetation Science*, **12**, 340–349.

Bongers, F., Parren, M.P.E., Swaine, M.D. and Traoré, D. (2005) Forest climbing plants of West Africa, introduction, in *Forest climbing Plants of West Africa* (eds F. Bongers, M.P.E. Parren and D. Traoré), CAB International, Wallingford, pp. 5–18.

Bongers, F., Popma, J., Meave del Castillo, J. and Carabias, J. (1988) Structure and floristic composition of the lowland rain forest of Los Tuxtlas, Mexico. *Vegetatio*, **74**, 55–80.

Bongers, F., Schnitzer, S.A., & Traoré, D. (2002) The importance of lianas and consequences for forest management in West Africa. *Bioterre: Revue International des Sciences de la Vie et de la Terre*, Special issue, 59–70.

Colwell, R.K. (2006) Estimates, statistical estimation of species of species richness and shared species from samples. Version 8. Persistent URL, purl.oclc.org/estimates.

CONANP (Comisión Nacional de Áreas Nacionales Protegidas) (2007) *Un nuevo ciclo de vida*, Logros 2007. Secretaría de Medio Ambiente y Recursos Naturales, Comisión Nacional de Áreas Naturales Protegidas, México.

Den Outer, R.W. (1972) *Tentative Determination Key to 600 Trees, Shrubs and Climbers from the Ivory Coast, Africa, Mainly Based on Characters on the Living Bark, Besides the Rhytidome and the Leaf. IV. Climbers*, Agricultural University, Wageningen, the Netherlands.

DeWalt, S.J., Schnitzer, S.A., Chave, J. et al. (2010) Annual rainfall and seasonality predict pan-tropical patterns of liana abundance and basal area. *Biotropica*, **42**(3), 309–317.

Diego, N. and Lozada, L. (2006) *Thoreauea guerrerensis* (Apocynaceae, Apocynoideae), una nueva especie de Guerrero, México. *Novon*, **16**(3), 332–335.

Durán, R. and Olmsted, I. (1987) *Listado florístico de la reserve de Sian Ka'an*, Amigos de Sian Ka'an, Quintana Roo.

Durán, R., Simá, P. and Juan-Qui, M. (1999a) *Listado florístico de Ría Celestún*, Centro de Investigación Científica de Yucatán A.C, Yucatán.

Durán, R., Simá, P., Juan-Qui, M. et al. (1999b) *Listado florístico de la reserva de Ría Lagartos*, Centro de Investigación Científica de Yucatán A.C, Yucatán.

Espejo, S.A. (2012) El endemismo en las Liliopsida mexicanas. *Acta Botanica Mexicana*, **100**, 195–257.

Ferrusquía-Villafranca, I. (1993) Geology of Mexico: a synopsis, in *Biological Diversity of Mexico: Origins and Distribution* (eds T.P. Ramamorthy, R. Bye and A.F.J. Lot), Oxford University Press, New York.

Fryxell, P.A. (1988) Malvaceae of Mexico. *Systematic Botany Monographs*, **25**, 1–522.

Gentry, A.H. (1982) Patterns of neotropical plant species diversity. *Evolutionary Biology*, **15**, 1–84.

Gentry, A.H. (1991a) The distribution and evolution of climbing plants, in *The Biology of Vines* (eds F.E. Putz and H.A. Mooney), Cambridge University Press, Cambridge, pp. 3–49.

Gentry, A.H. (1991b) Breeding and dispersal systems of lianas, in *The Biology of Vines* (eds F.E. Putz and H.A. Mooney), Cambridge University Press, Cambridge, pp. 393–423.

González-Espinosa, M., Meave del Castillo, J.A., Lorea-Hernández, F.G. et al. (eds) (2011) *The Red List of Mexican Cloud Forest Trees*, Fauna and Flora International, Cambridge.

Gotelli, N.J. and Colwell, R.K. (2001) Quantifying biodiversity: procedures and pitfalls in the measurement and comparison of species richnes. *Ecology Letters*, **4**, 379–391.

Hágsater, E., Soto-Arenas, M.A., Salazar-Chávez, G.A. et al. (2005) *Las orquídeas de México*, Instituto Chinoín, México, D.F.

Ibarra-Manríquez, G. and Martínez-Ramos, M. (2002) Liana community landscape variation in a neotropical rain forest. *Plant Ecology*, **160**, 91–112.

Ibarra-Manríquez, G., Martínez-Ramos, M. and Oyama, K. (2001) Seedling functional types in a lowland rain forest in Mexico. *American Journal of Botany*, **88**, 1801–1812.

Ibarra-Manríquez, G., Sánchez-Garfias, B. and González-García, L. (1991) Fenología de lianas y árboles anemócoros en una selva cálido-húmeda de México. *Biotropica*, **23**, 242–254.

Ibarra-Manríquez, G. and Sinaca, C.S. (1995) Lista florística de la Estación de Biología Tropical "Los Tuxtlas", Veracruz, México. *Revista de Biología Tropical*, **43**, 75–115.

Ibarra-Manríquez, G. and Sinaca, C.S. (1996a) Estación de Biología Tropical "Los Tuxtlas", Veracruz, México: lista florística comentada (Mimosaceae a Verbenaceae). *Revista de Biología Tropical*, **44**, 41–60.

Ibarra-Manríquez, G. and Sinaca, C.S. (1996b) Estación de Biología Tropical "Los Tuxtlas", Veracruz, México: lista florística comentada (Violaceae a Zingiberaceae). *Revista de Biología Tropical*, **44**, 427–448.

Ibarra-Manríquez, G., Villaseñor, R., J.L. and Durán, R. (1995) Riqueza y endemismo del componente arbóreo de la Península de Yucatán, México. *Boletín de la Sociedad Botánica de México*, **57**, 49–78.

Ibarra-Manríquez, G., Villaseñor, J.L., Durán, R., & Meave, J. (2002) Biogeographical analysis of the tree flora of the Yucatan Peninsula. *Journal of Biogeography*, **29**, 17–29.

Jiménez, R.J., Vega, F.K. and Cruz, D.R. (2011) *Balsas* (Sapindaceae), género nuevo de la Cuenca del río Balsas en el estado de Guerrero, México. *Novon*, **21**(2), 196–200.

Jongkind, C.C.H. (2005) Checklist of climber species in Upper Guinea, in *Forest Climbing Plants of West Africa* (eds F. Bongers, M.P.E. Parren and D. Traoré), CAB International, Wallingford.

Jongkind, C.C.H. and Hawthorne, W.D. (2005) A botanical synopsis of the lianes and other forest climbers, in *Forest Climbing Plants of West Africa* (eds F. Bongers, M.P.E. Parren and D. Traoré), CAB International, Wallingford.

Juárez-Jaimes, V., Alvarado-Cárdenas, L.O. and Villaseñor, R.J.L. (2007) La familia Apocynaceae *sensu lato* en México, diversidad y distribución. *Revista Mexicana de Biodiversidad*, **78**, 459–482.

Koleff, P., Urquiza-Haas, T. and Contreras, B. (2012) Prioridades de conservación de los bosques tropicales de México: reflexiones sobre su estado de conservación y manejo. *Ecosistemas*, **21**(1–2), 6–20.

Lott, E.J. (2002) Lista anotada de las plantas vasculares de Chamela-Cuixmala, in *Historia Natural de Chamela* (eds F.A. Noguera, R.J.H. Vega, A.N. García-Aldrete and M. Quesada-Avendaño), Instituto de Biología, UNAM, México, pp. 99–136.

Lott, E.J., Bullock, S.H. and Solís-Magallanes, A.J. (1987) Floristic diversity and structure of upland and arroyo forest of coastal Jalisco. *Biotropica*, **19**, 228–235.

Martínez, E., Ramos, A.C.H. and Chiang, F. (1994) Lista florística de la Lacandona, Chiapas. *Boletín de la Sociedad Botánica de México*, **54**, 99–177.

Martínez, E., Sousa, S.M. and Ramos, A.C.H. (2001) *Región de Calakmul, Campeche. Listados florísticos de México XXII*, Instituto de Biología. UNAM, México.

Martínez-Meléndez, J., Pérez-Farrera, M.A. and Farrera-Sarmiento, O. (2008) Inventario florístico del Cerro El Cebú y zonas adyacentes en la Reserva de la Biosfera El Triunfo (Polígono V), Chiapas, México. *Boletín de la Sociedad Botánica de México*, **82**, 21–40.

Meave, J.A., Romero-Romero, M.A., Valle-Doménech, A. et al. (2008) Plant diversity assessment in the Yaxchilán Natural Monument, Chiapas, Mexico. *Boletín de la Sociedad Botánica de México*, **83**, 53–76.

Mittermeier, R.A. (1988) Primate diversity and the tropical forest, in *Biodiversity* (ed E.O. Wilson), National Academy Press, Washington, DC.

Morales, J.F. (2005) Estudios en las Apocynaceae Neotropicales XV. Sinopsis del género *Thoreauea* (Apocynoideae, Echiteae), con una nueva especie de Veracruz, México. *Brittonia*, **57**(3), 258–263.

Padilla-Velarde, E., Cuevas-Guzmán, R., Ibarra-Manríquez, G. and Moreno-Gómez, S. (2006) Riqueza y biogeografía de la flora arbórea del estado de Colima, México. *Revista Mexicana de Biodiversidad*, **77**, 271–295.

Palacio-Prieto, J.L., Bocco, G., Velásquez, A. et al. (2000) La condición actual de los recursos forestales en México: resultados del inventario Forestal Nacional 2000. *Investigaciones Geográficas, Boletín del Instituto de Geografía, UNAM*, **43**, 183–203.

Pérez-Cueto, E. (1995) Menispermaceae. *Flora de Veracruz*, **87**, 1–43.

Pérez-Farrera, M.A., Martínez-Camilo, R., Martínez-Meléndez, N. et al. (2012) Listado florístico del Cero Quetzal (Polígono III), de la Reserva de la Biosfera El Triunfo, Chiapas. *Botanical Sciences*, **90**, 113–142.

Pérez-García, E.A., Meave, J.A., Villaseñor, J.L. et al. (2010) Vegetation heterogeneity and life-strategy diversity in the flora of the heterogeneous landscape of Nizanda, Oaxaca, Mexico. *Folia Geobotanica*, **45**, 143–161.

Pérez-Salicrup, D. and de Meijere, W. (2005) Number of lianas per tree and number of trees climbed by lianas, at Los Tuxtlas, Mexico. *Biotropica*, **37**(1), 153–156.

Pineda-García, F., Arredondo-Amezcua, L. and Ibarra-Manríquez, G. (2007) Riqueza y diversidad de especies leñosas del bosque tropical caducifolio de El Tarimo, Cuenca del Balsas, Guerrero. *Revista Mexicana de Biodiversidad*, **78**, 129–139.

Ribeiro, L.d.S.J.E., Hopkins, M.J.G., Vicentini, A. et al. (1999) *Flora da Reserva Ducke. Guia de identificação das plantas vasculares de uma floresta de terra-firme na Amazônia Central*, Instituto Nacional de Pesquisas da Amazônia (INPA) and Department for International Development (DFID), Manaus.

Ricker, M. and Hernández, H.M. (2010) Tree and tree-like species of Mexico, gymnosperms, monocotyledons, and tree ferns. *Revista Mexicana de Biodiversidad*, **81**, 27–38.

Rzedowski, J. (1978) *Vegetación de México*, Ed. Limusa, México, D.F.

Rzedowski, J. (1991a) Diversidad y orígenes de la flora fanerogámica de México. *Acta Botanica Mexicana*, **14**, 3–21.

Rzedowski, J. (1991b) El endemismo de la flora fanerogámica de México, una apreciación analítica preliminar. *Acta Botanica Mexicana*, **15**, 47–64.

Rzedowski, J. (1993) Diversity and origins of the Phanerogamic Flora of Mexico, in *Biological Diversity of Mexico: Origins and Distribution* (eds T.P. Ramamorthy, R. Bye and A.J.F. Lot), Oxford University Press, New York, pp. 129–144.

Rzedowski, J. and Calderón de Rzedowski, G. (1993) Bignoniaceae. *Flora del Bajío y de Regiones Adyacentes*, **22**, 1–44.

Schnitzer, S.A. and Bongers, F. (2002) The ecology of lianas and their role in forests. *Trends in Ecology and Evolution*, **17**, 223–230.

Schnitzer, S.A. and Bongers, F. (2011) Increasing liana abundance and biomass in tropical forests: emerging patterns and putative mechanisms. *Ecology Letters*, **14**, 397–406.

Solórzano, S., Ibarra-Manríquez, G. and Oyama, K. (2002) Liana diversity and reproductive attributes in two tropical forests in Mexico. *Biodiversity and Conservation*, **11**, 197–212.

StatSoft, Inc. (1998) STATISTICA for Windows. Tulsa, Oklahoma.

Trejo, I. and Dirzo, R. (2000) Deforestation of seasonally dry tropical forest: a national and local analysis in Mexico. *Biological Conservation*, **94**, 133–142.

Trejo, I. and Dirzo, R. (2002) Floristic diversity of Mexican seasonally dry tropical forests. *Biodiversity and Conservation*, **11**, 2063–2084.

Vázquez, G.J.A., Cuevas, G.R., Cochrane, T.S. et al. (1995) Flora de Manantlán. *Sida, Botanical Miscellany*, **13**, 1–312.

Villaseñor, J.L. (2003) Diversidad y distribución de las Magnoliophyta de México. *Interciencia*, **28**(3), 160–167.

Villaseñor, J.L. (2004) Los géneros de plantas vasculares de la flora de México. *Boletín de la Sociedad Botánica de México*, **75**, 105–135.

Williams, J.K. (2002) *Thoreauea* (Apocynaceae, Apocynoideae), a new genus from Oaxaca, Mexico. *Lundellia*, **5**, 47–58.

WCMC (World Conservation Monitoring Centre) (1992) *Global Biodiversity: Status of the Earth's Living Resources*, Chapman and Hall, London.

Chapter 9

CLIMBING PLANT DIVERSITY IN AUSTRALIA: TAXONOMY, BIOGEOGRAPHY AND FUNCTIONAL TRAITS

Rachael V. Gallagher

Macquarie University, Sydney, Australia

Ecology of Lianas, First Edition. Edited by Stefan A. Schnitzer, Frans Bongers, Robyn J. Burnham, and Francis E. Putz.

Companion Website: www.wiley.com/go/schnitzer/lianas

OVERVIEW

The Australian continent houses a diverse array of vegetation types, all of which contain a number of plant species that climb. Although woody lianas dominate in closed-canopy forests in tropical and subtropical regions to the north, they represent a relatively minor proportion of Australia's climbing plant flora. The majority of climbing plant species in Australia are herbaceous or semi-woody vines and scramblers found in both closed (e.g., rain forest, evergreen forest) and open vegetation types (e.g., savanna, woodlands). They exhibit a wide variety of morphological and physiological adaptations that allow them to form persistent populations in some of Australia's most extreme environments, including deserts and alpine zones.

This chapter provides an description of the taxonomic diversity, biogeographic patterns and functional traits of Australia's climbing plant flora at both regional and continental scales. I define a climbing plant as one which uses the structural investment in stems and branches made by a host species to place its leaves in adequate light for photosynthesis. This definition encompasses species with a wide range of climbing strategies and growth habits and is not restricted to woody lianas. The taxonomic and functional trait information presented here has been gathered together from Floras, published literature, field surveys and herbarium records. I explore continent-level patterns of climbing plant taxonomic diversity and biogeographic distribution, and compare the traits and ecological strategies of tropical and temperate climbers of Australia.

INTRODUCTION: TAXONOMIC DIVERSITY OF AUSTRALIAN CLIMBING PLANTS

Over 800 species in the Australian flora are classified as vines, lianas or scramblers, or as possessing a climbing or scandent habit in major published Floras. These 834 species are drawn from just over one-third of Australia's vascular plant families (91 of 254 families) and represent 286 individual genera. As a percentage of total species richness recorded in the Australian Plant Name Index ($n = 19{,}530$ native species; see http://www.anbg.gov.au/apni/index.html), the climbing habit is present in around 4% of the Australian vascular plant flora. This percentage is a relatively small fraction of the flora and the scarcity of climbers may result from the paucity of habitats, such as rain forests, in Australia, which are typically associated with high climbing plant diversity in other regions of the world (i.e., less than 2% of Australia's vegetation is classified as rain forest). However, phylogenetically, climbing is a widespread adaptation in the Australian flora. The phylogenetic relationships between climbing plant families and the spread of genera within each family in Australia are depicted in Fig. 9.1. This phylogenetic reconstruction highlights that almost half of all families (46%; 42 families) contain only a single genus with climbers present, and that climbing is found in both monocot and eudicot lineages and in basal groups such as the Austrobaileyaceae.

Two families – Apocynaceae and Fabaceae – contain the greatest number of both genera and species with climbers in the Australian flora. While the Fabaceae has a higher number of genera with climbers present ($n = 35$), there are actually more individual species which climb within the family Apocynaceae ($n = 135$ species across 28 genera) than in the Fabaceae ($n = 123$ species). Woody lianas feature prominently in both families, making up around one-third of the species. Notably, many climbing plant families that are renowned for supporting a high diversity of woody lianas in other regions of the globe have few representatives in the Australian flora. Most striking is the absence of large radiations in the Bignoniaceae, which contains only three genera with species that climb in Australia (*Neosepicaea*, *Pandorea*, *Tecomanthe*) representing 11 species (see Australian Plant Census; http://www.anbg.gov.au/chah/apc/index.html). All Australian Bignoniaceae that climb are woody lianas, some of whose distributions extend into Papua New Guinea. However, *Tecomanthe* and *Neosepicea* species occur only in tropical and subtropical rain forests and gallery forests in tropical locations further north, whereas Australia's eight *Pandorea* species are widespread across the continent, extending into forests in drier, inland areas as well as featuring prominently in closed forest habitats throughout Australia.

Parsonsia (Family: Apocynaceae) is the most speciose genus of woody lianas in Australia. Thirty-seven species are currently recognized, the majority of which are endemic. These 37 species represent around one-quarter of the entire taxonomic diversity of this genus, which also occurs throughout Southeast Asia, New Guinea, New Caledonia, New Zealand, and Polynesia. Species occur throughout most vegetation

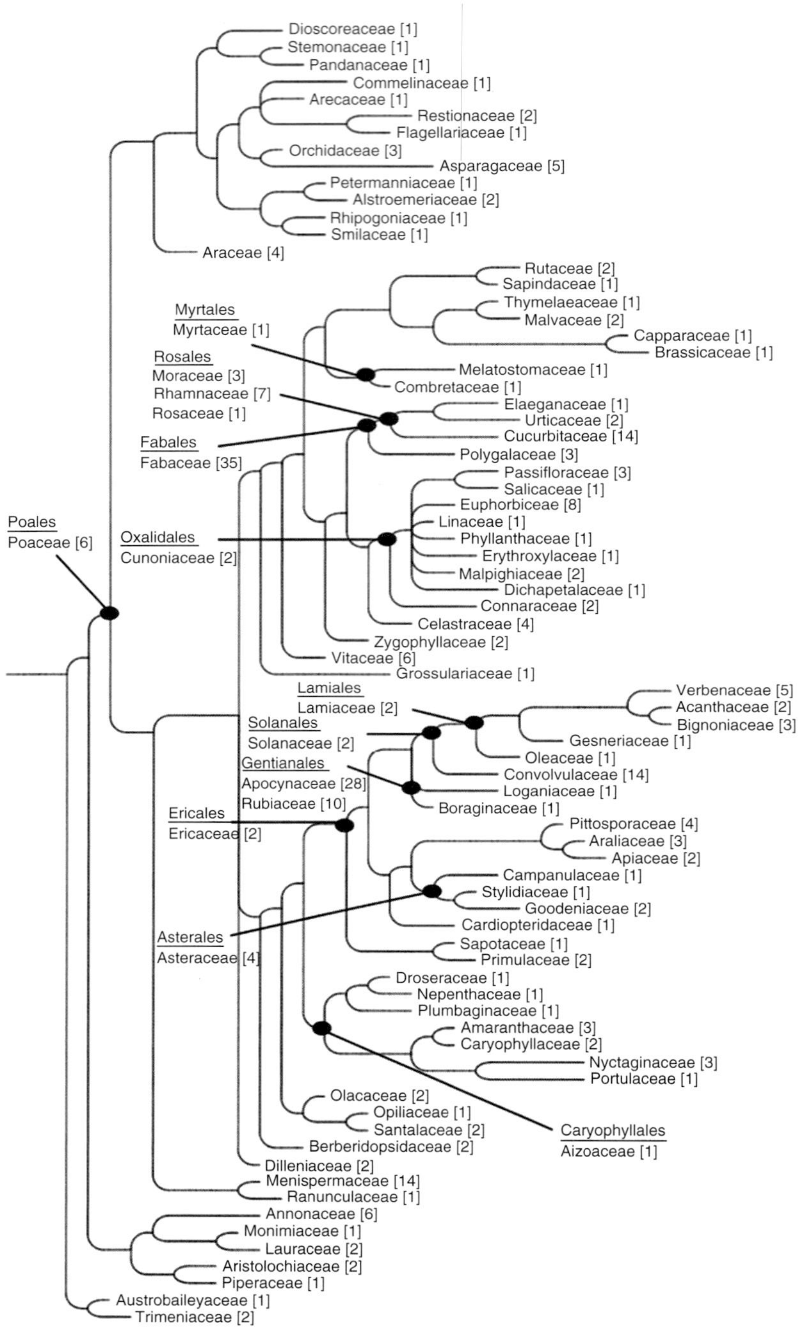

Fig. 9.1 Hypothesized phylogenetic relationships within the Australian climbing plant flora. This tree is based on relationships at the genus level depicted in a supertree constructed by the Angiosperm Phylogeny Group. Some 91of the 254 Australian vascular plant families contain at least one genus with climbers. Numbers in parentheses indicate the total number of genera in each family containing climbing plants. Dots and lines are used to indicate the placement of unresolved families within orders at the relevant node in the tree.

types in Australia, but reach a peak in abundance and diversity in tropical and subtropical regions on the eastern coast (Fig. 9.2).

BIOGEOGRAPHIC PATTERNS IN AUSTRALIAN CLIMBERS

All the major ecoregions of Australia include at least some species of climbing plant (Fig. 9.2). This includes vegetation formations as diverse as grasslands and shrublands in the arid interior of the continent, where trailing herbaceous perennials such as *Ipomoea* and *Convolvulus* dominate, through to various rain forest types (subtropical, littoral, tropical, cool temperate) spread primarily along the north-east coast of the continent. As a percentage of total land area, eucalypt-dominated plant communities are Australia's most widespread, covering over 60% of the continent; these also contain the highest proportion of climbing plant species of any vegetation type ($n = 608$ species). However, it is in the rain forests, which cover only a small fraction of the land area in Australia (<2%), where a remarkable diversity of climbing plants co-exist. Woody lianas become more dominant and species rich in the tropical regions, reflecting patterns identified across the globe (Gentry 1991; Schnitzer & Bongers 2011; Gallagher & Leishman 2012a; Burnham & Romero-Saltos, Chapter 5 in this volume; DeWalt et al., Chapter 11 in this volume; Schnitzer et al., Chapter 7 in this volume), but lianas are also an important physiognomic element in cool temperate, littoral and dry rain forests (i.e. those receiving only marginal rainfall with increasingly more species adapted to arid conditions) which occur further south, and inland (Webb 1959; Forster & Queensland Herbarium 1991; Butler et al. 2007; Gallagher & Leishman 2012b). In addition, lianas are a vital floristic element in vine thickets of inland and northern Australia, which are characterized by the presence of emergent deciduous tree species and vines such as *Cayratia clematidea*, *Clematis microphylla*, *Jasminum lineare*, and various *Parsonsia* species.

Range size and climatic breadth

The majority of Australian climbers occupy relatively limited ranges and high diversity is found in spatially restricted vegetation types such as tropical & subtropical moist broadleaf forests (Fig. 9.2). Figure 9.3 shows the number of 5 km × 5 km grid cells across Australia where each of 834 climbing plant species has been collected and catalogued in national herbariums. Using this measure as an indicator of range size and breadth, we see that even the most widespread of

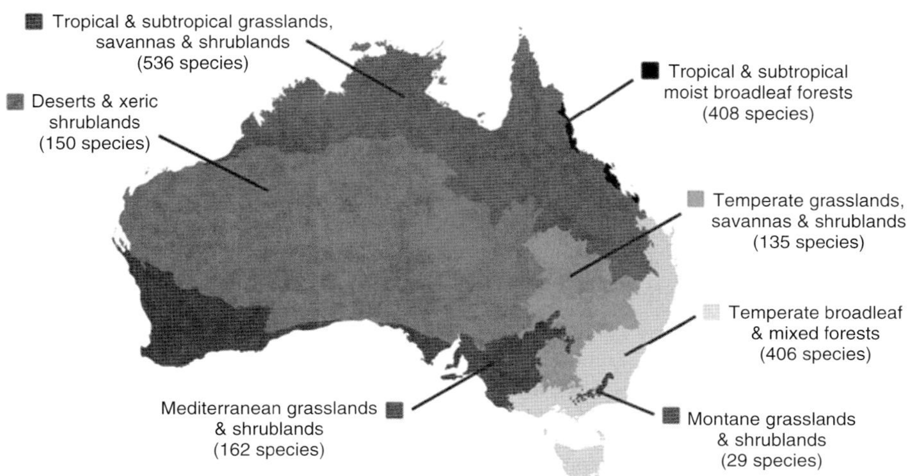

Fig. 9.2 Terrestrial ecoregions of Australia and their climbing plant diversity. The eight ecoregions represented are derived from the WWF Terrestrial Ecoregions Project. Species diversity data were compiled from collection records in Australia's Virtual Herbarium.

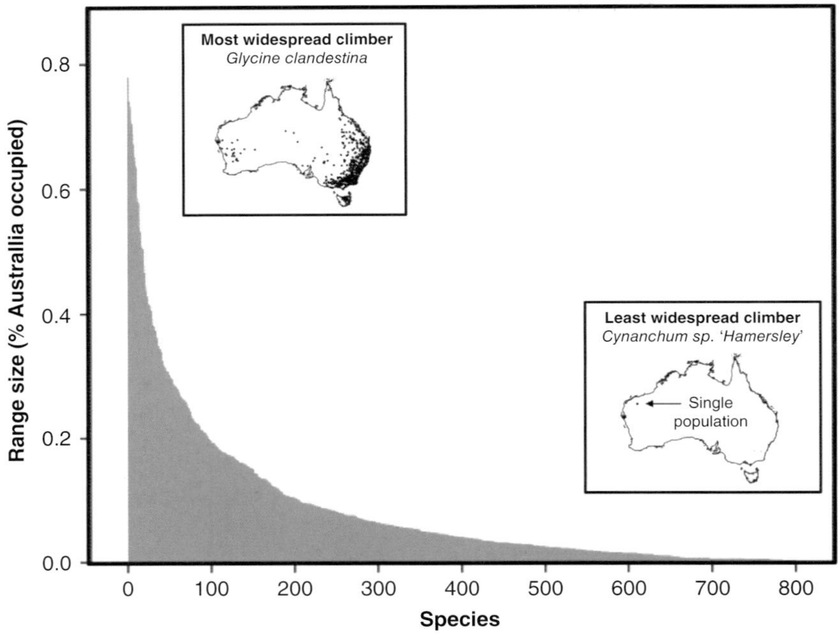

Fig. 9.3 Range sizes of Australian climbing plant species. Range size is estimated from the percentage of 5 km × 5 km grid cells occupied by each species across Australia.

climbers (e.g., *Glycine clandestina, Kennedia prostrata*) occupy less than 1% of grid cells across Australia. However, despite these seemingly small estimates of the ranges occupied by these species, it is important to note that collections for these plants have been made across a wide variety of vegetation types that span much of the biogeographic and climatic space available throughout Australia. For instance, *Glycine clandestina* (Fabaceae) occurs in 26 of the 29 recognized major vegetation groupings in Australia (see National Vegetation Information System; http://www.environment.gov.au/erin/nvis/about.html). Herbarium collections, however, may be subject to spatial bias, with less effort being dedicated to collecting in inland regions of Australia. Nevertheless, collections of this kind offer the most comprehensive record of species distributions currently available and are important for assessing patterns of diversity, if interpreted cautiously.

Taken collectively, climbing plants occupy almost all of the available climatic habitats – defined by mean annual temperatures and precipitation – across Australia (Fig. 9.4). There is a latitudinal gradient in woody liana diversity in Australia with more species being found in tropical regions than in temperate regions (Gallagher et al. 2011). In addition, a longitudinal gradient in total climber diversity exists as one moves inland from the coastal margin. Climates characterized by colder mean annual temperatures (<10 °C) typically have a smaller proportion of climbers than do areas in warmer regions within similar rainfall ranges. For example, only 28 species of climbing plant have been collected in montane vegetation in the Australian Alps region located in the high-elevation areas of the south-east. In these cooler environments, physiological processes, such as water transport, may be more challenging to maintain for climbers compared to co-occurring trees and shrubs as a result of differences in wood anatomy and vessel size (Ewers et al. 1991). Rainfall is highly variable across Australia with monsoonal climate conditions bringing pronounced wet and dry seasons to the northern portion of the continent, while southern regions experience rainfall year round. Rainfall in the southern and inland zones is affected by strong continental conditions, with the arid interior experiencing dramatically lower totals than coastal regions. This rainfall pattern reflects the

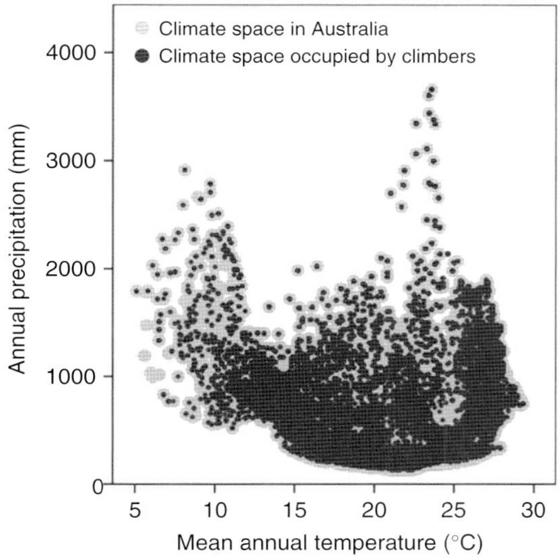

Fig. 9.4 The extent of climate space occupied by climbing plants across Australia, defined by mean annual precipitation (mm) and temperature (°C).

diversity of climbing plants throughout the continent, with the highest numbers of species being found away from the inland area in coastal regions and the tropics where rainfall is predictable and relatively abundant.

FORM AND FUNCTION IN AUSTRALIAN NATIVE CLIMBING PLANTS

Functional traits are the physiological, morphological, and phenological adaptations that describe how species "make a living" in the constraints of their physical environment (McGill et al. 2006). Traits such as seed mass, leaf size and shape, or photosynthetic rate have rapidly become an important empirical device in plant ecology over the last two decades, primarily because they can be used to quantify and compare the ecological strategies of species (Westoby et al. 2002). Functional traits also provide a vital link between observations at the species and whole plant level and key ecosystem functions, such as nutrient cycling and rates of above-ground net primary productivity (Garnier et al. 2007; Cornwell et al. 2008). Trait-based approaches are particularly useful for identifying how species from different biogeographic regions cope with and respond to prevailing abiotic conditions, such as those from tropical and temperate zones. In this section, the traits of climbing plants from tropical and temperate regions of Australia are used to compare ecological strategies and discuss the abiotic selection pressures that have contributed to the dominance of different climbers in each region.

The Tropic of Capricorn divides Australia roughly in half, with a monsoonal climate in the north and temperate seasonal climate in the south, and there is a high diversity of climbing plants in both regions. North of the tropic, average temperatures range between 21 and 27 °C and to the south between 2 and 21 °C; however, the tropics are dominated by open savannas, monsoon forest, wetlands and tropical rain forests, and temperate areas are dominated by eucalypt woodlands and hummock grasslands (George et al. 1982). Individual comparisons of the seed mass, leaf size, and growth habit (woody vs. herbaceous) of tropical and temperate climbers in Australia reveal clear differences between regions, while other traits such as dispersal mode remain relatively stable.

Seed mass

In seed plants, maternal investment in reproduction is allocated to one of two key strategies: the creation of few, large seeds or many, small seeds. Species that produce a small number of large, well-provisioned seeds have a higher chance of surviving environmental hazards during germination and establishment, whereas those producing a large number of small seeds are more suited to colonizing open habitats where light and nutrient resources are not limiting (Smith & Fretwell 1974). Both of these strategies for provisioning offspring are stable evolutionarily because maternal reproductive investment is integrated over the lifetime of a species (Moles et al. 2004). That is, the apparent advantage of increased fecundity associated with small-seeded species is balanced by the longer life spans and larger canopies associated with large-seeded species. In this way, seed size also provides an important link between investment in reproduction and patterns of vegetative growth and is central to our understanding of plant fecundity and seedling establishment (Howe & Smallwood 1982; Leishman 2001).

Seed mass tends to decline with increasing latitude across all plant growth forms, and Australian climbers are no exception. In tropical regions of Australia, climbers have seeds that are, on average, 22-times larger than the seeds of temperate climbers (mean tropical seed mass = 425.9 mg; mean temperate seed mass = 18.9 mg; $n = 205$ species; Gallagher et al. 2011). This tendency towards larger seeds in tropical climbers was also found in analyses of independent contrasts across nodes in a phylogenetic tree of Australian climbing plants. That is, larger seeded species have consistently been found to be associated with tropical habitats throughout the evolutionary history of Australian climbing plants. In addition, a previous study of seed size across multiple growth forms showed that latitude still explains a significant proportion of the variation in seed mass once vegetation type has been accounted for (Moles et al. 2007). The degree of canopy closure, as measured by leaf area index, may be a key determinant of the prevalence of large, well-provisioned seeds of these tropical vegetation types. Previous work in old-growth forests in Brazil showed that lianas with large seeds are more likely to colonize the types of low light microsites characteristic of late, successional closed canopy forests (Gerwing 2004). These larger seeds typically have a greater provision of endosperm tissue which can be used to prolong the amount of time seedlings have to transition between autotrophic and heterotrophic stages of establishment.

Dispersal mode

Seed size and dispersal mode are closely connected. Species with larger seeds (>100 mg) tend to be dispersed by vertebrate frugivores, while smaller seeds are more likely to be dispersed by the wind with the aid of aerodynamic appendages, such as wings, or pappuses (Hughes et al. 1994). The mode of dispersal, be it biotic (i.e., zoochory – via animals through ingestion, adhesion or carrying of the seed, particularly by ants), abiotic (i.e., anemochory – picked up by wind, hydrochory – buoyant seeds carried by water) or unassisted (i.e., seeds fall directly below maternal plants via gravity) provides information about the distance seeds can travel in the landscape and their germination requirements (Leishman et al. 2000). For instance, seeds of some Australian climbing plants (e.g., *Morinda jasminoides*, *Eustrephus latifolius*) require passage through the gut of vertebrates to stimulate germination (Bell 1999). In some cases, the mutualistic relationship between vertebrate disperser and plant is highly specialized such as that seen between climbing palms in the genus *Calamus* and the Australian ratite rain forest bird, the southern cassowary (*Casuarius casuarius*). *Calamus* species have relatively large seeds (e.g. *C. caryotoides* = 97 mg, *C. moti* = 107 mg; see Kew Seed Information Database) which are encapsulated in fleshy fruits that provide nutritious reward to the southern cassowary in return for dispersal. The wide gape-size of the cassowary places it among the few vertebrates capable of swallowing the seed, which is then transported across the home range of the bird (which is approximately 0.5–2.5 km^2) in the closed forests of the tropical north of Australia (Stocker & Irvine 1983).

Comparisons across species indicate that climbing plants from tropical regions, both in Australia and across the globe, tend to possess a higher proportion of wind-dispersed seeds than do tree or shrub species (Howe & Smallwood 1982). For example, in the neotropical dry forests of Costa Rica, 71% of the lianas species surveyed employed wind dispersal compared with 8–31% for sub-canopy and canopy trees. Similarly in the neotropical wet forests in Ecuador, 33% of liana species were wind dispersed, compared to 4–5% of trees (Howe & Smallwood 1982). A global compilation of dispersal syndromes for climbers indicates that 36% use wind dispersal as their primary mode of seed transport ($n = 952$ species; Gallagher & Leishman 2012a); however, this finding was not contrasted with the percentage of wind-dispersed species in other growth forms. In Australia, wind dispersal is the primary dispersal mode for 31% of tropical and 37% of temperate climbers in a comparison of 263 species, and dispersal mode did not differ significantly between regions in analyses across present-day species or across phylogenetically independent contrasts. This finding is particularly surprising given the poor representation of large, wind-dispersed climbing plant families in tropical Australia, such as the Bignoniaceae. Although a number of wind-dispersed species in the Australian tropics possess winged seed appendages (primarily those in the family Fabaceae), silky comas are also a prominent dispersal trait in species within the Apocynaceae.

Unlike patterns seen in other growth forms, the proportion of wind-dispersed climbing plants tends to be similar in tropical and temperate regions (Moles et al. 2007; Gallagher & Leishman 2012a). In other growth forms wind dispersal becomes less prevalent in tropical settings where vertebrate dispersal becomes more prominent. This pattern poses the key ecological question of why wind dispersal might be a selectively favorable strategy for moving the seeds of climbers around the landscape in tropical regions. Several important factors may be operating. First, wind dispersal may be a superior strategy for launching seeds from the top of the canopy where climbers, primarily woody lianas, overtop their hosts. It is likely that winds are stronger across the top of the canopy than within it, making seeds with aerodynamic appendages more likely to be dispersed from emergent canopy positions. Second, the high incidence of wind dispersal in tropical climbers may be linked to changes in the abiotic environment brought about by the onset of the dry season and the ability of lianas to outcompete free-standing hosts during this time (Schnitzer, Chapter 1 in this volume). For instance, the dry season brings with it an increase in strong winds (Strahler & Strahler 1989) and a reduction in heavy rainfall – both advantageous conditions for the movement of wind-borne seeds. It is proposed that lianas can grow longer into the dry season than can free-standing species by accessing deep repositories of soil water (Schnitzer 2005). This ability to persist in dry conditions and the favorability of dry season conditions for wind-dispersed seeds may work in concert, leading to a higher proportion of wind-dispersed climbing species being selected in tropical areas, particularly deciduous forests. Finally, small, wind-dispersed seeds are more successful at colonizing the high-light microsites characteristic of canopy gaps and early successional environments where climbers are commonly found (Howe & Smallwood 1982). The association between small seeds and colonization in more open environments arises because seedlings can more rapidly transition from autotrophic to heterotrophic growth where light is not limiting. Canopy gaps of this kind also provide trellis support for liana seedlings.

Leaf size

The leaves of temperate climbing plants in Australia are, on average, four times smaller than the leaves of tropical climbers (mean tropical leaf size = 56.9 cm^2; mean temperate leaf size = 12.7 cm^2; Gallagher et al. 2011). This shift to larger leaves in tropical climbers is also seen across phylogenetically independent contrasts testing for associations between leaf size and region, and in a global analysis of leaf size in climbing plants (Gallagher & Leishman 2012a). Large leaves offer a number of strategic ecological advantages to species, but can generally only be maintained where water is not limiting (Parkhurst & Loucks 1972). These advantages include greater surface area for light capture during photosynthesis, as well as a competitive advantage for monopolizing light resources in the canopy. Competition for light – which mediates interactions between species, particularly in closed canopy forests (DeWalt et al. 2000; Falster & Westoby 2003) – is likely to play an important role in determining the size and shape of leaves in tropical settings. Large leaves may increase the ability of climbers, in particular canopy lianas, to shade out host leaves and form a continuous light-gathering surface over the top of the canopy. This may be a particularly effective strategy in climbers because leaf size is less likely to be constrained by branching architecture and self-shading in a similar way to co-occurring trees (Hikosaka et al. 1994; Ackerly & Donoghue 1998). The leaves of climbers are typically arrayed linearly along a horizontal leading shoot which may be advantageous for locating and colonizing light-gaps (Putz & Holbrook 1991).

The pronounced reduction in leaf size between tropical and temperate climbers in Australia likely reflects a gradient in year-round soil water availability, which is typically higher in tropical forests than in temperate forests. Therefore, there may exist reduced selection pressure for leaf strategies that conserve water in the tropics and as water becomes more limiting, away from tropical regions, leaf size may reduce both within and among species in order to maintain optimal leaf temperatures for photosynthesis. While leaf sizes are significantly smaller in climbers from temperate compared to tropical regions in Australia, the tendency for leaves to reduce in size with increasing latitude has been reported for a range of plant growth forms and is not restricted to climbing plants. However, climbers have also consistently been shown to have larger leaves, on average, than co-occurring free-standing species in forests across Australia (Webb 1959; Gallagher & Leishman 2012b). Larger leaves in climbers may result from the lower costs, in terms of carbon invested in woody tissue, in moving water from the soil to the canopy. Unlike free-standing species, whose stems must provide water transport and structural support, the stems of lianas need only provide the former, and as a result water transport efficiency tends to be notably higher (Ewers et al. 1991). There is some evidence that this reduced allocation to support tissues may also allow for an increased investment in leaf material in terms of leaf mass fraction (Pickup et al. 2005) and leaf mass per area for a given leaf size (Gallagher & Leishman 2012b).

Growth habit

Woody lianas reach their peak in abundance and diversity in tropical Australia, making up 68% of 188 climbing plant species in this region. The remaining 32% of these tropical climbers are herbaceous. The richness of woody species in the Australian tropics is significantly larger than the richness in temperate Australia, where less than half the species are woody (48% of 162 species). Colder temperatures have previously been associated with a reduction in the proportion of woody lianas along latitudinal and altitudinal gradients across the globe (Webb 1968; Molina-Frearner et al. 2004; Jiménez-Castillo et al. 2007). As temperatures drop, the ability of the wide and long vessels characteristic of woody lianas to resist or recover from cavitation caused by freezing temperatures rapidly decreases (Ewers et al. 1991). Thus, cold temperatures may be unfavorable to woody liana climbers in more temperate regions that experience occasional freezing temperatures.

NON-NATIVE CLIMBING PLANTS IN AUSTRALIA AND THEIR IMPACT

Climbers can be aggressive invaders, and there is a wide range of introduced, non-native climbing plants that have become naturalized or invasive in the Australian landscape since European settlement in the late eighteenth century. Harris et al. (2007) estimated that 179 non-native climbers from 40 plant families have now become established and showed that these species had three primary introduction pathways. The majority (61%) of these climbers were introduced intentionally for the horticultural industry and nursery trade, mainly chosen for their showy ornamental flowers and rapid growth rates (Groves et al. 2005). The remaining 39% of non-native climbers in Australia were introduced for agricultural purposes, as either crop or fodder species (14%), or were accidentally introduced either as contaminants of traded goods or their origin was of an unknown source (25%). Relatively few of the non-native climbing species that have become established in Australia are woody lianas (6%), with the bulk of species being classified as herbaceous perennials or scramblers.

Of the 40 plant families from which non-native climbers in Australia are drawn, around one quarter (11 families) are not represented in the native species pool. Interestingly, these families contain some of the most invasive and widespread non-native climbers in Australia, such as the herbaceous perennial *Anredera cordifolia* (Ten.) Steenis (Madeira vine, Mignonette vine) in the family Basellaceae and the climbing cactus *Pereskia aculeata* Mill. (Cactaceae). The dominance of these species in Australia supports the idea that phylogenetic isolation from native species may play an important role in determining the success of introduced species in novel habitats (Darwin's Naturalization Hypothesis; see Diez et al. 2008). There is also evidence that in tropical and subtropical areas of Australia, non-native vines are functionally distinct from native vines (Murphy et al. 2006; Osunkoya et al. 2010a; Osunkoya et al. 2010b). Non-native vines pose particularly significant management challenges for the vegetation they invade. Impacts include smothering of host vegetation, which can reduce light levels for seedlings and saplings on the forest floor and may lead to the breakage of host tree branches and ultimately increased mortality (Horvitz et al. 1998; Batianoff & Butler 2002; Llorens & Leishman 2008).

The ability of non-native vines to dramatically alter abiotic conditions after invasion in Australian plant communities has resulted in them being described as 'transformer' weeds (Vivian-Smith & Panetta 2002). Management of non-native climbers in Australia is particularly important in subtropical and tropical rain forests, especially following cyclone damage (Grice & Setter 2003; Murphy et al. 2008) and in riparian zones where nutrient loads have increased and water availability is not limiting (Osunkoya et al. 2010b). Importantly, spatial models which predict areas of climatically suitable habitat for non-native climbers in Australia have consistently indicated that few of these species have reached the full extent of their suitable habitat under current average climate conditions (Gallagher et al. 2010). Thus, these species may have the capacity to spread across a wider area if factors critical to the invasions process, in particular, propagule availability (both seed and stem fragments), are not limiting.

CONCLUSION AND FUTURE DIRECTIONS

Climbing plants are a remarkable example of evolutionary adaptation. By exploiting the structural investment of host species, they reach the light at a fraction of the cost, in terms of carbon, of other growth forms. This innovative adaptation has radiated throughout the Australia flora and is now found in a diverse range of plant families that occupy all vegetation types and climatic habitats found throughout the continent. However, liana abundance and diversity vary between vegetation types and with environmental factors, which accompany gradients in latitude and longitude. For example, climber diversity is greatest in rain forest habitats; however, rain forests occupy a very small fraction of the Australian continent. In cooler, montane vegetation and arid, inland environments, climbing plant diversity is markedly lower than more climatically benign habitats close to the coast. This pattern reflects the importance of climatic factors – in particular, cold temperatures in montane locations, and limited rainfall in the arid interior – in restricting the distribution of climbers throughout Australia.

Accompanying the considerable taxonomic diversity of Australian climbing plants is a wide variety of functional traits and ecological strategies, which allow climbers to compete, grow, and reproduce. Some key traits vary with latitude. For example, adaptations such as large seeds and leaves, and a woody growth habit are characteristic of tropical Australian climbers, and become less prevalent with increasing latitude. These ecological adaptations to the differing environments in northern and southern regions of Australia make for a distinct climbing plant flora in the tropical and temperate zones of the continent. However, a key finding is that there appears to be no evidence of a latitudinal gradient in the prevalence of wind dispersal in climbers – a pattern that is distinct from other growth forms.

REFERENCES

Ackerly, D.D. & Donoghue, M.J. (1998) Leaf size, sapling allometry, and Corner's rules: phylogeny and correlated evolution in maples (*Acer*). *The American Naturalist*, **152**, 767–791.

Batianoff, G. & Butler, D. (2002) Assessment of invasive naturalized plants in south-east Queensland. *Plant Protection Quarterly*, **17**, 27–34.

Bell, D.T. (1999) Turner Review No. 1. The process of germination in Australian species. *Australian Journal of Botany*, **47**, 475–517.

Butler, D., Green, R.J., Lamb, D. et al. (2007) Biogeography of seed-dispersal syndromes, life-forms and seed sizes among woody rain-forest plants in Australia's subtropics. *Journal of Biogeography*, **34**, 1736–1750.

Cornwell, W.K., Cornelissen, J.H.C., Amatangelo, K. et al. (2008) Plant species traits are the predominant control on litter decomposition rates within biomes worldwide. *Ecology Letters*, **11**, 1065–1071.

DeWalt, S.J., Schnitzer, S.A. & Denslow, J.S. (2000) Density and diversity of lianas along a chronosequence in a central Panamanian lowland forest. *Journal of Tropical Ecology*, **16**, 1–19.

Diez, J.M., Sullivan, J.J., Hulme, P.E. et al. (2008) Darwin's naturalization conundrum: dissecting taxonomic patterns of species invasions. *Ecology Letters*, **11**, 674–681.

Ewers, F., Fisher, J. & Fichtner, K. (1991) Water flux and xylem structure in vines, in *The Biology of Vines* (eds F.E. Putz & H.A. Mooney), Cambridge University Press, New York, pp. 127–160.

Falster, D.S. & Westoby, M. (2003) Leaf size and angle vary widely across species: what consequences for light interception? *New Phytologist*, **158**, 509–525.

Forster, P.I. & Herbarium, Q. (1991) *Vineforest Plant Atlas for South-East Queensland with Assessment of Conservation Status*, Queensland Herbarium, Brisbane.

Gallagher, R.V., Hughes, L., Leishman, M.R. & Wilson, P.D. (2010) Predicted impact of exotic vines on an endangered ecological community under future climate change. *Biological Invasions*, **12**, 4049–4063.

Gallagher, R.V. & Leishman, M.R. (2012a) A global analysis of trait variation and evolution in climbing plants. *Journal of Biogeography*, **39**, 1757–1771.

Gallagher, R.V. & Leishman, M.R. (2012b) Contrasting patterns of trait-based community assembly in lianas and trees from temperate Australia. *Oikos*, **121**, 2026–2035.

Gallagher, R.V., Leishman, M.R. & Moles, A.T. (2011) Traits and ecological strategies of Australian tropical and temperate climbing plants. *Journal of Biogeography*, **38**, 828–839.

Garnier, E., Lavorel, S., Ansquer, P. et al. (2007) Assessing the effects of land-use change on plant traits, communities and ecosystem functioning in grasslands: a standardized methodology and lessons from an application to 11 European sites. *Annals of Botany*, **99**, 967–985.

Gentry, A.H. (1991) The distribution and evolution of climbing plants, in *The Biology of Vines* (eds F.E. Putz & H.A. Mooney), Cambridge University Press, New York, pp. 3–49.

George, A.S., Orchard, A.E., Hewson, H.J. & Thompson, H.S. (1982) *Flora of Australia*, Australian Government Publishing Service, Canberra.

Gerwing, J.J. (2004) Life history diversity among six species of canopy lianas in an old-growth forest of the eastern Brazilian Amazon. *Forest Ecology and Management*, **190**, 57–72.

Grice, A.C. & Setter, M.J. (2003) Cooperative Research Centre for Tropical Rain Forest Ecology and Management, in *Weeds of Rain Forests and Associated Ecosystems*, Cairns, Australia.

Groves, R.H., Lonsdale, M. & Boden, R. (2005) *Jumping the Garden Fence: Invasive Garden Plants in Australia and Their Environmental and Agricultural Impacts*, WWF-Australia, Canberra.

Harris, C.J., Murray, B.R., Hose, G.C. & Hamilton, M.A. (2007) Introduction history and invasion success in exotic vines introduced to Australia. *Diversity and Distributions*, **13**, 467–475.

Hikosaka, K., Terashima, I. & Katoh, S. (1994) Effects of leaf age, nitrogen nutrition and photon flux density on the distribution of nitrogen among leaves of a vine (*Ipomoea tricolor* Cav.) grown horizontally to avoid mutual shading of leaves. *Oecologia*, **97**, 451–457.

Horvitz, C.C., Pascarella, J.B., McMann, S. et al. (1998) Functional roles of invasive non-indigenous plants in hurricane-affected subtropical hardwood forests. *Ecological Applications*, **8**, 947–974.

Howe, F. & Smallwood, J. (1982) Ecology of seed dispersal. *Annual Review of Ecology, Evolution, and Systematics*, **13**, 201–228.

Hughes, L., Dunlop, M., French, K., et al. (1994). Predicting dispersal spectra: a minimal set of hypotheses based on plant attributes. *Journal of Ecology*, **82**, 933–950.

Jiménez-Castillo, M., Wiser, S.K. & Lusk, C.H. (2007) Elevational parallels of latitudinal variation in the proportion of lianas in woody floras. *Journal of Biogeography*, **34**, 163–168.

Leishman, M.R. (2001) Does the seed size/number trade-off model determine plant community structure? An assessment of the model mechanisms and their generality. *Oikos*, **93**, 294–302.

Leishman, M.R., Wright, I.J., Moles, A.T. & Westoby, M. (2000) The evolutionary ecology of seed size, in *Seeds: The Ecology of Regeneration in Plant Communities* (ed M. Fenner), CABI Publishing, New York, pp. 31–57.

Llorens, A.M. & Leishman, M.R. (2008) Climbing strategies determine light availability for both vines and associated structural hosts. *Australian Journal of Botany*, **56**, 527–534.

Lord, J., Egan, J., Clifford, T. et al. (1997) Larger seeds in tropical floras: consistent patterns independent of growth form and dispersal mode. *Journal of Biogeography*, **24**, 205–211.

McGill, B.J., Enquist, B.J., Weiher, E. & Westoby, M. (2006) Rebuilding community ecology from functional traits. *Trends in Ecology and Evolution*, **21**, 178–185.

Moles, A.T., Ackerly, D.D., Tweddle, J.C. et al. (2007) Global patterns in seed size. *Global Ecology and Biogeography*, **16**, 109–116.

Moles, A.T., Falster, D.S., Leishman, M.R. & Westoby, M. (2004) Small-seeded species produce more seeds per square metre of canopy per year, but not per individual per lifetime. *Ecology*, **92**, 384–396.

Molina-Freaner, F., Gámez, R.C., Tinoco-Ojanguren, C. & Castellanos, A.E. (2004) Vine species diversity across environmental gradients in northwestern Mexico. *Biodiversity & Conservation*, **13**, 1853–1874.

Murphy, H., Metcalfe, D.J., Bradford, M.G. et al. (2008) Recruitment dynamics of invasive species in rain forest habitats following Cyclone Larry. *Australian Ecology*, **33**, 495–502.

Murphy, H.T., Westcott, D.A. & Metcalfe, D.J. (2006) Functional diversity of native and invasive plant species in tropical rain forests, in *Proceedings of the 15th Annual Australian Weeds Conference* (eds C. Preston, J.H. Watts & N.D. Crossman), Weed Society of South Australia, Adelaide, pp. 199–202.

Osunkoya, O.O., Bayliss, D., Panetta, F.D. & Vivian-Smith, G. (2010a) Leaf trait co-ordination in relation to construction cost, carbon gain and resource-use efficiency in exotic invasive and native woody vine species. *Annals of Botany*, **106**, 371–380.

Osunkoya, O.O., Bayliss, D., Panetta, F.D. & Vivian-Smith, G. (2010b) Variation in ecophysiology and carbon economy of invasive and native woody vines of riparian zones in South-Eastern Queensland. *Australian Ecology*, **35**, 636–649.

Parkhurst, D.F. & Loucks, O.L. (1972) Optimal leaf size in relation to environment. *Journal of Ecology*, **60**, 505–537.

Pickup, M., Westoby, M. & Basden, A. (2005) Dry mass costs of deploying leaf area in relation to leaf size. *Functional Ecology*, **19**, 88–97.

Putz, F.E. & Holbrook, M. (1991) Biomechanical studies of vines, in *The Biology of Vines* (eds F.E. Putz & H.A. Mooney), Cambridge University Press, Cambridge, pp. 73–98.

Schnitzer, S.A. (2005) A mechanistic explanation for global patterns of liana abundance and distribution. *The American Naturalist*, **166**, 262–276.

Schnitzer, S.A. & Bongers, F. (2011) Increasing liana abundance and biomass in tropical forests: emerging patterns and putative mechanisms. *Ecology Letters*, **14**, 397–406.

Smith, C.C. & Fretwell, S.D. (1974) The optimal balance between size and number of offspring. *The American Naturalist*, **108**, 499–506.

Stocker, G. & Irvine, A. (1983) Seed dispersal by cassowaries (*Casuarius casuarius*) in North Queensland's rain forests. *Biotropica*, **15**, 170–176.

Strahler, A.N. & Strahler, A.H. (1989) *Elements of Physical Geography: Study Guide*, John Wiley and Sons Inc., Brisbane.

Vivian-Smith, G. & Panetta, D.J. (2002) Going with the flow: dispersal of invasive vines in coastal catchments, in *Coast to Coast 2002: Source to Sea: Conference Proceedings, Australia's National Coastal Conference* (ed B.G. Thom), Cooperative

Research Centre for Coastal Zone, Estuary and Waterway Management, Indooroopilly, Queensland, pp. 492–494.

Webb, L.J. (1959) A physiognomic classification of Australian rain forests. *Journal of Ecology*, **47**, 551–570.

Webb, L.J. (1968) Environmental relationships of the structural types of Australian rain forest vegetation. *Ecology*, **49**, 296–311.

Westoby, M., Falster, D.S., Moles, A.T. et al. (2002) Plant ecological strategies: some leading dimensions of variation between species. *Annual Review of Ecology and Systematics*, **33**, 125–159.

Chapter 10

PATTERNS OF LIANA SUCCESSION IN TROPICAL FORESTS

Susan G. Letcher

Purchase College (SUNY), Purchase, NY, USA

Ecology of Lianas, First Edition. Edited by Stefan A. Schnitzer, Frans Bongers, Robyn J. Burnham, and Francis E. Putz.

Companion Website: www.wiley.com/go/schnitzer/lianas

OVERVIEW

Lianas are particularly abundant in tropical forests and understanding the relationship between lianas and disturbance is a key step in understanding the dynamism and resilience of tropical forests systems. Lianas contribute only a small amount of the biomass in most forests, but they have disproportionate impacts on forest dynamics due to their competitive abilities and their capacity to alter forest structure. Despite the importance of lianas, our understanding of how the liana community responds to small-scale and large-scale disturbances lags behind our understanding of tree community responses.

This chapter summarizes current research on liana succession and suggests directions for future study. In response to the small-scale disturbances caused by treefall gaps, lianas rapidly colonize space and can inhibit the establishment of shade-tolerant trees, altering the successional trajectory of forest patches. After large-scale land clearing and forest regrowth, lianas respond differently than do trees: liana biomass recovers to pre-disturbance levels more slowly than tree biomass, and liana species richness stays the same or declines over time in contrast to the increased species richness observed in the tree community.

The extent to which lianas compete with trees and alter trajectories of secondary succession is largely unknown. The species composition of the liana community undergoes shifts during succession, and the traits of lianas that are prevalent at different points in succession suggest that lianas, like trees, occupy a continuum from small-seeded, light-demanding, fast-growing pioneer species to larger-seeded, shade-tolerant, slower-growing species characteristic of mature forest. Early successional lianas frequently have seedlings that lack a self-supporting phase, while late successional lianas have freestanding seedlings and saplings.

To fully appreciate the role of lianas during succession, comparative studies are needed from more ecoregions, particularly tropical dry forests where lianas tend to be abundant. Manipulative experiments and long-term demographic studies will also shed light on the ecology of lianas during succession. Trait-based approaches will permit a more nuanced understanding of lianas' responses to environmental variation and their impacts on trees. The conservation and appropriate management of tropical forests depend on understanding their dynamics, in which the liana community plays a vitally important role.

INTRODUCTION

Tropical forests are dynamic systems, marked by mosaics of disturbance and recovery (Aubréville 1938; Richards 1952; Clark 1990; Chazdon 2003). The disturbances vary in spatial extent and intensity: mudslides, fires, and floods that remove vegetation entirely (Foster et al. 1998; Turner et al. 1998); hurricanes and typhoons that fell some trees and defoliate others (Boose et al. 1994); and small-scale disturbance events such as tree- and branch-falls that leave ephemeral gaps in the forest (Brokaw 1982; Martínez-Ramos et al. 1988). Increasingly, human-mediated disturbances are affecting tropical forest structure and dynamics. A combination of large-scale forest clearance and subsequent demographic shifts in tropical countries have led to the re-establishment of forest on previous agricultural lands (Mather 1992; Chazdon 2003; FAO 2010); currently, degraded forests and secondary forests make up more than 60% of the remaining tropical forest on earth (FAO 2010). Even where forests have not been directly damaged or cleared by human activity, anthropogenic activity affects tropical forest dynamics. Higher temperatures and lower rainfall lead to lower wood production (Clark 2004; Clark et al. 2010), and the higher tree mortality observed in large-scale permanent plot studies in the past several decades (Phillips 1998; Wright 2005; Schnitzer, Chapter 30 in this volume) has been linked to climate change.

Because of the increase in forest disturbance, understanding the dynamics of successional forests in the tropics is urgently needed. Both stochastic and deterministic processes govern the regeneration of vegetation after disturbance (Chazdon 2008). Identifying emergent patterns in tropical forest succession could have implications for forest management and restoration (Chazdon et al. 2007; Chazdon 2008). The mosaic of successional habitats resulting from disturbances at different scales promotes the maintenance of diversity in tropical forests (Vandermeer et al. 2000; Schnitzer & Carson 2001), and understanding forest dynamism is crucial for conservation (Wright 2005). Lianas are particularly responsive to disturbance (Schnitzer & Carson 2001; Dalling et al. 2012; Ledo & Schnitzer 2014), and thus understanding the relationship between lianas and disturbance will be an important step in understanding tropical forest dynamics as a whole.

Early studies on lianas focused mainly on their physiology and morphology (Darwin 1865) and classification (Schenck 1892). The first studies on lianas *in situ* and their roles in forest communities focused on their silvicultural impacts. Lianas decrease tree growth (Fox 1968; Lowe & Walker 1977; Clark & Clark 1990), cause bole deformations (Stevens 1987; Hegarty 1991), compete with trees both above- and belowground (Barker & Pérez-Salicrup 2000; Pérez-Salicrup & Barker 2000; Pérez-Salicrup 2001; Schnitzer et al. 2005; Chen et al. 2008), and interfere with the regeneration of the forest canopy and the re-establishment of harvestable timber trees (Foggie 1960; Nicholson 1965; Fox 1968). When liana-laden trees fall, they take other trees with them, leading to larger disturbances (Fox 1968; Appanah & Putz 1984; but see also Parren & Bongers 2001). The negative impacts of lianas on trees, particularly their impacts on marketable wood volume, have led many foresters to advocate liana cutting and/or poisoning as a silvicultural practice prior to logging operations (Jones 1950; Foggie 1960; Nicholson 1965; Fox 1968; Appanah & Putz 1984). Fewer studies have addressed the positive roles of lianas in forests (see Schnitzer et al., Chapter 1 in this volume). However, lianas seal off forest edges, provide cover, and ameliorate the microclimate for regenerating tree seedlings (Kochummen & Ng 1977). Due to their rapid growth, high nutrient content in tissues (Kazda & Salzer 2000; Kazda, Chapter 22 in this volume), and high leaf litter production (Putz 1983), lianas play an important role in tropical forest carbon and nutrient cycling (Powers, Chapter 14 in this volume; van der Heijden et al., Chapter 13 in this volume). Their lateral growth is important for the translocation of nutrients (Tang et al. 2012; Powers, Chapter 14 in this volume). Lianas also provide food, pathways, and protected foraging sites for forest animals, including seed dispersers (Emmons & Gentry 1983; Gerwing & Vidal 2002; Arroyo-Rodríguez et al., Chapter 27 in this volume; Lambert & Halsey, Chapter 26 in this volume; Michel et al., Chapter 25 in this volume; Yanoviak, Chapter 24 in this volume).

In recent decades, the amount of research on lianas has expanded exponentially (Schnitzer et al., Chapter 1 in this volume). A new appreciation of the ecology of lianas is emerging, and with it, a new understanding of the complicated role of lianas in tropical forest dynamics. Here, I focus on three recent avenues of liana research: (1) the role of lianas in small-scale disturbances such as treefall gaps; (2) the changes in the liana community during secondary succession after deforestation; and (3) the life history traits of lianas that dominate in particular successional stages. Treefall gaps and large-scale clearing represent endpoints on a continuum of disturbance intensity, and life history traits provide a framework for understanding plant responses to disturbance (Grime 1977). Because the available data from tropical moist and wet forests vastly exceed those from tropical dry forests, this review focuses primarily on moist and wet forests.

LIANAS AND SUCCESSION IN GAPS

Treefall gaps are the most common and smallest-scale type of disturbance affecting tropical forests (Brokaw 1982). Using the criteria of Brokaw (1982), in which a gap is defined as an opening in the canopy that extends to within 2 m of the forest floor, at least 150 m^2 in area, the average times between gap formation at a given point on the forest floor range from 80–136 years in tropical rainforests worldwide (Brokaw 1982; Denslow 1987), meaning that between 0.7–2.9% of forest area will be converted to gaps in any given year (Clark 1990). Gaps appear to play less of a role in the dynamics of tropical dry forests, due to the shorter physiognomy and generally lower turnover rates of dry forest trees and the higher degree of deciduousness (Dickinson et al. 2001).

After gap formation, vegetation closes in and the canopy height increases (Denslow 1987). The regrowth of vegetation in canopy gaps does not always follow the same trajectory, however. Gap size affects the resulting microclimate and the rate of colonization following a disturbance, with the smallest gaps usually regenerating fastest (Brokaw 1982; Denslow 1987; Barton et al. 1989). Schnitzer et al. (2000) identified three distinct pathways of gap-phase regeneration in small gaps (25–75 m^2): depending on the identity of the earliest colonists in the gap, the first phase of regeneration may be dominated by palms, trees, or lianas. If trees win, the canopy is rapidly re-established (Schnitzer et al. 2000; Schnitzer & Bongers 2005). If palms win, it may take longer for trees to emerge from their deep shade (Schnitzer et al. 2000). If lianas win, they may form dense tangles that outcompete trees for aboveground and belowground resources, physically damage saplings, and stall canopy formation for a decade or more (Schnitzer et al. 2000; Schnitzer & Carson 2010; Paul & Yavitt 2011). Liana-dominated gaps leave tangles of lianas on the forest floor even after canopy closure (Campbell & Newbery 1983; Balée & Campbell 1990; Schnitzer et al. 2000).

Lianas have numerous adaptations that allow them to colonize gaps rapidly (reviewed in Paul & Yavitt 2011). Adult lianas that fall with the tree or branch that creates a gap can resprout, even after major stem damage (Putz 1984a; Fisher & Ewers 1991). Lianas have much faster rates of lateral stem extension than trees do, and liana shoots frequently colonize gaps laterally (Peñalosa 1984; Yorke et al. 2013). In the 1980s and 1990s, approximately 7.5% of canopy gaps on Barro Colorado Island followed the stalled, liana-filled pathway (Schnitzer et al. 2000). Experimental removal of lianas from gaps in this forest significantly increased the recruitment, growth, and species richness of shade-tolerant tree species (Schnitzer & Carson 2010).

Even in cases where lianas do not dominate a canopy gap, they may affect the competitive success of different tree guilds. Putz (1984a) and Schnitzer et al. (2000) found a strong positive correlation between the abundance of lianas and the proportion of pioneer tree species in gaps, and a strong negative correlation of liana abundance with shade-tolerant tree abundance. Other studies have noted the negative relationship between liana abundance and shade-tolerant tree performance (Lowe & Walker 1977; Putz et al. 1984; Stevens 1987). In an eight-year experiment, liana removal in treefall gaps increased the growth, recruitment, and diversity of shade-tolerant tree species, while the growth, recruitment, and survival of pioneer trees were not significantly affected (Schnitzer & Carson 2010). Pioneer trees can avoid and shed lianas via rapid diameter growth rate, frequent shedding of leaves and/or branches, flexible stems that sway in the wind, and shallow crowns that make it more difficult for lianas to colonize a tree and stay attached (Putz 1980; Putz 1984b). Some important pioneer genera also have symbiotic relationships with ants that remove vines; *Cecropia* is defended by *Azteca* ants in the Neotropics (Janzen 1973; Putz 1980) and *Macaranga* is defended by *Crematogaster* in the Paleotropics (Putz 1980).

The identity of the lianas that colonize a gap may also make a difference to the growth and survival of tree seedlings. Fast-growing liana species had stronger inhibitory effects on tree growth than slow-growing lianas, particularly for shade-tolerant tree species (Toledo-Aceves & Swaine 2008). Lianas with flexible low-density stems have a greater negative impact on tree growth than lianas with harder wood (Garrido-Pérez & Gerold 2009). The competitive interactions between lianas and trees are mediated by the environmental conditions in a gap: the physiological and morphological changes in tree seedlings in response to competition with lianas differ in high light and low light conditions (Chen et al. 2008).

Frequent disturbances may lead to higher liana abundance and diversity in forests (Webb 1958; Schnitzer & Bongers 2002; Paul & Yavitt 2011). At a landscape scale, lianas are more abundant in disturbed sites such as cyclone-impacted forests (Webb 1958) or areas prone to windthrows (Jones 1950; Putz & Chai 1987). Liana forests, where c. 86% of trees are colonized by lianas (Pérez-Salicrup et al. 2001), cover as much as 100,000 km^2 of the Amazon Basin (Pires 1973). Balée and Campbell (1990) found evidence of human activity in the liana forests between the Xingu and Tocantins rivers in Brazil, suggesting that liana forests could be a signature of intensive shifting cultivation by pre-Colombian civilizations. On a smaller scale in mature forests, lianas are more abundant in recent treefall gaps than in the rest of the forest (Putz 1984a), and areas in the forest with a higher incidence of gaps have more lianas (Ibarra-Manríquez & Martínez-Ramos 2002). Indeed, liana distribution in the Barro Colorado Island Forest Dynamics Plot was significantly associated with disturbance (Dalling et al. 2012; Ledo & Schnitzer 2014), and gaps appear to maintain the diversity of the vast majority of liana species, independent of stem density, at this site (Ledo & Schnitzer 2014; Schnitzer et al., Chapter 7 in this volume). Lianas are also positively associated with forest fragmentation (Laurance et al. 2001). However, as forests become increasingly fragmented and degraded, there may be a threshold beyond which liana density declines with increasing fragmentation (Arroyo-Rodríguez & Toledo-Aceves 2009).

Studies from permanent forest plots across the tropics suggest that lianas are increasing in abundance and biomass, with a number of potentially alarming consequences (reviewed in Schnitzer & Bongers 2011; Schnitzer, Chapter 30 in this volume). As noted above, shade-tolerant trees are inhibited by lianas in gaps, leading to slower growth and lower diversity of these species where lianas proliferate, while pioneer trees are largely unaffected (Schnitzer & Carson 2010). Because of their light wood and high allocation to leaf rather than stem tissue, lianas and pioneer trees store much less carbon than shade-tolerant trees (Schnitzer & Bongers 2002). Greater liana domination in gaps would reduce the recruitment of shade-tolerant tree species, thus altering the species composition of tropical forests, reducing diversity, and reducing landscape-level carbon storage (Schnitzer & Bongers 2011).

PATTERNS OF LIANA SUCCESSION IN SECONDARY FOREST

Forests are currently being logged, cleared, and converted to agricultural systems across large portions of the globe (Chazdon 2003; FAO 2010), and there is an increasing frequency of disturbance in old-growth forests (Wright 2005). Secondary forests recovering in abandoned agricultural areas provide important refuges for biodiversity in tropical landscapes (Chazdon et al. 2009). Numerous studies have investigated the recovery of biomass, species richness, and species composition of trees in tropical secondary forests (reviewed in Finegan 1996; Chazdon et al. 2007; Chazdon 2008), but few studies have focused directly on the liana community during succession. The most relevant studies are summarized in Table 10.1. The majority have been conducted in wet or moist forest in the neotropics. Certain common features of liana regeneration emerge from these studies, though more work is necessary to confirm the generality of these observations.

Biomass trajectories

Tree biomass generally recovers fairly rapidly during tropical wet forest succession, with sites reaching 70–100% of old-growth forest biomass in as little as 20–30 years (reviewed in Chazdon et al. 2007; Letcher & Chazdon 2009b), while tree stem density is much less predictable (Chazdon et al. 2007). In tropical dry forests, tree biomass accumulation is generally slower than in wet forests due to abiotic constraints (Chazdon et al. 2007).

Lianas make up a small fraction of the biomass in mature forests, compared to trees (Gerwing & Farias 2000; Gehring et al. 2004; van der Heijden et al., Chapter 13 in this volume), and their pattern of biomass change during succession is different. The stem density of lianas peaks at an early to intermediate phase of succession (DeWalt et al. 2000; Kuzee & Bongers 2005; Letcher & Chazdon 2009a; Madeira et al. 2009), with numerous small liana stems giving way to fewer but larger stems as the forest matures (DeWalt et al. 2000; Letcher & Chazdon 2009a). Some studies have found that liana biomass increases more slowly during succession than tree biomass (Letcher & Chazdon 2009a), while others have found roughly constant liana biomass over time (DeWalt et al. 2000). Because lianas allocate a large fraction of their biomass to more ephemeral tissues such as leaves rather than stem tissues, their standing crop biomass may take longer to accumulate (Schnitzer & Bongers 2002; Gehring et al. 2004). The slow accumulation or lack of accumulation of liana biomass over time may be related to changes in forest stature. Lianas are more successful at reaching the canopy in gaps or low-stature forests (Putz 1984a). As the canopy closes, the lianas that have reached it already can continue to accumulate biomass, but fewer new liana stems successfully reach the canopy (Putz 1984a; Letcher & Chazdon 2009a). In a comparison of forest structure in patches with different canopy heights in an intact eastern Amazonian forest, liana biomass was three times higher in the low-stature than high-stature forest, while the overall biomass of woody species was three times higher in the high-stature forest (Gerwing & Farias 2000).

Species richness trajectories

Tree species richness increases over time during forest succession, though it generally takes longer to recover than biomass (reviewed in Finegan 1996; Chazdon et al. 2007; Letcher & Chazdon 2009b). The highest liana species richness, on the other hand, is usually reported in early to intermediate stages of succession (DeWalt et al. 2000; Kuzee & Bongers 2005; Letcher & Chazdon 2009a; Madeira et al. 2009). Because of the confounding of individual density and species density, it is important to use appropriate statistical techniques to compare species richness when sampling effort (i.e., number of individuals sampled) differs among the treatments being compared (Gotelli & Colwell 2001). The decline in liana species richness observed during succession may be a statistical artifact of the decline in stem abundance. To accurately assess changes in species richness during succession, large samples are necessary in every age category. Letcher and Chazdon (2009a) found no change in liana species richness during succession using rarefaction techniques, while

Table 10.1 Studies on liana succession in tropical secondary forests.

Source	Region	Habitat type	Plot ages measured	Variables assessed
DeWalt et al. (2000)	Panama	Moist forest	20-100 years post-agricultural succession; old-growth	Abundance, diversity, climbing mode
Kuzee and Bongers (2005)	Côte d'Ivoire	Moist forest	2 and 20 years post-abandonment in cacao plantations (canopy left intact); old growth	Abundance, diversity, climbing mode
Gehring et al. (2005)	Central Amazonia	Wet forest	2–25 years after shifting cultivation; old-growth	Biomass
Tovar-Pérez (2007)	Western Mexico	Dry forest	3–20 years post-agricultural succession, old-growth	Abundance, diversity
Madeira et al. (2009)	Southeastern Brazil	Dry forest	structural criteria used instead of age	Abundance
Roeder et al. (2010)	Central Amazonia	Wet forest	17–25 years post-agricultural succession; old-growth	Abundance, diversity
Letcher and Chazdon (2009a)	Northeastern Costa Rica	Wet forest	11–44 years post-agricultural succession; old-growth	Abundance, diversity, species composition, biomass
Letcher and Chazdon (2012)	Northeastern Costa Rica	Wet forest	11–44 years post-agricultural succession; old-growth	Climbing mode, seed size, dispersal mode, seedling growth habit
Barry et al. (in press)	Central Panama	Moist forest	1–30 years post-agricultural succession	Abundance, diversity

Note: This table does not include studies conducted in temperate or subtropical forests (Yuan et al. 2009; Ladwig 2009).

non-parametric species richness estimators suggested a slight decrease in liana species richness over time. DeWalt et al. (2000) and Barry et al. (in press) also found a decline in liana species richness during succession using Fisher's α. In the relatively few studies that have examined the habitat associations of lianas during succession, more species tend to be associated with disturbed, early-successional habitats than with mature, closed-forest habitats (Putz 1984a; Kuzee & Bongers 2005; Letcher & Chazdon 2012; Dalling et al. 2012; Vargas 2012; Ledo & Schnitzer 2014).

Species composition patterns

It is unknown whether the tree species composition of secondary forests will converge with the species composition of old-growth forests, and this convergence may depend on land use history and landscape context (Mesquita et al. 2001; reviewed in Chazdon 2008; Letcher & Chazdon 2009b). In the few studies that have directly examined changes in liana species composition, there is an indication of species turnover during succession (DeWalt et al. 2000; Kuzee & Bongers 2005; Tovar Pérez 2007; Letcher 2008), particularly when stem abundance is taken into account. In a dry forest in Mexico, the liana communities of young secondary forest (3–8 yrs), older secondary forest (20 yrs) and old-growth forest showed very low similarity (Tovar Pérez 2007). In six degraded wet forest sites in Côte d'Ivoire, different liana species were found to be dominant in different forest ages (Kuzee & Bongers 2005), though the high number of rare species in this data set suggests that more sampling is necessary to fully characterize the liana species richness in the region.

In a chronosequence of 10 moist forest sites from 20 years old to old-growth, DeWalt et al. (2000) found directional changes in liana species composition, converging with old-growth forest composition, based on an NMDS ordination of species abundances. Letcher (2008) found a similar result based on NMDS ordination of pairwise similarity in liana species composition in a chronosequence of 30 sites ranging from 10 years post-abandonment to old growth forest (Fig. 10.1). Liana species composition showed directional turnover between forests of different ages, with the highest similarity to old-growth forest composition in the oldest secondary forests. However, the changes in the liana community were less pronounced than changes in the tree community; as suggested by the higher stress in Fig. 10.1B and the fewer significant differences between age categories for lianas than for trees (Table 10.2). Successional changes in the liana community in this landscape were highly correlated with those of the non-liana community (Mantel test, $\rho = 0.61$, $P = 0.0019$). A similar pattern of floristic congruence between trees and lianas has been found in the south-western and north central Amazon (Macía et al. 2007; Landeiro et al. 2012). Lianas in the Bignoniaceae had one of the highest levels of floristic congruence with other taxa, suggesting a possible role for lianas in biomonitoring and rapid vegetation inventories (Landeiro et al. 2012).

Of the 125 liana species in the dataset of Letcher (2008), 40 species had $\geq$20 individuals in the dataset, allowing an assessment of their specialization to particular forest age categories. Using indicator values analysis (Dufrêne & Legendre 1997) implemented in PC-ORD 4 (McCune & Mefford 1999), five of these species were identified as young forest indicators and five were identified as old-growth forest indicators. The relative abundance of the first group declined over succession, while the second group increased (Fig. 10.2). Although lianas in general are more diverse and abundant in young, disturbed, low-stature forests, particular species do show affinities for different successional stages (Putz 1984a; Kuzee & Bongers 2005; Tovar-Pérez 2007; Letcher 2008; Vargas 2012).

LIFE HISTORY TRAITS OF LIANAS DURING FOREST SUCCESSION

Life history traits provide a useful framework for understanding the ecological strategies of plant species (Grime 1977; Gilbert et al. 2006; Gallagher, Chapter 9 in this volume). Given the demonstrated differences between the liana communities of early successional and late successional forest, it is interesting to consider the life history traits that mediate liana establishment and survival in forests of different successional stages. Studies on the liana communities of intact forests have revealed a range of ecological strategies (Putz 1984a; Gerwing 2004; Gallagher et al. 2011; Manzané-Pinzón 2012). Putz (1990) found average annual diameter growth increments ranging from 0.23 to 5.82 mm in liana species on Barro Colorado Island, and suggested that the faster-growing lianas are analogous to pioneer trees in their aggressive occupation of space. Although liana diameter growth rates and biomass increments are generally at least an order of magnitude lower than those of trees (Putz 1990; Letcher & Chazdon 2009a), their rate of leaf area accumulation per unit diameter is much higher (Putz 1983; Bullock 1990).

As non-self-supporting plants, lianas face different selective pressures than trees do (Letcher & Chazdon 2009a, 2012; Paul & Yavitt 2011). Lianas of different species build a wide array of climbing structures (Hegarty 1991; DeWalt et al. 2000; reviewed in Schnitzer & Bongers 2002), and each type of structure is biomechanically constrained to attach to supports of particular diameters (Darwin 1865; Putz & Chai 1987; Rowe et al. 2006). Since tendril-climbing lianas depend on smaller-diameter supports than branch-twining lianas (Putz & Chai 1987), several authors have predicted that tendril climbers should decline over time during succession as the availability of small-diameter supports diminishes (Putz & Holbrook 1991; DeWalt et al. 2000). Lianas with specialized climbing mechanisms declined in abundance over 13 years in a mature rain forest in Gabon (Caballé & Martin 2001), a finding that the authors attributed to the closure of gaps and the low availability of small-diameter supports in their plots during the second census. DeWalt et al. (2000) found a decline in the abundance of tendril-climbing lianas in a moist forest chronosequence, and Kuzee and Bongers (2005) found a slight but non-significant decline in tendril climbing liana abundance with forest age in moist forest in the Côte d'Ivoire. However, other studies have shown no change in a tropical wet region (Letcher & Chazdon 2009a) or even an increase in tendril climbing lianas during succession in a temperate region (Ladwig 2009).

Lianas are sensitive to the same life history tradeoffs that trees face with respect to survival in shade vs.

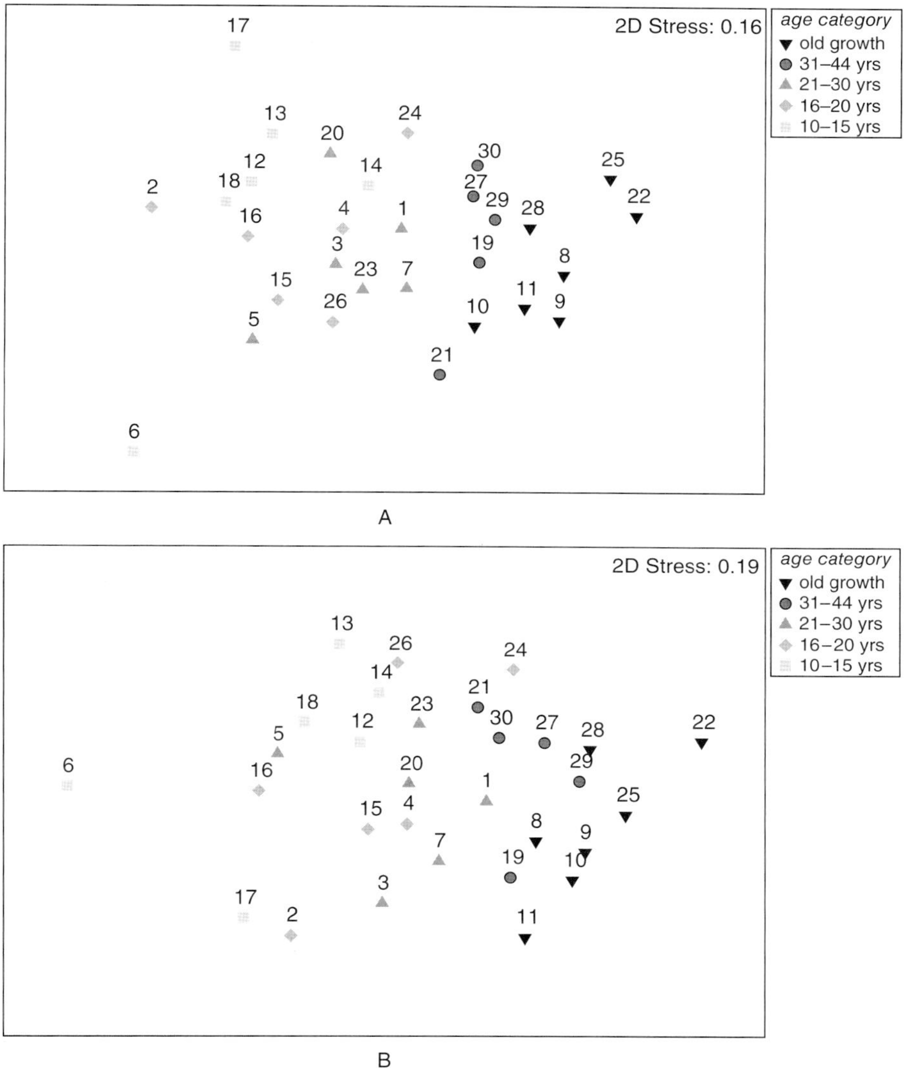

Fig. 10.1 Non-metric multidimensional scaling plot of the Chao-Jaccard similarity between trees (A) and lianas (B) in 30 sites in northeastern Costa Rica. Methods follow Letcher & Chazdon (2009b); numbers beside data points correspond to site numbers in Letcher & Chazdon (2009b). (Source: Letcher 2008. Reproduced with permission.)

rapid growth in high light (Gilbert et al. 2006). Letcher and Chazdon (2012) identified an important axis of variation in the liana community during succession: early successional lianas tend to have small seeds and seedlings that lose their self-supporting capacity at <1 m height (see also Gallagher, Chapter 9 in this volume). The species that dominate in old-growth forests tend to be large-seeded and have seedlings capable of maintaining a self-supporting growth form until they reach a larger size. The cotyledon reserves of a large-seeded liana species, in addition to providing the seedling with greater survival capacity in deep shade, allow the seedling to grow taller and sustain itself as a self-supporting stem in the forest understory

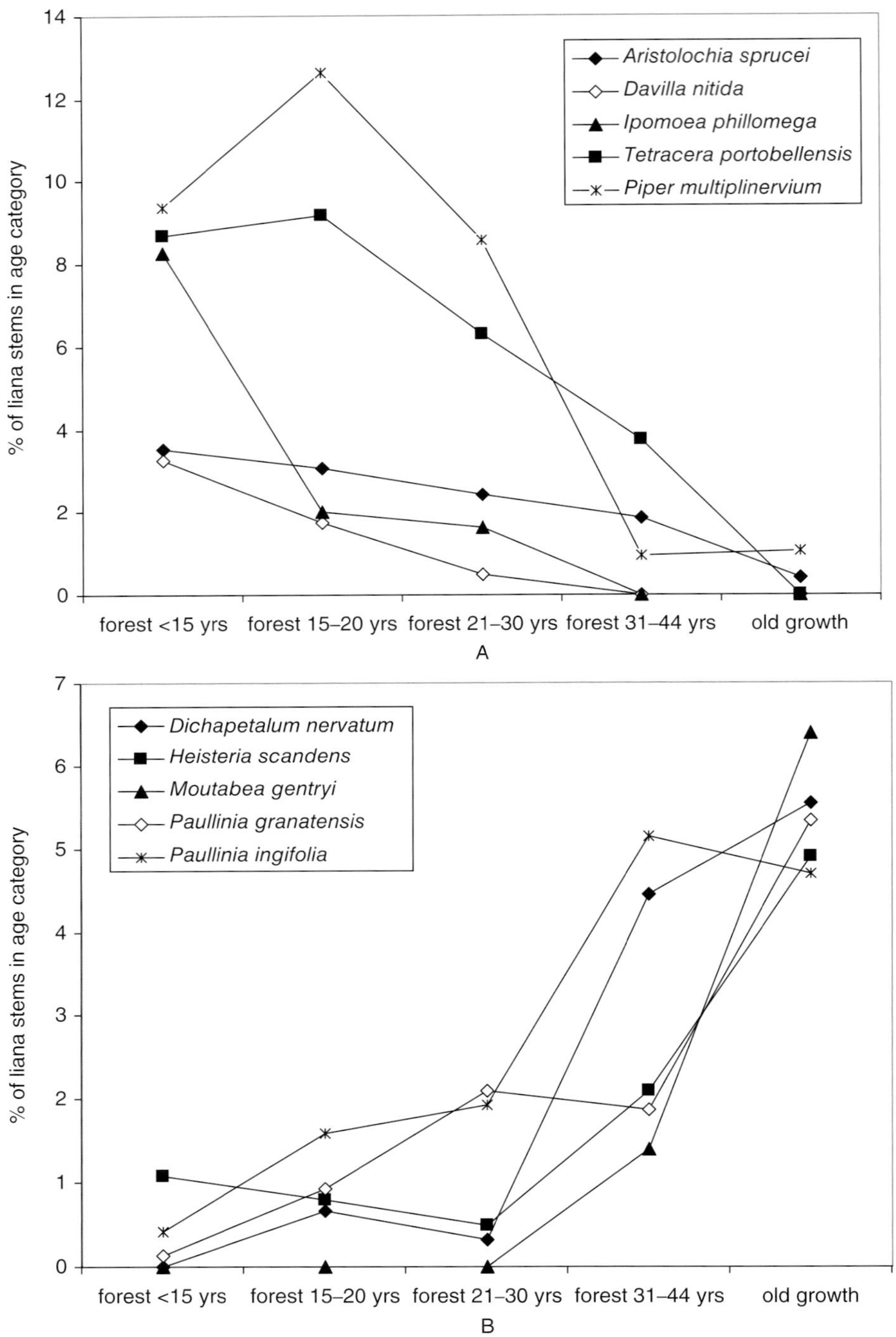

Fig. 10.2 The relative abundance of lianas identified as indicators of young secondary forest (A) and old-growth forest (B) in a chronosquence of 30 sites in northeastern Costa Rica. Note that the y axis scales differ between the panels. (Source: Letcher 2008. Reproduced with permission.)

Table 10.2 Results of a permutation test (ANOSIM; Clarke and Warwick 1994) evaluating the significance of pairwise comparisons among forest age categories in the NMDS plots (Figure 10.1). Each value is the P-value for the pairwise comparison between the two categories. Significant values (<0.05) are bold. Methods follow Letcher (2008).

Trees (Fig. 10.1A)		Forests 11–15 yrs	Forests 16–20 yrs	Forests 21–30 yrs	Forests 31–44 yrs
	Forest 16–20 yrs	0.113			
	Forest 21–30 yrs	**0.045**	0.147		
	Forest 31–44 yrs	**0.004**	**0.004**	**0.013**	
	Old growth	**0.002**	**0.001**	**0.001**	**0.039**
Lianas (Fig. 10.1B)	Forest 16–20 yrs	0.920			
	Forest 21–30 yrs	**0.030**	0.641		
	Forest 31–44 yrs	**0.004**	0.110	**0.028**	
	Old growth	**0.001**	**0.001**	**0.002**	0.321

Source: Letcher 2008. Reproduced with permission.

where supports are scarce (Putz 1984a; Letcher & Chazdon 2012). Liana species with self-supporting seedlings also tend to have a slower relative growth rate, higher leaf area, and higher stem cross-sectional area than climbing seedlings (Manzané-Pinzón 2012). The species identified by Letcher (2008) as indicators of young and old forest (Fig. 10.2) mainly conform to these predictions. All of the species identified as young forest specialists have diaspores <15 mm wide, and three of the five have non-self-supporting seedlings (Letcher 2008). The five species identified as old-growth specialists all have diaspore sizes >15 mm and seedlings capable of remaining self-supporting well past 1 m height.

In the tree community during tropical forest succession, the earliest colonists are usually small-seeded, fast-growing pioneer species. They eventually become overtopped by the slower-growing, shade-tolerant species that form the mature forest canopy (Finegan 1996; Lugo & Zimmerman 2002; Chazdon 2008). Lianas can avoid this overtopping, since they maintain their position at the top of the canopy by scrambling above tree branches (Letcher & Chazdon 2009a). This persistence may explain why the changes in the liana community are less pronounced than those of the tree community during succession (Fig. 10.1, Table 10.2). However, reaching the canopy from the forest floor after canopy closure is much more challenging for lianas than for trees (Putz 1984a), and so eventually the liana community of older forests does become distinct from that of young forests as the young-forest colonists die out and shade-tolerant lianas take over.

CONCLUSION AND FUTURE RESEARCH

Recent advances in liana ecology have refined our understanding of lianas and their roles in forest succession, yet many important questions remain. Large-scale comparative studies in different ecoregions will permit us to test the generality of patterns in liana succession and the intensity of lianas' impacts on trees in different contexts. In particular, more research is needed in tropical dry forests. The three topics reviewed here suggest a number of directions for future study.

Lianas in gaps

Gap dynamics have always been present in tropical forests, but recent changes in forest dynamism suggest that gap formation may become more frequent (Phillips 1998; Wright 2005; Schnitzer, Chapter 30 in this volume). Long-term studies of permanent plots and long-term manipulation experiments will be necessary to address the question of whether lianas are increasing in gaps and how their abundance affects tree dynamics (e.g., Schnitzer & Carson 2010). What factors affect the competitive success of lianas in treefall gaps and disturbed forests, and where are the ecological thresholds that tip the balance toward lianas? Does the identity of the liana species that colonize a gap affect the ability of shade-tolerant trees to compete? How will changes in liana abundance and biomass in mature tropical forests affect tree growth and mortality – and carbon storage – at a landscape scale, and ultimately at a global scale? A better understanding of how lianas respond to small-scale and mesoscale disturbances could also guide silvicultural interventions.

Lianas in secondary succession

The landscape-scale disturbances that result in secondary succession are interesting from an ecological standpoint and relevant for management and conservation, since secondary forests are an increasingly important component of the tropical landscape (Chazdon 2008). Liana removal experiments during succession will be useful for determining whether lianas shape successional trajectories after landscape-scale disturbance in the same way that they alter gap-phase regeneration trajectories. More studies of liana communities during succession will reveal whether there are emergent patterns in the way that lianas respond to successional change; for instance, why is it that tendril-climbing lianas decline in abundance during succession in some forests and increase in others (DeWalt et al. 2000; Ladwig 2009)? What are the ecological mechanisms that mediate species turnover during succession in the liana community? Does liana cutting in secondary forests speed the recovery of tree biomass and species richness, or does the presence of lianas buffer trees in young secondary forests from harsh microclimates (Kochummen & Ng 1977)?

Liana life history traits

Life history traits provide an important framework for grouping plants based on their responses to environmental change and their competitive ability (Grime 1977). Lianas come from an immense diversity of evolutionary backgrounds and have a wide range of growth strategies, climbing modes, dispersal modes, phenologies (Ibarra-Manríquez et al. 1991), seed germination strategies (Roeder et al. 2013), seedling morphologies (Ibarra-Manríquez et al. 2001), and physiological tolerances, etc. Building more comprehensive databases of traits will be the first step in developing functional classifications of lianas (Gallagher, Chapter 9 in this volume). As our knowledge of liana life history traits grows in sophistication, it becomes increasingly important to identify the sets of traits that most strongly determine the outcome of competitive interactions between trees and lianas in different environments. Do the life history tradeoffs evident in early liana establishment persist into adult growth patterns; i.e., do lianas that start out as fast-growing, weedy species always maintain a light-requiring physiology, or do ontogenetic shifts occur in liana growth patterns? Which aspects of liana morphology and physiology determine the intensity of their negative effects on tree growth? Can we improve silvicultural interventions by selectively targeting particular species or functional groups of lianas?

Despite the disproportionate impact of lianas in shaping forest dynamics, much remains to be learned about patterns of liana succession. Ultimately, understanding the role of lianas in forest dynamics will depend on an integration of these three research topics. Studying how lianas recolonize after large and small disturbances may provide insight into ecological tipping points and forest resilience. A better understanding of liana life history traits and functional types will provide a more nuanced approach to studying the ecological and silvicultural impacts of lianas.

ACKNOWLEDGMENTS

I am grateful to M. Peña-Claros, F. Bongers, and two anonymous reviewers for constructive feedback. Discussions with R.L. Chazdon, P. Foster, and S.A. Schnitzer contributed to the structure of the manuscript. Funding for the original research presented here came from the National Science Foundation (Graduate Fellowship to SGL and NSF DEB 0424767 to R. L. Chazdon), the University of Connecticut, the Ronald Bamford Endowment, the UCONN Center for Conservation and Biodiversity, the Organization for Tropical Studies. I thank M. Luna, A. Thrall, B. Lewis and J. Foley for field assistance, and the Organization for Tropical Studies and the Instituto Nacional de Biodiversidad for logistical support. Data from this study are archived at http://www.salvias.net.

REFERENCES

Appanah, S. & Putz, F.E. (1984) Climber abundance in virgin Dipterocarp forest and the effect of pre-felling climber cutting on logging damage. *Malaysian Forester*, **47**, 335–342.

Arroyo-Rodríguez, V. & Toledo-Aceves, T. (2009) Impact of landscape spatial pattern on liana communities in tropical rainforests at Los Tuxtlas, Mexico. *Applied Vegetation Science*, **12**, 340–349.

Aubréville, A. (1938) Regeneration patterns in the closed forest of Ivory Coast, in *World Vegetation Types* (ed S.R. Eyre), Macmillan, London Reprinted in R.L. Chazdon & T.C. Whitmore (eds.) (2002). *Foundations of Tropical Forest Biology: Classic Papers with Commentary*, pp. 523–538. University of Chicago Press, Chicago, pp. 41–55.

Balée, W. & Campbell, D.G. (1990) Evidence for the successional status of liana forest (Xingu River Basin, Amazonian Brazil). *Biotropica*, **22**, 36–47.

Barker, M.G. & Pérez-Salicrup, D.R. (2000) Comparative water relations of mature mahogany (*Swietenia macrophylla*) trees with and without lianas in a subhumid, seasonally dry forest in Bolivia. *Tree Physiology*, **20**, 1167–1174.

Barry, K.E., Schnitzer, S.A., van Breugel, M., & Hall, J.S. (in press). Rapid liana colonization and community development along a secondary forest chronosequence. *Biotropica*.

Barton, A.M., Fetcher, N. & Redhead, S. (1989) The relationship between treefall gap size and light flux in a Neotropical rain forest in Costa Rica. *Journal of Tropical Ecology*, **5**, 437–439.

Boose, E.R., Foster, D.R. & Fluet, M. (1994) Hurricane impacts to tropical and temperate forest landscapes. *Ecological Monographs*, **64**, 369–400.

Brokaw, N. (1982) Treefalls: frequency, timing, and consequences, in *The Ecology of a Tropical Forest: Seasonal Rhythms and Long-term Changes* (eds E.G. Jr. Leigh, A.S. Rand & D.M. Windsor), Smithsonian Institution Press, Washington, DC, pp. 101–108.

Bullock, S.H. (1990) Abundance and allometrics of vines and self-supporting plants in a tropical deciduous forest. *Biotropica*, **22**, 106–109.

Caballé, G. & Martin, A. (2001) Thirteen years of change in trees and lianas in a Gabonese rainforest. *Plant Ecology*, **152**, 167–173.

Campbell, E.J.F. & Newbery, D. (1983) Ecological relationships between lianas and trees in lowland rain forest in Sabah, East Malaysia. *Journal of Tropical Ecology*, **9**, 469–490.

Chazdon, R.L. (2003) Tropical forest recovery: legacies of human impact and natural disturbances. *Perspectives in Plant Ecology, Evolution and Systematics*, **6**, 51–71.

Chazdon, R.L. (2008) Chance and determinism in tropical forest succession, in *Tropical Forest Community Ecology* (eds W. Carson & S.A. Schnitzer), Wiley-Blackwell Publishing, Oxford, pp. 384–408.

Chazdon, R.L., Letcher, S.G., van Breugel, M. et al. (2007) Rates of change in tree communities in secondary Neotropical forests following major disturbances. *Philosophical Transactions of the Royal Society B: Biological Sciences*, **362**, 273–289.

Chazdon, R.L., Peres, C.A., Dent, D. et al. (2009) The potential for species conservation in tropical secondary forests. *Conservation Biology*, **23**, 1406–1417.

Chen, Y.-J., Bongers, F., Cao, K.-F. & Cai, Z.-Q. (2008) Above- and below-ground competition in high and low irradiance: tree seedling responses to a competing liana *Byttneria grandifolia*. *Journal of Tropical Ecology*, **24**, 517–524.

Clark, D.A. (2004) Sources or sinks? The responses of tropical forests to current and future climate and atmospheric composition. *Philosophical Transactions of the Royal Society of London, Series B: Biological Sciences*, **359**, 477–491.

Clark, D.B. (1990) The role of disturbance in the regeneration of neotropical moist forests, in *Reproductive Biology of Tropical Forest Plants* (eds K.S. Bawa & M. Hadley), pp. 219–315 *UNESCO Man and the Biosphere Series*, Vol. 7. Parthenon Publishing, Carnforth.

Clark, D.B. & Clark, D.A. (1990) Distribution and effects on tree growth of lianas and woody hemiepiphytes in a Costa Rican tropical wet forest. *Journal of Tropical Ecology*, **6**, 321–331.

Clark, D.B., Clark, D.A. & Oberbauer, S.F. (2010) Annual wood production in a tropical rain forest in NE Costa Rica linked to climatic variation but not to increasing CO_2. *Global Change Biology*, **16**, 747–759.

Clarke, K.R. & Warwick, R.M. (1994) *Change in Marine Communities: An Approach to Statistical Analysis and Interpretation*, Plymouth Marine Laboratory, Plymouth, UK.

Dalling, J.W., Schnitzer, S.A., Baldeck, C. et al. (2012) Resource-based habitat associations in a neotropical liana community. *Journal of Ecology*, **100**, 1174–1182.

Darwin, C.R. (1865) On the movements and habits of climbing plants. *Journal of the Linnean Society of London (Botany)*, **9**, 1–118.

Denslow, J.S. (1987) Tropical rainforest gaps and tree species diversity. *Annual Review of Ecology and Systematics*, **18**, 431–451.

DeWalt, S.J., Schnitzer, S.A. & Denslow, J.S. (2000) Density and diversity of lianas along a chronosequence in a central Panamanian lowland forest. *Journal of Tropical Ecology*, **16**, 1–19.

Dickinson, M.B., Hermann, S.M. & Whigham, D.F. (2001) Low rates of background canopy-gap disturbance in a seasonally dry forest in the Yucatan Peninsula with a history of fires and hurricanes. *Journal of Tropical Ecology*, **17**, 895–902.

Dufrêne, M. & Legendre, P. (1997) Species assemblages and indicator species: the need for a flexible asymmetric approach. *Ecological Monographs*, **67**, 345–366.

Emmons, L.H. & Gentry, A.H. (1983) Tropical forest structure and the distribution of gliding and prehensile-tailed vertebrates. *American Naturalist*, **121**, 513–524.

FAO (Food and Agriculture Organization of the United Nations). (2010) *Global Forest Resources Assessment*. FAO Forestry Paper 163, Rome, FAO.

Finegan, B. (1996) Pattern and process in neotropical secondary rain forests: the first 100 years of succession. *Trends in Ecology and Evolution*, **11**, 119–124.

Fisher, J.B. & Ewers, F.W. (1991) Structural responses to stem injury in vines, in *The Biology of Vines* (eds F.E. Putz & H.A. Mooney), Cambridge University Press, Cambridge, pp. 99–126.

Foggie, A. (1960) Natural regeneration in the humid tropical forest. *Caribbean Forester*, **2**, 73–81.

Foster, D.R., Knight, D.H. & Franklin, J.F. (1998) Landscape patterns and legacies resulting from large, infrequent forest disturbances. *Ecosystems*, **1**, 497–510.

Fox, J.E.D. (1968) Logging damage and the influence of climber cutting prior to logging in the lowland dipterocarp forest in Sabah, Malaysia. *Malaysian Forester*, **36**, 80–112.

Gallagher, R.V., Leishman, M.R. & Moles, A.T. (2011) Traits and ecological strategies of Australian tropical and temperate climbing plants. *Journal of Biogeography*, **38**, 828–839.

Garrido-Pérez, E.I. & Gerold, G. (2009) Land-use history and the origins and effects of lianas on tree communities: the case of secondary forests in the Northeastern Yucatan Peninsula. *Erdkunde*, **63**, 211–227.

Gehring, C., Denich, M. & Vlek, P.L. (2005) Resilience of secondary forest regrowth after slash-and-burn agriculture in Central Amazonia. *Journal of Tropical Ecology*, **21**, 519–527.

Gehring, C., Park, S. & Denich, M. (2004) Liana allometric biomass equations for Amazonian primary and secondary forest. *Forest Ecology and Management*, **195**, 69–83.

Gentry, A.H. (1991) The distribution and evolution of climbing plants, in *The Biology of Vines* (eds F.E. Putz & H.A. Mooney), Cambridge University Press, Cambridge, pp. 3–49.

Gerwing, J.J. (2004) Life history diversity among six species of canopy lianas in an old-growth forest of the eastern Brazilian Amazon. *Forest Ecology and Management*, **190**, 57–72.

Gerwing, J.J. & Farias, D.L. (2000) Integrating liana abundance and forest stature into an estimate of total aboveground biomass for an eastern Amazonian forest. *Journal of Tropical Ecology*, **16**, 327–335.

Gerwing, J.J. & Vidal, E. (2002) Changes in liana abundance and species diversity eight years after liana cutting and logging in an eastern Amazonian forest. *Conservation Biology*, **16**, 544–548.

Gilbert, B., Wright, S.J., Muller-Landau, H.C. et al. (2006) Life history tradeoffs in tropical trees and lianas. *Ecology*, **87**, 1281–1288.

Gotelli, N.J. & Colwell, R.K. (2001) Quantifying biodiversity: procedures and pitfalls in the measurement and comparison of species richness. *Ecology Letters*, **4**, 379–391.

Grime, J.P. (1977) Evidence for the existence of three primary strategies in plants and its relevance to ecological and evolutionary theory. *American Naturalist*, **111**, 1169–1194.

Hegarty, E.E. (1991) Vine-host interactions, in *The Biology of Vines* (eds F.E. Putz & H.A. Mooney), Cambridge University Press, Cambridge, pp. 493–501.

Hubbell, S.P. (1998) The maintenance of diversity in a neotropical tree community: conceptual issues, current evidence, and challenges ahead, in *Forest Biodiversity Research, Monitoring and Modeling* (eds F. Dallmeier & J.A. Comiskey), UNESCO, Paris, pp. 17–44.

Ibarra-Manríquez, G. & Martínez-Ramos, M. (2002) Landscape variation of liana communities in a Neotropical rain forest. *Plant Ecology*, **160**, 91–112.

Ibarra-Manríquez, G., Martínez-Ramos, M. & Oyama, K. (2001) Seedling functional types in a lowland forest in Mexico. *American Journal of Botany*, **88**, 1801–1812.

Ibarra-Manríquez, G., Sánchez-Garfias, B. & González-García, L. (1991) Fenología de lianas y arboles anemócoros en una selva calido-humeda de México. *Biotropica*, **23**, 242–254.

Janzen, D.H. (1973) Dissolution of mutualism between *Cecropia* and its *Azteca* ants. *Biotropica*, **5**, 15–28.

Jones, E.W. (1950) Some aspects of natural regeneration in the Benin rain forest. *Empire Forestry Review*, **29**, 108–124.

Kazda, M. & Salzer, J. (2000) Leaves of lianas and self-supporting plants differ in mass per unit area and in nitrogen content. *Plant Biology*, **2**, 268–271.

Kochummen, K.M. & Ng, F.S.P. (1977) Natural plant succession after farming in Kepong. *Malayan Forester*, **40**, 61–78.

Kuzee, M.E. & Bongers, F. (2005) Climber abundance, diversity, and colonisation in degraded forests of different ages in Côte d'Ivoire, in *Forest Climbing Plants of West Africa: Diversity, Ecology, and Management* (eds F. Bongers, M.P.E. Parren & D. Traoré), CAB International, Wallingford, pp. 73–91.

Ladwig, L.M. (2009) Ecology and impacts of lianas in regenerating forests. Master's thesis, Eastern Illinois University.

Landeiro, V.L., Bini, L.M., Costa, F.R.C. et al. (2012) How far can we go in simplifying biomonitoring assessments? An integrated analysis of taxonomic surrogacy, taxonomic sufficiency and numerical resolution in a megadiverse region. *Ecological Indicators*, **23**, 366–373.

Laurance, W.F., Pérez-Salicrup, D.R., Delamônica, P. et al. (2001) Rain forest fragmentation and the structure of Amazon liana communities. *Ecology*, **82**, 105–116.

Ledo, A. & Schnitzer, S.A. (2014) Disturbance, not negative density dependence or habitat specialization maintains liana diversity in a tropical forest. *Ecology*, in press.

Letcher, S.G. (2008) Structural, floristic, and phylogenetic dynamics of tropical forest stands during succession. Doctoral dissertation, University of Connecticut.

Letcher, S.G. & Chazdon, R.L. (2009a) Lianas and self-supporting plants during tropical forest succession. *Forest Ecology and Management*, **257**, 2150–2156.

Letcher, S.G. & Chazdon, R.L. (2009b) Rapid recovery of biomass, species richness, and species composition in a forest chronosequence in northeastern Costa Rica. *Biotropica*, **41**, 608–617.

Letcher, S.G. & Chazdon, R.L. (2012) Life history traits of lianas during tropical forest succession. *Biotropica*. doi: 10.1111/j.1744-7429.2012.00865.x

Lowe, R.G. & Walker, P. (1977) Classification of canopy, stem, crown status and climber infestation in natural tropical forest in Nigeria. *Journal of Applied Ecology*, **14**, 897–903.

Lugo, A.E. & Zimmerman, J.K. (2002) Ecological life histories, in *Tropical Tree Seed Manual* (ed J.A. Vozzo), USDA

Agricultural Handbook Number 721, USDA Forest Service, Washington, DC, pp. 191–213.

Macía, M.J., Ruokolainen, K., Tuomisto, H. et al. (2007) Congruence between floristic patterns of trees and lianas in a southwest Amazonian rain forest. *Ecography*, **30**, 561–577.

Madeira, B.G., Espírito-Santo, M.M., D'Ângelo Neto, S. et al. (2009) Changes in tree and liana communities along a successional gradient in a tropical dry forest in south-eastern Brazil. *Plant Ecology*, **201**, 291–304.

Manzané-Pinzón, E.J. (2012) *Freestanding and support-seeker liana seedlings: spatial distribution, life history and physiological traits in tropical forests of Central Panama*, University of Miami, Doctoral dissertation.

Martínez-Ramos, M., Alvarez-Buylla, E., Sarukhan, J. & Pinero, D. (1988) Treefall age determination and gap dynamics in a tropical forest. *Journal of Ecology*, **76**, 700–716.

Mather, A.S. (1992) The forest transition. *Area*, **24**, 367–379.

McCune, B. & Mefford, M.J. (1999) *PC-ORD: Multivariate Analysis of Eecological data, Version 4*, MJM Software Design, Gleneden Beach, Oregon.

Mesquita, R.C.G., Ickes, K., Ganade, G. & Williamson, G.B. (2001) Alternative successional pathways in the Amazon Basin. *Journal of Ecology*, **89**, 528–537.

Nicholson, D.I. (1965) A review of natural regeneration in the dipterocarp forest of Sabah. *Malayan Forester*, **28**, 4–25.

Parren, M. & Bongers, F. (2001) Does climber cutting reduce felling damage in southern Cameroon? *Forest Ecology and Management*, **141**, 175–188.

Paul, G.S. & Yavitt, J.B. (2011) Tropical vine growth and the effects on forest succession: a review of the ecology and management of tropical climbing plants. *Botanical Review*, **77**, 11–30.

Peñalosa, J. (1984) Basal branching and vegetative spread in two tropical rain forest lianas. *Biotropica*, **16**, 1–9.

Pérez Salicrup, D.R. (2001) Effect of liana-cutting on tree regeneration in a liana forest in Amazonian Bolivia. *Ecology*, **82**, 389–396.

Pérez-Salicrup, D.R. & Barker, M.G. (2000) Effect of liana removal on water potential and growth of adult *Senna multijuga* (Caesalpinioideae) trees in a Bolivian tropical forest. *Oecologia*, **124**, 469–475.

Pérez-Salicrup, D.R., Sork, V.L. & Putz, F.E. (2001) Lianas and trees in a liana forest of Amazonian Bolivia. *Biotropica*, **33**, 34–47.

Phillips, O.L. (1998) Increasing tree turnover in tropical forests as measured in permanent plots, in *Forest Biodiversity Research, Monitoring, and Modeling: Conceptual Background and Old World Case Studies* (eds F. Dallmeier & I.A. Comiskey), UNESCO, Paris, pp. 221–246.

Pires, J.M. (1973) Tipos de vegetação da Amazônia, in *O Museu Goeldi no Ano do Sesquicentenario* (ed M.F. Simões), Museu Paraense Emilio Goeldi, Belem, Brazil, pp. 179–202.

Putz, F.E. (1980) Lianas vs. trees. *Biotropica*, **12**, 224–225.

Putz, F.E. (1983) Liana biomass and leaf área of a 'tierra firme' forest in the Rio Negro basin, Venezuela. *Biotropica*, **15**, 185–189.

Putz, F.E. (1984a) The natural history of lianas on Barro Colorado Island, Panama. *Ecology*, **65**, 1713–1724.

Putz, F.E. (1984b) How trees avoid and shed lianas. *Biotropica*, **16**, 19–23.

Putz, F.E. (1990) Liana stem diameter growth and mortality rates on Barro Colorado Island, Panama. *Biotropica*, **22**, 103–105.

Putz, F.E. & Chai, P. (1987) Ecological studies of lianas in Lambir National Park, Sarawak, Malaysia. *Journal of Ecology*, **75**, 523–531.

Putz, F.E. & Holbrook, N.M. (1991) Biomechanical studies of vines, in *The Biology of Vines* (eds F.E. Putz & H.A. Mooney), Cambridge University Press, Cambridge, pp. 73–97.

Putz, F.E., Lee, H.S. & Goh, R. (1984) Effects of post-felling silvicultural treatments on woody vines in Sarawak. *Malaysian Forester*, **47**, 214–226.

Richards, P.W. (1952) *The Tropical Rain Forest*, Cambridge University Press, Cambridge.

Roeder, M., Ferraz, I.D.K. & Hölscher, D. (2013) Seed and germination characteristics of 20 Amazonian liana species. *Plants*, **2**, 1–15.

Roeder, M., Hölscher, D. & Ferraz, I.D.K. (2010) Liana regeneration in secondary and primary forests of central Amazonia. *Plant Ecology and Diversity*, **3**, 165–174.

Rowe, N.P., Isnard, S., Gallenmüller, F. & Speck, T. (2006) Diversity of mechanical architectures in climbing plants: an ecological perspective, in *A Mechanical Approach to the Ecology of Animals and Plants* (eds A. Herrel, T. Speck & N.P. Rowe), CRC Press, Boca Raton, FL, pp. 35–60.

Schenck, H. (1892) Beiträge zur Biologie und Anatomie der Lianen im besondere in Brasillien einheimischen Arten. *Botanische Mitteilungen aus den Tropen*, **4**, 1–253.

Schnitzer, S.A. & Bongers, F. (2002) The ecology of lianas and their role in forests. *Trends in Ecology and Evolution*, **17**, 223–230.

Schnitzer, S.A. & Bongers, F. (2005) Lianas and gap-phase regeneration: implications for forest dynamics and species diversity, in *Forest Climbing Plants of West Africa: Diversity, Ecology, and Management* (eds F. Bongers, M.P.E. Parren & D. Traoré), CAB International, Wallingford, pp. 59–72.

Schnitzer, S.A. & Bongers, F. (2011) Increasing liana abundance and biomass in tropical forests: emerging patterns and putative mechanisms. *Ecology Letters*, **14**, 397–406.

Schnitzer, S.A. & Carson, W.P. (2001) Treefall gaps and the maintenance of species diversity in a tropical forest. *Ecology*, **82**, 913–919.

Schnitzer, S.A. & Carson, W.P. (2010) Lianas suppress tree regeneration and diversity in treefall gaps. *Ecology Letters*, **13**, 849–857.

Schnitzer, S.A., Dalling, J.W. & Carson, W.P. (2000) The impact of lianas on tree regeneration in tropical forest canopy gaps:

evidence for an alternative pathway of gap-phase regeneration. *Journal of Ecology*, **88**, 655–666.

Schnitzer, S.A., Kuzee, M.E. & Bongers, F. (2005) Disentangling above- and below-ground competition between lianas and trees in a tropical forest. *Journal of Ecology*, **93**, 1115–1125.

Stevens, G.C. (1987) Lianas as structural parasites: the *Bursera simaruba* example. *Ecology*, **68**, 77–81.

Tang, Y., Kitching, R.L. & Cao, M. (2012) Lianas as structural parasites: a re-evaluation. *Chinese Science Bulletin*, **57**, 307–312.

Toledo-Aceves, T. & Swaine, M.D. (2008) Effect of lianas on tree regeneration in gaps and forest understorey in a tropical forest in Ghana. *Journal of Vegetation Science*, **19**, 717–728.

Tovar Pérez, A. (2007) Patrones fenológicos y composición de especies de lianas presentada en tres estadios de sucesión vegetal en un bosque tropical seco de Chamela-Cuixmala, Jalisco, México. Undergraduate thesis, Universidad Michoacana de San Nicolás de Hidalgo.

Turner, M.G., Baker, W.L., Peterson, C.J. & Peet, R.K. (1998) Factors influencing succession: lessons from large, infrequent natural disturbances. *Ecosystems*, **1**, 511–523.

Vandermeer, J., de la Cerda, I.G., Boucher, D. et al. (2000) Hurricane disturbance and tropical tree species diversity. *Science*, **290**, 788–791.

Vargas, W. (2012) Los bosques secos del Valle del Cauca, Colombia: una aproximación a su flora actual. *Biota Colombiana*, **13**, 102–164.

Webb, L.J. (1958) Cyclones as an ecological factor in tropical lowland rain-forest, North Queensland. *Australian Journal of Botany*, **6**, 220–228.

Wright, S.J. (2005) Tropical forests in a changing environment. *Trends in Ecology and Evolution*, **20**, 553–560.

Yorke, S.R., Schnitzer, S.A., Mascaro, J. et al. (2013) Increasing liana abundance and basal area in a tropical forest: the contribution of long-distance clonal colonization. *Biotropica*, **45**, 317–324.

Yuan, C.-M., Liu, W.-Y., Tang, C.Q. & Li, X.-S. (2009) Species composition, diversity, and abundance of lianas in different secondary and primary forests in a subtropical mountainous area, SW China. *Ecological Research*, **24**, 1361–1370.

Chapter 11

BIOGEOGRAPHICAL PATTERNS OF LIANA ABUNDANCE AND DIVERSITY

Saara J. DeWalt,[1] Stefan A. Schnitzer,[2,3] Luciana F. Alves,[4] Frans Bongers,[5] Robyn J. Burnham,[6] Zhiquan Cai,[7] Walter P. Carson,[8] Jérôme Chave,[9] George B. Chuyong,[10] Flavia R. C. Costa,[11] Corneille E. N. Ewango,[5,12] Rachael V. Gallagher,[13] Jeffrey J. Gerwing,[14] Esteban Gortaire Amezcua,[15] Terese Hart,[16] Guillermo Ibarra-Manríquez,[17] Kalan Ickes,[1] David Kenfack,[18] Susan G. Letcher,[19] Manuel J. Macía,[20] Jean-Remy Makana,[16] Agustina Malizia,[21] Miguel Martínez-Ramos,[17] Joseph Mascaro,[22] Chellam Muthumperumal,[23] S. Muthuramkumar,[24] Anselmo Nogueira,[25] Marc P. E. Parren,[26] Narayanaswamy Parthasarathy,[27] Diego R. Pérez-Salicrup,[17] Francis E. Putz,[28] Hugo G. Romero-Saltos,[29] M. Sridhar Reddy,[30] Moses Nsanyi Sainge,[31] Duncan Thomas,[32] and Juliano van Melis[33]

Ecology of Lianas, First Edition. Edited by Stefan A. Schnitzer, Frans Bongers, Robyn J. Burnham, and Francis E. Putz.

Companion Website: www.wiley.com/go/schnitzer/lianas

[1]Clemson University, USA
[2]University of Wisconsin-Milwaukee, USA
[3]Smithsonian Tropical Research Institute, Panama
[4]Instituto Ciências agrárias (IAC), Departamento de Botânica, Campinas, Brasil
[5]Wageningen University and Research Centre, The Netherlands
[6]University of Michigan, USA
[7]Chinese Academy of Sciences, China
[8]University of Pittsburgh, USA
[9]CNRS and Université Paul Sabatier, France
[10]University of Buea, Cameroon
[11]Instituto Nacional de Pesquisas da Amazônia – INPA, Brazil
[12]Wildlife Conservation Society (WCS)-DR Congo and Centre de Formation et de Recherche en Conservation Forestière (CEFRECOF), Reserve de Faune à Okapis, DR, Congo
[13]Macquarie University, Australia
[14]Portland State University, USA
[15]Botanist and Consultant, Quito, Ecuador
[16]Wildlife Conservation Society, DR, Congo
[17]Universidad Nacional Autónoma de México, México
[18]Center for Tropical Forest Science- Smithsonian Institution Global Earth Observatory, USA
[19]Purchase College (SUNY), USA
[20]Universidad Autónoma de Madrid, Spain
[21]Instituto de Ecología Regional (IER), Argentina
[22]American Association for the Advancement of Science, USA
[23]Madurai Kamaraj University, India
[24]V.H.N.S.N. College, India
[25]Universidade de São Paulo, Brazil
[26]Tropenbos International Congo-Basin Programme, Cameroon
[27]Pondicherry University, India
[28]University of Florida, USA
[29]Yachay Tech University, Liudad del conocimiento, Urcuquí, Ecuador
[30]Yogi Vemana University, India
[31]Tropical Plant Exploration Group, Cameroon
[32]Washington State University, USA
[33]University of Campinas, Brazil

OVERVIEW

Liana abundance and species diversity are higher in tropical than temperate forests, but substantial variation exists in liana community structure within tropical forests, particularly among biogeographical regions and along gradients in altitude, precipitation, and edaphic characteristics. We used liana surveys from across the world, which we compiled into the Global Liana Database (GLD), to examine patterns in liana abundance and diversity among continents, between islands and continental regions, and along climatic gradients. We focused particularly on the change in tropical liana abundance and diversity with mean annual precipitation and dry season length, both of which appear to be strong predictors of liana community structure.

Tropical sites in the GLD provide a wide environmental gradient, spanning gradients in mean annual rainfall of 850–7000 mm and seasonality (dry season length) of 0–7 months. The GLD presents a higher resolution alternative to the global dataset of 0.1 ha transects that Alwyn Gentry compiled in the 1980s and 1990s because the replication at each site is much greater, thus yielding a better estimate of liana abundance and diversity per site.

Using the GLD, we found that neotropical and African forests have almost twofold greater liana density than forests in Asia, and that African liana diversity is 50% greater than sites in South America and 140% greater than sites in Central America and Asia, after accounting for climatic conditions. Within each biogeographical region, liana density declines with increasing mean annual precipitation and increases with seasonality in precipitation. Liana diversity, in contrast, shows a tendency for a hump-shaped relationship with mean annual precipitation but no relationship with dry season length. Differences in liana density and diversity among biogeographical regions in the GLD mirror those found in Gentry's global dataset, but patterns of abundance and diversity in relation to mean annual precipitation and dry season length did not. Documentation of pantropical biogeographical patterns in liana abundance and diversity is a leap forward in the study of liana ecology because it provides the basis for the generation of hypotheses about pantropical drivers of liana proliferation and speciation.

INTRODUCTION

Tropical forests are well known for their high liana abundance and diversity relative to temperate forests (Gentry 1982; 1991). Lianas in mature lowland tropical forests are particularly abundant and diverse, but substantial variation exists in the structural importance and diversity of lianas among these forests (Gentry 1991; DeWalt & Chave 2004; Schnitzer 2005; DeWalt et al. 2010). For example, lianas compose 2–32% of woody stems and 9–35% of woody species, when comparing similarly sized stems (DeWalt & Chave 2004; van der Heijden & Phillips 2008, 2009; Schnitzer et al. 2012; Schnitzer et al., Chapter 7 in this volume). The abundance and diversity of lianas, like other plant growth forms, are affected by numerous processes, which lead to biogeographical differences among continents and islands, as well as variation along altitudinal, climatic, and edaphic gradients. Understanding the processes responsible for the abundance and distribution of organisms is a central goal in ecology. In this chapter, we examine pantropical patterns of liana abundance and species diversity and their correlates with climatic characteristics to gain insight into which processes are important for the distribution of tropical lianas.

Our analyses follow from the standard sampling protocol of liana diversity and abundance used by Alwyn Gentry in the 1980s and 1990s, which allowed for the first direct comparisons in these metrics for forests around the world (Gentry 1988; 1991). Specifically, this protocol enabled the first assessments of how liana abundance and species richness change with soil fertility and precipitation on regional and global scales (Gentry 1988; Clinebell et al. 1995; Schnitzer 2005; van der Heijden & Phillips 2008, 2009). Although the magnitude of Gentry's tremendous contributions should not be diminished, his transects were narrow (2 m × 50 m), and greater liana density and species richness are typically found in narrower rather than wider rectangles or in square plots, particularly for small plots (Schnitzer et al. 2006, unpublished data). In addition, Gentry used a fairly large size cutoff for lianas (2.5 cm diameter), and his transects encompassed small areas (10 transects each of 0.01 ha, equaling 0.1 ha at each site). Because lianas are typically clumped and most liana stems are in smaller size classes (Putz 1984; Campbell & Newbery 1993; Ibarra-Manríquez & Martínez-Ramos 2002; Dalling et al. 2012; Schnitzer et al. 2012), variance in liana density and species diversity is high among small plots. Not surprisingly, large differences have been found in liana abundance between the Gentry data and those gathered from larger plots, as well as wider plots with the same total sampled area as the Gentry transects (DeWalt & Chave 2004). Therefore, to test the correlates and possibly drivers of liana abundance and distributions, it is essential to use surveys from much larger sampling plots than those represented in the original Gentry datasets.

To this end, we have created the Global Liana Database (GLD) (http://www.LianaEcologyProject.com/global liana metadata). The GLD includes plot, species, and individual-level information for lianas surveyed in more than 30 tropical, subtropical, and temperate forest locations. Most of the sites include plots greater than 0.1 ha. In the GLD, we include only liana-centered studies designed to document liana abundance or diversity in fixed areas; excluded are tree-focused studies that primarily document liana loads and associations between liana abundance and tree demographics. An earlier version of the GLD, which included 24 sites, was used to examine relationships among density and basal area of lianas and a variety of abiotic, edaphic, and geographic factors (DeWalt et al. 2010).

To examine how climatic conditions are associated with liana density and diversity, we focus our analyses on old-growth continental forests (including continental-fragment islands such as Borneo) at <1050 m in elevation and with mean annual precipitation >850 mm yr^{-1}. These sites therefore include tropical dry broadleaf forest and tropical moist broadleaf forest. As of May 2013, the GLD included 29 such sites for which we can examine patterns of liana density. Species identifications are available for 20 of the 29 sites (4 in Central America and Mexico, 5 in South America, 2 in Africa, and 9 in Asia), which allow us to examine patterns of diversity (Fig. 11.1). We compare patterns generated by the GLD with those generated by the global Gentry dataset (Phillips & Miller 2002), which includes 11 sites in Africa including Madagascar, 65 in the neotropics (mostly in South America), 13 in Asia (including Taiwan, New Guinea, and Borneo), 1 in Australia, and 1 in New Caledonia. These liana surveys were previously reported in several publications (Gentry 1991; Clinebell et al. 1995; van der Heijden & Phillips 2008, 2009). To broaden the discussion of determinants of liana density and diversity beyond climatic factors, we also comment on differences between temperate and tropical forests and between continental and island regions. To this end, we discuss sites in the GLD from temperate forests in Australia, subtropical forest in Argentina, and an island in the Caribbean, and also review the available literature.

STANDARDIZATION OF LIANA SAMPLING

Without standardized sampling methods, it is difficult to compare liana species richness, composition, or even abundance across studies. Measuring liana diameter at different locations along the stem, treatment of the ramet-genet issue, plot shape, and inclusion or exclusion of hemi-epiphytes and rattans can all affect estimates of liana abundance and diversity (Schnitzer et al. 2006, 2012; Schnitzer et al., Chapter 7 in this volume). To standardize across studies, we followed the methods of DeWalt et al. (2010), who used the allometric equations from Schnitzer et al. (2006) to adjust diameter measurements made at points other than 130 cm along the stem from the last substantial roots. This point of measurement is the suggested point of measurement for liana surveys (Gerwing et al. 2006; Schnitzer et al. 2008). To include the most sites in the GLD, we used 2.5 cm as the minimum liana diameter. For studies that noted when single lianas had multiple rooted stems, we included only the largest diameter stem of each genet or "principal stem" for the density and diversity analyses (Gerwing et al. 2006; Schnitzer et al. 2008). We excluded species in the Araceae, Arecaceae, Marcgraviaceae, Poaceae, and all ferns. Members of the Araceae and Marcgraviaceae tend to be hemi-epiphytes but are sometimes included in liana studies (Gentry 1991; Rice et al. 2004; Gerwing et al. 2006). Arecaceae (e.g., rattans) and Poaceae (e.g., climbing bamboo) are monocots, do not have secondary thickening, and often have leaf sheaths around the stem that increase the apparent diameter. We included, however, the monocot families Smilacaceae and Dioscoreaceae, which are often included in liana studies because of their persistent fibrous stem and their presumed ecological equivalence to many liana species (Gerwing et al. 2006; Schnitzer et al. 2008).

It was not possible to standardize plot shape across the disparate studies. Liana individuals meeting the above criteria were simply counted and summed for each plot. For the estimate of liana diversity (see below), we included only individuals positively identified to species or morphospecies.

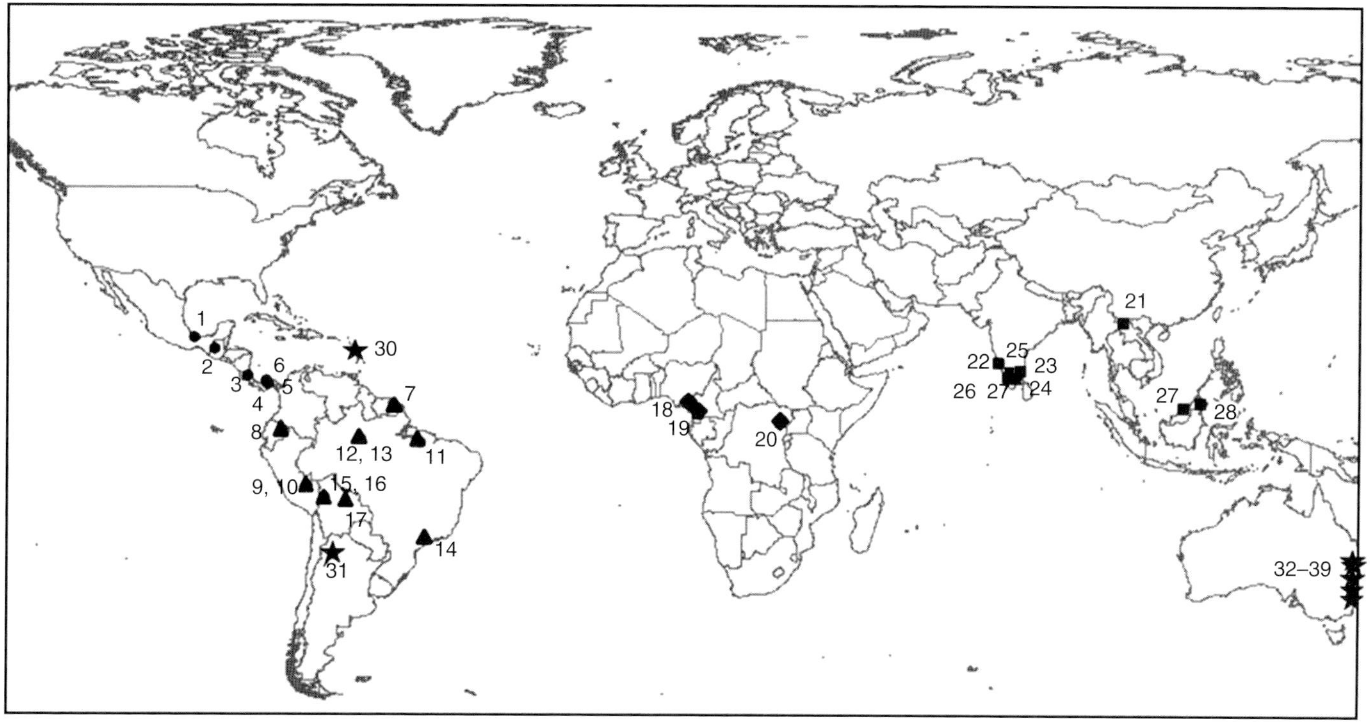

Fig. 11.1 Sites currently in the Global Liana Database and used in this study. Numbers correspond to site codes in Table 11.1. Sites are labeled with circles for Mexico and Central America, triangles for South America, diamonds in Africa, and squares in Asia. Shapes correspond to those used in the other figures. Stars denote sites not included in analyses but mentioned in the text.

Table 11.1 Density and diversity of lianas ≥2.5 cm diameter in sites included in the Global Liana Database. Density is mean number of lianas per ha at a site. Diversity is presented as Fisher's α. MAP is mean annual precipitation (mm yr^{-1}). DSL is the mean number of months per year with ≤100 mm/mo. Altitude is mean altitude (m above sea level) of plots within each site. Area is the total number of ha sampled across studies at each site. Means computed for each biogeographical region are not adjusted for differences in climate, as they are in Fig. 11.2. Source denotes the citation(s) for each study.

	Site	Location	Density (ind./ha)	Fisher's α	MAP (mm yr^{-1})	DSL (mo)	Altitude (m)	Area (ha)	Source
Central America									
1	Los Tuxtlas	Mexico	221	6.5	4725	1	100	1.00	3;22
2	Chajul	Mexico	622	14.9	3000	4	111	1.20	15
3	La Selva	Costa Rica	218	11.0	4000	0	103	1.48	7;16;23
4	Cocoli	Panama	463	–	1750	5	25	0.08	35
5	Barro Colorado Island	Panama	283	16.0	2600	4	79	47.00	7;9;37;38;40
6	Fort Sherman	Panama	300	–	3000	0	50	0.09	35
	Central America mean		*351*	*12.1*	*3179*	*2*	*78*		
South America									
7	Nouragues	French Guiana	417	–	3000	3	100	0.24	36
8	Yasuní	Ecuador	379	21.2	3000	0	243	4.00	4;17;34
9	Cocha Cashu	Peru	277	–	2099	3	357	1.40	5;7
10	Los Amigos	Peru	266	21.8	2800	3	303	1.00	5
11	Fazenda Sete	Brazil	1414	–	1700	6	100	0.29	14
12	Km41	Brazil	105	–	2650	2	75	0.60	7
13	Ducke	Brazil	220	–	2400	2	70	14.00	27
14	Ubatuba	Brazil	180	17.3	2200	3	375	1.00	1
15	Madidi	Bolivia	262	9.8	2335	3	620	1.30	18;19
16	NW Bolivia	Bolivia	275	–	2550	4	235	0.20	10
17	Oquiriquia	Bolivia	1283	7.6	1500	6	200	1.08	30
	South America mean		*462*	*15.5*	*2385*	*3*	*243*		
Africa									
18	Ebom	Cameroon	339	–	2035	5	409	26.00	29
19	Korup	Cameroon	320	35.7	5272	3	160	10.00	41
20	Ituri	D. R. Congo	464	16.3	1785	5	750	20.00	2;12;20
	Africa mean		*374*	*26.0*	*3031*	*4*	*440*		
Asia									
21	Baka	China	524	7.7	1050	7	750	0.50	6
22	Agumbe	India	121	4.8	7250	5	425	2.00	28
23	Coromandel Coast (Coastal)	India	353	5.5	1292	5	10	4.00	32
24	Coromandel Coast (Inland)	India	304	4.3	854	6	119	4.00	33
25	Eastern Ghats	India	210	4.2	1058	5	645	55.00	24
26	Valparai	India	113	10.4	3497	4	1048	4.40	25
27	Varagalaiar	India	126	9.1	1845	4	600	30.00	26
28	Lambir Hills	Borneo	122	–	2980	0	133	0.65	31;39
29	Sepilok	Borneo	136	7.1	2929	0	100	2.00	8
	Asia mean		*223*	*6.6*	*2528*	*4*	*426*		

Table 11.1 *(continued)*.

	Site	Location	Density (ind./ha)	Fisher's α	MAP (mm yr^{-1})	DSL (mo)	Altitude (m)	Area (ha)	Source
Island									
30	Dominica	Caribbean	76	1.7	–	–	492	3.50	11
Subtropical									
31	Tucumán	Argentina	323	2.1	1100	5	1000	6.00	21
32	Wooyung	Australia	333	–	1811	–	11	0.03	13
33	Iluka*	Australia	767	2.2	1522	–	20	0.03	13
Temperate									
34	Bongil Bongil*	Australia	1000	1.1	1679	–	15	0.03	13
35	Sea Acres*	Australia	967	1.2	1590	–	55	0.03	13
36	Myall Lakes*	Australia	800	1.2	1366	–	30	0.03	13
37	Bass Point	Australia	800	–	1407	–	23	0.03	13
38	Beecroft	Australia	867	–	1394	–	14	0.03	13
39	Mimosa	Australia	367	–	959	–	29	0.03	13

Note:* Fisher's α is calculated for lianas $\geq$1.0 cm diameter.

Sources:

1 Alves et al. (2012)
2 Bongers and Ewango (Chapter 3 in this volume)
3 Bongers et al. (1988)
4 Burnham (2002)
5 Burnham unpublished data
6 Cai et al. (2009)
7 DeWalt and Chave (2004)
8 DeWalt et al. (2006)
9 DeWalt et al. (2000)
10 DeWalt unpublished data
11 DeWalt and Ickes unpublished data
12 Ewango (2010)
13 Gallagher and Leishman (2012)
14 Gerwing unpublished data
15 Ibarra-Manríquez and Martínez-Ramos (2002)
16 Letcher and Chazdon (2009)
17 Macía (2008)
18 Macía (2011)
19 Macía et al. (2007)
20 Makana et al. (1998)
21 Malizia et al. (2010)
22 Martínez-Ramos unpublished data
23 Mascaro et al. (2004)
24 Muthumperumal and Parthasarathy (2010)
25 Muthuramkumar et al. (2006)
26 Muthuramkumar and Parthasarathy (2000)
27 Nogueira et al. (2011)
28 Padaki and Parathasarathy (2000)
29 Parren (2003)
30 Pérez-Salicrup et al. (2001)
31 Putz and Chai (1987)
32 Reddy and Parthasarathy (2003)
33 Reddy and Parthasarathy (2006)
34 Romero-Saltos et al. (2001)
35 Schnitzer (2005)
36 Schnitzer et al. (2006)
37 Schnitzer et al. (2012)
38 Schnitzer et al. (Chapter 7 in this volume)
39 Schnitzer unpublished data
40 Schnitzer and Carson unpublished data
41 Thomas et al. (Chapter 2 in this volume)

CALCULATING LIANA DIVERSITY

Sites may have greater species richness because of a greater underlying diversity of the species pool, greater abundance of individuals, or a combination of the two (Gotelli & Colwell 2011). It is not possible to standardize among sites by dividing observed species richness by the area sampled or the number of individuals sampled because of the non-linearity of species accumulation curves. Using the ratio of species richness to area in a small area will lead to substantial over-estimates of the number of species expected in large areas (Gotelli & Colwell 2001, 2011).

To examine broad patterns of liana species diversity with data from sites with different liana densities and areas sampled, we calculated Fisher's α, a commonly used diversity metric that is relatively unaffected by sample size (Magurran 2004). It is equal to the number of species expected to be represented by a single individual in a community; therefore, communities with many rare species are indicated by larger values of Fisher's α (Magurran 2004). Fisher's α was calculated only for plots with ≥25 individuals.

ANALYSES OF PATTERNS ASSOCIATED WITH CLIMATE AND BIOGEOGRAPHICAL REGION

We tested whether liana density and diversity differ among biogeographical regions (Central America, including Mexico, South America, Africa, and Asia) and if they vary with mean annual precipitation (MAP) or dry season length (DSL). Individual research groups provided the values for MAP and DSL. DSL was defined as the mean number of months per year with ≤100 mm of precipitation. Liana density was scaled to 1 ha for each plot/subplot. To control for sampling area, we computed Fisher's α and density for each 1-ha subplot in plots where ≥1 ha had been surveyed and then averaged the values. In addition, to avoid pseudoreplication, we calculated average Fisher's α and density among plots separated by <100 km if the climate and elevation were similar. In some cases, this meant averaging among several plots/studies conducted at a single site. We natural-log transformed the response variable of liana density to normalize residuals.

We first examined bivariate relationships between liana density or diversity and each of the two climatic variables (MAP and DSL) for each biogeographical region. For this purpose, we compared models that used the untransformed or natural-log transformed variables and fitted linear or quadratic polynomial regressions (i.e., $a \times \text{DSL}$ and $b \times \text{DSL}^2$ or $c \times \text{MAP}$ and $d \times \text{MAP}^2$, where a, b, c, and d are coefficients). We did not include interaction terms between biogeographical region and MAP or DSL because of the relatively low sample sizes for each region, particularly Africa. We selected the best-fitting bivariate model for each of MAP and DSL as the model with the lowest AIC value (data not shown). We then compared these bivariate models with ones containing both climatic variables (MAP and DSL) to determine the best-fitting overall model. All tests were conducted in R v. 3.0.1 (R Development Core Team 2009) using the "glm" function to obtain AIC values and the "lm" function for the F-statistics and multiple R^2 values.

BIOGEOGRAPHIC PATTERNS

Among tropical sites, the density of lianas ≥2.5 cm in diameter varied more than 10 times (range: 105–1414 lianas ha^{-1}; Table 11.1). African, Central American, and South American forests had more than twice the average liana density of Asian forests when calculated at mean values for MAP and DSL (Fig. 11.2A). Asian

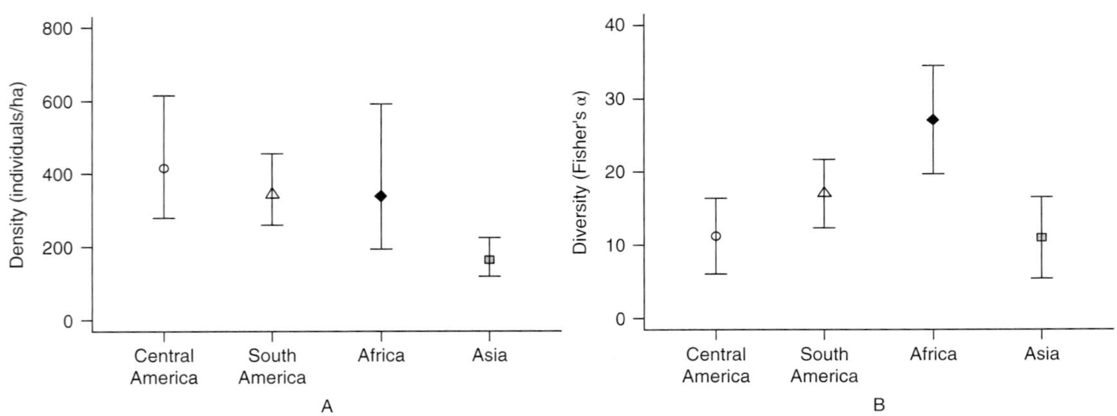

Fig. 11.2 Mean density (A) and diversity (B) of lianas ≥2.5 cm in diameter for the four biogeographical regions with MAP and DSL kept at their average values (LSmeans). Density was back-transformed. Diversity is measured as Fisher's α. Bars represent 95% confidence intervals. For density (A), there were 6 sites in Central America, 11 in South America, 3 in Africa, and 9 in Asia. For diversity (B), the number of sites included were 4 in Central America, 5 in South America, 2 in Africa, and 8 in Asia.

sites averaged 163 lianas ha^{-1}, whereas the other three biogeographical regions averaged 365 lianas ha^{-1}. Nonetheless, liana density differed greatly within biogeographical regions, particularly in South America where one site in Brazil (Gerwing & Vidal 2002) and one in Bolivia (Pérez-Salicrup et al. 2001) supported twice as many lianas as the other South American sites. Liana density did not differ among Africa, Central America, and South America even when these two high-density sites in South America were removed from the analysis.

Liana diversity, as measured by Fisher's α, varied almost nine times among sites (range: 4.2–35.7) and also differed among biogeographical regions. African sites had 59% greater diversity than South American ones, and 140% greater diversity than Central American and Asian sites at similar levels of MAP and DSL (Fig. 11.2B). Only two sites represented African forests, which is unlikely to be representative of all of Africa. The deficits in the data notwithstanding, the underlying diversity of the liana species pool appears greatest in Africa and South America.

Patterns of liana density and diversity among biogeographical regions in the GLD were similar to those reported by Gentry (1991). In Gentry's global analysis, African forests supported much higher liana densities than Australasian forests and more than neotropical forests. In addition, species richness was roughly equivalent in African and neotropical forests and much greater than in Asian sites (Gentry 1991). In both the GLD and Gentry datasets, Asian forests appear to contain substantially fewer lianas and lower liana diversity than other biogeographical regions. The GLD suggests that Central American sites have lower diversity than South American sites, when controlling for MAP and DSL. Thus far, Central and South American forests have not been directly compared in Gentry's dataset.

PATTERNS RELATED TO PRECIPITATION

Liana density

Liana abundance was negatively associated with mean annual precipitation and positively associated with dry season duration using the GLD dataset (Fig. 11.3; Table 11.2). Liana density declined most quickly with

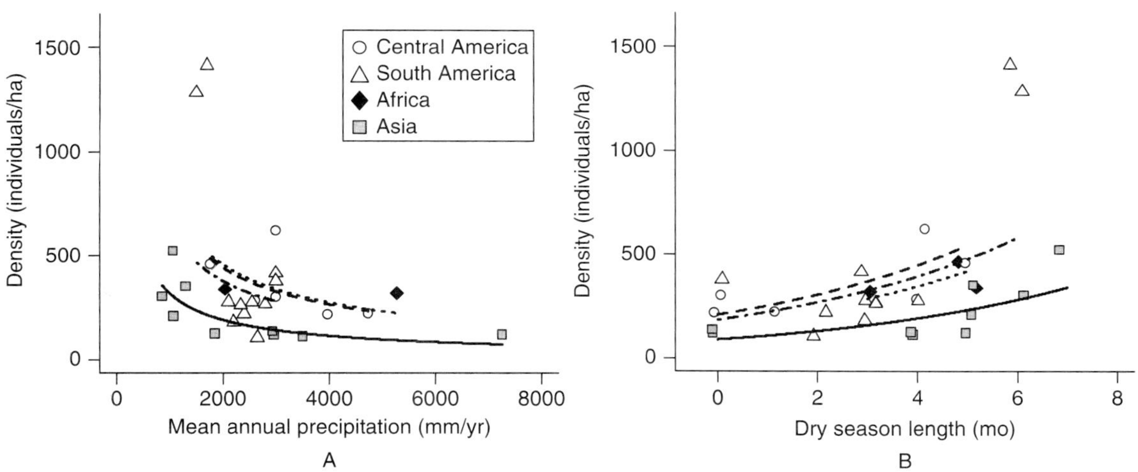

Fig. 11.3 Bivariate relationships between liana density and (A) mean annual precipitation and (B) dry season length for the four biogeographical regions. Each point represents the liana density for a site listed in Table 11.1. In (A), the point for Ituri (Africa) is not visible behind Cocoli (Central America) at 1750 mm yr^{-1}. The points in (B) are jittered so that they are visible. Lines represent the regression equations calculated from the best-fitting bivariate model (MAP or DSL) for Central America (dashed), South America (dashed and dotted), Africa (dotted), and Asia (solid). In (A), the regression lines for Central America and Africa overlap. In (A), density = *intercept* × $MAP^{-0.739}$ with *intercept* = 121,176 for Central America; 103,708 for South America; 125,713 for Africa; and 52,300 for Asia. In (B), density = *intercept* × exp(0.193[DSL]) with *intercept* = 207 for Central America; 182 for South America; 160 for Africa; and 88 for Asia.

Table 11.2 Model coefficients and *t*-values from the best-fitting models explaining log(density) and Fisher's α diversity of lianas. For each variable, we examined bivariate models for the MAP or DSL parameters as well as overall models potentially including both parameters. All models also included biogeographical region. Multiple R^2 values are provided for each model.

Variable	R^2	Coefficient	t^a	Parameter
(A) Density				
	0.45			MAP only
		−0.739	−3.44*	log(MAP)
	0.52			DSL only
		0.193	4.10**	DSL
	0.57			Overall
		−0.402	−1.72	log(MAP)
		0.140	2.56*	DSL
(B) Diversity				
	0.71			MAP only
		7.24e-03	2.07•	MAP
		−8.34e-07	−1.88•	MAP^2
	0.64			DSL only
		−0.76	−1.10	DSL
	0.71			Overall
		7.24e-03	2.07•	MAP
		−8.34e-07	−1.88•	MAP^2

Notes: •$0.05 < P < 0.1$, *$P < 0.01$, **$P < 0.001$
[a] *df* for each model parameter for (A) density are 1,24 for the MAP and DSL bivariate models, and 1,23 for the overall model. *df* for (B) diversity are 1,13 for the MAP and overall models, and 1,14 for the DSL model.

increases in precipitation at the lowest MAP values in the dataset. For each biogeographical region, liana density was predicted to be 2.7 times as high in seasonally dry forests with 5-month dry seasons than in rain forests with no dry months. The best-fitting overall model for density, which explained 57% of the variance, included biogeographical region, log(MAP), and DSL (Table 11.2). In addition, the length of the dry season explained more of the variation in liana density than mean annual precipitation.

The clear relationship between rainfall patterns and liana density in the GLD contrasts with previous reports based on the Gentry data. Using data solely from Gentry's neotropical dataset, neither Gentry (1982, 1991) nor van der Heijden and Phillips (2008) found an association between liana density and MAP or DSL. In contrast, Schnitzer (2005) found a significant negative relationship between liana density and MAP using all of Gentry's data from neotropical and paleotropical sites as well as his own data from the steep rainfall gradient across the isthmus of Panama. The lack of a relationship between liana abundance and climatic variables in Gentry's dataset appears, for some reason, restricted to the neotropics. In the GLD, the negative relationship between liana density and MAP and the positive relationship between liana density and DSL are apparent for both neotropical and paleotropical sites (Fig. 11.3A). Gentry's neotropical dataset covers longer gradients in MAP than the GLD (Gentry: 400–9000 mm yr^{-1}; GLD: 850–7250 mm yr^{-1}) and DSL (Gentry: 0–11 mo; GLD: 0–7 mo), and includes sites located on unusual soil types (e.g., floodplains and white sands) that may obscure the relationship with MAP and DSL. Further analyses of both datasets are needed to determine the source of the discrepancy (see below). In addition, the GLD requires more liana surveys in areas on the low and high ends of the gradients in MAP and DSL.

Liana diversity

Species diversity appeared to have a hump-shaped relationship with MAP but was not related to DSL in the GLD analysis (Fig. 11.4; Table 11.2). The best-fit model explained 71% of the variance in Fisher's α, but DSL did

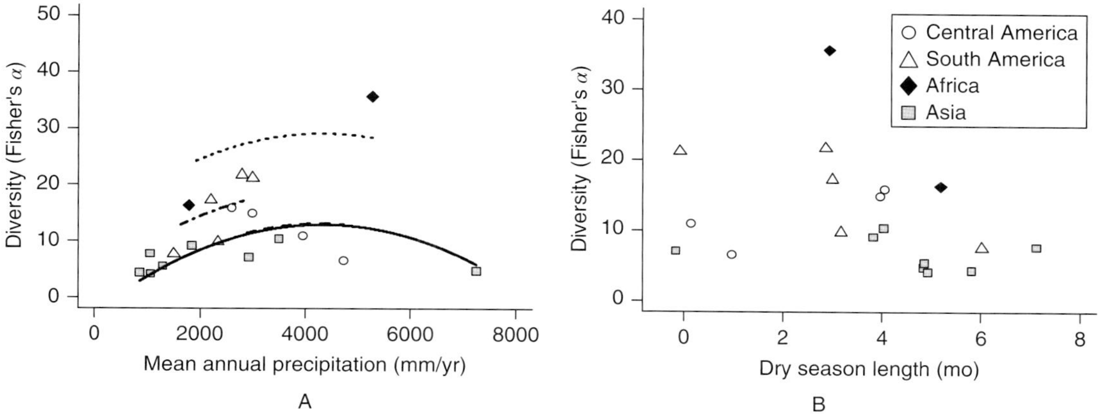

Fig. 11.4 Bivariate relationships between liana diversity (Fisher's α) of lianas $\geq$2.5 cm diameter and (A) mean annual precipitation and (B) dry season length for sites of Africa, Central America, South America, and Asia that contained at least 25 liana individuals. Each point represents the mean Fisher's α as calculated over individual plots at a site. Lines represent the regression equations calculated from the best-fitting bivariate model (MAP or DSL) for Central America (dashed), South America (dashed and dotted), Africa (dotted), and Asia (solid). In (A), the regression lines for Central America and Asia overlap. The points in (B) are jittered so that they are visible. In (A), Fisher's $\alpha = intercept + 0.0072(MAP) - 0.0000008(MAP^2)$ with $intercept = -2.53$ for Central America; 3.31 for South America; 13.4 for Africa; and -2.77 for Asia. Note that MAP and MAP^2 were only marginally significant in the model ($P = 0.06$ and 0.08, respectively; Table 11.2). In (B), the DSL parameter was not significantly different from 0 (Table 11.2).

not contribute significantly and MAP and MAP^2 were only marginally significant (Table 11.2). Interestingly, the Asian plots in the GLD spanned 100% of the range in the dataset of MAP (850–7250 mm yr^{-1}) and DSL (0–7 mo) but spanned only 20% of the range in Fisher's α (Asia: 4.2–10.4; GLD: 4.2–35.7).

Thus far, only the neotropical Gentry dataset has been used to study the relationship between liana species richness and climatic variables, and therefore an analysis using the pantropical dataset is awaiting. Climatic variables are more strongly associated with species richness in Gentry's neotropical dataset than in the GLD. When differences in density are accounted for, greater liana species richness was positively, linearly related to MAP and negatively related to DSL in Gentry's neotropical dataset (van der Heijden & Phillips 2009). In addition, DSL explained more variation in species richness than MAP (van der Heijden & Phillips 2009). In the GLD, there is an indication that Fisher's α diversity may be greatest at intermediate levels of rainfall but appears unrelated to dry season length. Therefore, in contrast with the Gentry dataset, neotropical diversity in the GLD dataset does not appear to increase linearly with MAP or negatively with DSL.

GLD versus the Gentry dataset

Why do the relationships between liana abundance and diversity and climatic variables differ between Gentry's dataset and those of the GLD? Because Gentry's data were conducted in very small areas and with narrow transects, they may be a poor estimate of liana density or diversity. Using the fairly homogeneous 50-ha plot on Barro Colorado Island (BCI), Panama, Schnitzer et al. (2012; unpublished data) reported that the coefficient of variation for liana density decreased considerably up to 1 ha, indicating that samples <1 ha are likely a poor representation of forest-wide liana density. The minimum sampling area for a robust estimate of species richness was even greater; >5 ha was necessary for a precise estimate of plot-level liana diversity based on the BCI 50-ha plot (Schnitzer et al. 2012). The large areas sampled for each site in the GLD relative to those in Gentry's dataset likely include areas of both high and low liana density and species richness. In contrast, the small number of replicates and limited total areas sampled for each site in the Gentry dataset may fail to provide an accurate and representative estimate of site-wide liana density and richness.

Alternatively, the limited number of sites in the GLD compared to the large number in Gentry's dataset, particularly in the neotropics, may obscure the true relationships. Nonetheless, none of these possibilities explain why climatic variables were strongly related to liana diversity but not density in Gentry's dataset for the neotropics, with the opposite pattern found in the GLD for the neotropics and paleotropics. Assuming that there are no major errors in either dataset, it could be that processes shaping liana communities differ among biogeographical regions. We can only test this hypothesis after more sites are added to the GLD, particularly in western and eastern Africa and at the ends of the gradients in climatic variables. These additions would then allow us to examine whether relationships with climatic variables differ among regions (i.e., a region × MAP or region × DSL interaction).

TROPICAL ISLANDS

Relatively few liana inventories are available for tropical islands, but the few surveys conducted in the Caribbean suggest that Puerto Rico, Jamaica, and Dominica support fewer lianas and lower species diversity than mainland tropical forests (Gentry 1991; Rice et al. 2004; DeWalt & Ickes, unpublished data; Florens et al. 2012). Across 17 0.25-ha plots that span gradients of precipitation and hurricane disturbance on the island of Dominica, the average liana density was 76 individuals ≥2.5 cm diameter ha^{-1}, with a Fisher's α of 1.7 (Table 11.1). Dominica has roughly one-third of the liana density and only one-fourth of the liana diversity compared to Los Tuxtlas, Mexico, a continental rain forest site at similar latitude.

On Caribbean islands, and perhaps other oceanic islands, it appears that non-liana climbers are abundant. On Puerto Rico, about half of the climbers surveyed in tropical wet forest were species of *Marcgravia* (Rice et al. 2004), which are considered "lianescent hemiepiphytes," "hemiepiphytic lianas," or "epiphytic lianas" because they establish as seedlings in the ground, climb with adventitious roots, and can become facultative epiphytes if their climbing stems degenerate. In some cases, epiphytic *Marcgravia* can produce new aerial roots and reestablish contact with the ground (DeWalt & Ickes, pers. obs.). On Dominica, species in two genera within the Marcgraviaceae, *Marcgravia* and *Schwartzia*, were particularly abundant (DeWalt & Ickes, pers. obs.), and, in some sites, almost every tree ≥10 cm diameter had a least one *Marcgravia umbellata* stem/root appressed to the trunk. These non-liana climbing species contribute to forest structure, but they are difficult to census using methods designed for lianas (e.g., Gerwing et al. 2006; Schnitzer et al. 2008).

TEMPERATE AND SUBTROPICAL FORESTS

Lianas are more diverse and typically more abundant in tropical than temperate forests, with subtropical forests in between (Gentry 1982, 1991; Schnitzer & Bongers 2002; Schnitzer 2005; Ladwig & Meiners, Chapter 15 in this volume). Unfortunately, the GLD currently includes few subtropical or temperate forests. The only samples from temperate forests are from Australia, and these plots are small (0.03 ha each; Gallagher & Leishman 2012), such that the density estimates for 1 ha are likely overinflated. The subtropical montane forest of Argentina and Australia support relatively high liana density (Argentina = 323 individuals ha^{-1}; Australia = 474 individuals ha^{-1}) but low diversity (Argentina Fisher's $\alpha = 2.1$; Australia Fisher's $\alpha = 2.2$ for lianas ≥1.0 cm at the Iluka site). Temperate forests of Australia had similar liana density and even lower diversity (800 individuals ha^{-1}; Fisher's $\alpha = 1.2$ for lianas ≥1.0 cm). Thus, even liana-rich and liana-dense temperate and subtropical forests tend to have only a fraction of the diversity of tropical forests. Gentry (1991) also found that temperate forests have few liana species. He further noted that temperate forests of the southern hemisphere, such as the Valdivian forests of Chile, tend to have double the liana density and species richness as northern hemisphere temperate forests. In North America, liana abundance appears to be positively related to disturbance and is particularly abundant along edges (Londré & Schnitzer 2006). However, liana densities vary substantially among North American broadleaf forest stands, even those located in close proximity to each other (Londré & Schnitzer 2006). In summary, other than disturbance, little is known about the climatic and other factors that drive liana density and diversity in temperate forests, and more sites in temperate, as well as subtropical, forests of the northern and southern hemispheres are needed in the GLD.

CONCLUSION AND FUTURE DIRECTIONS

The GLD is becoming a valuable dataset for determining biogeographic patterns of liana abundance and diversity. Furthermore, the GLD can be used to examine additional interesting relationships such as change in the liana community among forests and continents, and across abiotic gradients. For instance, we could examine how liana phylogenetic and functional composition differs among biogeographical regions and changes along climatic gradients. These kinds of analyses can be accomplished only with a large database, such as the GLD, that has species identifications, individual stem measurements, and relatively large plot sizes. In particular, we could test whether certain climbing types are more common under particular climatic regimes and verify that root climbers are more common in wetter, less seasonal temperate and tropical forests, as was found using Gentry's global dataset (Durigon et al. 2013). We could also use the GLD to test liana community-assembly mechanisms (*sensu* Webb et al. 2002). However, many more sites are needed in the GLD, particularly in Africa, where we have few sites and liana diversity is high (e.g., Bongers and Ewango, Chapter 3 in this volume; Thomas et al., Chapter 2 in this volume), and in Asia in areas outside of India, such as Peninsular Malaysia, Vietnam, and Cambodia.

As more sites are added to the GLD and information about trees also is added to the liana datasets, we can begin to examine whether lianas and trees co-vary in their patterns of abundance and diversity. Swaine and Grace (2007) found that liana species represent a greater percentage of the flora as annual rainfall decreases along a climatic gradient in Ghana. We could test whether lianas represent a greater proportion of the woody plant density or diversity in sites with lower MAP and greater DSL with plot-level surveys of both lianas and trees. Gentry's dataset suggests that the strength of the relationship between climatic variables and species richness in the neotropics is not as strong for lianas as it is for trees and shrubs (Gentry 1982; van der Heijden & Phillips 2009). We have yet to compile comparable diversity estimates for trees at sites for which we have liana diversity estimates to be able to conduct an independent assessment of this pattern.

Additional questions that we plan to address include: why do liana abundance and diversity vary with climatic factors (DeWalt et al. 2010; Šímová et al. 2011), and why are particular biogeographical regions especially liana-rich or liana-poor? Functional traits such as deeper roots, more efficient water transport, and larger xylem vessels may promote liana proliferation in drier areas in which trees are less competitive (Schnitzer 2005; Swaine & Grace 2007), though the ubiquity of such traits among the hundreds of liana species still needs to be tested. If there are contrasting patterns between liana and tree abundance across climatic gradients, as found by Schnitzer (2005), determination of the mechanisms for these differences may give insight into the factors that control liana abundance across broad abiotic gradients. Furthermore, if lianas continue to increase over time in abundance, productivity, and biomass in neotropical forests (reviewed by Schnitzer & Bongers 2011; Schnitzer, Chapter 30 in this volume; Schnitzer et al., Chapter 7 in this volume), then determination of the mechanisms responsible for liana distribution will help explain why lianas are increasing in tropical forests. It will also be critical to predict the forest types where lianas will play increasing ecological roles in the future. Building the GLD by adding more liana surveys with species identifications from tropical continental forests (in areas such as Amazonian Brazil, Queensland Australia, West Africa, East Africa, and mainland areas of Southeast Asia); tropical islands; and temperate forests will help to document these important patterns and may enable us to determine the evolutionary, biogeographical, and ecological processes that influence liana density and diversity.

ACKNOWLEDGMENTS

The Global Liana Database was begun as part of the Lianas and Tropical Forest Dynamics Working Group supported by the National Center for Ecological Analysis and Synthesis (NCEAS), a center funded by the National Science Foundation of the United States (NSF, Grant #EF-0553768), the University of California, Santa Barbara, and the State of California. Helene Muller-Landau contributed substantially to the creation and updating of the GLD. Data collection and management were funded by many organizations, including multiple US National Science Foundation grants, the HSBC Climate Partnership, the National Science Foundation of China, Wageningen University, Clemson University, the Department of Science and Technology of the Government of India, the World Wildlife Fund (Brazil), USAID, the Organization for Tropical Studies, the British Ecological Society,

CONACyT (Mexico), the European Commission, Fondo Mexicano para la Conservación de la Naturaleza, French Investissement d'Avenir grants of the ANR, BOLFOR, PAPIIT-UNAM, the University of Connecticut, the Center for Tropical Forest Science, the Smithsonian Tropical Research Institute, the University of Wisconsin-Milwaukee, International Cooperative Biodiversity Groups, Department of Biotechnology (DBT) of New Delhi, the Bioresources Development and Conservation Programme-Cameroon, Ministry of Environment and Forests-Cameroon, the Centre National de la Recherche Scientifique, UNESCO, Tropenbos International, the University of Miami, the Pontificia Universidad Católica del Ecuador, REPSOL-YPF, the Andrew W. Mellon Foundation, CONICET (Argentina), the Brazilian National Council for Science Development, and the Brazilian Long-Term Ecological Research Program.

REFERENCES

Alves, L.F., Assis, M.A., van Melis, J. et al. (2012) Variation in liana abundance and biomass along an elevational gradient in the tropical Atlantic Forest (Brazil). *Ecological Research*, **27**, 323–332.

Bongers, F., Popma, J., Meave del Castillo, J. & Carabias, J. (1988) Structure and floristic composition of the lowland rain forest of Los Tuxtlas, Mexico. *Vegetatio*, **74**, 55–80.

Burnham, R.J. (2002) Dominance, diversity and distribution of lianas in Yasuní, Ecuador: who is on top? *Journal of Tropical Ecology*, **18**, 845–864.

Cai, Z.-Q., Chen, Y.J., Wen, B. et al. (2009) Liana communities in three tropical forests in Xishuangbanna, South-west China. *Journal of Tropical Forest Science*, **21**, 252–264.

Campbell, E.J.F. & Newbery, D.M. (1993) Ecological relationships between lianas and trees in lowland rain forest in Sabah, East Malaysia. *Journal of Tropical Ecology*, **9**, 469–490.

Clinebell, R.R. II., Phillips, O.L., Gentry, A.H. et al. (1995) Prediction of Neotropical tree and liana species richness from soil and climatic data. *Biodiversity and Conservation*, **4**, 56–90.

Dalling, J.W., Schnitzer, S.A., Baldeck, C. et al. (2012) Resource-based habitat associations in a neotropical liana community. *Journal of Ecology*, **100**, 1174–1182.

DeWalt, S.J. & Chave, J. (2004) Structure and biomass of four lowland neotropical forests. *Biotropica*, **36**, 7–19.

DeWalt, S.J., Ickes, K., Nilus, R. et al. (2006) Liana habitat associations and community structure in a Bornean lowland tropical forest. *Plant Ecology*, **186**, 203–216.

DeWalt, S.J., Schnitzer, S.A., Chave, J. et al. (2010) Annual rainfall and seasonality predict pan-tropical patterns of liana density and basal area. *Biotropica*, **42**, 309–317.

DeWalt, S.J., Schnitzer, S.A. & Denslow, J.S. (2000) Density and diversity of lianas along a chronosequence in a central Panamanian tropical forest. *Journal of Tropical Ecology*, **16**, 1–19.

Durigon, J., Durán, S.M. & Gianoli, E. (2013) Global distribution of root climbers is positively associated with precipitation and negatively associated with seasonality. *Journal of Tropical Ecology*, **29**, 357–360.

Ewango, C.E.N. (2010) The liana assemblage of a Congolian rainforest: diversity, structure and dynamics. Ph.D. dissertation, Wageningen University.

Florens, F.B.V., Baider, C., Martin, G.M.N. & Strasberg, D. (2012) Surviving 370 years of human impact: what remains of tree diversity and structure of the lowland wet forests of oceanic island Mauritius? *Biodiversity and Conservation*, **21**, 2139–2167.

Gallagher, R.V. & Leishman, M.R. (2012) Contrasting patterns of trait-based community assembly in lianas and trees from temperate Australia. *Oikos*, **121**, 2026–2035.

Gentry, A.H. (1982) Patterns of neotropical plant species diversity. *Evolutionary Biology*, **15**, 1–84.

Gentry, A.H. (1988) Changes in plant community diversity and floristic composition on environmental and geographical gradients. *Annals of the Missouri Botanical Garden*, **75**, 1–34.

Gentry, A.H. (1991) The distribution and evolution of climbing plants, in *The Biology of Vines* (eds F.E. Putz & H.A. Mooney), Cambridge University Press, Cambridge, pp. 3–49.

Gerwing, J.J., Schnitzer, S.A., Burnham, R.J. et al. (2006) A standard protocol for liana censuses. *Biotropica*, **38**, 256–261.

Gerwing, J.J. & Vidal, E. (2002) Changes in liana abundance and species diversity eight years after liana cutting and logging in an eastern Amazonian forest. *Conservation Biology*, **16**, 544–548.

Gotelli, N.J. & Colwell, R.K. (2001) Quantifying biodiversity: procedures and pitfalls in the measurement and comparison of species richness. *Ecology Letters*, **4**, 379–391.

Gotelli, N.J. & Colwell, R.K. (2011) Estimating species richness, in *Biological Diversity: Frontiers in Measurement and Assessment* (eds A.E. Magurran & B.J. McGill), Oxford University Press, Oxford, pp. 39–54.

Ibarra-Manríquez, G. & Martínez-Ramos, M. (2002) Landscape variation of liana communities in a neotropical rain forest. *Plant Ecology*, **160**, 91–112.

Letcher, S.G. & Chazdon, R.L. (2009) Lianas and self-supporting plants during tropical forest succession. *Forest Ecology and Management*, **257**, 2150–2156.

Londré, R.A. & Schnitzer, S.A. (2006) The distribution of lianas and their change in abundance in temperate forests over the past 45 years. *Ecology*, **87**, 2973–2978.

Macía, M.J. (2008) Woody plants diversity, floristic composition and land use history in the Amazonian rain forests of Madidi National Park, Bolivia. *Biological Conservation*, **17**, 2671–2690.

Macía, M.J. (2011) Spatial distribution and floristic composition of trees and lianas in different forest types of an Amazonian rainforest. *Plant Ecology*, **212**, 1159–1177.

Macía, M.J., Ruokolainen, K., Tuomisto, H. et al. (2007) Congruence between floristic patterns of trees and lianas in a southwest Amazonian rain forest. *Ecography*, **30**, 561–577.

Magurran, A.E. (2004) *Measuring Biological Diversity*, Blackwell Scientific Publications, Oxford.

Makana, J.-R., Hart, T.B. & Hart, J.A. (1998) Forest structure and diversity of lianas and understory treelets in monodominant and mixed stands in the Ituri Forest, Democratic Republic of the Congo, in *Forest Biodiversity Research, Monitoring and Modeling: Conceptual Background and Old World Case Studies* (eds F. Dallmeier & J.A. Comiskey), Parthenon Publishing, Washington, DC.

Malizia, A., Grau, H.R. & Lichstein, J.W. (2010) Soil phosphorus and disturbance influence liana communities in a subtropical montane forest. *Journal of Vegetation Science*, **21**, 551–560.

Mascaro, J., Schnitzer, S.A. & Carson, W.P. (2004) Liana diversity, abundance, and mortality in a tropical wet forest in Costa Rica. *Forest Ecology and Management*, **19**, 3–14.

Muthumperumal, C. & Parthasarathy, N. (2010) A large-scale inventory of liana diversity in tropical forests of South Eastern Ghats, India. *Systematics and Biodiversity*, **8**, 289–300.

Muthuramkumar, S., Ayyappan, N., Parthasarathy, N. et al. (2006) Plant community structure in tropical rain forest fragments of the Western Ghats, India. *Biotropica*, **38**, 143–160.

Muthuramkumar, S. & Parthasarathy, N. (2000) Alpha diversity of lianas in a tropical evergreen forest in the Anamalais, Western Ghats, India. *Diversity and Distributions*, **6**, 1–14.

Nogueira, A., Costa, F.R.C. & Castilho, C.V. (2011) Liana abundance patterns: the role of ecological filters during development. *Biotropica*, **43**, 442–449.

Padaki, A. & Parthasarathy, N. (2000) Abundance and distribution of lianas in tropical lowland evergreen forest of Agumbe, central Western Ghats, India. *Tropical Ecology*, **41**, 143–154.

Parren, M.P.E. (2003) Lianas and logging in West Africa. Ph.D. dissertation, Wageningen University.

Pérez-Salicrup, D.R., Sork, V.L. & Putz, F.E. (2001) Lianas and trees in a liana forest of Amazonian Bolivia. *Biotropica*, **33**, 34–47.

Phillips, O.L. & Miller, J. (2002) *Global Patterns of Plant Diversity: Alwyn H. Gentry's Forest Transect Data Set*, Missouri Botanical Garden, St. Louis, MO.

Putz, F.E. (1984) The natural history of lianas on Barro Colorado Island, Panama. *Ecology*, **65**, 1713–1724.

Putz, F.E. & Chai, P. (1987) Ecological studies of lianas in Lambir National Park, Sarawak, Malaysia. *Journal of Ecology*, **75**, 523–531.

R Development Core Team (2009) R: A language and environment for statistical computing. R Foundation for Statistical Computing, Vienna, Austria. Available at URL: http://www.R-project.org.

Reddy, M.S. & Parthasarathy, N. (2003) Liana diversity and distribution in four tropical dry evergreen forests on the Coromandel coast of south India. *Biodiversity and Conservation*, **12**, 1609–1627.

Reddy, M.S. & Parthasarathy, N. (2006) Liana diversity and distribution on host trees in four inland tropical dry evergreen forests of peninsular India. *Tropical Ecology*, **47**, 109–123.

Rice, K., Brokaw, N. & Thompson, J. (2004) Liana abundance in a Puerto Rican forest. *Forest Ecology and Management*, **190**, 33–41.

Romero-Saltos, H., Valencia, R. & Macía, M.J. (2001) Patrones de diversidad, distribución y rareza de plantas leñosas en el Parque Nacional Yasuní y la Reserva Étnica Huaorani, Amazonía ecuatoriana, in *Evaluación de productos forestales no maderables en la Amazonía noroccidental* (eds J.F. Duivenvoorden, H. Balslev, J. Cavelier et al.), IBED, Universiteit van Amsterdam, Amsterdam, the Netherlands, pp. 131–162.

Schnitzer, S.A. (2005) A mechanistic explanation for global patterns of liana abundance and distribution. *American Naturalist*, **166**, 262–276.

Schnitzer, S.A. & Bongers, F. (2002) The ecology of lianas and their role in forests. *Trends in Ecology and Evolution*, **17**, 223–230.

Schnitzer, S.A. & Bongers, F. (2011) Increasing liana abundance and biomass in tropical forests: emerging patterns and putative mechanisms. *Ecology Letters*, **14**, 397–406.

Schnitzer, S.A., DeWalt, S.J. & Chave, J. (2006) Censusing and measuring lianas: a quantitative comparison of the common methods. *Biotropica*, **38**, 581–591.

Schnitzer, S.A., Mangan, S.A., Dalling, J.W., et al. (2012) Liana abundance, diversity, and distribution on Barro Colorado Island, Panama. *PLOS ONE* 7(12): e52114.

Schnitzer, S.A., Rutishauser, S. & Aguilar, S. (2008) Supplemental protocol for censusing lianas. *Forest Ecology and Management*, **255**, 1044–1049.

Šímová, I., Storch, D., Keil, P. et al. (2011) Global species-energy relationship in forest plots: role of abundance, temperature and species climatic tolerances. *Global Ecology and Biogeography*, **20**, 842–856.

Swaine, M.D. & Grace, J. (2007) Lianas may be favoured by low rainfall: evidence from Ghana. *Plant Ecology*, **192**, 271–276.
van der Heijden, G.M.F. & Phillips, O.L. (2008) What controls liana success in neotropical forests? *Global Ecology and Biogeography*, **17**, 372–383.
van der Heijden, G.M.F. & Phillips, O.L. (2009) Environmental effects on neotropical liana species richness. *Journal of Biogeography*, **36**, 1561–1572.
Webb, C.O., Ackerly, D.D., McPeek, M.A. & Donoghue, M.J. (2002) Phylogenies and community ecology. *Annual Review of Ecology and Systematics*, **33**, 475–505.

Part III

Liana–Tree Competition: Community and Ecosystem Level Effects

Chapter 12

ABOVE- AND BELOWGROUND COMPETITION BETWEEN LIANAS AND TREES

Tarin Toledo-Aceves

Instituto de Ecología, A.C., Veracruz, Mexico

Ecology of Lianas, First Edition. Edited by Stefan A. Schnitzer, Frans Bongers, Robyn J. Burnham, and Francis E. Putz.

Companion Website: www.wiley.com/go/schnitzer/lianas

OVERVIEW

Competition is one of the cornerstone processes in the study of ecology, and competition between lianas and trees has become an active area of tropical forest ecology. Lianas compete with trees for both above- and belowground resources, reducing tree physiological performance, growth, fecundity, and even survival. Lianas compete aboveground by displaying their leaves on top of tree crowns, intercepting the light and effectively reducing light availability for the plants below. Lianas compete belowground for limited soil water and nutrients. The relative importance of above- *vs.* belowground competition likely varies depending on the relative availability of resources (which varies spatially and temporally), as well as tree life-history stages. For example, when light is particularly abundant, such as in a treefall gap, lianas may compete more for belowground resources (water and nutrients). When a canopy gap appears, lianas can rapidly colonize and dominate for many years, suppressing tree seedling and sapling regeneration and delaying forest succession. Liana competitive effects during the gap regeneration phase may have a strong influence on the structure and composition of the forest. In contrast, in the forest understory, where light is limited, lianas may compete with trees more for light.

Lianas compete with trees throughout all tree life stages. For tree seedlings and saplings, lianas can negatively affect patterns of biomass allocation, growth and recruitment. For adult trees, lianas may compete for both above- and belowground resources, and liana infestation can strongly reduce tree performance. However, as more studies of competition are carried out, support for the importance of belowground negative effects has increased. Several studies now demonstrate intense belowground competitive effects of lianas on trees at the community and ecosystem levels. Experimental evaluations show that the competitive effects of lianas on trees can vary according to tree and liana species identity and that the interaction between lianas and trees is temporally dynamic. Because lianas are increasing in abundance, biomass, and productivity in many tropical forests, the intensity of competition between lianas and trees is expected to increase over time, with consequent negative effects on forest diversity, structure, and function. Research on the synergistic effects of increased forest disturbance and liana dominance on the above- and belowground competition between lianas and trees is necessary for a better understanding of the competitive effect of lianas on trees.

INTRODUCTION: LIANAS *VS.* TREES

Competition for resources is a key process in community organization and exerts a profound influence on the dynamics, composition, and biodiversity of communities (e.g., Tilman 1982; Schoener 1983). Plants require an assemblage of essential resources, which, if limited, can reduce growth, fecundity, and even survival. Lianas and trees in tropical forests compete for essentially the same resources (light, water, and nutrients). However, the interaction between lianas and trees may be a special form of competition because lianas compete with trees not only for resources, but they also exploit their host tree's architecture to improve their own light interception. Competition for light can be strong between lianas and trees, but competition for other essential resources, such as water and nutrients, is also intense (Putz 1984a; Pérez-Salicrup & Barker 2000; Pérez-Salicrup et al. 2001; Schnitzer 2005; Schnitzer et al. 2005).

Lianas can have a detrimental impact on trees throughout all the life stages of a tree. For tree seedlings and saplings, lianas can alter the patterns of biomass allocation, as well as physiological performance, and can reduce growth and recruitment (Schnitzer et al. 2005; Toledo-Aceves & Swaine 2008a; Schnitzer & Carson 2010). Exploring the effects of liana competition on tree seedlings in a semi-deciduous tropical forest in Bolivia, Pérez-Salicrup et al. (2001) reported that the removal of lianas significantly increased tree seedling growth. In a seasonally moist lowland tropical forest in Panama, Schnitzer and Carson (2010) found that removing lianas increased tree sapling recruitment by 46%, which resulted in a significant increase in tree diversity. For adult trees, lianas can affect the physiological performance (Pérez-Salicrup & Barker 2000; Tobin et al. 2012), can reduce growth rates (Grauel & Putz 2004; Van der Heijden & Phillips 2009; Ingwell et al. 2010), fecundity (Stevens 1987; Wright et al. 2005) and jeopardize survival (Phillips et al. 2002; Ingwell et al. 2010), as well as cause an increased probability of falling (Putz 1984a; Garrido-Pérez et al. 2012). Lianas can also alter the ecosystem properties of tropical forests (e.g., Schnitzer et al. 2011; van der Heijden et al. 2013, Chapter 13 in this volume). Consequently, lianas have myriad detrimental effects on trees at all tree life stages and ultimately affect forest diversity, structure, and function. Furthermore, lianas are increasing in abundance, biomass, and productivity in many tropical forests (Phillips et al. 2002; Londré & Schnitzer 2006; Schnitzer & Bongers 2011; Schnitzer, Chapter 30 in this volume), and thus the intensity of competition between lianas and trees, which is explored below, is also likely to increase.

In this chapter, the focus is mainly on the competitive effects of lianas on trees. In the first section, the current evidence for liana–tree competition is explored and summarized, with important consequences at the community level. Evidence of liana competition for various resources is described and the empirical evidence for competitive effects exerted by lianas during post-disturbance tree regeneration is detailed. The study then evaluates the relative importance of above- *vs.* belowground competition on tree performance. Throughout the chapter, it is stated that liana–tree interactions are highly dynamic and that they are subject to considerable temporal variation. Finally, the chapter concludes with a summary of the implications of liana competition on tropical forest dynamics, and identifies areas of future research that would improve our understanding of the competitive interactions between trees and lianas.

COMPETITION BETWEEN LIANAS AND TREES

Liana infestation of a large proportion of adult trees now seems to be common in tropical forests. On Barro Colorado Island, Panama, Putz (1984) found that 43% of the trees (≥20 cm dbh) were infested with lianas. Twenty-five years later in the same forest, Ingwell et al. (2010) reported that lianas were present in the crowns of nearly 75% of the canopy trees (≥20 cm dbh). In an Amazonian forest in Bolivia, 86% of the trees (≥10 cm dbh) supported lianas (Pérez-Salicrup et al. 2001), while in the tropical forest of Los Tuxtlas, Mexico, 63% of trees (≥10 cm dbh) carried at least one liana (Pérez-Salicrup & de Meijere 2005). In Malaysia, over 50% of large trees (≥20 cm dbh) were liana-infested (Putz & Chai 1987; Campbell & Newbery 1993). Lianas can also be abundant in temperate forests; for example, Ladwig and Meiners (2010) reported the presence of at least one liana on 68% of the trees in secondary oak-hickory forest plots in New Jersey, USA (see also Ladwig and Meiners, Chapter 15 in this volume). High liana infestation of trees is also common at forest edges (Williams-Linera 1990; Laurance et al. 2001; Londré & Schnitzer 2006), and likely contributes to high tree mortality in these areas. Collectively, these studies provide a strong indication of the potential for intense liana–tree competition in forests throughout the world.

Over the past 30 years, there has been an explosion of studies documenting intense competition between lianas and tree seedlings, saplings, and adults (Table 12.1; see also Schnitzer et al., Chapter 1 in this volume). The following sections summarize the current evidence for the competitive effects of lianas on seedling, sapling and adult trees.

Liana effects on tree seedlings and saplings

Tree regeneration is of vital importance because established seedlings constitute the source of the future tree community. High seedling mortality, which is typical during the early regeneration phase, can have a profound influence on the composition and diversity of the forest (Swaine 1996). Survival through the seedling stage of the tree life cycle is critical for reproductive success and many studies provide evidence for a significant reduction in tree seedling and sapling growth as a result of competition with lianas. Putz (1984) examined the effects of lianas on tree regeneration on Barro Colorado Island in Panama and reported a significant negative relationship between the average height of saplings and the abundance of lianas found in gaps. Pérez-Salicrup et al. (2001) experimentally removed lianas and reported that, after 18 months, seedlings of the trees *Clarisia ilicifolia* and *Astronium fraxinifolium* grew taller and produced more leaves than seedlings in control plots, though their survival was not greater than in the control plots. In an experiment in a tropical forest in Ghana, Toledo-Aceves and Swaine (2008a) found that one liana seedling species suppressed tree sapling biomass by 32% in *Nauclea diderichii* and 50% in *Khaya anthotheca*, while the tree *Garcinia kola* was unaffected. Thus, experimental evidence demonstrates that the competitive effects of lianas on tree seedlings and saplings can be severe and can vary with tree species identity.

The effect of lianas on tree seedlings and saplings may also be dependent on the liana species involved. For example, in Ghana, Toledo-Aceves and Swaine (2007) demonstrated that saplings of the tree *Ceiba pentandra*

Table 12.1 Correlative and experimental studies on the effects of lianas on trees.

Vegetation type or condition	Tree stage	Effect of lianas on trees	Country	Reference
Experimental field	Adult	Reduction in growth	USA	Whigham (1984)
Tropical deciduous forest	Adult	Reduction in fruit production	Costa Rica	Stevens (1987)
Experimental garden	Sapling	Reduction in growth and leaf expansion	USA	Dillenburg et al. (1993a)
Shade house	Sapling	Reduction in leaf nitrogen concentration	USA	Dillenburg et al. (1993b)
Experimental garden	Sapling	Reduction in leaf photosynthesis and modification of biomass allocation patterns	USA	Dillenburg et al. (1995)
Lowland tropical forest	Sapling	Suppression of tree regeneration	Panama	Schnitzer et al. (2000)
Tropical dry-moist forest	Seedling	Growth and leaf production suppression	Bolivia	Pérez-Salicrup (2001)
Swamp forest	Seedling and adult	Growth and regeneration suppression	Panama	Grauel & Putz (2004)
Lowland tropical forest	Sapling	Growth suppression and modification of biomass allocation patterns. Poorly developed crowns, shorter and thicker stems	Ivory Coast	Schnitzer et al. (2005)
Tropical rain forest	Seedling	Growth suppression, crown reduction and increased internode length	Ghana	Toledo-Aceves & Swaine (2007)
Shade house	Seedling	Reduction of photosynthetic rate, leaf area, and growth. Alteration of biomass allocation patterns	China	Chen et al. (2008)
Tropical rain forest	Seedling – Sapling	Growth suppression	Ghana	Toledo-Aceves & Swaine (2008a)
Shade house	Seedling	Reduction in growth	Ghana	Toledo-Aceves & Swaine (2008b)
Secondary temperate forest	Juvenile and adult	Growth suppression	USA	Ladwig & Meiners (2009)
Tropical rain forest	Adult	Growth suppression	Peru	van der Heijden & Phillips (2009)
Lowland tropical forest	Adult	Growth suppression and increased mortality	Panama	Ingwell et al. (2010)
Lowland tropical forest	Sapling	Reduction of growth, recruitment success and richness	Panama	Schnitzer & Carson (2010)
Lowland tropical forest	Adult	Reduction of sap flow velocity	Panama	Tobin et al. (2012)
Lowland tropical forest	Adult	Reduction of sap flow velocity and growth	Panama	Alvarez-Cansino et al. (unpublished data)

responded differently when planted with different species of climbers in gaps. The large canopy liana *Combretum racemosum* caused a significant reduction in *C. pentandra* height and diameter, while competition with the nitrogen-fixing woody climber *Centrosema pubescens* was associated with a significant reduction in *C. pentandra* crown area and increased internode length. Lengthening of the internodes is a response that has been interpreted as a strategy for escaping from colonization by climbers (Putz 1984b). It is also possible that extension of the internodes could be caused by etiolation in response to shading by the climber. When planted with climbers, the mean crown area of *C. pentandra* was less than a third of that in the control. The reduction in crown area was also accompanied by a reduction in the leaf area. Crown traits can be strongly related to tree sapling growth and survival (Poorter & Werger 1999; Sterck et al. 2003); the biomass gain of *Ceiba* was significantly related to its crown area. Schnitzer et al. (2005) also found that lianas affected biomass allocation in saplings of *C. pentandra* in the Ivory Coast; saplings covered by lianas developed shorter, thicker stems and a poorly develop crown. These results illustrate the range of negative effects on trees produced by very different climbers. The mechanisms that produce these different effects (or lack of them) among climber species remain to be further elucidated, but it is clear that climber/tree species pair interactions can vary widely.

The competitive effects of lianas on tree regeneration could modify the establishment capacity of tree species, acting as a selective force that could reduce the relative abundance of some species relative to others, thereby altering forest community composition (Putz 1984; Schnitzer et al. 2000; Chittibabu & Parthasarathy 2001; Schnitzer & Bongers 2002; Toledo-Aceves & Swaine 2008a). Support for this idea was provided by Schnitzer and Carson (2010), who found that experimental removal of lianas in gaps in Panama produced increased tree growth (55%), recruitment (46%), and species richness (65%). Shade-tolerant tree species appear to suffer more from lianas than pioneer trees – a phenomenon that is particularly prevalent in forest canopy gaps (Putz 1984b; Clark & Clark 1990; Schnitzer et al. 2000; Schnitzer & Carson 2010).

One of the main attributes of pioneer species is a high growth rate under high light conditions, which confers the ability to outgrow other vegetation, including trees, herbs, shrubs, and climbers. As a result of this capability, pioneer trees in gaps may be able to escape from liana colonization more readily than other groups, such as the shade-tolerant trees (Schnitzer & Carson 2010). Thus, by having a greater effect on some tree species than others, lianas may play a key role in the dynamics of tropical forests by altering patterns of tree regeneration.

The effects of lianas in gaps may last for decades, and may persist even after the gap regenerates a high canopy (Putz 1984a; Schnitzer et al. 2000; Foster et al. 2008). Schnitzer et al. (2000) proposed that lianas thus cause gap phase regeneration to follow an alternative pathway that is dominated by lianas and pioneer trees. In the absence of lianas, however, gaps regenerate "normally" and are dominated by trees (Fig. 12.1). Indeed, there are numerous reports of lianas dominating gaps in African and American tropical forests (Putz 1984a; Babweteera et al. 2000; Parren 2003; Schnitzer et al. 2004; Schnitzer & Carson 2010).

Evaluation of the impact of lianas on tree regeneration along a gradient of canopy openness shows that the light environment modifies the liana–tree interaction. In the forest understory, plant growth is light-limited and the reduction in light caused by the forest canopy may outweigh reductions caused solely by lianas. In canopy gaps, however, increased resource availability results in higher liana density and biomass, which may result in greater competition intensity (see next section). For example, Toledo-Aceves and Swaine (2008a) evaluated the effect of the liana *Acacia kamerunensis* on seedlings of the trees *Nauclea diderichii*, *Khaya anthotheca* and *Garcia kola* in the forest understory, in small (39 m^2) and large (197 m^2) canopy gaps in a tropical forest in Ghana. They found that lianas only reduced the biomass of *N. diderichii* and *K. anthotheca* in the large gaps, but that there was no detectable effect on trees in the smaller gaps or in the forest understory. Ultimately, these data support the idea that competition intensity depends on the ratio of resource supply to demand, and that competition will always be strongest for the most limiting resources (Taylor et al. 1990; Reader et al. 1994).

Liana effects on adult trees

Through competition, lianas can significantly reduce the growth rate of adult trees (Van der Heijden & Phillips 2009; Ingwell et al. 2010; Schnitzer & Carson 2010), reproductive output (Stevens 1987; Kainer et al. 2006; Wright et al. 2005) and survival (Phillips

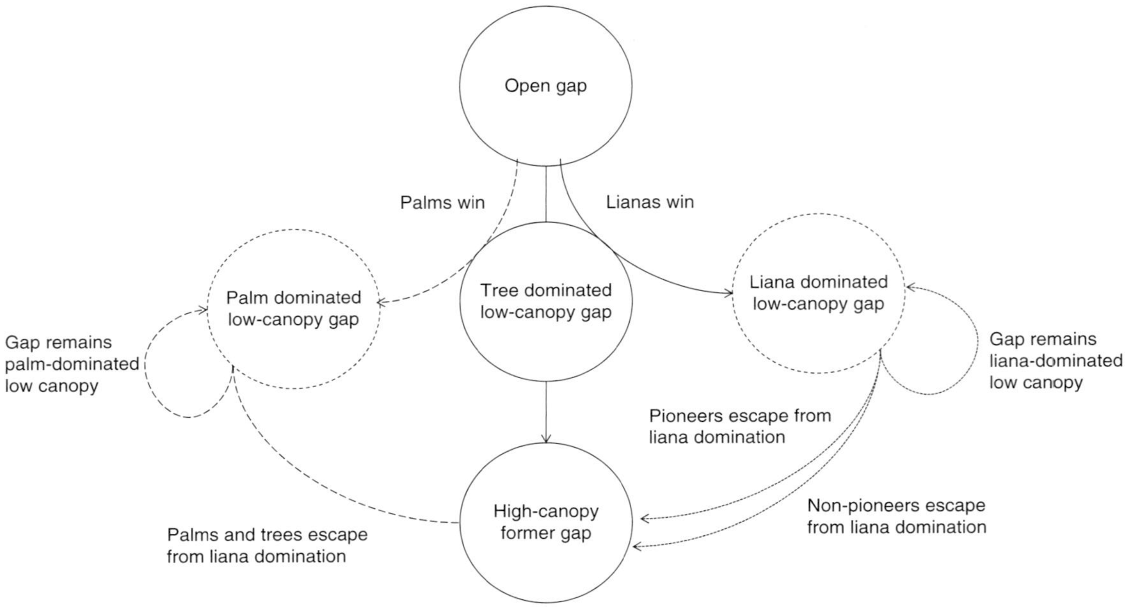

Fig. 12.1 Alternative forest gap regeneration path proposed by Schnitzer et al. (2000). In the typical gap phase regeneration model (middle), a gap is created and is dominated by rapid growing trees that will close the gap within three to six years of creation. In the liana-dominated gap pathway (right) tree seedlings and saplings are suppressed, delaying their growth and maintaining a low canopy state for an extended period (>13 years) (Schnitzer et al. 2000). The palm-dominated gap phase (left) follows a similar pattern. (Source: Schnitzer et al. 2000. Reproduced with permission of John Wiley & Sons, Ltd.)

et al. 2005; Ingwell et al. 2010), as well as affect the physiological performance (Pérez-Salicrup & Barker 2000; Tobin et al. 2012; Alvarez-Cansino et al. unpublished data) and increase the probability of treefall (Putz 1984a; Garrido-Pérez et al. 2012). In Amazonian Peru, van der Heijden and Phillips (2009) reported that by competing with their support trees, lianas reduced the growth of tree annual aboveground biomass by ~10% (see also van der Heijden et al., Chapter 13 in this volume). In French Guiana, Chave et al. (2001) reported that tree biomass was much reduced in a liana-dominated plot compared to the rest of the forest. Grauel and Putz (2004) cut lianas in *Prioria copaifera*-dominated forests in eastern Panama and found that, after five years, *P. copaifera* diameter had doubled in plots where all lianas had been cut, compared to control plots, regardless of the severity of liana infestation.

In central Panama, Alvarez-Cansino et al. (unpublished data) examined the response of seven tree species to liana cutting and reported that all species experienced increased growth and sap velocity. Likewise, Tobin et al. (2012) reported that lianas have a strong negative effect on water transport in canopy trees in a seasonal tropical moist forest in Panama. In a deciduous forest in Costa Rica, Stevens (1987) found a significant reduction in the reproduction of *Bursera simarouba* trees that were heavily loaded with lianas. In Amazonian Brazil, Kainer et al. (2006) reported that *Bertholletia excelsa* trees with lianas produced significantly fewer fruits than liana-free trees, and that 10 years following liana cutting, trees without lianas produced three times more fruits than trees with lianas (Kainer et al. 2014). Kainer and colleagues (2014) also found that the number of lianas rooted within 5 m of a host tree significantly explained fruit production, concluding that both above- and belowground competition were at work.

The detrimental effects of lianas on trees can increase tree mortality risk. In the Amazonian forest, Phillips

Canopy gap dynamics

Competition for light plays a central role in tropical forest dynamics since light is frequently the most limited resource for plants in the forest understory (Whitmore 1996). Creation of gaps in the canopy, by single or multiple tree falls, causes a sudden increase in light availability in the forest understory. Gap formation produces multiple effects; apart from the increase in irradiance, other resources that can be altered by gap formation include nutrients, soil moisture content, infection by pathogens and herbivory (Kitajima & Augspurger 1989; Denslow et al. 1998; Veenendaal et al. 1996). The ability of plants to exploit the change in resource availability in gaps is related to the degree of plasticity they exhibit mainly in response to changes in light. Plasticity of functional traits in response to resource availability allows plants to maximize the capture of resources that limit their performance (Bazzaz & Carlson 1982; Kitajima 1994). These ephemeral light-rich environments are vital for the maintenance of pioneer or light-demanding species (Brokaw 1985; Swaine & Whitmore 1988; Whitmore 1989). The high growth rates that characterize pioneers confer a competitive advantage in the rich environment created by treefall gaps.

In the typical model of gap-phase regeneration, the first stage is dominated by fast-growing species, including pioneer trees, herbs, shrubs, and climbers (Richards 1996). Liana seedlings represent a high proportion of the seedlings found in tropical forests, where they can account for around 20% of the total (Putz 1983, 1984a). According to Uhl et al. (1988), lianas could represent about 30% of all woody stems in gaps following four years of gap succession. In the second stage, when pioneer trees emerge from the seedling bank, their rapid growth can close the canopy in only 5–10 years, and the other groups subsequently tend to become more scarce. During this stage, non-pioneer light-demanding trees can also take advantage and become dominant. Eventually, more shade-tolerant species replace the canopy (Whitmore 1989). At this stage, pioneer seeds can remain dormant in the soil, while shade-tolerant species can regenerate under the forest canopy. Different factors influence this process, including the gap size (Brokaw 1985; Kennedy & Swaine 1992), the intensity of the disturbance, e.g., the removal of soil (Dickinson et al. 2000; Pinard et al. 2000), damage to the advanced regeneration (Uhl et al. 1988), soil characteristics (Ostertag 1998; Lewis & Tanner 2000), the seed bank (Dalling et al. 1998), the state of the surrounding matrix, the root and shoot resprouting and pathogens (Augspurger 1984), among others.

In natural gaps, advanced regeneration (seedlings established in the understory prior to gap creation) can often dominate regrowth (Brokaw & Busing 2000) and consists mainly of shade-tolerant species (about 97%; Uhl et al. 1988). In gaps created by timber extraction, however, advanced regeneration is damaged and the soil is disturbed and compacted by the use of heavy machinery (Pinard et al. 2000). Under these artificial conditions, gaps are dominated by pioneer vegetation (Denslow et al. 1990). In some cases, canopy opening is desirable to promote the regeneration of pioneers of commercial interest (Snook 1996; Toledo-Aceves et al. 2009). However, excessive canopy opening can lead to regeneration problems, mainly related to competition with non-tree species, and can also promote the dominance of lianas. Proliferation of lianas in these environments represents a significant threat to tree regeneration; tree seedlings and saplings in gaps are frequently covered by lianas, which can suppress tree growth and survival (Putz 1984a; Pérez-Salicrup 2001; Schnitzer & Bongers 2002; Toledo-Aceves & Swaine 2008a; Schnitzer et al. 2000). Thus, the negative effect of lianas on tree regeneration can be considerable, especially after logging in managed forests, where liana abundance can be exceptionally high (Putz 1984a; Parren 2003; Grauel & Putz 2004; Schnitzer et al. 2004).

et al. (2005) found evidence demonstrating that liana infestation strongly reduced adult tree survival; liana-infested trees were three times more likely to die than liana-free trees. Evaluating the long-term effects of lianas on trees in Panama, Ingwell et al. (2010) found a significant reduction in growth and increased mortality in trees with severe liana infestation: 42% of trees with more than 75% of the crown covered by lianas died in comparison to 21% mortality in trees without lianas.

Lianas can also cause mechanical stress to their hosts by adding weight to the tree crown. Liana infestation of the tree canopy can result in tree trunk and crown deformation, and trees may even become more susceptible to pathogens and breakage (Putz 1991). Trees can also be damaged or killed when a neighboring tree falls because lianas bind tree crowns together and, as they are pulled from a tree canopy, they can break branches and pull down neighboring trees (Putz 1991; Garrido-Pérez et al. 2012).

UNTANGLING ABOVE- AND BELOWGROUND LIANA EFFECTS ON TREES

Controlled experiments have demonstrated that lianas compete with trees for both above- and belowground resources at tree seedling, juvenile, and adult stages.

The relative contribution of above- and belowground competition between lianas and tree saplings was initially explored in a series of studies by Whigham (1984) and Dillenburg and collaborators (1993a, 1993b, 1995), in which the authors found that belowground competition with the lianas *Lonicera japonica* and *Parthenocissus quinquefolia* had a greater influence on the saplings of the tree *Liquidambar styraciflua* than aboveground competition. Even though the lianas reduced the amount of light received by the host tree, there was no clear relationship between this reduction and tree growth (Dillenburg et al. 1993b). In a disturbed tropical forest in the Ivory Coast, Schnitzer et al. (2005) found that, while competition with climbers for above- and belowground resources affected patterns of biomass allocation in tree saplings, belowground competition alone reduced tree sapling dry biomass by 83%. In a shade-house environment, Toledo-Aceves and Swaine (2008b) also found that belowground competition made a greater contribution than aboveground competition to the negative effect of the liana *Acacia kamerunensis* on tree sapling growth, regardless of tree species (Fig. 12.2). In this same study, shoot competition had no effect on seedling growth, whereas root competition reduced tree growth in biomass by 24%. Both root and shoot competition combined reduced growth in biomass by 42% relative to control trees.

In all of the studies mentioned above, water was not explicitly considered as a factor, even though water deficiency is a common limiting factor for tree seedlings and saplings in seasonal forests and is crucially important in terms of their growth and survival. In a seasonal forest in central Panama, the effect of liana cutting on large (>10 cm dbh) individuals of seven tree species was strong during the dry season (as measured by sap velocity), but was largely absent during the wet season, suggesting that much of the competition was for soil moisture (Alvarez-Cansino et al. unpublished data). In this same forest, Tobin et al. (2012) found a similar strong effect of lianas during the dry season and a much lower response of large trees to liana cutting during the wet season. These authors examined the change in water transport (measured as tree sap velocity) of trees from which they cut lianas or a comparable biomass of tree saplings. In response to the release from liana competition, canopy trees increased water transport by ~8%, while the removal of tree saplings had no significant effect (Tobin et al. 2012).

While lianas can interfere with tree water uptake and affect canopy tree performance (Schnitzer & Carson 2010; Tobin et al. 2012), there can be a variety of responses to lianas among tree species. Comparing the effect of lianas on water availability in a dry tropical forest in Bolivia, Pérez-Salicrup and Barker (2000) demonstrated that lianas interfered with water availability in adult trees of *Senna multijuga* during the dry season but did not affect the leaf water potential of the tree *Swietenia macrophylla* (Barker and Pérez-Salicrup, 2000). Thus, the contribution of the above- *vs.* belowground competitive effect of lianas on trees may also vary as a function of the particular species involved.

In an attempt to disentangle above- and belowground competition between lianas and adult trees (≥10 cm diameter) at the stand level, van der Heijden and Phillips (2009) reported that, by presumably competing mainly aboveground, lianas reduced tree annual aboveground biomass growth by ~10% in Amazonian Peru. In this evaluation, leaf area of the lianas present in the tree crown was used as a measure of aboveground competition and the basal area of lianas rooted within a radius of 1 metre from the tree was assumed to indicate belowground competition with the tree. However, the absence of an important contribution of belowground competition may be partly explained by the distance considered (1 m), which may have excluded important belowground liana competition. Van der Heijden and Phillips (2009) suggest that a possible explanation for their findings is that the competitive effect of lianas may change as trees gets older, as found by Dillenburg et al. (1995) and Toledo-Aceves and Swaine (2008a). Evaluation of the relative contribution of above- *vs.* belowground competition at stand level poses logistic difficulties that restrict our understanding of the interaction between lianas and trees.

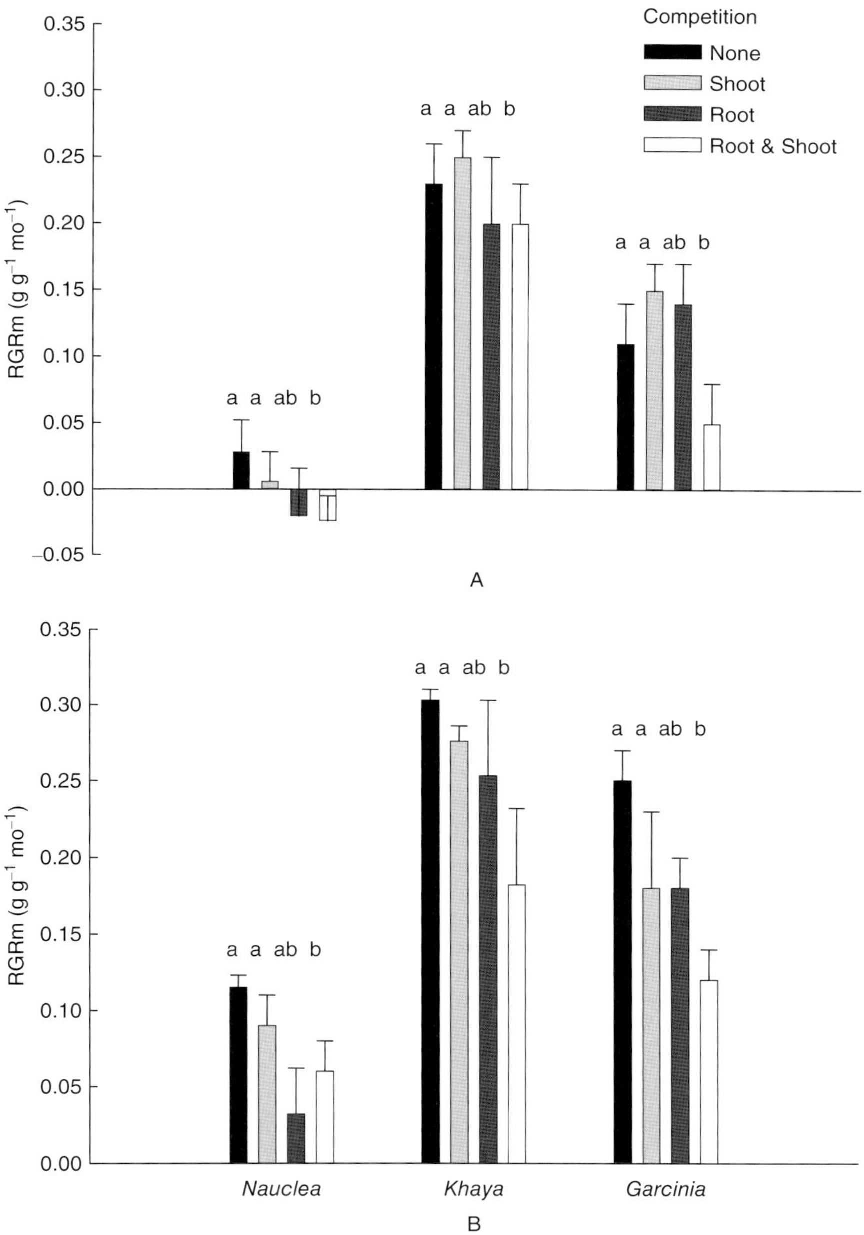

Fig. 12.2 Effect of above- and belowground competition with the liana *Acacia kamerunensis* on the relative growth rate in dry biomass (RGRm; mean + 1 SE; n = 6) of seedlings of the pioneer *Nauclea diderichii*, non-pioneer light-demanding *Khaya anthotheca* and the shade-tolerant *Garcinia afzelii*. Competition treatments: no competition, root competition, shoot competition and root and shoot competition, under (A) 7% photosynthetic active radiation (PAR); and (B) 27% PAR, in shade-houses after 12 months. Treatments that do not share a letter differed at $P < 0.05$, comparisons are between competition treatments within tree species and light treatments were pooled. (Source: Toledo-Aceves & Swaine 2008b. Reproduced with permission of Springer Science+Business Media.)

Collectively, the available evidence indicates that belowground competition plays a significant role in the interaction between lianas and trees in both intact forest and in treefall gaps. According to Schnitzer et al. (2005), the strong role of belowground competition from lianas reported in various studies could be the result of the predominantly high light conditions (Dillenburg et al.1993a, 1993b; Pérez-Salicrup & Barker 2000; Chen et al. 2008; Toledo-Aceves & Swaine 2008a) causing the competition to be mainly for belowground resources. In canopy gaps, this belowground competition may become aboveground competition as canopy gaps develop into a closed canopy (Schnitzer et al. 2005). Supporting this idea, Chen et al. (2008) reported that the relative importance of above- and belowground competition between the liana *Byttneria grandifolia* and saplings of the trees *Litsea dilleniifolia*, *Pometia tomentosa* and *Bauhinia variegate*, differed between two light levels (4% and 35% of full sunlight) in a shade-house experiment. In low light, the relative growth rate (RGR) of the three tree species was reduced more strongly by shoot competition (23–29% reduction) than by root competition (5–26%). Conversely, in high light, RGR was greatly reduced by root competition rather than shoot competition. Van der Heijden and Phillips (2009) found that trees growing in favorable conditions (i.e., high light and low belowground competition with neighboring trees) tended to be less affected by lianas compared to trees in low light conditions and with high competition with neighboring trees for belowground resources. Thus, the relative importance of above- *vs.* belowground competition is likely to vary depending on the relative availability of resources, a fact that may explain the disparate findings among studies. However, differences in the conditions in controlled environments, such as shade-houses and greenhouses, as well as the artificial conditions created by trench experiments, could limit our understanding of belowground interactions. To date, there are currently no fully factorial, *in situ* experimental studies examining the interaction between the light environment and water and nutrient availability on competition between trees and lianas.

IN WHAT WAY ARE LIANAS GOOD COMPETITORS?

The foliage and stems of lianas are conspicuous as lianas display their leaves on top of their tree hosts. This strategy allows lianas to effectively intercept light and also to reduce light availability for the plants below (Putz 1984a; Avalos & Mulkey 1999; Pérez-Salicrup et al. 2001). Compared to trees of similar stem diameter, the contribution of lianas to the foliar biomass of tropical forests is higher, as their leaf area ratio (LAR, leaf area per unit plant biomass) and Huber value (leaf area per active xylem area) are thought to be higher than that of trees (Putz 1983; Wyka et al. 2013). Thus, lianas have the advantage of being able to allocate more resources to the production of photosynthetic tissue and less to the development of support tissue, compared to trees (Pérez-Salicrup et al. 2001). Lianas also appear to have lower leaf mass per unit leaf area than that of their tree hosts (Kazda & Salzer 2000; Kazda & Mehltreter 2001; Kazda, Chapter 22, in this volume), which is advantageous in terms of capturing light at a lower photosynthetic tissue production "cost" (Asner & Martin, Chapter 21 in this volume).

Lianas may be particularly adept at competing for belowground resources because of their reputed well-developed root systems (Restom & Nepstad 2001), which give them the ability to efficiently acquire limited nutrients and water from the soil (Andrade et al. 2005). The ability to acquire water during seasonal droughts may favor lianas in seasonal forests, in contrast to many trees and shrubs (Barker & Perez-Salicrup 2000; Schnitzer 2005; Schnitzer & Bongers 2011; DeWalt et al., Chapter 11 in this volume). In general, climbers have higher water conductivity than other life forms (Hegarty 1989; Ewers et al. 1991), which is necessary for them to adequately support their high amount of leaf area.

In terms of the ability of lianas to colonize gaps, various attributes may allow them to colonize these environments and suppress tree regeneration. First, lianas have an advantage over trees through their capacity to colonize gaps from the seed rain, seed bank and pre-existing seedlings in the understory prior to gap formation (Putz 1984a). In addition, they can reach the gap from the surrounding forest (Peñalosa 1984) and produce new axes from resprouts more frequently than trees, considerably increasing their abundance in gaps (Appanah & Putz 1984a; Lieberman & Mingguang 1992; Schnitzer et al. 2000, 2012; Schnitzer et al., Chapter 7 in this volume). In comparison, trees typically use only the first of these three colonization methods. In addition, lianas grow faster than trees, particularly in high light environments because they can allocate more resources to growth in the high light gap environment and relatively less to the development of support tissue (Pérez-Salicrup et al. 2001). Lianas in gaps cause mechanical stress by adding weight to the tree crown, forcing trees to increase stem diameter at the expense of height (Schnitzer et al. 2005). These attributes may allow lianas to outcompete tree seedlings and saplings during regeneration. Moreover, lianas are resilient and can survive falling into a gap (Putz 1984a; Alvira et al. 2004); approximately 80–90% of adult lianas survive after being pulled into a gap when a tree is felled (Putz 1984a, Alvira et al. 2004; but see Kennard 1998). In the census of the BCI 50-ha plot, Schnitzer et al. (2012; Chapter 7 in this volume) found that nearly 30% of all rooted lianas were clones, and most of the clones appeared to be associated with treefall gaps and disturbance (see also Dalling et al. 2012; Ledo & Schnitzer 2014).

Effects of trees on lianas

Competition is defined as a negative-negative interaction between competitors. However, most studies have focused on the effects of climbers on trees; interaction between trees and climbers is generally considered to be asymmetrical, with climbers exerting the dominant effect. For climbers, the trees represent a physical support on which to grow but they also compete for resources and can have a detrimental effect on liana performance. In a shade-house experiment designed to evaluate the contribution of above- and belowground competition effects of the liana *Acacia kamerunensis* on tree saplings, Toledo-Aceves and Swaine (2008b) found that the liana inhibited the tree *Nauclea diderichii* by reducing the availability of belowground resources, while the tree significantly suppressed the growth of this liana species by interfering with its shoot function. The effects of tree species on liana performance are still largely unknown and require further experimental examination under conditions of varying resource availability.

LIANAS AND TREES: A DYNAMIC INTERACTION

The interaction between lianas and trees is temporally dynamic. Changes in the intensity of competition between lianas and trees indicate that trees may go through seasonal or annual cycles of high and low intensity competition from lianas. For example, the temperate tree *Liquidambar styraciflua* suffered a reduction in leaf photosynthetic capacity with liana root competition, but this effect disappeared after two growing periods (Dillenburg et al. 1995). Toledo-Aceves and Swaine (2008a) found similar results for the non-pioneer, light-demanding tree *K. anthotheca*, which was initially unaffected by the liana *Acacia kamerunensis* in terms of growth in leaf area, height and diameter after six months but was negatively affected after 15 months. In contrast, while seedlings of the pioneer tree *N. diderichii* were affected by the presence of the liana in terms of the growth rate in leaf area (32% reduction) and crown area (56% reduction) after six months, this effect was no longer apparent after one year, probably as a result of the tree out-competing the liana. In a shade-house experiment, the effect of root competition on tree growth was significant for the tree *K. anthotheca* after six months, but had disappeared after one year (Toledo-Aceves & Swaine 2008b). Tree attributes can change during ontogeny; for example, some trees can change from being very shade-tolerant at the seedling stage to exhibiting much faster growth when older (Poorter et al. 2005). Liana competition can also change seasonally, with lianas competing much more during the dry season than the wet season (Tobin et al. 2012; Alvarez-Cansino et al. unpublished data). Temporal changes in liana–tree competition is a new area of study and may lead to important insights into the predominant form of competition in any given forest.

Liana infestation of canopy trees is also dynamic; trees can undergo cycles of heavy and light crown infestation. Phillips et al. (2005) found that the liana populations are much more dynamic than their supports: turnover of large lianas was three times faster than that of the trees in a tropical forest in Peru. In Barro Colorado Island in Panama, Ingwell et al. (2010) reported that 10.9% of trees found to be severely infested by lianas in their crowns in 1996 were free of lianas by 2007, while 5.3% of the trees with no lianas in 1996 presented severe liana infestation in 2007. These authors proposed that the detrimental effects of lianas on trees contribute to the dynamic changes observed in liana communities.

CONCLUSION AND FUTURE DIRECTIONS

Lianas play an important role as competitors with trees for both above- and belowground resources throughout all the tree life stages. Competition studies between lianas and trees show that the nature of this interaction is specific to both liana and tree species involved. The competitive effect of lianas on trees can also vary as a function of the most limiting resource. Thus, the relative importance of above- *vs.* belowground competition may vary depending on the relative availability of light, nutrients, and water. Given that the magnitude of competitive effects can vary with the stress imposed by the environment, a comparative study of the interaction under contrasting conditions of soil fertility and water availability would provide a broader perspective of this process.

Competition between lianas and trees may also be intensified by abiotic stresses generated by increased fragmentation and disturbance. Fragmentation contributes to increased abundance of lianas, which can lead to high tree mortality in treefall gaps and forest edges as a result of physical stresses and high resource competition with lianas (Williams-Linera 1990; Laurance et al. 2001; Arroyo-Rodríguez & Toledo-Aceves 2009; Schnitzer & Carson 2010). Furthermore, under the current patterns of climate change, the increased presence of lianas and the incidence of drought in certain areas will likely increase the intensity of aboveground and belowground competition (see Schnitzer, Chapter 30 in this volume). If the outcome of increasing liana abundance and biomass is a higher turnover of trees and tree growth suppression, further research should concentrate on the synergistic effects of increased forest disturbance and drought on above- and belowground competition between lianas and trees at different stages, giving particular consideration to species-specific interactions.

ACKNOWLEDGMENTS

I would like to thank Stefan Schnitzer and his team for their helpful comments on an earlier version of this chapter. Thanks also go to Keith MacMillan for revision of the manuscript.

REFERENCES

Alvira, D., Putz, F.E. & Fredericksen, T.S. (2004) Liana loads and post-logging liana densities after liana cutting in a lowland forest in Bolivia. *Forest Ecology and Management*, **190**, 73–86.

Andrade, J.L., Meinzer, F.C., Goldstein, G. & Schnitzer, S.A. (2005) Water uptake and transport in lianas and co-occurring trees of a seasonally dry tropical forest. *Trees Structure and Function*, **19**, 282–289.

Appanah, S. & Putz, F.E. (1984) Climber abundance in virgin dipterocarp forest and the effect of pre-felling climber cutting on logging damage. *The Malaysian Forester*, **47**, 335–342.

Arroyo-Rodríguez, V. & Toledo-Aceves, T. (2009) Impact of landscape spatial pattern on liana communities in tropical rainforests at Los Tuxtlas, Mexico. *Applied Vegetation Science*, **12**, 340–349.

Augspurger, C.K. (1984) Seedling survival of tropical tree species interactions of dispersal distance, light-gaps, and pathogens. *Ecology*, **65**, 1705–1712.

Avalos, G. & Mulkey, S.S. (1999) Photosynthetic acclimation of the liana *Stigmaphyllon lindenianum* to light changes in a tropical dry forest canopy. *Oecologia*, **120**, 475–484.

Babweteera, F., Plumptre, A. & Obua, J. (2000) Effect of gap size and age on climber abundance and diversity in Budongo Forest Reserve, Uganda. *African Journal of Ecology*, **38**, 230–237.

Barker, M.G. & Pérez-Salicrup, D. (2000) Comparative water relations of mature mahogany (*Swietenia macrophylla*) trees with and without lianas in a subhumid, seasonally dry forest in Bolivia. *Tree Physiology*, **20**, 1167–1174.

Bazzaz, F.A. & Carlson, R.W. (1982) Photosynthetic acclimation to variability in the light environment of early and late successional plants. *Oecologia*, **54**, 313–316.

Brokaw, N.V. (1985) Gap-phase regeneration in a tropical forest. *Ecology*, **66**, 682–687.

Brokaw, N.V. & Busing, R.T. (2000) Niche versus chance and tree diversity in forest gaps. *Trends in Ecology and Evolution*, **15**, 183–188.

Cai, Z.Q. & Bongers, F. (2007) Contrasting nitrogen and phosphorus resorption efficiencies in trees and lianas from a tropical montane rain forest in Xishuangbanna, south-west China. *Journal of Tropical Ecology*, **23**, 115–118.

Campbell, E.J.F. & Newbery, D.M.C. (1993) Ecological relationships between lianas and trees in lowland rain forest in Sabah, East Malaysia. *Journal of Tropical Ecology*, **9**, 469–490.

Carter, G.A. & Teramura, A.H. (1988) Vine photosynthesis and relationships to climbing mechanics in a forest understory. *American Journal of Botany*, **75**, 1011–1018.

Chave, J., Riéra, B. & Dubois, M. (2001) Estimation of biomass in a Neotropical forest in French Guiana: spatial and temporal variability. *Journal of Tropical Ecology*, **17**, 79–96.

Chen, Y., Bongers, F., Cao, K. & Cai, Z. (2008) Above- and below-ground competition in high and low irradiance: tree seedling responses to a competing liana *Byttneria grandifolia*. *Journal of Tropical Ecology*, **24**, 517–524.

Chittibabu, C.V. & Parthasarathy, N. (2001) Liana diversity and host relationships in a tropical evergreen forest in the Indian Eastern Ghats. *Ecological Research*, **16**, 519–529.

Clark, D.B. & Clark, D.A. (1990) Distribution and effects on tree growth of lianas and woody hemiepiphytes in a Costa Rican tropical wet forest. *Journal of Tropical Ecology*, **6**, 321–331.

Dalling, J.W., Hubbell, S.P. & Silvera, K. (1998) Seed dispersal, seedling establishment and gap partitioning among tropical pioneer trees. *Journal of Ecology*, **86**, 674–689.

Dalling, J.W., Schnitzer, S.A., Baldeck, C. et al. (2012) Resource-based habitat associations in a neotropical liana community. *Journal of Ecology*, **100**, 1174–1182.

Denslow, J.S., Ellison, A.M. & Sanford, R.E. (1998) Treefall gap size effects on above- and below-ground processes in a tropical wet forest. *Journal of Ecology*, **86**, 597–609.

Denslow, J.S., Scultz, J.C., Vitousek, P.M. & Strain, B.R. (1990) Growth response of tropical shrubs to treefall gap environments. *Ecology*, **71**, 165–179.

Dickinson, M.B., Whigham, D.F. & Hermann, S.M. (2000) Tree regeneration in felling and natural treefall disturbances in a semideciduous forest in Mexico. *Forest Ecology and Management*, **134**, 137–151.

Dillenburg, L.R., Teramura, A.H., Forseth, I.N. & Whigham, D.F. (1995) Photosynthetic and biomass responses of *Liquidambar styraciflua* (Hamamelidaceae) to vine competition. *American Journal of Botany*, **82**, 454–461.

Dillenburg, L.R., Whigham, D.F., Teramura, A.H. & Forseth, I.N. (1993a) Effects of below- and aboveground competition from the vines *Lonicera japonica* and *Parthenocissus quinquefolia* on the growth of the tree host *Liquidambar styraciflua*. *Oecologia*, **93**, 48–54.

Dillenburg, L.R., Whigham, D.F., Teramura, A.H. & Forseth, I.N. (1993b) Effects of vine competition on availability of light, water, and nitrogen to a tree host (*Liquidambar styraciflua*). *American Journal of Botany*, **80**, 244–252.

Emmons, L.H. & Gentry, A.H. (1983) Tropical forest structure and the distribution of gliding and prehensile-tailed vertebrates. *American Naturalist*, **121**, 513–524.

Ewers, F.W., Fisher, J.B. & Fichtner, K. (1991) Vine physiology and development, in *The Biology of Vines* (eds F.E. Putz & H.A. Mooney), Cambridge University Press, Cambridge, pp. 127–160.

Foster, J.R., Townsend, P.A. & Zganjar, C.E. (2008) Spatial and temporal patterns of gap dominance by low-canopy lianas detected using EO-1 Hyperion and Landsat Thematic Mapper. *Remote Sensing Environment*, **112**, 2104–2117.

Garrido-Pérez, E.I., Durán, R. & Gerold, G. (2012) Liana-tree relationships: consequences for tree communities and tree evolution. *Interciencia*, **37**, 183–189.

Gentry, A.H. (1991a) The distribution and evolution of climbing plants, in *The Biology of Vines* (eds F.E. Putz & H.A. Mooney), Cambridge University Press, Cambridge, pp. 3–42.

Gentry, A.H. (1991b) Breeding and dispersal systems of lianas, in *The Biology of Vines* (eds F.E. Putz & H.A. Mooney), Cambridge University Press, Cambridge, pp. 393–423.

Grauel, W.T. & Putz, F.E. (2004) Effects of lianas on growth and regeneration of *Prioria copaifera* in Darien, Panama. *Forest Ecology and Management*, **190**, 99–108.

Hegarty, E.E. (1989) The climbers-Lianes and vines, in *Tropical Rain Forest Ecosystems* (eds H. Lieth & M.J.A. Werger), Elsevier, New York, pp. 339–353.

Ingwell, L.L., Wright, S.J., Becklund, K.K. et al. (2010) The impact of lianas on 10 years of tree growth and mortality on Barro Colorado Island, Panama. *Journal of Ecology*, **98**, 879–887.

Kainer, K.A., Wadt, L.H.O., Gomes-Sliva, D.A.P. & Capanu, M. (2006) Liana loads and their association with *Bertholletia excelsa* fruit and nut production, diameter growth and crown attributes. *Journal of Tropical Ecology*, **22**, 147–154.

Kainer, K.A., Wadt, L.H.O., & Staudhammer, C.L. (2014) Testing a silvicultural recommendation: Brazil nut responses 10 years after liana cutting. *Journal of Applied Ecology*, doi: 10.1111/1365–2664.12231.

Kazda, M. & Mehltreter, K. (2001) Contrasting leaf nutrition and leaf mass per unit area in lianas and trees from subtropical island Martin Garcia, Argentina, in *Life Forms and Dynamics in Tropical Forests* (eds G. Gottsberger & S. Liede), Gebrüder Bornträger, Berlin, pp. 131–139.

Kazda, M. & Salzer, J. (2000) Leaves of lianas and self-supporting plants differ in mass per unit area and nitrogen content. *Plant Biology*, **2**, 268–271.

Kennard, D.K. (1998) Biomechanical properties of tree saplings and free-standing lianas as indicators of susceptibility to logging damage. *Ecology*, **50**, 147–153.

Kennedy, D.N. & Swaine, M.D. (1992) Germination and growth of colonizing species in artificial gaps of different sizes in dipterocarp rain forest. *Philosophical Transactions of the Royal Society of London*, **335**, 357–366.

Kitajima, K. (1994) Relative importance of photosynthesis traits and allocation patterns as correlates of seedling shade tolerance of 13 tropical trees. *Oecologia*, **98**, 419–428.

Kitajima, K. & Augspurger, C.K. (1989) Seed and seedling ecology of a monocarpic tropical tree, Tachigalia versicolor. *Ecology*, **70**, 1102–1114.

Körner, C. (2006) Forests, biodiversity and CO_2: surprises are certain. *Biologist*, **53**, 82–90.

Kusumoto, B. & Enoki, T. (2008) Contribution of a liana species, *Mucuna macrocarpa* Wall., to litterfall production and nitrogen input in a subtropical evergreen broadleaved forest. *Journal of Forestry Research*, **13**, 35–42.

Ladwig, L. & Meiners, S. (2010) Liana host preference and implications for deciduous forest regeneration. *Journal of the Torrey Botanical Society*, **137**, 103–112.

Laurance, W.F., Pérez-Salicrup, D., Delamónica, P. et al. (2001) Rain forest fragmentation and the structure of Amazonian liana communities. *Ecology*, **82**, 105–116.

Ledo, A. & Schnitzer, S.A. (2014) Disturbance, not negative density dependence or habitat specialization maintains liana diversity in a tropical forest. *Ecology* in press.

Lewis, S.L. & Tanner, E.V.J. (2000) Effects of above- and belowground competition on growth and survival of rain forest tree seedlings. *Ecology*, **81**, 2525–2538.

Lieberman, D. & Mingguang, L. (1992) Seedling recruitment patterns in a tropical dry forest in Ghana. *Journal of Vegetation Science*, **3**, 375–382.

Londré, R.A. & Schnitzer, S.A. (2006) The distribution of lianas and their change in abundance in temperate forests over the past 45 years. *Ecology*, **87**, 2973–2978.

Ostertag, R. (1998) Belowground effects of canopy gaps in a tropical wet forest. *Ecology*, **79**, 1294–1304.

Parren, M. (2003) *Lianas and Logging in West Africa*, Tropenbos International, Wageningen.

Peñalosa, J. (1984) Basal branching and vegetative spread in two tropical rain forest lianas. *Biotropica*, **16**, 1–9.

Pérez-Salicrup, D.R. & Barker, M.G. (2000) Effect of liana cutting on water potential and growth of adult *Senna multijuga* (Caesalpinioideae) trees in a Bolivian tropical forest. *Oecologia*, **124**, 469–475.

Pérez-Salicrup, D.R. & de Meijere, W. (2005) Number of lianas per tree and number of trees climbed by lianas at Los Tuxtlas, Mexico. *Biotropica*, **37**, 153–156.

Pérez-Salicrup, D.R., Sork, V.L. & Putz, F.E. (2001) Lianas and trees in a liana forest of Amazonian Bolivia. *Biotropica*, **33**, 34–47.

Phillips, O.L., Vásquez Martínez, R., Arroyo, L. et al. (2002) Increasing dominance of large lianas in Amazonian forests. *Nature*, **418**, 770–774.

Phillips, O.L., Vasquez Martinez, R., Mendoza, A.M. et al. (2005) Large lianas as hyperdynamic elements of the tropical forest canopy. *Ecology*, **86**, 1250–1258.

Pinard, M.A., Barker, M.G. & Tay, J. (2000) Soil disturbance and post-logging forest recovery on bulldozer paths in Sabah, Malaysia. *Forest Ecology and Management*, **130**, 213–225.

Poorter, L., Bongers, F., Sterck, F.J. & Wöll, H. (2005) Beyond the regeneration phase: differentiation of height-light trajectories among tropical tree species. *Journal of Ecology*, **93**, 256–267.

Poorter, L. & Werger, M.J.A. (1999) Light environment, sapling architecture, and leaf display in six rain forest tree species. *American Journal of Botany*, **86**, 1464–1473.

Powers, J.S., Kalicin, M. & Newman, M. (2004) Tree species do not influence local soil chemistry in a species-rich Costa Rican rain forest. *Journal of Tropical Ecology*, **20**, 587–590.

Putz, F.E. (1983) Liana biomass and leaf area of a "Tierra Firme" forest in the Rio Negro Basin, Venezuela. *Biotropica*, **15**, 185–189.

Putz, F.E. (1984a) The natural history of lianas on Barro Colorado Island, Panama. *Ecology*, **65**, 1713–1724.

Putz, F.E. (1984b) How trees avoid and shed lianas. *Biotropica*, **16**, 19–23.

Putz, F.E. (1991) Silvicultural effects of lianas, in *The Biology of Vines* (eds F.E. Putz & H.A. Mooney), Cambridge University Press, Cambridge.

Putz, F.E. & Chai, P. (1987) Ecological studies of lianas in Lambir National Park, Sarawak, Malaysia. *Journal of Ecology*, **75**, 523–531.

Reader, R.J., Wilson, S.D., Belcher, J.W. et al. (1994) Plant competition in relation to neighbor biomass: an intercontinental study with *Poa pratensis*. *Ecology*, **75**, 1753–1760.

Restom, T.G. & Nepstad, D.C. (2001) Contribution of vines to the evapotranspiration of a secondary forest in eastern Amazonia. *Plant Soil*, **236**, 155–163.

Richards, P.W. (1996) *The Tropical Rain Forest: An Ecological Study*, Cambridge University Press, Cambridge.

Sánches, M.C. & Valio, I.F.M. (2002) Seedling growth of climbing species from a southeast Brazilian tropical forest. *Plant Ecology*, **159**, 51–59.

Schnitzer, S.A. (2005) A mechanistic explanation for global patterns of liana abundance and distribution. *The American Naturalist*, **166**, 262–276.

Schnitzer, S.A. & Bongers, F. (2002) The ecology of lianas and their role in forests. *Trends in Ecology and Evolution*, **17**, 223–230.

Schnitzer, S.A. & Bongers, F. (2011) Increasing liana abundance and biomass in tropical forests: emerging patterns and putative mechanisms. *Ecology Letters*, **14**, 397–406.

Schnitzer, S.A., Bongers, F. & Wright, S.J. (2011) Community and ecosystem ramifications of increasing lianas in neotropical forests. *Plant Signaling & Behavior*, **6**, 598–600.

Schnitzer, S.A. & Carson, W.P. (2010) Lianas suppress tree regeneration and diversity in treefall gaps. *Ecology Letters*, **13**, 849–857.

Schnitzer, S.A., Dalling, J.W. & Carson, W.P. (2000) The impact of lianas on tree regeneration in tropical forest canopy gaps: evidence for an alternative pathway of gap-phase regeneration. *Journal of Ecology*, **88**, 655–666.

Schnitzer, S.A., Kuzee, M. & Bongers, F. (2005) Disentangling above- and belowground competition between lianas and trees in a tropical forest. *Journal of Ecology*, **93**, 1115–1125.

Schnitzer, S.A., Mangan, S.A., Dalling, J.W. et al. (2012) Liana abundance, diversity, and distribution on Barro Colorado Island, Panama. *PLoS ONE*, **7**(12), e52114. doi: 10.1371/journal.pone.0052114

Schnitzer, S.A., Parren, M.P.E. & Bongers, F. (2004) Recruitment of lianas into logging gaps and the effects of pre-harvest climber cutting in a lowland forest in Cameroon. *Forest Ecology and Management*, **190**, 87–98.

Schoener, T.W. (1983) Field experiments on interspecific competition. *The American Naturalist*, **122**, 240–285.

Smith, A.P. (1973) Stratification of temperate and tropical forests. *American Naturalist*, **107**, 671–683.

Snook, L.K. (1996) Is there a future for mahogany? Catastrophic disturbance, logging and the ecology of mahogany (*Swietenia macrophylla* King): grounds for listing a major tropical timber species in CITES. *Botanical Journal of the Linnean Society*, **122**, 35–46.

Sterck, F.J., Martinez-Ramos, M., Dyer-Leal, G. et al. (2003) The consequences of crown traits for the growth and survival of tree saplings in a Mexican lowland rainforest. *Functional Ecology*, **17**, 194–200.

Stevens, G.C. (1987) Lianas as structural parasites: the *Bursera simaruba* example. *Ecology*, **68**, 77–81.

Swaine, M.D. (ed) (1996) *The Ecology of Tropical Forest Tree Seedlings*, UNESCO, Paris.

Swaine, M.D. & Whitmore, T.C. (1988) On the definition of ecological species groups in tropical rain forests. *Vegetatio*, **75**, 81–86.

Taylor, D.R., Aarssen, L.W. & Loehle, C. (1990) On the relationship between r/K selection and environmental carrying capacity: a new habitat templet for plant life-history strategies. *Oikos*, **58**, 239–250.

Tilman, D. (1982) *Resource Competition and Community Structure*, Monographs in population biology 17, Princeton University Press, Princeton, NJ.

Tobin, M.F., Wright, A.J., Scott, A.M. & Schnitzer, S. (2012) Lianas have a greater competitive effect than trees of similar biomass on tropical canopy trees. *Ecosphere*, **3**, 1–11.

Toledo Aceves, T., Purata Velarde, S. & Peters, C. (2009) Regeneration of commercial tree species in a logged forest in Selva Maya, Mexico. *Forest Ecology and Management*, **258**, 2481–2489.

Toledo-Aceves, T. & Swaine, M.D. (2007) Effect of three species of climber on the performance of *Ceiba pentandra* seedlings in gaps in a tropical forest in Ghana. *Journal of Tropical Ecology*, **23**, 45–52.

Toledo-Aceves, T. & Swaine, M.D. (2008a) Effect of lianas on tree regeneration in gaps and forest understorey in a tropical forest in Ghana. *Journal of Vegetation Science*, **19**, 717–728.

Toledo-Aceves, T. & Swaine, M.D. (2008b) Above- and below-ground competition between the liana *Acacia kamerunensis* and tree seedlings in contrasting light environments. *Plant Ecology*, **196**, 233–244.

Uhl, C., Clark, K., Dezzeo, N. & Maquirino, P. (1988) Vegetation dynamics in Amazonian treefall gaps. *Ecology*, **69**, 751–763.

Van Breugel, M., van Breugel, P., Jansen, P.A. et al. (2012) The relative importance of above- versus belowground competition for tree growth during early succession of a tropical moist forest. *Plant Ecology*, **213**, 25–34.

Van der Heijden, G.M.F. & Phillips, O.L. (2009) Liana infestation impacts tree growth in a lowland tropical moist Forest. *Biogeosciences*, **6**, 2217–2226.

Van der Heijden, G.M.F., Schnitzer, S.A., Powers, J.S. & Phillips, O.L. (2013) Liana impacts on carbon cycling, storage and sequestration in tropical forests. *Biotropica*, **45**, 682–692.

Veenendaal, E.M., Swaine, M.D., Agyeman, V.K. et al. (1996) Differences in plant and soil water relations in and around a forest gap in West Africa during the dry season may influence seedling establishment and survival. *Journal of Ecology*, **84**, 83–90.

Whigham, D. (1984) The influence of vines on the growth of *Liquidambar styraciflua* L. (sweetgum). *Canadian Journal of Forest Research*, **14**, 37–39.

Whitmore, T.C. (1989) Canopy gaps and the two major groups of forest trees. *Ecology*, **70**, 536–538.

Whitmore, T.C. (1996) A review of some aspects of tropical rain forest seedling ecology with suggestions for further enquiry, in *The Ecology of Tropical Forest Tree Seedlings* (ed M.D. Swaine), UNESCO, Paris, pp. 3–39.

Williams-Linera, G. (1990) Vegetation structure and environmental conditions of forest edges in Panama. *Journal of Ecology*, **78**, 356–373.

Wright, S.J., Jaramillo, M.A., Pavon, J. et al. (2005) Reproductive size thresholds in tropical trees: variation among individuals, species and forests. *Journal of Tropical Ecology*, **21**, 307–315.

Wyka, T.P., Oleksyn, J., Karolewski, P. & Schnitzer, S.A. (2013) Phenotypic correlates of the lianescent growth form: a review. *Annals of Botany*, **112**, 1667–1681.

Chapter 13

IMPACTS OF LIANAS ON FOREST-LEVEL CARBON STORAGE AND SEQUESTRATION

Geertje M.F. van der Heijden,[1,2] *Oliver L. Phillips,*[3] *and Stefan A. Schnitzer*[1,2]

[1]University of Wisconsin-Milwaukee, Milwaukee, WI, USA
[2]Smithsonian Tropical Research Institute, Panama
[3]University of Leeds, Leeds, UK

Ecology of Lianas, First Edition. Edited by Stefan A. Schnitzer, Frans Bongers, Robyn J. Burnham, and Francis E. Putz.

Companion Website: www.wiley.com/go/schnitzer/lianas

OVERVIEW

Mature tropical forests sequester large quantities of atmospheric carbon (CO_2), which they store as plant biomass. These forests, however, are changing, which may affect their capacity for carbon storage and sequestration and thus alter global carbon budgets. Changes include an increase in liana abundance and biomass over recent decades in neotropical forests. Lianas reduce the growth and survival of their host trees, which can decrease the net carbon uptake by tropical forests. Lianas also compete much more intensely with shade-tolerant, densely wooded trees than with fast-growing, light-wooded trees. Thus, lianas may shift tree species composition towards faster-growing tree species, which store relatively little carbon, and may therefore reduce the amount of carbon being stored in tropical forests. Lianas themselves, however, have low wood volume, and they store and sequester only a small proportion of the biomass in tropical forests. As lianas increase in tropical forests, they will effectively replace carbon-dense trees, but the greater stocks of lianas will not compensate for the liana-induced loss in net carbon sequestration and storage by trees. The increase in lianas will exacerbate their effects on carbon sequestration and stocks, significantly reducing the carbon balance of tropical forests, with potentially far-reaching consequences for global atmospheric CO_2 levels and associated climate change.

INTRODUCTION

Tropical forests worldwide contribute more than half of the global terrestrial biodiversity, constitute the largest pool of global terrestrial biomass carbon stocks, and contribute to a third of the global terrestrial net primary productivity (Field et al. 1998; Pan et al. 2011). Consequently, any modification to the productivity and dynamics in these forests may have global consequences, such as increasing the concentration of atmospheric greenhouse gases and depletion of the world's biodiversity. Recent evidence indicates that tropical forests are now changing, presumably due to a combination of higher atmospheric CO_2 concentrations, changes in climatic conditions, and intensification of human interferences, such as logging and hunting (e.g., Lewis et al. 2009a; Phillips et al. 2009). One of these changes is an increase in the abundance and biomass of lianas in neotropical forests (Phillips et al. 2002; Schnitzer & Bongers 2011; Schnitzer, Chapter 30 in this volume).

Lianas are an important and conspicuous component of tropical forests, where they compete intensely with trees and likely contribute to the carbon dynamics of tropical forests. Lianas compete strongly with their tree hosts for belowground resources, such as nutrients and water, as well as aboveground for light (see Toledo-Acevez, Chapter 12 in this volume). For instance, there is now considerable evidence both from correlative and experimental studies that lianas suppress individual tree growth (Putz 1984a; Whigham 1984; Clark & Clark 1990; Dillenburg et al. 1993a, 1993b, 1995; Pérez-Salicrup & Barker 2000; Gerwing 2001; Grauel & Putz 2004; Schnitzer et al. 2005; Campanello et al. 2007; Toledo-Aceves & Swaine 2008; Grogan & Landis 2009; van der Heijden & Phillips 2009; Schnitzer & Carson 2010), and increase tree mortality by competition for resources or by increasing the risk of wind- and rain-related breakage and tree-fall (Putz 1984a; Phillips et al. 2005; Grogan & Landis 2009; Ingwell et al. 2010). By depriving their host trees of resources, lianas can also severely reduce tree fecundity as both the probability of producing fruits and the quantity of fruit production diminish when trees become infested by lianas (Stevens 1987; Wright et al. 2005; Kainer et al. 2006, 2007, 2014; Fonseca et al. 2009; Nabe-Nielsen et al. 2009; Klimas et al. 2012). When controlling for biomass, competition from lianas appears to be far more intense than competition from trees (Tobin et al. 2012). However, not all tree species are equally affected by lianas (Putz 1984a, 1984b; Campbell & Newbery 1993; Balfour & Bond 1993; Pérez-Salicrup et al. 2001; Alvira et al. 2004; van der Heijden et al. 2008). Slow-growing and shade-tolerant tree species carry lianas more often and apparently suffer more from liana infestation than faster-growing species (van der Heijden et al. 2008; van der Heijden & Phillips 2009); lianas may therefore indirectly favor or even promote faster-growing species and, by doing so, affect tree species composition (Putz 1984b; Schnitzer et al. 2000; van der Heijden & Phillips 2009; van der Heijden et al. 2008; Schnitzer & Carson 2010). Thus, lianas pose a considerable risk to trees – decreasing their growth and increasing their mortality – and may have a particularly strong effect on slow-growing, shade-tolerant trees.

The strong effect of lianas on tree growth and survival has direct implications for tropical forest-level carbon storage and growth of carbon stocks. Nearly all of the aboveground biomass in tropical forests is contained in trees and, by reducing tree growth and survival, lianas may play an important role in forest-level biomass (and therefore carbon) dynamics. Several observational studies have already shown that the tree biomass of forests with a high abundance of lianas is considerably lower compared to forests with lower liana pressures (Chave et al. 2001; Laurance et al. 2001; Malhi et al. 2006). Over the last few decades, lianas have been increasing in abundance and biomass in neotropical forests (e.g. Phillips et al. 2002; Schnitzer & Bongers 2011; Schnitzer, Chapter 30 in this volume). If this trend continues, the impact of lianas on forest carbon storage and sequestration will intensify in the future, which may weaken the ability of tropical forests to store and sequester carbon. However, few studies have provided a comprehensive estimate of the effect that lianas may have on forest carbon (van der Heijden & Phillips 2009; van der Heijden et al. 2013; Schnitzer et al. in press).

The liana-induced reduction in carbon stored in tropical forests is likely caused by: (1) reduced tree growth; (2), increased tree mortality; and (3) a shift in tree species composition due to different susceptibilities of trees to liana infestation and competition. In this chapter, we outline these three distinct effects of lianas on the carbon balance and how increasing liana abundance and biomass might influence the carbon balance of tropical forests in the future. We also examine the extent to which lianas may compensate for these liana-induced reductions in carbon storage and sequestration. Our goal is to provide a comprehensive assessment of the potential role of lianas in tropical forest carbon dynamics, which, ultimately, may prove to be one of the most important and wide-ranging effects of lianas on tropical forest ecosystems.

LIANA-INDUCED EFFECTS ON CARBON SEQUESTRATION

Tropical forests store 285 ± 64 Pg C worldwide in above-ground biomass (Feldpausch et al. 2012), which is over 30% of the Earth's terrestrial carbon (e.g., Cao & Woodward 2002). The exchanges of carbon between this long-term pool of carbon and the atmosphere dictate the carbon balance of tropical forests. The carbon balance specifically focuses on whether fluxes into this carbon pool equal fluxes out of this carbon pool. Any differences in those fluxes provide valuable information on whether tropical forests act as a carbon sink, i.e., carbon stocks increase over time, or carbon source, i.e., carbon stocks decrease over time.

Aboveground carbon stocks of tropical forests change over time due to the difference in carbon gains by tree growth and carbon losses due to tree branch loss and tree mortality and the subsequent branch and stem decomposition. If carbon gains exceed carbon losses, forest carbon stocks continue to grow. However, carbon stocks will decrease when carbon losses exceed carbon gains. Currently, carbon stocks in tropical forests worldwide are thought to be growing by 1.3 Pg C yr^{-1}, which is equivalent to 0.49 Mg ha^{-1} (Lewis et al. 2009b; Pan et al. 2011) and corresponds to 50% of the global terrestrial carbon sink (Denman et al. 2007).

Lianas reduce tropical forest carbon uptake (sequestration) by competing with trees and thereby reducing tree growth and constraining the amount of carbon trees are able to sequester. There is now overwhelming empirical evidence for the severe negative effect that lianas can have on tree growth and survival (Putz 1984a; Clark & Clark 1990; Dillenburg et al. 1993b, 1995; Gerwing 2001; Grauel & Putz 2004; Schnitzer et al. 2005; Campanello et al. 2007; Chen et al. 2008; Toledo-Aceves & Swaine 2008; Grogan & Landis 2009; van der Heijden & Phillips 2009; Ingwell et al. 2010; Schnitzer & Carson 2010; Schnitzer et al. in press). Adult trees from which lianas were experimentally removed experience a two- to three-fold increase in diameter growth compared to liana-infested control trees (Gerwing 2001; Pérez-Salicrup & Barker 2000; Grauel & Putz 2004; Campanello et al. 2007; Grogan & Landis 2009). Moreover, the liana-induced effects on tree growth intensify when trees carry larger liana loads (Putz 1984a; Clark & Clark 1990; Grogan & Landis 2009; van der Heijden & Phillips 2009; Ingwell 2010). For example, Putz (1984a) showed that the growth of *Luehea seemannii* trees on Barro Colorado Island, Panama, decreased significantly with increasing liana basal area in the tree crown, with annual basal area increment of trees with light liana infestations being approximately twice as fast as that of trees with heavy liana infestations. Similarly, a study in Peru indicated that lianas could reduce tree growth by as much as 84%, depending on local growing conditions and functional traits of the tree (van der Heijden & Phillips 2009). Up to 80% of the trees in neotropical forests are carrying lianas (Toledo-Aceves, Chapter 12 in this volume), with ~30% of these trees having more than 75% of their crown covered by lianas (van der

Heijden et al. 2010; Ingwell et al. 2010). Therefore, lianas may considerably reduce the growth and carbon uptake of heavily infested trees.

Lianas also affect the forest carbon balance by increasing tree mortality, which means additional carbon stocks are being lost from the system by decomposition. Several studies have now indicated that trees infested with lianas are more likely to die (Putz 1984a; Phillips et al. 2005; Grogan & Landis 2009; Ingwell et al. 2010). In western Amazonia, large-liana-infested large trees (≥50 cm diameter), which store the majority of forest carbon, were three times more likely to die than large trees without large canopy lianas (Phillips et al. 2005). Tree mortality risks rise with increasingly severe infestation of lianas (Grogan & Landis 2009; Ingwell et al. 2010). For instance, compared to trees with no or only light liana infestation (<50% of the crown area covered with lianas), the mortality rate of *Swietenia macrophylla* trees in the eastern Amazon doubled for individuals with 50–75% and tripled for those with >75% of their crown area covered by lianas (Grogan & Landis 2009). In central Panama, nearly 75% of the trees had lianas in their crowns, and 42% of the trees with severe liana infestation (>75% of their crown covered by lianas) had died within 10 years (Ingwell et al. 2010).

To date, only two published studies have focused specifically on estimating the effect of lianas on the uptake of carbon by trees at a forest level. Van der Heijden and Phillips (2009) assessed liana infestation levels for trees ≥10 cm diameter in an intensely-studied forest site in Amazonian Peru. They evaluated the effect of lianas on tree growth while accounting for competition for light, and potential competition for nutrients and water with neighboring trees, and wood density, by combining all these factors in a multi-species model. The model predicted tree growth in the presence of lianas (as measured) and in a situation in which none of the trees were competing with lianas (by setting the liana competition part in the model to zero). The predicted growth rates were used to calculate the increment in aboveground biomass per plot for each of the scenarios, with the difference between the scenarios being a forest-level estimate of the effect of lianas (Fig. 13.1). Their results indicated that competition with lianas reduced aboveground tree biomass increment by, on average, 0.51 Mg dry weight ha^{-1} yr^{-1}. Since approximately 50% of the woody tissue of trees is composed of carbon, lianas reduced carbon gain by tree growth by 0.25 Mg C ha^{-1} yr^{-1} in this forest, equivalent to a relative reduction in tree carbon increment of ~10%. Adding mortality to estimates of liana-induced reduction in carbon uptake, which were not included in the calculations by van der Heijden & Phillips (2009), will increase the effect of lianas on the net forest-level carbon sequestration.

A liana removal study by Schnitzer et al. (in press) experimentally quantified the effect of lianas on the net carbon uptake of forests, i.e., the extent to which lianas may reduce the growth of carbon stocks. Over an eight-year period, they recorded the recruitment, diameter, mortality, and identity of all trees and lianas (≥1.3 m tall) in 17 newly formed gaps in a late secondary forest in Panama (Schnitzer & Carson 2010). They removed lianas from eight gaps and these gaps were kept free of lianas for the duration of the study. The remaining nine gaps were used as controls, where lianas were present. After eight years, the cumulative aboveground biomass increment of the liana-free gaps was on average 0.16 Mg ha^{-1} higher than that of the control gaps (controlling for gap size and initial tree biomass; Fig. 13.2). An estimated 84% of this increase in carbon was due to a greater tree growth in the gaps where lianas were removed, whereas the remaining 16% was attributed to the decrease in tree mortality. Based on these average effects and using a conservative mortality rate of 1% and the assumption that gaps are closed after eight years, i.e., in total 8% of the forest area is in a form of gap state each year; the authors estimated that the effect of lianas in gaps alone currently reduce forest-level biomass increment by 0.159 Mg ha^{-1} yr^{-1}. This liana-induced reduction of net carbon uptake in gaps is equivalent to 0.080 Mg C ha^{-1} yr^{-1}, corresponding to a ~10% reduction of the total net carbon gain of these forests. Using a less conservative 2% annual mortality rate doubled the effects of lianas to nearly 20% reduction of the total net carbon gain of these forests. Expanding this liana-removal approach to include gap and non-gap forest sites will undoubtedly increase the estimate of liana-induced carbon reduction in tropical forests.

Increases in liana stem density and biomass in tropical forests will likely exacerbate liana effects on tree growth and mortality, as the liana loads carried by trees increase (Schnitzer, Chapter 30 in this volume). From 1980 to 2000 in Amazonia, the abundance of lianas relative to that of trees increased by up to 4.6% a year (Phillips et al. 2002). This trend is matched by studies on Barro Colorado Island, Panama (BCI), for example, which have indicated that the relative contribution

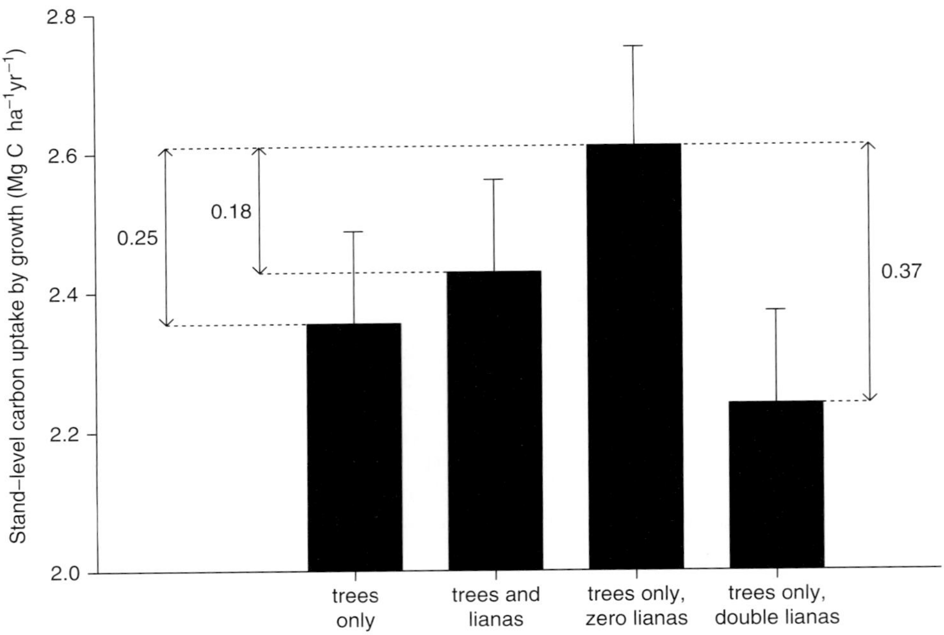

Fig. 13.1 Mean predicted aboveground rates of carbon uptake by trees (Mg C ha^{-1} yr^{-1}) for trees only (leftmost bar), carbon uptake for the combination of trees and lianas (including both the competitive effect of lianas on trees and their contribution to carbon uptake; bar second from left), trees competing without lianas (bar third from left), and trees competing with double the currently observed liana infestation (right-most bar). Arrows and values indicate the difference in carbon uptake (Mg C ha^{-1} yr^{-1}) between the different scenarios (adapted from van der Heijden & Phillips 2009). The results are based upon a multi-species model predicting the effect of lianas on individual tree growth taking functional traits and individual growing conditions into account, and include all dicot trees ≥10 cm diameter occurring in five different 1-ha plots in Tambopata, Peru, measured in the first census (2003) and which were still alive in the second census (2006). The median growth rates of individual trees per scenario were converted to biomass estimates and summed for all trees occurring in a single plot. The bars indicate the mean (±s.e.) of the five plots. Carbon content was assumed to be 50% of biomass, both for trees and for lianas. For more information, see van der Heijden & Phillips (2009). (Source: Adapted from van der Heijden & Phillips 2009. Reproduced with permission.)

of lianas to overall leaf and flower production had increased by 55% and 50%, respectively, over the same time period (Wright et al. 2004; Wright & Calderon 2006). The proportion of trees carrying lianas in their crowns on BCI has increased from 37% in 1967 to almost 75% in 2007 (Ingwell et al. 2010), with severe liana infestation (>75% of the crown covered by lianas) increasing 65% from 1996–2007. Liana stem density on BCI has also increased considerably, leaving little doubt that lianas are increased at this site (Schnitzer et al. 2012).

Using what we know about liana impacts and current forest structural and dynamical trends, we can now develop a simple "back-of-the-envelope" first estimate of the likely impact of lianas over the first quarter of the twenty-first century. The relative basal area of large lianas increased notably in many western Amazonian locations and has doubled over the previous 20 years (Phillips et al. 2002). Van der Heijden and Phillips (2009) used a tree growth model to suggest that if this rate of lianas increase were to continue, a further increase in the liana-induced stand-level reduction in carbon gains by tree growth from 0.25 Mg C ha^{-1} yr^{-1} (now) to 0.37 Mg C ha^{-1} yr^{-1} (mid-2020s) is to be expected. The difference is equivalent to ~25% of the total average carbon uptake of tropical forests (currently 0.49 Mg ha^{-1} yr^{-1}; Lewis et al. 2009b). However, as increasing liana pressure is also expected to result in a higher proportion of trees carrying lianas and higher tree mortality rates (Ingwell et al. 2010), this

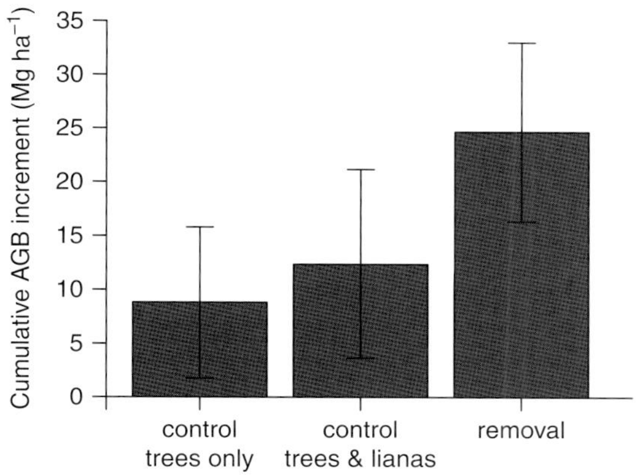

Fig. 13.2 Mean cumulative aboveground biomass (AGB) increment (kg m^{-2}) plus 90% confidence interval eight years after liana removal for plots where liana where removed (far right bar, $n = 8$) and control plots ($n = 9$) where lianas were still present in Gigante Peninsula, Panama. For the control gaps, the mean cumulative AGB increment of just trees (far left bar) and that of both trees and lianas (middle bar) is shown. Results were generated using linear mixed effects (LME) model. Gaps were approximately 1 year old when lianas were removed. (Source: Modified from Schnitzer et al. (in press).)

estimate is likely to be conservative (van der Heijden & Phillips 2009). Similarly, tree mortality in many tropical old-growth forests has increased by approximately 2–3% per year over the last three decades (Phillips & Gentry 1994; Phillips et al. 2004), with no obvious sign of a let-up (e.g., enhanced mortality reported in Phillips et al. 2009, 2010). At this rate of tree mortality, assuming that the effects of lianas in gaps remains constant, in a ten-year time span there would be a further reduction in net carbon uptake, i.e., growth of carbon stocks, by trees in gaps alone from 0.08 Mg C ha^{-1} yr^{-1} to 0.10 Mg C ha^{-1} yr^{-1} (Schnitzer et al. in press). By the year 2025, the liana effect in gaps could reach 0.12 Mg C ha^{-1} yr^{-1}. The difference between these estimates would be equivalent to ~10% of the total average carbon uptake in tropical forests worldwide.

LIANA-INDUCED EFFECTS ON TREE SPECIES COMPOSITION

Lianas may also be able to reduce carbon stocks by altering tree community composition, because tree species differ in their susceptibility to lianas (Putz 1984a, 1984b; Clark & Clark 1990; Balfour & Bond 1993; Campbell & Newbery 1993; Schnitzer et al. 2000; Pérez-Salicrup et al. 2001; Alvira et al. 2004; van der Heijden et al. 2008; Ingwell et al. 2010). In particular, fast-growing and light-wooded tree species tend to have lower levels of liana infestation, whereas dense-wooded, shade-tolerant tree species, which store the vast majority of forest carbon, have much higher liana infestation rates and suffer a proportionally greater reduction in liana-induced growth, survival, and recruitment (Putz 1984b; Clark & Clark 1990; Schnitzer et al. 2000; van der Heijden et al. 2008; van der Heijden & Phillips 2009; Ingwell et al. 2010; Schnitzer & Carson 2010). Indeed, fast-growing trees have wood densities that are, on average, 15% lower and have potential growth rates about three times higher than shade-tolerant trees that suffer high liana infestations (van der Heijden et al. 2008).

By imposing a much stronger effect on slower-growing shade-tolerant tree species than on fast growing trees, lianas may alter the outcome of competition among tree species (Schnitzer et al. 2000; Schnitzer & Carson 2010). The differential effect of lianas has been shown in a liana-removal study in Panama, which tested whether lianas comparatively affected recruitment and diversity of shade-tolerant trees more than that of pioneer trees (Schnitzer & Carson 2010). After lianas had been removed in treefall gaps for eight years, shade-tolerant tree diameter growth, density, and diversity increased considerably in gaps where lianas where removed compared to control gaps, whereas pioneer trees remained relatively unaffected. These results may indicate that by suppressing the growth and recruitment of shade-tolerant tree species, lianas may alter the floristic composition of tropical forests (Schnitzer & Carson 2010). These liana-driven changes in tree composition may have important consequences for the carbon balance of tropical forests because dense-wooded, shade-tolerant tree species may be replaced by liana and tree species with lower density wood.

There is now evidence that the tree composition of tropical forests is changing – an effect that may, in part, be due to lianas. For example, in Amazonian Peru, where liana infestation has been examined in detail and where lianas have increased in abundance over recent decades (Phillips et al. 2002), the incidence of liana infestation varied significantly across tree taxa (van der Heijden et al. 2008). Since around 1990, the stand-level mean wood density

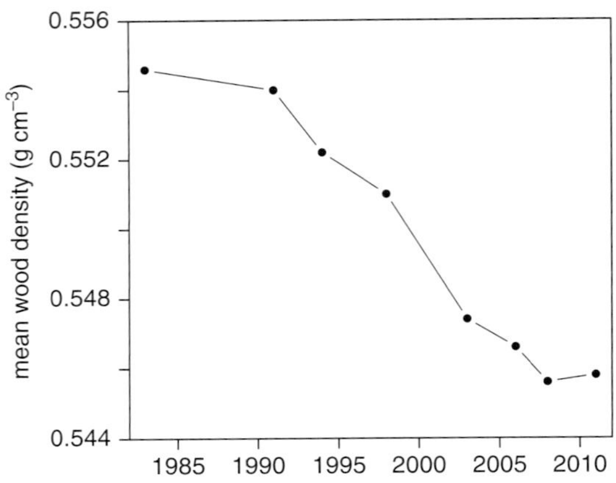

Fig. 13.3 Mean wood density (±0.02 SE) across five 1-ha plots in Tambopata, Peru, from 1983–2011. Mean wood density of five plots decreased over time ($P < 0.01$: Kendal's tau) – a 2% reduction (data extracted from Lopez-Gonzalez et al. 2011, 2012). The mean trend was similar to that of four of the plots, all of which significantly decreased in wood density by about 2–3% over time. The wood density of one plot did not change over time. (Source: Reproduced from van der Heijden et al. (2013). Reproduced with permission of John Wiley & Sons, Ltd.)

has decreased, indicating an overall relative increase of more fast-growing tree species (Fig. 13.3; van der Heijden et al. 2013). In addition, data from a large cluster of plots in central Amazonia also showed that the basal area or density of many genera of faster-growing trees increased whereas there was no such decrease in slower-growing trees (Laurance et al. 2004). These changes in floristic composition have been attributed to a rise in atmospheric CO_2 concentrations, which are thought to be responsible for increases in tree growth, recruitment, and mortality and to which faster-growing species may respond more strongly (Phillips & Gentry 1994; Körner 1998; Laurance et al. 2004; Phillips et al. 2004). The directional changes in tree species composition toward fast-growing trees with low-density wood may also be driven, in part, by the negative effects of lianas.

Liana-driven tree compositional changes would further reduce the ability of tropical forests to store carbon. A study simulating the effect of several biodiversity scenarios on the carbon storage capacity for a 50-ha forest in BCI, showed that a liana-induced loss of slower-growing tree taxa could reduce the carbon storage capacity by trees of this forest by as much as 34% (Bunker et al. 2005). If the increase in liana abundance and biomass persists and these simulation results are representative of the wider tropics, the potential liana-driven reduction of the carbon currently stored in tropical forest trees could reach ~36 Pg C in Amazonia and ~97 Pg C worldwide (~30% of the total forest carbon).

DO LIANAS COMPENSATE FOR THE LIANA-INDUCED REDUCTION IN TREE CARBON GAINS AND STOCKS?

Liana carbon increment and potential increase in biomass stocks must compensate, to some extent, for the liana-induced reduction in tree biomass storage and sequestration. However, the low wood volume and relatively small size of lianas may mean that the compensatory effect for the liana-induced reduction in tree biomass may be correspondingly small. Lianas forgo constructing a large stem and, instead, use the architecture of trees to access the canopy. By doing so, lianas can invest the resources otherwise necessary for support tissue into height growth, increased leaf area, and stem and presumably root elongation without large investment in woody support tissue (Putz 1983; Gehring et al. 2004; Gerwing 2004; Cai et al. 2007; Selaya et al. 2007). While lianas can contribute up to 40% of the leaf area of tropical forests (Putz 1983; Hegarty & Caballé 1991), they tend to compose less than 5% of the total biomass of these forests (Putz 1983; Nascimento & Laurance 2002; DeWalt & Chave 2004). For example, on the BCI 50-ha plot, lianas composed 25% of the woody stems (lianas and trees > 1 cm diameter), 35% of the woody species, but less than 3% of the basal area (Schnitzer et al. 2012; Schnitzer et al., Chapter 7 in this volume). Additionally, lianas are dynamic components of tropical forests, with turnover rates as much as three times faster than those of trees (Phillips et al. 2005), suggesting that a high proportion of the carbon sequestered by liana biomass growth is released back into the atmosphere much sooner than that of trees.

To date, few studies have attempted to quantify the compensation of lianas for the liana-induced decreases in tree biomass increment and stocks. Liana biomass growth has been estimated to compensate for 0.07 Mg ha^{-1} yr^{-1}, equivalent to 30%, of the liana-induced reduction in carbon uptake by tree growth in an old-growth forest in Peru (van der Heijden & Phillips

2009). In addition, in a secondary forest in Panama, Schnitzer et al. (in press) found that net carbon gain by lianas in tree-fall gaps compensated for only 24% of the liana-induced reduction in tree net carbon gain. In a fragmentation study in Brazil, the compensatory effect of lianas was even smaller: increases in liana biomass in forest edges only compensated for 8% of the tree biomass lost 10–17 years after fragmentation (Laurence et al. 1997). Increasing liana abundance and biomass in tropical forests are therefore indeed unlikely to offset the associated reduction in tree carbon uptake and storage caused by the increase in competitive effects of lianas.

CONCLUSION AND FUTURE DIRECTIONS

Lianas are currently reducing carbon sequestration and carbon storage in tropical forests: (1) through impacts on tree growth and mortality; (2) by causing a shift in tree species composition towards faster-growing species with low-density wood; and (3) by replacing carbon-dense trees with lianas, which sequester and store far less carbon than the trees they replace. Increasing liana abundance and biomass enhance the above-mentioned effects and are therefore projected to cause even more severe losses in carbon sequestration and storage in these forests. Increasing liana effects may therefore reduce or even reverse the carbon sink function of tropical forests and severely reduce the tropical forest carbon storage capacity and may therefore have significant consequences for global atmospheric CO_2 levels and hence climate change.

To accurately quantify the total effect of lianas on forest carbon sequestration and storage and to validate previous observational studies, large-scale experimental studies are necessary. A long-term experimental study in a secondary forest in central Panama is currently being conducted by S.A. Schnitzer, J. Powers, G.M.F. van der Heijden and colleagues. They are following more than 10,000 trees and 7,500 lianas ($\geq$1 cm) in eight 80 × 80 m control plots and more than 10,000 trees ($\geq$1 cm) in eight 80 × 80 m plots where lianas have been experimentally removed, with the aim of using this experiment to quantify the effects of lianas on tree and forest-level carbon dynamics, as well as on tree community composition. Preliminary results one year following liana cutting indicate a rapid biomass increment caused by both higher biomass gains due to tree growth and lower biomass losses due to tree mortality in the removal plots compared to the control plots. The prediction is that the increase in biomass of the plots without lianas will quickly exceed the sum of the biomass of trees and lianas in the control plots – possibly within three years. Additional experiments of this kind in other tropical forests will allow researchers to accurately quantify the effects of lianas in tropical forests, and these studies will provide data that can be used to more accurately predict the effects of increasing liana abundance and biomass on carbon sequestration and storage in tropical forests.

ACKNOWLEDGMENTS

The Tambopata fieldwork was made possible through grants from the Explorer's Club, the Coalbourn Trust, and the Alberta Mennega Foundation, financial support from the University of Leeds and logistical support from the Instituto National para Recursos Naturales (INRENA) and Peruvian Safaris S.A. We thank Tatiana Boza Espinoza for help with liana data collection and Abel Monteagudo and Rodolfo Vasquez for their contribution to tree species determination at Tambopata. Support for O.L. Phillips was provided by a European Research Council Advanced Grant, "Tropical forests in the changing Earth system." Research and support for S.A. Schnitzer and G.M.F van der Heijden in Panama were supported by a NSF grants DEB-0845071, DEB-1019436, and the University of Wisconsin–Milwaukee Research Growth Initiative program.

REFERENCES

Alvira, D., Putz, F.E. & Fredericksen, T.S. (2004) Liana loads and post-logging liana densities after liana cutting in a lowland forest in Bolivia. *Forest Ecology and Management*, **190**, 73–86.

Balfour, D.A. & Bond, W.J. (1993) Factors limiting climber distribution and abundance in a southern African forest. *Journal of Ecology*, **81**, 91–100.

Bunker, D.E., DeClerk, F., Bradford, J.C. et al. (2005) Species loss and aboveground carbon storage in a tropical forest. *Science*, **310**, 1029–1031.

Cai, Z.-Q., Poorter, L., Cao, K.-F. & Bongers, F. (2007) Seedling growth strategies in *Bauhinia* species: comparing lianas and trees. *Annals of Botany*, **100**, 831–838.

Campanello, P.I., Garibaldi, J.F., Gatti, M.G. & Goldstein, G. (2007) Lianas in a subtropical Atlantic forest: host preference and tree growth. *Forest Ecology and Management*, **242**, 250–259.

Campbell, E.J.F. & Newbery, D.M. (1993) Ecological relationships between lianas and trees in lowland rain forest in Sabah, East Malaysia. *Journal of Tropical Ecology*, **9**, 469–490.

Cao, M. & Woodward, F.I. (2002) Net primary and ecosystem production and carbon stocks of terrestrial ecosystems and their responses to climate change. *Global Change Biology*, **4**, 185–198.

Chave, J., Riera, B. & Dubois, M.-A. (2001) Estimation of biomass in a neotropical forest of French Guiana: spatial and temporal variability. *Journal of Tropical Ecology*, **17**, 79–96.

Chen, Y.-J., Bongers, F., Cao, K.-F. & Cai, Z.-Q. (2008) Above- and below-ground competition in high and low irradiance: tree seedling responses to a competing liana *Byttneria grandifolia*. *Journal of Tropical Ecology*, **24**, 517–524.

Clark, D.B. & Clark, D.A. (1990) Distribution and effects on tree growth of lianas and woody hemi-epiphytes in a Costa Rican tropical wet forest. *Journal of Tropical Ecology*, **6**, 321–331.

Denman, K.L., Brasseur, G., Chidtaisong, A. et al. (2007) Couplings between changes in the climate system and biochemistry, in *Climate Change 2007: The Physical Science Basis* (eds S. Solomon, D. Qin, M. Manning et al.), Cambridge University Press, Cambridge.

DeWalt, S.J. & Chave, J. (2004) Structure and biomass of four lowland Neotropical forests. *Biotropica*, **36**, 7–19.

Dillenburg, L.R., Teramura, A.H., Forseth, I.N. & Whigham, D.F. (1995) Photosynthetic and biomass allocation responses of *Liquidambar styracifula* (Hamamelidaceae) to vine competition. *American Journal of Botany*, **82**, 454–461.

Dillenburg, L.R., Whigham, D.F., Teramura, A.H. & Forseth, I.N. (1993a) Effects of vine competition on availability of light, water, and nitrogen to a tree host (*Liquidambar styraciflua*). *American Journal of Botany*, **80**, 244–252.

Dillenburg, L.R., Whigham, D.F., Teramura, A.H. & Forseth, I.N. (1993b) Effect of below- and aboveground competition from the vines *Lonicera japonica* and *Parthenocissus quinquefolia* on the growth of the tree host *Liquidambar styraciflua*. *Oecologia*, **93**, 48–54.

Feldpausch, T.R., Lloyd, J., Lewis, S.L. et al. (2012) Integrating height into tropical biomass estimates. *Biogeosciences*, **9**, 3381–3403.

Field, C.B., Behrenfeld, M.J., Randerson, J.T. & Falkowski, P. (1998) Primary production of the biosphere: integrating terrestrial and oceanic components. *Science*, **281**, 237–240.

Fonseca, M.G., Vidal, E. & Maes dos Santos, F.A. (2009) Intraspecific variation in the fruiting of an Amazonian timber tree: implications for management. *Biotropica*, **41**, 179–185.

Gehring, C., Park, S. & Denich, M. (2004) Liana allometric biomass equations for Amazonian primary and secondary forest. *Forest Ecology and Management*, **195**, 69–83.

Gerwing, J.J. (2001) Testing liana cutting and controlled burning as silvicultural treatments for a logged forest in the eastern Amazon. *Journal of Applied Ecology*, **38**, 1264–1276.

Gerwing, J.J. (2004) Life history diversity among six species of canopy lianas in an old-growth forest of the eastern Brazilian Amazon. *Forest Ecology and Management*, **190**, 57–72.

Grauel, W.T. & Putz, F.E. (2004) Effects of lianas on growth and regeneration of *Prioria copaifera* in Darien, Panama. *Forest Ecology and Management*, **190**, 99–108.

Grogan, J. & Landis, R.M. (2009) Growth history and crown vine coverage are principal factors influencing growth and martality rates of big-leaf mahogany *Swietenia macrophylla* in Brazil. *Journal of Applied Ecology*, **46**, 1283–1291.

Hegarty, E. & Caballé, G. (1991) Distribution and abundance of vines in forest communities, in *The Biology of Vines* (eds F.E. Putz & H.A. Mooney), Cambridge University Press, Cambridge, pp. 313–335.

Ingwell, L.L., Wright, S.J., Becklund, K.K. et al. (2010) The impact of lianas on 10 years of tree growth and mortality on Barro Colorado Island, Panama. *Journal of Ecology*, **98**, 879–887.

Kainer, K.A., Wadt, L.H.O., Gomes-Sliva, D.A.P. & Capanu, M. (2006) Liana loads and their association with *Bertholletia excelsa* fruit and nut production, diameter growth and crown attributes. *Journal of Tropical Ecology*, **22**, 147–154.

Kainer, K.A., Wadt, L.H.O. & Staudhammer, C.L. (2007) Explaining variation in Brazil nut fruit production. *Forest Ecology and Management*, **250**, 244–255.

Kainer, K.A., Wadt. L.H.O, & Staudhammer, C.L. (2014) Testing a silvicultural recommendation: Brazil nut responses 10 years after liana cutting. *Journal of Applied Ecology*, doi: 10.1111/1365-2664.12231.

Klimas, C.A., Kainer, K.A., Wadt, L.H.O. et al. (2012) Control of *Carapa guianensis* phenology and seed production at multiple scales: a five year study exploring the influences of tree attributes, habitat heterogeneity and climate cues. *Journal of Tropical Ecology*, **28**, 105–118.

Körner, C. (1998) Tropical forests in a CO_2-rich world. *Climatic Change*, **39**, 297–315.

Laurance, W.F., Laurance, S., Ferreira, L. et al. (1997) Biomass collapse in Amazonian forest fragments. *Science*, **278**, 1117–1118.

Laurance, W.F., Oliveira, A.A., Laurance, S.G. et al. (2004) Pervasive alteration of tree communities in undisturbed Amazonian forests. *Nature*, **428**, 171–175.

Laurance, W.F., Pérez-Salicrup, D., Delemônica, P. et al. (2001) Rain forest fragmentation and the structure of Amazonian liana communities. *Ecology*, **82**, 105–116.

Lewis, S.L., Lloyd, J., Sitch, S. et al. (2009a) Changing ecology of tropical forests: evidence and drivers. *Annual Review of Ecology, Evolution and Systematics*, **40**, 529–549.

Lewis, S.L., Lopez-Gonzalez, G., Sonke, B. et al. (2009b) Increasing carbon storage in intact African tropical forests. *Nature*, **457**, 1003–1007.

Lopez-Gonzalez, G., Lewis, S.L., Burkitt, M. & Phillips, O.L. (2011) Forestplots.net: a web application and research tool to manage and analyse tropical forest plot data. *Journal of Vegetation Science*, **22**, 582–597.

Lopez-Gonzalez, G., Lewis, S.L., Phillips, O.L., & Burkitt, M. (2012) Forest plots database. www.forestplots.net (accessed 12/09/2012).

Malhi, Y., Wood, D., Baker, T.R. et al. (2006) The regional variation of aboveground live biomass in old-growth Amazonian forests. *Global Change Biology*, **12**, 1107–1138.

Nabe-Nielsen, J., Kollman, J. & Peña-Claros, M. (2009) Effects of liana load, tree diameter and distances between conspecifics on seed production in tropical timber trees. *Forest Ecology and Management*, **257**, 987–993.

Nascimento, H. & Laurance, W. (2002) Total aboveground biomass in central Amazonian rainforests: a landscape-scale study. *Forest Ecology and Management*, **168**, 311–321.

Pan, Y., Birdsey, R.A., Fang, J. et al. (2011) A large and persistent sink in the world's forests. *Science*, **333**, 988–993.

Pérez-Salicrup, D. & Barker, M.G. (2000) Effect of liana cutting on water potential and growth of adult *Senna multijuga* (Ceasalpinioideae) trees in a Bolivian tropical forest. *Oecologia*, **124**, 469–475.

Pérez-Salicrup, D., Sork, V.L. & Putz, F.E. (2001) Lianas and trees in a liana forest of Amazonian Bolivia. *Biotropica*, **33**, 34–47.

Phillips, O.L., Aragao, L.E.O.C., Lewis, S.L. et al. (2009) Drought sensitivity of the Amazon rainforest. *Science*, **323**, 1344–1347.

Phillips, O.L., Baker, T.R., Arroyo, L. et al. (2004) Pattern and process in Amazon tree turnover, 1976–2001. *Philosophical Transactions of the Royal Society B. Biological Sciences*, **359**, 381–407.

Phillips, O.L. & Gentry, A.H. (1994) Increasing turnover through time in tropical forests. *Science*, **263**, 954–958.

Phillips, O.L., van der Heijden, G.M.F., Lewis, S.L. et al. (2010) Drought-mortality relationships for tropical forests. *New Phytologist*, **187**, 631–646.

Phillips, O.L., Vásquez Martínez, R., Arroyo, L., et al. (2002) Increasing dominance of large lianas in Amazonian forests. *Nature*, **418**, 770–774.

Phillips, O.L., Vásquez Martínez, R., Monteagudo Mendoza, A. et al. (2005) Large lianas as hyperdynamic elements of the tropical forest canopy. *Ecology*, **86**, 1250–1258.

Putz, F.E. (1983) Liana biomass and leaf area of a "Tierra Firme" forest in the Rio Negro Basin, Venezuela. *Biotropica*, **15**, 185–189.

Putz, F.E. (1984a) The natural history of lianas on Barro Colorado Island, Panama. *Ecology*, **65**, 1713–1724.

Putz, F.E. (1984b) How trees avoid and shed lianas. *Biotropica*, **16**, 19–23.

Schnitzer, S.A. & Bongers, F. (2011) Increasing liana abundance and biomass in tropical forests: emerging patterns and putative mechanisms. *Ecology Letters*, **14**, 397–406.

Schnitzer, S.A. & Carson, W.P. (2010) Lianas suppress tree regeneration and diversity in treefall gaps. *Ecology Letters*, **13**, 849–857.

Schnitzer, S.A., Dalling, J.W. & Carson, W.P. (2000) The impact of liana on tree regeneration in tropical forest canopy gaps: evidence for an alternative pathway of gap-phase regeneration. *Journal of Ecology*, **88**, 655–666.

Schnitzer, S.A., Kuzee, M.E. & Bongers, F. (2005) Disentangling above- and below-ground competition between lianas and trees in a tropical forest. *Journal of Ecology*, **93**, 1115–1125.

Schnitzer, S.A., Mangan, S.A., Dalling, J.W. et al. (2012) Liana abundance, diversity, and distribution on Barro Colorado Island, Panama. *PLoS One*, **7**, e52114.

Schnitzer, S.A., van der Heijden, G.M.F., Mascaro, J., & Carson W.P. (in press). Lianas reduce biomass accumulation in a tropical forest. *Ecology*.

Selaya, N.G., Anten, N.P.R., Oomen, R.J. et al. (2007) Above-ground biomass investments and light interception of tropical forest trees and lianas early in succession. *Annals of Botany*, **99**, 141–151.

Stevens, G.C. (1987) Lianas as structural parasites: the *Bursera simaruba* example. *Ecology*, **68**, 77–81.

Tobin, M.F., Wright, A.J., Mangan, S.A. & Schnitzer, S.A. (2012) Lianas have a greater competitive effect than trees of similar biomass on tropical canopy trees. *Ecosphere*, **3**, 20.

Toledo-Aceves, T. & Swaine, M.D. (2008) Above- and below-ground competition between the liana *Acacia kamerunensis* and tree seedlings in contrasting light environments. *Plant Ecology*, **196**, 233–244.

van der Heijden, G.M.F., Feldpausch, T.R., de la Fuente Herrero, A. et al. (2010) Calibrating the liana crown occupancy index in Amazonian forests. *Forest Ecology and Management*, **260**, 549–555.

van der Heijden, G.M.F., Healey, J.R. & Phillips, O.L. (2008) Infestation of trees by lianas in a tropical forest in Amazonian Peru. *Journal of Vegetation Science*, **19**, 747–756.

van der Heijden, G.M.F. & Phillips, O.L. (2009) Liana infestation impacts tree growth in a lowland tropical moist forest. *Biogeosciences*, **6**, 2217–2226.

van der Heijden, G.M.F., Schnitzer, S.A., Powers, J.S. & Phillips, O.L. (2013) Liana impacts on carbon cycling, storage and sequestration in tropical forests. *Biotropica*, **45**, 682–692.

Whigham, D.F. (1984) The influence of vines on the growth of *Liquidambar styraciflua* L. (Sweetgum). *Canadian Journal of Forest Research*, **14**, 37–39.

Wright, S.J. & Calderon, O. (2006) Seasonal, El Niño and longer-term changes in flower and seed production in a moist tropical forest. *Ecology Letters*, **9**, 35–44.

Wright, S.J., Calderón, O., Hernandéz, A. & Paton, S. (2004) Are lianas increasing in importance in tropical forests? A 17-year record from Panama. *Ecology*, **85**, 484–489.

Wright, S.J., Jaramillo, M.A., Pavon, J. et al. (2005) Reproductive size thresholds in tropical trees: variation among individuals, species and forests. *Journal of Tropical Ecology*, **21**, 307–315.

Chapter 14

RECIPROCAL INTERACTIONS BETWEEN LIANAS AND FOREST SOIL

Jennifer S. Powers

University of Minnesota, St. Paul, MN, USA

Ecology of Lianas, First Edition. Edited by Stefan A. Schnitzer, Frans Bongers, Robyn J. Burnham, and Francis E. Putz.

Companion Website: www.wiley.com/go/schnitzer/lianas

OVERVIEW

Lianas have their roots in soil and thus variation among forests in soil properties may affect liana community characteristics and function. Lianas are often a conspicuous component of tropical forests with traits that differ from those of trees, and they may affect soil processes including biogeochemical cycling and hydrology. Nevertheless, few studies have addressed whether liana species richness, diversity and abundance vary systematically with edaphic gradients, and even fewer studies have asked the reciprocal question of how lianas affect soil processes such as nutrient cycling or soil moisture availability. This chapter reviews the available data on how variation in soils affects lianas, and how lianas affect soils. There is not strong evidence that soil variation directly affects liana species richness or biomass at coarse or continental spatial scales, though some studies do show patterns at finer scales. By contrast, there are convincing data indicating that lianas have higher leaf litter fall quantities relative to their basal area or stem biomass and that their foliage has higher nutrient concentrations than the leaves of trees. Thus, lianas may have greater effects on nutrient cycling and trace gas emissions compared to trees relative to what their low basal area alone would suggest. At the ecosystem scale, lianas may also affect hydrological processes through competing with trees for water and accessing both shallow and deep sources of water. Because lianas are apparently increasing in abundance, studying how lianas affect ecosystem-scale biogeochemical processes and forest hydrology should be a research priority.

INTRODUCTION

For centuries, tropical forests have been known for their high species richness and staggering biological diversity. By contrast, tropical soils were long regarded as uniformly ancient, nutrient poor, and largely static. Only recently has the great heterogeneity of tropical soils been acknowledged (Richter & Babbar 1991; Townsend et al. 2008). Lianas are a conspicuous component of tropical forests, and it is likely that they interact with soils in many ways. Unfortunately, any discussion of the effects of soils on lianas or lianas on soils is necessarily incomplete, as few studies have focused exclusively on either topic. Nevertheless, given the apparent increases in liana abundance and biomass in many tropical forests (Schnitzer & Bongers 2011; Schnitzer, Chapter 30 in this volume) coupled with recent evidence for the ability of lianas to affect ecosystem carbon and nitrogen stocks and fluxes (Hickman et al. 2010; van der Heijden et al. 2013; Schnitzer et al. in press; van der Heijden et al., Chapter 13 in this volume), it suggests that examining the question of liana–soil relationships is timely, especially if it serves to identify major gaps in our knowledge and spur further research.

Why should we expect that lianas might affect soils and vice versa? As plants with roots in soil, it would be surprising if there were no direct effects of soil properties on any aspects of lianas such as community composition and structure, abundance, tissue elemental composition, and growth or productivity. Because lianas are also associated with host trees that serve as scaffolds, any effects of soil variation on lianas may also be indirect, i.e., mediated through host trees (Putz 1984; Schnitzer & Bongers 2002). Conversely, because lianas have distinctive features compared to trees such as low stem mass to leaf area ratios (Schnitzer, Chapter 30 in this volume), high water conductance (Santiago et al., Chapter 20 in this volume), and high nutrient levels (Asner & Martin, Chapter 21 in this volume; Kazda, Chapter 22 in this volume), they might affect key processes in the soil such as nutrient cycling, water availability, and trace gas emissions. In particular, these soil-related processes may vary as a function of lianas' detrital carbon inputs, their capacity to extract water from the soil, or the spatial locations of their crowns relative to their root zones. Three key questions to answer then are: (1) what evidence is there that soils affect lianas?; (2) what evidence is there that lianas affect soil moisture dynamics and biogeochemical cycling?; and (3) are liana–soil relationships quantitatively or qualitatively different from relationships between self-supporting trees and soil?

This chapter first reviews some of the ways that variation in soils within and among tropical forests may affect lianas. It then discusses how lianas may affect soils and key ecosystem properties and processes such as soil moisture and nutrient cycling. The chapter will contrast liana–soil relationships with tree–soil relationships and differentiate among what points are firmly established in the literature, which ideas are speculative, and which directions for future studies may be especially fruitful.

SOIL EFFECTS ON LIANAS

From white sands to red clays to black volcanic soils, there is tremendous variation in the physical and chemical properties of soils across the tropics (Richter & Babbar 1991). These properties may affect both liana community characteristics (e.g., the distribution and abundance of liana taxa) as well as liana function (e.g., primary production, nutrient content of liana tissues, etc). However, in interpreting edaphic-driven variation in liana communities and function among forests, it is important to remember that soil nutrient availability is often correlated with precipitation through weathering, and typically decreases as annual rainfall increases (Clinebell et al. 1995; Huston 2012). Liana abundance tends to peak in the drier tropics (Schnitzer 2005; DeWalt et al. 2010, Chapter 11 in this volume), which also tend to have higher nutrient availability, thus disentangling the mechanisms underlying any soil–liana abundance relationships can be challenging (Clinebell et al. 1995). However, the relationship between lianas and soil fertility is not evident (Gentry 1991; van der Heijden & Phillips 2008; Dalling et al. 2012). Moreover, the processes that affect liana abundance and diversity may differ, and liana diversity may increase with or have a unimodal relationship with soil moisture (Gentry 1991; van der Heijden & Phillips 2009; DeWalt et al., Chapter 11 in this volume). Given the wide range in soil properties within and among forests, it would be surprising if there were no effects of soils on any aspects of liana communities.

Soil effects on patterns of liana community composition, structure and abundance

Soil properties, such as particle-size distribution or texture, depth, topographic position, moisture availability, and nutrient availability may affect liana community composition and structure or biomass indirectly via distribution of host trees or directly via distributions of lianas themselves. A first step toward evaluating how the template of soil variation affects liana communities is assessing whether the data support strong relationships between trees and soil. There is ample evidence that some tropical tree species distributions vary across edaphic gradients at the landscape or meso-scale (Clark et al. 1998; Clark et al. 1999; John et al. 2007; Baldeck et al. 2013). However, there is little compelling evidence for continental-scale patterns in species richness that are related to soil variation (Wright 1992; Clinebell et al. 1995; Ter Steege et al. 2000). The case for edaphic effects on forest biomass stocks is even less well established. Early reviews did not find strong relationships between forest biomass and nutrient availability or soil order (Proctor 1983; Vitousek & Sanford 1986). Recent studies corroborate this view. Various studies have found no support for simple relationships between biomass and soil nutrients and/or physical characteristics in South America (Baraloto et al. 2011), positive relationships between soil phosphorus (P) and biomass in Dipterocarp forests in Borneo (Paoli et al. 2008), or more complicated relationships that depended on forest dynamics and environmental filtering of species with different traits (Gourlet-Fleury et al. 2011; Quesada et al. 2012). It should be noted, however, that though the data do not support a strong relationship between soil fertility and biomass stocks at the pan-tropical scale, there is a weak, positive relationship between soil P availability and the flux of aboveground net primary productivity in lowland tropical forests (Cleveland et al. 2011).

Are the relationships between soil properties and lianas as inconsistent as they appear to be for trees? A large number of studies have addressed the question of the extent to which edaphic variables are correlated with liana community composition or biomass and abundance in tropical and sub-tropical forests in Central and South America, Asia, and Africa (Table 14.1). Indeed, out of the possible ways that soils might affect lianas and vice versa, this topic has received the most study. Studies of liana–soil associations at the landscape or mesoscale can be divided broadly into three categories based on how soil variation is addressed (Table 14.1). First, many researchers have investigated whether there are associations between liana diversity, species composition, and/or abundance as a function of habitat type, which typically are defined by some combination of topographic position, soil parent materials, and/or hydrologic regime (e.g., periodic flooding). As many forest dynamic and edaphic parameters typically co-vary within habitat types (i.e., soil moisture and nutrient availability; Balfour & Bond 1993), these studies are important for highlighting the potential importance of edaphic factors in driving liana community composition and structure, but do not help identify which particular edaphic variables are responsible for observed patterns. A second line of inquiry concerns comparing the relative importance of edaphic versus forest structural or disturbance variables like gap abundances. Last, some studies have attempted to identify specific nutrients that are correlated to liana distributions.

Given the wide range of forest types, geographic regions, and rainfall regimes represented by the studies in Table 14.1, it would be surprising if a single pattern of liana–soil relationships emerged. At the continental scale, the Asian tropics likely have higher soil fertility than Africa and South America, because of high geologic activity (Huston 2012). Interestingly, the studies cited in Table 14.1 that report correlations between liana communities and/or biomass and indices of nutrient availability or soil fertility have tended to be conducted in Asia (Putz & Chai 1987; DeWalt et al. 2006; Nurfazliza et al. 2012), while many of the studies implicating forest dynamics, trellis availability, and/or soil moisture were conducted in Central or South America (Ibarra-Manríquez & Martínez-Ramos 2002; Homeier et al. 2010; Nogueira et al. 2011; Dalling et al. 2012). Furthermore, liana diversity and particularly density may be lower in Asian forests than in neotropical forests (DeWalt et al., Chapter 11 in this volume). Collectively these studies raise the untested hypothesis that the strength of soil nutrients as a determinant of liana community composition and structure may vary among continents as a function of underlying site fertility, i.e., stronger liana–soil correlations in sites with higher soil fertility. Alternatively, the lack of evidence for direct effects of edaphic variables on liana community composition does not preclude the possibility that soils affect forest dynamics or host tree distribution and thereby indirectly affect lianas, as some studies have suggested (Molina-Freaner & Tinoco-Ojanguren 1997; van der Heijden & Phillips 2008). Finally, it is possible that various spatial scales of studies and thoroughness of datasets vary by continent, and thus any patterns at the continental scale are explained by differential sampling effort. Undoubtedly, the sampling issue will be resolved if comparable datasets similar to the mapped 50-ha plot in Panama (Dalling et al. 2012; Schnitzer et al. 2012; Schnitzer et al., Chapter 7 in this volume) continue to be generated.

Soil effects on growth, productivity, and elemental composition of lianas

At the scale of individual trees or ecosystems, tree growth and net primary productivity of lowland tropical rainforests are often positively correlated with soil properties like phosphorus (P) or base cation availability (Cleveland et al. 2011; Baribault et al. 2012). Soil moisture and fertility may similarly affect liana growth, productivity, and elemental composition (e.g., foliar nitrogen (N) or P concentrations); however, few if any studies have examined these relationships directly in field settings. The data that do exist on this topic come from field fertilization or pot experiments with seedlings under relatively low light to mimic conditions in the forest understory. Two studies added fertilizer to seedlings that were planted or naturally growing in intact forest understories on Barro Colorado Island in Panama (Hattenschwiler 2002; Yavitt & Wright 2008). Both studies (which had one liana species in common) showed increased growth as a consequence of fertilization, but in Hättenschwiler's study of three liana species, the capacity to respond to increased nutrient availability varied with species and light environment.

In another study, researchers grew seedlings of trees and lianas from five species in the genus *Bauhinia* in a shade-house in mineral soil from a nearby forest with and without added N fertilizer under two light regimes (Cai et al. 2008). For both tree and liana species, relative growth rates were more affected by light than the fertilization treatment (Cai et al. 2008). Foliar N (%) increased in response to fertilization in the two tree species and in a shade-tolerant liana, but not in the two light-demanding liana species (Cai et al. 2008). Translating these results from seedlings to mature lianas in forest conditions is tenuous, but what they do suggest is that both lianas and trees vary in their capacity to respond to added nutrients (Lawrence 2003). Understanding whether lianas differ predictably compared to trees in response to added nutrients is a key priority, as anthropogenic nitrogen deposition is increasing throughout the tropics (Hietz et al. 2011).

LIANA EFFECTS ON SOILS AND ECOSYSTEM PROCESSES

Through their large leaf area, their potential capacity to extract water from deep soil layers, and other distinct morphological and functional properties, one could hypothesize that nutrient cycling and hydrology in a forest with lianas would look very different from a similar forest that lacked lianas. Although studies of the consequences of lianas on ecosystem processes

Table 14.1 Studies of liana community composition and structure across soil conditions. For each study we include site, mean annual precipitation, the focus of the study, the key soil parameters examined, and the published reference. Studies are grouped by how soil variation was addressed.

Site and location	Mean annual precipitation (MAP) (mm yr^{-1})	Goal, patterns, or findings	Key soil parameter implicated in liana distribution/abundance	Reference
		Studies of habitat types		
Lambir National Park, Sarawak, Malaysia	2950	Valley habitats (inceptisol) had greater liana density than ridge habitats (ultisol)	Soil fertility or forest disturbance may be higher in valley habitats	Putz & Chai 1987
Knysna and Alexandria Forests, South Africa	(MAP ranges with altitude from 600–1200 at Knysna and 700–900 at Alexandria)	Examined climber abundances over an altitudinal gradient and correlated it to abiotic factors and biotic factors like host tree architecture	Trellis availability more important that nutrients	Balfour & Bond 1993
Centro Ecológico de Sonora, Mexico	245	In this desert, vine species richness and Shannon diversity was highest along streams, intermediate in hillslopes, and lowest in plains areas	Higher water availability near streams may support higher densities of trees that act as trellises	Molina-Freaner & Tinoco-Ojanguren 1997
Chajul Tropical Biological Station, Mexico	3000	Liana abundance, species richness, and species identities varied among habitat types (alluvial terraces, floodplains, karst range, and low hills)	Soil moisture in addition to treefall dynamics, but not nutrients	Ibarra-Manríquez & Martínez-Ramos 2002
Sepilok Forest Reserve, Malaysia, Borneo	2929	Liana density, biomass, and diversity of stems ≥2.5 cm declined along with soil fertility and moisture in three habitats: alluvial > sandstone hill > keranga (heath forest)	Soil moisture and/or fertility	DeWalt et al. 2006
		Edaphic factors relative to disturbance and forest dynamics		
Sumaco Biosphere Reserve, Ecuador	2500–4000	Liana diameter (but not basal area or density) decreased along an elevation gradient	Soil C/N explained a small amount of variation in density, but stand structure was more important; no correlation between liana basal area and edaphic variables	Homeier et al. 2010

(*continued overleaf*)

Table 14.1 *(continued).*

Site and location	Mean annual precipitation (MAP) (mm yr^{-1})	Goal, patterns, or findings	Key soil parameter implicated in liana distribution/ abundance	Reference
Ducke Reserve, Amazonia, Brazil	2400	Forest structure, palm density, and sand % were correlated with liana density in small size classes ($1 \leq D \leq 2.4$ cm); soil variables had no correlation with density of larger lianas	Sand concentration, which correlates negatively with nutrients, but positively with soil moisture	Nogueira et al. 2011
Sumaco Biosphere Reserve, Ecuador	2500-4000	Liana diameter (but not basal area or density) decreased along an elevation gradient	Soil C/N explained a small amount of variation in density, but stand structure was more important; no correlation between liana basal area and edaphic variables	Homeier et al. 2010
Ailao Mountains National Nature Reserve, China	1931	Higher liana species richness in old growth forest, which also had higher soil moisture; distinct communities in different successional stages	Canopy height and openness were important, as well as soil moisture (other soil variables were not measured)	Yuan et al. 2009
Barro Colorado Island, Panama	2600	In a 50-ha plot of mapped trees and lianas, more liana species were associated with regions of low forest height, rather than topographic habitat or soil nutrients	Only weak associations with edaphic factors	Dalling et al. 2012, Ledo & Schnitzer 2014
		Specific soil variables		
Pasoh Forest Reserve, Malaysia	1733	Many liana species were differentially distributed along edaphic gradients	pH, ammonium, nitrate, phosphorus (P), potassium (K), and magnesium (Mg)	Nurfazliza et al. 2012
Parque Sierra de San Javier, subtropical montane forest, Argentina	1300-1500	In a 6-ha plot, liana composition and density increase in relation to soil P and recent treefall gaps	P was more important than cations, nitrogen, soil texture, or soil moisture	Malizia et al. 2010

are in their infancy, evidence is accumulating on the roles that lianas play in forest hydrologic dynamics, productivity and nutrient relations, and trace gas emissions.

Hydrologic dynamics

It is well established that the physiological, morphological, and anatomical characteristics of most liana species differ from those of trees (Isnard & Silk 2009; Rowe & Speck 2005; Rowe & Speck, Chapter 23 in this volume; Isnard & Feild, Chapter 17 in this volume). In general, liana vessels have larger diameters and greater lengths than those of trees (Ewers et al. 1997; Angyalossy et al. 2012; Jacobsen et al. 2012, Angyalossy et al., Chapter 19 in this volume), which should confer higher conductance and transpiration rates (Zhu & Cao 2009; Santiago et al., Chapter 20 in this volume). Lianas are also typically assumed to have greater rooting depths than trees, but only a few studies have tested this hypothesis directly by measuring rooting depths (Restom & Nepstad 2004) or indirectly by using isotopic tracers to compare the depth at which trees and lianas are extracting water from the soil profile (Moreira et al. 2000; Andrade et al. 2005).

Collectively, these hydraulic traits suggest that lianas may exert measurable effects on soil–water balances (Restom & Nepstad 2001). Most tests of this hypothesis are framed in terms of whether lianas can effectively compete with trees for soil water, especially in seasonal forests during times of low water availability. To answer this question, several studies have removed lianas from individual host trees or forest stands and then monitored soil moisture and/or physiological indices of plant water status (e.g., leaf water potential, conductance, or sap flow) (Barker & Pérez-Salicrup 2000; Pérez-Salicrup & Barker 2000; Pérez-Salicrup 2001; Campanello et al. 2007; Tobin et al. 2012). The studies that quantified soil moisture directly have yielded mixed results. Campanello et al. (2007) cut lianas and bamboo from three 1 ha plots in a subtropical semi-deciduous Atlantic forest of north-eastern Argentina. They measured the soil matric potential in surface soil (0–5 cm) using the filter paper method and found that soil water was higher in liana/bamboo removal plots compared to control plots under closed canopy microsites, but there were no differences in soil moisture between control and removal plots in gaps (Campanello et al. 2007). Thus, the effects of liana removal on soil water depended on microsite characteristics. By contrast, Pérez-Salicrup (2001) cut lianas from replicated 900 m^2 plots of transitional dry-moist forest in lowland Bolivia to study the effects on seedling growth and survivorship. In addition, he measured gravimetric soil moisture in the top 0–10 cm monthly, and at deeper layers in the soil profile (0.95–1.05 m depth) three times over the course of a year. He did not find any difference in gravimetric soil moisture between control and cut plots at any depth or time (Pérez-Salicrup 2001). Of course, failure to find differences in soil moisture between liana removal and control plots does not necessarily imply no effects of lianas, if lianas were accessing sources of water that were not accounted for in the experimental design (e.g., very deep water) or if tree–water relations change.

Studies that measured the water status of trees following liana cutting, which is a more direct indicator of plant competition, have also yielded variable results. Some studies have shown clear effects of increased sap flow (Tobin et al. 2012) or less negative leaf water potential (Pérez-Salicrup & Barker 2000) in trees with experimental liana removal relative to those with lianas. Another study failed to find differences in conductance and leaf water potential in large mahogany trees (*Swietenia macrophylla*) in a liana cutting experiment, perhaps because the trees had access to deep sources of water (Barker & Pérez-Peréz-Salicrup 2000). Collectively, these studies provide evidence that lianas can affect the tree–water balance and soil moisture, but the strength of these effects might vary with season, microsite (e.g., gap versus understory), tree size, and how strongly lianas are actually competing with trees for water.

Although useful for demonstrating liana–tree interactions, these studies were not designed to examine the consequences of liana presence or absence at the ecosystem scale. One of the few studies to address this question used sap flow measurements in lianas and trees to estimate transpiration during the wet and dry season, and compared these to estimates of total evapotranspiration in a young secondary forest in Brazil (Restom & Nepstad 2001). Lianas had higher transpiration rates expressed as a function of cross-sectional area, and thus accounted for ~8% of evapotranspiration despite comprising only 5.5% of basal area (Restom & Nepstad 2001). As liana biomass is increasing in tropical forests (Schnitzer & Bongers 2011; Schnitzer, Chapter 30 in this volume), more studies, such as this one that track the ecosystem-level consequence of liana removal on water balance and canopy conductance, are needed.

Other important questions about the role of lianas in forest hydrology remain unresolved. Hydraulic redistribution is defined as passive water flow through plant roots from moister to drier areas in the soil (Prieto et al. 2012). One common example of hydraulic redistribution is upward hydraulic lift, i.e., movement of water from wetter, deeper layers in the soil to drier, shallow soil layers during the night, when plants are not transpiring (Prieto et al. 2012). The magnitude and ecological significance of hydraulic redistribution vary among ecosystems (Neumann & Cardon 2012), but a recent review shows that average hydraulic redistribution as a percentage of average transpiration can exceed 10% for many ecosystem types including Brazilian savanna and Amazonian forest (Neumann & Cardon 2012). Because lianas are thought to have deep roots and retain leaves longer into the dry season than co-occurring deciduous trees in seasonally dry or moist forest, it is possible that they contribute disproportionately to hydraulic redistribution and thus may affect competitive relationships among juvenile trees and lianas in the seedling layer. The occurrence of hydraulic redistribution in trees varies among species with factors such as leaf habit and root system morphology (Scholz et al. 2008). Documenting whether liana species are capable of hydraulic redistribution and whether any liana-mediated hydraulic redistribution differs quantitatively from trees is a ripe avenue for future research.

Litterfall quantity, chemical composition, and spatial patterns

Lianas differ from trees in many ways including: large allocation to leaf biomass relative to support structures like stems, deep root systems, and nutrient-rich foliage (Gentry 1983; Hegarty & Caballé 1991; Asner & Martin 2012; Asner & Martin, Chapter 21 in this volume; Kazda, Chapter 22 in this volume; van der Heijden et al., Chapter 13 in this volume). These traits suggest that lianas may have disproportionate impacts on ecosystem processes relative to their stem biomass (Schnitzer & Bongers 2002). These characteristics could affect forest carbon and nutrient cycles through a number of mechanisms including the quantity, quality or composition, and spatial patterns of detrital inputs to the soil. Although there are far fewer studies that address the question of liana effects on ecosystem processes compared to those of trees or forest-wide patterns, the data that do exist show large and unequivocal contributions of lianas to forest productivity and ecosystem processes. Below we highlight how lianas contribute to forest productivity, evidence for higher nutrient concentrations than trees, and potential effects on the spatial patterns of nutrient cycling and soil chemistry.

Many studies have measured liana stem basal area and used these data with allometric equations to estimate aboveground biomass in different tropical forests (e.g., DeWalt & Chave 2004) or have collated values from the literature (e.g., Hegarty & Caballé 1991). Among the whole gamut of tropical forests, the basal area of lianas ranges from 0.5–3.5 m^2 ha^{-1}, while aboveground biomass estimates for lianas range from 2.4–46 Mg ha^{-1}, which represents a variable but small percentage of the aboveground biomass of the entire forest (from $\leq$1–7.2%) (Hegarty & Caballé 1991; DeWalt & Chave 2004; van der Heijden et al., Chapter 13 in this volume; Schnitzer et al., Chapter 7 in this volume). By contrast, Yong et al. (2012) reviewed studies that partitioned leaf litter fall into contributions from lianas versus trees and found that lianas account for 13–38% of leaf litter mass in a range of tropical forests from Africa, Central America, India, and Australia. Because lianas make up a large fraction of leaf litter in tropical forests, they could have correspondingly large contributions to nutrient cycling because the chemical composition and physical properties of liana litter differs from that of trees (Asner & Martin, Chapter 21 in this volume; Kazda, Chapter 22 in this volume), and these traits affect decomposition and nutrient mineralization rates (Cornwell et al. 2008).

Several studies have addressed whether there are systematic differences in foliar element concentrations and traits such as specific leaf area between life forms (Bigelow 1993; Kazda & Salzer 2000; Asner & Martin 2012; Yong et al. 2012; Asner & Martin, Chapter 21 in this volume; Kazda, Chapter 22 in this volume). The most comprehensive test to date compared 22 leaf traits between 563 liana species and 3322 tree species in a pan-tropical sampling scheme (Asner & Martin 2012; Asner & Martin, Chapter 21 in this volume). This study found large and consistent differences between leaves of lianas and trees across geographic regions, taxonomy, and environmental variables that can affect element concentrations and leaf traits

(Asner & Martin 2012). On a mass basis, liana leaves have higher concentrations of N, P, potassium (K), magnesium (Mg), manganese (Mn), boron (B), iron (Fe), zinc (Zn), and particularly calcium (Ca). Lianas also invest less in compounds for defense, consistent with their higher investment in chemicals for light capture and growth (on a mass basis) (Asner & Martin 2012; Asner & Martin, Chapter 21 in this volume). In sum, this benchmark study provided strong evidence for how green liana leaves differ systematically from those of trees.

To affect nutrient cycling through decomposition would require lianas also to have different chemical composition of senesced leaves and litter. There are limited data to suggest whether this is so, and some of the relevant studies have focused exclusively on trees. For example, in a study of 45 tree species on a nutrient-poor soil in French Guiana, green leaf and senesced leaf nutrients concentrations were positively correlated for some nutrients like N, but not others like P (Hättenschwiler et al. 2008). Some elements like Ca are not resorbed prior to senescence (Vitousek & Sanford 1986), and thus the large differences between liana and tree leaves should be conserved moving from live to dead leaves. Together these observations imply that lianas and trees should differ in litter nutrient concentrations. Two other studies have tested this hypothesis directly. Cai and Bongers (2007) quantified nutrient resorption and litter N and P in a taxonomically diverse sample of lianas and trees from Xishuangbanna, China, and found significantly higher concentrations of P but not N in litter from lianas compared to trees. By contrast, litter from the N_2-fixing legume species *Mucuna macrocarpa* in a subtropical evergreen forest in Japan had approximately twice the N concentrations than bulk leaf litter did (Kusumoto & Enoki 2008). In summary, while it seems like a fair assumption that the large differences in foliar chemistry between green liana and tree leaves should yield similar differences in litter nutrient concentrations, this question has yet to be resolved definitely and is a promising avenue for future work.

What are the potential consequences of large litter fall quantities and elevated foliar nutrient concentrations of lianas for nutrient cycling in tropical forest soils? Two studies have speculated that lianas affect spatial patterns of nutrients in forests, but came to opposite conclusions. Powers et al. (2004) studied whether four species of large canopy trees create their own zones of influence in terms of soil chemistry in a rain forest in Costa Rica. This work was based on a classic study that mapped the spatial patterns of soil chemical variables like pH and extractable N in transects that radiated outward from the tree bole to beyond the crown drip line of a single *Pinus contorta* tree growing on a sand dune (Zinke 1962). Zinke hypothesized that the systematic spatial patterns of soil properties under the pine tree crown was due to the decomposition of bark and needle litter inputs that differed predictably in quantity and chemical composition moving from the bole to the crown (1962) (Fig. 14.1A). The prediction of individual-tree generated spatial patterns in soil chemistry has been shown in many different ecosystems (Schlesinger et al. 1996; Finzi et al. 1998; Deschaseaux & Ponge 2001), but was not supported in the diverse tropical rain forest (Powers et al. 2004). Lianas may redistribute nutrients both vertically from the soil to the canopy and back as trees do, but also horizontally, as single lianas can connect multiple tree canopies and have rooting zones that are distant from their crown zones (Putz 1984). Powers et al. speculated that the input of nutrients from liana leaf litter fall effectively homogenizes litter nutrient concentrations on the forest floor and disrupts any individual-tree generated soil heterogeneity (Fig. 14.1B). Yong et al. (2012) used similar assumptions to arrive at a different conclusion. They tried to resolve the paradox of how large trees that are covered in lianas remain productive, and hypothesized that nutrient inputs from liana leaf litter enrich the soils of their host trees and thus ameliorate some of the negative effects of liana infestation, though they did not test this hypothesis with field data (Yong et al. 2012). Both of these studies are essentially thought experiments; more data are needed to determine whether the presence of lianas influences the amounts and spatial or temporal patterns of soil nutrient availability.

Trace gas emissions

Soils are a major source of atmospheric trace gases including carbon dioxide (CO_2), nitrous oxide (N_2O), and methane (CH_4) (Conrad 1996; Schlesinger & Andrews 2000). As many of these compounds are potent greenhouse gases, there is much interest in understanding the factors that regulate their production and flux rates. Many of the processes that control the release of these gases are ultimately under biological control, and a few recent papers have highlighted

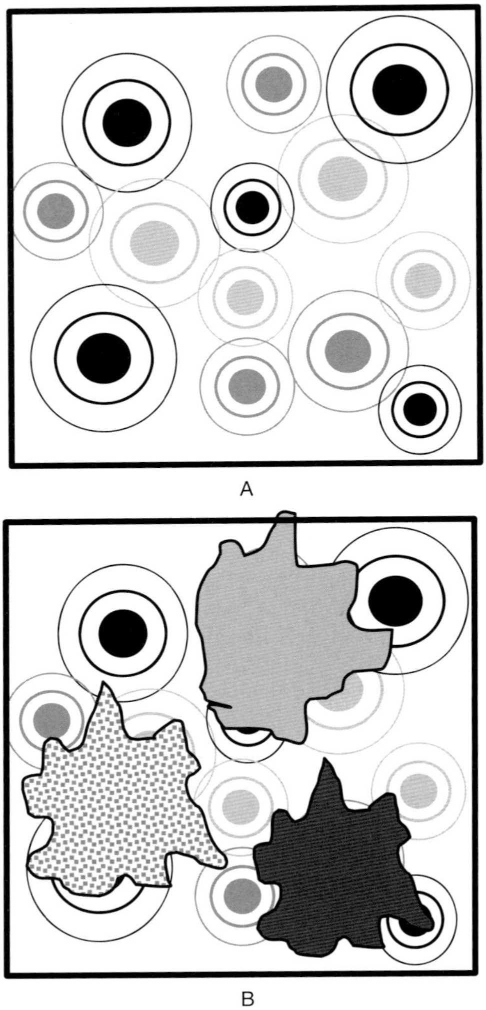

Fig. 14.1 Hypothesized overview of a forest with individual trees (filled in circles) and tree-generated patterns of soil heterogeneity depicted as concentric circles radiating outwards from each bole. Different shades indicate different species. Panel (A) depicts a forest lacking lianas, and (B) suggests how the presence of liana canopies might disrupt individual tree generated soil heterogeneity.

fascinating interactions between lianas and trace gas emissions from soils. Van Haren et al. (2010) studied whether soil CO_2 (the sum of root respiration and heterotrophic respiration from decomposer organisms) and N_2O effluxes (arising primarily from denitrification) depended upon tree species identity in a diverse tropical forest in the Brazilian Amazon. Including the basal area of nearby liana stems explained significant amounts of the variation in soil CO_2 effluxes for 6 of 10 tree species and N_2O effluxes for 2 of 6 species (van Haren et al. 2010). They speculate that increased carbon investment belowground by lianas accounts for the positive effect of lianas on trace gas emissions from soils, by providing more substrate for microbial respiration (van Haren et al. 2010).

A more direct demonstration of how vines may control trace gas emissions and the biogeochemical cycle of nitrogen is found in the work of Hickman et al. (2010), who studied kudzu (*Pueraria montana*), a non-native, N_2-fixing vine from the family Fabaceae that has invaded much of the south-eastern United States. They compared invaded and uninvaded sites in Georgia and found that kudzu presence dramatically increased soil N pool sizes and flux rates, as well as NO and N_2O emissions, which have negative impacts on air quality (Hickman et al. 2010). Taken together, these studies indicate that the distinctive traits that differentiate lianas from trees (e.g., high carbon allocation belowground relative to aboveground, large capacities to respond to elevated atmospheric CO_2) may allow them to have disproportionate impacts on soil biogeochemical processes and trace gas emissions relative to their biomass.

CONCLUSION AND FUTURE DIRECTIONS

The questions posed at the beginning of this chapter focused on reciprocal interactions between lianas and soils, and whether these interactions differ from relationships between trees and soils. Relationships and effects of lianas and soils are only now being teased apart and we clearly have a long way to go before robust generalizations can be made. Of potential interactions, the best studied is the question of whether variation in soils affects liana species distributions and/or biomass. At continental scales, there is little evidence to suggest direct linkages between liana density, biomass, or species richness with soil properties (Clinebell et al. 1995; van der Heijden & Phillips 2008, 2009; Dalling et al. 2012); however, at landscape or mesoscales some studies (particularly in the Asian tropics) have found significant patterns (Table 14.1). Surprisingly few studies have sought to determine how variation in soils affects liana growth or nutrient relations, so generalizations at this point are premature. The few studies on the effect of lianas on forest ecosystem functioning in terms of hydrology, nutrient cycling, and trace gas emissions have yielded tantalizing results that hint at potentially large effects that do not scale with their biomass (perhaps unlike trees). Priorities for future research include ecosystem-scale studies of how lianas affect biogeochemical cycling and hydrology both above- and belowground.

ACKNOWLEDGMENTS

I thank Geertje van der Heijden, Joey Reid, Peter Tiffin, and two anonymous reviewers for comments on previous drafts of this chapter. I also thank Stefan Schnitzer for lively discussions that stimulated the ideas in this chapter. Finally, I gratefully acknowledge the financial support from the National Science Foundation (DEB-1019441).

REFERENCES

Andrade, J.L., Meinzer, F.C., Goldstein, G. & Schnitzer, S.A. (2005) Water uptake and transport in lianas and co-occurring trees of a seasonally dry tropical forest. *Trees-Structure and Function*, **19**(3), 282–289.

Angyalossy, V., Angeles, G., Pace, M.R. et al. (2012) An overview of the anatomy, development and evolution of the vascular system of lianas. *Plant Ecology & Diversity*, **5**(2), 167–182.

Asner, G.P. & Martin, R.E. (2012) Contrasting leaf chemical traits in tropical lianas and trees: implications for future forest composition. *Ecology Letters*, **15**(9), 1001–1007.

Baldeck, C.A., Harms, K.E., Yavitt, J.B. et al. (2013) Soil resources and topography shape local tree community structure in tropical forests. *Proceedings of the Royal Society B: Biological Sciences*, **280**(1753).

Balfour, D.A. & Bond, W.J. (1993) Factors limiting climber distribution and abundance in a southern African forest. *Journal of Ecology*, **81**(1), 93–100.

Baraloto, C., Rabaud, S., Molto, Q. et al. (2011) Disentangling stand and environmental correlates of aboveground biomass in Amazonian forests. *Global Change Biology*, **17**(8), 2677–2688.

Baribault, T.W., Kobe, R.K. & Finley, A.O. (2012) Tropical tree growth is correlated with soil phosphorus, potassium, and calcium, though not for legumes. *Ecological Monographs*, **82**(2), 189–203.

Barker, M.G. & Pérez-Salicrup, D. (2000) Comparative water relations of mature mahogany (Swietenia macrophylla) trees with and without lianas in a subhumid, seasonally dry forest in Bolivia. *Tree Physiology*, **20**(17), 1167–1174.

Bigelow, S.W. (1993) Leaf nutrients in relation to stature and life form in tropical rain forest. *Journal of Vegetation Science*, **4**, 401–408.

Cai, Z.Q. & Bongers, F. (2007) Contrasting nitrogen and phosphorus resorption efficiencies in trees and lianas from a tropical montane rain forest in Xishuangbanna, south-west China. *Journal of Tropical Ecology*, **23**, 115–118.

Cai, Z.Q., Poorter, L., Han, Q. & Bongers, F. (2008) Effects of light and nutrients on seedlings of tropical Bauhinia lianas and trees. *Tree Physiology*, **28**(8), 1277–1285.

Campanello, P.I., Gatti, M.G., Ares, A. et al. (2007) Tree regeneration and microclimate in a liana and bamboo-dominated semideciduous Atlantic Forest. *Forest Ecology and Management*, **252**(1–3), 108–117.

Clark, D.B., Clark, D.A. & Read, J.M. (1998) Edaphic variation and the mesoscale distribution of tree species in a Neotropical rain forest. *Journal of Ecology*, **86**, 101–112.

Clark, D.B., Palmer, M.W. & Clark, D.A. (1999) Edaphic factors and the landscape-scale distributions of tropical rain forest trees. *Ecology*, **80**, 2662–2675.

Cleveland, C.C., Townsend, A.R., Taylor, P. et al. (2011) Relationships among net primary productivity, nutrients and climate in tropical rain forest: a pan-tropical analysis. *Ecology Letters*, **14**(9), 939–947.

Clinebell, R.R., Phillips, O.L., Gentry, A.H. et al. (1995) Prediction of neotropical tree and liana species richness from soil and climatic data. *Biodiversity and Conservation*, **4**, 56–90.

Conrad, R. (1996) Soil microorganisms as controllers of atmospheric trace gases (H-2, CO, CH_4, OCS, N_2O, and NO). *Microbiological Reviews*, **60**(4), 609.

Cornwell, W.K., Cornelissen, J.H.C., Amatangelo, K. et al. (2008) Plant species traits are the predominant control on litter decomposition rates within biomes worldwide. *Ecology Letters*, **11**, 1065–1071.

Dalling, J.W., Schnitzer, S.A., Baldeck, C. et al. (2012) Resource-based habitat associations in a neotropical liana community. *Journal of Ecology*, **100**, 1174–1182.

Deschaseaux, A. & Ponge, J.-F. (2001) Changes in the composition of humus profiles near the trunk base of an oak tree (Quercus petraea (Mattus.) Liebl.). *European Journal of Soil Biology*, **37**, 9–16.

DeWalt, S.J. & Chave, J. (2004) Structure and biomass of four lowland Neotropical forests. *Biotropica*, **36**.

DeWalt, S.J., Ickes, K., Nilus, R. et al. (2006) Liana habitat associations and community structure in a Bornean lowland tropical forest. *Plant Ecology*, **186**(2), 203–216.

DeWalt, S.J., Schnitzer, S.A., Chave, J. et al. (2010) Annual rainfall and seasonality predict pan-tropical patterns of liana density and basal area. *Biotropica*, **42**(3), 309–317.

Ewers, F.W., Carlton, M.R., Fisher, J.B. et al. (1997) Vessel diameters in roots versus stems of tropical lianas and other growth forms. *Iawa Journal*, **18**(3), 261–279.

Finzi, A.C., Canham, C.D. & Van Breemen, N. (1998) Canopy tree soil interactions within temperate forests: species effects on pH and cations. *Ecological Applications*, **8**, 447–454.

Gentry, A.H. (1983) Lianas and the "paradox" of contrasting latitudinal gradients in wood and litter production. *Tropical Ecology*, **24**, 63–67.

Gentry, A.H. (1991) The distribution and evolution of climbing plants, in *The Biology of Vines* (eds F.E. Putz & H.A. Mooney), Cambridge University Press, Cambridge, pp. 3–49.

Gourlet-Fleury, S., Rossi, V., Rejou-Mechain, M. et al. (2011) Environmental filtering of dense-wooded species controls above-ground biomass stored in African moist forests. *Journal of Ecology*, **99**(4), 981–990.

Hättenschwiler, S. (2002) Liana seedling growth in response to fertilisation in a neotropical forest understorey. *Basic and Applied Ecology*, **3**(2), 135–143.

Hättenschwiler, S., Aeschlimann, B., Coûteaux, M.-M. et al. (2008) High variation in foliage and leaf litter chemistry among 45 tree species of a neotropical rainforest community. *New Phytologist*, **179**, 165–175.

Hegarty, E.E. & Caballé, G. (1991) Distribution and abundance of vines in forest communities, in *The Biology of Vines* (eds F.E. Putz & H.A. Mooney), Cambridge University Press, Cambridge, pp. 313–335.

Hickman, J.E., Wu, S.L., Mickley, L.J. & Lerdau, M.T. (2010) Kudzu (Pueraria montana) invasion doubles emissions of nitric oxide and increases ozone pollution. *Proceedings of the National Academy of Sciences of the United States of America*, **107**(22), 10115–10119.

Hietz, H., Turner, B.L., Wanek, W. et al. (2011) Long-term change in the nitrogen cycle of tropical forests. *Science*, **334**, 664–666.

Homeier, J., Englert, F., Leuschner, C. et al. (2010) Factors controlling the abundance of lianas along an altitudinal transect of tropical forests in Ecuador. *Forest Ecology and Management*, **259**(8), 1399–1405.

Huston, M.A. (2012) Precipitation, soils, NPP, and biodiversity: resurrection of Albrecht's curve. *Ecological Monographs*, **82**, 277–296.

Ibarra-Manríquez, G. & Martínez-Ramos, M. (2002) Landscape variation of liana communities in a Neotropical rain forest. *Plant Ecology*, **160**(1), 91–112.

Isnard, S. & Silk, W.K. (2009) Moving with climbing plants from Charles Darwin's time into the 21st century. *American Journal of Botany*, **96**, 1205–1221.

Jacobsen, A.L., Pratt, R.B., Tobin, M.F. et al. (2012) A global analysis of xylem vessel length in woody plants. *American Journal of Botany*, **99**(10), 1583–1591.

John, R., Dalling, J.W., Harms, K.E. et al. (2007) Soil nutrients influence spatial distributions of tropical tree species. *Proceedings of the National Academy of Science*, **104**, 864–869.

Kazda, M. & Salzer, J. (2000) Leaves of lianas and self-supporting plants differ in mass per unit area and nitrogen content. *Plant Biology*, **2**, 268–271.

Kusumoto, B. & Enoki, T. (2008) Contribution of a liana species, Mucuna macrocarpa Wall., to litterfall production and nitrogen input in a subtropical evergreen broad-leaved forest. *Journal of Forest Research*, **13**(1), 35–42.

Lawrence, D. (2003) The response of tropical tree seedlings to nutrient supply: meta-analysis for understanding a changing tropical landscape. *Journal of Tropical Ecology*, **19**, 239–250.

Ledo, A. & Schnitzer, S. A. (2014) Disturbance, not negative density dependence or habitat specialization maintains liana diversity in a tropical forest. *Ecology*, in press.

Malizia, A., Grau, H.R. & Lichstein, J.W. (2010) Soil phosphorus and disturbance influence liana communities in a subtropical montane forest. *Journal of Vegetation Science*, **21**(3), 551–560.

Molina-Freaner, F. & Tinoco-Ojanguren, C. (1997) Vines of a desert plant community in central Sonora, Mexico. *Biotropica*, **29**(1), 46–56.

Moreira, M.Z., Sternberg, L.D.L. & Nepstad, D.C. (2000) Vertical patterns of soil water uptake by plants in a primary forest and an abandoned pasture in the eastern Amazon: an isotopic approach. *Plant and Soil*, **222**(1–2), 95–107.

Neumann, R.B. & Cardon, Z.G. (2012) The magnitude of hydraulic redistribution by plant roots: a review and synthesis of empirical and modeling studies. *New Phytologist*, **194**(2), 337–352.

Nogueira, A., Costa, F.R.C. & Castilho, C.V. (2011) Liana abundance patterns: the role of ecological filters during development. *Biotropica*, **43**(4), 442–449.

Nurfazliza, K., Nizam, M.S. & Supardi, M.N.N. (2012) Association of liana communities with their soil properties in a lowland forest of Negeri Sembilan, Peninsular Malaysia. *Sains Malaysiana*, **41**(6), 679–690.

Paoli, G.D., Curran, L.M. & Slik, J.W.F. (2008) Soil nutrients affect spatial patterns of aboveground biomass and emergent tree density in southwestern Borneo. *Oecologia*, **155**(2), 287–299.

Pérez-Salicrup, D.R. (2001) Effect of liana cutting on tree regeneration in a liana forest in Amazonian Bolivia. *Ecology*, **82**(2), 389–396.

Pérez-Salicrup, D.R. & Barker, M.G. (2000) Effect of liana cutting on water potential and growth of adult Senna multijuga (Caesalpinioideae) trees in a Bolivian tropical forest. *Oecologia*, **124**(4), 469–475.

Powers, J.S., Kalicin, M.H. & Newman, M.E. (2004) Tree species do not influence local soil chemistry in a species-rich Costa Rican rain forest. *Journal of Tropical Ecology*, **20**, 587–590.

Prieto, I., Armas, C. & Pugnaire, F.I. (2012) Water release through plant roots: new insights into its consequences at the plant and ecosystem level. *New Phytologist*, **193**(4), 830–841.

Proctor, J. (1983) Mineral nutrients in tropical forests. *Progress in Physical Geography*, **7**(3), 422–431.

Putz, F.E. (1984) The natural history of lianas on Barro Colorado Island, Panama. *Ecology*, **65**, 1713–1724.

Putz, F.E. & Chai, P. (1987) Ecological studies of lianas in Lambir National Park, Sarawak, Malaysia. *Journal of Ecology*, **75**(2), 523–531.

Quesada, C.A., Phillips, O.L., Schwarz, M. et al. (2012) Basin-wide variations in Amazon forest structure and function are mediated by both soils and climate. *Biogeosciences*, **9**(6), 2203–2246.

Restom, T.G. & Nepstad, D.C. (2001) Contribution of vines to the evapotranspiration of a secondary forest in eastern Amazonia. *Plant and Soil*, **236**(2), 155–163.

Restom, T.G. & Nepstad, D.C. (2004) Seedling growth dynamics of a deeply rooting liana in a secondary forest in eastern Amazonia. *Forest Ecology and Management*, **190**(1), 109–118.

Richter, D.D. & Babbar, L.I. (1991) Soil diversity in the tropics. *Advances in Ecological Research*, **21**, 315–389.

Rowe, N. & Speck, T. (2005) Plant growth forms: an ecological and evolutionary perspective. *New Phytologist*, **166**, 61–72.

Schlesinger, W.H. & Andrews, J.A. (2000) Soil respiration and the global carbon cycle. *Biogeochemistry*, **48**, 7–20.

Schlesinger, W.H., Raikes, J.A., Hartley, A.E. & Cross, A.F. (1996) On the spatial pattern of soil nutrients in desert ecosystems. *Ecology*, **77**, 364–374.

Schnitzer, S.A. (2005) A mechanistic explanation for global patterns of liana abundance and distribution. *American Naturalist*, **166**, 262–276.

Schnitzer, S.A. & Bongers, F. (2002) The ecology of lianas and their role in forests. *Trends in Ecology and Evolution*, **17**, 223–230.

Schnitzer, S.A. & Bongers, F. (2011) Increasing liana abundance and biomass in tropical forests: emerging patterns and putative mechanisms. *Ecology Letters*, **14**(4), 397–406.

Schnitzer, S.A., Mangan, S.A., Dalling, J.W. et al. (2012) Liana abundance, diversity, and distribution on Barro Colorado Island, Panama. *PLoS One*, **7**(12), e52114. doi: 10.1371/journal.pone.0052114

Schnitzer, S.A., van der Heijden, G.M.F., Mascaro, J. & Carson, W.P. In press. Lianas reduce biomass accumulation in a tropical forest. *Ecology*.

Scholz, F.G., Bucci, S.J., Goldstein, G. et al. (2008) Biophysical and life-history determinants of hydraulic lift in Neotropical savanna trees. *Functional Ecology*, **22**(5), 773–786.

Ter Steege, H., Sabatier, D., Castellanos, H. et al. (2000) An analysis of the floristic composition and diversity of Amazonian forests including those of the Guiana Shield. *Journal of Tropical Ecology*, **16**, 801–828.

Tobin, M.F., Wright, A.J., Mangan, S.A., & Schnitzer, S.A. (2012) Lianas have a greater competitive effect than trees of similar biomass on tropical canopy trees. *Ecosphere*, **3**(2), article 20: 1–11. 10.1890/ES11-00322.1

Townsend, A.R., Asner, G.P. & Cleveland, C.C. (2008) The biogeochemical heterogeneity of tropical forests. *Trends in Ecology and Evolution*, **23**, 424–431.

Van der Heijden, G.M.F., Schnitzer, S.A., Powers, J.S., & Phillips, O.L. (in review). Liana impacts on carbon cycle, storage and sequestration in tropical forests. *Biotropica*.

van der Heijden, G.M.F. & Phillips, O.L. (2008) What controls liana success in Neotropical forests? *Global Ecology and Biogeography*, **17**(3), 372–383.

van der Heijden, G.M.F. & Phillips, O.L. (2009) Environmental effects on Neotropical liana species richness. *Journal of Biogeography*, **36**(8), 1561–1572.

van der Heijden, G.M.F., Schnitzer, S.A., Powers, J.S. & Phillips, O.L. (2013) Liana impacts on carbon cycling, storage and sequestration in tropical forests. *Biotropica*, **45**, 682–692.

van Haren, J.L.M., de Oliveira, R.C., Restrepo-Coupe, N. et al. (2010) Do plant species influence soil CO_2 and N_2O fluxes in a diverse tropical forest? *Journal of Geophysical Research, Biogeosciences*, **115**.

Vitousek, P.M. & Sanford, R.L. (1986) Nutrient cycling in moist tropical forests. *Annual Review of Ecological Systems*, **17**, 137–167.

Wright, S.J. (1992) Seasonal drought, soil fertility and the species density of tropical forest plant communities. *Trends in Ecology & Evolution*, **7**(8), 260–263.

Yavitt, J.B. & Wright, S.J. (2008) Seedling growth responses to water and nutrient augmentation in the understorey of a lowland moist forest, Panama. *Journal of Tropical Ecology*, **24**, 19–26.

Yong, T., Kitching, R.L. & Cao, M. (2012) Lianas as structural parasites: a re-evaluation. *Chinese Science Bulletin*, **57**, 307–312.

Yuan, C.M., Liu, W.Y., Tang, C.Q. & Li, X.S. (2009) Species composition, diversity, and abundance of lianas in different secondary and primary forests in a subtropical mountainous area, SW China. *Ecological Research*, **24**(6), 1361–1370.

Zhu, S.D. & Cao, K.F. (2009) Hydraulic properties and photosynthetic rates in co-occurring lianas and trees in a seasonal tropical rainforest in southwestern China. *Plant Ecology*, **204**(2), 295–304.

Zinke, P.J. (1962) The pattern of influence of individual forest trees on soil properties. *Ecology*, **43**, 130–133.

Chapter 15

THE ROLE OF LIANAS IN TEMPERATE TREE COMMUNITIES

Laura M. Ladwig[1] *and Scott J. Meiners*[2]

[1]University of New Mexico, NM, USA
[2]Eastern Illinois University, IL, USA

Ecology of Lianas, First Edition. Edited by Stefan A. Schnitzer, Frans Bongers, Robyn J. Burnham, and Francis E. Putz.

Companion Website: www.wiley.com/go/schnitzer/lianas

OVERVIEW

Lianas occupy temperate forests worldwide and presumably perform important ecological functions; however, the ecological roles of lianas in temperate ecosystems remain poorly understood. In this chapter, we propose a scale-dependent framework of ecological drivers of lianas to organize existing information and suggest opportunities for further research on temperate liana ecology. At the coarsest scale, moisture limitations and cold temperatures shape regional distributions of lianas. Superimposed on broad climate-driven patterns are numerous species introductions that influence the spread, diversity, and density of lianas in temperate forests. Natural and anthropogenic disturbance, combined with the general opportunistic strategies of lianas, creates areas of high liana density across the temperate landscape. Throughout individual forest stands, processes at many spatial and temporal scales shape local liana density and composition. Within individual trees, host–liana interactions vary with species, location, and demographic age to shape both liana colonization patterns and their impacts on trees. Although temperate liana research is increasing, it still lacks a solid theoretical underpinning and lags greatly behind tropical studies. In the light of ongoing global climate change and increasing anthropogenic pressures, there is a pressing need for additional studies on the ecology of temperate lianas at multiple scales.

INTRODUCTION: TEMPERATE LIANAS

The global distribution of temperate forests is defined by precipitation and temperature, with temperate forests receiving 50–350 cm of precipitation annually and mean annual temperatures largely ranging from 4–20 °C. Lianas occur worldwide within the temperate forests of eastern North America, western Europe, eastern Asia, and smaller regions of temperate rainforest in southwest South America, eastern Asia, southeast Australia, northwest North America and New Zealand (Alaback 1991). Although liana density and diversity vary among temperate forests, in general, lianas comprise up to ~10% of woody temperate floras (Gentry 1991; Schnitzer & Bongers 2002). Lianas occur throughout temperate forest strata, from understory to upper canopy, and influence all forest ages, from young regenerating forests to mature old-growth stands (Londré & Schnitzer 2006; Rogers et al. 2008; Ladwig & Meiners 2010a, 2010b). Current knowledge on temperate liana ecology largely comes from North American systems, but throughout this chapter we include as much information as possible from other global regions in an attempt to provide a worldwide perspective of temperate lianas.

Lianas are often thought of as a tropical forest phenomenon; however, they are also an influential component of temperate forests. The primary differences between tropical and temperate lianas and their host communities revolve around climate, particularly water availability and the presence of a cold dormant season. These climactic differences alter liana physiology, necessitate additional dormancy adaptations, and constrain liana function within the forest. Liana diversity is higher in tropical forests, commonly reaching 25–35% of topical woody floras (Schnitzer & Bongers 2002; Schnitzer et al. 2012). For example, a relatively diverse Chilean temperate rainforest contained 13 liana spp/ha while a much greater 60 and 65 spp/ha occurred within tropical forests in La Selva, Costa Rica and Barro Island Colorado, Panama (Putz 1984; Mascaro et al. 2004; Gianoli et al. 2010; Yorke et al. 2013). Although generally much lower, temperate liana colonization and abundance can reach high levels comparable to tropical ecosystems. Within alluvial forests in Italy, 45–79% of trees contained lianas, with similar colonization occurring in several tropical and subtropical forests (Castagneri et al. 2013). Thus, while lianas are generally far more abundant and diverse in tropical forests, temperate forests also have the capacity for high liana colonization and abundance. Lianas likely influence forest community dynamics worldwide, and the differences between temperate and tropical environments necessitates a focused, separate discussion of temperate lianas.

In this chapter, we examine many aspects of liana interactions with temperate tree communities. The chapter follows a spatially hierarchical view of liana ecology to organize our current knowledge. Empirical data address the current understanding of the role of lianas in temperate forests and provide a basic framework of how temperate lianas fit into the surrounding community. Following the empirical data, we discuss current research needs and unanswered questions, and conclude by recapping the future direction of temperate liana ecology.

A HIERARCHICAL VIEW OF LIANA ECOLOGY

Temperate liana ecology is a relatively young area of study that has progressed somewhat haphazardly; therefore it lacks a solid conceptual background. To organize the disparate studies and perspectives regarding lianas in temperate forests, we propose a spatially hierarchical conceptual framework to organize temperate liana ecology (Fig. 15.1). The processes incorporated into this framework range from regional determinants of liana diversity and abundance at the coarsest (regional) scale, down to individual tree–liana interactions at the finest scale. Within this framework, processes functioning at coarser spatial scales may constrain processes occurring at finer spatial scales. Below, we use this framework to discuss the ecology of lianas in temperate systems and to highlight the most pressing research needs.

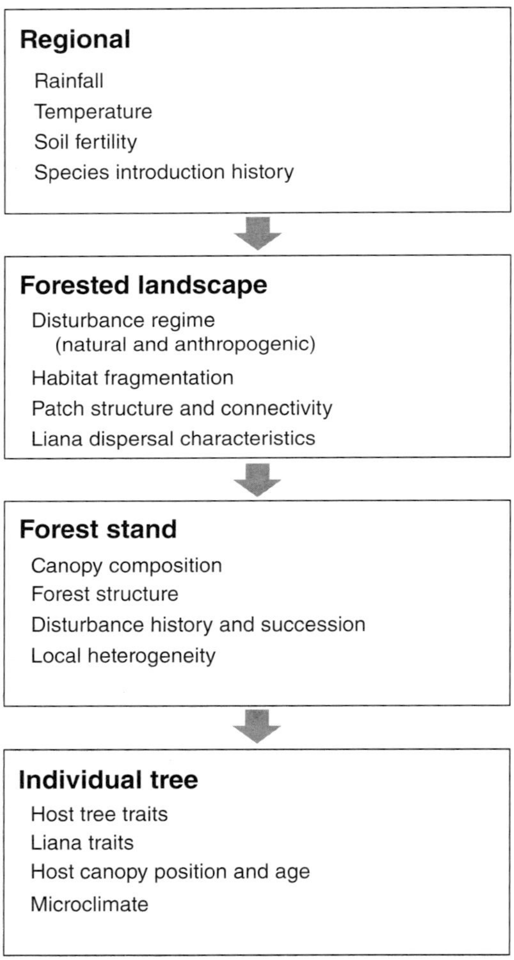

Fig. 15.1 Hierarchical conceptual diagram organizing the roles of lianas within temperate forests. Processes operating at coarser spatial scales (top) may constrain processes within finer spatial scales (progressively lower boxes).

REGIONAL CONTROLLERS

Liana abundance decreases with increasing latitude in the northern hemisphere, largely due to climate (Schnitzer 2005). As temperature and precipitation decrease across regional climate gradients, liana richness decreases worldwide, including within eastern Russia (Krestov 2003), China (Hu et al. 2010; Guo et al. 2012), and East Asia (Hu 2011). The distributions of many liana species are limited by water or temperature, yet lianas are highly adaptive and some species extend into arid regions (Molina-Freaner & Tinoco-Ojanguren 1997; Parsons 2005) and forests with extensive cold seasons (Zecca et al. 2012). Temperate regions with mild temperatures and consistent soil moisture, such as the south-eastern USA, appear more favorable to lianas and contain high temperate liana richness (Allen et al. 2007). At least 27 liana taxa occur in the Congaree floodplain in South Carolina (Allen et al. 1997), while only 13 liana species occur in more northern old secondary growth forests along southern Lake Michigan (Leicht-Young et al. 2010).

Lianas appear to respond to water availability and, as precipitation decreases, the percentage of lianas within the flora also typically decreases (Hu et al. 2010; Guo et al. 2012; but see also Swaine & Grace 2007). Across a decreasing annual rainfall gradient from 250 cm to 50 cm in China, the percentage of lianas in floras drops from 15% to 3% (Hu et al. 2010). To support the transpirational needs of disproportionately large canopies, lianas produce large xylem vessels to compensate for space restrictions imposed by narrow trunks (Angyalossy et al., Chapter 19 in this volume). Large vessels increase the efficiency of water movement, but also make the xylem vulnerable to drought-induced embolism; therefore, water conservation is necessary. To deal with water stress, some lianas decrease leaf area or grow within host canopies to take advantage of host canopy transpiration (Parsons 2006). Other temperate lianas are adapted to low soil moisture and can outperform non-liana neighboring or host plants. For example, *Wisteria sinensis*, a multi-stemmed liana native to China, maintained normal photosynthetic and transpiration rates until soil moisture reached 12% of full saturation, while potential host trees decreased photosynthesis rates much earlier at 41–71% of full saturation (Zhang et al. 2007). Differential adaptations for water conservation contribute to variation in liana distributions along climatic gradients.

Cold temperatures contribute to the latitudinal distribution of many plant species (Iverson & Prasad 1998), including lianas. However, lianas in particular are sensitive to cold temperatures compared to their host trees (Jiménez-Castillo & Lusk 2013). The large xylem vessels of lianas are problematic during cold temperatures, as ice formation can cavitate vessels and limit the ability of lianas to compete, ultimately limiting liana distributions (Ewers 1985; Sperry & Sullivan 1992). Across a temperature gradient in China, as mean January temperature drops from 15 °C to 5 °C, the percentage of lianas in the flora drops from 15% to 2% (Hu et al. 2010). Because of their susceptibility to cold temperatures, temperate lianas utilize various mechanisms to combat freeze damage to xylem. For example, *Vitis riparia* empties its xylem at the beginning of winter to prevent cavitation and uses root pressure to refill its xylem prior to spring bud break (Sperry et al. 1987; Tibbetts & Ewers 2000). *Celastrus orbiculatus* photosynthesizes until winter temperatures cavitate the xylem, then new xylem is produced in the spring (Tibbetts & Ewers 2000). Aside from several notable liana species that can tolerate freezing (Schnitzer 2005), cold temperatures generally limit liana distributions in temperate regions, decreasing liana species richness.

Superimposed on the climatically driven patterns of native liana diversity is the regional distribution of non-native lianas (Leicht-Young & Pavlovic, Chapter 29 in this volume). Lianas are common problematic non-native invasive species and comprise 9% of invasive plants species in eastern North America (Bargeron et al. 2003). Some notable examples include *Pueraria lobata* (kudzu), *Lonicera japonica* (Japanese honeysuckle), and *Hedera helix* (English ivy). Species introduced into regional species pools may fill unoccupied niches (Horvitz et al. 1998) or compete directly with both native liana and tree communities (Leicht-Young et al. 2010, 2011). Resident liana communities can aid the invasion of related non-natives via hybridization, as in the case of *Celastrus orbiculaus* and *C. scandens* in North America (Pooler et al. 2002). In general, non-native woody species, particularly lianas, are largely bird-dispersed and more diverse near urban areas (Aronson et al. 2007). Non-native species richness correlates with human population density and results in regional variation in non-native liana species pools (Pyšek et al. 2009; Gavier-Pizarro et al. 2010). This pattern likely reflects the continued influence of anthropogenic introduction of plant species (Mack 2000; Pyšek et al. 2003), as well as the ongoing fragmentation and modification of temperate forests (Londré & Schnitzer 2006).

LANDSCAPE-LEVEL PROCESSES

Contemporary research on the landscape ecology of temperate lianas began with accounts of liana dynamics after disturbance. Following ice and wind-storms in north-eastern USA, Siccama et al. (1976) observed that trees hosting lianas were more severely damaged than those without lianas. After hurricanes in south-eastern USA, lianas heavily colonized the most severely damaged forests (Allen et al. 2005) and hurricane damage also provided colonization opportunities for non-native lianas (Horvitz et al. 1998). While liana colonization does not necessarily depend on an initial disturbance and, in the apparent absence of recent disturbance, even mature forests are not resistant to liana colonization (Allen et al. 2007; Pavlovic & Leicht-Young 2011), liana abundances generally increase following disturbance (e.g., Buron et al. 1998; Londré & Schnitzer 2006).

Anthropogenic disturbances often increase forest edges, thus resulting in high liana densities (Buron et al. 1998; Londré & Schnitzer 2006). Over 45 years, liana abundance remained the same in the intact understory of 14 old-growth forests in Wisconsin, USA, but appeared to increase overall due to an increase in forest edges (Londré & Schnitzer 2006). Lianas were 4.5 times more abundant along forest edges than in interiors (Londré & Schnitzer 2006). While forest fragmentation can initially increase liana abundance, too much deforestation decreases liana habitat and can reduce liana abundance in the landscape. For example, as human land use decreased gallery forests along European river valleys, native *Vitis* populations declined (Schnitzler 1995). The combined effects of flood mitigation tactics and reduced forest habitat pushed *Vitis* populations along European river valleys to near extinction (Arnold et al. 2005, 2010). Depending on the patterns and magnitude of deforestation, lianas can flourish along newly established edges or flounder if remaining habitat is too small to maintain populations.

Disturbance frequently increases light availability, but light levels alone may not govern liana colonization in temperate forests. Many temperate lianas are shade-tolerant (Valladares et al. 2011) and in some temperate forests, light availability is related to growth and photosynthetic rates (Gianoli et al. 2012), but unrelated to liana abundance. Carrasco-Urra and Gianoli (2009) reported large differences in light infiltration in mature and edge forest sites, yet after accounting for size and abundance of host trees, light availability did not relate to liana density. In a mesic hardwood forest, canopy openness did not correlate with understory liana abundance or size, indicating that, contrary to expectations, understory individuals were not concentrated in high light areas (Collins & Wein 1993). Additionally, similar liana abundance and composition occurred in treefall gaps, secondary, and old-growth temperate rainforests in Chile (Gianoli et al. 2010). As light availability does not always drive temperate liana expansion following disturbance, additional factors may help explain the increase in lianas associated with disturbance.

A key component in the colonization of disturbed habitats is the ability of species to disperse into available sites. Many tropical lianas rely heavily on clonal expansion (Schnitzer et al. 2012) and this strategy is also likely to be important in temperate systems (Larson et al. 2002). Clonal spreading can be extensive within forests (Yorke et al. 2013); however, longer-distance seed dispersal would be necessary to colonize unconnected forested patches within a fragmented landscape. Following seedling establishment, many lianas can expand clonally, compensating for low initial rates of colonization or limited seed production (Sakai et al. 2002; Deiller et al. 2003; Schnitzer et al. 2012).

To address the comparative vagility of lianas in landscapes, we determined seed dispersal modes for North American genera of lianas, herbaceous vines, and trees using the United States Department of Agriculture's PLANTS database (www.plants.usda.gov). Comparing lianas and trees allowed us to evaluate how seeds of lianas and their tree hosts move throughout the landscape, while comparing seed movement of lianas and herbaceous vines captured the dispersal of these potentially competing climbing growth forms. Overall,

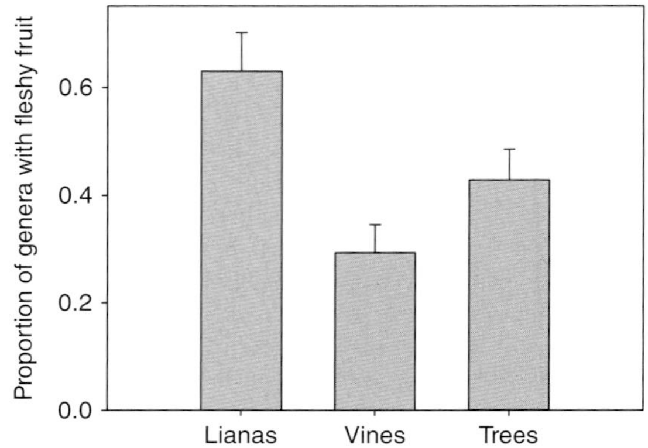

Fig. 15.2 Proportion of temperate climbers and trees (angiosperms only) with fleshy fruit based on genus level data from the USDA PLANTS database. The database search resulted in 75 vine, 46 liana, and 168 tree genera.

we found that temperate lianas were predominantly fleshy fruited. Lianas were more than twice as likely to have fleshy fruits than herbaceous vines (Fig. 15.2), illustrating the dependence of lianas on avian vectors compared to their herbaceous counterparts. Lianas also had 50% more fleshy fruits than trees, indicating that host and structural parasites use different mechanisms to move across the landscape. While focusing on fleshy fruits specifically excluded other types of vertebrate dispersal, epizoochorpy is often functionally difficult to separate from other dispersal modes. Vertebrate dispersed seeds are typically larger, a critical trait for establishing in shaded conditions of a forest understory (Westoby & Wright 2006). In contrast, herbaceous vines, many of which are annuals, often establish in more open conditions where the production of many small seeds with high abiotic dispersal ability is favored (Grime 2001). At a global scale, the proportion of lianas with animal-dispersed seeds tends to increase with latitude (Gallagher & Leishman 2012a; Gallagher, Chapter 09 in this volume), therefore animal dispersal may be more important in temperate forests than in the tropics. Mechanistically, the association between lianas and fleshy fruits may be functionally important in maintaining their abundance in the landscape.

Within temperate forests, animal-dispersed seeds may have a smaller seed shadow than wind-dispersed species (Clark et al. 1999), but there is the potential for directed, long-distance dispersal. Directed dispersal is important for lianas since they require a suitable host tree for success. Birds effectively disperse liana seeds as their selective usage of perch sites often represents structural supports for lianas. Therefore, lianas are often abundant along fencerows and forest edges, where birds congregate (Malmborg & Willson 1988; Londré & Schnitzer 2006). The tendency of birds to utilize perches in open habitats provides a vector for rapid, long-distance colonization by lianas (McDonnell & Stiles 1983; Robinson & Handel 1993).

LIANAS WITHIN FOREST STANDS

Liana abundance and composition in individual forest patches change over space and time. Highest liana densities typically occur along forest edges (Buron et al. 1998; Londré & Schnitzer 2006), reflecting the processes that drive landscape patterns of lianas. In addition to higher densities, individuals along forest edges grow taller than those in the forest interior (Ichihashi et al. 2010) and may also provide propagules for colonizing intact forest interiors and gaps. Following dispersal, local soil depressions may impact soil moisture and can alter liana colonization (Allen et al. 1997) as soil moisture and other microclimate variations may influence the survival of lianas in the forest understory (Collins & Wein 1993; West et al. 2010). Once established, many liana species exhibit a "sit and wait" strategy (Greenberg et al. 2001), persisting as small understory individuals (Ellsworth et al. 2004) until light openings appear and stunted lianas bolt into the canopy (Leicht & Silander 2006). The sit and wait strategy, plus the fast potential growth rates and plastic nature of lianas, necessitate the study of both canopy and understory individuals.

As tree composition changes through forest regeneration, the effects of lianas on trees also change. Lianas can recruit early in forest regeneration and influence the trajectory of succession (Fike & Niering 1999; Morrissey et al. 2009; Ladwig & Meiners 2010b). Even before trees become dominant components of successional communities, lianas can alter leaf and stem allocation in early successional herbs (Friedland & Smith 1982). Lianas may decrease the speed of succession through resource competition with trees. In extreme cases, lianas dominate woody communities to form a persistent scrub stage and in effect halt forest regeneration (Fike & Niering 1999; Royo & Carson 2006). Alternatively, lianas may preferentially colonize and compete with early successional tree species, which may release late successional trees and accelerate successional transitions (Ladwig & Meiners 2010b).

As forests develop, liana communities transition with changing host and resource availability. Within younger forests, liana communities largely consist of twiners, climbing the abundant young, small diameter trees (Putz 1995; Leicht-Young et al. 2010). As temperate forests age, total liana abundance may decrease (Robertson et al. 1994), while liana community composition shifts (Rogers et al. 2008). For example, during 50 years of forest regeneration in north-eastern USA, liana cover peaked during the transition from a herbaceous to woody-dominated community, presumably when hosts became abundant yet sufficient light remained (Ladwig & Meiners 2010b). Additionally, a functional separation between early and late successional lianas appeared during the transition from early to late successional trees (Fig. 15.3; Ladwig & Meiners 2010b). Early successional lianas, such as *Lonicera japonica* and *Toxicodendron radicans*, are more shade-intolerant and expand rapidly in open communities, following temporal patterns similar to shrubs in succession. In contrast, late successional lianas, including many species in the genus *Vitis*, colonize more slowly and spread throughout a site, but continue to expand once forest canopy closure occurs (Fig. 15.3). Thus, liana communities change through time, perhaps in conjunction with their host trees and potentially initiating cascading effects on surrounding forest entities.

Lianas contribute to various facets of forest diversity and productivity. The inclusion of lianas in forest inventories increases forest diversity, though not as much as in the tropics (Schnitzer & Bongers 2002; Schnitzer et al. 2012). Lianas can dominate the understory, averaging 78% cover under Maritime forest in coastal Virginia, USA (Naumann & Young 2007), and decrease understory herbaceous vegetation through shading and resource competition (Friedland & Smith 1982; Biggerstaff & Beck 2007). Some lianas can thrive under dense, shaded conditions (Leicht-Young et al. 2011) and may influence overall forest productivity by occurring throughout the forest strata, from the understory to the upper canopy; however, the extent of liana contribution to productivity in temperate forests remains unknown. Additionally, non-native lianas photosynthesize later into the autumn (Fridley 2012) and can alter carbon dynamics in the forest. Lianas also add structural complexity to the forest, which increases niche space for forest fauna and increases

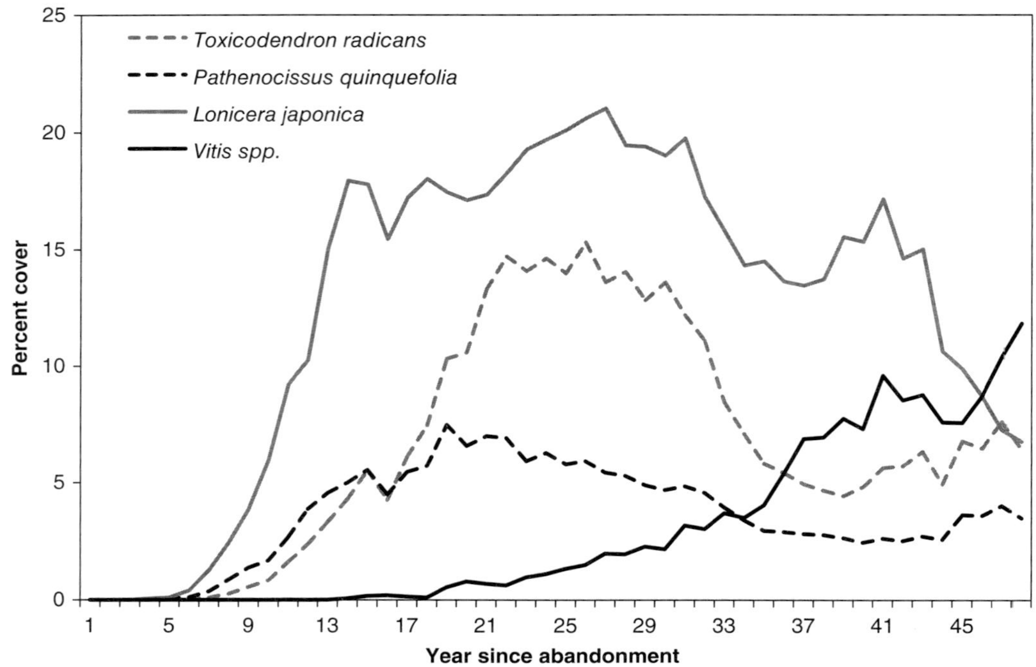

Fig. 15.3 Cover of the four dominant liana species within long-term forest regeneration plots of the Buell-Small Succession Study, located in the piedmont region of New Jersey, USA. Early successional liana species (*Lonicera japonica* and *Toxicodendron radicans*) increased earlier in forest succession and decreased as tree cover increased. Meanwhile, late successional species (*Vitis* spp.) entered the community later in succession and are continuing to increase within the forest. (Source: Ladwig & Meiners 2010b.)

ground-canopy connectivity (Michel et al., Chapter 25 in this volume, Yanoviak, Chapter 24 in this volume). Within bottomland hardwood forest, predatory snakes that prey on bird nests only climbed large *Quercus nuttallii* trees if a liana was present (Mullin & Cooper 2002). As many liana species are bird dispersed, they may provide seasonally important resources for birds (Stiles 1980). The foliage of some deciduous lianas also changes color early in the fall to serve as a 'fruit flag' to attract migrating birds (Stiles 1982). Lianas have the potential to influence many aspects of temperate forest function, most of which we do not fully understand.

LIANA INTERACTIONS WITH INDIVIDUAL TREES

Landscape and stand characteristics contribute to the abundance and diversity of lianas; however, not all trees in a forest are equally impacted. Lianas may exhibit host specificity based on tree species identity, size, or shape (Carter & Teramura 1988; Talley et al. 1996; Buron et al. 1998; Ladwig & Meiners 2010a). Host specificity relates to a combination of potential characteristics, including host architecture, bark texture, and degree of light attenuation, as well as the liana climbing mechanism (Carter & Teramura 1988), and may result in differential impacts of lianas on tree species (Ladwig & Meiners 2009). For example, *Vitis* spp. were more often associated with *Prunus*, *Ulmus*, and *Juglans* spp. trees within forests of mid-western USA (Morrissey et al. 2009) and, independent of host size, lianas more often climbed *Tepualia stipularis* and *Podocarpus nubigena* trees in a Chilean temperate rain forest (Muñoz et al. 2003). Twiners and tendril climbers require smaller diameter or trellis-like hosts while lianas with specialized attachment structures (e.g., root climbers) frequently climb larger trees with rough, firmly attached bark that can support the weight of the liana (Carter & Teramura 1988; Talley et al. 1996; Allen et al. 1997; Schnitzler & Heuzé 2006; Carrasco-Urra & Gianoli 2009; Morrissey et al. 2009; Leicht-Young et al. 2010). Within regenerating forests in New Jersey, the USA, *Lonicera japonica*, a twining liana, occurred most often on *Juniperus virginiana*, a tree with trellis-like branching, and *Cornus florida*, a small understory tree (Ladwig & Meiners 2010a). Therefore, liana host preference may correspond to specific liana climbing needs.

Liana host specificity can also relate to tree diameter, with larger trees often hosting more lianas (Carter & Teramura 1988; Leicht-Young et al. 2010; Ladwig & Meiners 2010a). Larger trees are often presumed to be older and therefore had a longer time to accumulate lianas. Although, when host tree and liana age were taken into account, age minimally impacted colonization and larger trees were mainly colonized due to greater surface area, not length of time present in the forest (Castagneri et al. 2013). Existing lianas can also act as a support for further liana colonization, therefore, once a tree contains one liana, it is more likely to be colonized by additional stems and species.

Lianas can occupy different regions within an individual host tree and the positioning of lianas throughout a host also influences how lianas impact that tree's performance. Some liana species occur predominately along the trunk, others within the lower canopy, and some occupy the upper canopy layer above the host foliage (Baars et al. 1998; Ladwig & Meiners 2010a; Blick & Burns 2011; Ichihashi & Tateno 2011). Separation of lianas within a tree could be a type of niche partitioning and allow multiple lianas to successfully grow on a single host. Lianas growing in uppermost tree canopies not only intercept more light, but also make host trees top-heavy and prone to fall, compared to lianas located lower on a host tree (Ichihashi & Tateno 2011). Lianas wrapped around branches can break small canopy twigs or physically constrict and alter the efficiency of vascular transport in larger limbs and trunks (Lutz 1943). Lianas also alter host tree resource allocation through both above- and belowground competition (Dillenburg et al. 1993, 1995). The presence of a host can also influence liana form, with *Toxicodendron diversilobum* climbing available hosts yet also growing shrub-like in the absence of a host (Gartner 1991a, 1991b). This flexibility in growth form may increase liana survival and reproduction, and enable the subsequent colonization of tree hosts.

Lianas compete with trees at many developmental stages and can influence tree growth and regeneration (cf., Toledo-Aceves, Chapter 12 in this volume). Lianas presumably reduce tree productivity but the impact of lianas on tree growth is difficult to quantify in the absence of experimental manipulations (e.g., Dillenburg et al. 1993; Schnitzer & Carson 2010). In a young oak-hickory forest in north-eastern USA, reduced tree growth was associated with high liana canopy cover, but unrelated to the number of stems or basal area of lianas along the trunk (Fig. 15.4; Ladwig & Meiners 2009). Presumably, canopy interactions between lianas and trees, including shading or physical damage to canopy branches, decreased tree growth. Additionally, decreased tree growth occurred only within dominant canopy trees (Fig. 15.4), therefore lianas appeared to differentially influence tree growth throughout the forest. Removal of lianas can also lead to increased host growth (Whigham 1984), demonstrating that lianas suppress tree performance.

The degree of liana colonization and the resulting impact on tree growth can change with time. Within *Quercus*-dominated forests in southern Italy, *Hedera helix* initially had a positive association with host tree growth that transitioned into a competitive relationship as forest regeneration progressed (Garfi & Ficarrotta 2003). Although it is difficult to separate causation from correlation without the appropriate experimental manipulations, differential decomposition rates of tree and liana leaves suggest litter from *Hedera helix* may increase soil fertility for host trees (Badre et al. 1998). What is unknown, however, is whether lianas also uptake more nutrients at the expense of their host trees, thus reducing both above- and belowground competition (e.g., Dillenburg et al. 1993, 1995).

Lianas not only impact the trees they climb, but can also influence neighboring trees. The tendency of lianas to connect neighboring canopies lowers tree flexibility, altering tree response to certain disturbances. As liana canopy density increases, the risk of mortality following severe storms also increases, damaging and reducing residence time for lianas in the canopy (Siccama et al. 1976; Allen et al. 1997). Furthermore, some temperate lianas possess allelopathic potential that may influence liana–host relations (Pisula & Meiners 2010). In laboratory bioassays examining the allelopathic potential of sun and shade leaves from dominant liana species (*Celastrus orbiculatus*, *Lonicera japonica*, *Toxicodendron radicans*, *Parthenocissus*

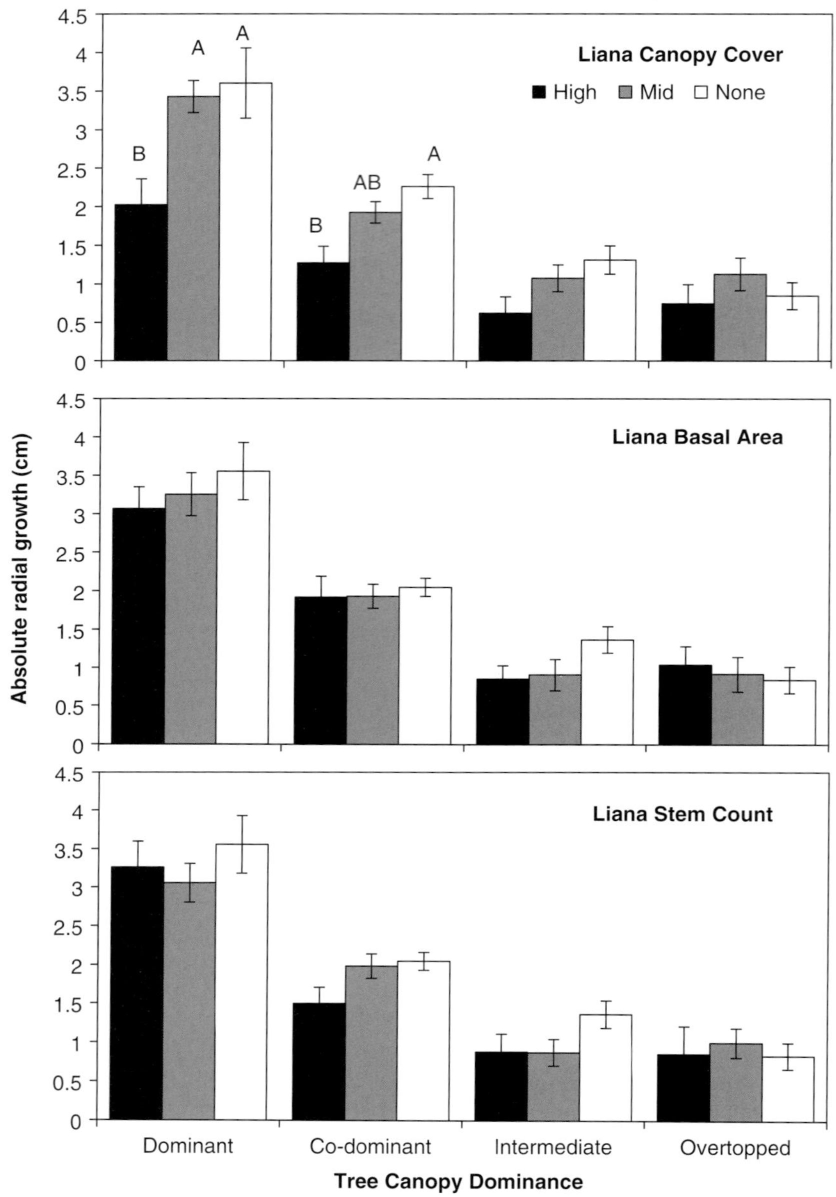

Fig. 15.4 Absolute radial growth (cm) of trees from 1999–2008 in relation to tree canopy dominance classes and liana loads. Letters indicate where significant differences in tree growth occurred between liana loads. Significantly less host tree growth was associated with high amounts of liana canopy cover in dominant and co-dominant trees. Canopy dominance classification was based on Smith (1986). Dominant crowns were above the general canopy and received full sun. Co-dominant crowns formed the canopy and received full sun, except along the edges. Intermediate crowns also reached the canopy, but had smaller crowns and received mainly indirect sun. Overtopped trees occurred below the canopy and received no direct sun. (Source: Ladwig & Meiners 2009. Reproduced with permission of Elsevier.)

quinquefolia, and *Vitis vulpina*), the greatest allelopathic potentials varied between species and ranged from 10 to 65% germination inhibition. Shade leaves of *T. radicans*, *L. japonica* and *P. quinquefolia* also had greater allelopathic potential than sun leaves, suggesting allelopathic activity may be induced by shading in some liana species (Ladwig et al. 2012), leading to differential impacts based on local conditions.

RESEARCH NEEDS AND UNANSWERED QUESTIONS

As temperate lianas have received less research attention in general, it is not surprising that many fundamental questions remain unanswered. Below we outline some of the issues that arose while examining the existing temperate liana literature.

Methodological issues

From a practical standpoint, there is a strong need to determine the optimal method to measure liana growth rates. Aging individuals is difficult as liana wood is often amorphous without distinct growth rings. Even when stems form distinguishable growth rings (Castagneri et al. 2013), individuals may consist of multiple stems of various ages. Additionally, lianas can sit and wait for extended times or grow remarkably fast, preventing estimates of size-age (DBH-biomass) relationships. Proper age assessments are necessary to quantify the speed at which lianas colonize regenerating forests, and the reciprocal nature of tree–liana interactions. Similarly, abiding by standard liana sampling protocols (e.g., Gerwing et al. 2006; Schnitzer et al. 2006, 2008) will allow unambiguous combination of datasets to address important questions on regional and pan-temperate patterns of liana geography.

Climate-liana linkages

Liana communities change across temperature and precipitation gradients, but it is unclear how liana species respond to climate change across their distribution. Based on common garden studies, growth and competitive ability of *Lonicera japonica* can vary between core and edge liana populations (e.g., Evans et al. 2013). However, more species need to be examined to generate a broad understanding of how lianas will respond to global change. Surveying liana species across the entire distributional range to assess changes in anatomy and behavior across climatic gradients can develop a mechanistic linkage with climate change (Schnitzer & Bongers 2011; Schnitzer, Chapter 30 in this volume). While soil fertility may have minimal effects on tropical liana distributions (e.g., Dalling et al. 2012; Ledo & Schnitzer in press; Powers, Chapter 14 in this volume), similar patterns may occur in temperate regions (Gallagher & Leishman 2012b) but influences of soil properties on temperate lianas are largely unexplored. Broader surveys that determine interactions and relative contributions of climate, soil, light, precipitation, and landscape structure will help project future shifts in liana dynamics under climate change scenarios.

Spatial dynamics of lianas in landscapes

Studies of liana distributions in temperate landscapes are few and largely descriptive. Studies that incorporate the role of corridors, location of source/sink populations and disperser behavior are needed to link patterns with landscape theory. Future research should combine multiple movement mechanisms, including species introductions, sexual and clonal expansion, human land alterations, and temporal changes during forest succession to provide a dynamic and comprehensive view of liana spatial dynamics at the landscape scale. Finally, detailed analyses of dispersal mode beyond the simple survey presented here will be necessary to separate phylogenetic influences and to determine the ecological implications of vertebrate dispersal for native and non-native temperate lianas.

Population dynamics of lianas

Research on changing liana abundances has typically examined population expansions (e.g., Allen et al. 2007; Schnitzer & Bongers 2011), largely neglecting the dynamics of population decline (but see Schnitzler 1995; Arnold et al. 2005, 2010). Determining when and how lianas naturally move into and out of a community will add to our understanding of the ecology of temperate lianas, and help manage problematic lianas. Furthermore, the inclusion of lianas at all demographic stages, from understory through to the upper canopy, will provide a more unified view of liana dynamics within forests.

Liana-forest stand linkages

Liana density and diversity vary greatly among temperate forest stands, but how this relates to stand density, diversity or function is poorly understood. For example, across a spectrum of northern temperate forests in the mid-western USA, both liana and tree species richness varied among sites, but were unrelated to each other (Leicht-Young et al. 2010). Thus, controls on liana richness appear to differ from those on canopy trees. We need research that integrates lianas with forest structure and history, wildlife population dynamics, trophic interactions and ecosystem processes. The potentially high abundance of lianas alone argues for their importance in temperate forests.

Liana–host interactions

A major limitation in liana research is the separation of correlation and causation in assessing liana impacts. It is largely impossible to separate whether lianas decrease tree growth or thrive on slow-growing individuals. Most experimental manipulations of temperate lianas evaluate interactions between young lianas and trees and it remains unclear whether these interactions will remain consistent through time. Furthermore, the belowground dynamics of temperate tree-liana competition remain virtually unknown (but see Dillenburg et al. 1993, 1995; Toledo-Aceves, Chapter 12 in this volume). Research relating the importance of above- and belowground competition to host and liana developmental stage is needed to place liana–tree interactions in an appropriate context.

CONCLUSION: THE FUTURE OF LIANAS IN TEMPERATE FORESTS

Changing landscapes and climates will inevitably influence temperate lianas and their host trees. Growing human populations increase ecological disturbances, which will likely increase liana abundance. Landscape changes will presumably include increased forest fragmentation and shifting forest locations, as some forests are cleared while other sites become reforested. Projections of future climates include substantial changes in temperature and precipitation regimes, both of which influence lianas at regional scales. Successful species will need to effectively disperse and establish within the new disturbance and climatic regimes. However, intrinsic physiological characteristics may favor lianas relative to other woody species under future conditions. For example, *Hedera helix*, a dominant liana in European forests, has more efficient water regulation than its host trees and is predicted to perform better in warmer, drier climates (Leuzinger et al. 2011). Under elevated CO_2, *H. helix* grows faster in the forest understory than other co-occurring woody species (Hättenschwiler & Körner 2003; Zotz & Körner 2006). *Toxicodendron radicans*, a widespread North American liana that causes dermatitis, becomes more toxic under increased CO_2 (Mohan et al. 2006). The response in the secondary chemistry of *T. radicans* suggests that climate change can generate subtle chemical shifts with currently unknown ecological implications. As the landscape becomes increasingly influenced by humans, it is important for liana ecology research to extend into more urban environments to better understand the potential roles of lianas in human-dominated systems (Francis 2011).

There is still much to be learned about the role of lianas in temperate forest communities. Within the spatially hierarchical framework that we use to organize temperate liana research, most aspects of temperate liana ecology are poorly understood. Most of our knowledge comes from simple descriptive surveys that are limited in determining mechanisms and separating pattern from process. Most important are perhaps some of the linkages among scales. For example, lianas are often notably more abundant at forest edges. While light availability is often assumed to drive this pattern, it has not been explicitly separated from dispersal processes. We also do not know whether the abundance of lianas at forest edges leads to greater overall abundance in fragmented landscapes or increases the probability of colonizing interior habitats. Some unknowns can be tackled by utilizing pre-existing datasets that were not initiated with lianas in mind but can be re-analyzed to fill knowledge gaps. Many unknowns will require new experiments designed with specific liana related questions in mind. Lianas are an important component of temperate forests, but far more research is necessary to discover their myriad contributions.

REFERENCES

Alaback, P.B. (1991) Comparative ecology of temperate rain-forests of the Americas along analogous climate gradients. *Revista Chilena de Historia Natural*, **64**, 399–412.

Allen, B.P., Pauley, E.F. & Sharitz, R.R. (1997) Hurricane impacts on liana populations in an old-growth southeastern bottomland forest. *Journal of the Torrey Botanical Society*, **124**, 34–42.

Allen, B.P., Sharitz, R.R. & Goebel, P.C. (2005) Twelve years post-hurricane liana dynamics in an old-growth southeastern floodplain forest. *Forest Ecology and Management*, **218**, 259–269.

Allen, B.P., Sharitz, R.R. & Goebel, P.C. (2007) Are lianas increasing in importance in temperate floodplain forests in the southeastern United States? *Forest Ecology and Management*, **242**, 17–23.

Arnold, C., Schnitzler, A., Douard, A. et al. (2005) Is there a future for wild grapevine (*Vitis vinifera* subsp silvestris) in the Rhine Valley? *Biodiversity and Conservation*, **14**, 1507–1523.

Arnold, C., Schnitzler, A., Parisot, C. & Maurin, A. (2010) Historical reconstruction of a relictual population of wild grapevines (*Vitis vinifera* ssp sylvestris, gmelin, hegi) in a floodplain forest of the upper Seine Valley, France. *River Research and Applications*, **26**, 904–914.

Aronson, M.F.J., Handel, S.N. & Clemants, S.E. (2007) Fruit type, life form and origin determine the success of woody plant invaders in an urban landscape. *Biological Invasions*, **9**, 465–475.

Baars, R., Kelly, D. & Sparrow, S.D. (1998) Liane distribution within native forest remnants in two regions of the South Island, New Zealand. *New Zealand Journal of Ecology*, **22**, 71–85.

Badre, B., Nobelis, P. & Trémolières, M. (1998) Quantitative study and modeling of the litter decomposition in a European alluvial forest: is there an influence of overstory tree species on the decomposition of ivy litter (*Hedera helix* L.)? *Acta Oecologia*, **19**, 491–500.

Bargeron, C.T., Moorhead, D.J., Douce, G.K., et al. (2003) Invasive plants of the eastern United States: Identification and control, The University of Georgia, USDA APHIS PPQ and USDA Forest Service Forest Health Technology Enterprise Team. www.invasive.org (accessed January 2009).

Biggerstaff, M.S. & Beck, C.W. (2007) Effects of method of English ivy removal and seed additions on regeneration of vegetation in a southeastern piedmont forest. *American Midland Naturalist*, **158**, 206–220.

Blick, R.A.J. & Burns, K.C. (2011) Liana co-occurrence patterns in a temperate rainforest. *Journal of Vegetation Science*, **22**, 868–877.

Buron, J., Lavigne, D., Grote, K. et al. (1998) Association of vines and trees in second-growth forest. *Northeastern Naturalist*, **5**, 359–362.

Carrasco-Urra, F. & Gianoli, E. (2009) Abundance of climbing plants in a southern temperate rain forest: host tree characteristics or light availability? *Journal of Vegetation Science*, **20**, 1155–1162.

Carter, G.A. & Teramura, A.H. (1988) Vine photosynthesis and relationships to climbing mechanisms in a forest understory. *American Journal of Botany*, **75**, 1011–1018.

Castagneri, D., Garbarino, M. & Nola, P. (2013) Host preference and growth patterns of ivy (*Hedera helix* L.) in a temperate alluvial forest. *Plant Ecology*, **214**, 1–9.

Clark, J.S., Silman, M., Kern, R. et al. (1999) Seed dispersal near and far: patterns across temperate and tropical forests. *Ecology*, **80**, 1475–1494.

Collins, B.S. & Wein, G.R. (1993) Understory vines: distributions and relation to environment on a southern mixed hardwood site. *Bulletin of the Torrey Botanical Club*, **120**, 38–44.

Dalling, J.W., Schnitzer, S.A., Baldeck, C. et al. (2012) Resource-based habitat associations in a neotropical liana community. *Journal of Ecology*, **100**, 1174–1182.

Deiller, A.F., Walter, J.M.N. & Tremolieres, M. (2003) Regeneration strategies in a temperate hardwood floodplain forest of the Upper Rhine: sexual versus vegetative reproduction of woody species. *Forest Ecology and Management*, **180**, 215–225.

Dillenburg, L.R., Teramura, A.H., Forseth, I.N. & Whigham, D.F. (1995) Photosynthesis and biomass allication responses of *Liquidambar styraciflua* (Hamamelidaceae) to vine competition. *American Journal of Botany*, **82**, 454–461.

Dillenburg, L.R., Whigham, D.F., Teramura, A.H. & Forseth, I.N. (1993) Effects of belowground and aboveground competition from the vines *Lonicera japonica* and *Parthenocissus quinquefolia* on the growth of the tree host *Liquidambar styraciflua*. *Oecologia*, **93**, 48–54.

Ellsworth, J.W., Harrington, R.A. & Fownes, J.H. (2004) Survival, growth and gas exchange of *Celastrus orbiculatus* seedlings in sun and shade. *American Midland Naturalist*, **151**, 233–240.

Evans, G.A., Kilkenny, F.F. & Galloway, L.F. (2013) Evolution of competitive ability within *Lonicera japonica*'s invaded range. *International Journal of Plant Sciences*, **174**, 740–748.

Ewers, F.W. (1985) Xylem structure and water conduction in conifer trees, dicot trees, and lianas. *IAWA Bulletin*, **6**, 309–317.

Fike, J. & Niering, W.A. (1999) Four decades of old field vegetation development and the role of *Celastrus orbiculatus* in the northeastern United States. *Journal of Vegetation Science*, **10**, 483–492.

Francis, R.A. (2011) Wall ecology: a frontier for urban biodiversity and ecological engineering. *Progress in Physical Geography*, **35**, 43–63.

Fridley, J.D. (2012) Extending leaf phenology and the autumn niche in deciduous forest invasions. *Nature*, **485**, 359–362.

Friedland, A.J. & Smith, A.P. (1982) Effects of vines on successional herbs. *American Midland Naturalist*, **108**, 402–403.

Gallagher, R.V. & Leishman, M.R. (2012a) A global analysis of trait variation and evolution in climbing plants. *Journal of Biogeography*, **39**, 1757–1771.

Gallagher, R.V. & Leishman, M.R. (2012b) Contrasting patterns of trait-based community assembly in lianas and trees from temperate Australia. *Oikos*, **121**, 2026–2035.

Garfi, G. & Ficarrotta, S. (2003) Influence of ivy (*Hedera helix* L.) on the growth of downy oak (*Quercus pubens* s.l.) in the Mont Caraci Natural Reserve (central-western Sicily). *Ecologia Mediterranea*, **29**, 5–14.

Gartner, B.L. (1991a) Is the climbing habit of poison oak ecotypic? *Functional Ecology*, **5**, 696–704.

Gartner, B.L. (1991b) Structural stability and architecture of vines vs shrubs of poison oak, *Toxicodendron diversilobum*. *Ecology*, **72**, 2005–2015.

Gavier-Pizarro, G.I., Radeloff, V.C., Stewart, S.I. et al. (2010) Housing is positively associated with invasive exotic plant species richness in New England, USA. *Ecological Applications*, **20**, 1913–1925.

Gentry, A.H. (1991) The distribution and evolution of climbing plants, in *The Biology of Vines* (eds F.E. Putz & H.A. Mooney), Cambridge University Press, Cambridge, pp. 3–49.

Gerwing, J.J., Schnitzer, S.A., Burnham, R.J. et al. (2006) Censusing lianas. *Biotropica*, **38**, 256–261.

Gianoli, E., Saldaña, A. & Jiménez-Castillo (2012) Ecophysiological traits may explain the abundance of climbing plant species across the light gradient in a temperate rainforest. *PLOS One*, **7**, e38831.

Gianoli, E., Saldana, A., Jiménez-Castillo, M. & Valladares, F. (2010) Distribution and abundance of vines along the light gradient in a southern temperate rain forest. *Journal of Vegetation Science*, **21**, 66–73.

Greenberg, C.H., Smith, L.M. & Levey, D.J. (2001) Fruit fate, seed germination and growth of an invasive vine: an experimental test of 'sit and wait' strategy. *Biological Invasions*, **3**, 363–372.

Grime, J.P. (2001) *Plant Strategies, Vegetation Processes, and Ecosystem Properties*, John Wiley & Sons, Chichester.

Guo, Y., Li, G., Kang, B. et al. (2012) The differential response of lianas and vines to rainfall gradients in Qinling Mountains, China. *Plant Ecology*, **213**, 1749–1755.

Hättenschwiler, S. & Körner, C. (2003) Does elevated CO_2 facilitate naturalization of the non-indigenous *Prunus laurocerasus* in Swiss temperate forests? *Functional Ecology*, **17**, 778–785.

Horvitz, C.C., Pascarella, J.B., McMann, S. et al. (1998) Functional roles of invasive non-indigenous plants in hurricane-affected subtropical hardwood forests. *Ecological Applications*, **8**, 947–974.

Hu, L. (2011) Distribution and diversity of climbing plants in temperate East Asia. *Biodiversity Science*, **19**, 567–573.

Hu, L., Li, M. & Li, Z. (2010) Geographical and environmental gradients of lianas and vines in China. *Global Ecology and Biogeography*, **19**, 554–561.

Ichihashi, R., Nagashima, H. & Tateno, M. (2010) Biomass allocation between extension- and leaf display-oriented shoots in relation to habitat differentiation among five deciduous liana species in a Japanese cool-temperate forest. *Plant Ecology*, **211**, 181–190.

Ichihashi, R. & Tateno, M. (2011) Strategies to balance between light acquisition and the risk of falls of four temperate liana species: to overtop host canopies or not? *Journal of Ecology*, **99**, 1071–1080.

Iverson, L.R. & Prasad, A.M. (1998) Predicting abundance of 80 tree species following climate change in the eastern United States. *Ecological Monographs*, **68**, 465–485.

Jiménez-Castillo, M. & Lusk, C.H. (2013) Vascular performance of woody plants in a temperate rain forest: lianas suffer higher levels of freeze-thaw embolism than associated trees. *Functional Ecology*, **27**, 403–412.

Krestov, P.V. (2003) Forest vegetation of easternmost Russia (Russia Far East), in *Forest Vegetation of Northeast Asia* (eds J. Kolbek, M. Srutek & E.E.O. Box), Springer, Berlin, pp. 93–180.

Ladwig, L.M. & Meiners, S.J. (2009) Impacts of temperate lianas on tree growth in young deciduous forests. *Forest Ecology and Management*, **259**, 195–200.

Ladwig, L.M. & Meiners, S.J. (2010a) Liana host preference and implications for deciduous forest regeneration. *Journal of the Torrey Botanical Society*, **137**, 103–112.

Ladwig, L.M. & Meiners, S.J. (2010b) Spatiotemporal dynamics of lianas during 50 years of succession to temperate forest. *Ecology*, **91**, 671–680.

Ladwig, L.M., Meiners, S.J., Pisula, N.L. & Lang, K.A. (2012) Conditional allelopathic potential of temperate lianas. *Plant Ecology*, **213**, 1927–1935.

Larson, K.C., Fowler, S.P. & Walker, J.C. (2002) Lack of pollinators limits fruit set in the exotic *Lonicera japonica*. *American Midland Naturalist*, **148**, 54–60.

Leicht, S.A. & Silander, J.A. Jr. (2006) Differential responses of invasive *Celastrus orbiculatus* (Celastraceae) and native *C. scandens* to changes in light quality. *American Journal of Botany*, **93**, 972–977.

Leicht-Young, S.A., Latimer, A.M. & Silander, J.A. Jr. (2011) Lianas escaping self-thinning: Experimental evidence of positive density dependence in temperate lianas *Celastrus orbiculatus* and *C. scandens*. *Perspectives in Plant Ecology, Evolution and Systematics*, **13**, 163–172.

Leicht-Young, S.A., Pavlovic, N.B., Frohnapple, K.J. & Grundel, R. (2010) Liana habitat and host preferences in northern temperate forests. *Forest Ecology and Management*, **260**, 1467–1477.

Leuzinger, S., Hartmann, A. & Körner, C. (2011) Water relations of climbing ivy in a temperate forest. *Planta*, **233**, 1087–1096.

Londré, R.A. & Schnitzer, S.A. (2006) The distribution of lianas and their change in abundance in temperate forests over the past 45 years. *Ecology*, **87**, 2973–2978.

Lutz, H.J. (1943) Injuries to trees caused by *Celastrus* and *Vitis*. *Bulletin of the Torrey Botanical Club*, **70**, 436–439.

Mack, R. (2000) Cultivation fosters plant naturalization by reducing environmental stochasticity. *Biological Invasions*, **2**, 111–122.

Malmborg, P.K. & Willson, M.F. (1988) Foraging ecology of avian frugivores and some consequences for seed dispersal in an Illinois woodlot. *Condor*, **90**, 173–186.

Mascaro, J., Schnitzer, S.A. & Carson, W.P. (2004) Liana diversity, abundance and mortality in a tropical wet forest in Costa Rica. *Forest Ecology and Management*, **190**, 3–14.

McDonnell, M.J. & Stiles, E.W. (1983) The structural complexity of an old field vegetation and the recruitment of bird-dispersed plant species. *Oecologia*, **56**, 109–116.

Mohan, J.E., Ziska, L.H., Schlesinger, W.H. et al. (2006) Biomass and toxicity responses of poison ivy (*Toxicodendron radicans*) to elevated atmospheric CO_2. *Proceedings of the National Academy of Sciences of the United States of America*, **103**, 9086–9089.

Molina-Freaner, F. & Tinoco-Ojanguren, C. (1997) Vines of a desert plant community in central Sonora, Mexico. *Biotropica*, **29**, 46–56.

Morrissey, R.C., Gauthier, M.M., Kershaw, J.A. Jr. et al. (2009) Grapevine (*Vitis* spp.) dynamics in association with manual tending, physiography, and host tree associations in temperate deciduous forests. *Forest Ecology and Management*, **257**, 1839–1846.

Mullin, S.J. & Cooper, R.J. (2002) Barking up the wrong tree: climbing performance of rat snakes and its implications for depredation of avian nests. *Canadian Journal of Zoology*, **80**, 591–595.

Muñoz, A.A., Chacon, P., Perez, F. et al. (2003) Diversity and host tree preferences of vascular epiphytes and vines in a temperate rainforest in southern Chile. *Australian Journal of Botany*, **51**, 381–391.

Naumann, J.C. & Young, D.R. (2007) Relationship between community structure and seed bank to describe successional dynamics of an Atlantic Coast maritime forest. *Journal of the Torrey Botanical Society*, **134**, 89–98.

Parsons, R.F. (2005) Desert vines: a comparison of Australia with other areas. *Journal of Biogeography*, **32**, 121–126.

Parsons, R.F. (2006) Notes on the origins of Australian desert vines. *Andansonia*, **28**, 405–412.

Pavlovic, N.B. & Leicht-Young, S.A. (2011) Are temperate mature forests buffered from invasive lianas? *Journal of the Torrey Botanical Society*, **138**, 85–92.

Pisula, N.L. & Meiners, S.J. (2010) Relative allelopathic potential of invasive plant species in a young disturbed woodland. *Journal of the Torrey Botanical Society*, **137**, 81–87.

Pooler, M.R., Dix, R.L. & Feely, J. (2002) Interspecific hybridizations between the native bittersweet, *Celastrus scandens*, and the introduced invasive species, *C. orbiculatus*. *Southeastern Naturalist*, **1**, 69–76.

Putz, F.E. (1984) The natural history of lianas on Barro Colorado Island, Panama. *Ecology*, **65**, 1713–1724.

Putz, F.E. (1995) Relay ascension of big trees by vines in Rock Creek Park, District of Columbia. *Castanea*, **60**, 167–169.

Pyšek, P., Křivánek, M. & Jarošík, V. (2009) Planting intensity, residence time, and species traits determine invasion success of alien woody species. *Ecology*, **90**, 2734–2744.

Pyšek, P., Sádlo, J., Mandák, B. & Jarošík, V. (2003) Czech alien flora and the historical pattern of its formation: what came first to central Europe? *Oecologia*, **135**, 122–130.

Robertson, D.J., Robertson, M.C. & Tague, T. (1994) Colonization dynamics of 4 exotic plants in a northern piedmont natural area. *Bulletin of the Torrey Botanical Club*, **121**, 107–118.

Robinson, G.R. & Handel, S.N. (1993) Forest restoration on a closed landfill: rapid addition of new species by bird dispersal. *Conservation Biology*, **7**, 271–278.

Rogers, D.A., Rooney, T.P., Olson, D. & Waller, D.M. (2008) Shifts in southern Wisconsin forest canopy and understory richness, composition, and heterogeneity. *Ecology*, **89**, 2482–2492.

Royo, A.A. & Carson, W.P. (2006) On the formation of dense understory layers in forests worldwide: consequences and implications for forest dynamics, biodiversity, and succession. *Canadian Journal of Forest Research*, **36**, 1345–1362.

Sakai, A., Nomiya, H. & Suzuki, W. (2002) Horizontal distribution of stolons of a temperate liana *Wisteria floribunda* DC. and its ecological significance. *Journal of Forest Research*, **7**, 125–130.

Schnitzer, S.A. (2005) A mechanistic explanation for global patterns of liana abundance and distribution. *The American Naturalist*, **166**, 262–276.

Schnitzer, S.A. & Bongers, F. (2002) The ecology of lianas and their role in forests. *Trends in Ecology and Evolution*, **17**, 223–230.

Schnitzer, S.A. & Bongers, F. (2011) Increasing liana abundance and biomass in tropical forests: emerging patterns and putative mechanisms. *Ecology Letters*, **14**, 397–406.

Schnitzer, S.A. & Carson, W.P. (2010) Lianas suppress tree regeneration and diversity in treefall gaps. *Ecology Letters*, **13**, 849–857.

Schnitzer, S.A., DeWalt, S.A. & Chave, J. (2006) Censusing and measuring lianas: a quantitative comparison of the common methods. *Biotropica*, **38**, 581–591.

Schnitzer, S.A., Mangan, S.A., Dalling, J.W. et al. (2012) Liana abundance, diversity, and distribution on Barro Colorado Island, Panama. *PLOS One*, **7**, e52114.

Schnitzer, S.A., Rutishauser, S. & Aguilar, S. (2008) Supplemental protocol for liana censuses. *Forest Ecology and Management*, **255**, 1044–1049.

Schnitzler, A. (1995) Community ecology of arboreal lianas in gallery forest of the Rhine Valley, France. *Acta Oecologica: International Journal of Ecology*, **16**, 219–236.

Schnitzler, A. & Heuzé (2006) Ivy (*Hedera helix* L.) dynamics in riverine forests: effects of river regulation and forest disturbance. *Forest Ecology and Management*, **236**, 12–17.

Siccama, T.G., Weir, G. & Wallace, K. (1976) Ice damage in a mixed hardwood forest in Connecticut in relation to *Vitis* infestation. *Bulletin of the Torrey Botanical Club*, **103**, 180–183.

Sperry, J.S., Holbrook, N.M., Zimmermann, M.H. & Tyree, M.T. (1987) Spring filling of xylem vessels in wild grapevine. *Plant Physiology*, **83**, 414–417.

Sperry, J.S. & Sullivan, J.E.M. (1992) Xylem embolism in response to freeze-thaw cycles and water-stress in ring-porous, diffuse-porous, and conifer species. *Plant Physiology*, **100**, 605–613.

Stiles, E.W. (1980) Patterns of fruit presentation and seed dispersal in bird-disseminated woody plants in the eastern deciduous forest. *The American Naturalist*, **116**, 670–688.

Stiles, E.W. (1982) Fruit flags: two hypotheses. *The American Naturalist*, **120**, 500–509.

Swaine, M.D. & Grace, J. (2007) Lianas may be favored by low rainfall: evidence from Ghana. *Plant Ecology*, **192**, 271–276.

Talley, S.M., Lawton, R.O. & Setzer, W.N. (1996) Host preferences of *Rhus radicans* (Anacardiaceae) in a southern deciduous hardwood forest. *Ecology*, **77**, 1271–1276.

Tibbetts, T.J. & Ewers, F.W. (2000) Root pressure and specific conductivity in temperate lianas: Exotic *Celastrus orbiculatus* (Celastraceae) vs. native *Vitis riparia* (Vitaceae). *American Journal of Botany*, **87**, 1272–1278.

Valladares, F., Gianoli, E. & Saldaña, A. (2011) Climbing plants in a temperate rainforest understory: searching for high light or coping with deep shade? *Annals of Botany*, **108**, 231–239.

West, N.M., Gibson, D.J. & Minchin, P.R. (2010) Microhabitat analysis of the invasive exotic liana *Lonicera japonica* Thunb. *Journal of the Torrey Botanical Society*, **137**, 380–390.

Westoby, M. & Wright, I.J. (2006) Land-plant ecology on the basis of functional traits. *Trends in Ecology and Evolution*, **21**, 261–268.

Whigham, D. (1984) The influence of vines on the growth of *Liquidambar styaciflua* L. (sweetgum). *Canadian Journal of Forest Research*, **14**, 37–39.

Yorke, S.R., Schnitzer, S.A., Mascaro, J. et al. (2013) Increasing liana abundance and basal area in a tropical forest: the contribution of long-distance clonal colonization. *Biotropica*, **45**, 317–324.

Zecca, G., Abbott, J.R., Sun, W. et al. (2012) The timing and the mode of evolution of wild grapes (*Vitis*). *Molecular Phylogenetics and Evolution*, **62**, 736–747.

Zhang, S., Xia, J., Zhou, Z. & Zhang, G. (2007) Photosynthesis response to various soil moisture in leaves of *Wisteria sinensis*. *Journal of Forestry Research*, **18**, 217–220.

Zotz, G., Cueni, N. & Körner, C. (2006) In situ growth stimulation of a temperate zone liana (*Hedera helix*) in elevated CO_2. *Functional Ecology*, **20**, 763–769.

Part IV

Liana Evolution

Chapter 16

CLIMBING PLANTS IN THE FOSSIL RECORD: PALEOZOIC TO PRESENT

Robyn J. Burnham

University of Michigan, Ann Arbor, MI, USA

Ecology of Lianas, First Edition. Edited by Stefan A. Schnitzer, Frans Bongers, Robyn J. Burnham, and Francis E. Putz.

Companion Website: www.wiley.com/go/schnitzer/lianas

OVERVIEW

As an iconic member of modern tropical forests, lianas fill a unique niche in warm, humid forests. The species that dominate the liana guild in tropical forests today are drawn from the full breadth of phylogenetic diversity of flowering plants. Their leaves and stems contribute a small percentage of the aboveground biomass of most extant forests, but they disproportionately influence the ecosystems in which they live. While not restricted to the tropics, paleobotanists interpret the presence of fossil climbers to suggest a tropical climate due to the current overrepresentation of climbers in tropical ecosystems (Manchester 2000; Doria et al. 2008). In addition, because lineages of climbers today are dominated by angiosperms, recognition of the climbing life form in pre-angiosperm periods gives a new perspective on the several extinct plant lineages that played the roles of today's angiosperms. Here I present a review of the quality and preservation of lianas in the fossil record and the importance of the major lineages of lianas over time using published occurrence data based on the fossil record. The Fossil Record of Climbers database (http://www-personal.umich.edu/~rburnham/) consists of individual fossil records of species, each occurrence from a limited geographic area. Multiple occurrence records can represent a species' spatial distribution during one time interval. Parallel to the occurrence-based record (abundance), I analyze the temporal record of species richness as represented by a single count of a species occurrence for each time period. This updated analysis of the fossil record demonstrates that climbing was an early innovation among land plants, and that it was achieved by many, but not all, land plant lineages. In spite of abundant occurrences, species diversity of climbers was relatively low and individual species were more widespread in Paleozoic wetlands than in the wetlands of the Early Cenozoic. Because fossil deposits are representative of wetland environments, this cross-time comparison is probably most similar to the lakesides, marshes, and deltas of extant landscapes, not drier (i.e., *terra firme*) forested habitats. The domination of the climbing habit by angiosperms today contrasts with a much broader phylogenetic spectrum of ancient non-angiosperm climbing vascular plants, most of which did not survive past the Permian-Triassic boundary.

INTRODUCTION

Given that climbing plants typically contribute only small amount of biomass to terrestrial ecosystems (see Putz 1983; Gerwing & Farias 2000; Pérez-Salicrup & Sork 2001; Wyka et al. 2013; van der Heijden et al., Chapter 13 in this volume), it is somewhat surprising that they produce much of a fossil record at all. After all, it is the morphological features of the plants that are recognizable by paleobotanists and if lianas, as "structural parasites," produce less biomass and fewer organs for fossilization, their fossil record should be poorer than for trees. Indeed, the record of fossil lianas is thin compared to abundant arborescent lycopods, sphenopsids, conifers, and angiosperms. It has taken over a century of concerted fossil collecting, archiving, creative searching, and cataloguing to compile an outline of the fossil history of climbing plants. In 2009, I published an overview of the fossil record of climbers through time, which focused on published evidence of Paleozoic and Mesozoic climbing plants. I was surprised to find that while angiosperms hyperdominate tropical liana composition today, very diverse climbing communities prior to the evolution of angiosperms were well documented, but little known outside of the field of paleobotany. At the time of that publication, slightly more than 1100 records of fossil plants were catalogued. I documented the presence of 10 major groups of land plants that included species of lianas, of which only two are extant. A third extant lineage that includes lianas, *Gnetum* of the Gnetales, is virtually unknown as a fossil (but see Guo et al. 2009).

Since the first version of the Fossil Record of Climbers database (FRC, available online at http://www-personal.umich.edu/~rburnham/) was published (Burnham 2009), I have since added more than 400 published occurrence records of climbing fossil plants. Greater effort has been dedicated to augmenting the formerly sparse records from the late Mesozoic and Cenozoic during which climbing angiosperms are known to be common, based largely on records of angiosperm families that are predominantly climbing (e.g., Menispermaceae, Vitaceae, Araceae, Smilacaceae). Records of Araceae and other potential hemiepiphytes are included, departing from the practice of excluding these taxa by most modern liana biologists, because hemiepiphytic records are likely to otherwise be uncounted among fossil habit summaries.

For the same reason, I also included a few Paleozoic species whose life form is not clear: they may have been epiphytes. Here, I present a summary of the current FRC database, including 40% more fossil plant records than were listed in 2009. I pose the following questions:

1. Are climbers substantially more diverse morphologically and in species richness in more recent fossil records (60 million years and younger) than in the Paleozoic?
2. Do the older (non-angiosperm) lineages of climbers coexist temporally with one another, or do they replace one another sequentially, with angiosperms representing the most recent example of this pattern?

THE COMPLEXITY OF READING AND APPLYING THE PLANT FOSSIL RECORD

Obvious and routine to paleobotanists but less widely recognized among modern systematists and ecologists are the limitations and advantages of using the fossil record to address biogeographical and ecological questions. The foremost limitation is the disarticulation of plants as they are incorporated into the rock record (Burnham 2008; Gastaldo & Demko 2011). Disarticulation complicates identification because a suite of characters derived from different organs is generally used to recognize extant plant species and genera. As leaves and stem are separated, information about the arrangement of leaves on stems is separated from information about leaf venation. Similarly, wood anatomy is separated from flower morphology. A simple character like the presence of thorns or spines is very hard to combine with the correct foliage, while the presence of grasping organs, like tendrils or adhesive roots, becomes a matter of a lucky find.

Another paleobotanical challenge relates to the settings in which fossilization occurs and how that contributes to differential preservation of plant organs. In permineralized (or petrified) organs, three-dimensional structure is maintained by the infilling of cells with calcium carbonate, silica, or more rarely with iron compounds (Shute & Cleal 1986). Serial sectioning of permineralized organs can reveal remarkable details of the internal anatomy of any organ, which can stimulate substantial revision of relationships and time of origin of major lineages (Beck 1960; Hilton et al. 2004; Rothwell et al. 2012). Structures that are three-dimensional in life (stems, seeds, fruits, branching structure) are best preserved by this mode of fossilization. Paleozoic sediments from Appalachia (eastern U.S.A.) and Great Britain are well known for coal-bearing strata within which calcium carbonate infiltrates and preserves cell structure, resulting in "coal balls," a source of anatomical detail for sphenophylls, sphenopsids, seed ferns, conifers, cordaitaleans, ferns, and lycophytes. Adpression (=compression/impression) fossils are the result of compression into an essentially two-dimensional form of any organ whose cells are not structurally replaced, such as a leaf, flattened seed, or piece of bark. The surface structure of a plant organ is best preserved in the finest-grained sediments, which create a mold of the surface. Any carbon remaining on the rock surface can be viewed directly or removed and surface details studied. Due to their sheer bulk and regular abscission, leaves may be the most commonly preserved plant organ. Foliage is often preserved after being waterlogged in calm water, sinking, and being covered with sediment in anoxic conditions. Adpression fossils may be fortuitously exposed and discovered several million years later. Internal anatomy of adpression fossils is rarely preserved: cells are flattened, leaving only the basic cell patterns and details of the epidermis; exceptions include rare cell replacement with iron compounds that preserve both internal anatomy and external structures (e.g., the exquisite detail of Permian leaves reported by Chaney et al. 2011). Interestingly, these two major modes of preservation (permineralization and adpression) are most often found and studied from plant-bearing sediments of different ages or different geographic areas, rather than from the same deposits. Local and regional sediment-rich areas, lateral to permineralized deposits, can preserve adpression fossils that reveal external details of the same species preserved in coal balls (see Taylor et al. 2009). Adpression fossils of the western interior of the United States are renowned for their high quality preservation of flattened fish, aquatic vertebrates, leaves, flowers, and insects of Cenozoic age (MacGinitie 1969; Leopold et al. 2008). Coal balls are rare-to-absent in the Cenozoic, leaving anatomical details only available from silicified (petrified) material. The effect of this dichotomy in study material and age is that evidence of lianoid anatomy, such as wide tracheids or vessels, long internodes, and abundant parenchyma (vitalized) in xylem or phloem rays are rarely present in the same fossil deposits as leaf-based morphological traits, such as a cordate base, pulvinus, etc. The published fossil record of lianas provides permineralization fossils that reveal stem anatomical details as well as adpressions from the Paleozoic, versus petrified seed, fruits, and rare wood remains along with adpressions from the Cenozoic.

Differences in the ease of identification of isolated plant organs introduce a further bias in the fossil record. As with the reconstruction of the habits of trees or herbs from the fossil record, it is more often the case that systematic affinities alone provide evidence for the liana habit in the Cenozoic record, whereas more direct evidence from anatomy, climbing mechanisms, or internode lengths are used to infer climbing habits in the Paleozoic. Cenozoic fossil records of lianas may be biased toward easily recognized families. For example, perhaps because seeds of Menispermaceae and Vitaceae are quite distinctive, the fossil records of these liana families are rich. Conversely, the leaves of some common extant vine families are difficult to distinguish without additional characters that are rarely available from the fossil record (e.g., Bignoniaceae, Dichapetalaceae, Icacinaceae, and Celastraceae). This taxonomic bias may skew the record toward those easily recognized taxa, regardless of the age of the fossils. In light of the preservation and identification issues outlined above, accounts of the plant fossil record should always be viewed as provisional.

The fossil record of climbers database

We compiled records of fossil climbing plants from books, journals, governmental open file reports, and internet resources. Literature was reviewed by the author or by a trained student assistant whose knowledge of the modern record of climbing plants allowed a search for fossil taxa that were most likely climbers. Characters on which the climbing habit was determined are listed in Table 16.1, accompanied by literature sources supporting the interpretation. The earlier compendium of climbers (Burnham 2009) focused on the Paleozoic and Mesozoic whereas for this

Table 16.1 Characters used to diagnose climbing habit and examples of supporting literature from the Fossil Record of Climbers. The database is available in static form at http://www-personal.umich.edu/~rburnham/. A current version may be available from the author on request.

Character for diagnosis of climbing habit	Example and/or rationale reference
Long internodes	Ray 1986, Galtier 1988, Dubuisson et al. 2003, Dunn et al. 2006, DiMichele et al. 2006
Small stem diameter to length ratio	Dunn et al. 2006, Harris et al. 2007, Ichihashi et al. 2009
Small stem diameter relative to supported foliage	Galtier 1988, Selaya & Anten 2008
Delayed apical foliage expansion and/or dense glandular trichomes	Baxter 1949, Hegarty 1991, Putz & Holbrook 1991, Krings et al. 2003, Ichihashi et al. 2009
Adventitious roots	Gentry 1991, Hegarty 1991, Speck 1994
Large petiole bases relative to stem diameter	Tomescu et al. 2001, Dunn et al. 2006
Hooks, spines, or grappling structures	Menninger 1970, Hegarty 1991, Teramura et al. 1991
Heterophylly	Batenburg 1981, Lee & Richards 1991, Krings & Kerp 2000, Krings et al. 2001, 2003
Anomalous wood anatomy: successive cambia, excessive parenchyma, multiple vascular tissue cycles	Taylor & Millay 1981, Carlquist 1991, Ewers et al. 1991, Caballé 1993, Mosbrugger & Roth 1996
Structural anatomy inconsistent with self-support	Li & Taylor 1998, Li et al. 1994, Speck 1994
Taxonomic affinities to other climber taxa	Gianoli 2004
Direct observation of climber wrapped on larger individuals in "snapshot" deposits	Oplustil et al. 2007; 2009

revision, adding records from the Cenozoic through the end of the Neogene to the data base was a priority. The most recent fossil record (Pleistocene–Holocene) still has not been thoroughly searched for records of climbing plants, though included are selected records encountered during searches. The logic of paying less attention to the most recent fossil record is that much of the Pleistocene and Holocene record is based on palynological remains, which are notoriously difficult to characterize to genus or species, and from which pollen is highly biased toward wind-pollinated lineages. Many extant taxa of lianas are insect- or bird-pollinated, and thus produce little pollen to contribute to the recent fossil record. Further, the questions we address here about the fossil record of lianas are not as relevant to the most recent history insofar as the questions relate to the evolution of major plant lineages and their relative contributions to the record of climbing plants. Because extant climbing lineages (ferns, angiosperms) are largely the same as those present in the Pleistocene and Holocene, little is gained by an exhaustive compilation of recent fossil liana records for this study. Of course, the recent fossil records can shed light on family- and genus-level diversification as well as taphonomic (burial) biases in certain lineages. We are continuously adding to the database, and periodically update the static version that is archived on our website (http://www-personal.umich.edu/~rburnham). Currently, the Fossil Record of Climbers database has the largest geographic data gaps in the Mediterranean and Middle East, Southeast Asia, India, Africa, and high latitudes of the eastern hemisphere.

In addition to data on literature source, taxonomic names, authority, and synonyms, records are coded for major plant lineage represented, the organ(s) preserved, specific locality, country or state of locality, major geological time period/sub-period when known, basis for scoring as a climbing plant, additional references to the same occurrence, and notes including source of age determinations, and conflicting opinions on taxonomic placement. Table 16.2 shows examples of database entries for the three major time periods (Cenozoic, Mesozoic, and Paleozoic) to demonstrate the kinds of data available and to highlight a sample of the major clades represented for different time periods. All time period categories, their durations, and number of records and species per time period are listed in Table 16.3.

Data on the abundance of species in a plant fossil assemblage are not often collected in an unbiased manner, and the issues involved with reconstruction of community abundance are complicated (DiMichele et al. 2004; Burnham 2008; Cascales-Miñana & Diez 2012; Cleal et al. 2012). The FRC database, as currently constructed, represents the occurrence of species at distinct geographic locations and the presence of species over time. It does not reflect abundance on a locale (per locality) basis. From the FRC data, we can re-create geographic distributions, regional diversity, and global species abundances for each time period. We entered a single species more than once for a single time period if it occurred in multiple countries or multiple states of the United States. Each species was entered within each time period for which it was represented by a distinct specimen. For example, *Lygodium kaulfussii* Heer is a widespread fossil climbing fern that occurred during the Eocene in Russia, China, Germany, England, and in eight states of the United States, so 12 counts of the species are recorded for the Eocene alone.

The multiple listings of a single species during a time period can be interpreted as evidence for a wide geographical distribution, but it also inflates liana diversity estimates over time. I have paired the multiple-record plots of climber richness/distributions (Figs. 16.1, 16.2) with single-count per time period plots in which a species is counted only once in any single time period. Figure 16.2 provides the known worldwide climber species richness for each time period without multiple records per species in a single time period.

For the angiosperm record, there are often cases in which a genus includes some modern species that are climbers, and other species that are self-supporting. In these cases, if the great majority (>70%) of the species climb, the record is included in the database. If <70% of the modern species climb, the fossil record is not included. For example, the genus *Dalbergia* is registered in the Index of Genera and Species of Fossilium Catalogus Plantae (van der Burgh 2008) by more than 60 fossil records, but the genus is known as both a tree/shrub or liana and the fossil record of leaf or fruit adpressions provide no information about habit. Conversely, because most species of *Davilla* (Dilleniaceae) climb, all known records are included for the genus. This consistent solution gives us a database from which we can estimate the importance of different climbers in the fossil record. It should nevertheless be noted that

Table 16.2 Fossil record of climbers database example entries for three time periods. Examples from Cenozoic, Mesozoic and Paleozoic strata highlight the presence of angiosperms and ferns. Data include geographic region, organ(s) represented, and time period details. Each taxon record has citation to geological context citation as well as additional references on time period verification and secondary citations of the same taxon from compendia. See similar examples in Burnham (2009).

Information category	Sample entry Miocene	Sample entry Cretaceous	Sample entry Permian
Literature citation	Graham, A. 1963 Systematic revision of the Sucker Creek and Trout Creek Miocene floras of southeastern Oregon. *Amer. J. Bot.*, **50**(9), 921–936.	Skog, J. 1993 The relationship of the fossil fern *Schizaeaopsis* Berry to modern genera in the Schizaeaceae. *Amer. Fern J.* **83**(1), 20–29.	Roßler, R. 2000 The late Palaeozoic tree fern *Psaronius* – an ecosystem unto itself. *Rev. Palaeobot. Palynol.*, **108**, 55–74.
Higher Taxonomic Group	Malpighiaceae	Schizaeaceae	Filicales (zygopterid)
Genus, Species, Author	*Hiraea knowltoni* (Berry) Graham	*Schizaeopsis macrophylla* Skog	*Ankyropteris brongniartii* (Renault) Mickle
Organ(s) Represented	Leaf, Fruit	Frond	Wood, Frond
Preservation type	Adpression	Adpression	Permineralization
Specific locality & region	Sucker Creek	Patuxtent Formation	Cheminitz
Country or state	Oregon	Maryland	Germany
Time period, sub-period	Miocene	Cretaceous, Aptian	Permian, Early
Evidence on which climbing interpretation is based	Modern genus is climber	Modern genus is climber	Literally caught in the act on tree of *Psaronius*
Synonomies	*Nyssa knowltoni, Ptelea miocenica*		*Senftenbergia* of Jennings & Eggert 1977
Additional references Notes	Berry, E.W. 1929. U.S. Geological Survey Professional Paper 154-H. Compared to modern *Hiraea quapara*		Grows on *Psaronius* and *Calamodendron* and has insect frass too

due to this approach, Cenozoic records of both filicalean and angiospermous climbers are underrepresented in the Fossil Record of Climbers database.

Fossil occurrence records of climbing plants

The data compiled expand the Fossil Record of Climbers database (Burnham 2009) from 1175 to 1623 occurrence records. The database now includes 1066 distinct species from over 385 million years, a mere shadow of the actual number of climbing species that must have been fossilized, and only a small portion of the number that have lived and climbed. The largest number of records from a time period (521) was from the Pennsylvanian Subperiod (Carboniferous Period), a time of both high sea level and generally warm climates with extensive deposits of terrestrial sediments world-wide (Cecil et al. 1985; Cecil 2003; Miller et al. 2005).

Table 16.3 Comparison of occurrence records with number of distinct species per time period in the fossil record of climbers database. Eleven major groups are distinguished here: sphenophylls, filicalales, medullosales, mariopteridales, lyginopteridales, callistophytales, caytoniales/peltasperms/corystosperms, gigantopterids, bennettitales, gnetales, and angiosperms. Time scale from Walker et al. (2012).

Time period		Time period duration (mya)	Number of records	Number of species	Number of major groups
Devonian	Paleozoic	359–419	1	1	1
Mississippian		323–359	55	44	5
Pennsylvanian		299–323	521	173	6
Early Permian		272–299	61	38	7
Mid & Late Permian		252–272	23	20	5
Early & Mid Triassic	Mesozoic	237–252	1	1	1
Late Triassic		201–237	14	14	3
Jurassic		145–201	5	5	1
Early Cretaceous		100–145	19	15	2
Late Cretaceous		66–100	92	80	2
Paleocene	Cenozoic	56–66	66	61	2
Eocene		34–56	363	297	2
Oligocene		23–34	81	68	2
Miocene		5–23	227	166	2
Pliocene		2.5–5	59	50	2
Quaternary		0–2.5	35	33	2
TOTAL			1623	1066	

While the Pennsylvanian Sub-period represents only 24 of the ~385 million years (~6%) surveyed (Walker et al. 2012; ICS 2013), nearly one-third (32.1%) of all plant fossil occurrences of climbers are from this time period.

The Paleozoic record

The oldest and possibly the most spectacular ancient climbing plants were recently discovered in the mid-Devonian (Givetian age) Gilboa forest of New York (Stein et al. 2012). These relics were preserved in fluvial sediments of the Moscow Formation, 398–385 million years ago. The scrambling plants were reconstructed as a member of the Aneurophytalean Progymnosperms, the first plants to produce a cambium that produced internal xylem and external phloem. They were provisionally assigned to the genus *Tetraxylopteris*. Large diameter (up to 15 cm diameter) stems appear to have scrambled along the forest floor, ascended large diameter fern-like trees of the genus *Eospermatopteris*, and in some cases probably fell from the canopy prior to sedimentation. No modified climbing organs have been recovered thus far. This recent discovery may spur re-evaluation of many occurrences of *Tetraxylopteris*, as it was not recognized as a climbing plant prior to this excavation of 1000 m^2 of Devonian forest floor. Calculations by Speck and Rowe (2003) suggested that certain species of the genus were not self-supporting, and some interpretations were of a three-dimensional interlocking shrub-height thicket in which a thicket of arching stems produced no modified organs for climbing, but rather scrambling up one another and any tree encountered (Hammond & Berry 2005). In this light, it seems reasonable to include the lax climbing growth form in the concept of lianescent growth during the Devonian. From this early record of surprisingly large-stemmed climbers, the breadth of phylogenetic diversity of climbers increases in the Early Carboniferous (Mississippian Subperiod) when early climbing ferns (e.g., *Psalixochlaena antiqua*) and Pteridosperms of various lineages: Lyginopterids, Mariopterids, and Sphenophyllaleans are first recorded. It is important to note that some Calamopityalean plants reconstructed as being incapable of self-support (Rowe et al. 1993) are not included in the FRC database because evidence for climbing is somewhat weak. Still, their small diameter stems and large leaves suggest

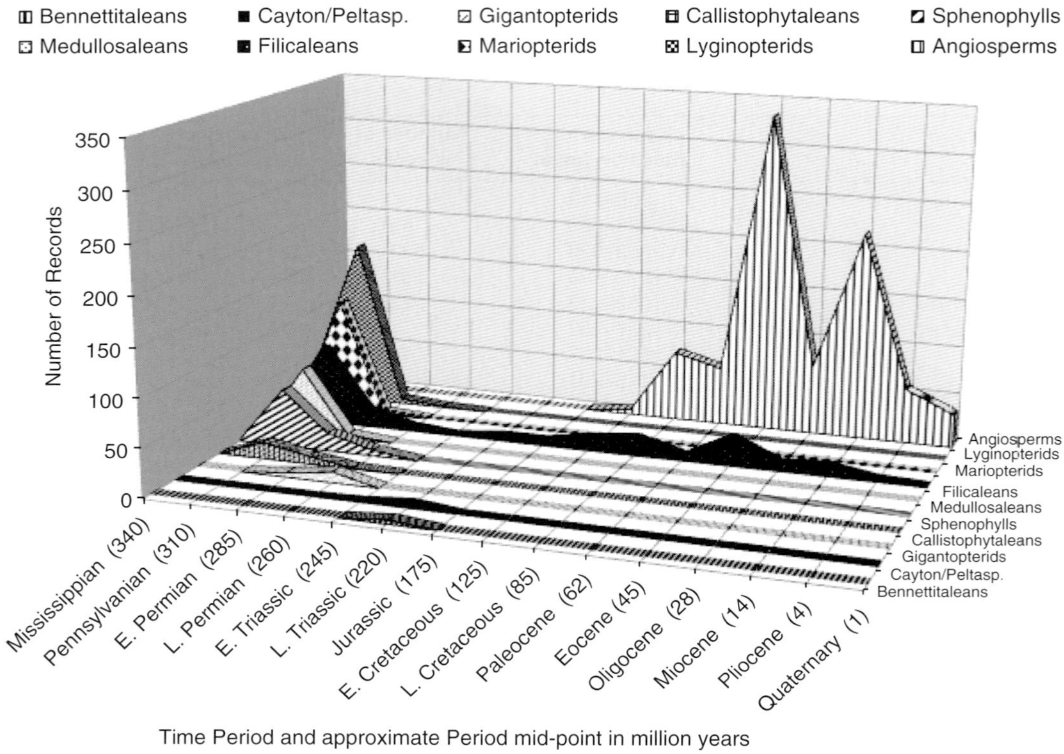

Fig. 16.1 1623 species records of fossil climbers plotted by major plant lineage. For clarity, single occurrences of two species are not shown: the Devonian Aneurophytalean *Tetraxylopteris* sp. (Stein et al. 2012) and the Early Cretaceous Gnetalean, *Khitania columnispicata* (Guo et al. 2009). Records for the Quaternary are incompletely surveyed and do not reflect the diversity represented.

that these plants were indeed lianescent (see DiMichele & Gastaldo 2008) and they appear to have increased in stem diameter distally, where support may have been available from surrounding vegetation. Lacking further support from data on attachment mechanisms, I excluded this early group of potential scramblers from the FRC. The lack of data on attachment mechanism is a common problem encountered in classifying extant climbers; for example, scramblers, such as members of the genus *Rubus* (Rosaceae), are difficult to classify neatly into a single growth-strategy because of their propensity to grow as a bush or a trailing liana, depending on conditions.

While the Mississippian climbers represent an explosion in diversity of lineages, there are relatively few occurrence records (only 55) from the Mississippian Sub-period. Lyginopterids (one of several seed fern groups) include the most speciose lineage of climbers in the Mississippian. Yet, there is representation by five of the six major groups that dominate the subsequent Pennsylvanian record. This pattern reflects the abundance of fossil plants known from the Mississippian Period assignable to taxon or life-form: not only are these older terrestrial lowland deposits less abundant worldwide, but the plants are far less confidently identified on the basis of the dispersed organs. While this may indeed be a wholly taphonomic artifact, it still reflects the facility with which the climber habit was adopted by a variety of seed-plant lineages.

The phylogenetic diversity of climbing plant lineages reaches a peak in the Pennsylvanian Sub-period (~300 million years ago), with at least six major lineages of lianescent plants (Sphenophyllaleans, Filicaleans, Medullosans, Mariopterids, Lyginopterids,

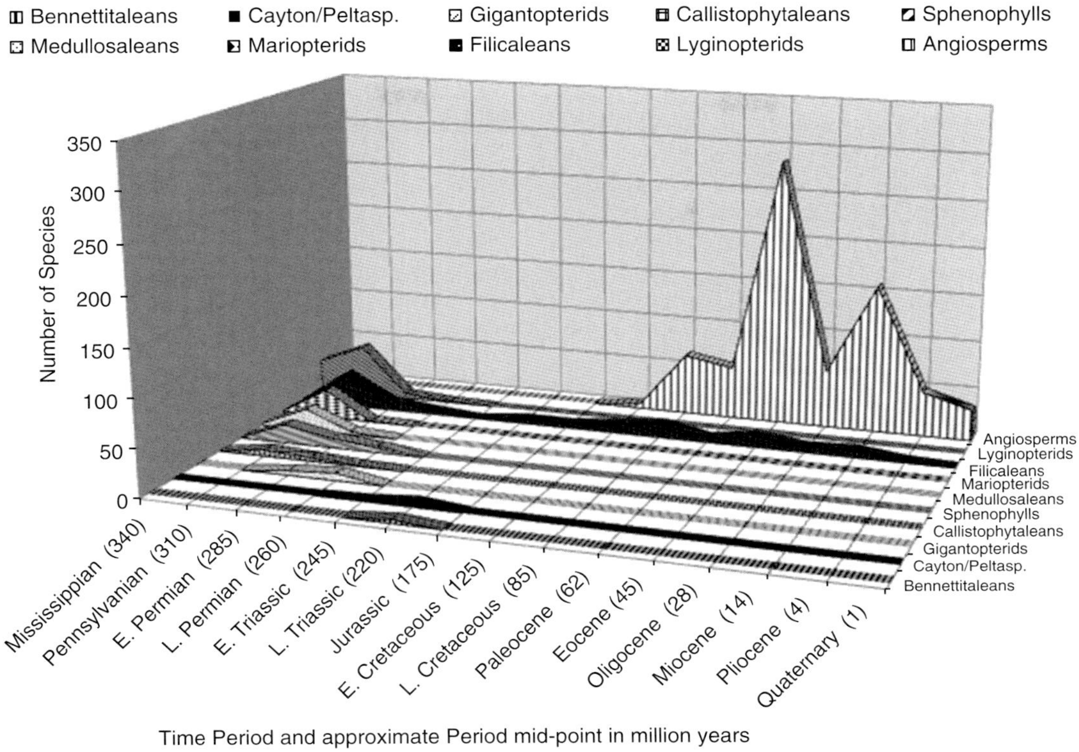

Fig. 16.2 1066 species records of fossil climbers plotted by major plant lineage. Each species is represented by a single occurrence per time period, thus this represents global diversity for each time period. For clarity, single occurrences of two species are not shown: the Devonian Aneurophytalean *Tetraxylopteris* sp. (Stein et al. 2012) and the Early Cretaceous Gnetalean, *Khitania columnispicata* (Guo et al. 2009). Records for the Quaternary are incompletely surveyed and do not reflect the diversity represented.

Callistophytaleans; Figs. 16.1 and 16.2). These diverse lineages include species with organs modified for climbing and climbing mechanisms that are recognized in extant angiosperms, such as tendrils, adhesive roots or tendrils, twining stems, prickles, and thorns. Spectacular preservation of whole ecosystems is rare, but a few Pennsylvanian Sub-period records are derived from descriptions of *in-situ* tree trunks with multiple climbing stems, with the whole forest buried in volcanic ash (see Oplustil et al. 2007, 2009; Libertin et al. 2009).

By the mid-Permian (~250 million years ago), a widespread decline in diversity and abundance of fossilized plants has been recognized repeatedly (Rees 2002; DiMichele et al. 2006; Stevens et al. 2011). The species richness decline was placed in regional context by Stevens and colleagues (2011), based on the plant fossil record in China. These authors speculated that the decline reflected continental movement, increased aridity, and increased seasonality. In China, the record of decline is well documented, but many plant groups appear to have survived there longer than in North America or Europe (Hilton et al. 2004; Seyfullah & Hilton 2011). Sediments of the late Permian in China harbor climbing plants otherwise last seen in the early Permian (various species of *Sphenophyllum*) or even the Pennsylvanian. This phenomenon is particularly noticeable for lianas, which decline precipitously in North America in the early Permian according to records of the FRC. In drier and more seasonal environments, fossil preservation would decline, so sampling bias is almost certainly involved. As noted above, the abundance of Pennsylvanian coal deposits, coal balls, and the associated overbank adpression fossils may give an unusually complete picture of wetland environments of this subperiod, whereas during much of the Mesozoic, fossil plants are preserved in clastic sediments of riversides and levees, rather than the mineralizing water of extensive swamps. If the species of Gigantopterids can be more confidently reconstructed, it may be that these large-leaved plants will help fill in the Permian record with a larger group of climbers (see DiMichele et al. 2011).

The Mesozoic record

During the Mesozoic (Triassic through Cretaceous) there is scant evidence of recognizable climbing plants. Less than 10% (131 of 1623) of the climber occurrence records occur in this 186 million-year period (251 to 65 million years ago). Of those, 92 records are from the Late Cretaceous and represent ferns and angiosperms. Two new major lineages of climbers appearing during the Mesozoic, Bennetitaleans and Caytonialeans, are represented by just a few records. These two groups are not well recognized as including lianas, and the records may be inaccurate with respect to plant habit. The possibility of climbing Gigantopterids is recorded in the FRC, known mostly from the Late Permian, but the records are too few to be interpretable until more data are available (Li & Taylor 1998, 1999; DiMichele et al. 2011). In the Mesozoic, even ferns (Filicaleans) have their lowest abundance of climbing records. In the Early Triassic there is, as yet, not a single record of climbing Filicalean in the database. This pattern, particularly for the Triassic, reflects the pattern for almost all plant life forms that suffered substantial diversity loss during the end-Permian mass extinctions (Rees 2002; McElwain et al. 2009). The presumed phenomenon of inhospitable habitats for climber growth, combined with increasingly clastic-based sedimentation, appears to have continued through the Jurassic (145–200 million years ago).

Records of climbing species increased again during the late Cretaceous (66–100 million years ago), as the number of wetland habitats preserved increased, the preserved rock volume increased, and the taxonomic recognition of isolated plant organs increased because of the similarity to modern forms. Another important factor during the Cretaceous, of course, is the evolution of the flowering plants. This evolutionary variant on the seed plant life cycle excels in the production of lianescent diversity today, and did so throughout the Cenozoic (66 million years ago to the present).

The Cenozoic record

During the Cenozoic all habits of angiosperms, in addition to the ferns, cycads, and coniferophytes that were common in the previous epochs, flourished and dominated the mesic warm habitats. As early as the Paleocene, angiosperms of large stature were distributed from the Arctic Circle (Greenwood & Basinger 1994; Moiseeva et al. 2009; Vavrek 2012) to Antarctica and Patagonia (Poole et al. 2000; Wilf et al. 2013). The Paleocene climber records in the FRC (66) are fewer than in the late Cretaceous (92), probably reflecting a combination of less intense study, a shorter time period (only 10 million years vs. 35 million years; Table 16.3), and a low point in diversity following the end-Cretaceous impact event (Nichols & Johnson 2008; Barreda et al. 2012). In contrast, the Eocene (34–56 million years ago) was a relatively long, warm, and well-studied geological period that harbored abundant well-preserved adpression fossils as well as permineralizations, petrifactions, and coalifications. From the Eocene, 334 individual records of flowering plants and more than 29 records of ferns are recognized in the FRC. This Eocene collection certainly under-represents the magnitude possible from the fossil record, given the large lacunae in various geographical areas in the database (e.g., Southeast Asia, the Middle East, Africa, Antarctica). The relatively short Oligocene Period (23–34 million years ago) is also represented by few climber records (Table 16.1, Fig. 16.1), again reflecting a shorter time period but also reflects decreased angiosperm abundance and diversity, related to the decline in global temperatures. Indeed, the transition from Eocene to Oligocene has been characterized as a greenhouse to icehouse transition (Prothero et al. 2009). The comparison of Oligocene records (81, mostly angiosperm) with the diverse and abundant Miocene records is stark: records for the Miocene (5–23 million years ago), based largely on European and North American fossils, include 217 angiosperms and 10 ferns. The proportion of climber diversity to tree diversity is not clearly known for any of these time periods because no similar compendium has been amassed for records of fossil trees. I propose that our knowledge of the true fossil liana diversity of Cenozoic angiosperms and ferns is limited because interpretation of the climbing habit for these plants is tied to correct identification of a nearest modern climbing relative. Unlike the Paleozoic record, support for the climbing habit in the Cenozoic is based on comparison with modern climbing species rather than directly observable characters that demonstrate a climbing habit (e.g., tendrils, hooks, lianescent wood, etc., but see Smith et al. 2013).

The diversification of angiosperm lianas cannot be separated from the diversification of the angiosperms. Liana diversification parallels the opportunities gained by all angiosperms as a consequence of their rapid reproductive cycles, biotic pollination and dispersal partners, diverse mycorrhizal associations, and diverse growth habits. Added to this suite of traits for lianas is the nearly unique attribute of anomalous secondary growth. Found in at least 30 families of climbing angiosperms (Angyalossy et al. 2012; Angyalossy et al., Chapter 19 in this volume), the inclusion of supernumerary cambia and excessive parenchymatous cells within vascular tissues may help explain the additional diversification within clades bearing an anomalous secondary group among lianas.

Occurrence records vs. species diversity of climbers in the fossil record

Evaluation of the fossil record based on occurrence records, instead of species records, potentially obscures important phylogenetic patterns that are evident at the species level. By allowing each species to be represented only once per time period, we can determine whether the repeated collection of widespread species is influencing our impression of clade diversity over time. Therefore, I removed all time-duplicate appearances of species from the database to compare the pattern of species richness over time (Fig. 16.2) to occurrences over time (Fig. 16.1), leaving 1066 species in the culled FRC, with each species recorded no more than once per time period. More than half of the non-angiosperm records were eliminated during this culling (413 of 767 occurrence records), whereas only 16% of angiosperm records were removed (144 of 856).

The overall pattern of species records is quite similar to the occurrence records, with the highest species richness among angiosperms, very low species richness in the Triassic and Jurassic, and a remarkable temporal segregation of the contributions by the early groups of land plants (various seed fern groups, true ferns, and seed plants) versus the modern domination by angiosperms. Only the Filicales (true ferns) bridge the temporal gap between the early multiple lineages of lianas and the angiosperm lianas.

Angiosperms contribute most of the species records, even though their presence in the fossil record is still not exhaustively included in the FRC database. Species of Filicales, as well as Sphenophylls, Lyginopteridaceae, Mariopteridaceae, and other seed ferns, are more spatially widespread in the fossil record than are angiosperms. Many species of non-angiosperms occurred across vast geographic areas, generating many records in the FRC occurrence database. For example, the sphenophyll, *Sphenophyllum oblongifolium* (Germar and Kaulfuss) Unger, is known from the United Kingdom, Canada, France, Germany, Morocco, Romania, Turkey, Mongolia, Virginia, West Virginia, Texas, New Mexico, Ohio, and Massachusetts. It is exceedingly rare for any species to be so widespread in the Cenozoic, with perhaps the exception of the climbing Schizeaceous fern, *Lygodium kaulfussii*, which has been found, within the Eocene alone, in 10 different states of the United States, Germany, England, China, and Kamchatka. The flowering plants represent 52% of all occurrence records, whereas they represent 67% of all species records, leaving a relatively small proportion of all climbing species to be apportioned among the nine other major groups of plants over 200 million years (Fig.16.2). Among extant angiosperms, high geographic species richness (alpha diversity) and species turnover (beta diversity) are widely recognized among most life-forms of Angiosperms (herbs, shrubs, branching trees, monocaulous trees like palms). These high diversity levels in angiosperms, compared to other groups, have been attributed to directed dispersal of pollen and seeds, short life cycles, and diverse chemical defenses (Burger 1981; Dodd et al., 1999; Crepet 2008, but see Gorelick 2001). These attributes, combined with narrow geographic ranges, set up the ideal conditions for increased diversity through speciation and damped extinction. Fossil angiosperm lianas are therefore no exception to this pattern of high alpha and beta diversity among angiosperms in general.

CONCLUSION

The Fossil Record of Climbers (FRC) database has grown more robust since its original publication (Burnham

2009), currently including more than 1600 records and more than 1000 species. It documents the clear presence of lianas from almost the earliest stages of land plants on Earth (Devonian, 398–385 million years ago; Stein et al. 2012). The FRC recounts the wide diversity of plant groups that assumed the climbing habit through time, and demonstrates that by the Late Cretaceous, phylogenetic diversity declined to one lineage of ferns and one diverse seed plant group, the angiosperms (Figs. 16.1, 16.2). That narrow phylogenetic diversity persists to the present, with only approximately 100 species of climbing ferns, a single Gnetalean climbing genus (*Gnetum*), and an overwhelming diversity of angiosperm lianas from at least 90 families (Gentry 1991). Lianas were present in readily fossilized habitats, such as swamps, disturbed areas, and lowland forests, which create excellent opportunities for repeated sampling of the fossil record. The major bias posed by the fossil record is the lack of data from areas not subject to deposition of fine sediments (often called extra-basinal sites). In the Paleozoic, deposits that have survived to the present mostly represent wet, low-lying, coastally-proximal floras. This pattern holds true for the late Mesozoic and Cenozoic, however, because younger Cenozoic deposits can retain some evidence of more continental floras (compared with the older, subsided and eroded deposits), there is a greater chance for extra-basinal taxa. Throughout the record, evidence for the climbing habit is derived from stem morphology, stem anatomy, reproductive structure anatomy, climbing organs, and taxonomic affinity to extant groups that are represented almost exclusively by lianas.

In the Paleozoic, several groups of land plants climbed, with examples among the Sphenophylls, Mariopterids, Lyginopterids, Filicaleans, and Callistophytales, among other lineages. Once these groups waned in diversity and all but Filicaleans went extinct, they were replaced only by the angiosperms, a group derived from a single lineage of seed plants but with far greater species richness (see Table 16.3). For comparison, among non-climbers today there are at least six major lineages representing woody plants: Angiosperms, Cycads, Conifers, Pteridophytes, Gnetophytes, and Ginkgophytes. Several lineages of seed ferns, gymnosperms, and arborescent lycophytes have become extinct but were well represented among woody non-climbing plants of the Paleozoic.

As a single evolutionary lineage, angiosperms are equivalent in morphological divergence to any other group of land plants that include lianas (Fig. 16.1, Table 16.3). Angiosperms currently occupy the habitat space that was occupied collectively by a minimum of nine other groups through 250 million years (Mississippian Sub-period to mid-Cretaceous). Angiosperms have held this position since their diversification in the early part of the Cretaceous, more than 100 million years ago (Magallón & Castillo 2009; Friis et al. 2010).

Neither this review nor the FRC addresses the abundance of lianas in ecosystems of the past because these data are almost non-existent. Fortunately, there are a few intensively censused fossil ecosystems (Burnham 1994; Johnson 2002; Ellis et al. 2003; Iglesias et al. 2007; Wing et al. 2012) from which a more complete picture of liana contributions to past ecosystems can be built. In addition, a few spectacular one-of-a-kind collections prove that these data are possible, even if limited in number (see Libertin et al. 2009; Oplustil et al. 2009). Data on paleoecological abundance of lianas are a priority for understanding their roles in ancient environments, which should be readily accessible from the abundance of river and levee deposits, as well as volcanic ash falls, in the fossil record.

ACKNOWLEDGMENTS

The database would not be possible without the extraordinary work of my colleagues in paleobotany over the past 200 years. I am particularly grateful to Hans Kerp for his discussions with me on the topic. I was assisted in locating the fossil references by the diligent searches of ReBecca Sonday, Bradley Sisson, and Jenna Dorey. In addition, I would like to acknowledge the helpful and thorough reviews by Paul F. Foster, Frans Bongers, Francis E. Putz, Stefan A. Schnitzer, and William A. DiMichele, the latter's work and fruitful discussion have been instrumental in following and interpreting the tangle of Paleozoic literature on lianas.

REFERENCES

Angyalossy, V., Angeles, G., Pace, M.R. et al. (2012) An overview of the anatomy, development and evolution of the vascular system of lianas. *Plant Ecology and Diversity*, **5**, 167–182.

Barreda, V.D., Cúneo, N.R., Wilf, P. et al. (2012) Cretaceous/Paleogene floral turnover in Patagonia: drop in diversity, low extinction, and a *Classopollis* spike. *PLoS ONE*, **7**(12), e52455.

Batenburg, L.H. (1981) Vegetative anatomy and ecology of *Sphenophyllum zwickaviense*, *S. emarginatum*, and other "compression species" of *Sphenophyllum*. *Review of Palaeobotany and Palynology*, **32**, 275–313.

Baxter, R.W. (1949) Some pteridosperm stems and fructifications with particular reference to the Medullosae. *Annals of the Missouri Botanical Garden*, **36**, 287–353.

Beck, C.B. (1960) The identity of *Archaeopteris* and *Callixylon*. *Brittonia*, **12**, 351–368.

Burger, W.C. (1981) Why are there so many kinds of flowering plants? *Bioscience*, **31**: 572, 577–581.

Burnham, R.J. (1994) Paleoecological and floristic heterogeneity in the plant fossil record: an analysis based on the Eocene of Washington. *United States Geological Survey Bulletin*, **2085B**, 1–36.

Burnham, R.J. (2008) Hide and go seek: what does presence mean in the fossil record? *Annals of the Missouri Botanical Garden*, **95**, 51–71.

Burnham, R.J. (2009) An overview of the fossil record of climbers: bejucos, sogas, trepadoras, lianas, cipós, and vines. *Revista Brasileira de Paleontologia*, **12**, 149–160.

Caballé, G. (1993) Liana structure, function and selection: a comparative study of xylem cylinders of tropical rainforest species in Africa and America. *Botanical Journal of the Linnean Society*, **113**, 41–60.

Carlquist, S. (1991) Anatomy of vine and liana stems: a review and synthesis, in *The Biology of Vines* (eds F.E. Putz and H. Mooney), Cambridge University Press, Cambridge, pp. 53–72.

Cascales-Miñana, B. & Diez, J.B. (2012) The effect of singletons and interval length on interpreting diversity trends from the palaeobotanical record. *Palaeontologia Electronica*, **15**, 6A.

Cecil, C.B. (2003) The concept of autocyclic and allocyclic controls on sedimentation and stratigraphy, emphasizing the climate variable, in *Climate Controls on Stratigraphy* (eds C.B. Cecil & N.T. Edgar), Society for Sedimentary Geology, Tulsa, OK, pp. 13–20.

Cecil, C.B., Stanton, R.W., Neuzil, S.G. et al. (1985) Paleoclimate controls on late Paleozoic sedimentation and peat formation in the central Appalachian basin (U.S.A.). *International Journal of Coal Geology*, **5**, 195–230.

Chaney, D.S., Mamay, S.H., DiMichele, W.A. & Kerp, H. (2011) *Auritifolia* gen. nov., probable seed plant foliage with Comioid affinities from the Early Permian of Texas, U.S.A. *International Journal of Plant Sciences*, **170**, 247–266.

Clark, D.B., Olivas, P.C., Oberbauer, S.F. et al. (2008) First direct landscape-scale measurement of tropical rain forest Leaf Area Index, a key driver of global primary productivity. *Ecology Letters*, **11**, 163–172.

Cleal, C.J., Uhl, D., Cascales-Miñana, B. et al. (2012) Plant biodiversity changes in Carboniferous tropical wetlands. *Earth-Science Reviews*, **114**, 124–155.

Crepet, W.L. (2008) The fossil record of Angiosperms: requiem or renaissance? *Annals of the Missouri Botanical Garden*, **95**, 3–33.

DiMichele, W.A., Behrensmeyer, A.K., Olszewski, T.D. et al. (2004) Long-term stasis in ecological assemblages: evidence from the fossil record. *Annual Reviews of Ecology, Evolution, and Systematics*, **35**, 285–322.

DiMichele, W.A. & Gastaldo, R.A. (2008) Plant paleoecology in deep time. *Annals of the Missouri Botanical Garden*, **95**, 144–198.

DiMichele, W.A., Looy, C.V. & Chaney, D.S. (2011) A new genus of Gigantopterid from the Middle Permian of the United States and China and its relevance to the Gigantopterid concept. *International Journal of Plant Sciences*, **172**, 107–119.

DiMichele, W.A., Phillips, T.L. & Pfefferkorn, H.W. (2006) Paleoecology of Late Paleozoic pteridosperms from tropical Euramerica. *Journal of the Torrey Botanical Society*, **133**, 83–118.

Dodd, M.E., Silvertown, J. & Chase, M.W. (1999) Phylogenetic analysis of trait evolution and species diversity variation among angiosperm families. *Evolution*, **53**, 732–744.

Doria, G., Jaramillo, C.A. & Herrera, F. (2008) Menispermaceae from the Cerrejón Formation, middle to late Paleocene, Colombia. *American Journal of Botany*, **95**, 954–978.

Dubuisson, J.-Y., Schneider, H. & Hennequin, S. (2008) Epiphytism in ferns: diversity and history. *Comptes Rendus Biologies*, **332**, 120–128.

Dunn, M.T., Mapes, G., & Rothwell, G.W. (2006) The Fayetteville flora of Arkansas (USA): a snapshot of terrestrial vegetation patterns within a clastic swamp at Late Mississippian time. In *Geological Society of America Special Paper* 399 (eds. S.F. Greb & W.A. DiMichele, pp. 127–137.

Ellis, B., Johnson, K.R. & Dunn, R.E. (2003) Evidence for an in situ early Paleocene rainforest from Castle Rock, Colorado. *Rocky Mountain Geology*, 73–100.

Ewers, F.W., Fisher, J.B. & Fichtner, K. (1991) Water flux and xylem structure in vines, in *The Biology of Vines* (eds F.E. Putz & H. Mooney), Cambridge University Press, Cambridge, pp. 127–160.

Friis, E.M., Pedersen, K.R. & Crane, P.R. (2010) Diversity in obscurity: fossil flowers and the early history of angiosperms. *Philosophical Transactions of the Royal Society B*, **365**, 369–382.

Galtier, J. (1988) Morphology and phylogenetic relationships of early Pteridosperms, in *Origin and Evolution of Gymnosperms* (ed C.B. Beck), Columbia University Press, New York, pp. 135–176.

Gastaldo, R.A. & Demko, T.M. (2011) The relationship between continental landscape evolution and the plant-fossil record: long term hydrologic controls on preservation. In *Taphonomy: Process and Bias through Time, Topics in Geobiology* (eds. P.A. Allison & D.J. Bottjer), **32**, 249–285.

Gentry, A.H. (1991) The distribution and evolution of climbing plants, in *The Biology of Vines* (eds F.E. Putz & H. Mooney), Cambridge University Press, Cambridge, pp. 3–49.

Gerwing, J.J. & Farias, D.L. (2000) Integrating liana abundance and forest stature into an estimate of total aboveground biomass for an eastern Amazonian forest. *Journal of Tropical Ecology*, **16**, 327–335.

Gianoli, E. (2004) Evolution of a climbing habit promotes diversification in flowering plants. *Proceedings of the Royal Society B: Biological Sciences*, **271**(1552), 2011–2015.

Gorelick, R. (2001) Did insect pollination cause increased seed plant diversity? *Biological Journal of the Linnean Society*, **74**, 407–427.

Greenwood, D.R. & Basinger, J. (1994) The paleoecology of high-latitude Eocene swamp forests from Axel Heiberg Island, Canadian High Arctic. *Review of Palaeobotany and Palynology*, **81**, 38–97.

Guo, S.-X., Sha, J.-G., Bian, L.-Z. & Qiu, Y.-L. (2009) Male spike strobiles with *Gnetum* affinity from the Early Cretaceous in western Liaoning, Northeast China. *Journal of Systematics and Evolution*, **47**, 93–102.

Hammond, S.E. & Berry, C.M. (2005) A new species of *Tetraxylopteris* (Aneurophytales) from the Devonian of Venezuela. *Botanical Journal of the Linnean Society*, **148**, 275–303.

Harris, C., Murray, B.R., Hose, G.C. & Hamilton, M.A. (2007) Introduction history and invasion success in exotic vines introduced to Australia. *Diversity and Distributions*, **13**, 467–475.

Hegarty, E.E. & Caballé, G. (1991) Distribution and abundance of vines in forest communities, in *The Biology of Vines* (eds F.E. Putz & H. Mooney), Cambridge University Press, Cambridge, pp. 313–335.

Hilton, J., Wang, S.-J., Galtier, J. et al. (2004) An Upper Permian permineralized plant assemblage in volcaniclastic tuff from the Xuanwei Formation, Guizhou Province, southern China, and its palaeofloristic significance. *Geological Magazine*, **141**, 661–674.

Ichihashi, R., Nagashima, H. & Tateno, M. (2009) Morphological differentiation of current-year shoots of deciduous and evergreen lianas in temperate forests in Japan. *Ecological Research*, **24**, 393–403.

Iglesias, A., Wilf, P., Johnson, K.R. et al. (2007) A Paleocene lowland macroflora from Patagonia reveals significantly greater richness than North American analogs. *Geology*, **35**, 947–950.

International Commission on Stratigraphy (2013) International Chronostratigraphic Chart v2013/01. http://stratigraphy.org/index.php/ics-chart-timescale

Johnson, K.R. (2002) Megaflora of the Hell Creek and lower Fort Union Formations in the western Dakotas: vegetational response to climate change, the Cretaceous-Tertiary boundary event, and rapid marine transgression. *Geological Society of America Special Papers*, **361**, 329–391.

Krings, M. & Kerp, H. (2000) A contribution to the knowledge of pteridosperm genera *Pseudomariopteris* Danzé-Corsin nov. emend. and *Helenopteris* nov. gen. *Review of Palaeobotany and Palynology*, **111**, 145–195.

Krings, M., Kerp, H., Taylor, T.N. & Taylor, E.L. (2001) Reconstruction of *Pseudomariopteris busquetti*, a vine-like Late Carboniferous-Early Permian pteridosperm. *American Journal of Botany*, **88**, 767–776.

Krings, M., Kerp, H., Taylor, T.N. & Taylor, E.L. (2003) How Paleozoic vines and lianas got off the ground: on scrambling and climbing Carboniferous Early Permian Pteridosperms. *The Botanical Review*, **69**, 204–224.

Labandeira, C.C. & Currano, E.D. (2013) The fossil record of plant-insect dynamics. *Annual Reviews of Earth and Planetary Science*, **41**, 287–311.

Lee, D.W. & Richards, J.H. (1991) Heteroblastic development in vines, in *The Biology of Vines* (eds F.E. Putz & H.A. Mooney), Cambridge University Press, Cambridge, pp. 205–244.

Leopold, E.B., Manchester, S.R. & Meyer, H.W. (2008) Phytogeography of the late Eocene Florissant flora reconsidered. *Geological Society of America Special Papers*, **435**, 53–70.

Li, H. & Taylor, D.W. (1998) *Aculeovinea yunguiensis* gen. et sp. nov., a new taxon of gigantopterid axis from the Upper Permian of Guizhou province, China. *International Journal of Plant Sciences*, **159**, 1023–1033.

Li, H. & Taylor, D.W. (1999) Vessel-bearing stems of *Vasovinea tianii* gen. et sp. nov. (Gigantopteridales) from the Upper Permian of Guizhou Province, China. *American Journal of Botany*, **86**, 1563–1575.

Li, H., Tian, B., Taylor, E.L. & Taylor, T.N. (1994) Foliar anatomy of *Gigantonoclea guizhouensis* (Gigantopteridales) from the Upper Permian of Guizhou Province, China. *American Journal of Botany*, **81**, 678–689.

Libertin, M., Psenicka, J., Dakova, F. et al. (2009) A palaeoecological model for a vegetated early Westphalian intramontane valley (Intra-Sudetic Basin, Czech Republic). *Review of Palaeobotany and Palynology*, **155**, 175–203.

MacGinitie, H.D. (1969) The Eocene Green River flora of northwestern Colorado and northeastern Utah. *University of California Publications in Geological Sciences*, **83**.

Magallón, S. & Castillo, A. (2009) Angiosperm diversification through time. *American Journal of Botany*, **96**, 349–365.

Manchester, S.R. (2000) Late Eocene fossil plants of the John Day Formation, Wheeler County, Oregon. *The Ore Bin*, **62**, 49–63.

McElwain, J.E., Wagner, P.J. & Hesselbo, S.P. (2009) Fossil plant relative abundances indicate sudden loss of Late Triassic biodiversity in East Greenland. *Science*, **324**, 1554–1556.

Menninger, E.A. (1970) *Flowering Vines of the World: An Encyclopedia of Climbing Plants*, Hearthside Press Inc., New York.

Miller, K.G., Kominz, M.A., Browning, J.V. et al. (2005) The Phanerozoic record of global sea-level change. *Science*, **312**, 1293–1298.

Moiseeva, M.G., Herman, A.B. & Spicer, R.A. (2009) Late Paleocene flora of the northern Alaska peninsula: the role of transberingian plant migrations and climatic change. *Paleontological Journal*, **43**, 1298–1308.

Mosbrugger, V. & Roth, A. (1996) Biomechanics in fossil plant biology. *Review of Palaeobotany and Palynology*, **90**, 195–207.

Nichols, D.J. & Johnson, K.R. (2008) *Plants and the K–T Boundary*, Cambridge University Press, Cambridge.

Oplustil, S., Psenicka, J., Libertin, M. et al. (2009) A Middle Pennsylvanian (Bolsovian) peat-forming forest preserved in situ in volcanic ash of the Whetstone Horizon in the Radnice Basin, Czech Republic. *Review of Palaeobotany and Palynology*, **155**, 234–274.

Oplustil, S., Psenicka, J., Libertin, M. & Simunek, Z. (2007) Vegetation patterns of Westphalian and Lower Stephanian mire assemblages preserved in tuff beds of the continental basins of Czech Republic. *Review of Palaeobotany and Palynology*, **143**, 107–154.

Pérez-Salicrup, D.R. & Sork, V.L. (2001) Lianas and trees in a liana forest of Amazonian Bolivia. *Biotropica*, **33**, 34–47.

Poole, I., Gottwald, H. & Francis, J.E. (2000) Illiciaceae, an element of Gondwanan polar forests? Late Cretaceous and early Tertiary woods of Antarctica. *Annals of Botany*, **86**, 421–432.

Prothero, D.R., Draus, E., & Burns, C. (2009) Magnetostratigraphy and tectonic rotation of the Eocene-Oligocene Makah and Hoko River Formations, Northwest Washington, USA. *International Journal of Geophysics* 2009, Article ID 930612.

Putz, F.E. (1983) Liana biomass and leaf area of the "Tierra Firme" forest in the Rio Negro basin, Venezuela. *Biotropica*, **15**, 185–189.

Putz, F.E. & Holbrook, N.M. (1991) Biomechanical studies of vines, in *The Biology of Vines* (eds F.E. Putz & H.A. Mooney), Cambridge University Press, Cambridge, pp. 73–97.

Ray, T.S. (1986) Growth correlations within the segment in the Araceae. *American Journal of Botany*, **73**, 993–1001.

Rees, P.M. (2002) Land-plant diversity and the end-Permian mass extinction. *Geology*, **30**, 827–830.

Rothwell, G.W., Mapes, G., Stockey, R.A. & Hilton, J. (2012) The seed cone *Eathiestrobus* gen. nov.: fossil evidence for a Jurassic origin of Pinaceae. *American Journal of Botany*, **99**, 708–720.

Rowe, N.P., Speck, T. & Galtier, J. (1993) Biomechanical analysis of a Palaeozoic gymnosperm stem. *Proceedings of the Royal Society of London, Series B, Biological Sciences*, **252**, 19–28.

Rydin, C. & Friis, E.M. (2010) A new early Cretaceous relative of Gnetales: Siphonospermum simplex gen. et sp. nov. from the Yixian formation of northeast China. *BMC Evolutionary Biology*, **10**, 183.

Seyfullah, L.J. & Hilton, J. (2011) Callistophytalean pteridosperms from Permian aged floras of China. *Palaeontology*, **54**, 287–302.

Selaya, N.G. & Anten, N.P.R. (2008) Differences in biomass allocation, light interception and mechanical stability between lianas and trees in early secondary tropical forest. *Functional Ecology*, **22**, 30–39.

Shute, C.H. & Cleal, C.J. (1986) Palaeobotany in museums. *Geological Curator*, **4**, 553–559.

Smith, S.Y., Little, S.A., Cooper, R.L. et al. (2013) A ranunculalean liana stem from the Cretaceous of British Columbia, Canada: *Atli morinii* gen. et sp. nov. *International Journal of Plant Sciences*, **174**, 818–831.

Speck, T. (1994) A biomechanical method to distinguish between self-supporting and non-self supporting fossil plants. *Review of Palaeobotany and Palynology*, **81**, 65–82.

Speck, T. & Rowe, N.P. (2003) Hydraulics and mechanics of plants: novelty, innovation and evolution, in *The Evolution of Plant Physiology* (eds I. Poole & A.R. Hemsley), Elsevier Academic Press, Kew, pp. 297–325.

Speck, T. & Rowe, N.P. (2005) Plant growth forms: an ecological and evolutionary perspective. *New Phytologist*, **166**, 61–72.

Stein, W.E., Berry, C.M., Hernick, L.V. & Mannolini, F. (2012) Surprisingly complex community discovered in the mid-Devonian fossil forest at Gilboa. *Nature*, **483**, 78–81.

Stevens, L.G., Hilton, J., Bond, D.P.G. et al. (2011) Radiation and extinction patterns in Permian floras from North China as indicators for environmental and climate change. *Journal of the Geological Society*, **168**, 607–619.

Taylor, T.N. & Millay, M.A. (1981) Morphologic variability of Pennsylvanian lyginopterid seed ferns. *Review of Palaeobotany and Palynology*, **32**, 27–62.

Taylor, T.N., Taylor, E.L. & Krings, M. (2009) *Paleobotany: The Biology and Evolution of Fossil Plants*, 2nd edn, Academic Press/Elsevier, Amsterdam.

Teramura, A.H., Gold, W.G. & Forseth, I.N. (1991) Physiological ecology of mesic, temperate woody vines, in *The Biology of Vines* (eds F.E. Putz & H.A. Mooney), Cambridge University Press, Cambridge, pp. 245–285.

Tomescu, A.M.F., Rothwell, G.W. & Mapes, G. (2001) *Lyginopteris royalii* sp. nov. from the Upper Mississippian of North America. *Review of Palaeobotany and Palynology*, **116**, 159–173.

Van der Burgh, J. (2008) *Index of Angiosperm Leaf Species Names, A–Z, 1823–2006*, Backhuys Publishers, Leiden.

Vavrek, M.J., Evans, D.C., Braman, D.R. et al. (2012) A Paleogene flora from the upper Bonnet Plume Formation of northeast Yukon Territory, Canada. *Canadian Journal of Earth Sciences*, **49**, 547–558.

Walker, J.D., Geissman, J.W., Bowring, S.A., & Babcock, L.E., compilers (2012) Geologic Time Scale v. 4.0: *Geological Society of America*, doi: 10.1130/2012.CTS004R3C.

Wilf, P., Cúneo, N.R., Escapa, I.H. et al. (2013) Splendid and seldom isolated: the paleobiogeography of Patagonia. *Annual Review of Earth and Planetary Sciences*, **41**, 561–603.

Wing, S.L., Strömberg, C.A.E., Hickey, L.J. et al. (2012) Floral and environmental gradients on a Late Cretaceous landscape. *Ecological Monographs*, **82**, 23–47.

Wyka, T.P., Oleksyn, J., Karolewski, P., & Schnitzer, S.A. (2013) Phenotypic correlates of the lianescent growth form: a review. *Annals of Botany*. doi:10.1093/aob/mct236.

Chapter 17

THE EVOLUTION OF ANGIOSPERM LIANESCENCE: A PERSPECTIVE FROM XYLEM STRUCTURE-FUNCTION

Sandrine Isnard[1] *and Taylor S. Feild*[2]

[1]IRD, UMR AMAP, Laboratoire de Botanique et d'Écologie Végétale Appliquées, Nouméa, Nouvelle-Calédonie
[2]James Cook University, Townsville, Queensland, Australia

Ecology of Lianas, First Edition. Edited by Stefan A. Schnitzer, Frans Bongers, Robyn J. Burnham, and Francis E. Putz.

Companion Website: www.wiley.com/go/schnitzer/lianas

INTRODUCTION

The evolution of lianas has punctuated the history of land plants, with the angiosperm lineages representing the most recent stage of liana exploration (see Burnham, Chapter 16 in in this volume). As lianas evolve, the ecology of plant communities changes, in particular the tenor of and response to disturbance in forests as well as how energy and resource cycling in ecosystems (see van der Heijden et al., Chapter 13 in this volume).

During the last two decades, we have learned that lianas are essential operators in tropical forest dynamics and function (see Schnitzer et al, Chapter 01 in this volume), to the extent that lianas act as ecosystem transformers. Liana abundance is correlated with treefall gaps, larger-scale forest disturbances, as well as human perturbation (cutting, selective logging) (Schnitzer & Bongers 2002). They are often classified as early successional or gap-dependent pioneer species (Dewalt et al. 2000), despite a wide range of species being shade-tolerant (Putz 1984; Putz & Chai 1987). Lianas act as disturbance generators, by increasing host tree mortality and pulling neighboring trees down during treefall (the domino effect). They can arrest forest succession and alter successional trajectories by their ability to colonize and proliferate in gaps. As such, lianas can alter tree composition by altering gap-phase regeneration. Liana evolution might thus have enhanced forest turnover and reduced carbon sequestration (van der Heijden et al., Chapter 13 in this volume). With increasing forest perturbation worldwide, an increase in the relative importance of lianas appears inevitable, as liana expansion will affect many aspects of forest dynamics (see Schnitzer, Chapter 30 in this volume). Thus, understanding the functional biology of lianas is of critical importance in forecasting the fate of tropical forests in changing environments.

Water transport and the cost of plant body construction are two properties largely determined by the xylem structure-function. These features lie at the heart of how lianas transform ecosystems functions. Of key importance is that lianas have slender stems that are capable of moving lots of water and supplying a large canopy of leaves. Lianas can contribute up to 40% of leaf productivity of tropical forest, however, they represent less than 5% of total aboveground biomass (Hegarty & Caballé 1991). Such a pattern of resource allocation is associated with efficient nutrient cycling to enhance light-stimulated carbon uptake and growth response under high light conditions.

The model of lianas as fast-growing, disturbance-loving plants emerges largely from the function of eudicot and magnoliid angiosperms. However, lianescence has evolved in other clades (Carlquist 1991; Feild & Wilson 2012; Isnard et al. 2012; Feild and Isnard 2013) with older divergence ages (Magallón & Castillo 2009). New discoveries about xylem structure and function in climbers among extant lineages diverging near the root of angiosperms show that these climbers function differently from the common view of fast-growing, light-loving lianas.

Here we look at some specific properties of ecology and function, derived from functional aspects of stem hydraulics, which appear to be restricted to lianas. We chart the broad picture of xylem structure and function of angiosperm lianescence framed by what is known about the comparative ecophysiologies of eudicot/magnoliid and monocot lianas from temperate/tropical zones. We next review the liana hydraulic paradigm illustrated by these angiosperm climbers. We then demonstrate that lianas have explored heteroxylly (defined as the amount of xylem structural diversity), an innovation critical to the evolution of several key hydraulic, mechanical, and storage functions linked to the climbing habit. Finally, we explore how other climber lineages differ from generally accepted views about how lianas function hydraulically.

EVOLUTIONARY SUCCESS OF MESANGIOSPERMS LIANAS

The climbing habit has a long evolutionary history (see Burnham, Chapter 16 in this volume) and arose independently in the angiosperms, gymnosperms, and ferns. The only vessel-bearing lineage of extant gymnosperm, the Gnetales, comprise about 33 climbing species in the genus *Gnetum* (Won & Renner 2006). Some fern groups, such as *Trichomanes* (Dubuisson et al. 2003) or the Lycopsids (Rowe & Speck 1997, 1998), or the extinct seeds ferns Medullosales (Dunn et al. 2003; Krings & Kerp 2006; Wilson & Fisher 2011) provide examples of the climbing habit in tracheid-bearing species.

Among extant plants, climbers occur in most major lineages of Mesangiosperms (Fig. 17.1). The Mesangiosperms are a large clade that includes the vast majority of the angiosperms: magnoliids, Chloranthaceae, monocots, Ceratophyllaceae, and eudicots (Cantino et al. 2007). They diversified rapidly during the Cretaceous and the clade is posited as representing the "big bang" of angiosperm evolution (Soltis et al. 2008). In Mesangiosperms, lianas are mostly represented by the eudicot clade, monocots, and the magnoliids (Fig. 17.1). More than 60% of eudicot and magnoliid plant orders (30 out of 48 orders) (Stevens 2001 onwards) and 130 families (Gentry 1991) (about 30% of angiosperm families, APGIII) have at least one climber species. Such widespread phylogenetic diversity reflects the repeated and successful evolution of the climbing habit in vessel-based eudicot and magnoliid (see Gianoli, Chapter 18 in this volume). In monocotyledons, the climbing habit is also well represented, particularly in palms, which constitute one of the largest climbing families in Asia.

Less considered in the literature, though of evolutionary interest, the climbing habit evolved in other clades of angiosperms that diverged before the rise of the Mesangiosperms (Magallón & Castillo 2009). As such, the early diverging clade of Austrobaileyales, in which conduit elements are transitional between tracheids and vessels (Carlquist 1999), includes several climbing species in four genera: *Austrobaileya*, *Schisandra*, *Kadsura*, and *Trimenia*.

AN OVERVIEW OF THE MESANGIOSPERM LIANA PARADIGM

Major hydraulic attributes of lianas

Xylem structure and hydraulic performance

Long-distance water transport in most angiosperm lianas occurs through the lumen of non-living xylem vessels (Fig. 17.2). In vessel-bearing angiosperms, lumen and pit (total area of pits on a conduit) resistance each generally account for approximately half of the total resistance to water flow (Wheeler et al. 2005). Vessel diameter is important because the Hagen-Poiseuille equation predicts that hydraulic conductance (K_H) is proportional to the fourth power of conduit radius. Thus, a small increase in conduit diameter imparts a considerable increase in hydraulic flow (Tyree & Zimmermann 2002). Xylem conduit length is also important because it influences the hydraulic resistance that arises from flow between vessels. Flow between vessels is resistive because it occurs through non-perforated endwall cells that mark the endpoints of vessels. Longer conduits mean a diminished resistance that is attributable to vessel endwalls (Sperry et al. 2006).

Numerous studies of angiosperm liana hydraulic performance have defined a paradigm: lianas transport water fast and cheaply as compared to free-standing plants (Ewers et al. 1990, 1991a; Gartner 1991; Chui & Ewers 1992). The paradigm emerged from comparative functional and structural measurements of stem xylem made on several clades of Mesangiosperms.

Comparative hydraulic performance measures: lianas versus non-climbers

Hydraulic tests of water transport capacity conducted on tropical and temperate eudicot and magnoliids lianas have demonstrated that their stem xylem hydraulic transport efficiencies, defined as the amount of hydraulic conductivity for a given amount of stem cross-sectional area (K_S, sapwood area specific hydraulic conductivity), are among the highest range of values (Fig. 17.3A). Mesangiosperm lianas exhibit a mean value of K_s, about three times higher than self-supporting angiosperms. Consistent with possessing enormous K_S, xylem vessels of liana stems often possess greater maximum vessel diameters and lengths longer than trees and shrubs species (Figs. 17.3C, 17.3D). The enormous K_S values of liana stems result from water flow through a few wide diameter vessels (Gartner et al. 1990; Carlquist 1991). For instance, in the hook climber *Bauhinia fassoglensis* (Fabaceae), 14% of the vessels of the widest diameter class contributed 95% of the total theoretical hydraulic conductance as calculated by the Hagen-Poiseuille theory of ideal pipes (Ewers et al. 1991a). In the twiner *Lonicera japonica* (Caprifoliaceae), only 0.6% of the total number of vessels belonged to the widest diameter class and contributed 27.7% of the total theoretical conductivity (Chui & Ewers 1992). In the two temperate lianas *Celastrus orbiculatus* (Celastraceae) and *Vitis riparia* (Vitaceae) the largest vessel diameter categories ($\geq$160 μm) which account for only 8% of the total number of vessels, are responsible for about 63% of the theoretical conductivity. Thus, most of the water transport in lianas occurs though a small proportion of the stem cross-section, while the remaining wood tissues consist of narrow vessels, tracheids, fibers, and parenchyma.

In addition, perforation plates in vessel-based climbers are predominantly simple (Carlquist 1991). The significance of simple perforation plates is that they offer minimal resistance to water flow (Christman &

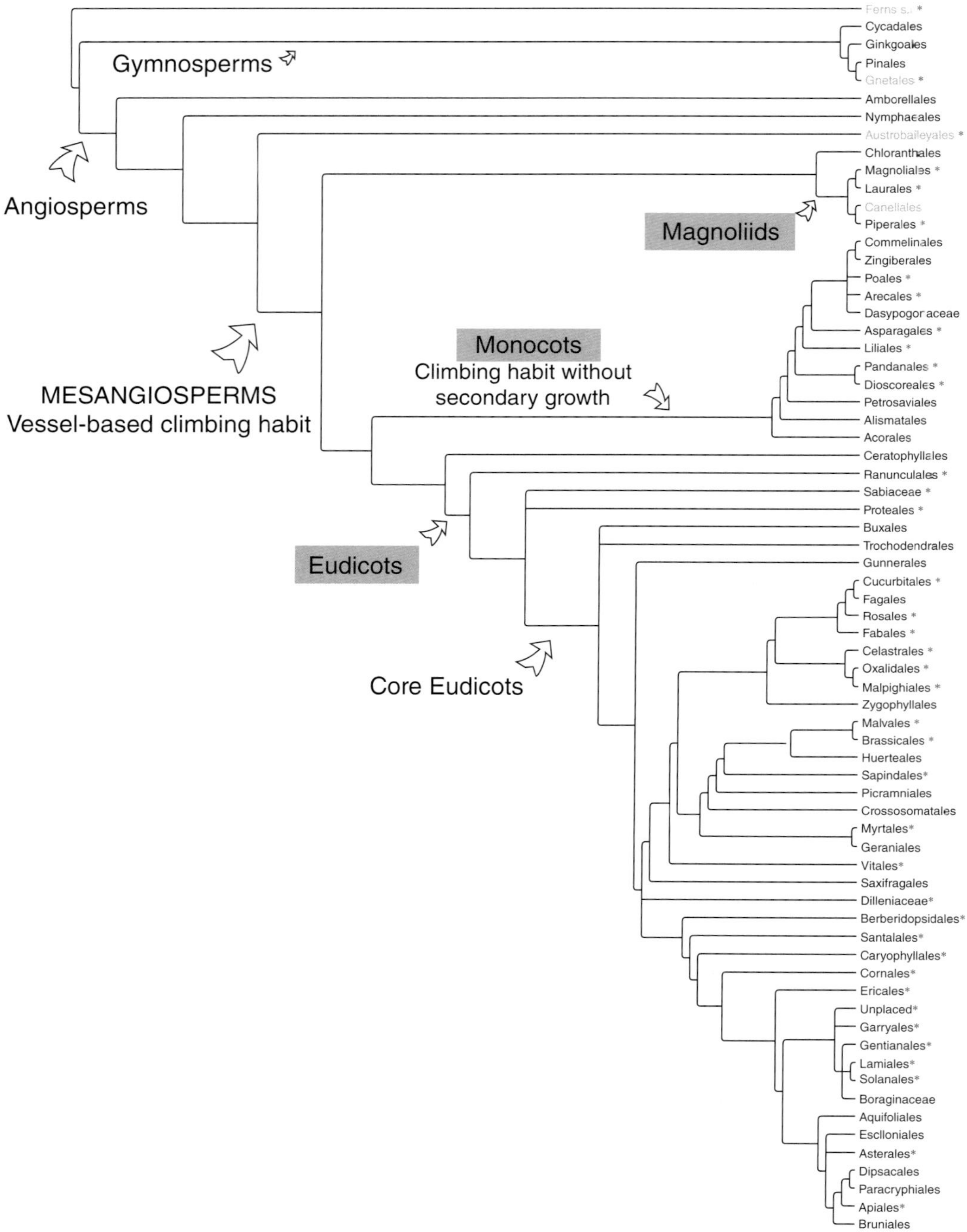

Fig. 17.1 A phylogeny of plants (APGIII, Stevens 2001 onward) illustrating the distribution of the climbing habit in different lineages. Lineages that contain climbing species are annotated with an asterisk (*). Clades that contain most of the diversity of lianas are highlighted (Magnoliids, Monocots and Eudicots). These clades belonging to the Mesangiosperms represent the paradigm of vessel-based lianas (see text). The climbing habit also occurs more sporadically in angiosperms lineages with tracheid/vessel transition (Austrobaileyales) or as a single known species in the tracheid-based Winteraceae (Canellales).

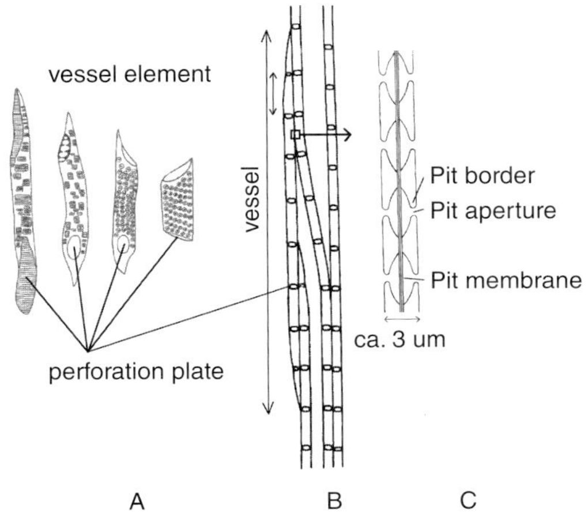

Fig. 17.2 Structure of conduits elements. When water flows through the vessels and from one vessel to another, it encounters several sectors of resistance: the lumen of vessel walls, the perforation plates joining vessels' elements, and finally the pith of interconnected conductive elements. (A) Evolution of vessel elements: decrease in length, reduction in inclination of end walls, change from scalariform to simple perforation plates, and from opposite to alternate pit arrangement (after Bailey & Tupper 1918). The various types of perforations plate (ends of vessel elements) differentially affect water transport along vessels, simple perforation plates inducing much less water flow resistance than scalariform perforation plate (Tyree & Zimmermann 2002). (B) Schematic representation of a vessel network. Vessels vary in length and consist of a series of vessel elements stacked end-to-end interconnected by perforation plates. (C) Pits are thin, porous depressions in the primary wall where secondary wall material has not been deposited (Sperry et al. 2006), and represent a high resistance to water flow (after Tyree & Zimmermann 2002). (Sources: Fig. 17.2A, adapted from Bailey & Tupper 1918. Reproduced with permission. Fig. 17.2B, from Zimmermann & McDonough 1978. Reproduced with permission of Elsevier. Fig. 17.2C, from Zimmermann & McDonough 1978. Reproduced with permission of Elsevier).

Sperry 2010). Several comparative studies have shown that eudicot lianas from lineages with free-standing taxa that possess scalariform perforation plates (e.g., Dilleniaceae, Lardizabalaceae, Icacinaceae), bear simple plated vessels (Carlquist 1991). With respect to lateral wall pitting, we mentioned above that the longer the conduit, the smaller resistance attributable to pit membrane (Sperry et al. 2006). By their very long vessels lianas could approach the "saturating length" (Sperry & Hacke 2004), where the conductivity is equal to the Hagen-Poiseuille lumen conductivity and thus lateral pitting is of minimal influence on axial water flow.

In terms of hydraulic cost, a long-noted feature of liana stems is that they possess very low ratios of stem cross-sectional area to distal leaf area (Huber value, H_V) versus free-standing plants (Ewers & Fisher 1991a; Tyree & Zimmermann 2002). Thus, wood investment in hydraulic transport per stem cross-sectional area is cheaper in lianas than in other growth forms. For instance, a 6-mm-wide liana stem of *Bauhinia fassoglensis* bears the same amount of distal leaf area (ca. 2.45 m^2) as an 18 mm diameter stem in the tree *Bauhinia blakeana* (Ewers & Fisher 1991a). Broadly, the ratio of stem-cross-section to leaf area is about one to two orders of magnitude lower in lianas as compared to self-supporting trees or shrubs (Fig. 17.3F). Due to the large proportion of non-structural tissues (phloem, rays, parenchyma) compared to fibers and vessels in most lianoid stems (Carlquist 1991), the carbon investment in stem tissues for a given leaf area is less in lianas than in self-supporting species.

However, in terms of the capacity of the stem hydraulic supply to deliver water to transpiring leaves, which is quantified by leaf area-specific hydraulic conductivity (K_L), liana hydraulic capacities broadly parallel those of trees (Tyree & Zimmermann 2002; Feild & Balun 2008; Zhu & Cao 2009) (Fig. 17.3B). As K_L is broadly correlated with leaf photosynthetic and transpirational capacities, the gas exchange capacities of lianas and trees also broadly overlap (Fig. 17.3E). Thus, the large K_S values of lianas offset their very low H_v to produce similar K_L. As a consequence, the typical ecological success of Mesangiosperms lianas in disturbed open areas, leading to arrested succession in some cases, is caused by a large shift in increased xylem hydraulic efficiency rather than possession of especially high capacities for carbon gain relative to competing free-standing plants.

Hydraulic safety in lianas to drought-induced cavitation

Due to the reliance on a smaller number of wide and long vessels, the water transport systems of liana stems are often hypothesized to be more prone to losses in xylem hydraulic conductance when xylem emboli occur (Ewers et al. 1991a). However, actual xylem vulnerability data (i.e., a measure of percentage loss of

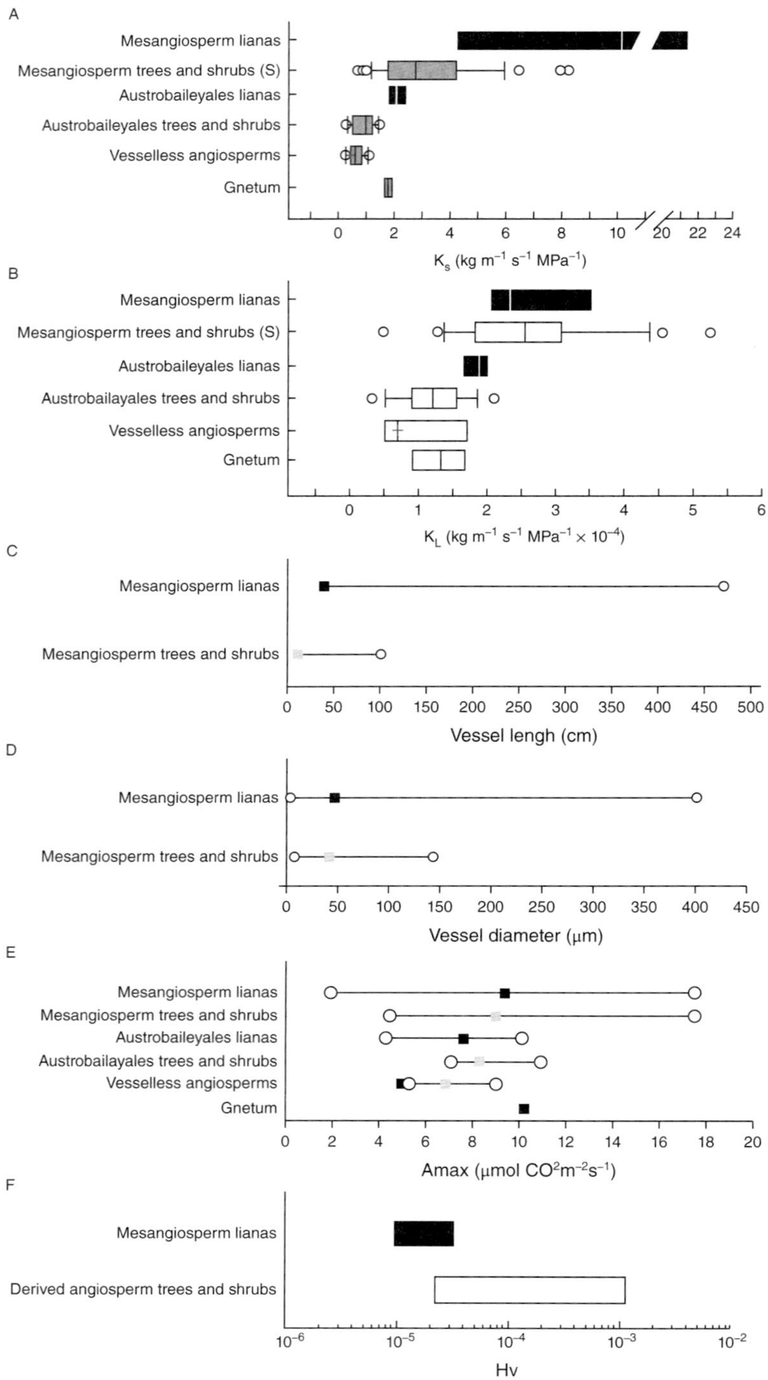

Fig. 17.3 Comparative hydraulic performances and vessel dimensions for lianas and self-supporting species with varying phylogenetic affinities. (A) Sapwood area-specific hydraulic conductivity (K_S). (B) Leaf area-specific conductivity (K_L). (A, B) The black cross represents the vessel-less *Tasmania cordata* (after Feild & Wilson 2012; Feild et al. 2012). (C) Vessel lengths. (D) Vessel diameters. (E) Maximum photosynthetic rate (Amax). (F) Range of Huber values (after Tyree & Zimmermann 2001). (A, B) S, simple perforation plate, Sc, long scalariform perforation plates with 30–120 scalariform pits per plates. (C–E) square, mean, white circles, minimum and maximum values. Vessel dimensions come from published sources (Ewers et al. 1990; Tomlinson et al. 2001; Chui & Ewers 1992), as well as photosynthetic data (Feild & Franks 2003; Feild & Chatelet 2012; Zhu & Cao 2009; Castellanos 1991; Cai et al. 2009). (Source: Feild & Wilson 2012. Reproduced with permission of University of Chicago Press.)

hydraulic conductivity due to embolism over a range of xylem water potentials) that would test such a hypothesis for lianas are limited. Measuring stem hydraulic performance of lianas is difficult as their generally long vessels present challenges for commonly used procedures to generate vulnerability curves (Cochard et al. 2005). When long vessels are present, they are open along the measured segment and tend to cavitate at anomalously low tensions. A high percentage loss of conductivity might thus indicate the presence of open vessels instead of vulnerability to cavitation. The standard bench drying method (Tyree & Sperry 1989) remains the most reliable method to measure liana vulnerability to drought-induced embolism since it allows the measurement of segments whose length can be longer than the longest vessels' length.

Using air drying methods in a comparison of co-occurring lianas and trees species in seasonally dry tropical forest, Zhu et al. (2009) showed that the liana xylem was significantly more vulnerable to cavitation as compared to co-occurring tree species (higher P50 values, above -1.5 MPa for stem segments between 80–110 cm length; the P50 is a measure of xylem water potential corresponding to 50% loss of stem conductivity. This threshold is used as a standard for comparative studies.) The percentage loss of conductivity also increased faster with decreasing water potential than in tree species, indicating a narrower safety margin. Very high vulnerability to cavitation was reported for two temperate lianas *Celastrus orbiculatus* and *Vitis riparia* (P50 ≥0.2 MPa), where hydraulic conductivity was measured on short stem segments of 10 cm length (Tibbetts & Ewers 2000). The paucity of available data and potential artifacts of open vessels during vulnerability measurements leave open the question of how vulnerable lianas are to drought-induced embolism.

Another hydraulic paradigm of eudicot lianas is that they possess great ability to repair embolized vessels. Root pressure is often cited as the mechanism that lianas employ to refill their vessels after embolism (Putz 1983). Water is taken into the root osmotically and forced up though the stem at positive pressure to quickly push gas out of vessels or more slowly to dissolve gas (Sperry et al. 1987). Indeed, root pressures occur in several climbers among monocots and eudicots (Cochard et al. 1994; Ewers et al. 1997; Fisher et al. 1997; Cobb et al. 2007). However, in a survey of 29 lianas species, Ewers et al. (1997) found that only three species had positive pressure. Root pressure may not be the only mechanism for embolism reversal in lianas. Recent studies have indicated that living cells adjacent to the embolized vessels may create a driving gradient that draws water into the lumen of the embolized vessels. Such a mechanism has been proposed to occur while adjacent functioning vessels are under tension (Holbrook & Zwieniecki 1999). The abundant parenchyma in liana wood and common wide rays, composed of abundant paratracheal parenchyma (Carlquist 1991), may potentially aid in refilling embolized xylem cells.

On the other hand, lianas are suggested to suffer little from drought water stress since several physiological and structural features would insure the stability of water transport and avoid embolism. First, lianas can transport more water over a less steep gradient in water potential from roots to leaves (Andrade et al. 2005), despite often very long hydraulic path lengths, because they have high specific conductivity (K_S). Several studies have also found that lianas possess strong stomatal control (Tay et al. 2007; Cai et al. 2009), though the underlying control remains unclear and might be caused by vapor pressure deficit (VDP) or leaf water potential (Ψleaf). Liana water storage capacity, as indicated by their vast numbers of living parenchyma cells and perhaps also wide diameter vessels, could significantly regulate evaporative demand, though there have been few direct measurements of stem xylem capacitance. The fast-growing weedy climber Kudzu (*Pueraria montana*), capable of growing under divergent radiation regimes, exhibits constant leaf water potential over a wide range of transpiration rates, suggesting that water storage tissues can buffer water stresses (Forseth & Teramura 1987). In this species, a strong stomatal control in response to vapor pressure deficit (VPD) was observed, coupled with rapid leaf orientation in response to irradiance. Another potential physiological attribute of lianas might lie in their potentially deep root systems, as lianas would be able to tap deep soil water and consequently avoid embolism during the dry season (Restom & Nepstad 2004; Andrade et al. 2005). Additionally, layers of apotracheal parenchyma that often surround the walls of wide vessels in many lianas might prevent embolism spread in adjacent vessels.

Patterns of carbon allocation in lianas

An important structural dimension of liana physiology is that lianas are more cheaply constructed and exhibit more efficient resource allocation as compared to free-standing taxa. This interpretation is based on evidence that lianas have low leaf construction and defense costs per unit of photosynthetic area (low leaf mass per area, LMA) (Cai et al. 2009; Asner & Martin 2012), which is related to shorter leaf lifetimes and faster leaf turnover rates (Castellanos 1991). In the *leaf economic spectrum*, such a leaf investment strategy is related to quick investment of nutrient and dry mass (Wright et al. 2004). Lianas exhibit a high mass-based concentration of leaf metabolic components that promote photosynthetic light capture and growth (e.g., N, P, chlorophyll), suggesting that they support higher rates of N and P cycling by concentrating these elements in the leaves (Asner & Martin 2012). Consequently, N and P requirements are lower in lianas than in trees per unit of photosynthetic surface area (Asner & Martin 2012), and this might result from the lower carbon investment in wood structure as lianas use external mechanical support to access light.

The pattern of carbon allocation might be the main functional advantage of liana growth habit. A higher relative growth rate than in free-standing species is achieved thanks to their ability to allocate more carbon to photosynthetically active tissues and root growth instead of stems or leaf structures and defense (Forseth & Teramura 1987; Cernusak et al. 2008). A high photosynthetic rate per mass might not be the key factor, instead efficient nutrient cycling in lianas enhances light-stimulated carbon uptake and growth response in high-light conditions.

CAMBIAL VARIANTS AND THE FUNDAMENTAL ROLE OF HETEROXYLLY IN MESANGIOSPERM LIANA DIVERSIFICATION

Vessel-less wood is mostly composed of tracheids, due to the small conductance ability of individual tracheids. In contrast, the presence of wider vessels allows for more hydraulic conductance to be realized with investment in fewer conduits, freeing up wood to include other cell types such as fibers, parenchyma cells, and rays (Sperry et al. 2007; Hudson et al. 2010; Feild & Wilson 2012).

The evolution of wood structural diversity – termed "heteroxylly" – is believed to represent a major leap in angiosperm diversification. Heteroxylly increases the number of developmental options available for specialization of hydraulic, mechanical, and storage functions within the xylem. Recent work proposes a correlation between greater heteroxylly and an expanded diversity of growth forms and architectures during angiosperm evolution (Carlquist 2009; Feild et al. 2012). Lianas in particular have explored heteroxylly through a great diversity of cambial functioning.

A "typical" bifacial, or bidirectional, cambium produces secondary xylem (wood) toward the inside and secondary phloem toward the outside thus forming a continuous cylindrical sheath around the secondary xylem (Evert 2006; Spicer & Groover 2010). In many angiosperm lianas, however, cambial activity deviates sharply from such a typical development to produce a broad range of stem bauplans (Fig. 17.4). Multiple cambia or differential production of secondary tissues produce various anatomical configurations such as dispersed xylem or xylem in successive rings, interxylary phloem, non-circular stems, and furrowed or crenate wood outlines (Schenck 1892, Obaton 1957; Carlquist 1991; Caballé 1993).

These cambial "variants," found in many climbers, represent a central feature of the liana habit and occur across the whole angiosperms' phylogeny (Spicer & Groover 2010; Angyalossy et al. 2012). The repeated evolution of cambial variants across the Mesangiosperm phylogeny provides compelling evidence of adaptive significance.

Cambial variants may provide both hydraulic and mechanical advantages over a uniform cambium. In Bignoniaceae, cambial variants are thought to have played an important role in the diversification of the family, through specialization of regular and variant phloem in distinct storage and transport function (Pace et al. 2009, 2011). The greatest diversity of cambial variants occurs in the core eudicots where the diversity of lianas reaches its apogee. But the other lineages among Mesangiosperms, such as the basal eudicot Ranunculales (Ranunculaceae, Menispermaceae, and Lardizabalaceae) and the magnoliids Piperales (Piperaceae, Aristolochiaceae) also exhibit cambial

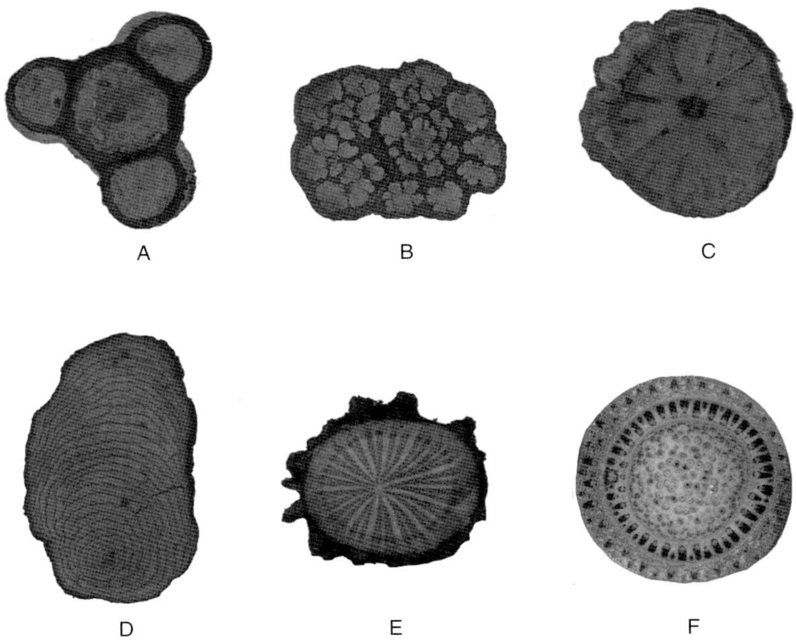

Fig. 17.4 Advanced heteroxylly in lianas. Cambial variants in Mesangiosperms lianas (A–E) and scattered vascular bundles embedded in a parenchyma matrix in the monocotyledon rattan (F). (A) peripheral xylem in *Serjania brachycarpa* (Sapindaceae). (B) dispersed wood in *Aegiphila sp.* (Lamiaceae). (C) furrowed wood in *Fridericia patellifera* (Bignoniceae), (D) multiple concentric cambia in Menispermaceae. (E) wide rays in *Aristolochia ovalifolia* (Aristolochiaceae). (F) scattered vascular bundles and two thick leaf sheaths surrounding the stem in *Calamus sp.* (Arecaceae).

variants (Fig. 17.4). In Piperales, both *Aristolochia* and *Piper* stems exhibit a "cable construction" with tall and wide rays representing a large cross-sectional area in climbing species (Carlquist 1993, Isnard et al. 2012, Wagner et al. 2012) (Fig. 17.4E). In *Piper* climbers, stems possess a polystelic organization, comprising several rings of vascular bundles, with peripheral bundles undergoing significant secondary growth (Isnard et al. 2012; Wagner et al. 2012). Thus in magnoliids, similarly to eudicots, the evolution of the climbing habit is linked to the evolution of a peculiar stem bauplan.

Large rays in Piperales derive from regions of interfascicular cambium that are incapable of producing fusiform initials (i.e., longitudinally aligned cells). Such anatomical organization appears to be highly conserved across the Piperales, and has been proposed to represent an "unbreakable trait" (Isnard et al. 2012; Wagner et al. 2012). Strong conservatism seems to govern secondary meristematic activity in diverse climbing lineages. Because many cambial variants are taxon-specific, stem anatomy can be used as a reliable feature to identify various plant families or genera (Caballé 1993) (e.g., phloem wedges in Bignonieae (Pace et al. 2009; Pace et al. 2011), successive concentric cambia in Menispermaceae (Carlquist 2007)).

Mechanical function of variant cambia

The mechanical function of cambial variants is probably most evident when large amounts of parenchyma, coupled with the compartmentalization of the vascular system, generate the so-called "cable construction." These innovations provide greater stem flexibility and protection of the vascular system, where sectors of wide and long vessels that are mechanically brittle are jacketed in softer living tissues (Putz & Holbrook 1991). Lianas with large amounts of parenchyma resist loss of water conduction under severe twisting (Putz & Holbrook 1991). Ewers et al. (1991a) further showed that regardless of the type of cambial variant, the different wood segments are interconnected in three dimensions, allowing for further continuity of the vascular system following injury. The redundant and connected vascular system thus ensures efficient and robust water transport to the leaves through multiple pathways. Such vascular organization follows a basic rule of engineering which predicts that the level of redundancy should increase with increasing risk of fatal system failure (Schenk et al. 2008). Lianas in particular are subjected to mechanical strain during their life in the canopy, during treefall or separation from the host support.

A mechanical attribute of cambial variant is consequently the ability of the stem to sustain and/or survive large deformation in bending or twisting without losing hydraulic function. Large amounts of parenchyma and multiple cambia also allow: (1) splitting of the stem following bending or twisting strain along a soft parenchyma matrix, thus preserving the hydraulic integrity; and (2) wound healing though proliferation of parenchyma or new cambia (Caballé 1986; Fisher & Ewers 1991).

Cambial variants and secondary phloem

Variant secondary growth may also serve to increase the relative amount of secondary phloem (Ewers & Fisher 1991a; Pace et al. 2009), which plays a crucial role in the long-distance carbohydrate transport for both storage and growth. A large proportion of the stem cross-sectional area is dedicated to phloem in lianas as compared to trees and shrubs. Ewers et al. (1991a) showed that the reduction of the phloem surface in the reduced stem cross-sectional area of *Bauhinia* climbers, compared to congeneric trees and shrubs, was less than the reduction of xylem surface. Included phloem, interxylary phloem, large phloem wedge, and phloem in bands, all illustrate the diversity of phloem organization resulting from cambial variants. In addition to shifts in phloem patterning within the stem, phloem conduits in lianas also appear to be tuned toward greater transport efficiency at the conduit scale. Phloem in lianas is distinguished by wide diameter sieve-tube elements, and simple sieve plates with large pores (Carlquist 1991). In Bignonieae, two distinct types of phloem coexist, and cells associated with these two phloem types are known to be anatomically different. Regular phloem is suggested to have specialized in storage while variant phloem has specialized in conducting photosynthates (Pace et al. 2011).

The storage function

The large amounts of parenchyma found in many lianas may be adaptively significant for both carbohydrate and water storage (Ewers & Fisher 1991a; Carlquist 2007; Spicer & Groover 2010). These functions have received little study in lianas, however, and are mostly conjectural. Capacitance has been suggested in species capable of maintaining constant water potential under divergent radiation regimes (Asner & Martin 2012). Storage capacity is probably a crucial aspect of liana ecophysiology and is suggested to confer a competitive advantage on lianas over trees during the dry season (Schnitzer 2005). Further research should address the effective role of water and carbohydrate storage with respect to water stress avoidance and the proliferation of lianas under high light environments and during seasonal droughts.

THE LIANA HABIT BEYOND VESSEL-BASED EUDICOTS AND MAGNOLIIDS

The highly heteroxylous vascular system and xylem structure found in the vast majority of Mesangiosperm lianas appear to be a prerequisite to meeting both high hydraulic demand of a large canopy and the mechanical function of a stem subject to environmental perturbation. This raises the question of the hydraulic and mechanical functions of climbers in lineages lacking secondary growth or possessing "transitional" conduit structures.

The climbing habit in monocots

The climbing habit has evolved repeatedly in different monocot lineages (Arecaceae, Liliales, Pandanales, Poales) in spite of their lack of secondary growth.

However, monocotyledon liana stems possess many features that mimic the heteroxyllous structure described above. These include scattered vascular bundles embedded in a parenchyma matrix ("cable construction"), marked differentiation between protoxylem and notably larger metaxylem vessels, and a fibrous sheath around vascular bundles offering protection to vascular tissues against mechanical stresses (Fig. 17.4F) (Carlquist 1991; Isnard & Silk 2009).

Monocot rattan palms can exhibit giant metaxylem vessels up to 300 μm in diameter and 3.5 m in length (Tomlinson et al. 2001). Rattans (climbing palms) provide a striking example of how hydraulic and mechanical functions need to be studied concomitantly to understand the evolution of climber functional performance. The vascular features of rattans are unusual compared to those of arborescent palms (Tomlinson et al. 2001; Tomlinson and Spangler 2002), suggesting xylem adaptation is related to water transport. This might be especially true given the high hydraulic demand on a thin and very long climbing stems, up to 100 m, without secondary growth. However, long-distance water transport in rattans, in spite of their enormous vessels, appears hydraulically constrained at two points. First, the interconnection of their large axial vessels occurs through transverse vessels of very narrow diameter, and, second, a layer of living cells separates their axial vessels from the tracheids leading to the leaf (Tomlinson et al. 2001).

Cobb (2006) explored the question of efficiency and resistance of such vascular construction and came to surprising conclusions. First, transverse vessels are so abundant that their narrowness would not significantly affect steady-state water transport, even though a slightly higher resistance to water flow was observed. The layer of parenchyma, where water is forced to flow to reach the leaves, forms such a large area along the rattan stem that the resistance to water flow from the stem to the leaves was minimal. Vascular peculiarities were also not related to reversal from embolism. Thus steady-state water transport efficiency does not benefit nor acutely suffer from the unusual vascular anatomy of rattans (Cobb 2006). The adaptive significance of such vascular construction in rattans is proposed to be mechanical, via making the water transport system less prone to mechanical strain caused by twisting or bending of the stem, inherent to the climbing habit (Cobb 2006). The cane of rattans has been shown to exhibit a high degree of stem flexibility (Isnard & Rowe 2008) and is extensively used in handicraft and construction because of their great flexibility. In rattans, stem mechanical construction has been shown to be broadly analogous to, though developmentally divergent from, eudicot lianas (Isnard & Rowe 2008). Stem flexibility arises after the senescence and shedding of the outer leaf sheath, which leaves behind the flexible stem (Isnard & Rowe 2008). Thus, in addition to wide vessels, to meet the hydraulic demands of a narrow stem, climbing palms have evolved a set a mechanical adaptations, demonstrating that the evolution of the climbing habit involves more than an efficient water system.

The climbing habit in Austrobaileyales

The lianoid habit is surprisingly common among species of the early diverging angiosperm clade Austrobaileyales. Austrobaileylean climbers are represented by the monotypic lineage *Austrobaileya*, two species of *Trimenia*, and all known species of *Schisandra* and *Kadsura* that compose the Schisandraceae (~50 species). Comparisons of the xylem anatomy of *Schisandra* and *Kadsura* with their free-standing closest relatives *Illicium* indicated that the differences in wood anatomy between climbers and free-standing plants were largely quantitative (Carlquist 1999; Feild & Isnard 2013). Xylem shifts associated with lianescence in the Schisandraceae included larger pits on tracheids, wider tracheid lumens, wider and shorter vessels elements, as well as broader multiseriate rays than in *Illicium*. Some *Schisandra* lianas also exhibited the development of vessels bearing simple perforation plates, whereas vessels found in free-standing taxa possessed highly resistive scalariform perforation plates consisting of numerous pits and frequent remnants of pit membranes (Carlquist 1999; Feild & Isnard 2013). In the genus *Trimenia* where vessels have been suggested to be "semi-tracheids," with prominent pit membrane remnants in the perforation (Carlquist 2012), self-supporting species have more numerous bars per perforation plate than do the climbing species. Other Austrobaileylean lianas appear to reflect the same pattern. For instance, *Austrobaileya scandens* possesses large diameter vessels and wide rays (Fig. 17.5B) (Carlquist 2001), but vessel elements are of a hydraulically resistive nature, based on their gradually tapered end-walls with perforation plates bearing numerous scalariform pits (Feild et al. 2003; Christman & Sperry 2010). A similar vessel anatomy also occurs in *Trimenia* lianas (Fig. 17.5A) (*T. moorei* and *T. macurura*, T. S. Feild, unpublished data). Thus Austrobaileyales lianas have a conduit structure that facilitates water flow, compared to self-supporting close relatives, but nevertheless retains some xylem characteristics that limit the hydraulic efficiency of their conduits.

Functionally, Austrobaileyalean lianas possess up to about 50% greater specific conductivity (*Ks*) than self-supporting closely-related species (Fig. 17.3A). But wood hydraulic capacity in these basal angiosperms is in the lower range of values, along with vessel-less angiosperms and conifers (Feild & Wilson 2012) (Fig. 17.3). Leaf area-specific conductivity (K_L) of Austrobaileyales lianas also exhibits a low and narrow range of values, linked with low maximal stomatal conductances (gs_{max}) and CO_2 assimilation rates (A_{max}, (Feild et al. 2003, 2011; Feild & Isnard 2013). The gas exchange capacities of lianas and trees in Austrobaileyales broadly overlap, as observed in Mesangiosperms, and are among the lower values (Fig. 17.3E). Similar to what is observed in Mesangiosperms, changes in growth habit mainly involved change in hydraulic capacity and xylem structure but not higher photosynthetic rates per mass.

Finally, Austrobaileyales lianas exhibit an increase in hydraulic performance compared with their free-standing relatives, but nevertheless show low xylem hydraulic capacities that are correlated with relatively low capacity for photosynthetic gas exchange when compared with Mesangiosperm climbers.

At least one Austrobaileylean liana (*Schisandra glabra*) occurs as a large-bodied climber, climbing up to 25 m in the forest canopy and exhibits extensive lateral exploration of understory and subcanopy spaces. The stems perform bending and torsional flexibility associated with low toughness (Feild & Isnard 2013). Compared with a non-climbing relative from similar habitats (*Illicium floridanum*), the climbing *Schisandra glabra* exhibits an average of 30% lower stem sapwood area to leaf area (H_v) (Feild & Isnard 2013). Despite lacking any cambial variant, this species exhibits lianoid anatomical features including wide rays, wide vessels, and a thick cortex. These results show that high mechanical flexibility, heteroxylly, and thin stems relative to leaf area might have preceded a shift toward high hydraulic efficiency in early angiosperm lianas.

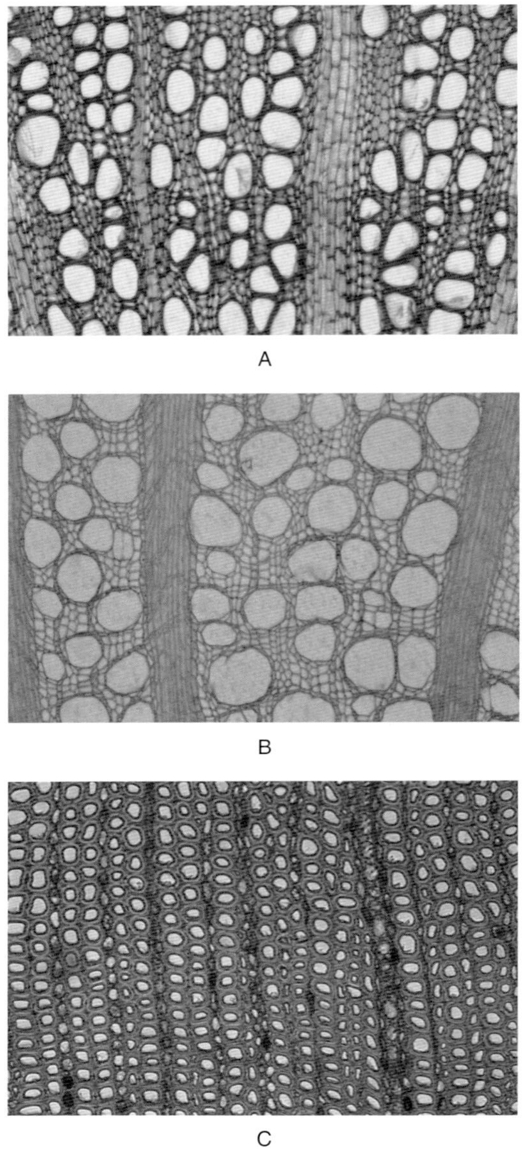

Fig. 17.5 Wood anatomy in Austrobaileyales and Canellales lianas. Large diameter vessels and wide rays in the lianas *Trimenia maclura* (A) and *Schisandra glabra* (B) (Austrobaileyales). (C) Tracheid-based homoxyllous wood in the liana *Tasmannia cordata* (Winteraceae).

The climbing habit in a vessel-less angiosperm (Winteraceae)

Recent discovery of an extant vessel-less homoxyllous angiosperm liana, *Tasmannia cordata* (Winteraceae) (Fig. 17.5C) from the highland cloud forests of New Guinea (Feild et al. 2012), could compromise the heteroxylly hypothesis as a prerequisite for climbing. But *T. cordata* exhibits very strong wood compared to what has been found in most eudicot lianas (Rowe et al. 2006): its wood is as much as six times stronger than its congeneric free-standing relatives (Feild et al. 2012). Additionally, *Tasmannia cordata* remains a small-sized climber and stems never become large (mean diameter for mature stems is about 17 mm and maximal length is ~5 m) (Feild et al. 2012). The ability of *T. cordata* to resprout from the base may give individual plants a better chance of survival after the death of a single stem from mechanical or hydraulic damage. Hydraulic capacity (K_s and K_L) as well as organization of stem anatomy do not differ from its self-supporting close relatives. Wood hydraulic capacity in these vessel-less angiosperms is nested in the lower range of values (Figs. 17.3A, 17.3B). *Tasmania cordata* combines ecophysiological features of shade-tolerant species and is restricted to wet and understory habitats. The hydraulic permissiveness of its habitat (high humidity and cool temperature) that reduces the evaporative demand and water stress, along with intermittent freeze-thaw conditions, may argue for limited competition with larger vessel-diameter climbers (Feild et al. 2012).

These hydraulic and mechanical features seem to run contrary to the climbing habit, and they contribute to the reason why *T. cordata* is the only known tracheid-based angiosperm climber. With a shade-adapted ecophysiology, very low stem specific conductivity, and leaf-area specific hydraulic conductivity, *Tasmannia cordata* may represent a case of an "evolutionary dead end" (Feild et al. 2012).

Hydraulic limitation of lianas in early diverging angiosperm

The dominant angiosperm liana strategy found in Mesangiosperms can be defined by their opportunism, fast growth, and high light-demanding ecophysiology related to hydraulically efficient xylem structure and highly heteroxylous vasculature which allows mechanical laxness, ecophysiological advantage (phloem transport and storage), and mechanical protection of the hydraulic system.

It is consequently surprising that all the angiosperm lianas studied outside this large and diverse clade, as well as the recently discovered vessel-less angiosperm climber *Tasmania cordata*, combine ecophysiological features of shade-tolerant species and seem to be restricted to wet and understory habitats. The main hydraulic-related feature shared by these wet-demanding species is the hydraulic limitation of their xylem conduit and their low photosynthetic capacity. Xylem conduits include tracheids (*Tasmannia*), and vessels with prominent pit membrane remnants in the perforation, conduit structures both known to limit water transport efficiency. In Schisandraceae, exhibiting the largest diversity of climbers among basal angiosperms, the vessels are wide and short and have been suggested to have predominantly simple perforations (Carlquist 1999). However, stem-specific conductivity (K_s) remains in the lowest range of values along with vessel-less angiosperms, probably due to the many irregularities of perforation plates, the remnants of pit membranes (Carlquist 1999), that increase resistance to water flow. We suggest that the evolution of lianoid features does not necessarily signal the evolution of a highly opportunistic, light-demanding climbing life history. Instead, the life-history strategies of the climbing habit need to be examined in light of the dual mechanical and hydraulic function of the xylem structure that determines how lianas are able to climb and transport water over long distances (Isnard & Silk 2009). At least one species of *Schisandra* (*S. glabra*) exhibits mechanical flexibility in both bending and torsion and the high degree of mechanical specialization during ontogeny in this species is comparable with what is observed in eudicot and magnoliids lianas (Feild & Isnard 2013). These results suggest that mechanical laxness might precede wood hydraulic performance during liana evolution. More studies on Austrobaileylean lianas would clarify this point.

The climbing habit outside of the angiosperm clade

Gnetum

The climbing habit has evolved in the gymnosperm vessel-bearing genus *Gnetum*, which comprises ~30 species of woody twiners and two free-standing species (Won & Renner 2006). Comparative xylem-related features of lianas versus free-standing species in *Gnetum* show trends similar to those characterizing the Mesangiosperms lianas. These include: (1) a skewed distribution of vessel length and trends toward larger vessel diameters (Fisher & Ewers 1995); (2) a reduction of stem stiffness during development (Rowe et al. 2006); (3) a cambial variant producing successive cambia (Carlquist 2007); and (4) higher K_s and HV values than self-supporting relatives. However, similar to what has been shown in climbers outside Mesangiosperms, *Gnetum* species exhibit low hydraulic capacities (K_s and K_L), and are nested in the lower range of values (Figs. 17.2A, 17.2B). *Gnetum* species from Papua New Guinea exhibit low photosynthetic and hydraulic capacities and have ecophysiological performances similar to co-occurring obligate shade-adapted angiosperms (Feild & Balun 2008). Ecologically, *Gnetum* species largely inhabit shaded understories of wet forest and swamp gallery forest, but few species are reported in more open and disturb habitats. The low hydraulic capacity of *Gnetum* climbers suggests that they only occasionally exploit high light environments, unlike the ecophysiologically adapted, light-demanding, disturbance-loving climbers. Still, the gas exchange capacity (A_{max}) of *Gnetum parviflorum*, from tropical seasonal forest, shows the same range of values found in Mesangiosperms (Fig. 17.3E). The complete picture of the ecological diversification of *Gnetum* climbers thus remains unclear and the ecophysiological performance of some species could exceed values provided by available studies.

Climbing ferns

Two distantly related tracheid-based lineages, the ferns and the extinct seeds ferns, illustrate that climbing habit can be decoupled from possessing vessels but not conduit gigantism. In the fossil record, several species of medullosan seed plants were probably climbers (Mosbrugger 1990; Krings et al. 2003). The tracheid structure of medullosan species, including large pit membrane areas and a high ratio of tracheid diameter to wall thickness, suggests that their vascular system provided little structural support to the whole plant, with support instead found in sheathing leaf bases. The recent finding that the cortex tissues of medullosan species exhibit very low lignin content, supports the idea that they were unable to grow free-standing (Wilson & Fisher 2011). The stem–leaf base structure of medullosan species somehow recall rattan palms, where leaf sheaths provide mechanical support and the senescence of sheathing leaf bases might release the flexible stem. Medullosan species include a set of lianoid features, comprising multiple vascular cylinders included in a thick cortex, a secondary vascular system in contact with axial parenchyma cells, anastomoses of vascular segments throughout the stem length, narrow stems, and tracheids among the longest and the widest known among seed plants, up to 200 μm in diameter (Wilson 1924; Wilson & Fisher 2011). The hydraulic conductivity of medullosan xylem per unit area is presumably relatively high and the large stem area dedicated to water transport suggests that they may have had a per-area conductivity comparable to vessel-bearing eudicots (Wilson et al. 2008).

In extant ferns (e.g. Lygodiaceae), the wide and long conduit which accounts for a high hydraulic conductivity, compensates for the small ratio of xylem to leaf area (Pittermann et al. 2011). Such hydraulic performance in ferns is suggested to arise because the xylem serves uniquely the hydraulic function while mechanical support is provided by the presence of a hypodermal sterome. Such hydraulic capacity and the separation of mechanical and hydraulic function, analogous to what occurs in the xylem of woody angiosperms, probably open an avenue to the climbing habit. Ferns tracheids exhibit some degree of flexibility thereby tolerating bending and torsional stress which may facilitate the climbing habit (Pittermann et al. 2011). In the climber *Lygodium flexuosum*, the central conduits are separated by parenchyma cells, an unusual stele organization for ferns, and the largest tracheids are distributed toward the center of the stele, a location that experiences minimal tensile and compressive stresses (Niklas 1992; Pittermann et al. 2011). This anatomical organization recalls heteroxylly in both structure and function because mechanics and hydraulic functions are distributed across two tissue systems – metaxylem conduits for hydraulics and the sclerenchyma tissues and fibers for mechanical support.

As in vessel-based plants, some tracheid-based plants exhibit a diversity of tracheid morphology and function, with maximum conduit diameters specific to the climbing species (Wilson et al. 2008; Pittermann et al. 2011). In both ferns and medullosan plants, tracheids tend to be larger and longer than those of conifers (Carlquist & Schneider, 2001), where the climbing habit is not known to occur. Small-sized tracheids in conifers and vessel-less angiosperms seem to rule out the climbing habit because homoxylous wood, with small dimension tracheids, is thought to disable the separation of mechanical and hydraulic functions.

CONCLUSION

Most of the variety of lianas occurs in the eudicots, monocots and magnoliids clades of Mesangiosperms. Angiosperm lianescence is characterized by hydraulic performance, cheap construction cost, and efficient nutrient cycling correlated with light-stimulated growth. These functions are related to the high frequency of wide and long vessels and predominantly simple perforation plates. Climbers beyond the Mesangiosperms show conduit structure that facilitate water flow, compared to their close self-supporting relatives, but at the same time they retain some conduit characteristics that limit their hydraulic efficiency.

Among angiosperms, lianas belonging to Austrobaileyales possess some lianoid features, but do not exhibit capabilities to compete in open and highly evaporative habitats, where many of the eudicot and magnoliids climbers excel. The Austrobaileyales taxa exhibit an ecophysiology of shaded-tolerant species and seem to be mostly restricted to wet and understory habitats. A similar pattern seems to occur in *Gnetum*. This suggests that the evolution of lianoid traits is not necessarily correlated with the evolution of highly opportunistic (competitive advantage) lianas. The vascular systems of lianas outside the Mesangiosperms is based on simple xylem, ray, and phloem construction; they lack the highly specialized vascular system of more derived eudicot or magnoliids and consequently do not experience a significant hydraulic and mechanical leap and do not benefit from the functional advantages of high light-demanding and opportunistic lianas.

Most climbing lineages have explored heteroxylly-like innovations to meet hydraulic, mechanical, and storage requirements inherent to the climbing strategy. Anatomical organization in climbers lacking secondary growth (monocots, ferns) somehow recalls heteroxylly in both structure and function because mechanics and hydraulic functions are distributed across different tissues. Heteroxylly is not necessarily expressed through an obvious cambial variant, but in most climbers it is often dramatically expressed through compartmentalization of large amounts of flexible tissues (wide vessels and most often parenchyma) juxtaposing stiff fibers and narrow vessels. As a good illustration of this, monocots lacking secondary growth, and thus cambial variants, nevertheless have evolved in a large variety of climbers because their anatomical structure with vascular bundles embedded in a soft matrix mimics the heteroxyllous structure.

In opportunistic, disturbance-loving Mesangiosperms lianas, a highly heteroxyllous vascular system appears to be a prerequisite to meet both the high hydraulic demand of the large canopy and the mechanical function of a stem subject to environmental perturbation. The limited ecophysiological and mechanical performance of the vessel-less *T. cordata* supports a hypothesis of the key role played by vessels and advances heteroxylly in angiosperms lianas' evolution and diversification.

The major advantage of being a liana probably lies in their pattern of carbon allocation resulting in a higher growth rate than free-standing species, while the major evolutionary constraints lie in their prerequisite efficient hydraulic system and stem flexibility. Heteroxylly and conduit structure are cornerstones of the highly opportunistic climbing habit, while cambial variants are the ultimate expression of heteroxylly that may have opened the way to the diversification and ecological success of the climbing habit in many Mesangiosperms clades. Hydraulic advantage over other growth forms may have permitted many Mesangiosperm lianas to be competitive in environments where evaporative demand is relatively high, such as disturbed habitats and seasonally dry tropical forests.

REFERENCES

Andrade, J.L., Meinzer, F.C., Goldstein, G. & Schnitzer, S.A. (2005) Water uptake and transport in lianas and co-occurring trees of a seasonally dry tropical forest. *Trees*, **19**, 282–289.

Angyalossy, V., Angeles, G., Pace, M.R. et al. (2012) An overview on the anatomy, development and evolution of the vascular system of lianas. *Plant Ecology and Diversity*, **5**.

Asner, G.P. & Martin, R.E. (2012) Contrasting leaf chemical traits in tropical lianas and trees: implications for future forest composition. *Ecology Letters*, **15**, 1001–1007.

Caballé, G. (1986) *Sur la biologie des lianes ligneuses en forêt gabonaise*. Ph.D. Thesis, Université Montpellier 2, Montpellier, France.

Caballé, G. (1993) Liana structure, function and selection: a comparative study of xylem cylinders of tropical rainforest species in Africa and America. *Botanical Journal of the Linnean Society*, **113**, 41–60.

Cai, Z.-Q., Schnitzer, S. & Bongers, F. (2009) Seasonal differences in leaf-level physiology give lianas a competitive advantage over trees in a tropical seasonal forest. *Oecologia*, **161**, 25–33.

Cantino, P.D., Doyle, J.A., Graham, S.W. et al. (2007) Towards a phylogenetic nomenclature of Tracheophyta. *Taxon*, **56**, 1E–44E.

Carlquist, S. (1991) Anatomy of vine and liana stems: a review and synthesis, in *The Biology of Vines* (eds F.E. Putz & H.A. Mooney), Cambridge University Press, Cambridge, pp. 53–71.

Carlquist, S. (1993) Wood and bark anatomy of Aristolochiaceae, systematic and habital correlations. *IAWA Journal*, **14**, 341–357.

Carlquist, S. (1999) Wood and bark anatomy of Schisandraceae: implications for phylogeny, habit, and vessel evolution. *Aliso*, **18**, 45–55.

Carlquist, S. (2001) Observations on the vegetative anatomy of Austrobaileya: habital, organographic and phylogenetic conclusions. *Botanical Journal of the Linnean Society*, **135**, 1–11.

Carlquist, S. (2007) Successive cambia revisited: ontogeny, histology, diversity, and functional significance. *Journal of the Torrey Botanical Society*, **134**, 301–332.

Carlquist, S. (2009) Xylem heterochrony: an unappreciated key to angiosperm origin and diversifications. *Botanical Journal of the Linnean Society*, **161**, 26–65.

Carlquist, S. (2012) How wood evolves: a new synthesis. *Botany*, **90**, 901–940.

Carlquist, S. & Schneider, E.L. (2001) Vessels in ferns: structural, ecological, and evolutionary significance. *American Journal of Botany*, **88**, 1–13.

Castellanos, A.E. (1991) Photosynthesis and gas exchange of vines, in *The Biology of Vines* (eds F.E. Putz & H.A. Mooney), Cambridge University Press, Cambridge, pp. 127–160.

Cernusak, L.A., Winter, K., Aranda, J. & Turner, B.L. (2008) Conifers, Angiosperm trees, and lianas: growth, whole-plant water and nitrogen use efficiency, and stable isotope composition ($d^{13}C$ and $d^{18}O$) of seedlings grown in a tropical environment. *Plant Physiology*, **148**, 642–659.

Christman, M.A. & Sperry, J.S. (2010) Single-vessel flow measurements indicate scalariform perforation plates confer higher flow resistance than previously estimated. *Plant, Cell and Environment*, **33**, 431–433.

Chui, S.-T. & Ewers, F.W. (1992) Xylem structure and water transport in a twiner, a scrambler, and a shrub of *Lonicera* (Caprifoliaceae). *Trees*, **6**, 216–224.

Cobb, A.R. (2006) *Water relations of rattans*, Ph.D. Thesis Harvard University, Cambridge, USA.

Cobb, A.R., Choat, B. & Holbrook, N.M. (2007) Dynamics of freeze-thaw embolism in *Smilax rotundifolia* (Smilacaceae). *American Journal of Botany*, **94**, 640–649.

Cochard, H., Damour, G., Bodet, C. et al. (2005) Evaluation of a new centrifuge technique for rapid generation of xylem vulnerability curves. *Physiologia Plantarum*, **124**, 410–418.

Cochard, H., Ewers, F.W. & Tyree, M.T. (1994) Water relations of a tropical vine-like bamboo (*Rhipidocladum racemiflorum*): root pressures, vulnerability to cavitation & seasonal changes in embolism. *Journal of Experimental Botany*, **45**, 1085–1089.

Dewalt, S.J., Schnitzer, S.A. & Denslow, J.S. (2000) Density and diversity of lianas along a chronosequence in a central Panamanian lowland forest. *Journal of Tropical Ecology*, **16**, 1–19.

Dubuisson, J.-Y., Hennequin, S., Rakotondrainibe, F. & Schneider, H. (2003) Ecological diversity and adaptive tendencies in the tropical fern *Trichomanes* L. (Hymenophyllaceae) with special reference to climbing and epiphytic habits. *Botanical Journal of the Linnean Society*, **142**, 41–63.

Dunn, M.T., Krings, M., Mapes, G. et al. (2003) Medullosa steinii sp. nov., a seed fern vine from the Upper Mississippian. *Review of Palaeobotany and Palynology*, **124**, 307–324.

Evert, R.F. (2006) *Esau's Plant Anatomy*, John Wiley & Sons, Inc., Hoboken, NJ.

Ewers, F.W., Cochard, H. & Tyree, M.T. (1997) A survey of root pressures in vines of tropical lowland forest. *Oecologia*, **110**, 191–196.

Ewers, F.W. & Fisher, J.B. (1991a) Why vines have narrow stems: histological trends in *Bauhinia* (Fabaceae). *Oecologia*, **88**, 233–237.

Ewers, F.W., Fisher, J.B. & Chiu, S.T. (1990) A survey of vessel dimensions in stems of tropical lianas and other growth forms. *Oecologia*, **84**, 544–552.

Ewers, F.W., Fisher, J.B., & Fichtner, K. (1991b) Water flux and xylem structure in vines. In *The Biology of Vines* (eds. F.E. Putz & H.A. Mooney). Cambridge University Press, Cambridge.

Feild, T.S. & Balun, L. (2008) Xylem hydraulic and photosynthetic function of *Gnetum* (Gnetales) species from Papua New Guinea. *New Phytologist*, **177**, 665–675.

Feild, T.S., Chatelet, D.S., Balun, L. et al. (2012) The evolution of angiosperm lianescence without vessels – climbing mode and wood structure-function in *Tasmannia cordata* (Winteraceae). *New Phytologist*, **193**, 229–240.

Feild, T.S., Franks, P.J. & Sage, T.L. (2003) Ecophysiological shade adaptation in the basal angiosperm, *Austrobaileya scandens* (Austrobaileyaceae). *International Journal of Plant Sciences*, **164**, 313–24.

Feild, T.S. & Isnard, S. (2013) Climbing habit and ecophysiology of *Schisandra glabra* (Schisandraceae): Implications for the early evolution of angiosperm lianescence. *International Journal of Plant Sciences*, **174**, 1121–1133.

Feild, T.S., Upchurch, G.R., Chatelet, D.S. et al. (2011) Fossil evidence for low gas exchange capacities for Early Cretaceous angiosperm leaves. *Paleobiology*, **37**, 195–213.

Feild, T.S. & Wilson, J.P. (2012) Evolutionary voyage of angiosperm vessel structure-function and its significance for early angiosperm success. *International Journal of Plant Sciences*, **173**, 596–609.

Fisher, J.B. & Ewers, F.W. (1991) Structural responses to stem injury in vines. In *The Biology of Vines* (eds. F.E. Putz & H.A. Mooney), pp. 99–124. Cambridge University Press, Cambridge.

Fisher, J.B. & Ewers, F.W. (1995) Vessel dimensions in liana and tree species of *Gnetum* (Gnetales). *American Journal of Botany*, **82**, 1350–1357.

Fisher, J.B., Ewers, F.W. & Lopez-Portillo, J. (1997) Survey of root pressure in tropical vines and woody species. *International Journal of Plant Sciences*, **158**, 44–50.

Forseth, I.N. & Teramura, A.H. (1987) Field photosynthesis, microclimate and water relations of an exotic temperate liana, *Pueraria lobata*, kudzu. *Oecologia*, **71**, 262–267.

Gartner, B.L. (1991) Stem hydraulic properties of vines vs. shrubs of western poison oak, Toxicodendron diversilobum. *Oecologia*, **87**, 180–189.

Gartner, B.L., Bullock, S.H., Mooney, H.A. et al. (1990) Water transport properties of vine and tree stems in a tropical deciduous forest. *American Journal of Botany*, **77**, 742–749.

Gentry, A.G. (1991) The distribution and evolution of climbing plants, in *The Biology of Vines* (eds F.E. Putz & H.A. Mooney), Cambridge University Press, Cambridge.

Hegarty, E.E. & Caballé, G. (1991) Distribution and abundance of vines in forest communities, in *The Biology of Vines* (eds F.E. Putz & H.A. Mooney), Cambridge University Press, Cambridge.

Holbrook, N.M. & Zwieniecki, M.A. (1999) Embolism repair and xylem tension: do we need a miracle? *Plant Physiology*, **120**, 7–10.

Hudson, P.J., Razanatsoa, J. & Feild, T.S. (2010) Early vessel evolution and the diverisification of wood function: Insights from Malagasy Canellales. *American Journal of Botany*, **97**, 80–93.

Isnard, S., Prosperi, J., Wanke, S. et al. (2012) Growth form evolution in Piperales and its relevance for understanding the angiosperm diversification: an integrative approach combining plant architecture, anatomy and biomechanics. *International Journal of Plant Sciences*, **173**, 610–639.

Isnard, S. & Rowe, N.P. (2008) The mechanical role of the leaf sheath in rattans. *New Phytologist*, **177**, 643–652.

Isnard, S. & Silk, W.K. (2009) Moving with climbing plants from Charles Darwin's time into the 21st century. *American Journal of Botany*, **96**, 1205–1221.

Krings, M. & Kerp, H. (2006) Neuropteris attenuata, a narrow-stemmed, leaning or lianescent seed fern from the Upper Pennsylvanian of Lower Saxony, Germany. *Flora – Morphology, Distribution, Functional Ecology of Plants*, **201**, 233–239.

Krings, M., Kerp, H., Taylor, T. & Taylor, E. (2003) How Paleozoic vines and lianas got off the ground: on scrambling and climbing carboniferous-early permian pteridosperms. *The Botanical Review*, **69**, 204–224.

Magallón, S. & Castillo, A. (2009) Angiosperm diversification through time. *American Journal of Botany*, **96**, 349–365.

Mosbrugger, V. (1990) *Tree Habit in Land Plants: A Functional Comparison of Trunk Constructions with a Brief Introduction into the Biomechanics of Trees*, Lecture Notes in Earth Sciences 28, Springer Verlag, Berlin.

Niklas, K.J. (1992) *Plant Biomechanics: An Engineering Approach to Plant Form and Function*, University of Chicago Press, Chicago.

Obaton, M. (1957) *Les lianes ligneuses à structure anormale des forêts denses d'Afrique occidentale*, Ph.D. Thesis, Faculté des Sciences, Paris, France.

Pace, M.R., Lohmann, L.G. & Angyalossy, V. (2009) The rise and evolution of the cambial variant in Bignonieae (Bignoniaceae). *Evolution and Development*, **11**, 465–479.

Pace, M.R., Lohmann, L.G. & Angyalossy, V. (2011) Evolution of disparity between the regular and variant phloem in Bignonieae (Bignoniaceae). *American Journal of Botany*, **98**, 602–618.

Pittermann, J., Limm, E., Rico, C. & Christman, M.A. (2011) Structure–function constraints of tracheid-based xylem: a comparison of conifers and ferns. *New Phytologist*, **192**, 449–461.

Putz, F.E. (1983) Liana biomass and leaf area of a "Tierra Firme" forest in the Rio Negro Basin, Venezuela. *Biotropica*, **15**, 185–189.

Putz, F.E. (1984) The natural history of lianas on Barro Colorado Island, Panama. *Ecology*, **65**, 1713–1724.

Putz, F.E. & Chai, P. (1987) Ecological studies of lianas in Lambir National Park, Sarawak, Malaysia. *Journal of Ecology*, **75**, 523531.

Putz, F.E. & Holbrook, N.M. (1991) Biomechanical studies of vines, in *The Biology of Vines* (eds F.E. Putz & H.A. Mooney), Cambridge University Press, Cambridge.

Restom, T.G. & Nepstad, D.C. (2004) Seedling growth dynamics of a deeply rooting liana in a secondary forest in eastern Amazonia. *Forest Ecology and Management*, **190**, 109–118.

Rowe, N.P., Isnard, S., Gallenmüller, F. & Speck, T. (2006) Diversity of mechanical architectures in climbing plants: an ecological perspective, in *Ecology and Biomechanics: A Mechanical Approach to the Ecology of Animals and plants* (eds A. Herrel, T. Speck & N.P. Rowe), CRC Press LLC, Boca Raton, FL, pp. 35–59.

Rowe, N.P. & Speck, T. (1997) Biomechanics of *Lycopodiella cernua* and *Huperzia squarrosa*: implications for inferring growth habits of fossil small-bodied lycopsids. *Mededelingen Nederlands Instituut voor Toegepaste Geowetenschappen TNO*, **58**, 293–302.

Rowe, N.P. & Speck, T. (1998) Biomechanics of plant growth forms: the trouble with fossil plants. *Review of Palaeobotany and Palynology*, **102**, 43–62.

Schenck, H. (1892) Beiträge zur Biologie und Anatomie der Lianen, in besonderen der in Brasilien einheimische Arten.

I Teil. *Beiträge zur Anatomie der Lianen*. G. Fischer, Jena, Germany.

Schenk, H.J., Espino, S., Goedhart, C.M. et al. (2008) Hydraulic integration and shrub growth form linked across continental aridity gradients. *Proceedings of the National Academy of Sciences of the United States of America*, **105**, 11248–11253.

Schnitzer, S.A. (2005) A mechanistic explanation for global patterns of liana abundance and distribution. *The American Naturalist*, **166**, 262–276.

Schnitzer, S.A. & Bongers, F. (2002) The ecology of lianas and their role in forests. *Trends in Ecology and Evolution*, **17**, 223–230.

Soltis, D.E., Bell, C.D., Kim, S. & Soltis, P.S. (2008) Origin and early evolution of Angiosperms. *Annals of the New York Academy of Sciences*, **1133**, 3–25.

Sperry, J.S. & Hacke, U.G. (2004) Analysis of circular bordered pit function I. Angiosperm vessels with homogenous pit membranes. *American Journal of Botany*, **91**, 369–385.

Sperry, J.S., Hacke, U.G., Feild, T.S. et al. (2007) Hydraulic consequences of vessel evolution in Angiosperms. *International Journal of Plant Sciences*, **168**, 1127–1139.

Sperry, J.S., Hacke, U.G. & Pittermann, J. (2006) Size and function in conifer tracheids and angiosperm vessels. *American Journal of Botany*, **93**, 1490–1500.

Sperry, J.S., Holbrook, N.M., Zimmermann, M.H. & Tyree, M.T. (1987) Spring filling of xylem vessels in wild grapevine. *Plant Physiology*, **83**, 414–417.

Spicer, R. & Groover, A. (2010) Evolution of development of vascular cambia and secondary growth. *New Phytologist*, **186**, 577–592.

Stevens, P.F. (2001 onwards). *Angiosperm Phylogeny Website, Version 12*, July 2012 [and more or less continuously updated since].

Tay, A., Abdullah, A., Awang, M. & Furukawa, A. (2007) Midday depression of photosynthesis in *Enkleia malaccensis*, a woody climber in a tropical rainforest. *Photosynthetica*, **45**, 189–193.

Tibbetts, T.J. & Ewers, F.W. (2000) Root pressure and specific conductivity in temperate lianas: exotic *Celastrus orbiculatus* (Celastraceae) vs. native *Vitis riparia* (Vitaceae). *American Journal of Botany*, **87**, 1272–1278.

Tomlinson, P.B., Fisher, J.B., Spangler, R.E. & Richer, R.A. (2001) Stem vascular architecture in the rattan palm *Calamus* (Arecaceae-Calamoideae-Calaminae). *American Journal of Botany*, **88**, 797–809.

Tomlinson, P.B. & Spangler, R.E. (2002) Developmental features of the discontinuous stem vascular system in the rattan palm *Calamus* (Arecaceae-Calamoideae-Calamineae). *American Journal of Botany*, **89**, 1128–1141.

Tyree, M.T. & Sperry, J.S. (1989) Vulnerability of xylem to cavitation and embolism. *Annual Review of Plant Physiology and Plant Molecular Biology*, **40**, 19–36.

Tyree, M.T. & Zimmermann, M.H. (2002) *Xylem Structure and the Ascent of Sap*, Springer Verlag, Berlin.

Wagner, S.T., Isnard, S., Rowe, N.P., et al. (2012) Escaping the lianoid habit: evolution of shrub-like growth forms in *Aristolochia* subgenus *Isotrema* (Aristolochiaceae). *American Journal of Botany*.

Wheeler, J.K., Sperry, J.S., Hacke, U.G. & Hoang, N. (2005) Inter-vessel pitting and cavitation in woody Rosaceae and other vesselled plants: a basis for a safety versus efficiency trade-off in xylem transport. *Plant, Cell and Environment*, **28**, 800–812.

Wilson, C.L. (1924) Medullary bundle in relation to primary vascular system in Chenopodiaceae and Amaranthaceae. *Botanical Gazette*, **78**, 175.

Wilson, J.P. & Fisher, W.W. (2011) Geochemical support for a climbing habit within the Paleozoic seed fern genus *Medullosa*. *International Journal of Plant Sciences*, **172**, 586–598.

Wilson, J.P., Knoll, A.H., Holbrook, N.M. & Marshall, C.R. (2008) Modeling fluid flow in *Medullosa*, an anatomically unusual Carboniferous seed plant. *Paleobiology*, **34**, 472–493.

Won, H. & Renner, S.S. (2006) Dating dispersal and radiation in the Gymnosperm *Gnetum* (Gnetales) – clock calibration when outgroup relationships are uncertain. *Systematic Biology*, **55**, 610–622.

Wright, I.J., Reich, P.B., Westoby, M. et al. (2004) The worldwide leaf economics spectrum. *Natur*, **428**, 821–827.

Zhu, S.-D. & Cao, K.-F. (2009) Hydraulic properties and photosynthetic rates in co-occurring lianas and trees in a seasonal tropical rainforest in southwestern China. *Plant Ecology*, **204**, 295–304.

Fig. 1.2 Lianas blanketing trees on Barro Colorado Island, Panama. (Source: C. Ziegler. Reproduced with permission.)

Fig. 1.3 Three-toed sloth in central Panama using a liana to ascend to the forest canopy. (Source: Photo credit: S.A. Schnitzer. Reproduced with permission.)

Ecology of Lianas, First Edition. Edited by Stefan A. Schnitzer, Frans Bongers, Robyn J. Burnham, and Francis E. Putz.

Companion Website: www.wiley.com/go/schnitzer/lianas

Fig. 4.2 Photographs of selected liana species from four forest types (WEF, SEF, SDF and DEF) of peninsular India

WEF: A) *Gnetum ula* (Gnetaceae – Gymnosperm), exhibiting cauliflory and drupes (mammal dispersed). B) *Elaeagnus indica* with silvery/coppery shining leaves and drupes for vertebrate dispersal. C) Stout thorns of *Zanthoxylum tetraspermum* (Rutaceae)

SEF: D) Armed straggler, *Pterolobium hexapetalum* (Caesalpiniaceae) with curved thorn from bullate protuberance. E) *Jasminum angustifolium* (Oleaceae), a stem twiner with scented white flower. F) *Lantana camara* (Verbenaceae), an invasive species

SDF: G) *Grewia rhamnifolia* (Tiliaceae). H) *Abrus precatorius* (Papilionaceae), dehising legumes exposing reddish black seeds. I) *Capparis sepiaria* (Capparaceae)

DEF: J) *Ventilago madraspatana* (Rhamnaceae) that occur in all four forest types of peninsular India. K) Thorny straggler *Acacia caesia* occurs in forest gaps and edges (Mimosaceae). L) Drupes of *Hugonia mystax* (Linaceae) are vertebrate dispersed.

Fig. 19.2 Macroscopic transverse sections of tracheophyte liana stems, figures not to scale. (A) *Salpichlaena volubilis* (Blechnaceae, Monilophytes) stem showing wide tracheary elements, twining rachis. (B) *Gnetum paniculatum* (Gnetaceae, Gymnosperm) stem showing wide vessels and successive cambia. (C) *Aristolochia gigantea* (Aristolochiaceae, Magnoliid) stem showing thick periderm, wide vessels and secondary vascular cylinder marked by very wide rays alternating with the axial elements of the xylem and phloem. Wide rays are derived from the interfascicular cambium. (D) *Desmoncus orthacanthos* (Arecaceae, Monocots) main stem (st) and two surrounding leaf sheaths (ls). (E) *Dioclea rufescens* (Leguminosae, Eudicot Rosid) stem with very wide vessels (up to 500 μm). (F) *Thunbergia grandiflora* (Acanthaceae, Eudicot Lamiid) stem showing wide vessels and islands of included phloem (arrows) produced by interfascicular cambium.

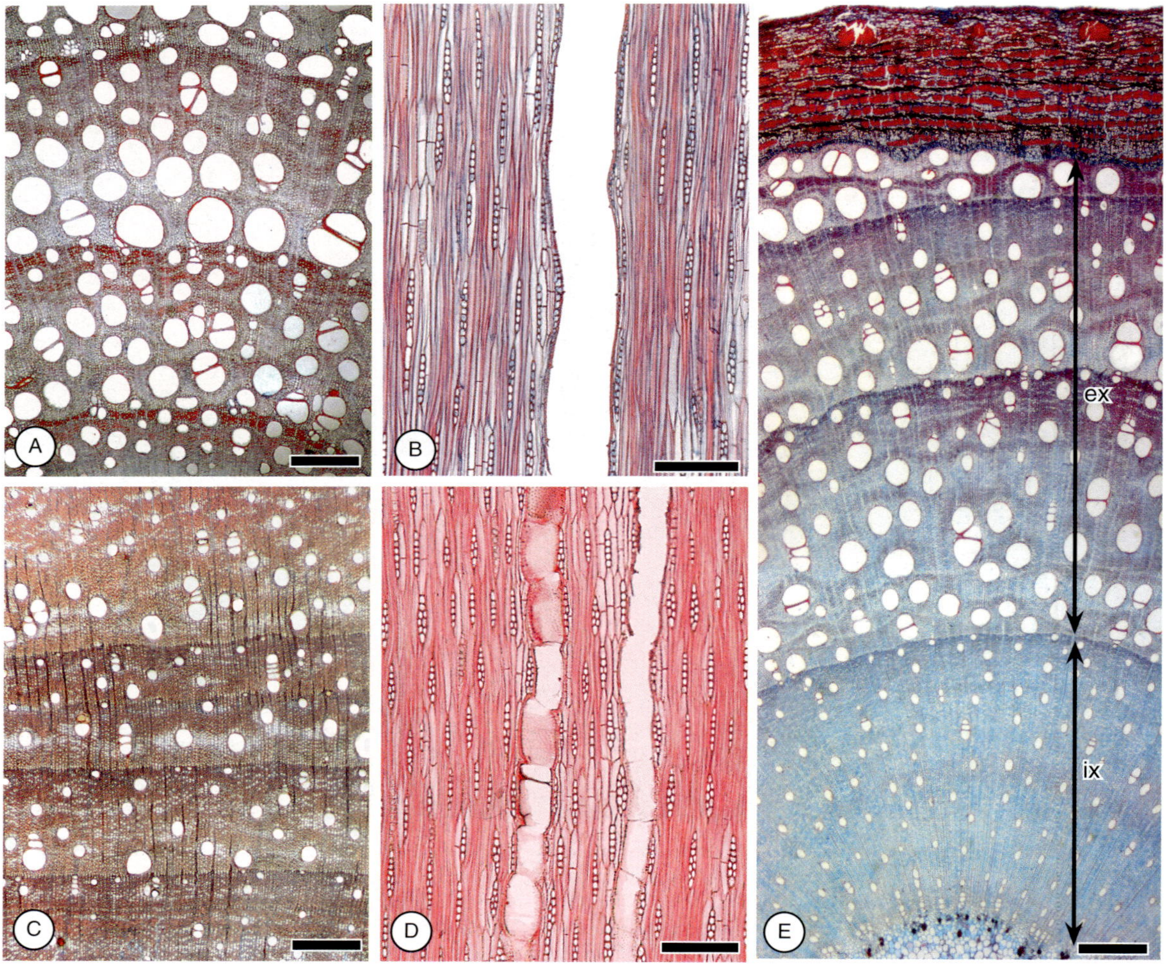

Fig. 19.3 Anatomical comparison between wood of a liana and a tree of *Dalbergia frutescens* (Leguminosae). (A, B, E), liana; (C, D), tree. (A) Wood transverse section, wide and narrow vessels combined, abundant banded paratracheal parenchyma. (B) Wood tangential section, showing lack of clear storied structure (rays irregularly storied), rays taller than those of the trees. (C) Wood transverse section, narrow vessels, aliform paratracheal parenchyma, forming short confluences. (D) Wood tangential section, showing storied structure (rays, axial parenchyma, vessel elements), narrow and short rays. (E) Stem transverse section from pith to cambium. Note the difference between the inner, dense, secondary xylem of self-supporting phase (ix) and the subsequent xylem with lianescent xylem syndrome (ex). Note: Scale bars: A, C, E = 500 μm, B, D = 200 μm.

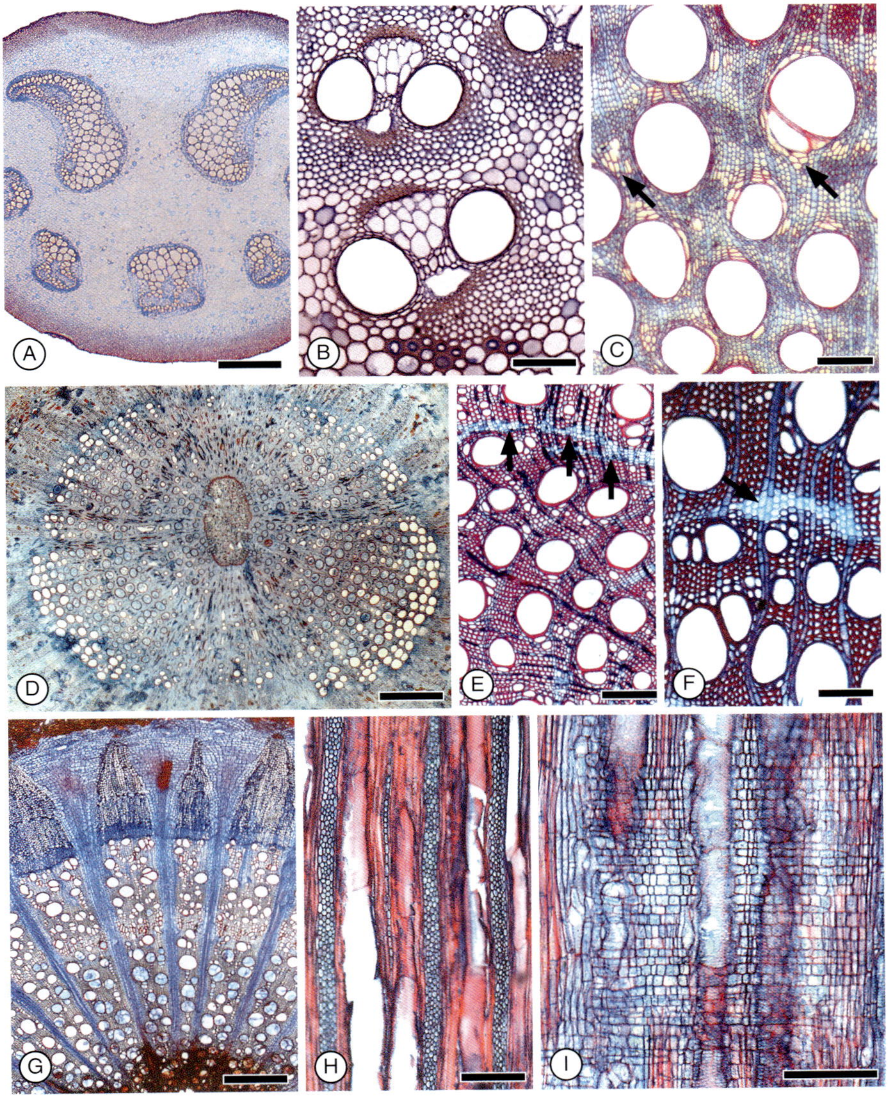

Fig. 19.4 Lianescent vascular syndrome. (A–G) Transverse section (TS). (A–C) Wide tracheary elements. (A) *Salpichlaena volubilis* (Blechnaceae, Monilophyte). (B) *Merostachys neesii* (Poaceae, Monocot). (C) *Gouania mollis* (Rhamnaceae Eudicot, narrow vessels (arrows) combined with wide vessels). (D) *Cissus sulcicaulis* (Vitaceae, Eudicot), the secondary xylem is composed of a matrix of non-lignified parenchyma where the vessels and their surrounding fibers are embedded. (E) *Heteropterys nordestina* (Malpighiaceae, Eudicot), non-lignified axial parenchyma in the secondary xylem (arrows). (F) *Dolichandra unguis-cati* (Bignoniaceae, Eudicot), rays are normally lignified, except when crossing islands of non-lignified axial parenchyma (arrow) in the secondary xylem. (G) *Cissampelos andromorpha* (Menispermaceae, Eudicot), wide and non-lignified rays. (H) *Bignonia campanulata* (Bignoniaceae, Eudicot), tall and wide rays, as seen in longitudinal tangential section (LT). (I) *Lundia corymbifera* (Bignoniaceae, Eudicot), heterogeneous rays, with procumbent, square and upright cells mixed, as seen in longitudinal radial section (LR). Note: Scale bars: A = 500 μm, B = 100 μm, C, E, H, I = 200 μm, D = 3 mm, F = 50 μm G=800 μm.

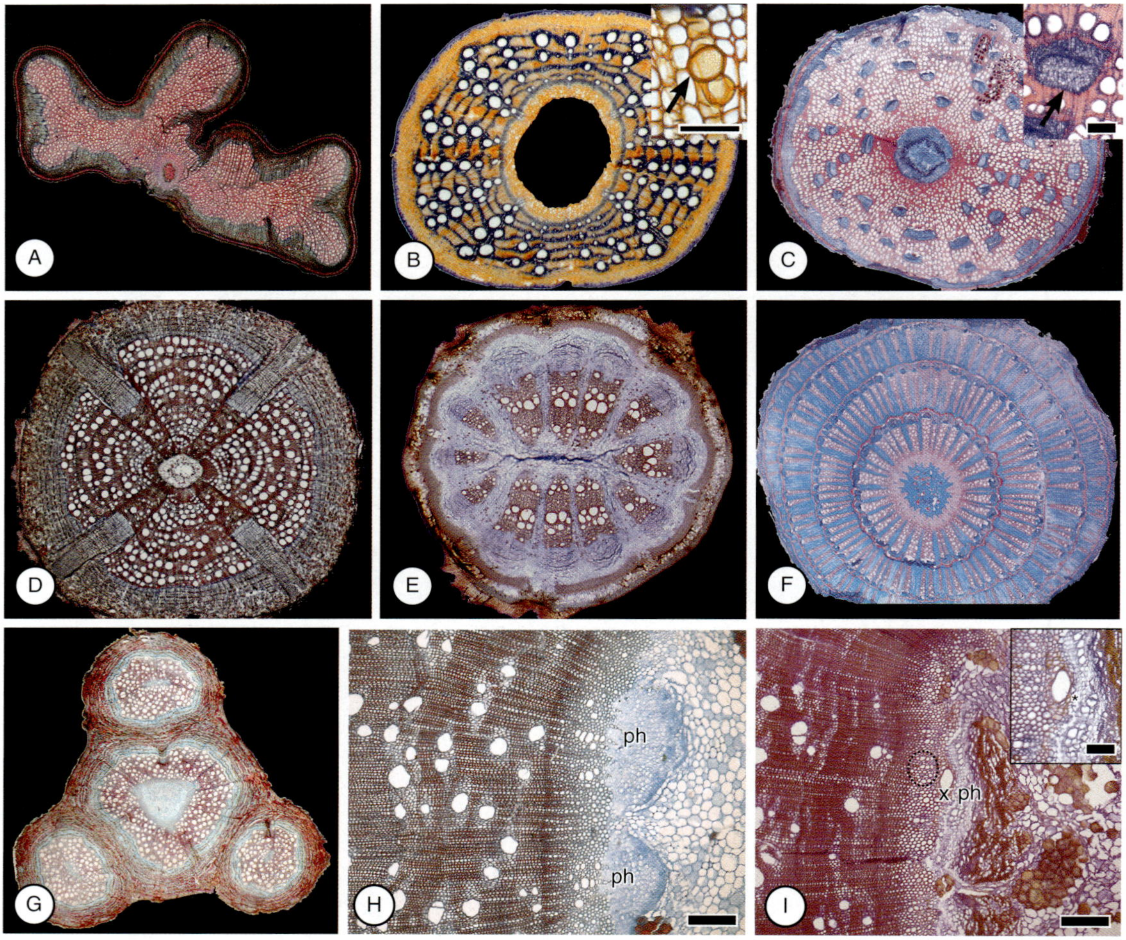

Fig. 19.8 Cambial variant types, as seen in transverse section (TS). (A) *Heteropterys nordestina* (Malpighiaceae), asymmetrical stem. (B) *Thunbergia grandiflora* (Acanthaceae), interxylary phloem without cambial inclusion (detail: sieve tube with simple sieve plate, arrow). (C) *Strychnos millepunctata* (Loganiaceae), interxylary phloem with cambial inclusion (detail: included cambium, arrow). (D) *Tanaecium pyramidatum* (Bignoniaceae), xylem furrowed by phloem wedges. (E) *Aristolochia gigantea* (Aristolochiaceae), axial vascular elements in segments alternating with wide vascular rays. (F) *Chondrodendron platyphyllum* (Menispermaceae), successive cambia. (G) *Serjania lethalis* (Sapindaceae), compound stem. (H) *Maripa glabra* (Convolvulaceae), intraxylary phloem (ph) produced by a perimedullar cambium. (I) *Telitoxicum duckei* (Menispermaceae), intraxylary phloem (ph) and xylem (x) produced by a perimedullar cambium shown in the detail (asterisk). Note primary xylem before the neoformed intraxylary xylem (dotted circle). Note: Scale bars: B = detail 50 μm, C = detail = 100 μm, H-I = 300 μm, I = detail = 100 μm.

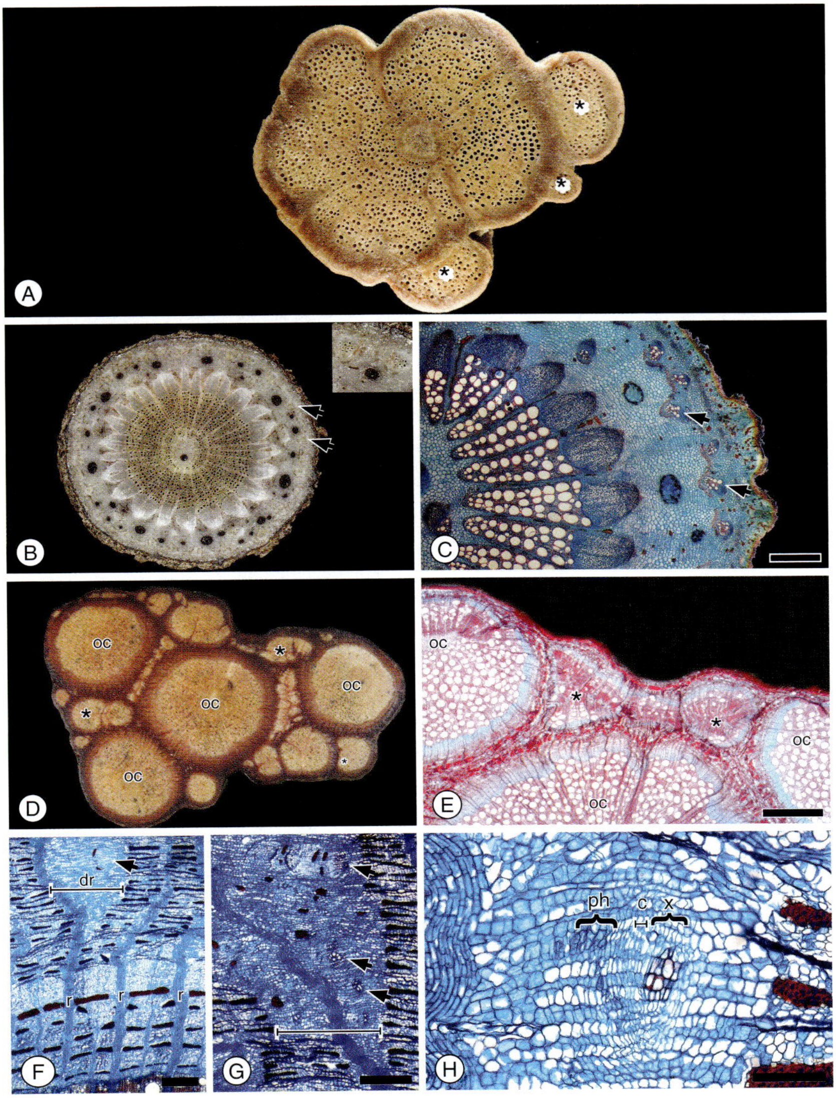

Fig. 19.9 Cambial variant types, as seen in transverse section (TS). (A–C) External vascular cylinders. (A) *Dalechampia ficifolia* (Euphorbiaceae), lateral vascular cylinders (asterisks) derived from neoformations. (B, C) *Manekia obtusa* (Piperaceae), secondary growth of external primary vascular bundle (detail: two external vascular cylinders). (D–H) Neoformations of vascular cylinders associated with another type of cambial variant. (D, E) Compound stem of *Serjania laruotteana* (Sapindaceae) with neoformed vascular cylinders (asterisks) between four original cylinders (oc). (F–H) *Amphilophium crucigerum* (Bignoniaceae), new vascular portions (arrows) derived from new cambia, perpendicular to original cambium, at dilated portion (dr) of phloem rays (r) in the nonconducting phloem. Note in H the detail of one new vascular portion with its phloem (ph) and xylem (x) derived from a neoformed cambium (c). Note: Scale bars: C = 1 mm, E = 250 μm, F–G = 500 μm, H = 150 μm.

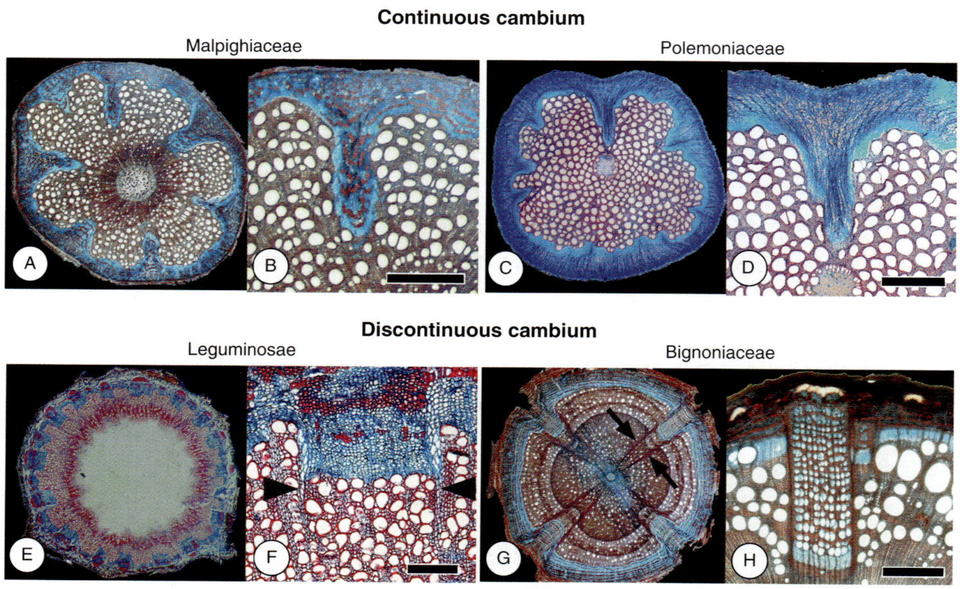

Fig. 19.10 Four examples of families with furrowed xylem (phloem arcs/wedges), as seen in transverse section (TS). (A–D) Continuous cambium. (A, B) *Mascagnia sepium* (Malpighiaceae). (C, D) *Cobaea scandens* (Polemoniaceae). (E–H) Discontinuous cambium (cambia included). (E–F) *Mimosa velloziana* (Leguminosae), arrows indicate limiting rays. (G) *Fridericia chica* (Bignoniaceae), arrows indicate limiting rays. (H) *Tynanthus cognatus* (Bignoniaceae), detail of a phloem wedge. Note: Scale bars: B, D = 500 μm; F = 350 μm; H = 450 μm.

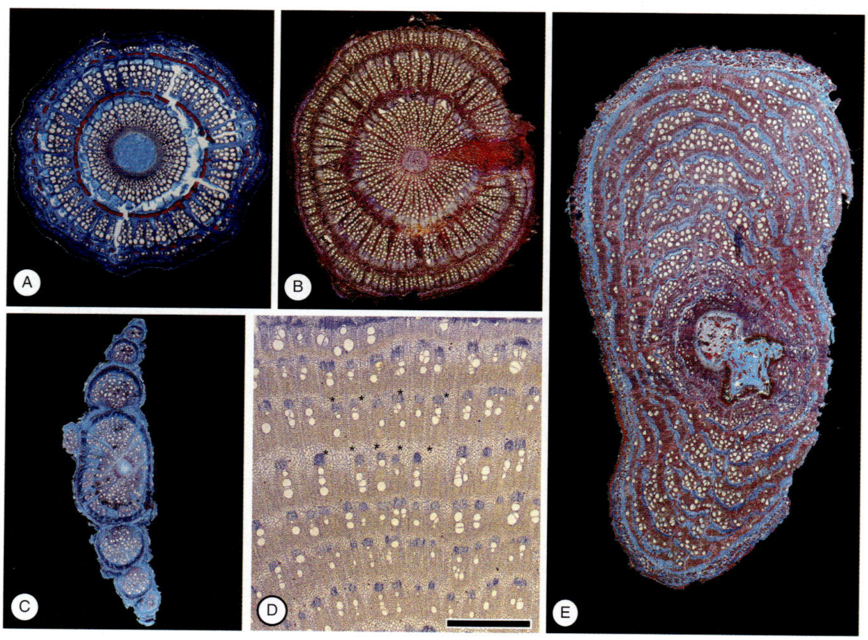

Fig. 19.11 Successive cambia of different types, seen in transverse section (TS). (A, B) Successive cambia resulting in a round stem. (C, E) Successive cambia resulting in flattened stem. (D) Conjunctive tissue (asterisks) one of the products of the successive cambia. (A) *Gnetum paniculatum* (Gnetaceae). (B) *Doliocarpus dentatus* (Dilleniaceae). (C) *Rhynchosia phaseoloides* (Leguminosae). (D) *Bouganvillea spectabilis* (Nyctaginaceae). (E) *Telitoxicum duckei* (Menispermaceae).

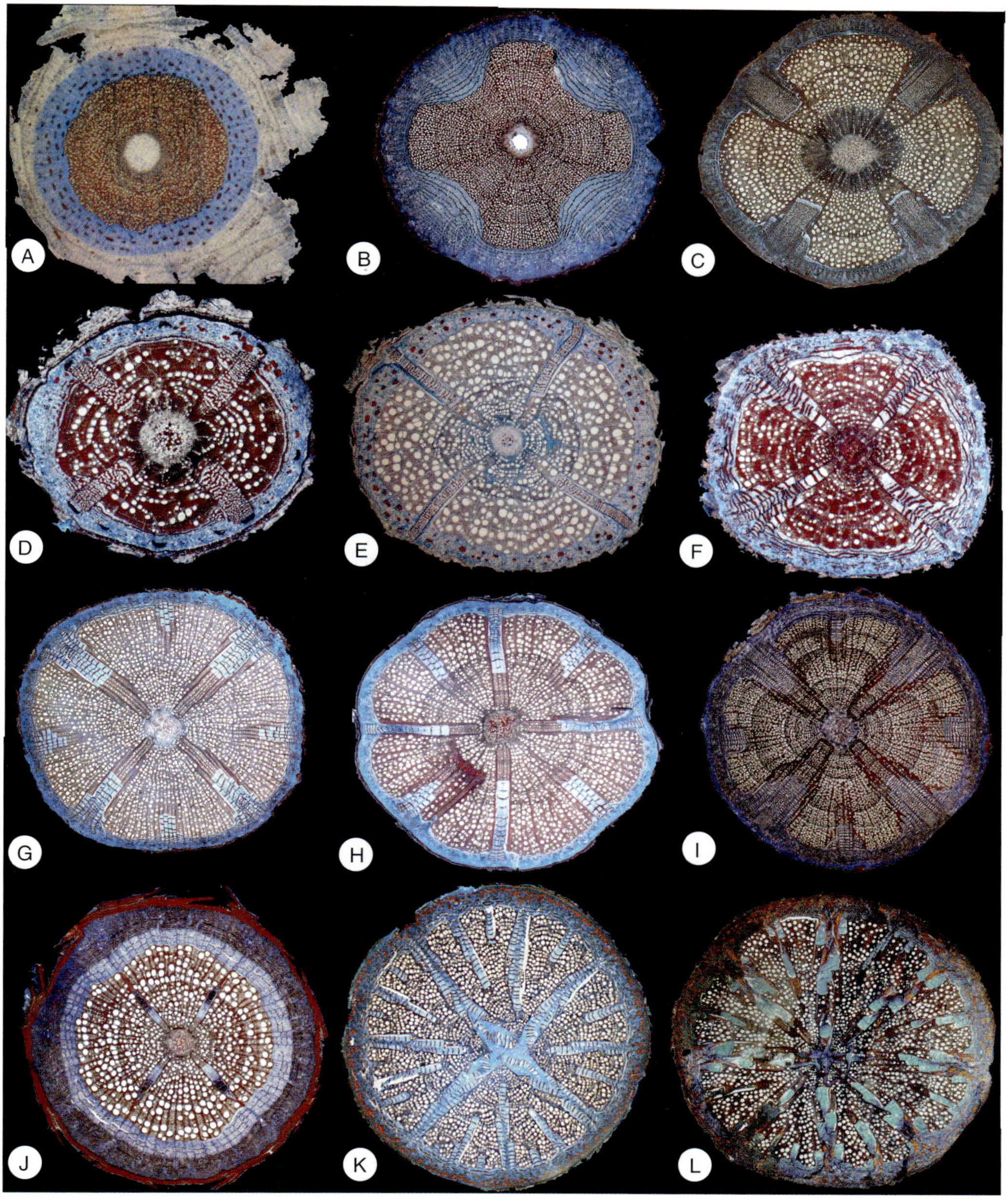

Fig. 19.12 Diversity of stem architectures in Bignoniaceae lianas, as seen in transverse section (TS). (A) Liana of tribe Tecomeae. (B–L) Lianas of tribe Bignonieae. (A) *Podranea ricasoliana*, stem with regular secondary growth. (B) *Perianthomega vellozoi*, stem with 4 arcs. (C–F) Stems with 4 phloem wedges. (C) *Fridericia samydoides*. (D) *Lundia longa*. (E) *Pleonotoma tetraquetra*. (F) *Manaosella cordifolia*. (G–I) Stems with multiple of 4 phloem wedges. (G) *Mansoa difficilis*. (H) *Pyrostegia venusta*. (I) *Adenocalymma divaricatum*. (J) *Amphilophium crucigerum*, stem with included phloem wedges. (K, L) Stems with multiple dissected phloem wedges. (K) *Dolichandra unguis-cati*. (L) *Dolichandra unguiculata*.

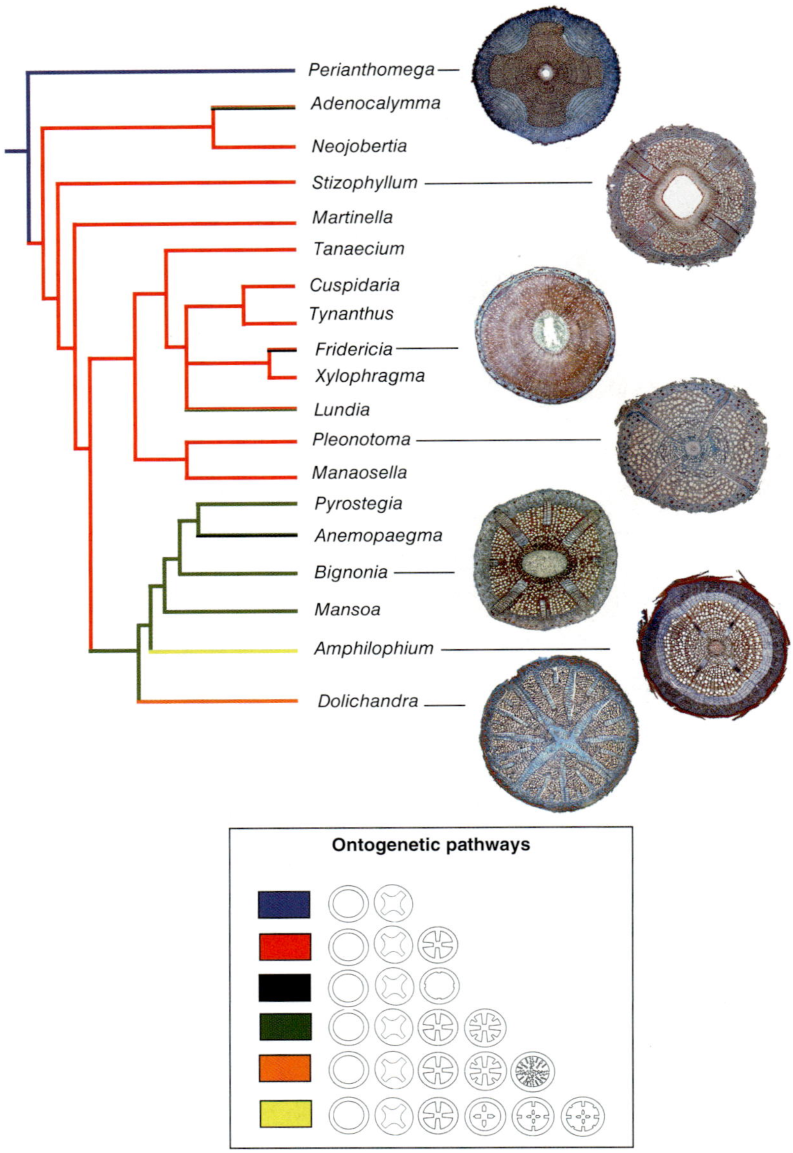

Fig. 19.13 Ancestral character state reconstruction of the ontogenetic pathways of tribe Bignonieae (Bignoniaceae). In Bignonieae the presence of four phloem arcs (blue) is ancestral in the family (synapomorphic), and different stem anatomical architectures have evolved from it, four phloem wedges (red) correspond to the majority of the tribe, multiple of four phloem wedges (green), multiple dissected (black) and included phloem wedges (yellow) have evolved in the terminal nodes of the phylogeny.

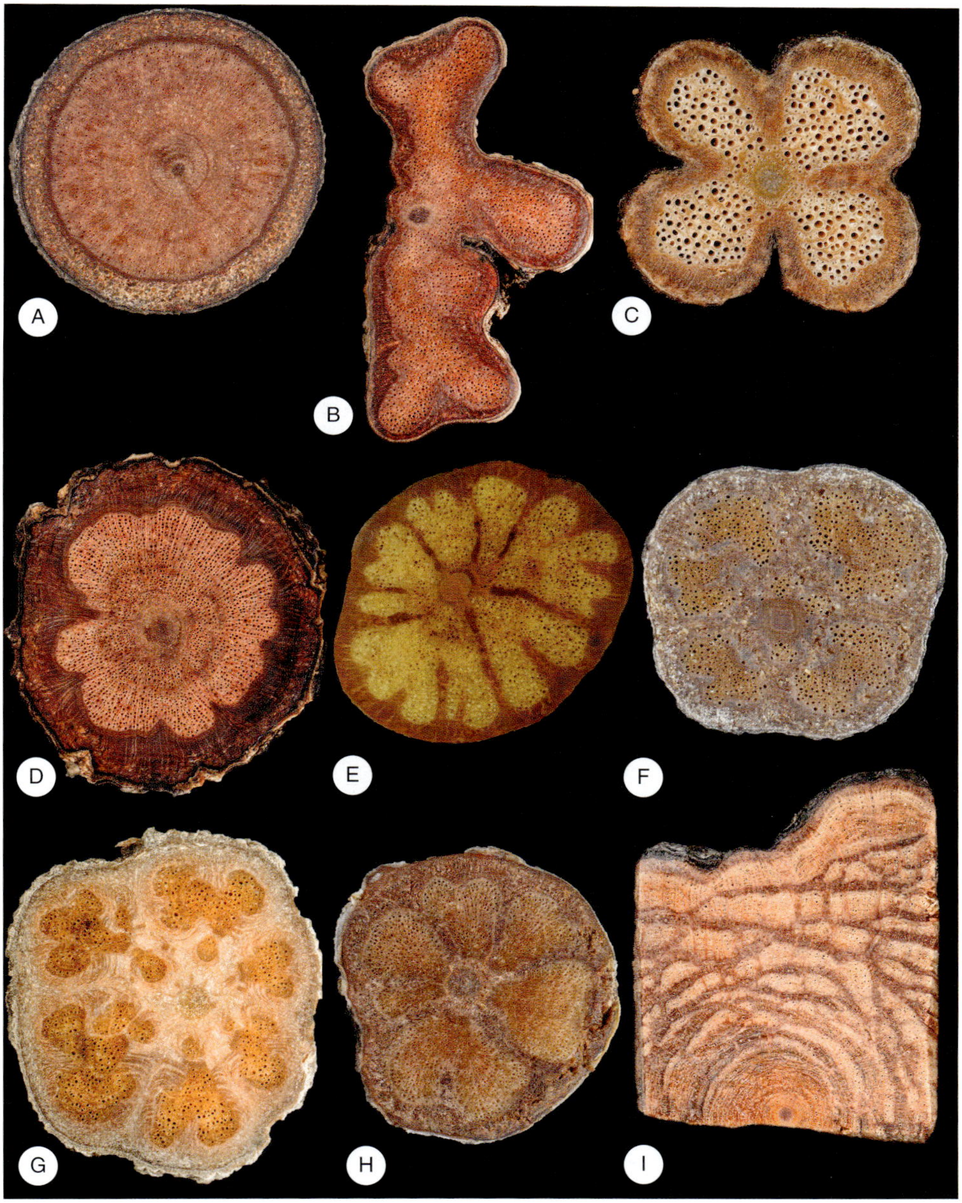

Fig. 19.14 Diversity of stem architectures in Malpighiaceae lianas, as seen in transverse section (TS). (A) *Heteropterys nitida* stem with regular secondary growth. (B) *Heteropterys nordestina* stem lobed with asymmetrical conformation. (C) *Stigmaphyllon blanchetii*, stem lobed with symmetrical conformation. (D) *Heteropterys megaptera*, stem with shallow phloem wedges. (E) *Mascagnia sepium*, stem with deep phloem wedges. (F–H) Stems with dissected xylem. (F) *Mezia mariposa*. (G) *Alicia anisopetala*. (H) *Heteropterys cordifolia*. (I) *Banisteropsis nummifera*, stem with successive cambia.

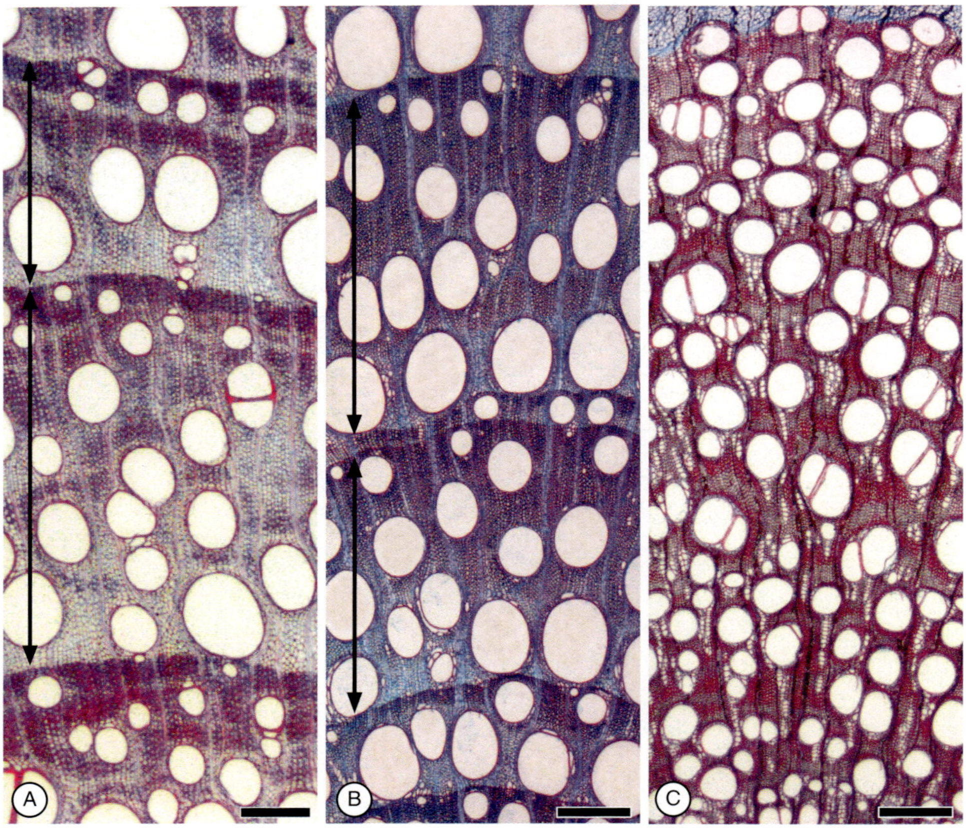

Fig. 19.16 Growth rings in secondary xylem of lianas, as seen in transverse section (TS). (A, B) Annual growth rings present (double-headed arrows). Annual growth rings marked by marginal parenchyma band and wide vessels in the earlywood and narrow vessels and radially flattened fibers in the latewood. (A) *Dalbergia frutescens* (Leguminosae). (B) *Stizophyllum riparium* (Bignoniaceae). (C) Growth rings absent in *Serjania laruotteana* (Sapindaceae). Scale bars: 400 μ m.

Fig. 23.1 Growth forms of lianas. (A) *Bauhinia guianensis* (Fabaceae), ribbon-shaped flexible stem surviving tree fall damage. (B) Self-supporting treelet of *Strychnos* sp. (Loganiaceae), the plant is still mechanically stable and has not yet started to produce hooks. (C) Understory self-supporting shoots of *Strychnos*, stems are at the point of instability and producing paired branches with sensitive hooks. (D) Three meter-long searchers of an unidentified liana extending from a twining stem in the understory; such narrow stems have high Young's moduli in order to retain their orientation and to span gaps between potential supports. (E) Multiple sprouting of robust self-supporting shoots of *Strychnos* sp. arising from fallen stem. (F) Sensitive hooks of *Strychnos* form closed spring-like hooks without contact with host branches; these thicken and toughen on contact with host. (G) Twining stem of *Smilax* sp., the species produces sharp spines on twining axes that undoubtedly assist in anchoring the twining stem to the host. (H) Wide-angle branches of canopy-climbing *Manihot quinquepartita* (Euphorbiaceae); these branches form irreversible connections with canopy branches. (I) Wide-angled petioles of *Croton nuntians* interlocking with host branches. (J) Self-supporting searcher stem of *Croton nuntians* in forest understory.

Fig. 23.2 Attachment structures in lianas. (A–D) *Galium aparine*; (A) tangle of shoots. (B) Lateral view of a leaf; hooks on the upper (adaxial) surface are curved towards the leaf tip (right), hooks on the underside of the leaf (abaxial surface) are curved towards the leaf base (left). (C) Adaxial hook. (D) Abaxial hook (C, D fluorescence micrographs, hooks stained with acridine orange). (E) Complex trichome on the underside of a leaf of *Croton pullei* (SEM); these structures increase friction with supporting structures. (F) Thorn on a shoot of *Asparagus setaceus* curved towards the shoot base (right). (G–I) *Parthenocissus tricuspidata*, (G) tendril with nine attachment pads on plaster. (H) Longitudinal section through an attachment pad on wood, at the interface between the wood and the underside of the pad, a thin layer of glue is visible. (I) Underside of an attachment pad (SEM). (J) Tendril of *Amphilophium crucigerum* with glueless attachment pads anchored in tree bark. (K–N) *Hedera helix*, (K) different hierarchical morphological levels of the attachment system: (K1) root cluster at the underside of an internode; (K2) three roots (SEM); (K3) root hairs (SEM). (L) Individual root hair demonstrating the orientation of cellulose microfibrils (SEM). (M) Stained single root hair in differential interference contrast, the darker lines on the root hair indicate cellulose microfibril orientation. (N) Bar plot diagram of mean orientation of cellulose microfibrils versus the relative position on the longitudinal axis of the root hair. (Sources: A, C, D from Bauer et al. 2011; H from Steinbrecher et al. 2011; K, L, M, N from Melzer et al. 2010. Reproduced with permission.)

Fig. 24.1 Dead lianas were conspicuous in the canopy of liana removal plots within weeks of cutting their stems in the understory. (Source: S.P. Yanoviak. Reproduced with permission.)

Fig. 24.2 Liana cutting increased light penetration to the understory in the months following the manipulation. (Source: S.P. Yanoviak. Reproduced with permission.)

Fig. 27.2 Black howler monkeys (*Alouatta pigra*, left side) and Geoffroyi's spider monkeys (*Ateles geoffroyi*, right side) using lianas as supporting structures during locomotion (A, B), resting (C, D) and feeding (E, F) in the Lacandona rain forest, Mexico. (Source: A. González-Zamora. Reproduced with permission.)

Chapter 18

EVOLUTIONARY IMPLICATIONS OF THE CLIMBING HABIT IN PLANTS

Ernesto Gianoli

Universidad de La Serena, La Serena, Chile
Universidad de Concepción, Concepción, Chile

Ecology of Lianas, First Edition. Edited by Stefan A. Schnitzer, Frans Bongers, Robyn J. Burnham, and Francis E. Putz.

Companion Website: www.wiley.com/go/schnitzer/lianas

OVERVIEW

This chapter discusses the evolutionary relevance of climbing plants from several lines of evidence based on bibliographic surveys, meta-analyses of published literature, and population genetics data. I first provide an updated assessment of the prevalence and taxonomic distribution of climbing plant species, showing that over one-third of all seed plant families and three-quarters of all dicot orders have climbers. These data imply that the climbing habit has evolved more times than it was originally believed, thus stressing the importance of the climbing habit in plant evolution. I discuss whether there is an association between specialized climbing mechanisms and species richness within families. I then address whether the climbing habit is an evolutionary key innovation that has allowed climbing clades to become more species-rich than their non-climbing sister groups. This pattern appears to have no biogeographic bias, i.e., climbing taxa are not more represented in the tropics than non-climbing taxa. One ecological hypothesis for this macroevolutionary pattern is increased exposure to pollinators and herbivores. Evolutionary and ecological differences between woody and herbaceous vines do not appear to correlate with different effects of the climbing habit on respective lineages. Finally, using genetic differentiation among populations as a surrogate for speciation, I test the key innovation of the climbing habit at an ecological scale. I compare population differentiation in climbing vs. non-climbing species in temperate forests of southern South America and show that temperate climbers have higher levels of genetic differentiation among populations than do non-climbers. The key innovation of climbing, evaluated at the macroevolutionary level, may also be found on an ecological scale. Finally, I identify key questions that should be addressed in future research on the evolution of climbing plants.

INTRODUCTION: EVOLUTION AND TAXONOMIC DISTRIBUTION OF THE CLIMBING HABIT

Trees, shrubs, herbs, and climbers represent the principal plant growth strategies that have evolved over time (Rowe & Speck 2005). Proliferation of species within plant families may be related to the diversity of growth forms displayed (Ricklefs & Renner 1994). The climbing habit has appeared numerous times, in independent events over the Phanerozoic eon (Burnham 2009; Burnham, Chapter 16 in this volume). Pteridosperms, an extinct group of gymnosperms, evolved the climbing life form during the Carboniferous period, ca. 300 million years ago (Krings et al. 2003). Most of the mechanisms of attachment that Darwin (1875) recognized (twining, tendrils, hooks, and adhesive roots) were present in the Carboniferous fossil record (Krings et al. 2003; Burnham 2009). Within angiosperms, there have been three origins of the climbing habit in Rhamnaceae (Richardson et al. 2000), two origins in each of Apocynaceae, Fabaceae, and Hernandiaceae (Gianoli 2004), and several origins within Calamoidea (Arecaceae) (Isnard & Rowe 2008). Gentry (1991a) reported that the total number of seed plant families including climbing species was at least 133. Schnitzer and Bongers (2002) suggest that about 60% of all dicotyledonous plant orders have at least one climbing species.

This chapter gives an update on the prevalence of climbers across plant orders and families to assess their evolutionary importance. I have supplemented Gentry's family list by conducting a systematic survey of the Angiosperm Phylogeny Website (APW; Stevens 2012), a volume on angiosperm classification by Cronquist (1981), and directed searches in Google Scholar and ISI Web of Science databases. Because I followed the classification criteria and taxonomic categories of the Angiosperm Phylogeny Group (APG; 1998, 2003, 2009), some of the earlier climber families listed by Gentry (1991a) have been subsumed into larger families. For instance, Asclepiadaceae and Hippocrateaceae are now included in Apocynaceae and Celastraceae, respectively, and all three Bombacaceae, Sterculiaceae and Tiliaceae are included in Malvaceae. Despite these family losses, the updated list of plant families including at least one climbing plant species is larger than before. I found 171 families (Table 18.1), distributed among ferns (9), gymnosperms (2), basal angiosperms (3), magnoliids (8), monocots (22), and eudicots (127). Seed plant families with climbing representatives number 162, over one-third of all seed plant families (35.6%). The APG phylogeny places these 162 families in 75.7% of all dicot orders. At the family level, only 38.7% of all dicot families include climbing species, but this is still larger than among the monocots, in which 24.7% of all families include climbing species (Table 18.1). The total number of orders and families are derived from APG (Stevens 2012). The updated list of plant families with climbers (Table 18.1) includes ca. 30% more families than did Gentry's list (Gentry

Table 18.1 Plant families including at least one climbing plant species (orders or families with climbing species/total number of orders or families). All types of climbers were considered, including scrambling species. Information was obtained from Gentry (1991a), systematic searches in APG (Stevens 2012), Cronquist (1981), Google Scholar and ISI Web of Science. Taxonomic categories follow the Angiosperm Phylogeny Group (APG; 1998, 2003, 2009). Order and family numbers were taken from the Angiosperm Phylogeny Website (APG, Stevens 2012).

FERNS
Orders: 4/11
Families: 9/31
Hymenophyllales
Hymenophyllaceae
Schizaeales
Lygodiaceae
Schizaeaceae
Polypodiales
Blechnaceae
Dryopteridaceae
Lomariopsidaceae
Oleandraceae
Polypodiaceae
Gleicheniales
Gleicheniaceae

GYMNOSPERMS
Orders: 1/4
Families: 2/13
Gnetales
Ephedraceae
Gnetaceae

BASAL ANGIOSPERMS
Orders: 1/3
Families: 3/7
Austrobaileyales
Austrobaileyaceae
Schisandraceae
Trimeniaceae

MAGNOLIIDS
Orders: 3/4
Families: 8/18
Laurales
Hernandiaceae
Lauraceae
Monimiaceae
Siparunaceae
Magnoliales
Annonaceae
Myristicaceae
Piperales
Aristolochiaceae
Piperaceae

MONOCOTS
Orders: 9/11
Families: 22/89
Alismatales
Araceae
Arecales
Arecaceae
Asparagales
Amaryllidaceae
Asparagaceae
Orchidaceae
Xanthorrhoeaceae
Commelinales
Commelinaceae
Dioscoreales
Dioscoreaceae
Liliales
Alstroemeriaceae
Liliaceae
Petermanniaceae
Philesiaceae
Rhipogonaceae
Smilacaceae
Pandanales
Cyclanthaceae
Pandanaceae
Stemonaceae
Poales
Bromeliaceae
Cyperaceae
Flagellariaceae
Poaceae
Zingiberales
Marantaceae

EUDICOTS
Orders: 28/37
Families: 126/328
Apiales
Apiaceae
Araliaceae
Griseliniaceae
Pittosporaceae
Aquifoliales
Aquifoliaceae
Cardiopteridaceae
Asterales
Asteraceae
Campanulaceae
Goodeniaceae
Rousseaceae
Stylidiaceae
Berberidopsidales
Berberidopsidaceae
Brassicales
Capparaceae
Caricaceae
Pentadiplandraceae
Salvadoraceae
Tovariaceae
Tropaeolaceae
Caryophyllales
Aizoaceae
Amaranthaceae
Ancistrocladaceae
Asteropeiaceae
Barbeuiaceae
Basellaceae
Cactaceae
Caryophyllaceae
Dioncophyllaceae
Droseraceae
Drosophyllaceae
Nepenthaceae
Nyctaginaceae
Phytolaccaceae
Plumbaginaceae
Polygonaceae
Portulacaceae
Talinacee
Celastrales
Celastraceae

Table 18.1 *(continued)*

Cornales	Schlegeliaceae	**Ranunculales**
Cornaceae	Scrophulariaceae	Lardizabalaceae
Hydrangeaceae	Verbenaceae	Menispermaceae
Loasaceae	**Malpighiales**	Papaveraceae
Cucurbitales	Achariaceae	Ranunculaceae
Begoniaceae	Calophyllaceae	**Rosales**
Coriariaceae	Chrysobalanaceae	Cannabaceae
Cucurbitaceae	Clusiaceae	Elaeagnaceae
Dilleniales	Dichapetalaceae	Moraceae
Dilleniaceae	Euphorbiaceae	Rhamnaceae
Dipsacales	Linaceae	Rosaceae
Caprifoliaceae	Lophopyxidaceae	Ulmaceae
Ericales	Malpighiaceae	Urticaceae
Actinidiaceae	Ochnaceae	**Santalales**
Cyrillaceae	Passifloraceae	Erythropalaceae
Ericaceae	Salicaceae	Loranthaceae
Lecythidaceae	Trigoniaceae	Olacaceae
Marcgraviaceae	Violaceae	Opiliaceae
Polemoniaceae	**Malvales**	Santalaceae
Primulaceae	Malvaceae	**Sapindales**
Theaceae	Thymelaeaceae	Anacardiaceae
Fabales	**Myrtales**	Meliaceae
Fabaceae	Combretaceae	Rutaceae
Polygalaceae	Lythraceae	Sapindaceae
Gentianales	Melastomataceae	**Saxifragales**
Apocynaceae	Myrtaceae	Aphanopetalaceae
Gelsemiaceae	Onagraceae	Saxifragaceae
Gentianaceae	Vochysiaceae	**Solanales**
Loganiaceae	**Oxalidales**	Convolvulaceae
Rubiaceae	Connaraceae	Montiniaceae
Lamiales	Cunoniaceae	Solanaceae
Acanthaceae	Elaeocarpaceae	**Vitales**
Bignoniaceae	Huaceae	Vitaceae
Gesneriaceae	Oxalidaceae	**Zygophyllales**
Lamiaceae	**Proteales**	Zygophyllaceae
Lentibulariaceae	Sabiaceae	**Unplaced - Asterids**
Oleaceae		Boraginaceae
		Icacinaceae

1991a). This suggests that the climbing habit may have evolved more times than originally believed. In this regard, the traditional view that climbers make a major contribution to plant taxonomic diversity only in tropical regions (Gentry 1991a) could be reconsidered. Recently compiled data indicate that when both herbaceous and woody vines are included, the number of climbing plant species in extratropical South America exceeds that of trees, shrubs and epiphytes (Durigon et al. 2014).

A fundamental question in evolution is why some groups are more species-rich than related groups (Sanderson & Donoghue 1996; Magallón & Sanderson 2001; Magallón & Castillo 2009). After reviewing the occurrence of climbing plants and their taxonomic distribution, Gentry (1991a) asked which features were those that led to the diversification success among some climbing taxa. He put forward the hypothesis that the more speciose liana families are those with more specialized climbing mechanisms. Climbing modes entail both simple and complex mechanisms of attachment (Putz & Holbrook 1991; Isnard & Silk 2009; Vaughn & Bowling 2010). Lianas with different climbing modes show differential distribution through forest succession, mainly because of the diameter of available trellises (Putz 1984; Putz & Chai 1987; Putz & Holbrook 1991; DeWalt et al. 2000; Carrasco-Urra & Gianoli 2009). At a global scale, however, no relationship was found between the proportion of the two main climbing mechanisms (twining and tendrils) and latitude (Gallagher & Leishman 2012).

A macroevolutionary pattern of association between climbing mode and diversification has not received quantitative attention. Gentry (1991a) pointed out that almost all the speciose families of climbers have specialized attachment mechanisms in addition to twining, while most of the less diversified families had only a single, simple attachment mechanism. For instance, 44% of all Panamanian climbing species belong to 10 families that have tendrils (Gentry 1985), arguably the most specialized climbing mechanism (Gentry 1991a; Vaughn & Bowling 2010). However, of the top ten most speciose climbing genera (Gentry 1991a, Table 1.6), only three (*Passiflora, Cissus, Smilax*) have tendrils. Moreover, of the 11 large (>100 species) New World families that are mostly climbers, only four families use tendrils as characteristic climbing mechanism (Gentry 1991a: Table 1.4). Thus, there is no direct relationship between speciation at the family level and the possession of tendrils. Additional research is needed to confirm or reject Gentry's assertion that specialized climbing mechanisms are associated with enhanced diversification of climbing plants.

THE CLIMBING HABIT: AN EVOLUTIONARY KEY INNOVATION

Lineages may exhibit evolutionary success, measured in terms of species richness, often explained via key innovations, defined as traits that promote diversification in a clade by allowing the exploitation of a previously unused or under-utilized resource (Simpson 1953; Heard & Hauser 1995; Hunter 1998; but see Davies et al. 2004). If speciation results from divergence in the same traits upon which natural selection operates (ecological speciation; Rundle & Nosil 2005), key innovations may directly influence taxonomic diversification (Hunter 1998). Key innovation hypotheses are tested by comparing species richness in pairs of closely related same-aged taxonomic groups (sister groups) with and without the candidate trait (Slowinski & Guyer 1993; Barraclough et al. 1998). Examples of key innovations in plants include latex and resin canals, which reduce herbivory and allow plant radiation (Farrell et al. 1991), floral nectar spurs in *Aquilegia* (Ranunculaceae) that promote pollinator specialization (Hodges 1997), fleshy fruits in tropical understory plants that promote speciation via seed dispersal by birds (Smith 2001), extrafloral nectaries in *Senna* (Fabaceae) that facilitate habitat colonization via ant–plant relationships (Marazzi & Sanderson 2010), and CAM photosynthesis in Bromeliaceae that allows colonization of arid environments by increased water efficiency (Quezada & Gianoli 2011). These cases illustrate how key innovations might operate on ecological time-scales by allowing taxa to enter into a new set of related ecological niches, where ecological specialization may proceed as a first step towards reproductive isolation (Heard & Hauser 1995).

Using the standard pairwise procedure for key innovation testing, Gianoli (2004) evaluated a hypothesis related to Gentry's (1991a) for climbing mechanisms, but focused on growth habit. It was hypothesized that climbers should be more diverse than their non-climbing relatives (Gianoli 2004), based on the number of times that the climbing habit has arisen (Gentry 1991a), the importance of lianas in tropical forest diversity (Schnitzer &Bongers 2002), and the increasing biomass of lianas in tropical forests (Phillips et al. 2002). Results showed that the climbing habit constitutes a key innovation in angiosperms: climbing plant taxa had higher species richness than non-climbing sister groups in 38 of 48 pairs of sister groups from 45 families of flowering plants (Gianoli 2004) (Fig. 18.1). Interestingly, a similar result was found for epiphytism: epiphyte genera were significantly richer in species than terrestrial genera in orchids and in non-orchids (Gravendeel et al. 2004).

Woody and herbaceous families of angiosperms differ in the relative importance of ecological controls on diversification (Eriksson & Bremer 1991, Tiffney & Mazer 1995). The pattern of increased diversification in climber lineages reported by Gianoli (2004) was evident when separate analyses were applied to herbaceous and woody climbers (Fig. 18.1). This was surprising because a stronger pattern was expected for herbaceous climbers based on the following ecological and evolutionary arguments. First, constraints for successfully reaching a support differ for

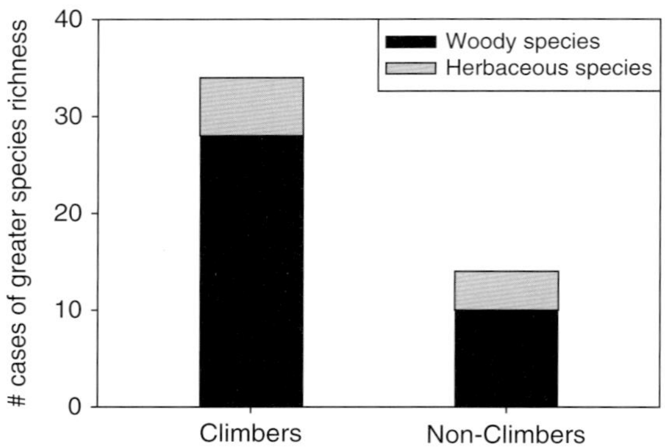

Fig. 18.1 Outcome of a series of paired comparisons of species richness between climbing plant lineages and their sister groups. Climbing taxa showed significantly greater species richness than their non-climbing sister groups, and this pattern was similar for both herbaceous and woody vines. Data and statistical analyses are in Gianoli (2004).

herbaceous versus woody vines (Gentry 1991a; Putz & Holbrook 1991). Woody climbers may be constrained because a minimum tree diameter is required to sustain a liana's mass. Pioneer tree species seldom bear woody vines in part because of their fast growth rates and flexible trunks (Putz 1984). Second, herbaceous families show higher speciation rates than woody families (Levin & Wilson 1976; Ricklefs & Renner 1994; Dodd et al. 1999). More research on the ecological causes of climbing plant diversification is needed to explain why evolutionary and ecological differences between woody and herbaceous vines do not translate into differential effects of the key innovation on the climbing habit. Herbaceous vines have received much less attention than lianas (woody vines) in ecological studies (Gianoli et al. 2010) and are often excluded from field sampling (Hegarty & Caballé 1991; Muthuramkumar & Parthasarathy 2000; Phillips et al. 2002). However, herbaceous vines contribute significantly to local abundance and diversity of climbers in tropical rainforests (Gentry & Dodson 1987) and account for a greater proportion of climbing plant species than lianas in temperate South America (Durigon et al. 2014).

Despite the accumulation of cases reporting evolutionary key innovations, few studies provide mechanistic explanations for the effects of novel traits on ecological and phenotypic expansion (Schluter 2000). It has been stressed that comparative tests and ecological arguments are complementary and necessary elements of a key innovation explanation (Heard & Hauser 1995). Climbers that find a suitable support generally have a greater biomass and reproductive output than those that grow unsupported (Putz 1984; Puntieri & Pyšek 1993; Gianoli 2002). However, does this divergence influence speciation? Key innovations are proposed to enhance species diversification via different pathways (Heard & Hauser 1995): (1) access to new adaptive zones free from competitors or natural enemies; (2) fitness increase leading one clade to replace another locally; or (3) enhancement of conditions for reproductive or ecological specialization. Here, I address the influence of the key innovation of climbing on the conditions for specialization in climbing plants.

The ecological opportunity needed for a trait to be considered a key innovation (Schluter 2000) may take the following form for a climbing plant species. First, in contrast to erect or prostrate species, which occupy a narrow range of the light gradient, climbing species can develop under a broad range of light conditions while they both are supported and unsupported individuals (Gianoli et al. 2010, 2012). In fact, long-lived vines may experience several changes in the light environment while climbing up, down, and across trees in the forest (Valladares et al. 2011). A similar hypothesis, the exploitation of a wider variety of habitats, may explain the greater species richness in herbaceous families compared to woody families (Ricklefs & Renner 1994). Second, a greater diversity of pollinators and fruit/seed dispersers may interact with climbers, promoting ecological specialization. This hypothesis is analogous to the prevailing explanation for the diversification of all angiosperms (Crepet & Niklas 2009). In support of this hypothesis, a study conducted in the temperate rainforest of southern Chile documented that the community of vines and epiphytes flowered longer (ca. 11 months) than the community of shrubs and trees (7–8 months) (Riveros & Smith-Ramírez 1995). Data from neotropical forests indicate that pollination by beetles, an exceedingly diverse group (Farrell 1998), and by large and mid-sized bees is more prevalent among climbers than among trees, shrubs, or herbs (Gentry 1991b). A study of woody vegetation across several wet tropical forests showed that the 20 liana species studied were pollinated by insects and not by wind or hummingbirds (Chazdon et al. 2003), while about 13% of the trees were pollinated by the wind or hummingbirds. Third, exposure to a greater array of herbivores, whose action may favor the presence and maintenance of defensive secondary compounds in climbers (Hegarty et al. 1991) could foster divergence, as has occurred in some plant lineages (Futuyma & Agrawal 2009). Conversely, a recent study comparing leaf chemical traits in 778 lianas and 6496 trees found that typical structure-defense compounds (lignin, phenols, and tannins) were on average 9% less common in lianas (Asner & Martin 2012).

The geographic distribution of the clades is rarely addressed in comparisons of species richness aimed at testing key innovations (but see von Hagen & Kadereit 2003; Davies et al. 2004, 2005). If clades possessing candidate traits are over-represented in the tropics, compared to sister groups, it is possible that the outcome of the test is related to the higher diversity of tropical taxa (Cox & Moore 2005) rather than to the putative key innovation. The greater species richness of herbaceous vines and lianas compared to trees in the tropics (Gentry 1991a) might have influenced the results of the climbing plant study by Gianoli (2004). Only in 3 out of 48 comparisons included in Gianoli's (2004) study was there a difference in geographic range between the climbing taxa and their sister groups. These included *Actinidia* (Actinidiaceae) vs. *Clematoclethra* (Actinidaceae) (tropical and subtropical vs. temperate), Hugonioideae (Linaceae) vs. Linoideae (Linaceae) (tropical vs. temperate and subtropical) and Smilacaceae vs. Liliaceae (tropical and subtropical vs. temperate and subtropical) (Gianoli 2004). In only one of these cases, *Actinidia* vs. *Clematoclethra*, the climbing taxon was more diverse, was distributed in the tropics, and had a sister group with temperate distribution. It is concluded, therefore, that it is extremely unlikely that biogeographic bias influenced the pattern of higher diversity of climbing taxa from my earlier study (Gianoli 2004).

THE KEY INNOVATION OF THE CLIMBING HABIT AT AN ECOLOGICAL SCALE

Although speciation cannot be evaluated at ecological time-scales, genetic differentiation among populations may be a testable surrogate for speciation (Avise 2000; Levin 2000). The hypothesis that a greater speciation rate explains the higher species richness in the tropics (Cox & Moore 2005) has been supported using population divergence within plant and vertebrate species (Eo et al. 2008; Martin & McKay 2004) as a proxy for speciation. Furthermore, it has been shown that ecological divergence is consistently and positively associated with reproductive isolation across plants and animals (Funk 1998; Filchak et al. 2000; Rundle & Nosil 2005; Funk et al. 2006; Peccoud et al. 2009).

I compared population differentiation in climbing vs. non-climbing taxa in temperate forests of southern South America. Using data from published studies and from my ongoing research, I evaluated the hypothesis that climbing plants have undergone greater among-population genetic differentiation than closely related trees or shrubs. Evidence available suggests that the hypothesis of more differentiation in climbing clades could be supported (Table 18.2). Overall, climbers ($n = 8$ species) showed an among-population genetic distance of 0.297 ± 0.073 (mean ± s.e.), while in the case of non-climbers ($n = 14$ species) genetic distance was 0.163 ± 0.023 (Table 18.2). These differences were statistically significant (t-test, t-value = 2.253, P = 0.036). The liana *Berberidopsis corallina* (Berberidopsidaceae) showed a strikingly high level of genetic differentiation between populations ($\Phi st = 0.548$; Etisham-Ul-Haq et al. 2001), much higher than any value reported for trees in this region (Table 18.2). Moreover, most studies conducted on trees compared populations that are more geographically distant than those included in the *B. corallina* study. It could be argued that the typical vegetative reproduction in climbers (Gerwing 2004) may have influenced the results, with lianas showing locally homogeneous populations because of undetected clonal individuals included in the sampling. However, evidence from the same region shows that climbing plants and trees show similar levels of within-population genetic diversity (Torres-Díaz et al. 2013; Gianoli et al., unpublished). Thus, the higher levels of genetic differentiation among populations of climbers, compared to non-climbers, are unlikely to be a consequence of low genetic diversity within climber populations.

The greater genetic differentiation observed in climbing plants is unlikely to be caused by a phylogenetic bias. That is, climbers and non-climbers in the study belong to very different phylogenetic lineages that, in turn, are associated with differential rates of diversification or differentiation (see Garland et al. 1992). For instance, the liana *Berberidopsis corallina* and the tree *Aextoxicon punctatum* (Table 18.2) belong to the sister families Berberidopsidaceae and Aextoxicaceae within Berberidopsidales (Stevens 2012). Likewise,

Table 18.2 Genetic differentiation among populations of woody and herbaceous vines (climbers) vs. trees, shrubs and herbs (non-climbers) in temperate forests of southern South America.

	Family	Genetic distance	Parameter	Molecular marker	Source
Climbers					
Berberidopsis corallina	Berberidopsidaceae	0.548	Φ_{ST}	RAPD	Etisham-Ul-Haq et al. 2001
Boquila trifoliolata	Lardizabalaceae	0.378	Φ_{ST}	AFLP	Gianoli et al., unpublished
Convolvulus chilensis	Convolvulaceae	0.307	G_{ST}	Allozymes	Gianoli et al. 2005
Elytropus chilensis	Apocynaceae	0.105	Φ_{ST}	AFLP	Gianoli et al., unpublished
Hydrangea serratifolia	Hydrangeaceae	0.093	Φ_{ST}	AFLP	Gianoli et al., unpublished
Lapageria rosea	Philesiaceae	0.440	G_{ST}	ISSR	Hoffens 2008
Luzuriaga radicans	Luzuriagaceae	0.037	Φ_{ST}	AFLP	Gianoli et al., unpublished
Tristerix corymbosus	Loranthaceae	0.465	Φ_{ST}	cpDNA	Amico and Nickrent 2009
Non-climbers					
Aextoxicon punctatum	Aextoxicaceae	0.182	Φ_{ST}	RAPD	Núñez-Ávila and Armesto 2006
Alstroemeria aurea	Alstroemeriaceae	0.144	G_{ST}	Allozymes	Souto and Premoli 2003
Araucaria araucana	Araucariaceae	0.287	G_{ST}	Allozymes	Ruiz et al. 2007
Embothrium coccineum	Proteaceae	0.202	F_{ST}	Allozymes	Souto and Premoli 2007
Eucryphia cordifolia	Cunoniaceae	0.214	G_{ST}	cpDNA	Segovia et al. 2012
Fitzroya crupressoides	Cupressaceae	0.144	Φ_{ST}	RAPD	Allnutt et al. 1999
Gomortega keule	Gomortegaceae	0.303	Φ_{ST}	ISSR	Herrera et al. 2005
Nothofagus alessandrii	Nothofagaceae	0.257	G_{ST}	Allozymes	Torres-Díaz et al. 2007
Nothofagus betuloides	Nothofagaceae	0.120	G_{ST}	Allozymes	Premoli 1997
Nothofagus dombeyi	Nothofagaceae	0.074	G_{ST}	Allozymes	Premoli 1997
Nothofagus nitida	Nothofagaceae	0.047	G_{ST}	Allozymes	Premoli 1997
Nothofagus obliqua	Nothofagaceae	0.052	F_{ST}	Allozymes	Marchelli and Gallo 2004
Pilgerodendron uviferum	Cupressaceae	0.186	Φ_{ST}	RAPD	Allnutt et al. 2003
Podocarpus salignus	Podocarpaceae	0.070	Φ_{ST}	RAPD	Allnutt et al. 2001

the liana *Luzuriaga radicans* and the perennial herb *Alstroemeria aurea* (Table 18.2) belong to the sister families Luzuriagaceae and Alstroemeriaceae within Liliales (Stevens 2012). Although more evidence is needed, available data suggest that the key innovation features of climbing habit, detectable at the macroevolutionary level, may also be traced at the ecological level. To my knowledge, this is one of the first times that an evolutionary key innovation detected by the standard protocol (meta-analysis of paired comparisons; Slowinski & Guyer 1993) is also supported at the scale of population divergence, which may underlie the speciation process (Heard & Hauser 1995; Rundle & Nosil 2005). Because other factors can influence both population divergence and speciation, these results need independent verification from numerous lineages of climber and non-climber pairs.

CONCLUSION AND FUTURE DIRECTIONS

Climbers are an important component of global plant diversity, and the climbing habit may give these plants evolutionary and ecological advantages over non-climbing close relatives. Further research on climbing plants should be aimed at addressing a number of important evolutionary issues. For instance, are some climbing mechanisms more efficient than others in facilitating ecological specialization (and hence population and species divergence)? Which ecological factors are more frequently involved in selection that leads to climbing plant differentiation? What features of lianas make them as prone to diversification as herbaceous vines, despite the intrinsic higher rate of diversification of herbaceous plants? Do lianas and trees differ in their within-population genetic diversity and among-population genetic differentiation? Finally, is the current pattern of increasing dominance of lianas in tropical forests related to enhanced diversification?

ACKNOWLEDGMENTS

I thank C. Torres-Díaz, E. Ruiz, R.S. Ríos, C. Salgado-Luarte, and J. Durigon for their collaboration. Comments by R.J. Burnham and A.T. Moles improved the text. Funding by FONDECYT 1100585 is acknowledged.

REFERENCES

Allnutt, T.R., Courtis, J.R., Gardner, M. & Newton, A.C. (2001) Genetic variation in wild Chilean and cultivated British populations of *Podocarpus salignus* D. Don (Podocarpaceae). *Edinburgh Journal of Botany*, **58**, 459–473.

Allnutt, T.R., Newton, A.C., Lara, A. et al. (1999) Genetic variation in *Fitzroya cupressoides* (alerce), a threatened South American conifer. *Molecular Ecology*, **8**, 975–987.

Allnutt, T.R., Newton, A.C., Premoli, A. & Lara, A. (2003) Genetic variation in the threatened South American conifer *Pilgerodendron uviferum* (Cupressaceae), detected using RAPD markers. *Biological Conservation*, **114**, 245–253.

Amico, G.C. & Nickrent, D.L. (2009) Population structure and phylogeography of the mistletoes *Tristerix corymbosus* and *T. aphyllus* (Loranthaceae) using chloroplast DNA sequence variation. *American Journal of Botany*, **96**, 1571–1580.

Angiosperm Phylogeny Group (1998) An ordinal classification for the families of flowering plants. *Annals of the Missouri Botanical Garden*, **85**, 531–553.

Angiosperm Phylogeny Group II (2003) An update of the Angiosperm Phylogeny Group classification for the orders and families of flowering plants: APG II. *Botanical Journal of the Linnean Society*, **141**, 399–436.

Angiosperm Phylogeny Group III (2009) An update of the Angiosperm Phylogeny Group classification for the orders and families of flowering plants: APG III. *Botanical Journal of the Linnean Society*, **161**, 105–121.

Asner, G.P. & Martin, R.E. (2012) Contrasting leaf chemical traits in tropical lianas and trees: implications for future forest composition. *Ecology Letters*, **15**, 1001–1007.

Avise, J. (2000) *Phylogeography: The History and Formation of Species*, Harvard University Press, Cambridge, MA.

Barraclough, T.G., Nee, S. & Harvey, P.H. (1998) Sister group analysis in identifying correlates of diversification. *Evolutionary Ecology*, **12**, 751–754.

Burnham, R.J. (2009) An overview of the fossil record of climbers: bejucos, sogas, trepadoras, lianas, cipós, and vines. *Revista Brasileira de Paleontologia*, **12**, 149–160.

Carrasco-Urra, F. & Gianoli, E. (2009) Abundance of climbing plants in a southern temperate rainforest: host-tree characteristics or light availability? *Journal of Vegetation Science*, **20**, 1155–1162.

Chazdon, R.L., Careaga, S., Webb, C. & Vargas, O. (2003) Community and phylogenetic structure of reproductive traits of woody species in wet tropical forests. *Ecological Monographs*, **73**, 331–348.

Cox, C.B. & Moore, P.D. (2005) *Biogeography: An Ecological and Evolutionary Approach*, Blackwell Publishing, Oxford.

Crepet, W.L. & Niklas, K.J. (2009) Darwin's second "abominable mystery": Why are there so many angiosperm species? *American Journal of Botany*, **96**, 366–381.

Cronquist, A. (1981) *An Integrated System of Classification of Flowering Plants*, Columbia University Press, New York.

Darwin, C.R. (1875) *The Movements and Habits of Climbing Plants*, John Murray, London.

Davies, T.J., Barraclough, T.J., Chase, M.W. et al. (2004) Darwin's abominable mystery: insights from a supertree of the angiosperms. *Proceedings of the National Academy of Sciences of the USA*, **101**, 1904–1909.

Davies, T.J., Savolainen, V., Chase, M.W. et al. (2005) Environment, area and diversification in the species-rich flowering plant family Iridaceae. *American Naturalist*, **166**, 418–425.

DeWalt, S.J., Schnitzer, S.A. & Denslow, J.S. (2000) Density and diversity of lianas along a chronosequence in a central Panamanian lowland forest. *Journal of Tropical Ecology*, **16**, 1–9.

Dodd, M.E., Silvertown, J. & Chase, M.W. (1999) Phylogenetic analysis of trait evolution and species diversity variation among angiosperm families. *Evolution*, **53**, 732–744.

Durigon, J., Miotto, S.T.S., & Gianoli, E. (2014) Distribution and traits of climbing plants in subtropical and temperate South America. *Journal of Vegetation Science*, in press.

Eo, S.Y., Wares, J.P. & Carroll, J.P. (2008) Population divergence in plant species reflects latitudinal biodiversity gradients. *Biology Letters*, **4**, 382–384.

Eriksson, O. & Bremer, B. (1991) Fruit characteristics, life forms, and species richness in the plant family Rubiaceae. *American Naturalist*, **138**, 751–761.

Etisham-Ul-Haq, M., Allnutt, T.R., Smith-Ramírez, C. et al. (2001) Patterns of genetic variation in *in* and *ex situ* populations of the threatened Chilean vine *Berberidopsis corallina*, detected using RAPD markers. *Annals of Botany*, **87**, 813–821.

Farrell, B.D. (1998) "Inordinate fondness" explained: why are there so many beetles? *Science*, **281**, 555–559.

Farrell, B.D., Dussourd, D.E. & Mitter, C. (1991) Escalation of plant defense: do latex and resin canals spur plant diversification? *American Naturalist*, **138**, 881–900.

Filchak, K.E., Roethele, J.B. & Feder, J.L. (2000) Natural selection and sympatric divergence in the apple maggot *Rhagoletis pomonella*. *Nature*, **407**, 739–742.

Funk, D.J. (1998) Isolating a role for natural selection in speciation: host adaptation and sexual isolation in *Neochlamisus bebbianae* leaf beetles. *Evolution*, **52**, 1744–1759.

Funk, D.J., Nosil, P. & Etges, W.J. (2006) Ecological divergence exhibits consistently positive associations with reproductive isolation across disparate taxa. *Proceedings of the National Academy of Sciences of the USA*, **103**, 3209–3213.

Futuyma, D.J. & Agrawal, A.A. (2009) Macroevolution and the biological diversity of plants and herbivores. *Proceedings of the National Academy of Sciences of the USA*, **106**, 18054–18061.

Gallagher, R.V. & Leishman, M.R. (2012) A global analysis of trait variation and evolution in climbing plants. *Journal of Biogeography*, **39**, 1757–1771.

Garland, T., Harvey, P.H. & Ives, A.R. (1992) Procedures for the analysis of comparative data using phylogenetically independent contrasts. *Systematic Biology*, **41**, 18–32.

Gentry, A.H. (1985) An ecotaxonomic survey of Panamanian lianas. *Monographs in Systematic Botany from the Missouri Botanical Garden*, **10**, 29–42.

Gentry, A.H. (1991a) The distribution and evolution of climbing plants, in *The Biology of Vines* (eds F.E. Putz & H.A. Mooney), Cambridge University Press, Cambridge, pp. 3–49.

Gentry, A.H. (1991b) Breeding and dispersal systems of lianas, in *The Biology of Vines* (eds F.E. Putz & H.A. Mooney), Cambridge University Press, Cambridge, pp. 393–423.

Gentry, A.H. & Dodson, C. (1987) Contribution of nontrees to species richness of a tropical rain forest. *Biotropica*, **19**, 149–156.

Gerwing, J.J. (2004) Life history diversity among six species of canopy lianas in an old-growth forest of the eastern Brazilian Amazon. *Forest Ecology and Management*, **190**, 57–72.

Gianoli, E. (2002) Maternal environmental effects on the phenotypic responses of the twining vine *Ipomoea purpurea* to support availability. *Oikos*, **99**, 324–330.

Gianoli, E. (2004) Evolution of a climbing habit promotes diversification in flowering plants. *Proceedings of the Royal Society of London B*, **271**, 2011–2015.

Gianoli, E., Saldaña, A. & Jiménez-Castillo, M. (2012) Ecophysiological traits may explain the abundance of climbing plant species across the light gradient in a temperate rainforest. *PLoS ONE*, **7**(6), e38831.

Gianoli, E., Saldaña, A., Jiménez-Castillo, M. & Valladares, F. (2010) Distribution and abundance of vines along the light gradient in a southern temperate rainforest. *Journal of Vegetation Science*, **21**, 66–73.

Gianoli, E., Suárez, L.H., Quezada, I.M., & Ruiz, E. (2005) Ecotipos y adaptación local en *Convolvulus chilensis* – Convolvulaceae. Proceedings of the XLVIII Annual Meeting of the Biological Society of Chile. *Biological Research*, **38**(4).

Gravendeel, B., Smithson, A., Silk, F.J.W. & Schuiteman, A. (2004) Epiphytism and pollinator specialization: drivers for orchid diversity? *Philosophical Transactions of the Royal Society of London B*, **359**, 1523–1535.

Heard, S.B. & Hauser, D. (1995) Key evolutionary innovations and their ecological mechanisms. *Historical Biology*, **10**, 151–173.

Hegarty, E.E. & Caballé, G. (1991) Distribution and abundance of vines in forest communities, in *The Biology of Vines* (eds F.E. Putz & H.A. Mooney), Cambridge University Press, Cambridge, pp. 313–336.

Hegarty, M.P., Hegarty, E.E. & Gentry, A.H. (1991) Secondary compounds in vines with an emphasis on those with defensive functions, in *The Biology of Vines* (eds F.E. Putz & H.A. Mooney), Cambridge University Press, Cambridge, pp. 287–310.

Herrera, R., Arias, M., Moya-León, M.A. et al. (2005) Genetic variation in a Chilean endangered endemic: *Gomortega keule* (Molina) Baillon. *Biodiversity and Conservation*, **14**, 2871–2881.

Hodges, S.A. (1997) Floral nectar spurs and diversification. *International Journal of Plant Sciences*, **158**, S81–S88.

Hoffens, K.A. (2008) *Diversidad genética en poblaciones de copihue (Lapageria rosea Ruiz et Pav.) determinada a través de marcadores moleculares*. PhD thesis, Universidad Austral de Chile, Valdivia.

Hunter, J.P. (1998) Key innovations and the ecology of macroevolution. *Trends in Ecology & Evolution*, **13**, 31–36.

Isnard, S. & Rowe, N.P. (2008) Mechanical role of the leaf sheath in rattans. *New Phytologist*, **177**, 643–652.

Isnard, S. & Silk, W.K. (2009) Moving with climbing plants from Charles Darwin's time into the 21st century. *American Journal of Botany*, **96**, 1205–1221.

Krings, M., Kerp, H., Taylor, T.N. & Taylor, E.N. (2003) How Paleozoic vines and lianas got off the ground: on scrambling and climbing Carboniferous- Early Permian pteridosperms. *The Botanical Review*, **69**, 204–224.

Levin, D.A. (2000) *The Origin, Expansion, and Demise of Plant Species*, Oxford University Press, Oxford.

Levin, D.A. & Wilson, A.C. (1976) Rates of evolution in seed plants: net increase in diversity of chromosome numbers and species numbers through time. *Proceedings of the National Academy of Sciences of the USA*, **73**, 2086–2090.

Magallón, S. & Castillo, A. (2009) Angiosperm diversification through time. *American Journal of Botany*, **96**, 349–365.

Magallón, S. & Sanderson, M.J. (2001) Absolute diversification rates in angiosperm clades. *Evolution*, **55**, 1762–1780.

Marazzi, B. & Sanderson, M.J. (2010) Large-scale patterns of diversification in the widespread legume genus *Senna* and the evolutionary role of extrafloral nectaries. *Evolution*, **64**, 3570–3592.

Marchelli, P. & Gallo, L.A. (2004) The combined role of glaciations and hybridization in shaping the distribution of genetic variation in a Patagonian southern beech. *Journal of Biogeography*, **31**, 451–460.

Martin, P.R. & McKay, J.K. (2004) Latitudinal variation in genetic divergence of populations and the potential for future speciation. *Evolution*, **58**, 938–945.

Muthuramkumar, S. & Parthasarathy, N. (2000) Alpha diversity of lianas in a tropical evergreen forest in the Anamalais, Western Ghats, India. *Diversity and Distributions*, **6**, 1–14.

Nosil, P., Harmon, L.J. & Seehausen, O. (2009) Ecological explanations for (incomplete) speciation. *Trends in Ecology & Evolution*, **24**, 145–156.

Núñez-Ávila, M.C. & Armesto, J.J. (2006) Relict islands of the temperate rainforest tree *Aextoxicon punctatum* (Aextoxicaceae) in semi-arid Chile: genetic diversity and biogeographic history. *Australian Journal of Botany*, **54**, 733–743.

Peccoud, J., Ollivier, A., Plantegenest, M. & Simon, J.C. (2009) A continuum of genetic divergence from sympatric host races to species in the pea aphid complex. *Proceedings of the National Academy of Sciences of the USA*, **106**, 7495–7500.

Phillips, O.L., Martínez, R.V., Arroyo, L. et al. (2002) Increasing dominance of large lianas in Amazonian forests. *Nature*, **418**, 770–774.

Premoli, A.C. (1997) Genetic variation in a geographically restricted and two widespread species of South American *Nothofagus*. *Journal of Biogeography*, **24**, 883–892.

Puntieri, J.G. & Pyšek, P. (1993) The effects of physical support and density on biomass production and size hierarchies of *Galium aparine* populations. *Oikos*, **66**, 279–284.

Putz, F.E. (1984) The natural history of lianas on Barro Colorado island, Panamá. *Ecology*, **65**, 1713–1724.

Putz, F.E. & Chai, P. (1987) Ecological studies of lianas in Lamber National Park, Sarawak, Malaysia. *Journal of Ecology*, **75**, 523–531.

Putz, F.E. & Holbrook, N.M. (1991) Biomechanical studies of vines. in *The Biology of Vines* (eds F.E. Putz & H.A. Mooney), Cambridge University Press, Cambridge, pp. 73–97.

Quezada, I.M. & Gianoli, E. (2011) Crassulacean acid metabolism photosynthesis in Bromeliaceae: an evolutionary key innovation. *Biological Journal of the Linnean Society*, **104**, 480–486.

Richardson, J.E., Fay, M.F., Cronk, Q.C.B. & Chase, M.W. (2000) A revision of the tribal classification of Rhamnaceae. *Kew Bulletin*, **55**, 311–340.

Ricklefs, R.E. & Renner, S.S. (1994) Species richness within families of flowering plants. *Evolution*, **48**, 1619–1636.

Riveros, M. & Smith-Ramírez, C. (1995) Patrones de floración y fructificación en bosques del sur de Chile, in *Ecología de los Bosques Nativos de Chile* (eds J.J. Armesto, C. Villagrán & M.K. Arroyo), Editorial Universitaria, Santiago, pp. 235–250.

Rowe, N. & Speck, T. (2005) Plant growth forms: an ecological and evolutionary perspective. *New Phytologist*, **166**, 61–72.

Ruiz, E., González, F., Torres-Díaz, C. et al. (2007) Genetic diversity and differentiation within and among Chilean populations of *Araucaria araucana* (Araucariaceae) based on allozyme variability. *Taxon*, **56**, 1221–1228.

Rundle, H.D. & Nosil, P. (2005) Ecological speciation. *Ecology Letters*, **8**, 336–352.

Sanderson, M.J. & Donoghue, M.J. (1996) Reconstructing shifts in diversification rates on phylogenetic trees. *Trends in Ecology & Evolution*, **11**, 15–20.

Schluter, D. (2000) *The Ecology of Adaptive Radiation*, Oxford University Press, Oxford.

Schnitzer, S.A. & Bongers, F. (2002) The ecology of lianas and their role in forests. *Trends in Ecology & Evolution*, **17**, 223–230.

Schnitzer, S.A. & Bongers, F. (2011) Increasing liana abundance and biomass in tropical forests: emerging patterns and putative mechanisms. *Trends in Ecology and Evolution*, **17**, 223–230.

Segovia, R.A., Pérez, M.F. & Hinojosa, L.F. (2012) Genetic evidence for glacial refugia of the temperate tree *Eucryphia cordifolia* (Cunoniaceae) in southern South America. *American Journal of Botany*, **99**, 121–129.

Simpson, G.G. (1953) *The Major Features of Evolution*, Columbia University Press, New York.

Slowinski, J.B. & Guyer, C. (1993) Testing whether certain traits have caused amplified diversification: an improved method based on a model of random speciation and extinction. *American Naturalist*, **142**, 1019–1024.

Smith, J.F. (2001) High species diversity in fleshy-fruited tropical understory plants. *American Naturalist*, **157**, 646–653.

Souto, C.P. & Premoli, A. (2003) Genetic divergence among natural populations of *Alstroemeria aurea* D. Don: a dominant clonal herb of the understory in subalpine *Nothofagus* forests. *Botanical Bulletin of Academia Sinica*, **44**, 329–336.

Souto, C.P. & Premoli, A. (2007) Genetic variation in the widespread *Embothrium coccineum* (Proteaceae) endemic to Patagonia: effects of phylogeny and historical events. *Australian Journal of Botany*, **55**, 809–817.

Stevens, P.F. (2012) *Angiosperm Phylogeny Website*. Version 12, July 2012. http://www.mobot.org/MOBOT/research/APweb/.

Tiffney, B.H. & Mazer, S.J. (1995) Angiosperm growth habit, dispersal and diversification reconsidered. *Evolutionary Ecology*, **9**, 93–117.

Torres-Díaz, C., Ruiz, E., González, F. et al. (2007) Genetic diversity in *Nothofagus alessandrii* (Fagaceae), and endangered endemic tree species of the coastal Maulino forest of central Chile. *Annals of Botany*, **100**, 75–82.

Torres-Díaz, C., Ruiz, E., Salgado-Luarte, C. et al. (2013) Within-population genetic diversity of climbing plants and trees in a temperate forest in central Chile. *Gayana Botanica*, **70**, 31–38.

Valladares, F., Gianoli, E. & Saldaña, A. (2011) Climbing plants in a temperate rainforest understory: searching for high light or coping with deep shade? *Annals of Botany*, **108**, 231–239.

Vaughn, K.C. & Bowling, A.J. (2010) Biology and physiology of vines. In *Horticultural Reviews*, Vol. 38 (ed. J. Janick), John Wiley & Sons, Inc., Hoboken, NJ. doi: 10.1002/9780470872376.ch1

Von Hagen, K.B. & Kadereit, J.W. (2003) The diversification of *Halenia* (Gentianaceae): ecological opportunity versus key innovation. *Evolution*, **57**, 2507–2518.

Part V

Liana Anatomy, Physiology, and Biomechanics

Chapter 19

LIANA ANATOMY: A BROAD PERSPECTIVE ON STRUCTURAL EVOLUTION OF THE VASCULAR SYSTEM

Veronica Angyalossy, Marcelo R. Pace, and André C. Lima

Instituto de Biociências, Universidade de São Paulo, São Paulo (SP), Brazil

Ecology of Lianas, First Edition. Edited by Stefan A. Schnitzer, Frans Bongers, Robyn J. Burnham, and Francis E. Putz.

Companion Website: www.wiley.com/go/schnitzer/lianas

OVERVIEW

The vascular system of lianas exhibits a number of shared features that result from convergent evolution and are independent of their taxonomic affinities. In this chapter we explore the vascular system of lianas in major lineages addressing different aspects of their anatomy, development, seasonal growth, and evolution. Since lianas are twining plants that climb in the search of light, various special features favor this habit and guarantee an efficient fluid conduction throughout their long stems. Both xylem and phloem of lianas are conspicuous for their wide conducting cells. In xylem it is also common to find a reduced volume of fibers associated with an increased amount of axial parenchyma. The flexibility required to twine seems to have been facilitated by the widespread presence of variant secondary growth, which derives from the activity of one or multiple cambia. The evolution of cambial variants in different lineages seems to have led to an increase in anatomical complexity in lineages in which variants occur. Cambial variants also affect xylem differentiation and seasonal response to environmental cues. Lianas may have active cambia for a very short period of time, from two months to up to a year, most likely a taxon-dependent feature. The vascular system of lianas provides a wealth of features to explore the link between habit, development, and the evolution of complex anatomies.

INTRODUCTION: THE UNIQUE VASCULAR SYSTEM OF LIANAS

The study of the vascular system of lianas has a long history. It flourished in the works of Heinrich Schenck (1892, 1893), who traveled from Germany to Brazil and compiled a treatise on both morphological (Schenck 1892) and anatomical aspects (Schenck 1893) of lianas. Most textbooks dealing with liana anatomy subsequently used Schenck's work and his detailed hand-line drawings as a basis for their treatments (Solereder 1908; Boureau 1957; Metcalfe & Chalk 1950). Most recently, Obaton (1960) on lianas of Africa, and Carlquist (1991, and references therein) on lianas of different continents, have introduced a large volume of new data, enriching the knowledge in this field. In this chapter we summarize general aspects of the vascular system of lianas, and provide new data that show how the vascular anatomy of lianas has evolved and diversified, and how lianas respond to environmental cues. A neotropical bias will be evident inasmuch as this is our research expertise, though aspects of paleotropical lianas are also covered based on the literature (Obaton 1960; Rajput & Rao 1999, 2003; Rajput et al. 2009; Patil et al. 2011). An Appendix of glossary terms is also given.

The lianescent habit evolved many times during the evolutionary history of plants (Gentry 1991; Burnham 2009; Spicer & Groover 2010; Isnard & Feild, Chapter 17 in this volume), being present in lycophytes, monilophytes (ferns), fossil pteridosperms, gymnosperms and angiosperms (Fig. 19.1). Several convergences evolved along with these multiple origins, including the development of modified organs specialized for climbing, such as tendrils, spines, or adventitious roots that enable vines to reach the canopy (Darwin 1865; Schenck 1892; Isnard & Silk 2009). Additionally, the development of a highly specialized vascular system has been key to the success of lianas, and this is the focus of this chapter. The vascular system of lianas is special since scandent plants must cope with very specific physical constraints inherent to their habit, including water transport along extremely long stems (Isnard & Feild, Chapter 17 in this volume), and the demand for flexibility to accommodate twining (or circumnutation). Lianas do not support their own body, but instead use external supports, usually other plants, as the means to reach the canopy and keep their bodies erect, a characteristic that gave them the epithet of "structural parasites" (Stevens 1987).

The fact that lianas are not self-supporting has enabled them, along their evolutionary trajectory, to favor conduction and flexibility at the expense of mechanical stiffness. This reliance on host support may explain why lianas have narrower stems than trees, even though they frequently sustain a leaf area comparable to that of trees (Ewers & Fisher 1991). The broad canopies of lianas, in turn, are only made possible by the development of a highly specialized vascular system with very wide vessels (or wider than those from closely related self-supporting species), low density, and intermixed soft and stiff tissues, features that ensure both optimal water conduction, flexibility, and toughness. Together these traits may be termed the "lianescent vascular syndrome," and they are common to most scandent plants.

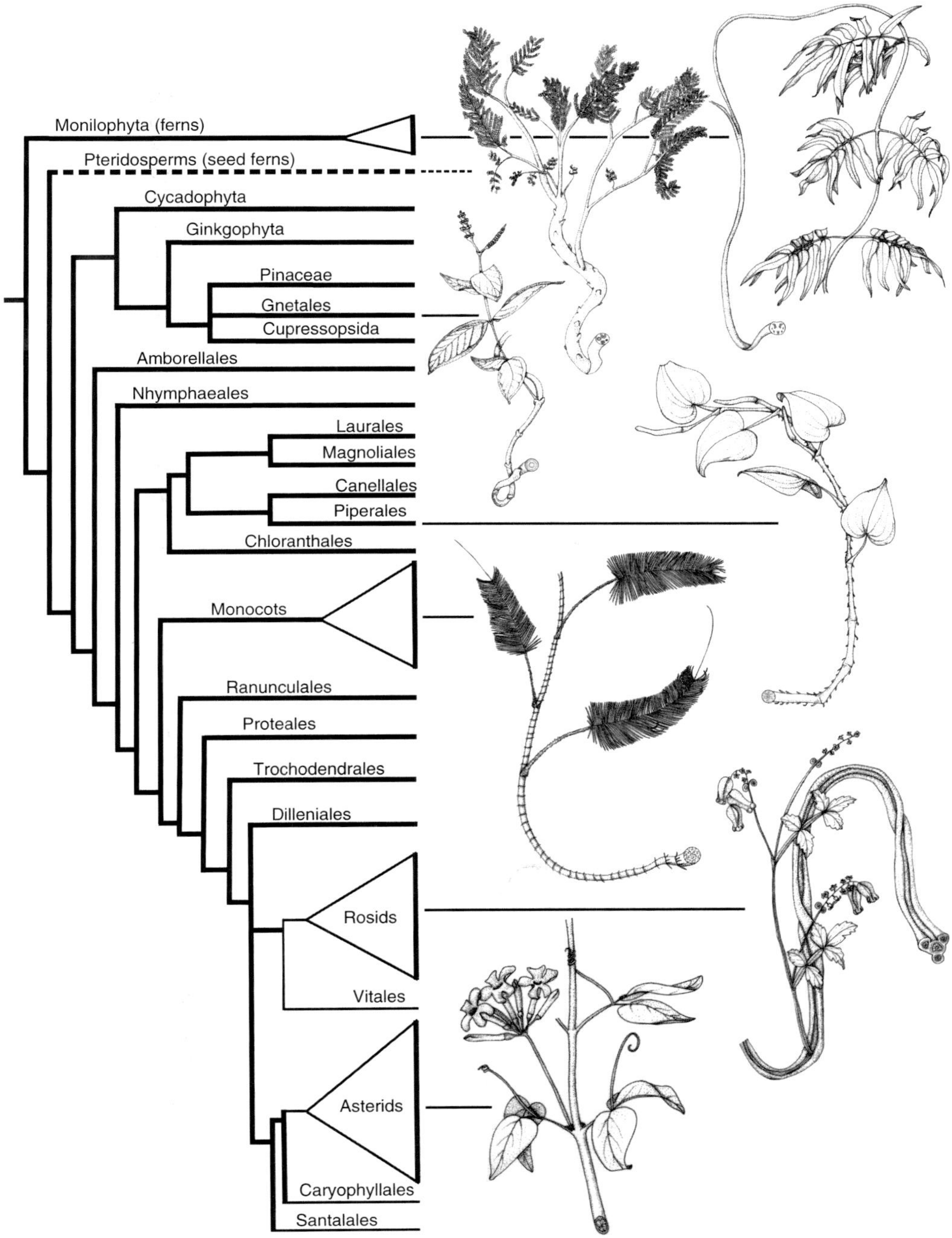

Fig. 19.1 Tracheophytes phylogeny, according to APweb (Stevens 2001 onwards, excluding lycophytes), showing examples of liana species that evolved independently in major lineages of vascular plants. Monilophyta: Blechnaceae, *Salpichlaena volubilis*. Pteridosperms: Medullosaceae, *Medullosa steinii* (fossil; dotted line). Gymnosperms: Gnetales, *Gnetum* sp. Angiosperms: Magnoliids, Piperales, Piperaceae, *Manekia obtusa*. Monocots: Arecales, Arecaceae, *Calamus* sp. Eudicots, Rosids, Sapindales, Sapindaceae, *Serjania* sp. Eudicots: Asterids, Lamiales, Bignoniaceae, *Lundia longa*. (Source: Illustrator L.W. Gussella. Reproduced with permission.)

THE LIANESCENT VASCULAR SYNDROME

Regardless of whether the vine is a fern (e.g., *Salpichlaena volubilis*, Fig. 19.2A), gymnosperm (e.g., *Gnetum* spp., Fig. 19.2B), magnoliid (e.g., *Piper nigrum* or *Aristolochia gigantea*, Fig. 19.2C), monocot (e.g. *Desmoncus orthacanthos*, Fig. 19.2D), or eudicot (various families, including Acanthaceae, Bignoniaceae, Leguminosae, Menispermaceae, and Sapindaceae, Figs. 19.2E, 19.2F), vines generally show a very similar vascular system. Xylem may be characterized by very wide conducting cells, fewer fibers, and the presence of soft tissues, usually parenchyma and/or phloem, interspersed in the stiff xylem and/or sclerenchyma (Schenck 1893; Obaton 1960; Carlquist 1985a). In lianas with secondary growth, the combination of wide and narrow vessels or vasicentric tracheids (termed vessel dimorphism by Carlquist 1981), the presence of non-lignified axial and ray parenchyma, high and wide rays, and cambial variants are also common. Together, these characters make up what can be termed the lianescent vascular syndrome.

In contrast, self-supporting plants typically show xylem with narrower vessels, no vessel dimorphism, less parenchyma, and more fibers compared with the secondary xylem of their lianescent counterparts. This can be seen, for example, in some species with a facultative scandent habit in the Euphorbiaceae (*Manihot*, Ménard et al. 2009) and the Leguminosae

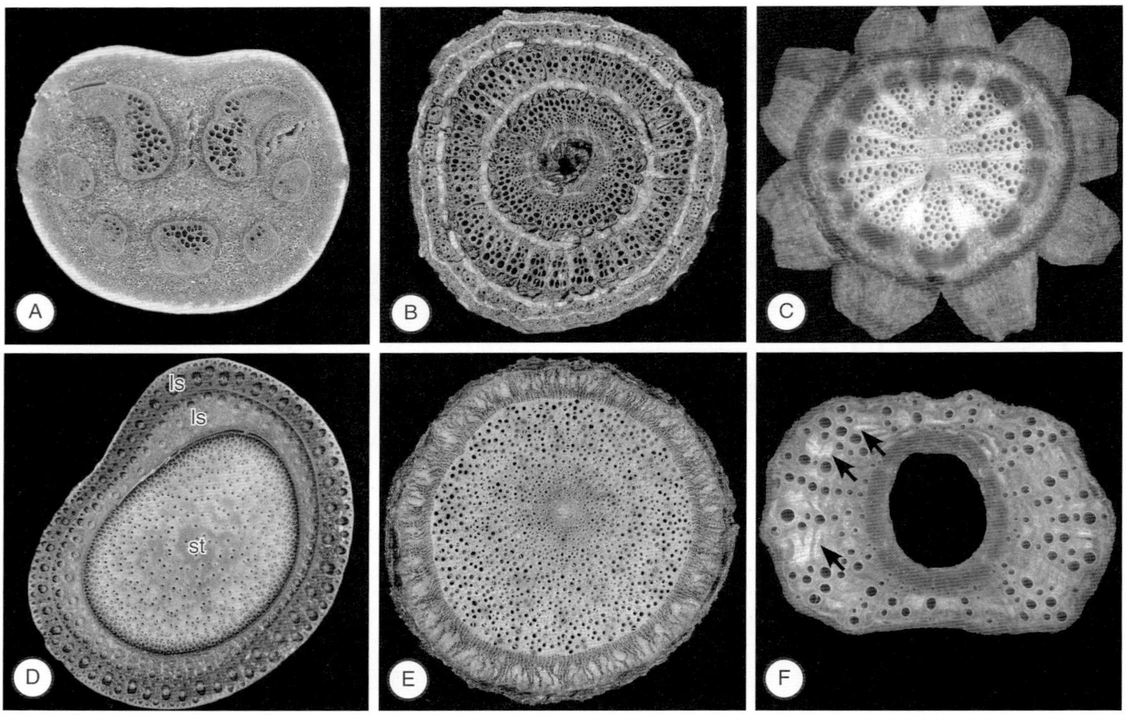

Fig. 19.2 Macroscopic transverse sections of tracheophyte liana stems, figures not to scale. (A) *Salpichlaena volubilis* (Blechnaceae, Monilophytes) stem showing wide tracheary elements, twining rachis. (B) *Gnetum paniculatum* (Gnetaceae, Gymnosperm) stem showing wide vessels and successive cambia. (C) *Aristolochia gigantea* (Aristolochiaceae, Magnoliid) stem showing thick periderm, wide vessels and secondary vascular cylinder marked by very wide rays alternating with the axial elements of the xylem and phloem. Wide rays are derived from the interfascicular cambium. (D) *Desmoncus orthacanthos* (Arecaceae, Monocots) main stem (st) and two surrounding leaf sheaths (ls). (E) *Dioclea rufescens* (Leguminosae, Eudicot Rosid) stem with very wide vessels (up to 500 μm). (F) *Thunbergia grandiflora* (Acanthaceae, Eudicot Lamiid) stem showing wide vessels and islands of included phloem (arrows) produced by interfascicular cambium. *See plate section for color representation of this figure.*

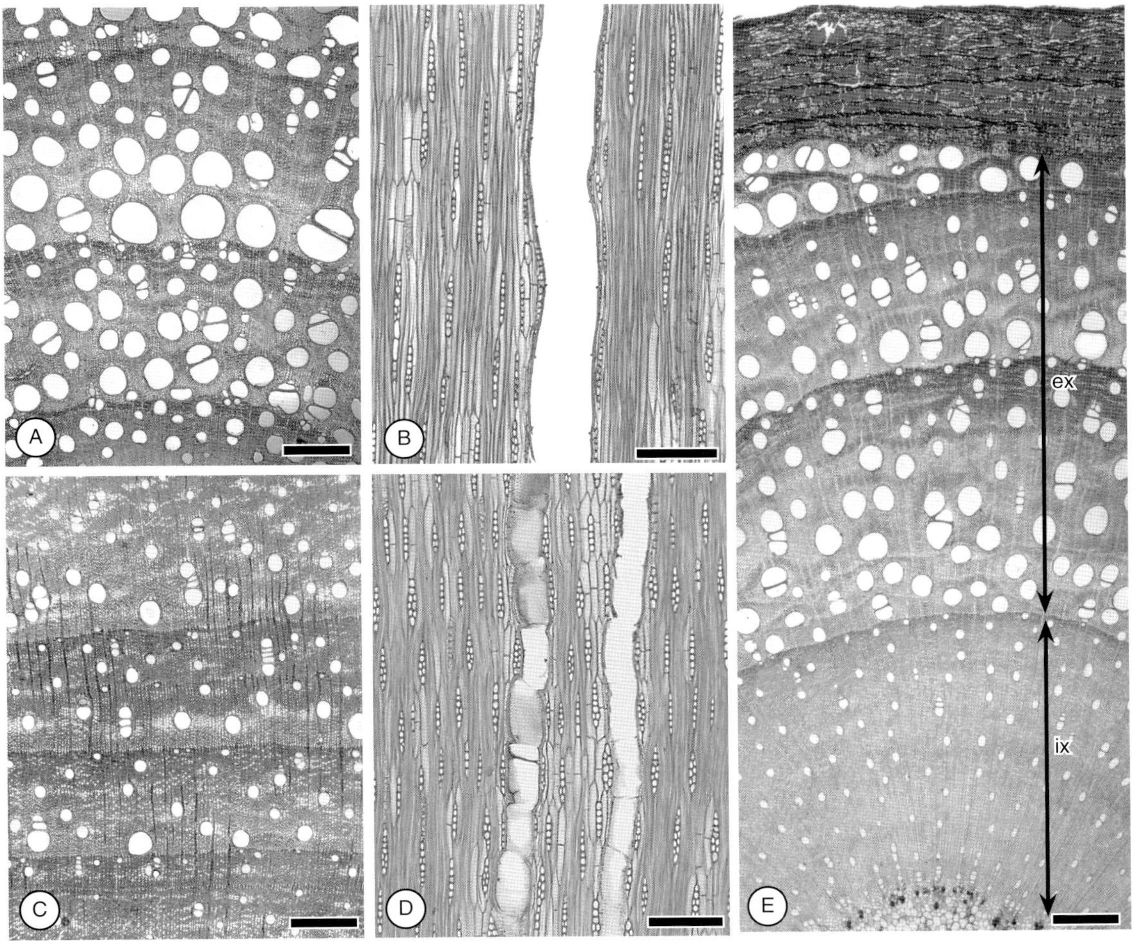

Fig. 19.3 Anatomical comparison between wood of a liana and a tree of *Dalbergia frutescens* (Leguminosae). (A, B, E), liana; (C, D), tree. (A) Wood transverse section, wide and narrow vessels combined, abundant banded paratracheal parenchyma. (B) Wood tangential section, showing lack of clear storied structure (rays irregularly storied), rays taller than those of the trees. (C) Wood transverse section, narrow vessels, aliform paratracheal parenchyma, forming short confluences. (D) Wood tangential section, showing storied structure (rays, axial parenchyma, vessel elements), narrow and short rays. (E) Stem transverse section from pith to cambium. Note the difference between the inner, dense, secondary xylem of self-supporting phase (ix) and the subsequent xylem with lianescent xylem syndrome (ex). Note: Scale bars: A, C, E = 500 μm, B, D = 200 μm. *See plate section for color representation of this figure.*

(*Dalbergia*, Figs. 19.3A–D). These differences are also true for lineages of shrubs that evolved from lianescent ancestors, e.g., Apocynaceae (Lahaye et al. 2005), Bignoniaceae (Pace et al. 2009), and Aristolochiaceae (Wagner et al. 2012). In addition, a number of lianas start their development as self-supporting shrubs or with stiff searching branches. At these stages, they have not developed the lianescent vascular syndrome and are characterized, instead, by a few narrow vessels and thick fibers, which may be modified as they are established on a support, switching to a lianescent vascular anatomy (Fisher & Ewers 1991; Rowe &

Speck 1996, 2004; Gallenmüller et al. 2001; Rowe et al. 2004; Angyalossy et al. 2012; Fig. 19.3E). Here, we discuss the anatomy and functional biology of this unique set of characters that develop both in herbaceous vines and lianas.

Vessel dimensions and morphology

Herbaceous vines and lianas are well known for possessing some of the widest (Figs. 19.2, 19.3A, 19.3E, 19.4A–C) and longest vessels in the plant kingdom (Klotz 1978; Ewers 1985; Ewers et al. 1990; Jacobsen et al. 2012). The widest vessel was found in the eudicot *Dioclea*, Leguminosae, with vessels up to 500 µm wide (measured in this study) and the longest vessel, i.e., a series of vessel elements, was found in *Amphilophium* (= *Pithecoctenium*) *crucigerum*, Bignoniaceae, at 7.73 m long (Ewers et al. 1990). In addition, lianas have xylem that remains conductive for a longer time than trees (Ewers & Fisher 1989; Gartner et al. 1990; Ewers et al. 1991, Holbrook & Putz 1996). Increases in vessel lumen raise water flow exponentially to the 4th power, as described by the Hagen-Poiseuille equation (Sperry et al. 2006). Thus, the wide vessels of lianas are extremely efficient in water conduction. In addition, the abundance of vessels in liana stems is often closer to their theoretical packing limit, i.e., the maximum number of conduits of a given width per square millimeter (McCulloh et al. 2010), compared with that of self-supporting plants growing under similar conditions (Crivellaro et al. 2012). Furthermore, in clades in which multiple perforation plates are common, the vessels of lianas usually bear a higher frequency of simple perforation plates than self-supporting species (e.g., Dilleniaceae, Carlquist 1991; Gnetaceae, Fisher & Ewers 1995; European species, Baas & Schweingruber 1987). Simple perforation plates are known to be less resistant to water flow than scalariform perforation plates (Christman & Sperry 2010). The large diameters and long lengths of vessels, together with the prevalence of simple perforation plates, allow lianas to construct a very efficient hydraulic system (Ewers & Fisher 1991; Schnitzer 2005; Isnard & Feild, Chapter 17 in this volume).

Wide vessels, however, have the shortcoming of being more susceptible to failure by embolism (Zimmermann & Jeje 1981; Ewers et al. 1990), a phenomenon that may explain why the diversity and abundance of lianas are reduced at higher latitudes (Gentry 1991) and altitudes (Jiménez-Castillo et al. 2007). However, even in tropical forests, embolisms can occur, especially during the dry season. This constraint might have been overcome, in part, by the ubiquitous presence of narrow vessels or vasicentric tracheids associated with the very wide vessels (vessel dimorphism; Figs. 19.4C, 19.4E–F). Vessel dimorphism guarantees water conduction when wide vessels embolize, by creating anastomoses within the water path (Carlquist 1985a). Narrow vessels act also as "vessel relays," creating "hydraulic bridges" between wide vessels and guaranteeing efficient radial water flow throughout the stem (Brodersen et al. 2013). This safety mechanism ensures conduction while root pressure (Ewers et al. 1997) or other mechanisms cope with vessel refilling. However, root pressure alone was shown not to be sufficient for vessel refilling in most lianas studied (Ewers et al. 1997). Other mechanisms may also participate in vessel refilling, such as the development of new tracheary elements to compensate for those lost by embolism (Sperry et al. 1987; Stiller & Sperry 2002), horizontal water redistribution from the secondary phloem to the secondary xylem (Salleo et al. 2004; Nardini et al. 2011; Sokołowska & Zagórska-Marek 2012), as well as growth of leaves and small branches in shade as to limit evaporative stress (Secchi & Zwieniecki 2012). Despite the mechanisms of action for embolism repair in lianas, it is a fact that most of their vascular system remains conductive for their entire lifespan (Ewers et al. 1991; Fisher & Ewers 1992).

Axial parenchyma

A higher abundance of axial parenchyma has been recorded in many lianas species in comparison to closely related trees and shrubs. Indeed, sometimes even the same species growing within different habits show these differences (e.g., *Machaerium floribundum*, Angyalossy et al. 2012; *Dalbergia frutescens*, Figs. 19.3A, 19.3C). This holds true also for the presence of non-lignified parenchyma (both axial and radial, Figs. 19.4D–G). Both features have been related to increased flexibility, storage capacity, and injury repair for liana stems (Carlquist 1985a, 1991; Fisher & Ewers 1991; Ewers et al. 1991). The lianas in the Bignoniaceae are an exception to this trend (Pace & Angyalossy 2013). They have less axial parenchyma than the self-supporting lineages in the family, with the

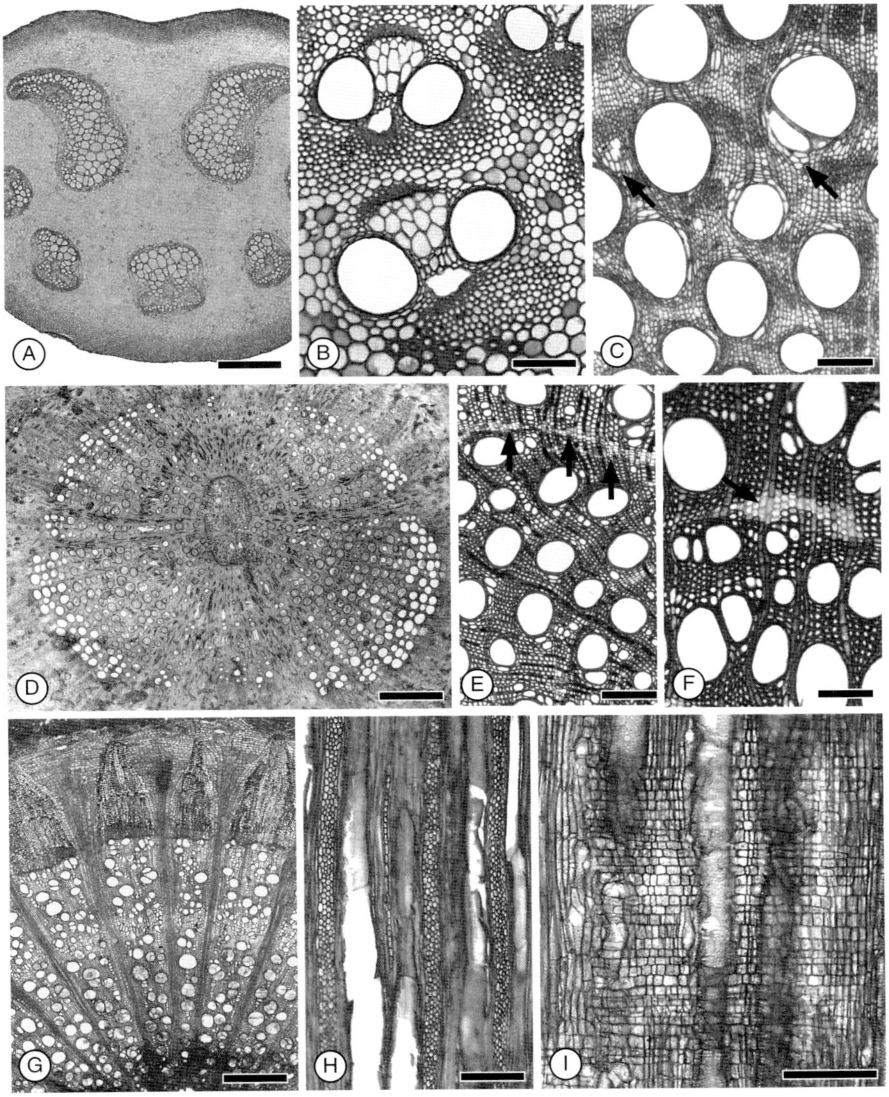

Fig. 19.4 Lianescent vascular syndrome. (A–G) Transverse section (TS). (A–C) Wide tracheary elements. (A) *Salpichlaena volubilis* (Blechnaceae, Monilophyte). (B) *Merostachys neesii* (Poaceae, Monocot). (C) *Gouania mollis* (Rhamnaceae Eudicot, narrow vessels (arrows) combined with wide vessels). (D) *Cissus sulcicaulis* (Vitaceae, Eudicot), the secondary xylem is composed of a matrix of non-lignified parenchyma where the vessels and their surrounding fibers are embedded. (E) *Heteropterys nordestina* (Malpighiaceae, Eudicot), non-lignified axial parenchyma in the secondary xylem (arrows). (F) *Dolichandra unguis-cati* (Bignoniaceae, Eudicot), rays are normally lignified, except when crossing islands of non-lignified axial parenchyma (arrow) in the secondary xylem. (G) *Cissampelos andromorpha* (Menispermaceae, Eudicot), wide and non-lignified rays. (H) *Bignonia campanulata* (Bignoniaceae, Eudicot), tall and wide rays, as seen in longitudinal tangential section (LT). (I) *Lundia corymbifera* (Bignoniaceae, Eudicot), heterogeneous rays, with procumbent, square and upright cells mixed, as seen in longitudinal radial section (LR). Note: Scale bars: A = 500 μm, B = 100 μm, C, E, H, I = 200 μm, D = 3mm, F = 50 μm G = 800 μm. *See plate section for color representation of this figure.*

flexibility and storage function probably being taken over by the ubiquitous presence of cambial variants, which typically mixes soft and stiff tissues with its phloem wedges, enhancing flexibility, and also by septate fibers, which resemble axial parenchyma in function, since they store starch (Harrar 1946; Pace et al. 2009, 2011; see Pace & Angyalossy 2013, for a discussion of this topic; see section on "furrowed xylem" in this chapter).

Non-lignified parenchyma is more common in lianas than in self-supporting species, and in some genera and/or species, such as *Cissus* (Vitaceae, Fig. 19.4D), *Heteropterys nordestina* (Malpighiaceae, Fig. 19.4E), *Stigmaphyllon* (Malpighiaceae), and *Dolichandra* (Bignoniaceae, Fig. 19.4F, Pace et al. 2009), non-lignified axial and radial parenchyma occupies spaces in which fibers, vessels, or lignified parenchyma normally are present. This replacement of one structure by another likely represents homeosis in plant anatomy (Sattler 1988), and is believed to be an important developmental mechanism in evolution (West-Eberhard 2003; Arthur 2011). Non-lignified parenchyma plays an important role in the development of the vascular system since non-lignified cells maintain the ability to divide, altering and transforming the secondary xylem after its formation. In *Dolichandra* (Bignoniaceae), the proliferation of non-lignified parenchyma is so intense that a multiple-dissected xylem is formed (Pace et al. 2009). Lianas of this genus were described as the most mechanically flexible in the entire group of 21 genera and more than 300 species (Bignonieae; Gentry 1980). Given this high division capacity, non-lignified parenchyma also plays an important role in injury repair (Carlquist 1985a; Dobbins & Fisher 1986; Fisher & Ewers 1991).

Wide and tall rays

Although some lianas, such as the Combretaceae, have generally narrow, uniseriate to biseriate rays, very wide and tall rays are more common in lianas (Figs. 19.4G, 19.4H), and they represent another feature correlated with the lianescent vascular syndrome. Anatomically, these rays are often five or more cells in width and can reach more than one millimeter in height in tangential section (Fig. 19.4H). In the families in which their composition has been described, they were also heterocellular (Klaassen 1999 for Sapindaceae; Tamaio 2006 for Menispermaceae; Rajput et al. 2012b for Leguminosae; Pace & Angyalossy 2013 for Bignoniaceae), marked by the presence of procumbent, square, and upright cells mixed throughout the rays (Fig. 19.4I). The presence of wide and tall rays increases the flexibility of the stem of lianas, allowing the twining around supports, and also avoiding damage to the vascular system when lianas' hosts fall or are dislodged (Schenck 1893; Carlquist 1985a, 1991).

In lianas of several lineages, the rays are remarkably wide and tall, such as in Dilleniaceae, Menispermaceae, and the entire order Piperales (Figs. 19.2C, 19.4G), and they originate from the interfascicular cambium that exclusively possesses ray initials (Esau 1967, 1977; Mennega 1982; Isnard et al. 2012; and personal observations). In the Piperales, the presence of wide and tall rays has even been suggested as acting as a preaptation (a process in which a feature becomes adaptive even though not primarily selected for that function, *sensu* Gould & Vrba, 1982) in the multiple evolution of the lianescent habit in this order (Isnard et al. 2012).

Conducting ray cells

Perforated ray cells (Figs. 19.5A–C) and sieve ray cells are conducting elements originated by ray initials. These are usually found connecting vessels or sieve tubes that are present at opposite sides of the rays (Fig. 19.5B). More information is available about perforated ray cells than sieve ray cells, due to the difficulties in working with phloem. Perforated ray cells occur frequently in lianas and are common even in families in which closely related self-supporting species lack them completely, such as in the Leguminosae and Bignoniaceae (Angyalossy et al. 2009, 2012; Pace & Angyalossy 2013). Specifically, in Bignoniaceae the lianescent habit evolved many times, accompanied each time by the evolution of perforated ray cells (Pace & Angyalossy 2013; Figs. 19.5D, 19.5E). Perforated ray cells are usually found connecting narrow and wide vessels, constituting by-passes when the wide vessels are embolized (Angyalossy et al. 2009). Future studies should address the possible role of perforated ray cells in vessel refilling, a mechanism thus far largely neglected.

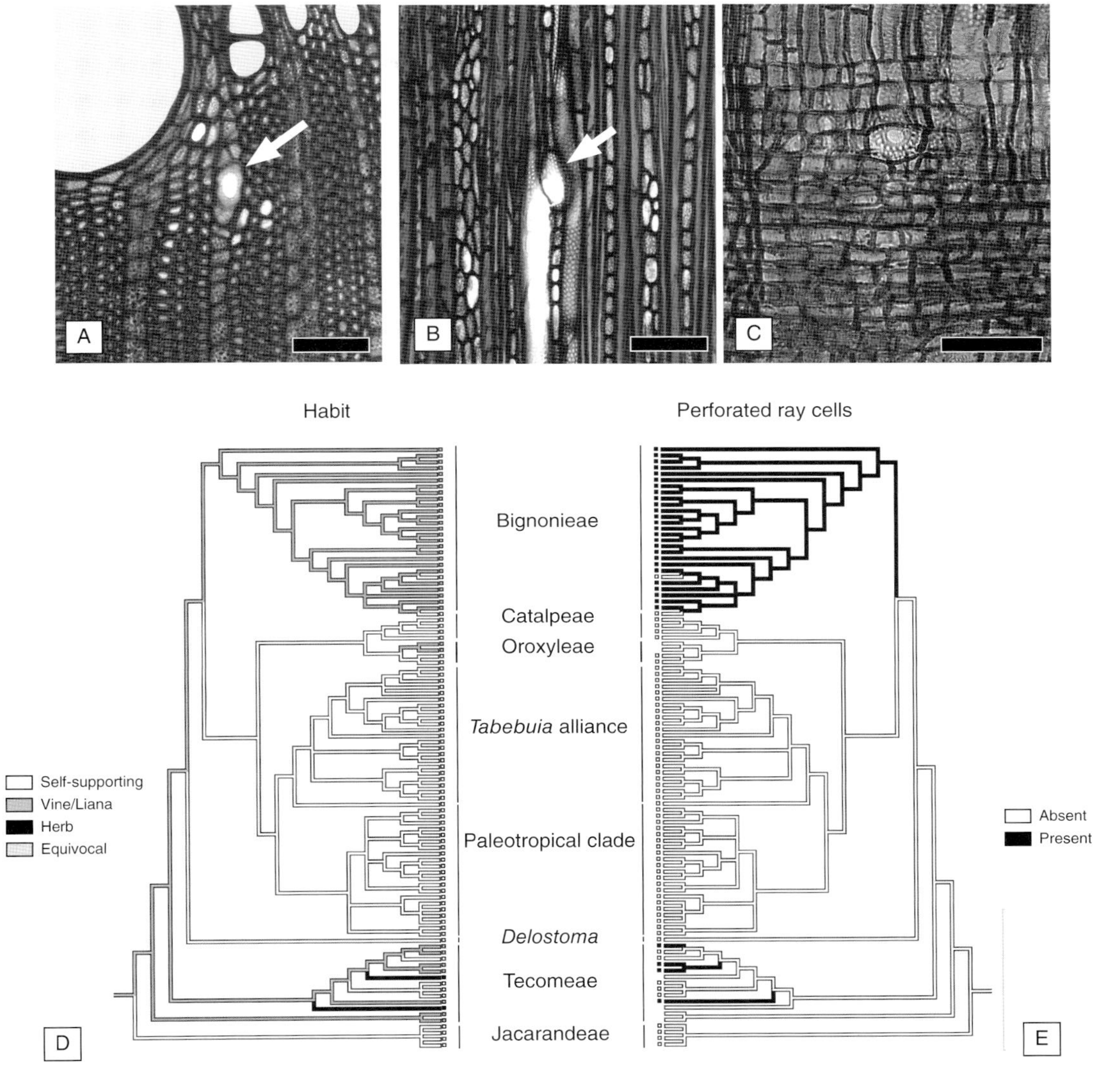

Fig. 19.5 Perforated ray cells in Bignoniaceae. (A) *Callichlamys latifolia*, perforated ray cell (arrow), in transverse section (TS). (B) *Stizophyllum riparium*, perforated ray cell (arrow) connecting two adjacent vessels, in longitudinal tangential section (LT). (C) *Campsis radicans*, perforated ray cell, in longitudinal radial section (LR). (D) Ancestral character state reconstruction of habits of Bignoniaceae. (E) Ancestral character state reconstruction of perforated ray cells in the wood of the Bignoniaceae. Note correlation between lianescent habit and presence of perforated ray cells. Note: Scale bars: A, C = 100 μm, B = 80 μm. (Source: Pace & Angyalossy 2013. Reproduced with permission of the University of Chicago Press.)

Phloem

In contrast to xylem, not much information is available on the possible correlations between the scandent habit and phloem anatomical modifications. Scientific literature suggests that lianas usually have very wide sieve tubes (Carlquist 1975, 1991; Pace et al. 2011), sometimes as wide as the small vessels found in the xylem of the same plant (Figs. 19.6A, 19.6B). Wide sieve tubes ensure higher photosynthate transport across the plant. Examples of wide sieve tubes are seen in both monocot and eudicot scandents. In *Calamus* palms (Fig. 19.6B) and in Bignoniaceae lianas, sieve tubes as wide as 70 μm can be found (Pace et al. 2011). In addition, these wide sieve tubes may bear more than one companion cell per sieve tube element, as seen in transverse section, possibly a response to the demands made by such a wide cell (Fig. 19.6C). Furthermore, in Bignoniaceae, where the lianas have both a regular and a variant phloem occurring side-by-side, it is notable that the sieve tubes of the phloem wedges are remarkably wider than those of the regular phloem, occupying a higher total area in that tissue, and possibly indicating that the variant phloem in these plants may have specialized in conduction, while the regular phloem has taken over the storage function (Pace et al. 2011), something that has been recently suggested also for plants with interxylary phloem (Carlquist 2013).

Cambial variants

Cambial variants (Carlquist 2001) are an alternative form of secondary growth that deviates from the most common secondary growth pattern found in vascular plants. Most vascular plants have a single vascular cambium, with fusiform and ray initials, which produce more secondary xylem to the inside than secondary phloem to the outside, in equal proportions around the organ circumference. Cambial variants are present in both self-supporting plants and lianas, though they are much more common in the latter (Putz & Mooney 1991; Angyalossy et al. 2012). Although not all lianas have cambial variants, in a quantitative report on the frequency of lianas

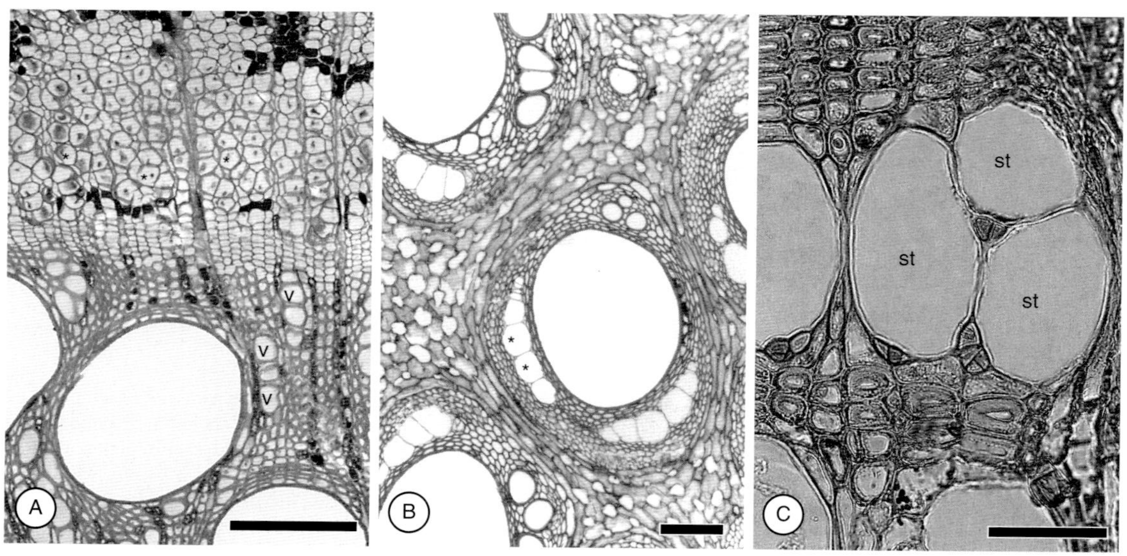

Fig. 19.6 The phloem of lianas (wide sieve tubes), as seen in transverse section (TS). (A) *Serjania lethalis* (Sapindaceae, Eudicot), sieve tubes (asterisks) as wide as the narrow vessels (v). (B) *Calamus trachycoleus* (Arecaceae, Monocot), wide sieve tubes (asterisks) in vascular bundles of a climbing palm. (C) *Tynanthus cognatus* (Bignoniaceae, Eudicot), the wide sieve tubes of Bignoniaceae lianas may have more than one companion cell. Note: Scale bars: A = 200 μm, B = 100 μm, C = 50 μm.

with variant secondary growth in the rain forests of Gabon (Africa), 56% of all lianas did have some sort of cambial variant (Caballé 1986). Cambial variants may represent another key feature in the evolution of lianas. Since most cambial variants differ in development and diversity, they are treated separately and in detail below.

DIVERSITY AND EVOLUTION OF THE CAMBIAL VARIANTS

Similar to the lianescent habit, cambial variants also evolved many times in the evolutionary trajectory of vascular plants (Fig. 19.7), being present in fossil pteridosperms (*Medullosa*, Dunn et al. 2003), gymnosperms (*Gnetum*, Fisher & Ewers 1995, Carlquist 1996; Fig. 19.2B), magnoliids (Aristolochiaceae, Carlquist 1991; Wagner et al. 2012; Fig. 19.2C), and eudicots (Apocynaceae, Bignoniaceae, Leguminosae, Malpighiaceae, Menispermaceae, Sapindaceae, among others; Schenck 1893; Carlquist 1991; Spicer & Groover 2010; Angyalossy et al. 2012; Fig. 19.2F). Cambial variants have been studied since the nineteenth century (Schenck 1893, Radlkofer 1875) and are especially notable for generating new architectures in organs in which they occur, forming ladder-like stems (*Phanera splendens*, Leguminosae), cables (tribe Paullinieae, Sapindaceae), and cross-vines (Bignonieae, Bignoniaceae), among other architectural forms. Cambial variants are commonly taxon-specific, such as the four, to multiples of four, equidistant phloem wedges of Bignonieae (Bignoniaceae; Lohmann 2006) or the successive cambia with wide rays of Menispermaceae (Mennega 1982; Carlquist 1996; Ortiz et al. 2007). This allows the identification of many individual plants to family or genus, even in the absence of leaves and flowers (Caballé 1993). Cambial variants have been suggested as playing a key role in providing higher flexibility to the stem of climbers (Putz & Holbrook 1991; Rowe et al. 2004), to actively participate in injury repair (Dobbins & Fisher 1986; Fisher & Ewers 1991), to store water and carbohydrates (Mooney & Gartner 1991; Carlquist 2001), to influence xylem development (Lima et al. 2010), and to provide a more efficient system for the conduction of photosynthates (Pace et al. 2011). Table 19.1 summarizes all cambial variants and the orders/families in which they occur.

Cambial variants are extraordinarily diverse, but, overall, they can be divided into two major groups: (1) those derived from a single cambium; and (2) those derived from multiple cambia (Carlquist 1991, 2001; Angyalossy et al. 2012). Within each category, new subgroups can be delimited. These categories are detailed below.

Cambial variants derived from a single cambium

Five different types represent this category. In the first type there is a cambium with regular activity, producing more xylem to the inside than phloem to the outside, but in different proportions around the stem's girth. This generates a stem with an irregular conformation, found, for example, in some Apocynaceae, Leguminosae, Malpighiaceae and Rubiaceae (Fig. 19.8A). In the second type there is a cambium which alternates its activity, sometimes producing xylem to the inside and phloem to the outside, and sometimes producing phloem in both directions (outside and inside), generating interxylary phloem, found, for instance, in some Acanthaceae, Combretaceae, and Malpighiaceae (Fig. 19.8B). In the third type there is a cambium that produces less xylem than phloem at certain portions of the stem, creating small phloem arcs that later become included with the coalescence of the cambium around the arcs. This generates interxylary phloem in which the included phloem strands possess a portion of cambium within them, as in some Loganiaceae (Fig. 19.8C). In the fourth type, the xylem is furrowed by the presence of phloem arcs/wedges, derived from portions of the cambium that decrease the production of xylem, but increase the production of phloem, found, for instance, in most lianas of Bignoniaceae (Fig. 19.8D) and in some Celastraceae, Icacinaceae, and Malpighiaceae. Finally, in the fifth type, axial elements of the xylem and phloem are present in segments alternating with very wide xylem and phloem rays (in wood anatomy also called xylem in plates by Pfeiffer, 1926, and others subsequently). In some groups these axial elements are derived exclusively from the fascicular cambium that produces only fusiform derivatives in the younger stems, while the interfascicular cambium produces only rays (Figs. 19.2C, 19.4G, 19.8E). Sometimes this variant can be combined with other ones, such as in the successive cambia of Menispermaceae (Fig. 19.8F).

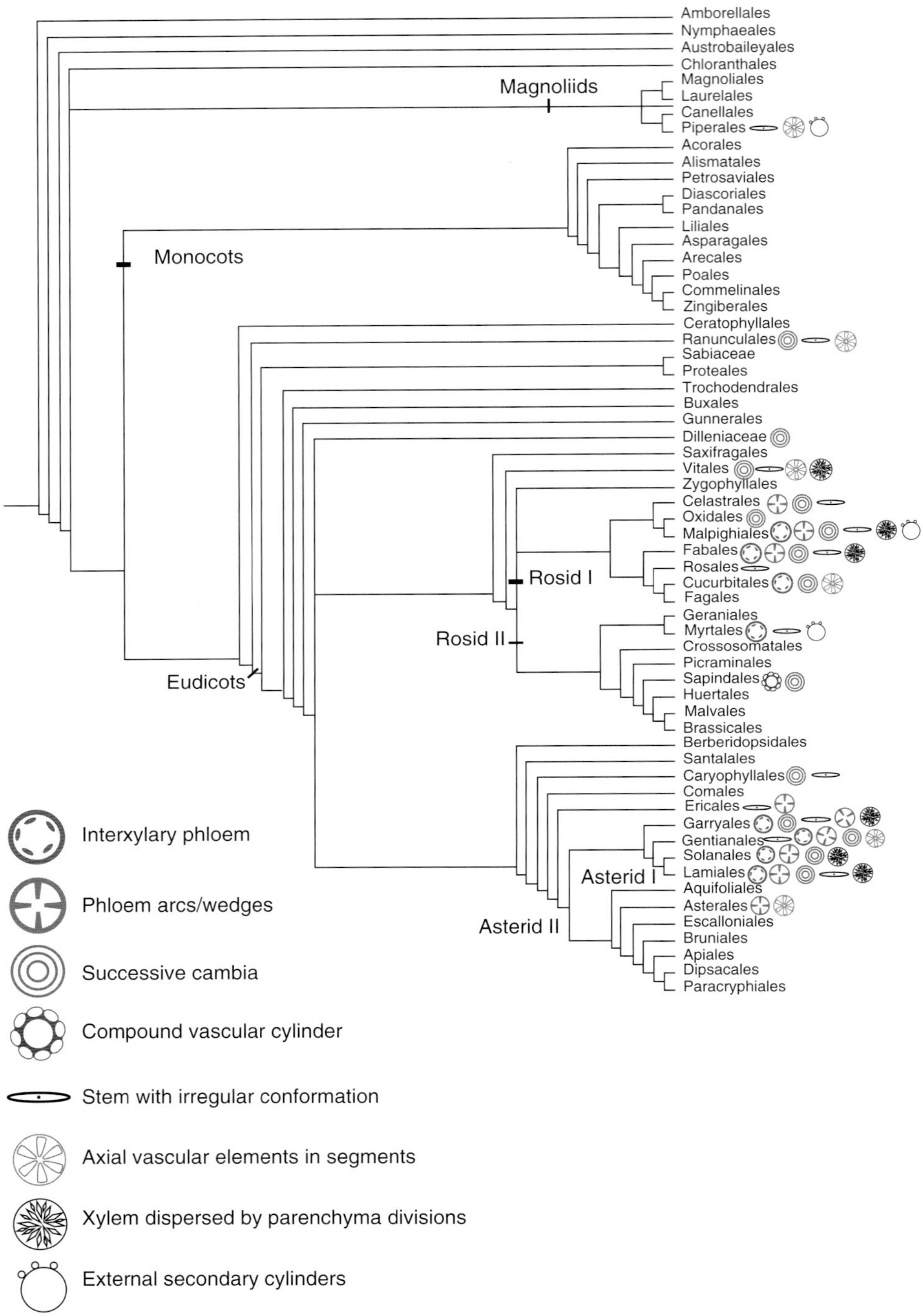

Fig. 19.7 Phylogeny of angiosperms (after Stevens 2001 onwards, APweb) illustrating distribution of cambial variants in the phylogeny. Note the higher concentration of cambial variants in the Eudicots, especially in the Rosids and the Asterids (inspired by Spicer & Groover 2010). For more detailed references on the distribution of cambial variants by family, see Table 19.1. (Source: Adapted from Angyalossy et al. 2012. Reproduced with permission of Taylor & Francis Ltd.)

Table 19.1 List of angiosperm orders and families of lianas that possess cambial variants, with their type specified and source information.

Order	Family	Cambial variant type	References	Additional references
Asterales	Asteraceae	Axial vascular elements in segments*. Phloem arc/wedges.	Schenck 1893, Pfeiffer 1926, Carlquist 1991	Carlquist 1992a, 1966
Caryophyllales	Amaranthaceae	Successive cambia.	Schenck 1893, Pfeiffer 1926, Carlquist 1991	Rajput & Rao 1999, 2000, Carlquist 2003a
Caryophyllales	Dionchophyllaceae	Successive cambia.	Carlquist 1991	–
Caryophyllales	Nyctaginaceae	Successive cambia.	Schenck 1893, Pfeiffer 1926, Carlquist 1991	Esau & Cheadle 1969, Carlquist 2004, Hernández-Ledesma et al. 2011
Caryophyllales	Polygonaceae	Stem with irregular conformation. Successive cambia. Intraxylary phloem.	Schenck 1893, Pfeiffer 1926, Carlquist 1991	Joshi 1936, Carlquist 2003b
Celastrales	Celastraceae *s.l.*	Phloem arc/wedges. Stem with irregular conformation. Successive cambia. Interxylary phloem.	Schenck 1893, Pfeiffer 1926, Obaton 1960, Carlquist 1991	Mennega 1997
Cucurbitales	Begoniaceae	Axial vascular elements in segments. Stem with irregular conformation	Schenck 1893	Carlquist 1985b
Cucurbitales	Cucurbitaceae	Successive cambia. Interxylary phloem. Axial vascular elements in segments	Schenck 1893, Pfeiffer 1926, Carlquist 1991	Carlquist 1992a, 1992b, Patil et al. 2011
Dilleniales	Dilleniaceae	Successive cambia. Axial vascular elements in segments.	Schenck 1893, Pfeiffer 1926, Carlquist 1991	Dickison 1967
Ericales	Marcgraviaceae	Stem with irregular conformation.	Carlquist 1991	
Fabales	Leguminosae	Successive cambia. Interxylary phloem. Stem with irregular conformation. Phloem arc/wedges. Xylem parts dispersed by parenchyma. Intraxylary phloem.	Schenck 1893, Pfeiffer 1926, Obaton 1960, Carlquist 1991	Kuo & Pate 1981, Nair et al. 1990, Nair 1993, Brandes and Barros 2008, Rajput et al. 2012b
Fabales	Polygalaceae	Successive cambia.	Schenck 1893, Pfeiffer 1926, Carlquist 1991	Caballé 1993, Rajput et al. 2012a

(*continued overleaf*)

Table 19.1 *(continued)*.

Order	Family	Cambial variant type	References	Additional references
Gentianales	Apocynaceae *s.l.*	Stem with irregular conformation. Successive cambia. Interxylary phloem. Axial vascular elements in segments. Phloem arc/wedges.	Schenck 1893, Pfeiffer 1926, Obaton 1960, Carlquist 1991	Patil & Rajput 2008
Gentianales	Gentianaceae	Interxylary phloem.	Pfeiffer 1926	Carlquist 1992a
Gentianales	Loganiaceae	Interxylary phloem.	Schenck 1893, Pfeiffer 1926, Obaton 1960, Carlquist 1991	Scott & Brebner 1889, Veenendal & Den Outer 1993, Rajput et al. 2010
Gentianales	Rubiaceae	Phloem arc/wedges. Stem with irregular conformation.	Schenck 1893, Pfeiffer 1926, Carlquist 1991	
Lamiales	Acanthaceae	Successive cambia. Interxylary phloem. Xylem parts dispersed by parenchyma.	Schenck 1893, Pfeiffer 1926, Obaton 1960, Carlquist 1991	Carlquist 1992a
Lamiales	Bignoniaceae	Phloem arc/wedges. Xylem parts dispersed by parenchyma.	Schenck 1893, Pfeiffer 1926, Dobbins 1969, 1970, 1971, 1981, Pace et al. 2009	Carlquist 1991, Gasson & Dobbins 1991, Dos Santos 1995, Ozório-Filho 2002, Pace et al. 2011.
Lamiales	Verbenaceae	Stem with irregular conformation.	Schenck 1893, Pfeiffer 1926, Carlquist 1991	–
Malpighiales	Malpighiaceae	Phloem arc/wedges. Xylem parts dispersed by parenchyma. Successive cambia. Interxylary phloem. Stem with irregular conformation.	Schenck 1893, Pfeiffer 1926, Obaton 1960, Carlquist 1991	–
Malpighiales	Passifloraceae	Successive cambia. Phloem arc/wedges.	Obaton 1960, Carlquist 1991	Hearn 2009
Malpighiales	Euphorbiaceae	External vascular cylinders.	Schenck 1893, Pfeiffer 1926	Ménard et al. 2009
Myrtales	Combretaceae	Interxylary phloem. Stem with irregular conformation. External vascular cylinders.** Intraxylary phloem.	Schenck 1893, Pfeiffer 1926, Obaton 1960, Carlquist 1991	Den Outer & Fundter 1976, Den Outer & Van Veenendaal 1995, Rajput et al. 2009, Zózimo et al. 2011
near Garryales	Icacinaceae	Successive cambia. Interxylary phloem. Stem with irregular conformation. Phloem arc/wedges. Xylem parts dispersed by parenchyma.	Pfeiffer 1926, Obaton 1960, Carlquist 1991	Bailey & Howard 1941, Lens et al. 2008, Utteridge et al. 2005

Oxalidales	Connaraceae	Successive cambia.	Obaton 1960, Carlquist 1991	
Piperales	Piperaceae	Axial vascular elements in segments*. Stem with irregular conformation. External vascular cylinders.**	Schenck 1893, Pfeiffer 1926, Obaton 1960, Carlquist 1991	Isnard et al. 2012
Piperales	Aristolochiaceae	Axial vascular elements in segments. Stem with irregular conformation.	Schenck 1893, Obaton 1960, Carlquist 1991	Isnard et al. 2012
Ranunculales	Ranunculaceae	Axial vascular elements in segments.	Schenck 1893, Pfeiffer 1926	Carlquist 1995, Isnard et al. 2003
Ranunculales	Menispermaceae	Successive cambia. Axial vascular elements in segments. Stem with irregular conformation. Intraxylary phloem.	Schenck 1893, Pfeiffer 1926, Obaton 1960, Carlquist 1991	Mennega 1982, Rajput & Rao 2003, Tamaio et al. 2009,
Rosales	Ulmaceae	Stem with irregular conformation.	Schenck 1893, Pfeiffer 1926	
Rosales	Moraceae	Stem with irregular conformation.	Schenck 1893, Pfeiffer 1926, Carlquist 1991	Koek-Noorman et al. 1984
Sapindales	Sapindaceae	Compound vascular cylinders. Divided and corded cylinder. Successive cambia.	Schenck 1893, Pfeiffer 1926, Obaton 1960, Carlquist 1991	Raldkofer 1932, Acevedo-Rodriguez 1993, Klaassen 1999, Araújo & Costa 2006, Tamaio & Somner 2010, Tamaio et al. 2011
Solanales	Convolvulaceae *s.l.*	Successive cambia. Interxylary phloem. Phloem arc/wedges. Xylem parts dispersed by parenchyma. Intraxylary phloem.	Schenck 1893, Pfeiffer 1926, Obaton 1960, Carlquist 1991	Scott 1891, Mikesell & Schroeder 1984, Rajput et al. 2008, Patil et al. 2009, Patil et al. 2009, Ceja-Romero & Perez-Olvera 2010, Terrazas et al. 2011
Vitales	Vitaceae	Axial vascular elements in segments. Stem with irregular conformation. Successive cambia. Xylem parts dispersed by parenchyma.	Schenck 1893, Pfeiffer 1926, Obaton 1960, Carlquist 1991	June Adkinson 1913, Carlquist 1985a

Notes:
*Terminology according to Tamaio et al. (2009).
**Terminology according to this chapter, grouping: (1) new formation of secondary vascular cylinders as in Euphorbiaceae, Combretaceae, Sapindaceae; and (2) secondary growth of the external primary vascular bundle as in Piperaceae.

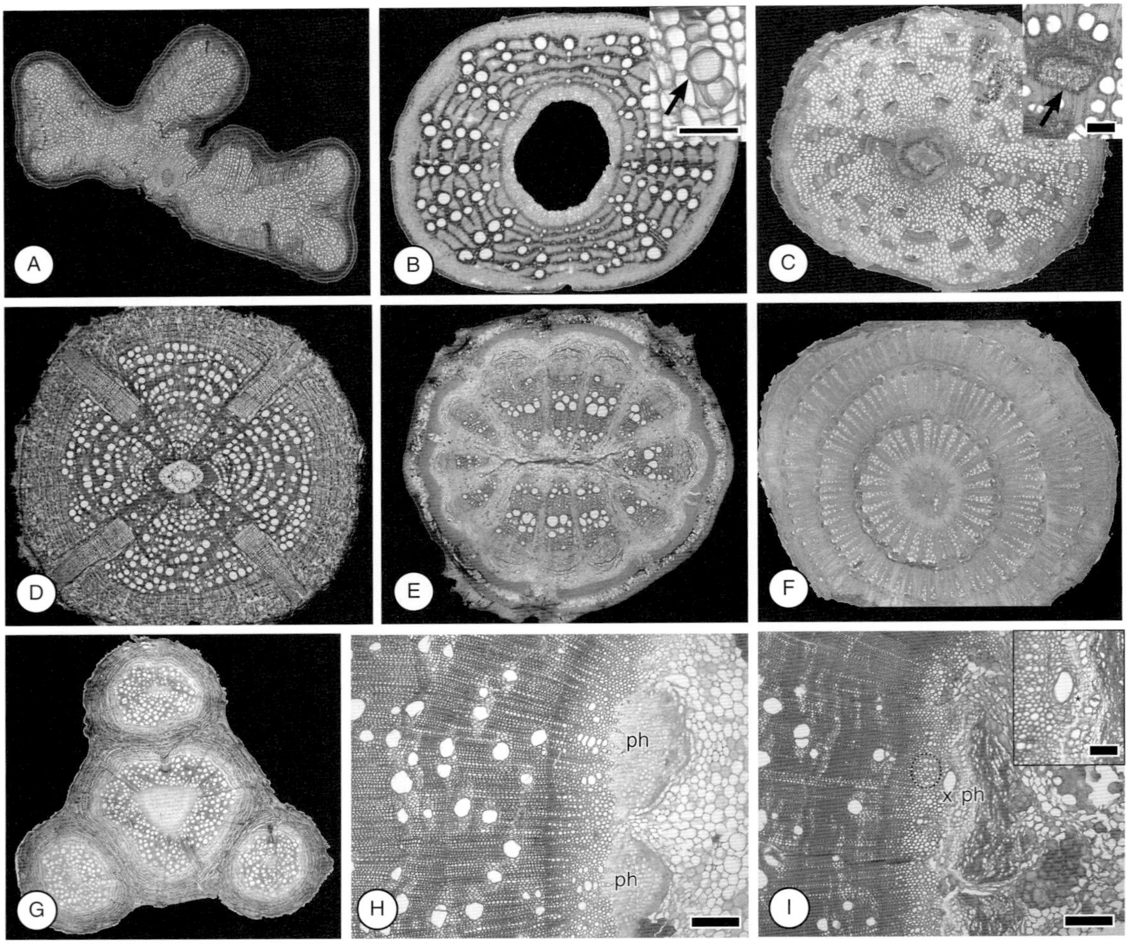

Fig. 19.8 Cambial variant types, as seen in transverse section (TS). (A) *Heteropterys nordestina* (Malpighiaceae), asymmetrical stem. (B) *Thunbergia grandiflora* (Acanthaceae), interxylary phloem without cambial inclusion (detail: sieve tube with simple sieve plate, arrow). (C) *Strychnos millepunctata* (Loganiaceae), interxylary phloem with cambial inclusion (detail: included cambium, arrow). (D) *Tanaecium pyramidatum* (Bignoniaceae), xylem furrowed by phloem wedges. (E) *Aristolochia gigantea* (Aristolochiaceae), axial vascular elements in segments alternating with wide vascular rays. (F) *Chondrodendron platyphyllum* (Menispermaceae), successive cambia. (G) *Serjania lethalis* (Sapindaceae), compound stem. (H) *Maripa glabra* (Convolvulaceae), intraxylary phloem (ph) produced by a perimedullar cambium. (I) *Telitoxicum duckei* (Menispermaceae), intraxylary phloem (ph) and xylem (x) produced by a perimedullar cambium shown in the detail (asterisk). Note primary xylem before the neoformed intraxylary xylem (dotted circle). Note: Scale bars: B = detail 50 μm, C = detail = 100 μm, H-I = 300 μm, I = detail = 100 μm. *See plate section for color representation of this figure.*

Cambial variants derived from multiple cambia

This category can be divided into four main types. The first type is characterized by successive cambia, in which new cambia arise successively by cell divisions of external portions of the secondary vascular system. Each new cambium produces secondary xylem to the inside and secondary phloem to the outside of the stem successively, found, for instance, in some Dilleniaceae, Gnetaceae, Menispermaceae and Leguminosae (Fig. 19.8F). The second type includes those species with a compound stem, which is formed by multiple vascular cylinders forming a single stem, creating a cable-like structure. The compound stem may originate in the primary growth, with the formation of 4–12 procambial cylinders that develop into discrete vascular bundles. Later on, a cambium is formed within each vascular bundle, each of which start secondary growth, forming 4–12 complete secondary vascular cylinders, such as in *Serjania caracasana* (Tamaio & Angyalossy 2009). In other species, the compound stem is formed later during secondary growth of the stem, with the dedifferentiation of pericycle parenchyma cells into new cambia that produce external vascular cylinders without pith (Tamaio & Somner 2010). The compound stem is found exclusively in the Sapindaceae (Fig. 19.8G). The third type consists of species exhibiting intraxylary phloem (Fig. 19.8H), and sometimes also intraxylary xylem (Fig. 19.8I), derived from a cambium formed between primary xylem and the pith. The last type is comprised of species showing external vascular cylinders that can be formed by (1) neoformation of secondary vascular cylinders, as in Euphorbiaceae (Fig. 19.9A) and Combretaceae, or (2) secondary growth of the external primary vascular bundles, as in Piperaceae (Figs. 19.9B, 19.9C).

Furthermore, a single plant can exhibit more than one type of cambial variant, a case found, for example, in the stems of Sapindaceae and Bignoniaceae, which in addition to the compound stems and furrowed xylem, respectively, can also exhibit neoformation of the vascular system in some genera. In the compound stems of lianescent Sapindaceae, new vascular cylinders are formed in older stems between the original vascular cylinders. New cambia arise by cell divisions of parenchyma cells of the nonconducting phloem of one or more of the original cylinders (Tamaio & Angyalossy 2009; Figs. 19.9D, 19.9E). In the Bignoniaceae, new vascular portions arise from new cambia, perpendicularly to the original cambium, derived from anticlinal cell divisions of the dilated phloem rays of the nonconducting phloem, resulting in new vascular portions (parallel to the rays of the original vascular system; Figs. 19.9F, 19.9H).

Another common trait of lianas is the parenchymatization of the stem associated with one type of cambial variant, as seen in the multiple-dissected wood of *Dolichandra* (see Fig. 9 in Pace et al. 2009). Parenchymatization occurs through the presence of non-lignified parenchyma in the stems, which may subdivide later on in the ontogeny of a species and create novel anatomical architectures.

DEVELOPMENT AND MULTIPLE ORIGINS OF CAMBIAL VARIANTS

Although different types of cambial variant can be considered typical of certain families (Caballé 1993), most variants are not exclusive to a single lineage, having evolved multiple times (Fig. 19.7). However, as expected in homoplasious evolution, even though some variations are included under the same category, they commonly differ in development and fine anatomy, a phenomenon explored in detail here for the lianas with furrowed xylem and successive cambia.

Furrowed xylem (phloem arcs/wedges)

Finding furrowed xylem in lianas is not unusual and although commonly related to lianas of Bignoniaceae – one of the top three most abundant liana families in the Neotropics (Gentry 1991) – it can be found in at least nine other orders/families of the eudicots (Fig. 19.7; labeled as phloem arcs/wedges), clearly the result of independent origins. Phloem wedges furrowing the xylem have evolved at least five times in the rosids (Celastraceae, Euphorbiaceae, Leguminosae, Malpighiaceae and Passifloraceae) and six times in the asterids (Apocynaceae, Asteraceae, Bignoniaceae, Convolvulaceae, Icacinaceae and Polemoniaceae). Although they appear similar, they include marked differences. First, in some, phloem wedges appear at regular intervals, e.g., Apocynaceae, Bignoniaceae, Leguminosae, and Malpighiaceae, while rather irregularly in other families. The Bignoniaceae are easily distinguished from the others by the constant number of four to multiples of four phloem wedges that

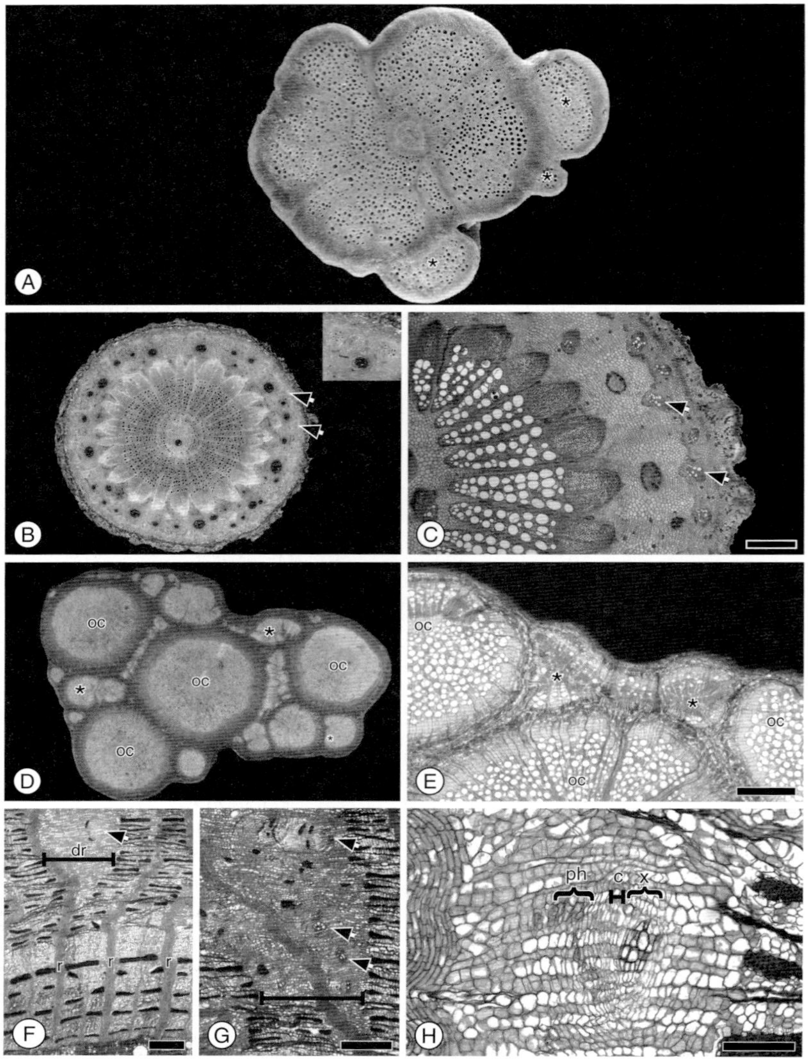

Fig. 19.9 Cambial variant types, as seen in transverse section (TS). (A–C) External vascular cylinders. (A) *Dalechampia ficifolia* (Euphorbiaceae), lateral vascular cylinders (asterisks) derived from neoformations. (B, C) *Manekia obtusa* (Piperaceae), secondary growth of external primary vascular bundle (detail: two external vascular cylinders). (D–H) Neoformations of vascular cylinders associated with another type of cambial variant. (D, E) Compound stem of *Serjania laruotteana* (Sapindaceae) with neoformed vascular cylinders (asterisks) between four original cylinders (oc). (F–H) *Amphilophium crucigerum* (Bignoniaceae), new vascular portions (arrows) derived from new cambia, perpendicular to original cambium, at dilated portion (dr) of phloem rays (r) in the nonconducting phloem. Note in H the detail of one new vascular portion with its phloem (ph) and xylem (x) derived from a neoformed cambium (c). Note: Scale bars: C = 1 mm, E = 250 μm, F–G = 500 μm, H = 150 μm. *See plate section for color representation of this figure.*

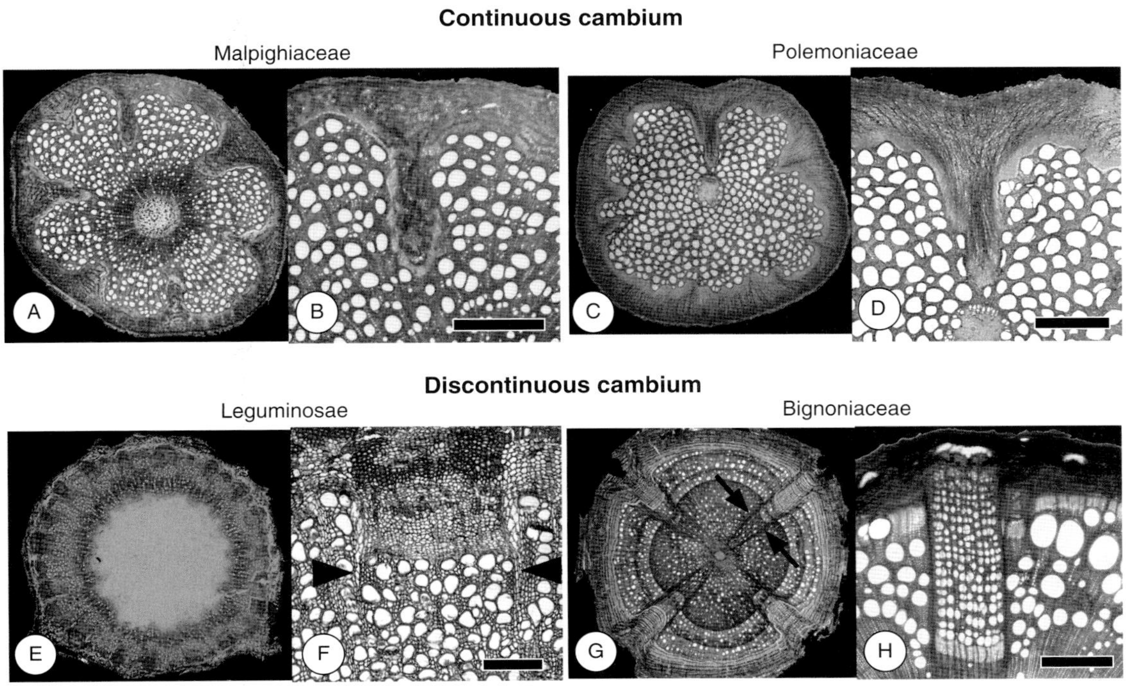

Fig. 19.10 Four examples of families with furrowed xylem (phloem arcs/wedges), as seen in transverse section (TS). (A–D) Continuous cambium. (A, B) *Mascagnia sepium* (Malpighiaceae). (C, D) *Cobaea scandens* (Polemoniaceae). (E–H) Discontinuous cambium (cambia included). (E–F) *Mimosa velloziana* (Leguminosae), arrows indicate limiting rays. (G) *Fridericia chica* (Bignoniaceae), arrows indicate limiting rays. (H) *Tynanthus cognatus* (Bignoniaceae), detail of a phloem wedge. Note: Scale bars: B, D = 500 μm; F = 350 μm; H = 450 μm. *See plate section for color representation of this figure.*

are regularly spaced (Schenck 1893; Dobbins 1971; Pace et al. 2009; Figs. 19.8D, 19.10G, 19.10H). This characteristic has been related to the decussate leaves typical of the Bignoniaceae (Dobbins 1970). In fact, the wedges are always found in alternation with the leaves (Schenck 1893; Pace et al. 2011) and traveling in a spiral throughout the stem.

From a broad perspective, the furrowed xylem (phloem wedges) cambial variant can be divided into two major groups, depending on the continuity of the cambium, as in Apocynaceae, Celastraceae, Convolvulaceae, Euphorbiaceae, Malpighiaceae, Passifloraceae, and Polemoniaceae (Figs. 19.10A–D), or its discontinuity, as in Bignoniaceae, Icacinaceae, and Leguminosae (Figs. 19.10E–H).

The development of phloem wedges involves at least two steps. First, in the onset of secondary growth a continuous cambium is installed with regular activity throughout its girth. Second, sooner or later depending on the species, grooves appear in some regions of the cambium. These correspond to zones that reduce the production of xylem and increase the production of phloem. Given this differential production of secondary tissues, phloem arcs/wedges are formed (Figs. 19.10B, 19.10D). In plants with a continuous cambium, this is the final stage of development.

In plants with a discontinuous cambium, on the other hand, there is a third step. In the places where the phloem wedges had first appeared, a cambium disruption with inclusion occurs. In Bignoniaceae and, most likely, other families where phloem wedges are accompanied by cambial disruption (Icacinaceae, Leguminosae), this phenomenon is correlated with the loss of anticlinal division capacity by the variant cambia (Pace et al. 2009). In the Bignoniaceae, the four equidistant portions of the cambium that change their activity lack anticlinal divisions, while the regular portions of the cambium go on increasing in girth by anticlinal divisions, leading to cambial disruption while growth proceeds. At the margins of the phloem wedges, wide and tall rays start to form.

The presence of wide and tall rays bordering the phloem wedges is typical of the furrowed xylem (phloem wedges) cambial variant, and it is especially evident in those families displaying cambial inclusion, although not exclusively. These wide rays were recognized by Schenck in 1893, who named them "limiting rays," given their marginal position (Schenck 1893; Figs. 19.10F, 19.10G). These wide rays were suggested as acting as a physiological barrier in Bignoniaceae, canalizing hormones originating from the leaves and promoting the development of the phloem wedges in this family (Dobbins 1990, 2005). However, this proposition does not seem to hold since the phloem wedges start their formation in Bignoniaceae before the appearance of limiting rays (Pace et al. 2009). It is therefore more likely that the limiting rays in the furrowed xylem cambial variant simply represent an architectural by-product (a spandrel, *sensu* Gould & Lewontin 1979). The limiting rays are present exactly at sharp notches in the limits of the phloem wedges, a zone of mechanically fragile tissue that could easily facilitate fracture propagation. Rays likely enlarge as a way to adjust for the differential growth rates between the regular and variant cambia and to avoid tissue fracture, similar to the dilating rays of the secondary phloem of various vascular plants that dilate to adjust to the increase in the circumference of the axis (Evert 2006). Ray divisions also allows the new cells produced by the included cambium to move outward while they differentiate, since they are within the phloem wedges and surrounded by stiff secondary xylem (Obaton 1960). Therefore, limiting rays in this cambial variant have probably not appeared as a result of adaptive selection, but as a by-product of architectural constraint in the dynamic of the tissues in the secondary body of these plants.

Although the variant cambium reduces the production of xylem, it is important to note that it never really ceases or becomes unidirectional, and both xylem and phloem continue to be produced throughout the plant's lifespan (Pace et al. 2011). In some Malpighiaceae, the presence of tilted axial xylem cells is noticeable, most likely because of the position of the cambium in the margins of the phloem wedges. The same pattern is noticeable in *Perianthomega*, the only genus of the Bignonieae tribe (Bignoniaceae) that possesses phloem arcs with a continuous cambium (Pace et al. 2009).

Successive cambia

Successive cambia are one of the most frequent types of cambial variant, having evolved independently at least eighteen times, in both gymnosperms (e.g., *Gnetum*, Fig. 19.11A) and angiosperms (Fig. 19.7, Table 19.1). Successive cambia refers to the appearance of multiple concentric cambia in the stem. It is especially frequent in the Caryophyllales (Amaranthaceae, Dionchophyllaceae, Nyctaginaceae, Fig. 19.11D), but it is also present in some Acanthaceae, Apocynaceae, Dilleniaceae (*Doliocarpus*, Fig. 19.11B), Celastraceae, Connaraceae, Convolvulaceae, Cucurbitaceae, Icacinaceae, Leguminosae (Fig. 19.11C), Malpighiaceae (some *Banisteriopsis*), Menispermaceae (Fig. 19.11E), Passifloraceae, Polygalaceae, Sapindaceae, and Vitaceae.

Similar to most cambial variants, at the onset of secondary growth the stems have regular growth with a single cambium of equal activity throughout the girth. Subsequently, successive cambia appear. These may be derived from: (1) divisions of the pericyclic cells, as reported for Convolvulaceae, Dilleniaceae and Menispermaceae (Eames & MacDaniels 1925; Tamaio et al. 2009; Terrazas et al. 2011; Fig. 19.11E); (2) cortex cells, such as reported for some Menispermaceae (Carlquist 2007) and Leguminosae (Rajput et al. 2012b; Fig. 19.11C); or even (3) parenchyma cells of the nonconducting phloem, as in some Leguminosae (Schenck 1893; Nair & Mohan-Ram 1990; Dias-Leme 1999; personal observations).

In the most common case, as in Caryophyllales, Dilleniaceae, and Menispermaceae, pericyclic or cortical cells divide and develop into a large meristematic zone. New successive cambia will periodically develop in the middle of this meristematic zone, with secondary phloem originating to the outside and secondary xylem to the inside. Part of the meristematic zone differentiates into parenchymatic tissue and/or sclerenchymatic tissue, termed conjunctive tissue (Fig. 19.11D). The outermost part of this meristematic tissue/cells, near the cortex or periderm, will form new successive cambia sequentially.

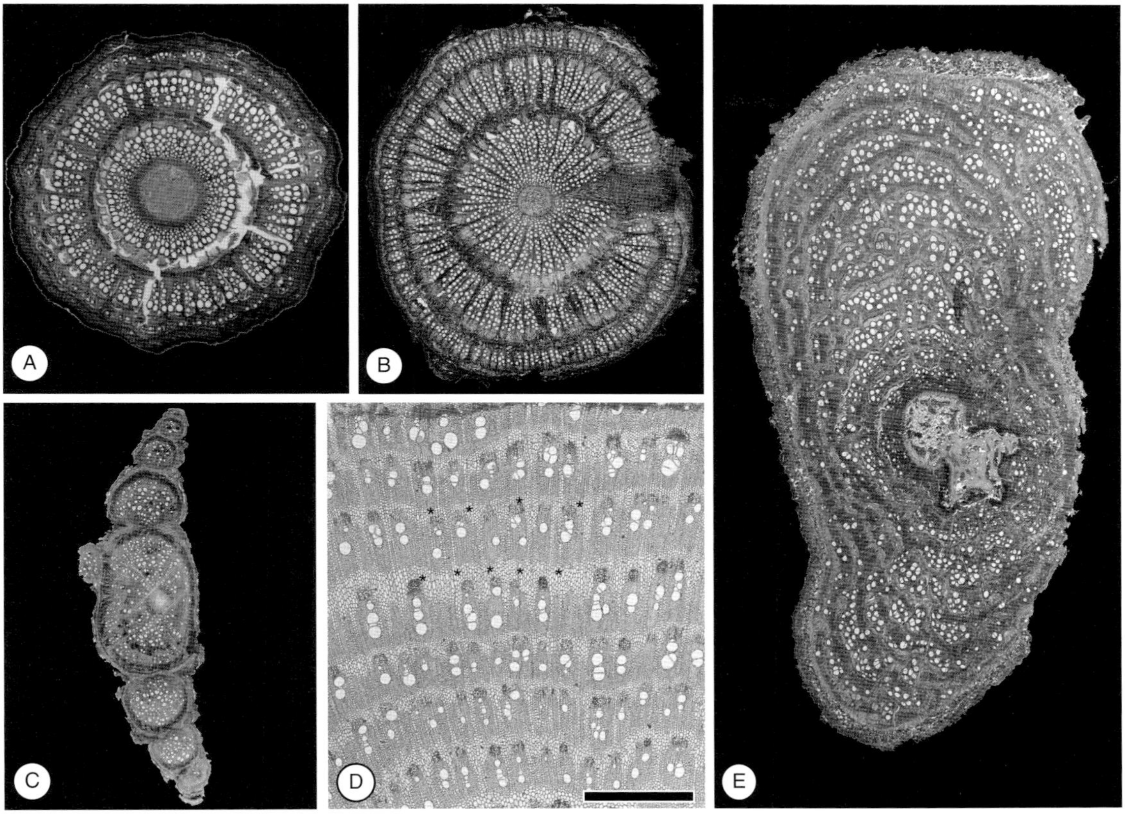

Fig. 19.11 Successive cambia of different types, seen in transverse section (TS). (A, B) Successive cambia resulting in a round stem. (C, E) Successive cambia resulting in flattened stem. (D) Conjunctive tissue (asterisks) one of the products of the successive cambia. (A) *Gnetum paniculatum* (Gnetaceae). (B) *Doliocarpus dentatus* (Dilleniaceae). (C) *Rhynchosia phaseoloides* (Leguminosae). (D) *Bouganvillea spectabilis* (Nyctaginaceae). (E) *Telitoxicum duckei* (Menispermaceae). *See plate section for color representation of this figure.*

In the legume species *Rhynchosia phaseoloides*, successive cambia arise from cells of the cortex (Schenck 1893; Rajput et al. 2012b; Fig. 19.11C). These cortical cells originate meristematic cells in radial files, which in the mid region differentiate into a vascular cambium and in the inner region a parenchymatic conjunctive tissue (Rajput et al. 2012b). The outermost part of this meristematic zone will give rise to new successive cambia. For some lianas species of the genus *Machaerium* (*M. tounateifolium*, Schenck 1893; *M. oblongifolium*, Dias-Leme 1999; personal observations), new cambia arise from the differentiation of axial parenchyma cells of the nonconductive secondary phloem. This axial phloem parenchyma does not produce a meristematic zone, but differentiates directly into a cambium. As a consequence, no conjunctive tissue is formed.

The successive cambia found in the nineteen families (Table 19.1) are not anatomically identical. For instance, in some, the successive cambia have more pronounced activity at certain portions than others, generating asymmetrical stems, typical of some Connaraceae, Leguminosae (Fig. 19.11C), and Menispermaceae (Fig. 19.11E). Other families exhibit successive cambia with more or less equal amounts of tissue through the stem girth, resulting in stems with a round external appearance, as in some Asteraceae, Dilleniaceae (Fig. 19.11B), Gnetaceae (Fig. 19.11A), Menispermaceae, and Nyctaginaceae (Fig. 19.11D).

CAMBIAL VARIANTS ENHANCE ANATOMICAL COMPLEXITY

Cambial variants result from a myriad of completely different developmental pathways. However, most of them seem to be initially derived from a stem with regular secondary growth. A remarkable exception is the compound stem of some Sapindaceae, which exhibit a different vascular structure initiated at primary growth (Araújo & Costa 2006; Tamaio & Angyalossy 2009).

More commonly, though, lianas seem to have evolved into gradually more complex forms, as is the case for species of the Bignonieae (Bignoniaceae). In these lianas, different anatomical architectures have evolved, but they are all linked by a sequence of gradual changes. The first anatomical variation to evolve in the tribe was the formation of four phloem arcs, and subsequently four phloem wedges, the most common pattern in the family (for details, see Pace et al. 2009; Fig. 19.12). These four phloem arcs/wedges are equally spaced (Figs. 19.12B–F), and between them, in the interwedges, the cambium maintains regular secondary growth. However, in three lineages within the Bignonieae (the "multiples of four clade," some *Adenocalymma* and *Lundia*), additional phloem wedges form between the first four, creating stems with 8, 16, 32, and 64 phloem wedges or more, also equally spaced (Figs. 19.12G–I). In another lineage, *Dolichandra*, sister to the "multiples of four clade," non-lignified rays and axial parenchyma are present in addition to the multiple phloem wedges (Figs. 19.12K, 19.12L, 19.13). This ray and axial parenchyma, in turn, subdivide during development, tearing the lignified cells of the secondary xylem into pieces due to parenchymatization. Both variations seem to increase the flexibility of the stem (Gentry 1980), but biomechanical tests are needed to confirm this perspective. Another lineage, *Amphilophium*, reunites six genera of earlier classifications (*Amphilophium*, *Distictella*, *Distictis*, *Glaziova*, *Haplolophium*, *Pithecoctenium*; Fischer et al. 2004; Lohmann & Taylor 2014) and develops normally four, or multiples of four, phloem wedges; however, these phloem wedges seem to disappear in adult stems. Careful anatomical analysis shows that the phloem wedges have not disappeared, but have been included in the secondary xylem (Fig. 19.12J), something detected by the interruption of the wood cells in four to multiple of four sites of the stem. At a certain stage in the development of *Amphilophium*, the regular cambium flanking the phloem wedges coalesces and includes the wedges. Apparently, the variant secondary phloem of *Amphilophium* remains conductive for a shorter period than that of other Bignonieae and this could be one reason why the regular cambium coalesces and includes the wedge in *Amphilophium*, but not in other lianas of Bignonieae. Interestingly, as seen in Fig. 19.13, the evolution of the cambial variant of Bignonieae is accompanied by an increase in anatomical complexity (more stages in their ontogenies) towards the terminal nodes of the phylogeny.

Other families in which multiple types of cambial variants occur, such as the Leguminosae, Malpighiaceae, and Sapindaceae, might also reproduce a similar evolutionary pattern by virtue of similar small-scale changes in the development of the stems. The Malpighiaceae, for instance, shows a remarkable diversity of cambial variants (Fig. 19.14), as well as lianas with regular secondary growth (Fig. 19.14A). The cambial variants include species with unequal production of xylem, generating lobed stems, that may be asymmetrical (Fig. 19.14B) or symmetrical (Fig. 19.14C), intraxylary phloem, furrowed xylem (Figs. 19.14D, 19.14E), and dissected xylem, caused by the proliferation of non-lignified parenchyma (Figs. 19.14F–H), as well as successive cambia (Fig. 19.14I). These variations may also be found in combination. How these variants have evolved within the family and whether or not they are developmentally interconnected is under evaluation (Pace & Angyalossy, in preparation).

SEASONALITY OF CAMBIAL ACTIVITY AND GROWTH RINGS IN LIANAS

There is still surprisingly little information on the seasonality of secondary growth of lianas, resulting from switches in cambial activity. In temperate lianas, such as *Celastrus scandens* (Celastraceae), *Parthenocissus inserta*, *Vitis riparia* and *Vitis vinifera* (Vitaceae), the cambium is active from one and a half to four months annually (Esau 1948; Davis & Evert 1970). Although the resumption of cambial activity every year was not well synchronized among the individuals – varying by one and a half months, from early to late spring – all species exhibited a dormant cambium before the onset of winter. Temperature has been highlighted as the most likely factor influencing dormancy and resumption of the cambium in temperate species (Kozlowski et al. 1991; Begum et al. 2007, 2008).

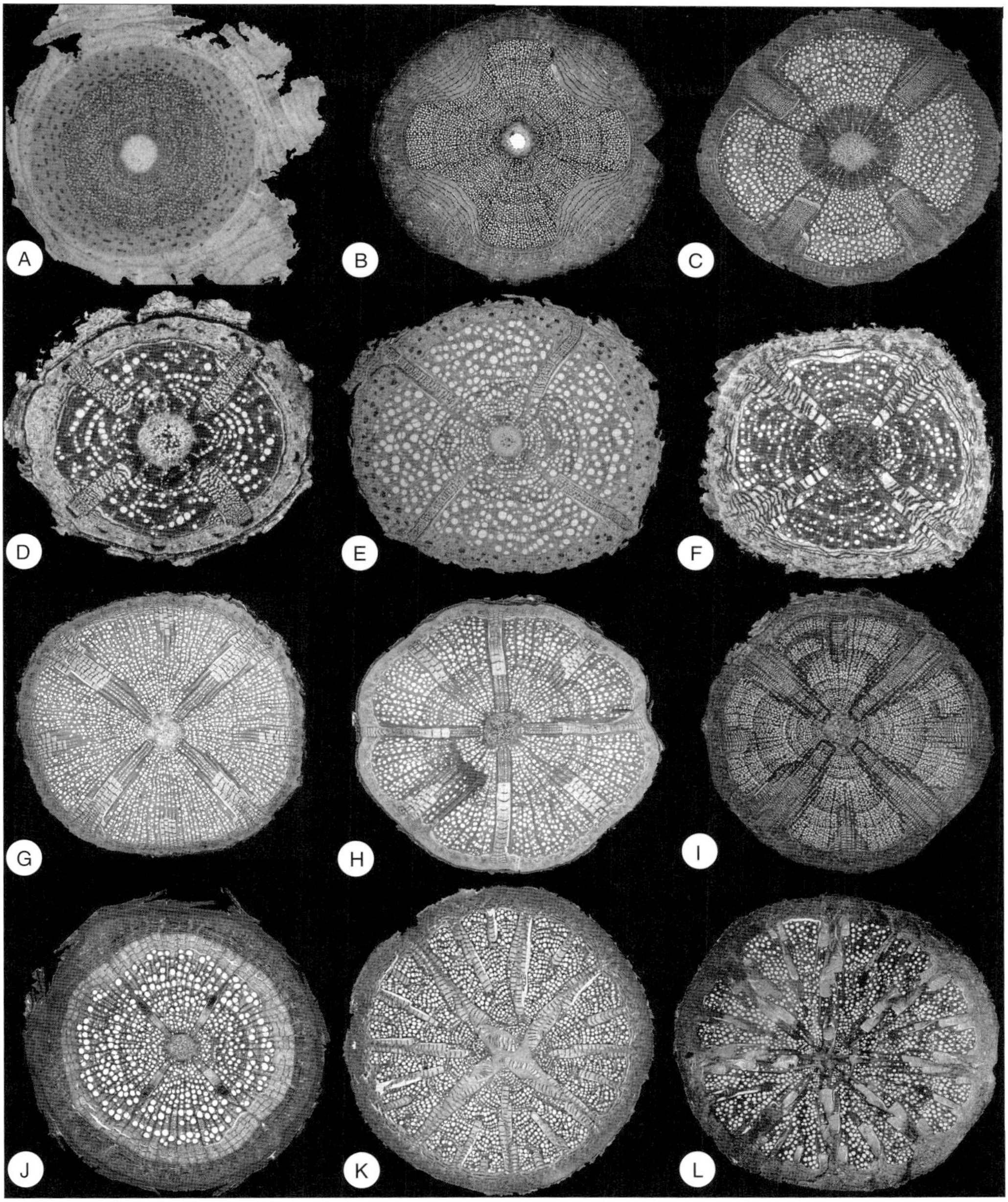

Fig. 19.12 Diversity of stem architectures in Bignoniaceae lianas, as seen in transverse section (TS). (A) Liana of tribe Tecomeae. (B–L) Lianas of tribe Bignonieae. (A) *Podranea ricasoliana*, stem with regular secondary growth. (B) *Perianthomega vellozoi*, stem with 4 arcs. (C–F) Stems with 4 phloem wedges. (C) *Fridericia samydoides*. (D) *Lundia longa*. (E) *Pleonotoma tetraquetra*. (F) *Manaosella cordifolia*. (G–I) Stems with multiple of 4 phloem wedges. (G) *Mansoa difficilis*. (H) *Pyrostegia venusta*. (I) *Adenocalymma divaricatum*. (J) *Amphilophium crucigerum*, stem with included phloem wedges. (K, L) Stems with multiple dissected phloem wedges. (K) *Dolichandra unguis-cati*. (L) *Dolichandra unguiculata*. *See plate section for color representation of this figure.*

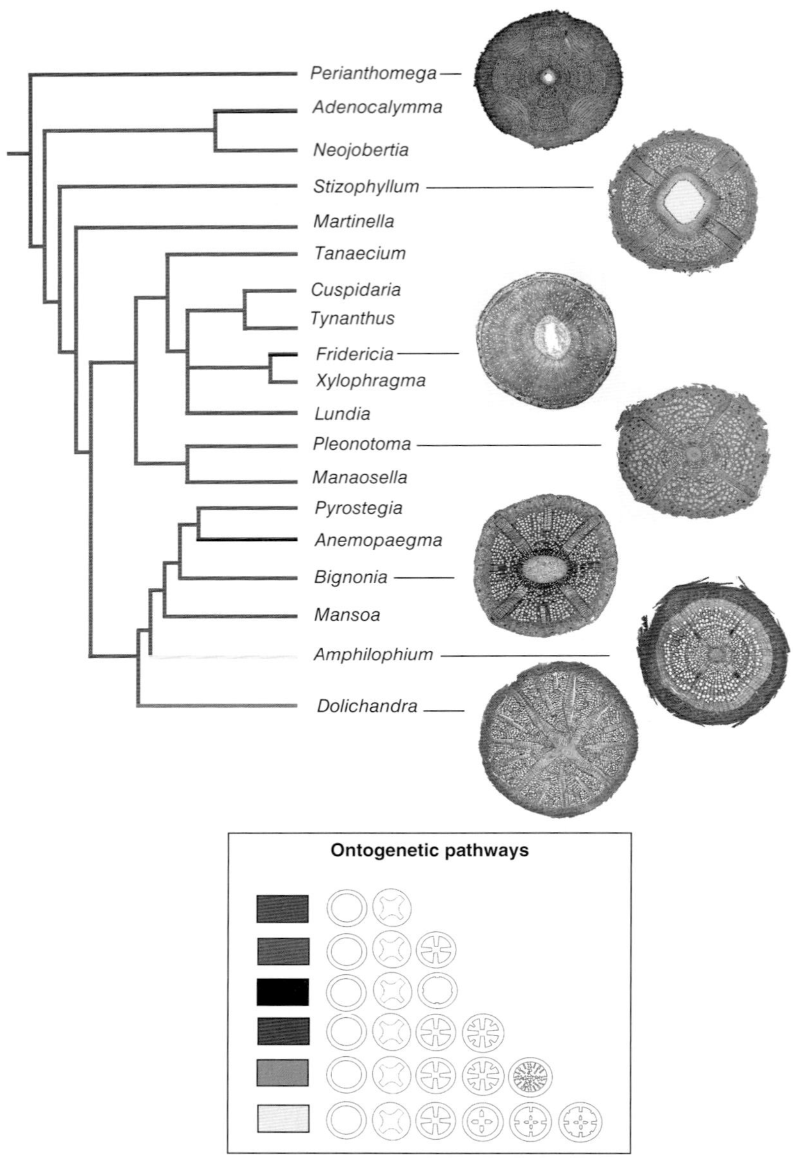

Fig. 19.13 Ancestral character state reconstruction of the ontogenetic pathways of tribe Bignonieae (Bignoniaceae). In Bignonieae the presence of four phloem arcs (blue) is ancestral in the family (synapomorphic), and different stem anatomical architectures have evolved from it, four phloem wedges (red) correspond to the majority of the tribe, multiple of four phloem wedges (green), multiple dissected (black) and included phloem wedges (yellow) have evolved in the terminal nodes of the phylogeny. *See plate section for color representation of this figure.*

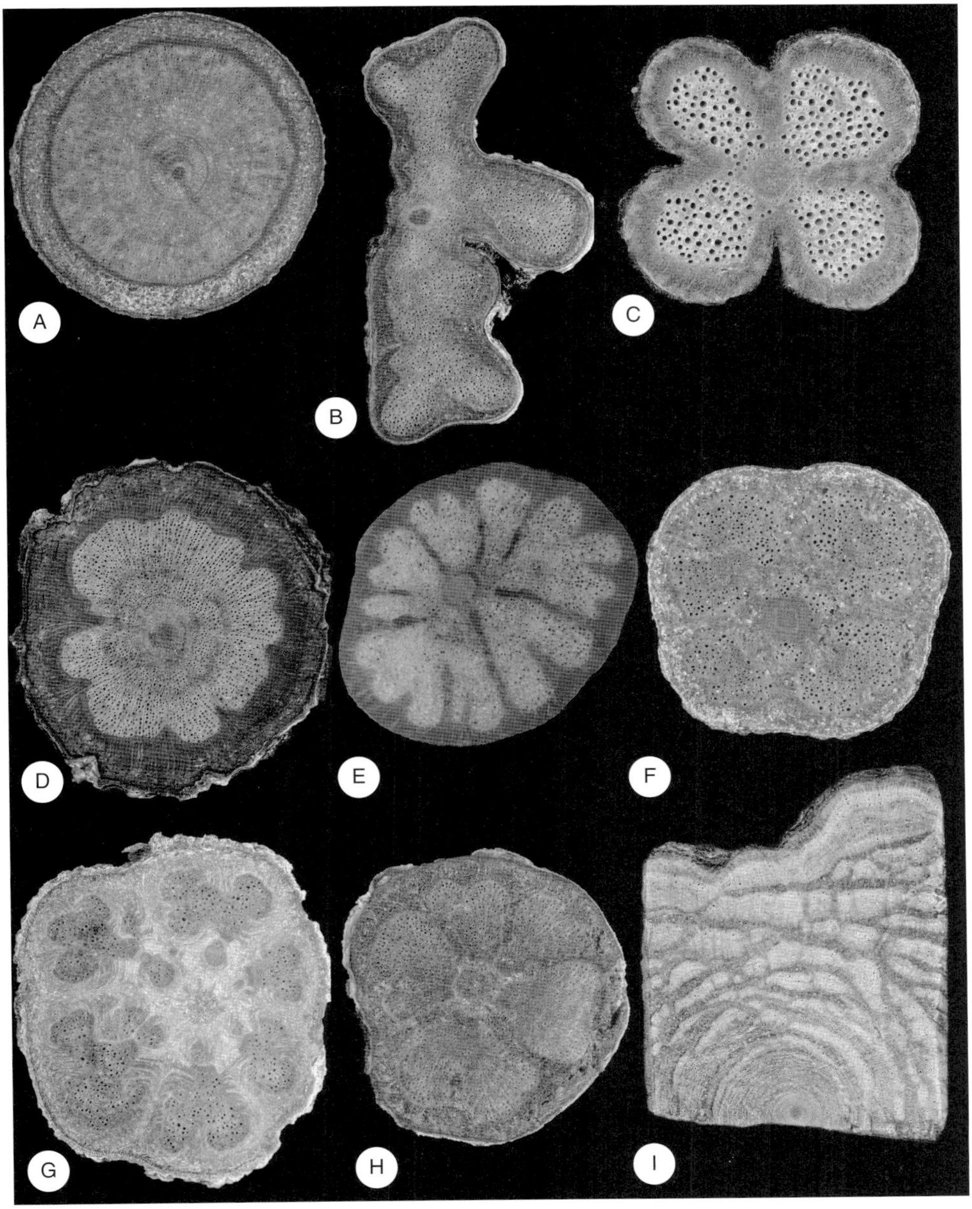

Fig. 19.14 Diversity of stem architectures in Malpighiaceae lianas, as seen in transverse section (TS). (A) *Heteropterys nitida* stem with regular secondary growth. (B) *Heteropterys nordestina* stem lobed with asymmetrical conformation. (C) *Stigmaphyllon blanchetii*, stem lobed with symmetrical conformation (D) *Heteropterys megaptera*, stem with shallow phloem wedges. (E) *Mascagnia sepium*, stem with deep phloem wedges. (F–H) Stems with dissected xylem. (F) *Mezia mariposa*. (G) *Alicia anisopetala*. (H) *Heteropterys cordifolia*. (I) *Banisteropsis nummifera*, stem with successive cambia. *See plate section for color representation of this figure.*

In contrast to temperate regions, the main factor affecting cambial activity in tropical tree species is the seasonal dry periods (Worbes 1995) or floods in floodplains (Worbes 1985; Schöngart et al. 2002). It was previously believed that growth rings of lianas were too irregular to have been produced annually (Schenck 1893); however, new analyses, including periodic sampling of stems (Léon-Gómez & Monroy-Ata 2005; Lima et al. 2010; Lima 2012), dendrochronology (Heuzé et al. 2009; Brandes et al. 2011), and cambial wounding (Brandes et al. 2011) have demonstrated that growth rings are indeed annual and positively correlated with precipitation in the tropics.

Periodic sampling analyzing the most abundant liana families of the Neotropics, including Bignoniaceae, Leguminosae, and Sapindaceae, from the Brazilian semi-deciduous seasonal forest in the city of São Paulo, shed light on the periodicity of growth in tropical lianas (Lima et al. 2010; Lima 2012). Studies of the cambial activity in four species (*Tynanthus cognatus* and *Styzophyllum riparium* of the Bignoniaceae; *Dalbergia frutescens* of the Leguminosae; and *Serjania laruotteana* of the Sapindaceae) have shown that cambial activity is strongly correlated with the rainy season for these four species (Fig. 19.15A). It is shown that the entire period of growth for most of the species and the most intense growth period of the only continuously growing species analyzed, i.e., *Serjania laruotteana*, are concentrated during the last half of the rainy season. Periodic sampling also showed different responses of cambial activity extent in the four studied lianas growing under the same climatic conditions. The different responses

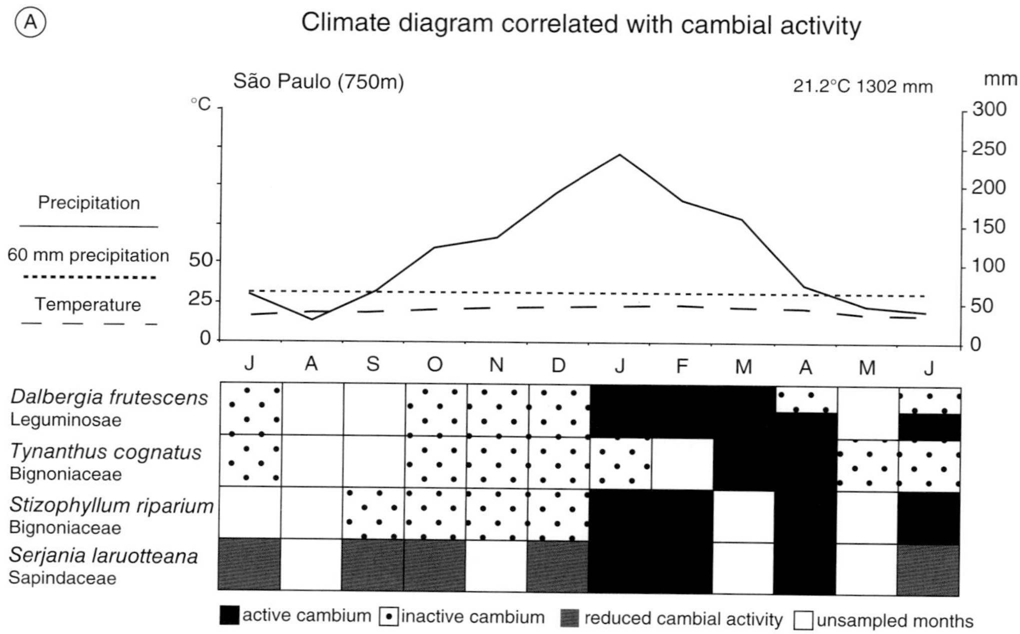

Fig. 19.15 (A) Climate diagram of São Paulo, Brazil, and the correlation with cambial activity of four species of lianas. Months in which the cambium was active are marked in black, inactive cambium is marked with dots. Gray indicates reduced activity. Black and dots together indicate periods in which the cambium was found both active and inactive in different specimens. White indicates unsampled months. (B) Diagram indicating cambial activity in the stem of the four studied liana species throughout the year. Active cambium is drawed by thick line and inactive by thin line. Note that the months depicted for each species are not the same, as the onset of cambial activity varied among them. The resumption of cambial activity in *Dalbergia frutescens* takes place in different regions of the stem in January and spread laterally until the end of the growth season in June. In *Tynanthus cognatus* and *Stizophyllum riparium* the cambial activity resumption takes place in the variant cambium inside the phloem wedges in March and January, respectively, and then spreads laterally to the regular cambium. *Serjania laruotteana* compound stem shows higher activity in the central cylinder and in portions of the peripheral cylinders of the cambium that face the central cylinder.

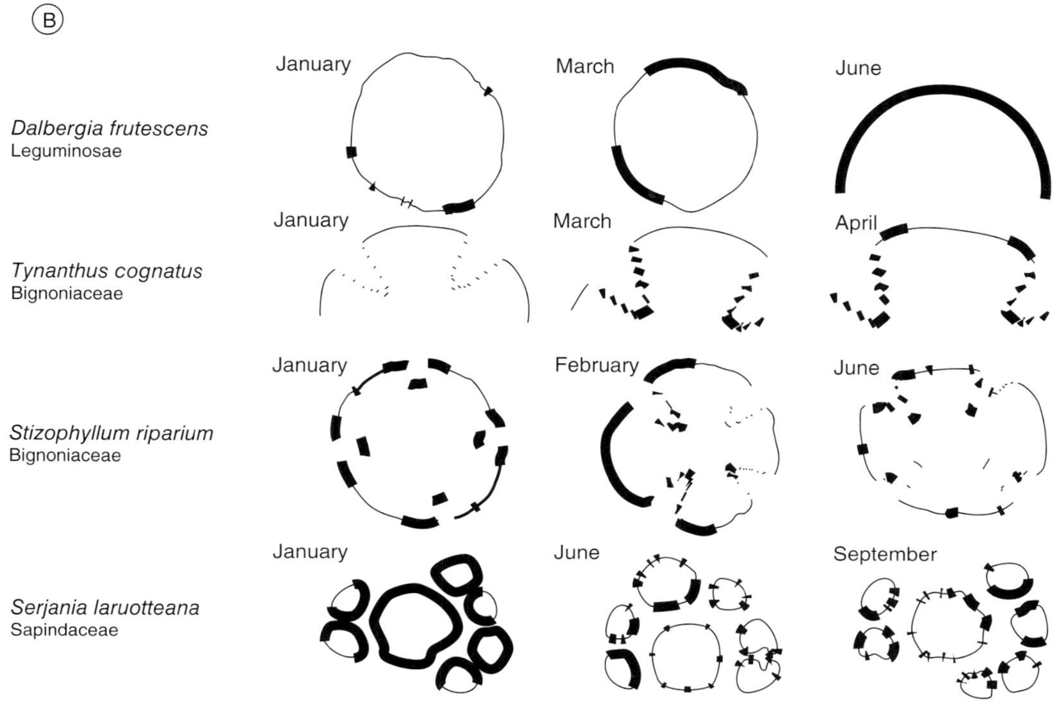

Fig. 19.15 *(continued).*

included a short cambial activity of only two months in *Tynanthus cognatus* (Lima et al. 2010), a longer period of six months in *Stizophyllum riparium* and *Dalbergia frutescens*, and continuous cambial activity in *Serjania laruotteana* (Lima 2012, Fig. 19.15A). Therefore, the different responses of cambial activities proved to be species-specific within a common environment. These idiosyncratic patterns of growth (Enquist & Leffler 2001) seen in different species may be controlled by the water relations of the plant, which is, in turn, influenced by both biotic and abiotic factors, such as the amount and seasonality of annual precipitation, the soil and wood water retention capacity, the size and depth of the root system, plant deciduousness, and seasonality of deciduousness (Borchert 1994a, 1994b).

Apart from temporal discontinuity, it is of significant interest that cambial activity is not spatially continuous throughout the stem circumference in all liana species studied. The cambial activity in *Dalbergia frutescens*, a species that lacks cambial variants, starts at different points scattered throughout the stem circumference and then spreads sideward, reaching all or most of the stem circumference by the end of the growing season (Fig. 19.15B).

The exact site where cambial activity resumes, or higher activity takes place, is not random, but is correlated with the cambial variant of the species. In *Tynanthus cognatus* and *Stizophyllum riparium* – species with furrowed xylem (phloem wedges, see Fig. 19.8D) – the resumption of cambial activity takes place first in the variant cambium located at the bottom of the phloem wedges, then spreads sideward to the regular portion of the cambium beside the phloem wedges, and then to the center of the regular portion of the cambium (Fig. 19.15B). Furthermore, in *Serjania laruotteana*, which has a compound stem and continuous cambial activity throughout the year, the higher activities are detected in the central cylinder and in portions of the cambium of the peripheral cylinders facing the central cylinder. Because of this activity, the piths are generally eccentric in the peripheral cylinders.

The seasonal cambium activity promotes a seasonal wood formation detectable by the presence of growth rings. In the studied species, annual growth rings occur in the three species that exhibit a dormant period of cambial activity (*Dalbergia frutescens* Fig. 19.16A; *Stizophyllum riparium* Fig. 19.16B; and *Tynanthus cognatus*); the only species with continuous growth during the year, *Serjania laruotteana*, lacks growth rings (Fig. 19.16C).

Several studies (Schenck 1893; Gasson & Dobbins 1991; Carlquist 1993; Brandes & Barros 2008; León 2009; Lima et al. 2010) shows that the most common earlywood growth markers found in the secondary xylem of lianas are marginal parenchyma, as found in Aristolochiaceae, Bignoniaceae (Fig. 19.16B), Combretaceae, Leguminosae (Fig. 19.16A), Rhamnaceae, Ulmaceae; and larger vessels, as found in Bignoniaceae (Fig. 19.16B), Leguminosae (Fig. 19.16A) and Verbenaceae. According to the same authors, the most common latewood markers are thick and radially flattened fibers, as found in Bignoniaceae, Combretaceae, Leguminosae, Rhamnaceae, Ulmaceae, Verbenaceae, and narrow vessels in Bignoniaceae (Lima et al. 2010). Less commonly, dilated rays and

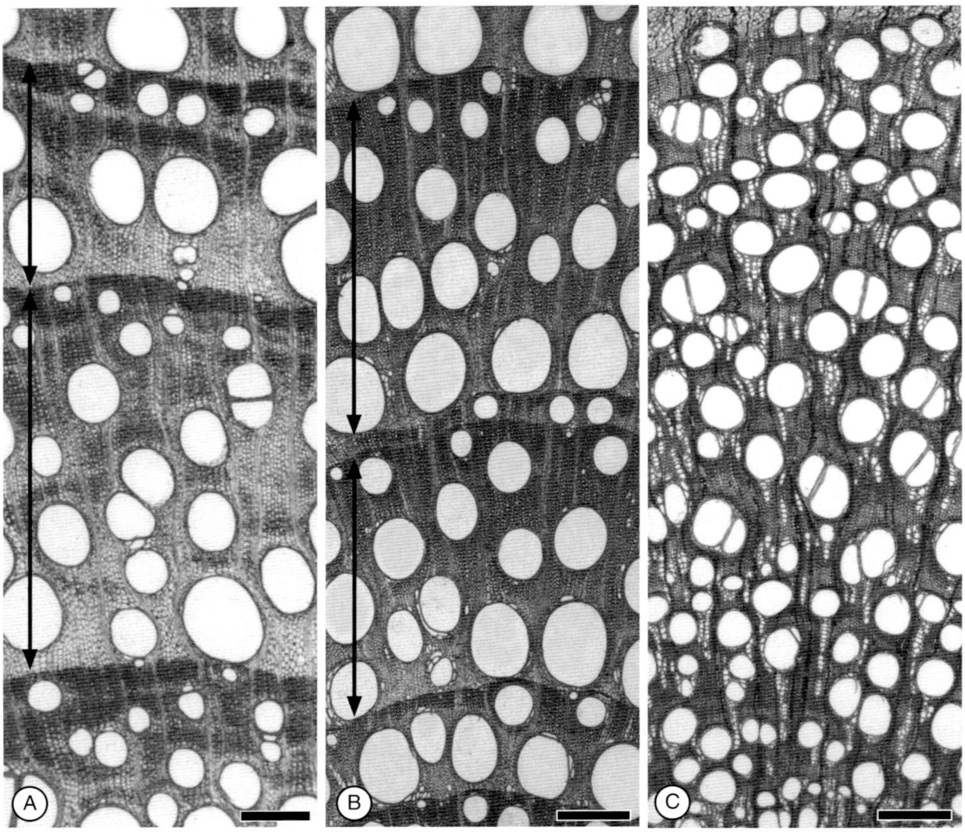

Fig. 19.16 Growth rings in secondary xylem of lianas, as seen in transverse section (TS). (A, B) Annual growth rings present (double-headed arrows). Annual growth rings marked by marginal parenchyma band and wide vessels in the earlywood and narrow vessels and radially flattened fibers in the latewood. (A) *Dalbergia frutescens* (Leguminosae). (B) *Stizophyllum riparium* (Bignoniaceae). (C) Growth rings absent in *Serjania laruotteana* (Sapindaceae). Scale bars: 400 μm.
See plate section for color representation of this figure.

radially shorter ray cells are also found as latewood markers in Bignoniaceae (Lima et al. 2010).

In addition, for the lianas of Bignoniaceae the latewood markers were shown to be the most reliable means of identifying growth rings, being found in all species that had growth rings, as opposed to earlywood markers that could be missing in several species (Lima et al. 2010).

CONCLUSION AND FUTURE DIRECTIONS

In this chapter, we have explored the anatomical characteristics of the vascular system of scandent plants, showing that they share common features that can be grouped under the term "lianescent vascular syndrome." These features include the presence of wide conducting cells of the xylem and phloem, a higher abundance of axial parenchyma, fewer fibers, broader rays, and cambial variants. New data on seasonal cambial activity and the formation of growth rings in Bignoniaceae, Leguminosae and Sapindaceae show that lianas respond differently to the presence of dry and wet seasons, depending on the taxon. Much more remains to be explored in liana anatomy, and the following are the most pressing topics:

1. The relationship between radial hydraulic water distribution from the secondary ray phloem towards the secondary ray xylem in taxonomically related liana and tree species to unravel the possible role of perforated ray cells in vessel refilling.
2. The seasonality of growth of lianas to confirm if (a) the observed timespan of cambial activity explored in the present study is shared by other lianas; and (b) whether different cambial variants induce different cambial activity.
3. The integration of ontogenetic and phylogenetic data of different families to learn whether optimization in photosynthate conduction of the cambial variant is a common feature in variant phloem of all lianas.
4. The integration of ontogenetic and phylogenetic studies to explore the mechanisms involved in the evolution of new architectures derived from cambial variants in the vascular system of lianas.

GLOSSARY

Anastomose: coalesce, get linked by anastomosis, like interconnected veins in a leaf join together.

Axial parenchyma: parenchyma cells in the axial system of the secondary vascular tissues, produced by the cambium.

Cambial variant: a term of convenience, use to indicate an alternative form of secondary growth different from the most common pattern, where a single cambium produces more secondary xylem to the inside than secondary phloem to the outside. Also known as anomalous secondary growth.

Fibers: in the secondary vascular tissue refers to axial elongated sclerenchyma cells with a lignified secondary wall.

Fusiform initials: elongated meristematic cells of the cambium that originate the elements of the axial system of the secondary xylem (vessel, tracheid, fiber, axial parenchyma) and secondary phloem (sieve tube elements, sieve cells, fibers, axial parenchyma).

Liana: any scandent plant that presents a conspicuous secondary growth.

Non-lignified axial parenchyma: axial parenchyma cells with non-lignified walls.

Perforated ray cell: cell within the rays that differentiates as vessel element and connect two adjacent vessels on each side of the ray.

Phloem wedges: a term of convenience that refers to the presence of wedges of phloem that penetrate into the secondary xylem. Also known as furrowed xylem.

Septate fiber: a fiber with a thin transverse wall (septum).

Rays: groups of cells of various shapes that are formed by the ray initials of the cambium, and that form a continuous radial system between phloem and xylem.

Ray initials: meristematic cells of the cambium that form the ray cells of both the secondary xylem and phloem.

Sieve ray cell: cell within the ray that differentiates as sieve tube element, usually connecting two adjacent sieve tubes at each side of the ray.

Vasicentric tracheids: tracheids adjacent to vessels in the secondary xylem.

Vessel dimorphism: a term of convenience coined by Carlquist (1981), which refers to the presence of very wide vessels co-occurring with very narrow vessels or tracheids in the secondary xylem, a feature very common in lianas.

Vine: any scandent plant.

*Most terms and definitions are based on Evert's (2006) *Esau's Plant Anatomy*.

ACKNOWLEDGMENTS

We are indebted to our colleagues for field collection support and plant identifications, S. Alcantara, R.F. Almeida, M. Alves, A. Amorim, L. Arroyo, R. Arruda, C.L. Bastos, G.C.T. Ceccantini, G.A. Damasceno-Junior, M.A. Farinaccio, M. Groppo, L.G. Lohmann, J. Lovo, A. Nogueira, J. Prado, A. Pott, G. Somner, M. Souza-Baena, N. Tamaio, B.L.P. Villagra, and A. Zuntini; A. Nogueira for incentives to produce this chapter; L.G. Lohmann for the supervision of the evolutionary aspects in Bignonieae developed during the MSc of the second author; G. Angeles for the discussion of some topics of the manuscript; N. Tamaio for providing anatomical slides of *Chondrodendron platyphyllum* (Menispermaceae) and *Strychnos millepunctata* (Loganiaceae); A.C.F. Barbosa for assistance in anatomical sectioning and stem macroscopic preparations; C.L. Bastos for laborious help with anatomical slide preparation and macroscopic images, V. Barão for assistance on stem macroscopic images, and J. Rando for advice on legumes' nomenclature. Financial support was provided by the Conselho Nacional de Desenvolvimento Científico e Tecnológico (CNPq, Brazil) to V. Angyalossy; and Fundação de Amparo à Pesquisa do Estado de São Paulo (FAPESP, Brazil) to M.R. Pace and A.C. Lima.

REFERENCES

Acevedo-Rodríguez, P. (1993) Systematics of *Serjania* (Sapindaceae) Part I: A revision of *Serjania* Sec. *Platycocccus*. *Memoirs of the New York Botanical Garden*, **67**, 1–93.

Angyalossy, V., Angeles, G., & Madero-Vega, C. (2009) The mirror effect on xylem and phloem radial conduction. *Proceedings of the Sixth Plant Biomechanics Conference, Cayenne*, pp. 156–162.

Angyalossy, V., Angeles, G., Pace, M.R. et al. (2012) An overview on the anatomy, development and evolution of the vascular system of lianas. *Plant Ecology & Diversity*, **5**, 167–182.

Araújo, G.U.C. & Costa, C.G. (2006) Cambial variant in the stem of *Serjania corrugata* (Sapindaceae). *IAWA Journal*, **27**, 269–280.

Arthur, W. (2011) *Evolution: A Developmental Approach*, Wiley-Blackwell, Chichester.

Baas, P. & Schweingruber, F.H. (1987) Ecological trends in the wood anatomy of trees, shrubs and climbers from Europe. *IAWA Bulletin n.s.*, **8**, 245–274.

Bailey, I.W. & Howard, R.A. (1941) The comparative morphology of the Icacinaceae I. Anatomy of the node and internode. *Journal of the Arnold Arboretum*, **22**, 125–132.

Begum, S., Nakaba, S., Bayramzadeh, V. et al. (2008) Temperature responses of cambial reactivation and xylem differentiation in hybrid poplar (*Populus sieboldii* × *P. grandidentata*) under natural conditions. *Tree Physiology*, **28**, 1813–1819.

Begum, S., Nakaba, S., Oribe, Y. et al. (2007) Induction of cambial reactivation by localized heating in a deciduous hardwood hybrid Poplar (*Populus sieboldii* × *P. grandidentata*). *Annals of Botany*, **100**, 439–447.

Borchert, R. (1994a) Soil and stem water storage determine phenology and distribution of tropical dry forest trees. *Ecology*, **75**, 1437–1449.

Borchert, R. (1994b) Water status and development of tropical trees during seasonal drought. *Trees*, **8**, 115–125.

Boureau, E. (1957) Structures libéro-ligneuses anormales, in *Anatomie végétale: l'appareil végétatif des phanérogames* (ed E. Boureau), Presses Universitaires de France, Paris, pp. 633–651.

Brandes, A.F.N. & Barros, C.F. (2008) Anatomia do lenho de oito espécies de lianas da família Leguminosae ocorrentes na Floresta Atlântica. *Acta Botanica Brasilica*, **22**, 465–480.

Brandes, A.F.N., Lisi, C.S. & Barros, C.F. (2011) Dendrochronology of lianas of the Leguminosae family from the Atlantic Forest, Brazil. *Trees*, **25**, 133–144.

Brodersen, C.R., Choat, B., Chatelet, D.S. et al. (2013) Xylem vessel relays contribute to radial connectivity in grapevine stems (*Vitis vinifera* and *V. arizonica*; Vitaceae). *American Journal of Botany*, **100**, 314–321.

Burnham, R.J. (2009) An overview of the fossil record of climbers: bejucos, sogas, trepadoras, lianas, cipós, and vines. *Revista Brasileira de Paleontologia*, **12**, 149–160.

Caballé, G. (1986) *Sur la biologie des lianes ligneuses en forêt gaonaise*. Doctoral dissertation, Université des Sciences et Technique du Languedoc, Montpellier.

Caballé, G. (1993) Liana structure, function and selection: a comparative study of xylem cylinders of tropical rainforest species in Africa and America. *Botanical Journal of the Linnean Society*, **113**, 41–60.

Carlquist, S. (1966) Wood anatomy of Compositae: a summary, with comments on factors controlling wood evolution. *Aliso*, **6**, 25–44.

Carlquist, S. (1975) *Ecological Strategies of Xylem Evolution*. University of California Press, Los Angeles.

Carlquist, S. (1981) Wood anatomy of Nepenthaceae. *Bulletin of the Torrey Botanical Club*, **108**, 324–330.

Carlquist, S. (1985a) Observations on functional wood histology of vines and lianas: vessel dimorphism, tracheids, vasicentric tracheids, narrow vessels, and parenchyma. *Aliso*, **11**, 139–157.

Carlquist, S. (1985b) Wood anatomy of Begoniaceae, with comments on raylessness, paedomorphosis, relationships, vessel diameter, and ecology. *Bulletin of the Torrey Botanical Club*, **112**, 59–69.

Carlquist, S. (1991) Anatomy of vine and liana stems: a review and synthesis, in *The Biology of Vines* (eds F.E. Putz & H.A. Mooney), Cambridge University Press, New York, pp. 53–71.

Carlquist, S. (1992a) Wood anatomy of sympetalous dicotyledon families: a summary, with comments on systematic relationships and evolution of the woody habit. *Annals of the Missouri Botanical Garden*, **79**, 303–332.

Carlquist, S. (1992b) Wood anatomy of selected Cucurbitaceae and its relationship to habit and systematics. *Nordic Journal of Botany*, **12**, 347–355.

Carlquist, S. (1993) Wood and bark anatomy of Aristolochiaceae, systematic and habit correlations. *IAWA Journal*, **14**, 341–357.

Carlquist, S. (1995) Wood anatomy of the Ranunculiflorae. *Plant Systematics and Evolution*, **9**, 11–24.

Carlquist, S. (1996) Wood, bark, and stem anatomy of Gnetales: a summary. *International Journal of Plant Sciences*, **157**, S58–S76.

Carlquist, S. (2001) *Comparative Wood Anatomy: Systematic, Ecological, and Evolutionary Aspects of Dicotyledon Wood*, 2nd edn, Springer, Lexington, MA.

Carlquist, S. (2003a) Wood and stem anatomy of woody Amaranthaceae s.s.: ecology, systematics and the problems of defining rays in dicotyledons. *Botanical Journal of the Linnean Society*, **143**, 1–19.

Carlquist, S. (2003b) Wood anatomy of Polygonaceae: analysis of a family with exceptional wood diversity. *Botanical Journal of the Linnean Society*, **141**, 25–51.

Carlquist, S. (2004) Lateral meristems, successive cambia and their products: a reinterpretation based on roots and stems of Nyctaginaceae. *Botanical Journal of the Linnean Society*, **146**, 129–143.

Carlquist, S. (2007) Successive cambia revisited: ontogeny, histology, diversity, and functional significance. *Journal of the Torrey Botanical Society*, **134**, 301–332.

Carlquist, S. (2013) Interxylary phloem: diversity and functions. *Brittonia*. DOI 10.1007/s12228-012-9298-128-012-9298-1

Ceja-Romero, J. & Perez-Olivera, C.D. (2010) Wood anatomy of the arboreal species of *Ipomoea* (Convolvulaceae). *Madera y Bosques*, **16**, 61–73.

Christman, M.A. & Sperry, J.S. (2010) Single-vessel flow measurements indicate scalariform perforation plates confer higher flow resistance than previously estimated. *Plant, Cell and Environment*, **33**, 431–443.

Crivellaro, A., McCulloh, K., Jones, F.A., & Lachenbruch, B. (2012) Anatomy and mechanical and hydraulic needs of woody climbers contrasted with subshrubs on the island of Cyprus. *IAWA Journal*, **33**, 355–373.

Darwin, C. (1865) On the movements and habits of climbing plants. *Journal of the Linnean Society*, London.

Davis, J.D. & Evert, R.F. (1970) Seasonal cycle of phloem development in woody vines. *Botanical Gazette*, **131**, 128–138.

Den Outer, R.W. & Fundter, J.M. (1976) The secondary phloem of some Combretaceae and the systematic position of *Strephonema pseudocola* A. Chev. *Acta Botanica Neerlandica*, **25**, 481–493.

Den Outer, R.W. & Van Veenendaal, W.L.H. (1995) Development of included phloem in the stem of *Combretum nigricans* (Combretaceae). *IAWA Journal*, **16**, 151–158.

Dias-Leme, C.L. (1999) *Árvores e lianas de Leguminosae Papilionoideae: anatomia do caule*. Doctoral dissertation, University of São Paulo, São Paulo.

Dickison, W.C. (1967) Comparative morphological studies in Dilleniaceae. 1. Wood anatomy. *Journal of the Arnold Arboretum*, **48**, 1–29.

Dobbins, D.R. (1969) Studies on the anomalous cambial activity in *Doxantha unguis-cati* (Bignoniaceae). I. Development of the vascular pattern. *Canadian Journal of Botany*, **47**, 2101–2106.

Dobbins, D.R. (1970) The influence of the leaves on anomalous cambial activity in the "cross-vine." *American Journal of Botany*, **57**, 735.

Dobbins, D.R. (1971) Studies on the anomalous cambial activity in *Doxantha unguis-cati* (Bignoniaceae). II. A case of differential production of secondary tissues. *American Journal of Botany*, **58**, 697–705.

Dobbins, D.R. (1981) Anomalous secondary growth in lianas of the Bignoniaceae is correlated with the vascular pattern. *American Journal of Botany*, **68**, 142–144.

Dobbins, D.R. (1990) The role of rays in anomalous cambial activity. *IAWA Bulletin n.s.*, **11**, 122–123.

Dobbins, D.R. (2005) *Development and Function of Included Phloem in Woody Lianas of the Bignoniaceae*. XVII International Botanical Congress Book of Abstracts, Vienna.

Dobbins, D.R. & Fisher, J.B. (1986) Wound responses in girdled stems of lianas. *Botanical Gazette*, **147**, 278–289.

Dos Santos, G. (1995) *Wood anatomy, chloroplast DNA, and flavonoids of the tribe Bignonieae (Bignoniaceae)*. Doctoral dissertation, University of Reading.

Dunn, M.T., Krings, M., Mapes, G. et al. (2003) *Medullosa steinii* sp nov., a seed fern vine from the Upper Mississippian. *Review of Palaeobotany and Palynology*, **124**, 307–324.

Eames, A.J. & MacDaniels, L.H. (1925) *An Introduction to Plant Anatomy*, 1st edn, McGraw-Hill Book Company, Inc., New York.

Enquist, B.J. & Leffler, A.J. (2001) Long-term tree ring chronologies from sympatric tropical dry-forest trees: individualistic responses to climatic variation. *Journal of Tropical Ecology*, **17**, 41–60.

Esau, K. (1948) Phloem structure in the grapevine, and its seasonal changes. *Hilgardia*, **18**, 217–296.

Esau, K. (1967) *Plant Anatomy*, 2nd edn, John Wiley and Sons, New York.

Esau, K. (1977) *Anatomy of Seed Plants*, 2nd edn, John Wiley & Sons, New York.

Esau, K. & Cheadle, V.I. (1969) Secondary growth in *Bougainvillea*. *Annals of Botany*, **33**, 807–819.

Evert, R.F. (2006) *Esau's Plant Anatomy: Meristems, Cells, and Tissues of the Plant Body – Their Structure, Function, and Development*, 3rd edn, John Wiley and Sons, Hoboken, NJ.

Ewers, F.W. (1985) Xylem structure and water conduction in conifer trees, dicot trees, and lianas. *IAWA Bulletin, n.s.*, **6**, 309–317.

Ewers, F.W., Cochard, H. & Tyree, M.T. (1997) A survey of root pressure in vines of a tropical lowland forest. *Oecologia*, **110**, 191–196.

Ewers, F.W. & Fisher, J.B. (1989) Variation in vessel length and diameter in stems of six tropical and subtropical lianas. *American Journal of Botany*, **76**, 1452–1459.

Ewers, F.W. & Fisher, J.B. (1991) Why vines have narrow stems: histological trends in *Bauhinia* (Fabaceae). *Oecologia*, **88**, 233–237.

Ewers, F.W., Fisher, J.B. & Chiu, S.T. (1990) A survey of vessel dimensions in stems of tropical lianas and other growth forms. *Oecologia*, **84**, 544–552.

Ewers, F.W., Fisher, J.B. & Fichtner, K. (1991) Water flux and xylem structure in vines, in *The Biology of Vines* (eds F.E. Putz & H.A. Mooney), Cambridge University Press, Cambridge, pp. 53–71.

Fischer, E., Theisen, I. & Lohmann, L.G. (2004) Bignoniaceae, in *The Families and Genera of Vascular Plants VII. Dicotyledons. Lamiales (Except Acanthaceae including Avicenniaceae)* (eds K. Kubitzki & J.W. Kadereit), Springer Verlag, Heidelberg, pp. 9–38.

Fisher, J.B. & Ewers, F.W. (1991) Structural responses to stem injury in vines, in *The Biology of Vines* (eds F.E. Putz & H.A. Mooney), Cambridge University Press, Cambridge, pp. 99–124.

Fisher, J.B. & Ewers, F.W. (1992) Xylem pathways in liana stems with variant secondary growth. *Botanical Journal of the Linnean Society*, **108**, 181–202.

Fisher, J.B. & Ewers, F.W. (1995) Vessel dimensions in liana and tree species of *Gnetum* (Gnetales). *American Journal of Botany*, **82**, 1350–1357.

Gallenmüller, F., Müller, U., Rowe, N. & Speck, T. (2001) The growth form of *Croton pullei* (Euphorbiaceae): functional morphology and biomechanics of a neotropical liana. *Plant Biology*, **3**, 50–61.

Gartner, B.L., Bullock, S.H., Mooney, H.A. et al. (1990) Water transport properties of vine and tree stems in a tropical deciduous forest. *American Journal of Botany*, **77**, 742–749.

Gasson, P. & Dobbins, D.R. (1991) Wood anatomy of the Bignoniaceae, with a comparison of trees and lianas. *IAWA Bulletin n.s.*, **12**, 389–417.

Gentry, A.H. (1980) Bignoniaceae: Part I (Crescentia and Tourrettieae). In *Flora Neotropical Monograph* 25 (1), The New York Botanical Garden, New York.

Gentry, A.H. (1991) The distribution and evolution of climbing plants, in *The Biology of Vines* (eds F.E. Putz & H.A. Mooney), Cambridge University Press, Cambridge, pp. 3–49.

Gould, S.J. & Lewontin, R.C. (1979) The spandrels of San Marco and the Panglossian paradigm: a critique of the adaptationist programme. *Proceedings of the Royal Society of London B*, **205**, 581–598.

Gould, S.J. & Vrba, E.S. (1982) Exaptation: a missing term in the science of form. *Paleobiology*, **8**, 4–15.

Harrar, E.S. (1946) Notes on starch grains in septate fiber-tracheids. *Tropical Woods*, **85**, 1–9.

Hearn, D.J. (2009) Developmental patterns in anatomy are shared among separate evolutionary origins of stem succulent and storage root-bearing growth habits in *Adenia* (Passifloraceae). *American Journal of Botany*, **96**, 1941–1956.

Hernández-Ledesma, P., Terrazas, T. & Flores-Olvera, H. (2011) Comparative stem anatomy of *Mirabilis* (Nyctaginaceae). *Plant Systematics and Evolution*, **292**, 117–132.

Heuzé, P., Dupouey, J. & Schnitzler, A. (2009) Radial growth response of *Hedera helix* to hydrological changes and climatic variability in the Rhine floodplain. *River Research Applications*, **25**, 393–404.

Holbrook, N.M. & Putz, F.E. (1996) Physiology of tropical vines and hemiepiphytes: plants that climb up and plants that climb down. In: S.S. Mulkey, R.L. Chazdon, A.P. Smith, editors. *Tropical forests plant ecophysiology*. New York (New York): Chapman & Hall. p. 363–393.

Isnard, S., Prosperi, J., Wanke, S. et al. (2012) Growth form evolution in Piperales and its relevance for understanding angiosperm diversification: an integrative approach combining plant architecture, anatomy, and biomechanics. *International Journal of Plant Sciences*, **173**, 610–639.

Isnard, S. & Silk, W.K. (2009) Moving with climbing plants from Charles Darwin's time into the 21st century. *American Journal of Botany*, **96**, 1205–1221.

Isnard, S., Speck, T. & Rowe, N.P. (2003) Mechanical architecture and development in *Clematis*: implications for canalised evolution of growth forms. *New Phytologist*, **158**, 543–559.

Jacobsen, A.L., Pratt, R.B., Tobin, M.F. et al. (2012) A global analysis of xylem vessel length in woody plants. *American Journal of Botany*, **99**, 1583–1591.

Jiménez-Castillo, M., Wiser, S.K. & Lusk, C.H. (2007) Elevational parallels of latitudinal variation in the proportion of lianas in woody floras. *Journal of Biogeography*, **34**, 163–168.

Joshi, A.C. (1936) The anatomy of *Rumex* with special reference to the morphology of the internal bundles and the origin of internal phloem in the Polygonaceae. *American Journal of Botany*, **23**, 362–369.

June Adkinson, A.M. (1913) Some features of the anatomy of the Vitaceae. *Annals of Botany*, **27**(1), 133–139.

Klaassen, R.K.W.M. (1999) Wood anatomy of the Sapindaceae. *IAWA Journal*, Suppl. 2.

Klotz, L.H. (1978) Observations on diameters of vessels in stems of palms. *Principes*, **22**, 99–106.

Koek-Noorman, J., Topper, S.M.C. & ter Welle, B.J.H. (1984) The systematic wood anatomy of the Moraceae (Urticales) III. Tribe Ficeae. *IAWA Bulletin, n.s.*, **5**, 330–334.

Kozlowski, T.T., Kramer, P.J. & Pallardy, S.G. (1991) *The Physiological Ecology of Woody Plants*, Academic Press, New York.

Kuo, J. & Pate, J.S. (1981) Intraxylary (medullary) phloem in *Macrotyloma uniflorum* Lam. (Fabaceae). *Annals of Botany*, **48**, 403–406.

Lahaye, R., Civeyrel, L., Speck, T. & Rowe, N.P. (2005) Evolution of shrub-like growth forms in the lianoid subfamily Secamonoideae (Apocynaceae s.l.) of Madagascar: phylogeny, biomechanics, and development. *American Journal of Botany*, **92**, 1381–1396.

Lens, F., Kårehed, J., Baas, P. et al. (2008) The wood anatomy of polyphyletic Icacinaceae s.l., and their relationship within asterids. *Taxon*, **57**, 525–552.

León, W.J.H. (2009) Anatomia de la madera de 20 especies de lianas del lote boscoso del Dorado-Tumeremo (estado Bolívar, Venezuela). *Hernstia*, **19**, 137–170.

Léon-Gómez, C. & Monroy-Ata, A. (2005) Seasonality in cambial activity of four lianas from a Mexican lowland tropical rainforest. *IAWA Journal*, **26**, 111–120.

Lima, A.C. (2012) *Sazonalidade da atividade cambial em lianas*. Master's dissertation. University of São Paulo.

Lima, A.C., Pace, M.R. & Angyalossy, V. (2010) Seasonality and growth rings in lianas of Bignoniaceae. *Trees*, **24**, 1045–1060.

Lohmann, L.G. (2006) Untangling the phylogeny of neotropical lianas (Bignonieae, Bignoniaceae). *American Journal of Botany*, **93**, 304–318.

Lohmann, L.G. & Taylor, C.M. (2014) A new generic classification of tribe Bignonieae (Bignoniaceae). *Annals of the Missouri Botanical Garden*, **99**, 348–489.

McCulloh, K., Sperry, J.S., Lachenbruch, B. et al. (2010) Moving water well: comparing hydraulic efficiency in twigs and trunks of coniferous, ring-porous, and diffuse-porous saplings from temperate and tropical forests. *New Phytologist*, **186**, 439–450.

Ménard, L., McKey, D. & Rowe, N.P. (2009) Developmental plasticity and biomechanics of treelets and lianas in *Manihot* aff. *quinquepartita* (Euphorbiaceae): a branch-angle climber of French Guiana. *Annals of Botany*, **103**, 1249–1259.

Mennega, A.W. (1982) Stem structure of the New World Menispermaceae. *Journal of the Arnold Arboretum*, **63**, 145–172.

Mennega, A.W. (1997) Wood anatomy of the Hippocrateoideae (Celastraceae). *IAWA Journal*, **18**, 331–368.

Metcalfe, C.R. & Chalk, L. (1950) *Anatomy of the Dicotyledons*, Clarendon Press, Oxford.

Mikesell, J. & Schroeder, A.C. (1984) Internal phloem development in *Parbitis nil* Chois. (Convolvulaceae). *Botanical Gazette*, **145**, 196–203.

Mooney, H.A. & Gartner, B.L. (1991) Reserve economy of vines, in *The Biology of Vines* (eds F.E. Putz & H.A. Mooney), Cambridge University Press, Cambridge, pp. 161–179.

Nair, M.N.B. (1993) Structure of stem and cambial variant in *Spatholobus roxburghii* (Leguminosae). *IAWA Journal*, **14**, 191–204.

Nair, M.N.B. & Mohan Ram, H.Y. (1990) Structure of wood and cambial variant in the stem of *Dalbergia paniculata* Roxb. *IAWA Bulletin n.s.*, **11**, 379–391.

Nardini, A., Lo Gullo, M.A. & Salleo, S. (2011) Refilling embolized xylem conduits: is it a matter of phloem unloading? *Plant Science*, **180**, 604–611.

Obaton, M. (1960) Les lianes ligneuses à structure anormale des forêts denses d'Afrique Occidentale. *Annales des Sciences Naturelles Botanique et Biologie Végétale*, **12**, 1–20.

Ortiz, R.D.C., Kellogg, E.A. & Werff, H.V.D. (2007) Molecular phylogeny of the moonseed family (Menispermaceae): implications for morphological diversification. *American Journal of Botany*, **94**, 1425–1438.

Ozório-Filho, H.L. (2002) *Variação cambial em Bignoniaceae: padrão anatômico e desenvolvimento do caule*. Master's dissertation, Universidade de São Paulo.

Pace, M.R. & Angyalossy, V. (2013) Wood evolution: a case study in the Bignoniaceae. *International Journal of Plant Sciences*, **174**, 1014–1048.

Pace, M.R., Lohmann, L.G. & Angyalossy, V. (2009) The rise and evolution of the cambial variant in Bignonieae (Bignoniaceae). *Evolution & Development*, **11**, 465–479.

Pace, M.R., Lohmann, L.G. & Angyalossy, V. (2011) Evolution of disparity between the regular and variant phloem in Bignonieae (Bignoniaceae). *American Journal of Botany*, **98**, 602–618.

Patil, V.S., Marcati, C.R. & Rajput, K.S. (2011) Development of intra- and extraxylary secondary phloem in *Coccinia indica* (Cucurbitaceae). *IAWA Journal*, **32**, 475–491.

Patil, V.S., Rao, K.S. & Rajput, K.S. (2009) Development of intraxylary phloem and internal cambium in *Ipomoea hederifolia* (Convolvulaceae). *Journal of the Torrey Botanical Society*, **136**, 423–432.

Pfeiffer, H. (1926) Das abnorme Dickenwachstum, in *Handbuch der Pflanzenanatomie*, Gebrüder Bornträger Verlag, Berlin.

Putz, F.E. & Holbrook, N.M. (1991) Biomechanical studies in vines, in *The Biology of Vines* (eds F.E. Putz & H.A. Mooney), Cambridge University Press, Cambridge, pp. 73–97.

Putz, F.E. & Mooney, H.A. (1991) *The Biology of Vines*, Cambridge University Press, Cambridge.

Radlkofer, L. (1875) *Monographie der Sapindaceen – Gattung Serjania*, Verlag der K. B. Akademie, München.

Radlkofer, L. (1932) Sapindaceae. In *Das Planzenreich* (ed. A. Engler), vol. IV. 165 (Heft 98a), pp. 414–425. Verlag von H.R. Engelmann (J. Cramer), Weinheim.

Rajput, K.S., Fiamengui, M.B. & Marcati, C.R. (2010) Stem anatomy of *Strychnos bicolor* Prog. (Loganiacceae) from Brazilian Cerrado. *Phytomorphology*, **60**, 48–56.

Rajput, K.S., Fiamengui, M.B. & Marcati, C.R. (2012a) Stem anatomy and development of successive cambia in the Neotropical liana *Securidaca rivinifolia* (Polygalaceae). *IAWA Journal*, **33**, 391–402.

Rajput, K.S., Nunes, O.M., Brandes, A.F.N. & Tamaio, N. (2012b) Development of successive cambia and pattern of secondary growth in the stem of the Neotropical liana *Rhynchosia phaseoloides* (SW.) DC. (Fabaceae). *Flora*, **207**, 607–614.

Rajput, K.S., Patil, V.S. & Rao, K. (2009) Development of included phloem of *Calycopteris floribunda* Lamk. (Combretaceae). *Journal of the Torrey Botanical Society*, **136**, 302–312.

Rajput, K.S. & Rao, K. (1999) Structural and developmental studies on cambial variant in *Pupalia lappacea* (Amaranthaceae). *Annales Botanici Fennici*, **36**, 137–141.

Rajput, K.S. & Rao, K.S. (2000) Secondary growth in the stem of some species of *Alternanthera* and *Achyranthes aspera* (Amaranthaceae). *IAWA Journal*, **21**, 417–424.

Rajput, K.S. & Rao, K. (2003) Cambial variant and xylem structure in the stem of *Cocculus hirsutus* (Menispermaceae). *IAWA Journal*, **24**, 411–420.

Rowe, N., Isnard, S. & Speck, T. (2004) Diversity of mechanical architectures in climbing plants: an evolutionary perspective. *Journal of Plant Growth Regulation*, **23**, 108–128.

Rowe, N.P. & Speck, T. (1996) Biomechanical characteristics of the ontogeny and growth habit of the tropical liana *Condylocarpon guianense* (Apocynaceae). *International Journal of Plant Sciences*, **157**, 406–417.

Salleo, S., Lo Gullo, M.A., Trifilò, P. & Nardini, A. (2004) New evidence for a role of vessel-associated cells and phloem in the rapid xylem refilling of cavitated stems of *Laurus nobilis* L. *Plant, Cell and Environment*, **27**, 1065–1076.

Sattler, R. (1988) Homeosis in plants. *American Journal of Botany*, **75**, 1606–1617.

Schenck, H. (1892) Beiträge zur Biologie und Anatomie der Lianen, im besonderen der in Brasilien einheimischen Arten. I Teil. Beiträge zur Biologie der Lianen, in *Botanische Mitteilungen aus den Tropen*, Verlag von Gustav Fischer, Jena.

Schenck, H. (1893) Beiträge zur Biologie und Anatomie der Lianen im besonderen der in Brasilien einheimischen Arten. II Teil, Beiträge zur Anatomie der Lianen, in *Botanische Mitteilungen aus den Tropen*, Verlag von Gustav Fisher, Jena.

Schnitzer, S.A. (2005) A mechanistic explanation for global patterns of liana abundance and distribution. *The American Naturalist*, **166**, 262–276.

Schöngart, J., Piedade, M.T.F., Ludwingshausen, S. et al. (2002) Phenology and stem-growth periodicity of tree species in Amazonian floodplain forests. *Journal of Tropical Ecology*, **18**, 581–597.

Scott, D.H. (1891) On some points in the anatomy of *Ipomoea versicolor* Meissn. *Annals of Botany*, **3**, 173–180.

Scott, D.H. & Brebner, G. (1889) On the anatomy and histogeny of the genus *Strychnos*. *Annals of Botany*, **3**, 275–304.

Secchi, S. & Zwieniecki, M.A. (2012) Analysis of xylem sap from functional (non-embolized) and non-functional (embolized) vessels of *Populus nigra* – chemistry of refilling. *Plant Physiology Preview*, **160**, 955–964.

Sokołowska, K. & Zagórska-Marek, B. (2012) Symplasmic, long-distance transport in xylem and cambial regions in branches of *Acer pseudoplatanus* (Aceraceae) and *Populus tremula* × *P. tremuloides* (Salicaceae). *American Journal of Botany*, **99**, 1745–1755.

Solereder, H. (1908) *Systematic Anatomy of the Dicotyledons*, Clarendon Press, Oxford.

Sperry, J.S., Hacke, U.G. & Pittermann, J. (2006) Size and function in conifer tracheids and angiosperm vessels. *American Journal of Botany*, **93**, 1490–1500.

Sperry, J.S., Holbrook, N.M., Zimmermann, M.H. & Tyree, M.T. (1987) Spring filling of xylem vessels in wild grapevine. *Plant Physiology*, **83**, 414–417.

Spicer, R. & Groover, A. (2010) Evolution of development of vascular cambia and secondary growth. *New Phytologist*, **186**, 577–592.

Stevens, G.C. (1987) Lianas as structural parasites: the *Bursera simaruba* example. *Ecology*, **68**, 77–81.

Stevens, P.F. (2001 onwards). *Angiosperm Phylogeny Website*. Version 12, July 2012 (and more or less continuously updated since). Available at: http://www.mobot.org/MOBOT/research/APweb.

Stiller, V. & Sperry, J.S. (2002) Cavitation fatigue and its reversal in sunflower (*Helianthus annuus* L.). *Journal of Experimental Botany*, **53**, 1155–1161.

Tamaio, N. (2006) *Anatomia do caule e da raiz em Menispermaceae*. PhD dissertation, University of São Paulo.

Tamaio, N. & Angyalossy, A. (2009) Variação cambial em Serjania caracasana (Sapindaceae): enfoque na adequação terminológica. *Rodriguésia*, **60**, 651–666.

Tamaio, N., Neves, M.F., Brandes, A.F.N. & Vieira, R.C. (2011) Quantitative analyses establish the central vascular cylinder as the standard for wood-anatomy studies in lianas having compound stems (Paullinieae: Sapindaceae). *Flora*, **206**, 987–996.

Tamaio, N. & Somner, G. (2010) Development of corded vascular cylinder in *Thinouia restingae* Ferruci & Somner (Sapindaceae: Paullinieae). *Journal of the Torrey Botanical Society*, **137**, 319–326.

Tamaio, N., Vieira, R.C. & Angyalossy, V. (2009) Origin of successive cambia on stem in three species of Menispermaceae. *Revista Brasileira de Botânica*, **32**, 839–848.

Terrazas, T., Aguilar-Rodriguez, S. & Tinoco-Ojanguren, C. (2011) Development of successive cambia, cambial activity, and their relationship to physiological traits in *Ipomea arborescens* (Convolvulaceae) seedlings. *American Journal of Botany*, **98**, 765–774.

Utteridge, T., Nagamasu, H., Teo, S.P. et al. (2005) *Sleumeria* (Icacinaceae): A new genus from Northern Borneo. *Systematic Botany*, **30**, 635–643.

Veenendaal, W.L.H. & Den Outer, R.W. (1993) Development of included phloem and organization of the phloem network in the stem of *Strychnos millepunctata* (Loganiaceae). *IAWA Journal*, **14**, 253–265.

Wagner, S.T., Isnard, S., Rowe, N.P. et al. (2012) Escaping the lianoid habit: evolution of shrub-like growth forms in *Aristolochia* subgenus *Isotrema* (Aristolochiaceae). *American Journal of Botany*, **99**, 1609–1629.

West-Eberhard, M.J. (2003) *Developmental Plasticity and Evolution*, Oxford University Press, New York.

Worbes, M. (1985) Structural and other adaptations to long term flooding by trees in Central Amazonia. *Amazoniana*, **9**, 459–484.

Worbes, M. (1995) How to measure growth dynamics in tropical trees: a review. *IAWA Journal*, **16**, 337–351.

Zimmermann, M.H. & Jeje, A.A. (1981) Vessel-length distribution in stems of some American woody plants. *Canadian Journal of Botany*, **59**, 1882–1892.

Zózimo, E., Tamaio, N. & Vieira, R.C. (2011) Development of intraxylary phloem in the stem of *Combretum rotundifolium* (Combretaceae). *IAWA Journal*, **32**, 14–24.

Chapter 20

PHYSIOLOGICAL IMPLICATIONS OF THE LIANA GROWTH FORM

Louis S. Santiago,[1,2] *Sarah C. Pasquini,*[1] *and Mark E. De Guzman*[1]

[1]University of California, Riverside, CA, USA
[2]Smithsonian Tropical Research Institute, Balboa, Ancon, Republic of Panama

Ecology of Lianas, First Edition. Edited by Stefan A. Schnitzer, Frans Bongers, Robyn J. Burnham, and Francis E. Putz.

Companion Website: www.wiley.com/go/schnitzer/lianas

OVERVIEW

Lianas represent a unique woody growth form in tropical forests and display several intriguing patterns of distribution and abundance, including high abundance early in succession, increasing abundance in some undisturbed forests, and peaks in abundance in seasonally dry tropical forest. Whereas many of the proposed mechanisms for the unique patterns of liana abundance are physiologically based, the available empirical data are relatively scarce. Therefore, we use the demonstrated physiological affinities of lianas to evaluate mechanisms for liana abundance patterns in tropical forest. This chapter is organized along increasing levels of plant morphological complexity, first, addressing leaves, then moving to the supply-and-demand coordination between stem and leaf water transport, and, finally, the whole plant scale.

The emerging pattern of greater photosynthesis in lianas than trees provides possible explanations for the victory of lianas early in forest succession and in seasonally dry tropical forest where light and other resources are relatively abundant. In addition, the ability of many lianas to support leaf physiology with efficient water transport is also consistent with peculiarities of their distribution and abundance. Perhaps not surprisingly, the belowground processes remain the most elusive for forming specific hypotheses on comparative physiology of lianas and trees. Further studies are needed to pin down the physiological mechanisms of liana abundance and distribution. Studies that are specifically designed to test particular hypotheses are urgently needed, as well as careful species selection based on stratified or phylogenetically based sampling designs or plant community-wide campaigns.

INTRODUCTION

Lianas are a unique growth form that reach their crescendo in tropical forests, but also occur in a variety of other ecosystems. The lianoid growth form, characterized by structural parasitism and a slender stem supporting a relatively large canopy of leaves (Putz 1983; Gerwing & Farias 2000; Isnard & Feild, Chapter 17 in this volume), is thought to have evolved as an extreme point on the tradeoff spectrum between stem strength, on one hand, and rapid vertical growth, on the other. Indeed, Rowe and Speck (2005) point out that early in succession, even among trees, there are individuals that barely have enough stem strength to remain vertical if the surrounding vegetation is removed. This observation highlights the intense selection for rapid vertical growth early in the race for light in young tropical vegetation, as well as the manner in which stem biomechanical strength can be reduced under strong selective pressure for rapid growth. In either instance, this tradeoff constrains numerous physiological aspects of lianas. Because of the need to reduce allocation to stems to maximize vertical growth, heteroxylly, the production of vessels of contrasting sizes, including large vessels that have high water transport capacity but occupy little volume, has been heralded as a necessary evolutionary breakthrough facilitating the liana growth form (Carlquist 1991; Ewers & Fisher 1991; but see Feild et al. 2012). Furthermore, leaves that are nutrient-rich and carry a high photosynthetic capacity relative to trees are thought to facilitate the early successional behavior of many lianas and have been implicated in the success of the liana growth form (Asner & Martin 2012; Asner & Martin, Chapter 21 in this volume). Similarly, deep roots with the potential to dependably supply water to a large-vesseled xylem that is potentially vulnerable to xylem cavitation, have also been proclaimed as a critical part of the lianoid strategy (Restom & Nepstad 2004; Schnitzer 2005). Thus, this chapter addresses the physiological implications of the liana growth form with an emphasis on how being a liana leads to certain physiological outcomes. Comparisons are made with trees to encompass the competitive framework that structures the co-dominance of woody growth forms in tropical forest.

Physiological processes of lianas are important because they form the basis for matter and energy exchange between lianas and the atmosphere and soil. Therefore, the constraints and particularities of the liana growth form may cause lianas to cycle nutrients and water differently than trees. For example, differences in stem structure and density between trees and lianas have large impacts on the aboveground standing carbon stock of tropical forests because per individual, liana stems contain much less carbon than tree stems (Chave et al. 2001). Furthermore, because lianas have shoots and leaves that can extend many meters from the main trunk (Putz 1984), lianas have great potential to distribute nutrients obtained within the rooting zone of one individual over large distances through the forest through litterfall, potentially muting the effects of individual trees on soil conditions below their canopy (Powers, Chapter 14 in this volume). If lianas do indeed vary consistently in their rooting depth compared to trees, they may access and move different water sources than trees to the atmosphere through the transpiration stream (Powers, Chapter 14 in this volume). Finally, because lianas have a greater propensity for wind-dispersed seeds than trees (Gentry 1991), they contribute substantially less to the base of animal food webs, a tendency that can trickle up to higher tropic levels. Therefore, lianas should not be viewed merely as skinny feeble versions of trees, but rather as unique organisms with a distinct physiology and predictable effects on ecosystem processes.

There is increased interest in how lianas influence ecosystem processes because of several unique and dynamic aspects of their abundance in tropical forests. First, lianas are particularly abundant early in forest succession (DeWalt et al. 2000; Letcher & Chazdon 2009; Madeira et al. 2009; Letcher, Chapter 10 in this volume). Not all liana species exhibit early successional life-history characteristics (Gilbert et al. 2006). Yet, lianas tend to do exceptionally well in the high light environments of fragmented forests (Laurance et al. 2001; Arroyo-Rodríguez & Toledo-Aceves 2009) and treefall gaps (Putz 1984; Schnitzer et al. 2000; Schnitzer & Carson 2001; Schnitzer et al., Chapter 7 in this volume), with high relative abundance compared to trees and compared to later successional ecosystems, if not per species, then certainly per stem. Second, several reports, mostly from the neotropics, indicate that the relative abundance of lianas is increasing (Phillips et al. 2002; Wright et al. 2004b; Chave et al. 2008; Schnitzer & Bongers 2011; Schnitzer, Chapter 30 in this volume). At least some of the reports of increasing liana abundance pertain to undisturbed or mature forest, and therefore represent a phenomenon that goes well beyond the effects of disturbance (Yorke et al. 2013). This pattern has been attributed to a variety of explanations including increasing rates of natural disturbance, changing land-use, and increasing atmospheric CO_2, but further studies are needed to determine the mechanism and the extent of this phenomenon (Schnitzer & Bongers 2011; Schnitzer, Chapter 30 in this volume). A third unique aspect of liana abundance is that whereas most forms of life increase with increasing water availability, lianas appear to reach their peak relative abundance in seasonally dry tropical forest and decrease in relative abundance as mean annual precipitation increases (Schnitzer 2005; DeWalt et al., Chapter 11 in this volume). This pattern has been attributed to the idea that lianas allocate relatively less resources to stem support compared to trees, and allocate relatively more of these resources belowground to enable deep roots and reliable water sources that support the dry season growth advantage over trees that has been observed in seasonally dry tropical forest (Restom & Nepstad 2004; Andrade et al. 2005; Schnitzer 2005). Thus, the dry season growth advantage hypothesis for liana abundance in seasonally dry tropical forest is consistent with the observed decline in relative abundance of lianas with increasing precipitation. Overall, the proposed mechanisms for the unique patterns of liana abundance are physiologically based, yet remain perched upon a frighteningly scarce set of empirical data. This chapter attempts to pull together demonstrated physiological affinities of lianas and use this information to evaluate physiological mechanisms for liana abundance patterns in tropical forest.

Tropical forests in general are increasing in relevance because they contribute a disproportionately large amount to global productivity, carbon cycling, and carbon storage (Wright 2010; van der Heijden et al., Chapter 13 in this volume). Because lianas appear to be part of the dynamic manner in which tropical forests seem to be responding to environmental change, it is now more critical than ever to understand the physiologically mediated consequences of changing liana abundance. As such, this chapter focuses on woody species, which are the major lianoid forms of tropical forest, and contribute most to the cycling of matter and energy. Consequently, herbaceous species, though important in the very early stages of tropical forest succession, are not addressed. However, we do include the woody hemi-epiphytes, such as species in the tropical genera *Ficus* and *Clusia*, which begin life as epiphytes rooted in the canopy, and grow down to root in the soil, terminating with narrow-stemmed, liana-like overall whole-plant morphology (Patiño et al. 1995; Holbrook & Putz 1996a, 1996b). Although hemiepiphytes differ from lianas in some life-history characteristics, for the purposes of the type of physiological variables discussed in this chapter, once they are rooted in soil, they are essentially the same as other soil-rooted lianas, so hereafter we make no further distinction. Finally, this chapter is organized along increasing levels of plant morphological complexity, first, addressing leaves, the characteristic organ of higher plants, then moving to shoots, including the supply-and-demand coordination between stem and leaf water transport, and finally addressing the whole plant scale, where data on overall morphology and water use are accumulating.

PHYSIOLOGICAL CHARACTERISTICS OF LEAVES

High rates of maximum photosynthesis have been described as one of the characteristics enabling lianas to succeed following disturbance, yet relatively little data exist comparing tree and liana photosynthesis. For example, canopy trees showed greater maximum photosynthetic rates than canopy lianas in wet forest (3100 mm precipitation year^{-1}) at the San Lorenzo Canopy Crane in Panama (Santiago & Wright 2007). However, in the canopy of seasonally dry tropical forest (1367 mm precipitation year^{-1}) of Xishuangbanna, China, several studies have demonstrated greater photosynthetic capacity in lianas than trees (Cai et al. 2009; Zhu & Cao 2009, 2010). It is clear that site-based studies such as these can result in contrasting patterns, based on local conditions. Therefore, for a greater understanding of how liana and tree leaf-level physiology compare overall, we assembled all known data on leaf photosynthetic traits for lianas and trees in a comparative context (Fig. 20.1). We began assembling traits such as specific leaf area (leaf area divided by leaf mass; SLA), maximum photosynthetic CO_2 assimilation rate per leaf area (A_{max}) and respiration per leaf area (R_{area}) from tropical trees and lianas using the Glopnet database (Wright et al. 2004a), which contains 5 lianas and 1102 tree species. We then complemented this with more recent published sources (Santiago & Wright 2007; Cai et al. 2009; Zhu & Cao 2009, 2010), as well as data obtained directly from researchers or collected in the field (S. Pasquini, unpublished data). These data indicate that among comparisons of nearly 500 species, SLA and R_{area} are the leaf traits that most differ between lianas and trees, with lianas generally exhibiting thinner leaves with greater respiration costs, but not greater photosynthetic rates per leaf area (Fig. 20.1). A major limitation of this analysis is that many of these data were collected on individual species at sites that had trees, but may or may not have had lianas present. Nonetheless, these data are consistent with differences in allocation to leaf characteristics that might allow lianas to take advantage of high resource availability following disturbance and increasing atmospheric CO_2 concentration (Granados & Körner 2002).

Another recent study that addresses liana and tree leaf physiology was conducted empirically on leaf chemical traits grouped into three categories: (1) light capture and growth; (2) structure and defense; and (3) metabolism and maintenance. Asner and Martin (2012; Chapter 21 in this volume) studied the canopy leaves of 778 lianas and 6494 trees across 48 forests and clearly demonstrate that elements favoring light capture and growth, such as nitrogen and phosphorus, are 14.5% higher; metabolism and maintenance chemicals, such as potassium and calcium, are 16.7% higher; and structural and defense chemicals,

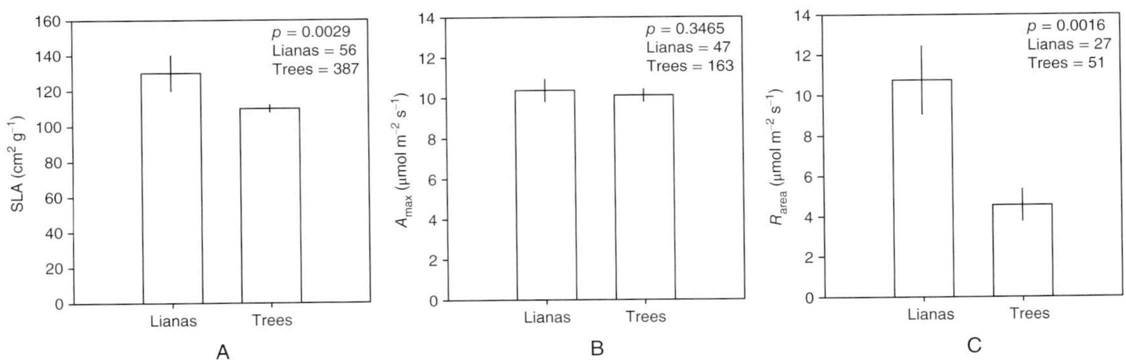

Fig. 20.1 Compiled means (±1 SE) of (A) Specific leaf area (SLA), (B) Maximum photosynthetic rate per leaf area (A_{max}) and (C) Respiration rate per leaf area (R_{area}) for lianas and trees. Numbers of liana and tree species are reported and results of statistical comparisons between liana and tree means by one-way ANOVA are reported as p-values. (Source: Data from Wright et al. (2004a), Santiago & Wright (2007), Zhu & Cao (2008, 2009), and Sarah Pasquini (unpublished data). Reproduced with permission.)

such phenols as and tannins, are 9% lower in lianas compared to trees. They also documented that the peak difference between liana and tree leaf chemistry occurred at approximately 2500 mm of annual precipitation and 25 °C, and that as precipitation increased or temperature decreased, liana leaf chemistry often converged with that of trees (Asner & Martin 2012). The observation that liana photosynthesis is greater than that of trees in seasonally dry tropical forest and that this advantage is lost in wetter forest is entirely consistent with the seemingly contrasting reports from Xishuangbanna, China (Cai et al. 2009; Zhu & Cao 2009, 2010), and San Lorenzo, Panama (Santiago & Wright 2007), and strongly supports the hypothesis of dry season growth advantage (Schnitzer 2005).

One emerging aspect of liana physiology is the question of whether lianas, as putatively early successional in nature, take rapid advantage of increased nutrient availability as would be expected of species that represent an *r*-selected life history. This question is especially relevant for seedlings because currently we do not understand if increasing liana abundance in tropical forests is the result of greater recruitment, or enhanced competition in later life stages. Cai et al. (2008) studied the responses of liana and tree seedlings of the genus *Bauhinia* to a range of light and nitrogen availability. They hypothesized that lianas would be more responsive than trees to environmental increases in resource availability. They found that leaf N concentration and A_{max} increased in response to increased light and soil N availability in the light-demanding tree species and the shade-tolerant liana, but surprisingly not in the two light-demanding lianas (Cai et al. 2008). Furthermore, they observed that though light and N had significant interactive effects on most physiological parameters, the consequences for relative growth rate differed little among species. Overall, they concluded that, contrary to predictions, lianas were no more responsive to variation in light and N availability than trees. In another study of the effects of nutrients on liana and tree seedling physiology, chlorophyll fluorescence-based techniques were used to evaluate photosynthetic capacity in an ongoing fully factorial nitrogen, phosphorus and potassium fertilization experiment at the Barro Colorado Nature Monument in Panama. After 14 years of fertilization, and after controlling for light availability with canopy photographs, it was determined that although liana photosynthetic capacity was no more responsive to nutrient addition than trees, both trees and lianas increased photosynthetic capacity with phosphorus addition, including an interaction with potassium (S. Pasquini, unpublished data). The most striking result was that in all analyses, no matter how the data are viewed, lianas seedlings always had greater photosynthetic capacity than tree seedlings, consistent with the idea that high photosynthetic rates contribute to liana success following disturbance.

HYDRAULIC COORDINATION BETWEEN THE STEM AND THE LEAF

A hallmark of liana construction is the slender stem containing a heteroxylous anatomy with some very large diameter vessels that promote high water transport capacity (Ewers & Fisher 1991; Angyalossy et al., Chapter 19 in this volume; Isnard & Feild, Chapter 17 in this volume). However, these slender stems also normally support a distal leaf area that is considerably greater than trees when normalized per unit stem area (Putz 1983; Ewers & Fisher 1991; Gerwing & Farias 2000; Zhu & Cao 2009). Therefore, lianas usually have a lower Huber value than trees (H_v; cross-sectional stem area divided by distal leaf area; Isnard & Feild, Chapter 17 in this volume), raising questions regarding the supply of water from stems to leaves and coordination in the physiological and morphological underpinnings of such arrangements. This question was first addressed anatomically by researchers measuring vessel diameter distributions and calculating potential rates of hydraulic conductivity (K_H; a measure of water transport capacity) based on the Hagen-Poiseuille equation. Indeed, large vessels in lianas produce high calculated rates of K_H (Ewers et al. 1997; Gutierrez et al. 2009). However, calculations of theoretical K_H can differ from actual rates during transpiration in live plants because of factors such as low conductance of pit membranes at vessel ends, an unrepresentative sample of measured vessels, non-circular geometry of the vessel cross-sections, and a nonlinear shape of vessels in the stem, making the actual length of the transport pathway greater than the measured stem length (Tyree & Ewers 1991; Santiago et al. 2004). Therefore, although relatively few measurements of K_H on liana stems exist, these data are critical for evaluating how stem water transport capacity is coordinated with leaf-scale patterns of carbon gain and water loss in lianas and trees.

Several key studies now exist that make it possible to evaluate stem hydraulics as a defining characteristic of liana physiology. In a seasonally dry tropical deciduous forest (748 mm precipitation year^{-1}) in Chamela, Mexico, ten liana species averaged significantly higher stem-specific hydraulic conductivity (K_S; water transport capacity normalized per sapwood or stem cross-sectional area) than ten tree species (Gartner et al. 1990). Furthermore, among free-standing and hemi-epiphytic *Ficus* species from Panama, Patiño et al. (1995) found that comparisons of K_S among species was difficult because these values changed with diameter, and stem diameter varied largely among species. Nonetheless, they report that hemi-epiphytes exhibited lower leaf area-specific hydraulic conductivity (K_L; water transport capacity normalized per distal leaf area) and lower H_v compared to free-standing species, indicating that hemi-epiphytes support a greater distal leaf area per stem area and that a given amount of hydraulic supply is spread among a greater area of leaves (Patiño et al. 1995). However, they also compiled data from other known studies of lianas and co-occurring trees available at the time and showed that the highest values for K_S were found in lianas, which they reasoned needed wide vessels to maintain efficient water transport to compensate for narrow stems. Lower H_v in lianas relative to trees has been reported in numerous other studies through a variety of metrics (Putz 1983; Gerwing & Farias 2000; Restom & Nepstad 2001; Feild & Balun 2008, Zhu & Cao 2009). Patterns of K_L between lianas and trees, on the other hand, show more variation. For example, the studies of Feild and Balun (2008) and Zhu and Cao (2009) show lower K_L in lianas relative to trees. However, Patiño et al. (1995) indicate that even though they found lower K_L in hemi-epiphytic species than in free-standing species, when they included data from other studies, overall, lianoid species showed high K_L relative to most tree species. K_L is a critical factor for evaluating efficient use of resources between trees and lianas. Santiago and Wright (2007) hypothesized that lianas exhibited relatively low leaf gas exchange rates because stem water is divided among a greater distal leaf area than in trees, or in other words, lianas should have lower K_L. We assembled preliminary results from a study in seasonally dry tropical forest (1800 mm precipitation year^{-1}) at the Parque Metropolitano Canopy Crane in Panama on six canopy liana species and six canopy tree species (DeGuzman, unpublished data), and combined these data with studies reporting these multiple hydraulic metrics (Feild & Balun 2008; Zhu & Cao 2009, van der Sande et al. in press), and found greater K_S and marginally greater K_L in lianas, but no statistical difference in H_v (Figs. 20.2A–C). These data suggest that in general, greater water transport capacity in lianas promotes higher leaf photosynthetic potential than trees as has been observed in some studies (Zhu & Cao 2009, 2010; Asner & Martin 2012; Asner & Martin, Chapter 21 in this volume).

Solving the water transport problem in narrow stems with large vessels by lianas carries a particular functional limitation. Although large vessels are efficient at transporting water because flows increase to the fourth power with an increase in radius, large vessels are also associated with greater risk of drought-induced xylem cavitation, when water demand exceeds supply and the negative pressure of the water exceeds the surface tension of water at the border of a gas-filled pit membrane (Sperry et al. 1988). Thus, it is not the size of the vessel *per se* that leads to a greater cavitation risk, but rather the size of pores in the pit membranes, and larger vessels are thought to have larger pores or a simply greater number of pores which would increase the chances of one giving away (Christman et al. 2009). In contrast, in smaller-diameter vessels, the smaller pores in pit membranes have the benefit of minimizing cavitation risk, but carry the cost of reduced water transport (Hacke et al. 2006). It is important to keep in mind that although lianas have some very large diameter vessels (Carlquist 1991), their heteroxylous tendencies also produce numerous small vessels which may function to maintain water transport when large vessels become cavitated (Holbrook & Putz 1996a). Nonetheless, the presence of few very large vessels has a substantial positive impact on the K_S of liana stems, but can also have a negative impact on resistance to cavitation. Cavitation resistance is commonly measured as the water potential at which 50% loss of hydraulic conductivity occurs (Ψ_{50}), where less negative values indicate greater vulnerability to drought-induced cavitation in xylem vessels. A compilation of existing data supports this hypothesis, with lianas averaging greater vulnerability to cavitation than trees (Fig. 20.2D). However, relatively little data comparing lianas and trees exist for the trade-off between K_S and Ψ_{50}, commonly known as the "safety versus efficiency" trade-off. The study by Zhu and Cao (2009) on canopy lianas and trees, and new data on canopy lianas and trees from Panama (M. DeGuzman, unpublished data), indicate that lianas and trees follow the same relationship of drought

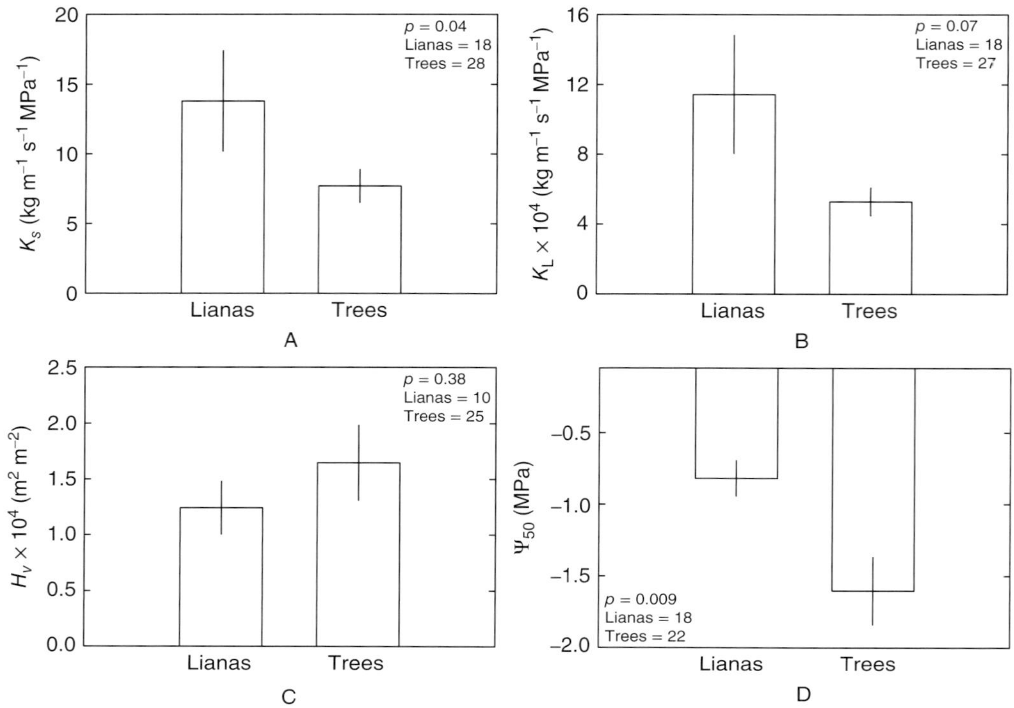

Fig. 20.2 Mean (±1 SE) of (A) Stem-specific hydraulic conductivity (K_S), (B) Leaf area-specific hydraulic conductivity (K_L), (C) Huber value, cross-sectional stem area divided by distal leaf area (H_v), and (D) Cavitation resistance measured as the water potential at which 50% loss of hydraulic conductivity occurs (Ψ_{50}), of trees and lianas. Numbers of liana and tree species are reported and results of statistical comparisons between liana and tree means by one-way ANOVA are reported as *p*-values. (Source: Data from Feild & Balun (2008), Zhu & Cao (2008, 2009), van der Sande et al. (in press), and Mark De Guzman (unpublished data). Reproduced with permission.)

resistance and maximum stem-specific conductivity, but that lianas show a tendency for greater values for K_S and a less negative Ψ_{50} (Fig. 20.3).

WHOLE-PLANT MORPHOLOGY AND TRANSPIRATION

At the whole-plant level, it is clear that lianas represent a distinct overall morphology and that the trade-offs implicit in the liana growth form lead to particular physiological outcomes. For example, though deep roots have been stated as a putative mechanism allowing lianas to reach their peak abundances in seasonally-dry tropical forest, relatively little data exist on this. Restom and Nepstad (2004) provide convincing data in this respect, as they actually excavated roots of the liana *Davilla kunthii*, and showed clearly that roots of this species extend to 10 m depth in Amazonian forest, yet they do not provide comparative data with trees. Beyond that, the literature is scarce, and most reports comparing rooting depth of lianas and trees in tropical forest utilize comparisons of stable isotopic composition of xylem water from tree and liana stems and the soil profile (Dawson 1993). For example, Jackson et al. (1995) used hydrogen stable isotopic composition to estimate the depth of water uptake of eight trees, one palm, and one liana species on Barro Colorado Island, Panama, in 1992, and determined that the liana species was accessing water sources that were comparable to the deepest sources of trees in their study (~1.2 m). In contrast, in 1997–1998, lianas were estimated to access relatively shallow water sources (0–60 cm) at the beginning of the dry season,

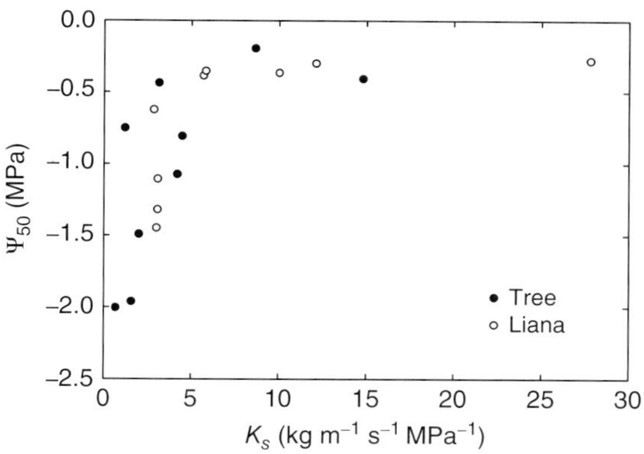

Fig. 20.3 Cavitation resistance measured as the water potential at which 50% loss of hydraulic conductivity occurs (Ψ_{50}) as a function of water transport capacity measured as maximum stem-specific hydraulic conductivity (K_S) for six liana and six tree species from the canopy of a seasonally dry tropical forest measured from the canopy crane, Parque Metropolitano, Panama (M. DeGuzman, unpublished data) and three liana and three tree species from Zhu & Cao (2009).

and increase water uptake to greater depths (>100 cm) as the dry season progressed (Andrade et al. 2005). In seasonally dry tropical forest (1800 mm precipitation year^{-1}) in eastern Amazonia, isotopic analysis of sap water from trees and lianas during the dry season showed that trees acquired labeled water at a deeper level in the soil profile compared to that acquired by lianas, although the depth of water uptake by lianas varied greatly on a seasonal basis (Moreira et al. 2000). Such isotopic techniques address the average depth of water uptake or the depth of water uptake at a snapshot in time, but not the entire extent of the rooting system. Numerous studies using these techniques indicate that water uptake is a dynamic process that varies seasonally, with many species tapping deeper water sources as the dry season progresses (Meinzer et al. 1999; Andrade et al. 2005). Thus, though enhanced dry season growth by lianas strongly implies dependence on deep roots (Schnitzer 2005), the existing results are not all in line with this idea and further studies that are specifically designed to address this phenomenon are desirable.

Studies of whole-plant water use with sap flow techniques have been increasing in recent years, and offer one of the few methods for analyzing whole-plant physiological processes. One of the earliest sap flow studies on lianas was conducted at Chamela, Mexico, by Fichtner and Schulze (1990) using prototype steady-state heating methods. Although this study provided no comparisons with co-occurring tree species, they found large variation in xylem water flow, and quite high rates in some species (Fichtner & Schulze 1990). In secondary Amazonian forest, Restom and Nepstad (2001) found that lianas transpire more than trees with stems of similar diameter, and with similar leaf crown exposure to sunlight, and that lianas experienced a smaller reduction in transpiration from the wet to the dry season than did trees. Most interestingly, they reported that during the dry season, lianas represented 8% of total transpiration, while only representing 5.5% of total basal area (Restom & Nepstad 2001). Similarly, in seasonally dry tropical forest at the Parque Metropolitano Canopy Crane in Panama, lianas exhibited similar transpiration rates relative to trees in a secondary forest, even though their stems were much narrower in diameter (Phillips et al. 2002). In contrast, in mature forest on Barro Colorado Island, Panama, sap flow in co-occurring lianas and trees were comparable at similar stem diameters, and lianas and trees conformed to the same linear relationships between daily sap flow and stem diameter (Andrade et al. 2005), suggesting that high sap flow rates in lianas relative to trees in secondary forest are curtailed as tropical forests mature. These data, which indicate greater whole-plant metabolism in lianas than in trees early in succession, are consistent with the high abundance of lianas in secondary forest, as well as the possibility that increasing disturbance is one of the factors favoring observed increases in liana abundance (Schnitzer, Chapter 30 in this volume).

CONCLUSION AND FUTURE DIRECTIONS

Overall, the available data comparing the physiological propensity of lianas relative to trees provide several illuminating patterns that could help explain curiosities in the abundance and distribution of lianas worldwide. The pattern of overall greater photosynthetic potential in lianas relative to trees that is emerging across the tropics (Asner & Martin 2012, this volume) provides possible explanations for the triumph of lianas early in forest succession (Letcher, Chapter 10 in this volume) where resource availability is high, in seasonally dry tropical forest where light availability is higher than in wetter forest, and in a higher CO_2 world. The ability of many lianas, in many situations, to provide a rapid and efficient water supply to support the physiological activity of leaves is also consistent with their peculiar patterns of distribution and abundance, though the cost of high water transport capacity in terms of increased risk of drought-induced cavitation could become a deciding factor if climate change reduces precipitation inputs in tropical forests as some models predict (Phillips et al. 2010). Additionally, emerging data on whole-plant physiological processes are consistent with leaf- and stem-level studies, and provide an important level of organization for evaluating hypotheses developed in more detailed physiological studies. Finally, and perhaps not surprisingly, belowground processes remain the black box in understanding the overall spectrum of trade-offs that structure the balance of costs and benefits of the liana growth form (see also Schnitzer et al., Chapter 1 in this volume).

Additional studies are necessary to pin down the physiological mechanisms of liana abundance and distribution. First, whereas meta-analyses are excellent for testing comparative hypotheses and putting conventional wisdom to the test of rigor, studies that are specifically designed to test hypotheses of increasing liana abundance are badly needed. Second, because lianas may show contrasting physiological tendencies in young versus old forest, or in seasonally-dry versus ever-wet forests, studies along gradients, or that are broad enough to capture the diversity of tropical environments will be highly valued in the future. Finally, because physiological measurements are time-consuming, but give much greater mechanistic detail than easier-to-measure functional traits, many physiological studies cannot accommodate hundreds of species and suffer from the doubt of species selection. Therefore, careful species selection based on stratified or phylogenetically based sampling designs or plant community-wide campaigns will be essential to maintain our understanding of tropical forests in a dynamic world.

REFERENCES

Andrade, J.L., Meinzer, F.C., Goldstein, G. & Schnitzer, S.A. (2005) Water uptake and transport in lianas and co-occurring trees of a seasonally dry tropical forest. *Trees: Structure and Function*, **19**, 282–289.

Arroyo-Rodríguez, V. & Toledo-Aceves, T. (2009) Impact of landscape spatial pattern on liana communities in tropical rainforests at Los Tuxtlas, Mexico. *Applied Vegetation Science*, **12**, 340–349.

Asner, G.P. & Martin, R.E. (2012) Contrasting leaf chemical traits in tropical lianas and trees: implications for future forest composition. *Ecology Letters*, **15**, 1001–1007.

Cai, Z.Q., Poorter, L., Han, Q. & Bongers, F. (2008) Effects of light and nutrients on seedlings of tropical *Bauhinia* lianas and trees. *Tree Physiology*, **28**, 1277–1285.

Cai, Z.Q., Schnitzer, S.A. & Bongers, F. (2009) Seasonal differences in leaf-level physiology give lianas a competitive advantage over trees in a tropical seasonal forest. *Oecologia*, **161**, 25–33.

Carlquist, S. (1991) Anatomy of vine and liana stem: A review and synthesis, in *The Biology of Vines* (eds F.E. Putz & H.A. Mooney), Cambridge University Press, Cambridge.

Chave, J., Olivier, J., Bongers, F. et al. (2008) Above-ground biomass and productivity in a rain forest of eastern South America. *Journal of Tropical Ecology*, **24**, 355–366.

Chave, J., Riera, B. & Dubois, M.A. (2001) Estimation of biomass in a neotropical forest of French Guiana: spatial and temporal variability. *Journal of Tropical Ecology*, **17**, 79–96.

Christman, M.A., Sperry, J.S. & Adler, F.R. (2009) Testing the 'rare pit' hypothesis for xylem cavitation resistance in three species of *Acer*. *New Phytologist*, **182**, 664–674.

Dawson, T.E. (1993) Water sources of plants as determined from xylem-water isotopic composition: perspectives on plant competition, distribution, and water relations, in *Stable Isotopes and Plant Carbon-Water Relations* (eds J.R. Ehleringer, A.E. Hall & G.D. Farquhar), Academic Press, San Diego, pp. 465–496.

DeWalt, S.J., Schnitzer, S.A. & Denslow, J.S. (2000) Density and diversity of lianas along a chronosequence in central Panamanian lowland forest. *Journal of Tropical Ecology*, **16**, 1–9.

Ewers, F.W., Carlton, M.R., Fisher, J.B. et al. (1997) Vessel diameters in roots versus stems of tropical lianas and other growth forms. *Iawa Journal*, **18**, 261–279.

Ewers, F.W. & Fisher, J.B. (1991) Why vines have narrow stems: histological trends in *Bauhinia* (Fabaceae). *Oecologia*, **88**, 233–237.

Feild, T.S. & Balun, L. (2008) Xylem hydraulic and photosynthetic function of *Gnetum* (Gnetales) species from Papua New Guinea. *New Phytologist*, **177**, 665–675.

Feild, T.S., Chatelet, D.S., Balun, L. et al. (2012) The evolution of angiosperm lianescence without vessels: climbing mode and wood structure-function in *Tasmannia cordata* (Winteraceae). *New Phytologist*, **193**, 229–240.

Fichtner, K. & Schulze, E.D. (1990) Xylem water flow in tropical vines as measured by a steady state heating method. *Oecologia*, **82**, 355–361.

Gartner, B.L., Bullock, S.H., Mooney, H.A. et al. (1990) Water transport properties of vine and tree stems in a tropical deciduous forest. *American Journal of Botany*, **77**, 742–749.

Gentry, A.H. (1991) The distribution and evolution of climbing plants, in *The Biology of Vines* (eds F.E. Putz & H.A. Mooney), Cambridge University Press, Cambridge, pp. 3–49.

Gerwing, J.J. & Farias, D.L. (2000) Integrating liana abundance and forest stature into an estimate of total aboveground biomass for an eastern Amazonian forest. *Journal of Tropical Ecology*, **16**, 327–335.

Gilbert, B., Wright, S.J., Muller-Landau, H.C. et al. (2006) Life history trade-offs in tropical trees and lianas. *Ecology*, **87**, 1281–1288.

Granados, J. & Körner, C. (2002) In deep shade, elevated CO_2 increases the vigor of tropical climbing plants. *Global Change Biology*, **8**, 1109–1117.

Gutierrez, M., San Miguel-Chavez, R. & Terrazas, T. (2009) Xylem conductivity and anatomical traits in diverse lianas and small tree species from a tropical forest of Southwest Mexico. *International Journal of Botany*, **5**, 279–286.

Hacke, U.G., Sperry, J.S., Wheeler, J.K. & Castro, L. (2006) Scaling of angiosperm xylem structure with safety and efficiency. *Tree Physiology*, **26**, 689–701.

Holbrook, N.M. & Putz, F.E. (1996a) Physiology of tropical vines and hemiepiphytes: plants that climb up and plants that climb down, in *Tropical Forest Plant Ecophysiology* (eds S.S. Mulkey, R.L. Chazdon & A.P. Smith), Chapman & Hall, New York, pp. 363–394.

Holbrook, N.M. & Putz, F.E. (1996b) Water relations of epiphytic and terrestrially-rooted strangler figs in a Venezuelan palm savanna. *Oecologia*, **106**, 424–431.

Jackson, P.C., Cavelier, J., Goldstein, G. et al. (1995) Partitioning of water resources among plants of a lowland tropical forest. *Oecologia*, **101**, 197–203.

Laurance, W.F., Perez-Salicrup, D., Delamonica, P. et al. (2001) Rain forest fragmentation and the structure of Amazonian liana communities. *Ecology*, **82**, 105–116.

Letcher, S.G. & Chazdon, R.L. (2009) Lianas and self-supporting plants during tropical forest succession. *Forest Ecology and Management*, **257**, 2150–2156.

Madeira, B., Espirito-Santo, M.M., Neto, S.D. et al. (2009) Changes in tree and liana communities along a successional gradient in a tropical dry forest in south-eastern Brazil. *Plant Ecology*, **201**, 291–304.

Meinzer, F.C., Andrade, J.L., Goldstein, G. et al. (1999) Partitioning of soil water among canopy trees in a seasonally dry tropical forest. *Oecologia*, **121**, 293–301.

Moreira, M.Z., Sternberg, L.d.S.L., & Nepstad, D. (2000) Vertical patterns of soil water uptake by plants in a primary forest and an abandoned pasture in the eastern Amazon: an isotopic approach. *Plant and Soil*, **222**, 95–107.

Patiño, S., Tyree, M.T. & Herre, E.A. (1995) Comparison of hydraulic architecture of woody plants of differing phylogeny and growth form with special reference to freestanding and hemi-epiphytic *Ficus* species from Panama. *New Phytologist*, **129**, 125–134.

Phillips, O.L., Martinez, R.V., Arroyo, L. et al. (2002) Increasing dominance of large lianas in Amazonian forests. *Nature*, **418**, 770–774.

Phillips, O.L., van der Heijden, G., Lewis, S.L. et al. (2010) Drought-mortality relationships for tropical forests. *New Phytologist*, **187**, 631–646.

Putz, F.E. (1983) Liana biomass and leaf area of a tierra firme forest in the Rio Negro Basin, Venezuela. *Biotropica*, **15**, 185–189.

Putz, F.E. (1984) The natural history of lianas on Barro Colorado Island, Panama. *Ecology*, **65**, 1713–1724.

Restom, T.G. & Nepstad, D.C. (2001) Contribution of vines to the evapotranspiration of a secondary forest in eastern Amazonia. *Plant and Soil*, **236**, 155–163.

Restom, T.G. & Nepstad, D.C. (2004) Seedling growth dynamics of a deeply rooting liana in a secondary forest in eastern Amazonia. *Forest Ecology and Management*, **190**, 109–118.

Rowe, N. & Speck, T. (2005) Plant growth forms: an ecological and evolutionary perspective. *New Phytologist*, **166**, 61–72.

Santiago, L.S., Goldstein, G., Meinzer, F.C. et al. (2004) Leaf photosynthetic traits scale with hydraulic conductivity and wood density in Panamanian forest canopy trees. *Oecologia*, **140**, 543–550.

Santiago, L.S. & Wright, S.J. (2007) Leaf functional traits of tropical forest plants in relation to growth form. *Functional Ecology*, **21**, 19–27.

Schnitzer, S.A. (2005) A mechanistic explanation for global patterns of liana abundance and distribution. *American Naturalist*, **166**, 262–276.

Schnitzer, S.A. & Bongers, F. (2011) Increasing liana abundance and biomass in tropical forests: emerging patterns and putative mechanisms. *Ecology Letters*, **14**, 397–406.

Schnitzer, S.A. & Carson, W.P. (2001) Treefall gaps and the maintenance of species diversity in a tropical forest. *Ecology*, **82**, 913–919.

Schnitzer, S.A., Dalling, J.W. & Carson, W.P. (2000) The impact of lianas on tree regeneration in tropical forest canopy gaps: evidence for an alternative pathway of gap-phase regeneration. *Journal of Ecology*, **88**, 655–666.

Sperry, J.S., Donnelly, J.R. & Tyree, M.T. (1988) A method for measuring hydraulic conductivity and embolism in xylem. *Plant, Cell and Environment*, **11**, 35–40.

Tyree, M.T. & Ewers, F.W. (1991) The hydraulic architecture of trees and other woody plants. *New Phytologist*, **119**, 345–360.

van der Sande, M.T., Poorter, L., Schnitzer, S.A., & Markesteijn, L. (in press). Are lianas more drought-tolerant than trees? A test for the role of hydraulic architecture and other stem and leaf traits. *Oecologia*, DOI 10.1007/s00442-012-2563-x

Wright, I.J., Reich, P.B., Westoby, M. et al. (2004a) The worldwide leaf economics spectrum. *Nature*, **428**, 821–827.

Wright, S.J. (2010) The future of tropical forests. *Year in Ecology and Conservation Biology 2010*, **1195**, 1–27.

Wright, S.J., Calderon, O., Hernandez, A. & Paton, S. (2004b) Are lianas increasing in importance in tropical forests? A 17-year record from Panama. *Ecology*, **85**, 484–489.

Yorke, S.R., Schnitzer, S.A., Mascaro, J. et al. (2013) Increasing liana abundance and biomass in a tropical forest: the contribution of long distance clonal colonization. *Biotropica*, **45**, 317–324.

Zhu, S.D. & Cao, K.F. (2009) Hydraulic properties and photosynthetic rates in co-occurring lianas and trees in a seasonal tropical rainforest in southwestern China. *Plant Ecology*, **204**, 295–304.

Zhu, S.D. & Cao, K.F. (2010) Contrasting cost-benefit strategy between lianas and trees in a tropical seasonal rain forest in southwestern China. *Oecologia*, **163**, 591––599.

Chapter 21

CANOPY CHEMISTRY EXPRESSES THE LIFE-HISTORY STRATEGIES OF LIANAS AND TREES

Gregory P. Asner and Roberta E. Martin
Carnegie Institution for Science, Stanford, CA, USA

Ecology of Lianas, First Edition. Edited by Stefan A. Schnitzer, Frans Bongers, Robyn J. Burnham, and Francis E. Putz.

Companion Website: www.wiley.com/go/schnitzer/lianas

OVERVIEW

Lianas are canopy climbers that depend upon host trees for structural support needed to maximize solar radiation capture. How does the foliar chemistry of lianas differ from trees in tropical forests, and can the differences help to explain the successful life strategy of lianas under varying environmental conditions? Our global tropical forest sampling indicates that co-existing canopy lianas and trees indeed differ substantially in their chemical traits. On average, lianas systematically invest more resources to enhance light capture, growth, maintenance and metabolism in their foliage as compared to neighboring trees. In contrast, lianas invest less in chemical compounds to support leaf structure and defense. However, ecoregional differences in the foliar chemistry of lianas and trees are mediated by soils and climate. Soils depleted in rock-derived nutrients suppress the distinction between lianas and trees in growth-related chemical traits, while defense chemistries may diverge between the two growth forms under these conditions. Warmer temperatures also foster divergence in liana and tree chemical traits. Any future changes in the growth, abundance or biomass of lianas in tropical forests will thus likely be constrained by changes in climate as well as by soil fertility.

INTRODUCTION

Lianas are a prominent feature in many tropical forests. As climbing plants, lianas have a much-reduced need for the woody tissue investments required to stand freely, nevertheless they successfully gain access to the upper canopy of their host trees. In doing so, lianas are often able to harvest the maximum available solar radiation, so their presence can alter a cascading range of ecosystem processes (Putz 1984; Schnitzer & Bongers 2002). Lianas also profoundly affect the structure of forest canopies by modifying the subcanopy light environment, the partitioning of the forest biomass, and by smothering the crowns of host trees (Putz 1983; Gerwing & Farias 2000; Ferment et al. 2001).

Given the potential effects of lianas on the structure and functioning of tropical forests, as illustrated in many of the chapters in this volume (e.g., van der Heijden et al. Chapter 13 in this volume; Toledo-Aceves, Chapter 12 in this volume, Schnitzer, Chapter 30 in this volume), an important goal in tropical biology is to understand how lianas attain high abundance in some regions while remaining low in others (Gentry 1983; Dewalt et al. 2000; Schnitzer 2005). Some studies have emphasized the importance of disturbance and other ecosystem-level processes in promoting liana proliferation (Putz & Chai 1987; Schnitzer & Bongers 2011; Dalling et al. 2012; Ledo & Schnitzer 2014; Schnitzer, Chapter 30 in this volume). Others have focused on liana physiology and their underpinning chemical traits, (Kalacska et al. 2007; Cai et al. 2009; Sanchez-Azofeifa et al. 2009; Kazda, Chapter 22 in this volume; Santiago et al., Chapter 20 in this volume). The consensus seems to be that lianas can short-circuit the need to allocate nutrients and carbon to stem components, allowing for increased allocation to foliage for light capture and to roots for nutrient and water acquisition (Isnard & Feild, Chapter 17 in this volume). As such, there is also increasing agreement that lianas fare best in regions with abundant light (less cloud cover) and in drier conditions (Schnitzer 2005; DeWalt et al. 2010; DeWalt et al., Chapter 11 in this volume). The unique distribution of lianas may indirectly suggest that lianas are energy-limited, and that liana life strategies make them more successful in regions with ample solar energy inputs.

Beyond our growing understanding of basic life strategies of lianas, previous research indicates that the abundance and biomass of lianas may be increasing in some tropical forests (reviewed by Schnitzer and Bongers 2011; Schnitzer, Chapter 30 in this volume). Suggested explanations for liana proliferation include the changing forest evapotranspirative (ET) balance due to changes in rainfall and temperature, boosted rates of natural and anthropogenic disturbance, and atmospheric CO_2 enrichment (Schnitzer & Bongers 2011; Schnitzer, Chapter 30 in this volume). Each of these explanations must be supported by physiological responses in lianas to changing resource conditions that favor them over trees. For instance, we might expect to find systematic differences in growth-related leaf traits advantaging lianas over trees under conditions of increasing water stress from rising ET demand, enhanced light or nutrient availability associated with disturbance, or increasing atmospheric CO_2. We might also anticipate differences in foliar traits related to the metabolism that would aid in energetic efficiency.

In a recent study conducted at 48 tropical forest sites globally, we indeed documented a widespread and systematic difference in the foliar chemical composition of liana and tree growth forms (Asner & Martin 2012).

Our findings confirmed and expanded upon previous research from individual sites and regions (Castro-Esau et al. 2004; Castro-Esau et al. 2006; Salzer et al. 2006; Zhang et al. 2006; Sanchez-Azofeifa et al. 2009; Zhu & Cao 2010; Kazda, Chapter 22 in this volume). In particular, we found that lianas maintain an average 14.5% and 16.7% higher concentrations of chemicals responsible for light capture, carbon fixation, and foliar metabolism. These chemicals include photosynthetic pigments, nitrogen (N), and phosphorus (P), as well as the direct product of photosynthesis – soluble carbon (C). We also found that the light capture- and growth-related differences in lianas and trees are maximal under sunnier and warmer climates within the tropical forest biome (Fig. 21.1), a finding in broad agreement with the distribution of lianas documented by Schnitzer (2005), DeWalt et al. (2010), and DeWalt et al. (Chapter 11 in this volume). We thus suggest that liana proliferation might be more likely to occur in forests undergoing increased at-canopy solar irradiance via disturbance, as well as with changes in climate, as proposed by Schnitzer and Bongers (2011) and Schnitzer (Chapter 30 in this volume).

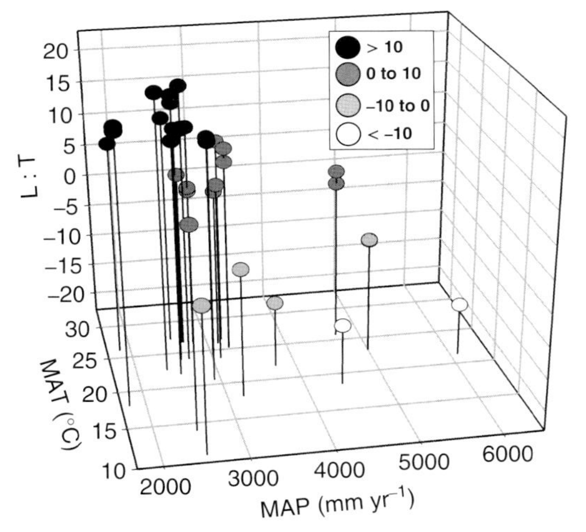

Fig. 21.1 Relative importance of mean annual precipitation (MAP) and temperature (MAT) in determining liana to tree ratio (L:T) of foliar light capture and growth chemical traits on a mass basis. Different shaded symbols (black to white) indicate the percentage difference in overall chemical allocation. These differences are calculated as [(L:T)x100]-100. (Source: Adapted from Asner & Martin 2012. Reproduced with permission of John Wiley & Sons Ltd.)

While growth-related leaf traits may help to understand changing abundances of lianas compared to trees, foliar structure and defense traits are also likely to play an important underlying role. Tropical leaf trait evolution has taken place under high herbivore and pathogen pressure (Janzen 1970; Coley 1983; Fine et al. 2004), so structure and defense compounds are key to tissue survival and longevity supporting whole-plant carbon gain. Yet the overall investment portfolio of a plant is also affected by resource availability, with light and rock-derived nutrients being two of the scarcer resources in humid tropical forests. As a result, trade-offs may need to be made to allocate finite resources to traits that promote growth in lieu of structure and defense, or vice versa (Dudt & Shure 1994; Dominy et al. 2003; Fine et al. 2006). In this context, we also recently found that canopy lianas average about 10% lower investment in foliar traits supporting structure-defense in comparison to co-existing canopy trees (Asner & Martin 2012). The measured traits included polyphenol, tannin, lignin and cellulose, as well as leaf mass per area (LMA). However, in contrast to the light capture and growth compounds, growth form-specific differences in structure-defense traits showed no trend with temperature, precipitation or solar radiation. These findings suggest that the evolution of chemical traits to protect leaves is not driven by climate, even if the light capture and growth traits are correlated with climate variation. So how do abiotic factors such as climate and soil fertility mediate the trade-off between light-capture, growth, structure and defense traits in lianas and trees in tropical forests?

Here we consider this question by extending our earlier work comparing three groups of foliar chemical compounds in canopy lianas and trees of tropical forests. A wide variety of chemical compounds are used in foliage to support multiple functional strategies, which we group into the categories: (1) light capture and growth; (2) structure and defense; and (3) maintenance and metabolism (Table 21.1). Light capture-growth traits include photosynthetic pigments such as chlorophylls and carotenoids, N and P, as well as the immediate products of photosynthesis such as soluble carbon (Evans 1989; Chapin III 1991; Niinemets et al. 1999). Structure-defense compounds include lignin and cellulose that support strength and longevity (Melillo et al. 1982), as well as phenols and tannins for chemical defense (Coley et al. 1993). Maintenance-metabolism elements (e.g., Ca, K, Mg, Zn, Mn, B, Fe) are those required to mediate many

Table 21.1 A portfolio of chemical compounds and elements provide a multi-dimensional indication of foliar investment in light capture and growth, structure and defense, and maintenance and metabolism. We use these chemical traits to compare the life-history strategies of lianas and trees in humid tropical forests.

Light capture and growth	Structure and defense	Maintenance and metabolism
Chlorophylls	Carbon (C)	Potassium (K)
Carotenoids	Hemi-cellulose	Calcium (Ca)
Nitrogen (N)	Cellulose	Magnesium (Mg)
Phosphorus (P)	Lignin	Zinc (Zn), Boron (B)
Water	Phenols	Manganese (Mn)
Soluble carbon	Tannins	Iron (Fe)

functions within the leaf, such as stomatal function and protein synthesis (Niinemets et al. 2001; Porder et al. 2005; John et al. 2007). In our previous work, we focused on these individual compounds and elements, and on the role of climate in maximizing the differences between lianas and trees (Fig. 21.1). Here we take a fresh look at global patterns based on general soil type as well as on an elevation gradient tightly linked to air temperature. We do so with the intention of understanding additional factors mediating differences in the liana and tree foliar chemical traits that underlie their distinct life strategies.

THE LEAF CHEMICAL DATABASE

We draw from a database of canopy foliar samples that includes 7274 individuals from tropical forest sites spanning Neotropical, Afrotropical (Madagascar), and Australasian ecoregions (Asner & Martin 2012). All of the specimens, along with taxonomic identifications and vouchers, can be viewed on the Spectranomics Project website (http://spectranomics.ciw.edu). Across the sites included here, mean annual precipitation ranges from 131–6096 mm yr^{-1}, elevation from 21–3700 m.s.l, and mean annual temperature of 8–27 °C (Table 21.2). The dataset includes 778 liana and 6496 tree specimens. The percentage of samples collected as lianas in any forest stand ranged from 4–41%, with a mean liana:tree sampling ratio of 15%. Lianas are distributed among 563 species, 216 genera in 69 families. Trees from corresponding locations were sampled from 3322 species, 807 genera in 140 families.

Due to the extreme effect of shade on foliar chemical concentrations and LMA (Poorter et al. 2000; Wright et al. 2007; Poorter et al. 2009), and its confusing impact on assessments of foliar traits among lifeforms or species (Asner et al. 2011), we very carefully controlled for full sunlight exposure in developing our canopy foliar dataset. Details on leaf collections, processing, LMA determination and chemical assays are provided on the Spectranomics Project website (http://spectranomics.ciw.edu).

THE POTENTIAL ROLE OF SOILS

We first consider differences in the soils found in the three major ecoregions included in our database (Table 21.2). The African sites are all located in Madagascar on soils dominated by highly dystrophic Entisols. These Malagasy soils are very low in plant-available N (Murray et al. 1985; Gade 1996; Rufino et al. 2006). In contrast, the neotropical and Australasian sites used here are dominated by Ultisols and Oxisols that are low in plant-available Ca and P, while N flows in relative abundance (Boto & Wellington 1984; Townsend et al. 2002, 2008; Gleason et al. 2009; Quesada et al. 2009).

The global differences in leaf traits among lianas and trees discussed earlier (Fig. 21.1) are mirrored to varying degrees within each ecoregion (Fig. 21.2). Differences among these ecoregions suggest a potential

Table 21.2 Ecoregional characteristics including mean annual precipitation (mm yr^{-1}), mean annual temperature (°C), and USDA soil order.

Ecoregion	Number of forest sites	MAP	MAT	Soil order
Madagascar	3	1700–3020	17.0–24.3	Entisols
Australasia	8	1615–3333	18.3–26.6	Ultisols-Oxisols
Neotropics	32	2000–6096	18.0–27.2	Ultisols-Oxisols

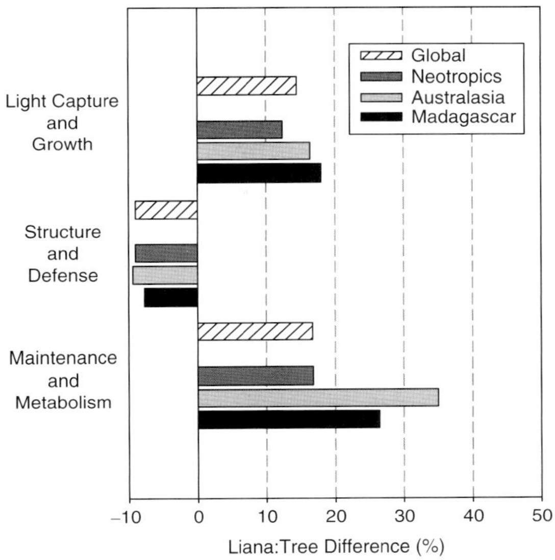

Fig. 21.2 Average differences in mass-based leaf chemical trait allocation between tropical canopy lianas and trees, by ecoregion and globally. Differences are calculated as [(L:T)x100]-100. (Source: Adapted from Asner & Martin 2012. Reproduced with permission of John Wiley & Sons Ltd.)

role of soils in mediating chemical trait differences between lianas and trees. For example, liana investment in light capture and growth chemicals is relatively greater than and more distinct from trees in the Malagasy sites. Much of the distinction is driven by high N and photosynthetic pigment concentrations in the Malagasy lianas, whereas liana:tree differences for these chemical traits are more subtle in the neotropics and Australasia (detailed data not shown). In contrast, defense trait differences between lianas and trees are the smallest in Madagascar. So in the region with the lowest soil N availability, relative investments in foliar growth chemicals are higher, and for defense are lower, in lianas as compared to trees. Perhaps lianas are adapted for maximum light capture and growth under conditions of N limitation as a life strategy to maximize competitiveness with trees. Such a response would follow general theories for plant stoichiometry and physiology under N-limiting conditions (Field & Mooney 1986; Chapin et al. 1987).

In contrast to Madagascar, the neotropical sites are depleted in soil Ca and P, making these the two most limiting nutrients. Such limitations appear to be expressed in liana foliage, which achieved only two-thirds of the differential boost in light capture and growth compounds over trees, as compared to the Malagasy lianas (Fig. 21.2). In other words, lianas and trees are closer to being the same in the neotropics than they are in Madagascar. On the other hand, low soil Ca and P also appear related to an increased divergence in liana and tree structural and defense compounds. While low Ca and P availability is related to boosted phenols, tannins, lignin, cellulose and LMA in neotropical tree foliage (Fine et al. 2006; Fyllas et al. 2009), we found no similar environmental effect on lianas (Asner & Martin 2012). This results in larger differences between the two growth forms (Fig. 21.2).

THE BREAKDOWN OF THE GLOBAL PATTERN

While such broad-brush comparisons suggest that soils mediate differences in foliar chemical traits between lianas and trees, the global-scale pattern remains clear. Yet how do the chemicals within each of the three broad trait categories contribute to life-strategy distinctions between lianas and trees? Much of the underlying difference rests in leaf mass per area, or LMA. On average, the LMA of trees is 20% higher than that of lianas (Fig. 21.3). LMA is a structural trait reflecting differences in construction strategies among plants (Poorter et al. 2009), and in turn, it is broadly representative of the overall chemical trait portfolio of the liana and tree growth forms. If LMA is high, investment in light capture and growth chemicals is often low, as are investments in chemicals supporting foliar maintenance and metabolism (Fig. 21.3B). Meanwhile, investment in structure and defense is also often high in this situation.

The pattern at this level of global integration follows the notion of an evolved leaf economics spectrum (Reich et al. 1997; Wright et al. 2004), particularly because lower LMA in lianas is matched by higher concentrations of photosynthetic pigments and N (Fig. 21.3B). Here we extend this to cations that regulate foliar metabolism – particularly Ca, Mg and K, which are found in higher concentration in lianas than in trees. And oppositely, the leaf economics spectrum is broadly expressed in lower structural and defense investment when nutrients (including cations) are more abundant in leaves.

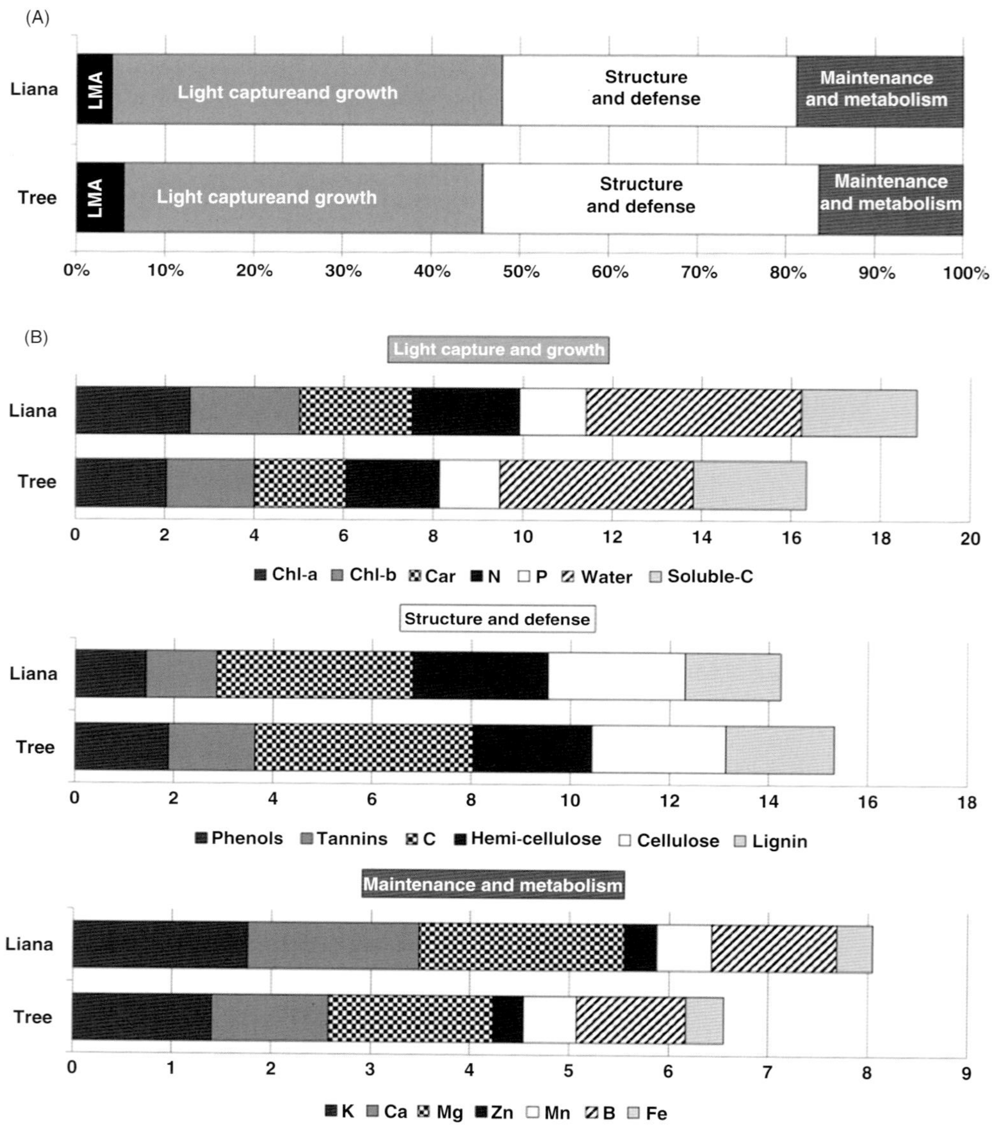

Fig. 21.3 (A) Broad categories of chemical trait strategies and LMA, each calculated as a percentage of the total chemical portfolio measured in canopy foliage of more than 7000 individual trees and lianas (Asner & Martin 2012). The scaled quantities of individual traits within each category are shown in (B). Each trait value is standardized and then relativized to the minimum value to facilitate comparison among traits and across growth forms ($Trait_{ij} = [(m_{ij}-M_j)/SD_j]/min_j$). The size of the bars indicates the amount of each chemical relative to other chemicals within each grouping (e.g., liana – light capture and growth.)

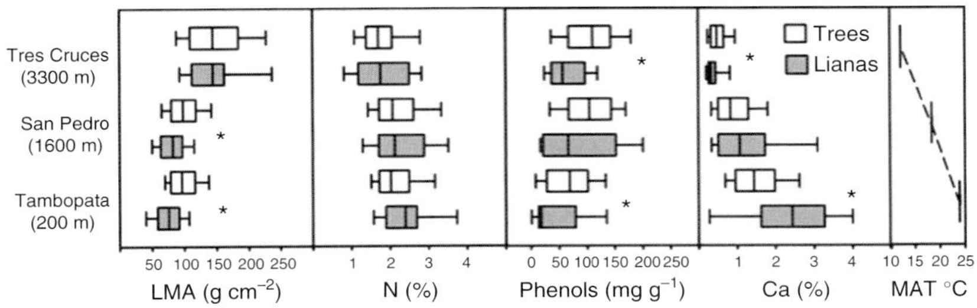

Fig. 21.4 Leaf chemical properties of 44 lianas (grey) and 612 trees (white) measured along an elevation gradient in the Peruvian Amazon. Mean annual temperature (MAT) is significantly correlated (standard least square linear regression; $p < 0.01$) with all traits within a given growth form with the exception of phenols in lianas. Regression r^2-values with LMA, N, phenols and Ca are 0.54, 0.12, 0.01, 0.51 in lianas and 0.28, 0.04, 0.09, and 0.36 in trees, respectively. Asterisk indicates statistical difference in median values between growth forms based on individuals (Wilcoxon rank test; $p < 0.05$.)

THE ROLE OF TEMPERATURE

As shown in Fig. 21.1, foliar chemical traits among co-existing lianas and trees diverge the most under warmer and drier conditions. However, this pattern seems only to hold for light capture and growth chemicals (e.g., N, pigments) and not so for structure and defense traits (Asner & Martin 2012). For example, in Fig. 21.4, we provide some new results from an elevation gradient stretching from the lowland Amazon to the treeline in the Peruvian Andes. Along this transect, mean annual temperature (MAT) changes from 24 °C at 200 m elevation to 8 °C at the highest site near 3300 m above sea level. Changes in MAT along this transect correspond to several trends in leaf structure, growth, defense and metabolism (Fig. 21.4). The pattern is clear for LMA – a trait that integrates the effects of multiple growth-related chemicals. LMA increases with elevation or with decreasing mean annual temperature. Additionally, LMA differences among lianas and trees are significant at the lower elevations, but the two growth forms converge in the coldest temperatures at higher altitude (Fig. 21.4). In contrast, there is a very weak elevational trend in N for either growth form, and we find no growth form-specific differences along the gradient (Fig. 21.4). Any potential trend in foliar N concentration with elevation is trumped by inter-specific variation at any point along the gradient. This pattern – of phylogeny swamping environmental controls – has been observed along gradients of soil fertility in the lowland forests of Amazonia (Fyllas et al. 2009; Asner & Martin 2011) and Borneo (Asner et al. 2012).

While this Andes-to-Amazon elevation gradient imparts little effect on foliar N, we do observe more pronounced trends in rock-derived nutrient concentrations, particularly among P and base cations. In the example presented here, foliar Ca concentrations decrease precipitously as elevation increases (Fig. 21.4). Moreover, at low and high elevation, lianas show divergent investment in Ca from that of co-existing tree canopies. This trend of growth form-specific differences embedded in large-scale environmental gradients is a sign that the chemical resource remains in high demand throughout the region. A repeating theme, demonstrated here and in our previous work, is that abundant nutrients rarely display growth form-specific divergence (e.g., lianas versus trees), whereas nutrients in the highest demand are often found in contrasting concentrations in lianas compared to trees. Lianas appear capable of taking maximum advantage of limiting nutrients, allocating relatively more to their foliage in warmer conditions. However, at altitude, this advantage diminishes in lianas, and growth form-specific differences in rock-derived nutrient concentrations become minimal or even reversed (Fig. 21.4).

CONCLUSION

Independent of floristic composition, climate, soil type, and other sources of environmental variation, mass-based concentrations of leaf chemicals promoting light capture and growth, as well as maintenance and metabolism, remain elevated in lianas as compared to trees in humid tropical forest canopies. In contrast, structure and defense compounds are, on average, lower in concentration in lianas compared to trees. However, there appears to be substantial biogeographic variation in the degree to which lianas and trees diverge in chemical strategy. This likely has much to do with phylogenetic trait evolution under contrasting environmental conditions such as soil type or fertility, but additional study is needed to more carefully tease apart the underlying controls.

The chemical strategy of strong growth and weak defense in liana foliage provides a physiological basis for interpreting observed increases in liana abundance and biomass in some tropical forests. Explanations for this phenomenon have focused on rising ET demand, natural and anthropogenic disturbance rates, and atmospheric CO_2 concentrations (Schnitzer & Bongers 2011; Schnitzer, Chapter 30 in this volume). Our previous results (Asner & Martin 2012), which included stable carbon isotopes sensitive to water uptake, suggested no systematic difference in the average water-use efficiency of lianas and trees (*sensu* Marshall & Zhang 1994; Beerling & Woodward 1995). This would mean that neither growth form possesses an underlying water-use advantage under either increasing ET or elevated CO_2. On the other hand, systematic differences in nutrient concentrations, particularly P and Ca, favor liana growth and metabolic responses to elevated CO_2 (Field & Mooney 1986; Niinemets et al. 1999). In addition, increasing disturbance frequency, such as via logging, fire and storm events (Nelson et al. 1994; Cochrane & Schulze 1999; Asner et al. 2005), could create conditions for added light and nutrient resources, thus advantaging lianas if they are capable of making the most of the liberated resource. In this context, the research community would benefit from experimental work to assess growth form-specific responses to nutrient fertilization across multiple gradients of climate and soil fertility in tropical forests.

ACKNOWLEDGMENTS

We thank R. Tupayachi, F. Sinca, P. Martinez, N. Jaramillo, L. Carranza, C. Anderson, M. Hocheime, P. Weiss, K. Smith, K. Ledesma, G. Powell, A. Bin Suhaili, J. Wright, A. Townsend, and many assistants who helped with logistics, field work and laboratory analyses. We thank F. Bongers, S. Schnitzer, and three reviewers for comments and support. The Carnegie Spectranomics Project is funded by the John D. and Catherine T. MacArthur Foundation.

REFERENCES

Asner, G.P., Knapp, D.E., Broadbent, E.N. et al. (2005) Selective logging in the Brazilian Amazon. *Science*, **310**, 480–482.

Asner, G.P. & Martin, R.E. (2011) Canopy phylogenetic, chemical and spectral assembly in a lowland Amazonian forest. *New Phytologist*, **189**, 999–1012.

Asner, G.P. & Martin, R.E. (2012) Contrasting leaf chemical traits in tropical lianas and trees: implications for future forest composition. *Ecology Letters*, **15**, 1001–1007.

Asner, G.P., Martin, R.E. & Suhaili, A. (2012) Sources of canopy chemical and spectral diversity in lowland Bornean forest. *Ecosystems*, **15**, 504–517.

Asner, G.P., Martin, R.E., Tupayachi, R. et al. (2011) Taxonomy and remote sensing of leaf mass per area (LMA) in humid tropical forests. *Ecological Applications*, **21**, 85–98.

Beerling, D.J. & Woodward, F.I. (1995) Leaf stable carbon isotope composition records increased water-use efficiency of C_3 plants in response to atmospheric CO_2 enrichment. *Functional Ecology*, **9**, 394–401.

Boto, K.G. & Wellington, J.T. (1984) Soil characteristics and nutrient status in a Northern Australian mangrove forest. *Estuaries*, **7**, 61–69.

Cai, Z.-Q., Schnitzer, S.A. & Bongers, F. (2009) Seasonal differences in leaf-level physiology give lianas a competitive advantage over trees in a tropical seasonal forest. *Oecologia*, **161**, 25–33.

Castro-Esau, K.L., Sanchez-Azofeifa, G.A. & Caelli, T. (2004) Discrimination of lianas and trees with leaf-level hyperspectral data. *Remote Sensing of Environment*, **90**, 353–372.

Castro-Esau, K.L., Sanchez-Azofeifa, G.A., Rivard, B. et al. (2006) Variability in leaf optical properties of Mesoamerican trees and the potential for species classification. *American Journal of Botany*, **93**, 517–530.

Chapin, F.S. III (1991) Integrated responses of plants to stress. *BioScience*, **41**, 29–36.

Chapin, F.S. III, Bloom, A.J., Field, C.B. & Waring, R.H. (1987) Plant responses to multiple environmental factors. *BioScience*, **37**, 49–57.

Cochrane, M.A. & Schulze, M.D. (1999) Fire as a recurrent event in tropical forests of the eastern Amazon: effects on forest structure, biomass, and species composition. *Biotropica*, **31**, 2–16.

Coley, P.D. (1983) Herbivory and defensive characteristics of tree species in a lowland tropical forest. *Ecological Monographs*, **53**, 209–234.

Coley, P.D., Kursar, T.A. & Machado, J.-L. (1993) Colonization of tropical rain forest leaves by epiphylls: effects of site and host-plant leaf lifetime. *Ecology*, **74**, 619–623.

Dalling, J.W., Schnitzer, S.A., Baldeck, C. et al. (2012) Resource-based habitat associations in a neotropical liana community. *Journal of Ecology*, **100**, 1174–1182.

DeWalt, S.J., Schnitzer, S.A. & Denslow, J.S. (2000) Density and diversity of lianas along a chronosequence in a central

Panamanian lowland forest. *Journal of Tropical Ecology*, **16**, 1–19.

Dominy, N.J., Lucas, P.W. & Wright, S.J. (2003) Mechanics and chemistry of rain forest leaves: canopy and understorey compared. *Journal of Experimental Botany*, **54**, 2007–2014.

Dudt, J.F. & Shure, D.J. (1994) The influence of light and nutrients on foliar phenolics and insect herbivory. *Ecology*, **75**, 86–98.

Evans, J.R. (1989) Photosynthesis and nitrogen relationships in leaves of C_3 plants. *Oecologia*, **78**, 9–19.

Ferment, A., Picard, N., Gourlet-Fleury, S. & Baraloto, C. (2001) A comparison of five indirect methods for characterizing the light environment in a tropical forest. *Annals of Forest Science*, **58**, 877–891.

Field, C. & Mooney, H.A. (1986) The photosynthesis - nitrogen relationship in wild plants, in *On the Economy of Plant Form and Function* (ed T.J. Givnish), Cambridge University Press, Cambridge, pp. 25–55.

Fine, P.V.A., Mesones, I. & Coley, P.D. (2004) Herbivores promote habitat specialization by trees in Amazonian forests. *Science*, **305**, 663–665.

Fine, P.V.A., Miller, Z.J., Mesones, I. et al. (2006) The growth-defense trade-off and habitat specialization by plants in Amazonian forests. *Ecology*, **87**, S150–S162.

Fyllas, N., Patiño, S., Baker, T. et al. (2009) Basin-wide variations in foliar properties of Amazonian forest: phylogeny, soils and climate. *Biogeosciences*, **6**, 2677–2708.

Gade, D.W. (1996) Deforestation and its effects in Highland Madagascar. *Mountain Research and Development*, **16**, 101–116.

Gentry, A.H. (1983) Lianas and the "paradox" of contrasting latitudinal gradients in wood and litter production. *Tropical Ecology*, **24**, 63–67.

Gerwing, J.J. & Farias, D.L. (2000) Integrating liana abundance and forest stature into an estimate of total aboveground biomass for an eastern Amazonian forest. *Journal of Tropical Ecology*, **16**, 327–335.

Gleason, S.M., Read, J., Ares, A. & Metcalfe, D.J. (2009) Phosphorus economics of tropical rainforest species and stands across soil contrasts in Queensland, Australia: understanding the effects of soil specialization and trait plasticity. *Functional Ecology*, **23**, 1157–1166.

Janzen, D.H. (1970) Herbivores and the number of tree species in tropical forests. *American Naturalist*, **104**, 501–528.

John, R., Dalling, J.W., Harms, K.E. et al. (2007) Soil nutrients influence spatial distributions of tropical tree species. *Proceedings of the National Academy of Sciences*, **104**, 864–869.

Kalacska, M., Bohlman, S., Sanchez-Azofeifa, G.A. et al. (2007) Hyperspectral discrimination of tropical dry forest lianas and trees: comparative data reduction approaches at the leaf and canopy levels. *Remote Sensing of Environment*, **109**, 406–415.

Ledo, A. & Schnitzer, S.A. (2014) Disturbance and clonal reproduction determine liana distribution and maintain liana diversity in a tropical forest. *Ecology*, in press.

Marshall, J.D. & Zhang, J. (1994) Carbon isotope discrimination and water-use efficiency in native plants of the north-central Rockies. *Ecology*, **75**, 1887–1895.

Melillo, J.M., Aber, J.D. & Muratore, J.F. (1982) Nitrogen and lignin control of hardwood leaf litter decomposition dynamics. *Ecology*, **63**, 621–626.

Murray, R.S., Coughlan, K.J. & Quirk, J.P. (1985) Nitrogen sorption isotherms and the microstructure of vertisols. *Australian Journal of Soil Research*, **23**, 137–149.

Nelson, B.W., Kapos, V., Adams, J.B. et al. (1994) Forest disturbance by large blowdowns in the Brazilian Amazon. *Ecology*, **75**, 853–858.

Niinemets, U., Ellsworth, D.S., Lukjanova, A. & Tobias, M. (2001) Site fertility and the morphological and photosynthetic acclimation of Pinus sylvestris needles to light. *Tree Physiology*, **21**, 1231–1244.

Niinemets, U., Tenhunen, J.D., Canta, N.R. et al. (1999) Interactive effects of nitrogen and phosphorus on the acclimation potential of foliage photosynthetic properties of cork oak, Quercus suber, to elevated atmospheric CO_2 concentrations. *Global Change Biology*, **5**, 455–470.

Poorter, L., Kwant, R., Hernandez, R. et al. (2000) Leaf optical properties in Venezuelan cloud forest trees. *Tree Physiology*, **20**, 519–526.

Poorter, H., Niinemets, U., Poorter, L. et al. (2009) Causes and consequences of variation in leaf mass per area (LMA): a meta-analysis. *New Phytologist*, **182**, 565–588.

Porder, S., Paytan, A. & Vitousek, P.M. (2005) Erosion and landscape development affect plant nutrient status in the Hawaiian Islands. *Oecologia*, **142**, 440–449.

Putz, F.E. (1983) Liana biomass and leaf area of a "tierra firme" forest in the Rio Negro basin, Venezuela. *Biotropica*, **15**, 185–189.

Putz, F.E. (1984) Natural history of lianas on Barro Colorado Island, Panama. *Ecology*, **6**, 1713–1724.

Putz, F.E. & Chai, P. (1987) Ecological studies of lianas in Lambir National Park, Sarawak, Malaysia. *Journal of Ecology*, **775**, 523–531.

Quesada, C., Lloyd, J., Schwarz, M. et al. (2009) Regional and large-scale patterns in Amazon forest structure and function are mediated by variations in soil physical and chemical properties. *Biogeosciences*, **6**, 3993–4057.

Reich, P.B., Walters, M.B. & Ellsworth, D.S. (1997) From tropics to tundra: global convergence in plant functioning. *Proceedings of the National Academy of Sciences*, **94**, 13730–13734.

Rufino, M.C., Rowe, E.C., Delve, R.J. & Giller, K.E. (2006) Nitrogen cycling efficiencies through resource-poor African crop-livestock systems. *Agriculture, Ecosystems & Environment*, **112**, 261–282.

Salzer, J., Matezki, S. & Kazda, M. (2006) Nutritional differences and leaf acclimation of climbing plants and the associated vegetation in different types of an Andean montane rainforest. *Oecologia*, **147**, 417–425.

Sanchez-Azofeifa, G.A., Castro, K., Wright, S.J. et al. (2009) Differences in leaf traits, leaf internal structure, and spectral reflectance between two communities of lianas and trees: implications for remote sensing in tropical environments. *Remote Sensing of Environment*, **113**, 2076–2088.

Schnitzer, S.A. (2005) A mechanistic explanation for global patterns of liana abundance and distribution. *American Naturalist*, **166**, 262–276.

Schnitzer, S.A. & Bongers, F. (2002) The ecology of lianas and their role in forests. *Trends in Ecology*, **17**, 223–230.

Schnitzer, S.A. & Bongers, F. (2011) Increasing liana abundance and biomass in tropical forests: emerging patterns and putative mechanisms. *Ecology Letters*, **14**, 397–406.

Townsend, A.R., Asner, G.P. & Cleveland, C.C. (2008) The biogeochemical heterogeneity of tropical forests. *Trends in Ecology and the Environment*, **23**, 424–431.

Townsend, A.R., Asner, G.P., Cleveland, C.C. et al. (2002) Unexpected changes in soil phosphorus dynamics along pasture chronosequences in the humid tropics. *Journal of Geophysical Research-Atmospheres*, **107**, 8067.

Wright, I.J., Ackerly, D.D., Bongers, F. et al. (2007) Relationships among ecologically important dimensions of plant trait variation in seven Neotropical forests. *Annals of Botany*, **99**, 1003–1015.

Wright, I.J., Reich, P.B., Westoby, M. et al. (2004) The worldwide leaf economics spectrum. *Nature*, **428**, 821–827.

Zhang, J., Rivard, B., Sanchez-Azofeifa, G.A. & Castro-Esau, K.L. (2006) Intra- and inter-class spectral variability of tropical tree species at La Selva, Costa Rica: implications for species identification using HYDICE imagery. *Remote Sensing of Environment*, **105**, 129–141.

Zhu, S.-D. & Cao, K.-F. (2010) Contrasting cost-benefit strategy between lianas and trees in a tropical seasonal rain forest in southwestern China. *Oecologia*, **163**, 591–599.

Chapter 22

LIANA–NUTRIENT RELATIONS

Marian Kazda

Institute of Systematic Botany and Ecology, Ulm University, Ulm, Germany

Ecology of Lianas, First Edition. Edited by Stefan A. Schnitzer, Frans Bongers, Robyn J. Burnham, and Francis E. Putz.

Companion Website: www.wiley.com/go/schnitzer/lianas

OVERVIEW

Lianas and trees differ considerably in their leaf-level traits. This chapter summarizes our investigations on leaf traits of lianas and trees based on paired samples of both growth forms. The data set comprises 374 pairs of leaves from lianas and associated supporting vegetation from eight different ecosystems from Africa, Latin America, and Europe. Three traits are evaluated: leaf mass per unit area (LMA), mineral nutrients, and aluminum concentrations in leaves from both growth forms. I also evaluate the importance of the interaction of these traits with disturbance on the growth of lianas. Our data indicate, in accordance with other published studies, that liana leaves are lower in LMA and higher in nitrogen and phosphorus concentrations compared to trees. These characteristics are a prerequisite for fast growth, but low LMA is disadvantageous in dry sites because of high evaporative demand. There, lianas strongly increase the leaf potassium concentrations at low LMA. On acidic sites in an Andean rain forest, leaf potassium was also higher in lianas compared to the self-supporting plants and no Al-accumulating lianas were found compared to 19 individuals regarded as hyper-accumulators among the self-supporting vegetation. Consistent with other recent studies, increasing disturbance in tropical forests is one of the driving factors for liana abundance. High nutrient availability after disturbance may bring about competitive advantages for lianas over trees in such environments. On the other hand, fast growth and build-up of leaf cover by lianas may prevent soil erosion and nutrient leaching from tropical soils. Due to high nutrient concentrations in their leaves, lianas have the potential to re-distribute nutrients in forests and may contribute to long-term productivity in highly dynamic tropical rain forests.

INTRODUCTION

After their seedling stage, climbers rely on external support to maximize their access to light. While self-supporting plants invest substantially in mechanical stability, lianas proportionately allocate less carbon to stem stability and can invest proportionately more in shoot length (Gartner 1991) at comparable stem densities (Putz 1990), or higher (Niklas 1994) compared to forest trees. The shift in carbon allocation in lianas allows climbers to grow particularly fast and reach the best-illuminated parts of the canopy quickly with small stem diameters (Mooney & Gartner 1991; Hegarty 1991; Kurzel et al. 2006).

Generally, plant species with inherently high growth rates tend to have greater photosynthetic capacity than slow-growing plant species. Plant organs of fast-growing species have shorter lifespans, lower dry matter content, and lower carbon concentrations (Aerts & Chapin 2000). Several investigations have shown that climbers have a lower leaf mass per unit area (LMA) compared to plants supporting them (Castellanos et al. 1989; Cornelissen et al. 1996; Kazda & Salzer 2000; Kazda & Mehltreter 2001; Salzer et al. 2006; Kazda et al. 2009). The latter study in a tropical rain forest in Madagascar showed that lianas had lower leaf carbon concentration than trees, and lianas also had also reduced carbon investment per crown volume, but increased leaf area ratio (LAR). High LAR is associated with high growth rates (Poorter & Remkes 1990), which require simultaneously high availability of both light and nutrients.

While investigating liana leaf parameters and comparing them with the self-supporting trees, one has to keep in mind the high plasticity of plants, especially in relation to the light environment (Asner & Martin, Chapter 21 in this volume). The assessment of leaf traits of lianas and trees referred in this chapter was based on paired samples of both growth forms. The t-test for paired samples does not analyse individual values but by subtracting the first score from the second for each pair of subjects, only those "pure" differences between trees and associated lianas emerge. As each "tree–liana pair" is growing under the same environmental conditions, this design excludes the variation in the data set that results from specific nutrient levels of individual subjects. The design is highly sensitive to differences in leaf parameters, such as LMA and nutrient levels. For analysis, leaves of lianas and trees had to grow together under identical conditions. The emphasis was put on simultaneously obtaining segments of both liana and tree twigs at the same site, including understory and canopy microsites. Through this approach, it is possible to evaluate liana leaf traits and to compare them directly with the associated self-supporting vegetation.

This chapter focuses first on leaf structure and leaf nitrogen, phosphorus, and potassium. However, highly weathered tropical soils are often low in available plant nutrients and have high levels of extractable aluminum due to low soil pH. Because high aluminum adversely affects plant growth, the chapter also evaluates the data on aluminum concentration in liana leaves.

Because the abundance of lianas increases dramatically after disturbance (Hegarty & Caballé 1991; De Walt et al. 2000; Laurance et al. 2001, Schnitzer et al., Chapter 7 in this volume), I attempt to link liana nutrient levels to disturbance, and discuss the importance of lianas for nutrient cycling in tropical forests.

LEAF MASS PER AREA AND NITROGEN CONCENTRATION

Data from our studies on pair-wise comparison of liana and tree nutrient levels (Kazda & Salzer 2000; Kazda & Mehltreter 2001; Salzer et al. 2006; Kazda et al. 2009) show that liana leaves consistently have low LMA and higher mass-based nitrogen concentrations (N_{mass}) compared with their host trees (see also Cai & Bongers 2007; Gallagher & Leishman 2012; Tang et al. 2012). The results compiled in Table 22.1 cover mainly tropical rain forests in Africa, the neotropics, seasonal forests (El Rodeo, Costa Rica), drought-tolerant vegetation (Island Martin Garcia, Argentina, Mediterranean islands in Croatia), and deciduous temperate forests in Germany. Despite the taxonomic differences, growth form exerts a major influence upon leaf structure (i.e., LMA) and leaf nutrient concentration.

The lower LMA in lianas compared to trees especially in highly illuminated top-canopy conditions has several positive effects, including: (1) decreased carbon requirements for leaf construction; (2) higher photosynthetic potential through lower internal resistance to CO_2 diffusion (the LMA effect, Poorter & Remkes 1990; Niinemets 1999); and (3) possibly lower load on the supporting tree and thus lower risk of fall (Ichihashi & Tateno 2011). Lower LMA provides a competitive advantage for canopy penetration as shown for invading species by Schieving & Poorter (1999). Furthermore, low LMA promotes high leaf area ratio, low Huber values, and thus high growth rate (Mooney & Gartner 1991) at low costs for space acquisition (Kazda et al. 2009). Furthermore, liana leaves have lower mass-based carbon concentration than tree leaves (Kazda et al. 2009, lianas 43.7%, trees 45.2%, $p < 0.001$, $n = 626$, sites see Table 22.1). Asner and Martin (2012; Chapter 21 in this volume), also report lower LMA and leaf carbon concentration in lianas in a study of 778 lianas of 563 species and 6496 trees of 3322 species from 48 tropical sites. These two traits present lianas leaves with low carbon investment per unit leaf area. Structural optimization like this can partially explain the effective canopy colonization by climbers, especially when linked to higher leaf nutrient concentrations.

Leaf traits have substantial consequences for carbon fixation and growth. Low LMA leads to lower resistance for CO_2 transport from the stomata to the palisade parenchyma, where most of the carbon fixation takes place (Poorter & Remkes 1990; Niinemets 1999). Lower resistance to CO_2 transport may be due to thin leaf tissue (i.e., shorter diffusion pathway) and to low leaf density (i.e. decreased diffusion resistance). High nitrogen concentration in leaves is a prerequisite for effective gas exchange rates (Ellsworth & Reich 1993; Reich et al. 1995). However, measurements of photosynthetic parameters in lianas have not yet shown a consistent picture. In a comparative study in a lowland tropical forest in Panama, lianas showed lower mass-based rates of gas exchange than trees (Santiago & Wright 2007), while no differences between lianas and trees were found in nutrient-rich ecosystems in southwest China (Han et al. 2011). Cai et al. (2009) found that lianas had a higher mass-based photosynthesis than trees, regardless of season, even though photosynthetic nitrogen use efficiency was significantly lower in lianas than in trees.

Across 44 species in a lowland tropical forest in Panama (Santiago & Wright 2007), the area-based relationship between leaf photosynthesis and nitrogen was significant for the full data set, but not when evaluated for trees and lianas separately. In a study comparing photosynthetic performance of the temperate forest liana *Clematis vitalba* to supporting vegetation (six trees and tall shrub species), no difference in carbon fixation rates per unit leaf area was found (Salzer, unpublished). For climbing species, a positive correlation between area-based nitrogen concentration and light-saturated photosynthesis (Fig. 22.1) indicates optimized photosynthetic potential. This relationship was insignificant for the supporting vegetation (trees), and also when analyzing only species pairs consisting of lianas and the shade-tolerant tree *Fagus sylvatica* (*Clematis*: $R^2 = 0.53$, $p < 0.001$, *Fagus*: $R^2 = 0.12$, $p = 0.17$, $n = 17$) (data from Salzer, unpublished).

The observed high growth rates of lianas (Mooney & Gartner 1991) especially height (Schnitzer 2005) may be explained by intensive exploitation of both aboveground and belowground resources (Schnitzer et al. 2005; Toledo-Aceves, Chapter 12 in this volume). High photosynthetic plasticity (Avalos & Mulkey 1999) and preferential carbon allocation in leaf area

Table 22.1 Ecosystems investigated, and number of plant families analyzed for mean values of LMA and N_{mass} by growth form (std. error in parenthesis). Differences between the simultaneously evaluated liana and tree leaves (10–40 leaves per sample, depending on leaf size) were tested by t-test for paired samples.

Ecosystem	No. of liana-tree pairs	No. of plant families		LMA ($g\,m^2$)		Leaf nitrogen ($mg\,g^{-1}$)	
		Trees	Lianas	Trees	Lianas	Trees	Lianas
Mountain rain forest Ecuador (Salzer et al. 2006)	83	15	13	163.0 (5.6)	*** 93.5 (5.5)	12.40 (0.60)	*** 15.72 (0.59)
Lowland tropical rain forest Gabon (Kazda & Salzer 2000)	51	15	10	97.9 (7.4)	*** 68.2 (6.6)	18.4 (0.78)	*** 24.1 (0.71)
Tropical rain forest Madagascar (Kazda et al. 2009)	67	21	15	173.4 (6.4)	*** 98.9 (6.1)	11.7 (0.68)	*** 15.7 (0.65)
Moist tropical forest Tirimbina, Costa Rica (Heinen et al. unpublished)	46	16	16	64.1 (4.6)	*** 43.5 (2.1)	23.6 (0.85)	*** 28.4 (1.1)
Dry tropical forest El Rodeo, Costa Rica (Heinen et al. unpublished)	30	13	13	70.8 (5.5)	*** 48.4 (2.6)	23.9 (1.5)	n.s. 27.3 (1.4)
Dry shrubland, Paraná delta Argentina (Kazda & Mehltreter 2001)	40	7	7	124.0 (9.6)	*** 73.0 (7.6)	26.2 (1.1)	n.s. 27.2 (0.8)
Mediterranean macchia Croatia (Kazda & Salzer, unpublished)	21	4	1	145.4 (10.7)	*** 109.2 (10.9)	13.1 (1.2)	*** 16.3 (1.1)
Temperate deciduous forests Southern Germany (Salzer, unpublished)	36	4	1	49.0 (3.2)	*** 39.9 (2.1)	26.36 (0.74)	*** 28.93 (3.80)
Overall	374	57	43	119.3 (3.2)	*** 75.3 (3.1)	18.09 (0.42)	*** 21.90 (0.40)

Note: ***$p < 0.001$, significance is indicated between the respective means.

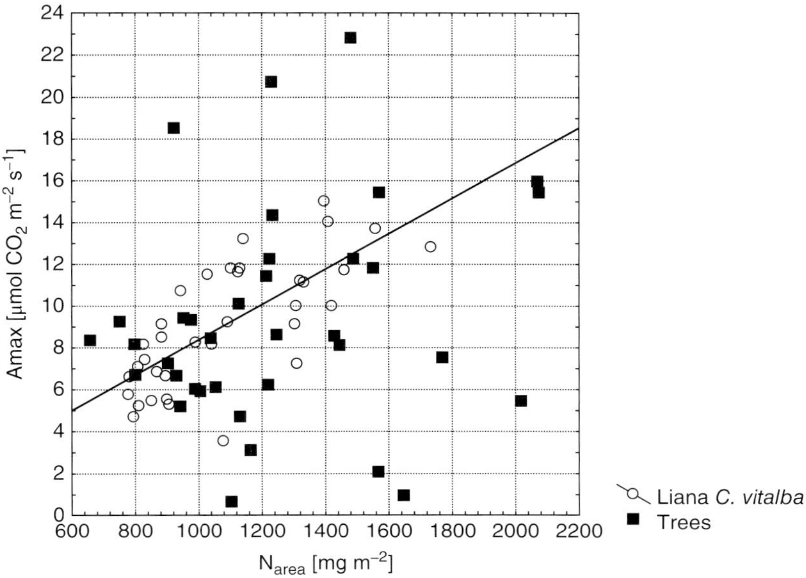

Fig. 22.1 Relationship between nitrogen per leaf area and maximum carbon fixation in *Clematis vitalba* ($y = -0.188 + 0.0085{*}x$, $R^2 = 0.53$, $p < 0.001$, $n = 35$) and in supporting trees ($n = 35$, n.s.) for fully developed leaves in deciduous forests in southern Germany. (Source: Data from Salzer (unpublished). Reproduced with permission.)

versus structural elements distinguish liana versus tree growth (Cai et al. 2009; Kazda et al. 2009) and are linked to higher rates of leaf expansion (Avalos et al. 1999). Variation in structural support in a forest can expose a climber's shoot to various light intensities, which requires high light acclimation potential. Ecophysiological experiments by Avalos and Mulkey (1999) confirmed high trait plasticity in lianas.

The above-mentioned liana traits are consistent with the optimized allocation of nitrogen for photosynthesis. Along an altitudinal gradient in an Andean montane rain forest, the relationship between LMA and area-based nitrogen concentration was twice as steep in climber leaves compared to leaves of their host trees (Salzer et al. 2006). Santiago and Wright (2007) reported a similar result for trees and lianas in a lowland forest in Panama. Smaller variation among N_{area} and a stronger relationship between N_{area} and LMA in climbers suggest better adjustment to light conditions than for their host trees (Salzer et al. 2006).

In a tropical rain forest in Madagascar, we found a pronounced difference in patterns of carbon, nitrogen, and other mineral nutrients between climbers and self-supporting trees growing together under different abiotic environments (Kazda et al. 2009). Understory lianas made a comparatively low per-volume investment in LMA and leaf carbon concentration, which resulted in leaf production at a much lower carbon cost per unit leaf area. Because leaves are spatially arranged, the leaf area density (LAD, $m^2\ m^{-3}$) multiplied by element concentrations in plant organs, gives carbon and nitrogen costs for growing space acquisition (Kazda et al. 2009). Under shady conditions, lianas invested 7.7 times less carbon and 5.5 times less nitrogen per volume of the shared growing space compared to trees. Once in the top canopy, this ratio approximated 3.6 and 2.7 for carbon and nitrogen, respectively. These volume-based carbon and nitrogen data demonstrate that lianas invest substantially less carbon and nitrogen per acquired growth space volume than trees, and that lianas preferentially invest more resources in the high-light environment of the canopy.

For the ecosystems in Gabon, Ecuador, Argentina, and Croatia (cf. Table 22.1), we found significantly higher phosphorus concentration in liana leaves compared to trees (liana 2.20 mg g^{-1}, trees 1.80 mg g^{-1}, $p < 0.001$, $N = 376$), which is in agreement with results by Cai and Bongers (2007), Santiago and Wright (2007), and Asner and Martin (2012; Chapter 21 in this volume). Investigations of forest composition in a subtropical montane forest in north-west Argentina revealed that total liana density was positively correlated with soil phosphorus (Malizia et al. 2010). Laurance et al. (2001) and van der Heijden and Phillips (2008) also found a positive correlation between liana biomass and species richness with soil phosphorus (see also Nurfazliza et al. 2012). In contrast, other studies have failed to find strong support for the variance in liana distribution with soil nutrients (e.g., Dalling et al. 2012; reviewed by Powers, Chapter 14 in this volume). Therefore, the effect of nitrogen and phosphorus availability on liana abundance is not yet resolved, and additional research is needed on how liana abundance changes across broad gradients in soil fertility and climate.

The concentrations of the macronutrients Ca and Mg as evaluated on the basis of ecosystems listed in Table 22.1, did not show any clear differences between climbers and their host trees. These two elements are related mainly to soil properties and are often in low supply in tropical rain forests, such as in Gabon or montane rain forest in Ecuador. In contrast, Asner and Martin (Chapter 21 in this volume) reported consistently higher levels of Ca in lianas than trees, but only in low-elevation forests. Potassium concentration, however, is significantly higher in lianas compared to their host trees (liana $17.7\,mg\,g^{-1}$, trees $11.9\,mg\,g^{-1}$, $p < 0.001$, $n = 728$, Table 22.1). Previous investigations also found higher mass-based potassium concentrations in lianas compared to trees (Santiago & Wright 2007; Asner & Martin 2011).

In summary, liana leaves from tropical rain forests have higher N_{mass} than leaves of the self-supporting plants (Ackerly 1992; Kusumoto & Enoki 2008; Asner & Martin 2012; cf. Table 22.1). However, lianas and trees do not differ regarding N_{mass} in ecosystems with more pronounced drought stress such as on the subtropical island Martin Garcia in Paraná delta or in El Rodeo, a dry tropical forest in Costa Rica (Table 22.1). In drier ecosystems, both growth forms displayed similar high variation in leaf nitrogen (N_{mass} about $26\,mg\,g^{-1}$). High leaf N concentrations in dry habitats may be part of a water conservation strategy (Wright et al. 2001). Increased N_{mass} indicates higher investment in photosynthetic enzymes, which lowers internal CO_2 concentration and thus maintains photosynthesis at lower stomatal conductance (Wright et al. 2001). Low LMA in liana leaves may result in lower diffusion resistance for CO_2, thus increasing carbon fixation, but may also increase the movement of H_2O vapor, thus increasing transpiration.

On dry sites, high photosynthetic water use efficiency and drought stress mitigation and compensation are prerequisites for plant performance. Thus, the question arises, why can lianas dominate in sites with predictable drought (e.g., Schnitzer 2005; DeWalt et al., Chapter 11 in this volume)? One explanation is that in seasonal tropical forests, trees can become dormant to conserve water while lianas remain active longer and may tap deeper sources of water (Schnitzer 2005; Santiago et al., Chapter 20 in this volume). A second could be a better osmotic adjustment through higher potassium concentrations.

THE ROLE OF POTASSIUM IN COMPENSATION FOR LOW LMA

Both lianas and trees display a negative relationships between LMA and leaf potassium (Fig. 22.2). For liana leaves at LMA below $50\,g\,m^{-2}$ the potassium concentration strongly increases (Fig. 22.2A). The temperate deciduous liana, *Clematis vitalba*, along with evergreen lianas from Tirimbina, Costa Rica, showed the lowest LMA and the highest potassium concentrations. For tree leaves, the relationship between LMA and leaf potassium is weaker (lianas: $R^2 = 0.33$, $p < 0.001$, trees: $R^2 = 0.16$, $p < 0.001$; Fig. 22.2B). It is obvious that LMA in liana leaves is always below $200\,g\,m^{-2}$, while the self-supporting vegetation in Madagascar and montane rain forests in Ecuador often exceed this value.

In two ecosystems with pronounced drought periods, a negative relationship between LMA and leaf potassium concentration was found in liana leaves. In the dry forest on the subtropical island, Martin Garcia, in the Paraná delta, Argentina, a significant negative correlation between LMA and potassium was found in liana leaves ($R^2 = 0.58$, $p < 0.001$), but not in tree leaves (Fig. 22.3A). Nitrogen and potassium were also significantly positively correlated in liana leaves ($R^2 = 0.43$, $p < 0.001$) but not for tree leaf LMA (data not shown), suggesting that both nutrients are simultaneously providing higher photosynthetic performance at better osmotic regulation (Kazda & Mehltreter 2001). For the dry tropical forest in El Rodeo, Costa Rica, a steep negative relationship was found between

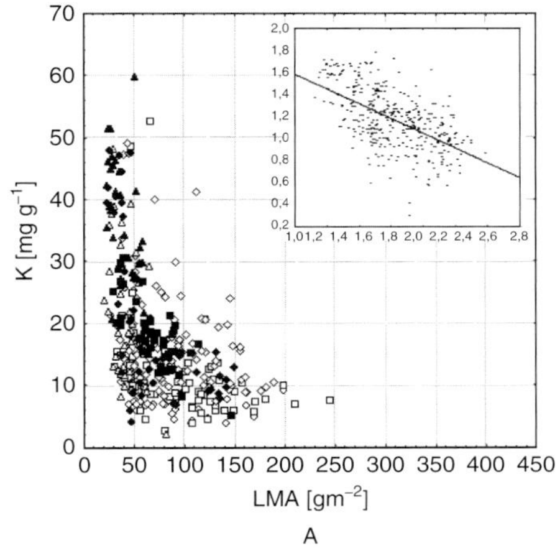

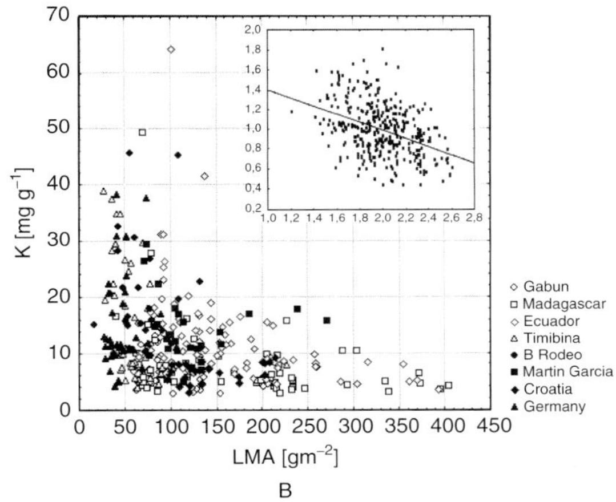

Fig. 22.2 Relationship between leaf mass per leaf area (LMA) and leaf potassium concentration for the whole data set (Table 22.1) for: (A) lianas ($n = 380$); and (B) trees ($n = 347$). Regression equations for log/log diagrams in inset: lianas: $\log(K) = 2.39 - 0.67 \log(LMA)$, $R^2 = 0.33$, $p < 0.001$; trees: $\log(K) = 1.82 - 0.42 \log(LMA)$, $R^2 = 0.17$, $p < 0.001$.

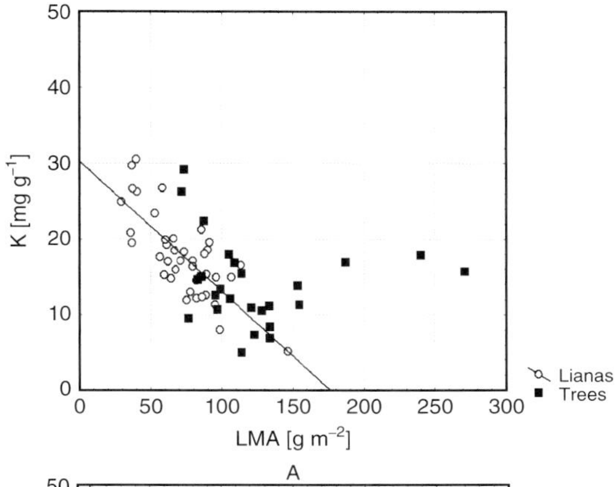

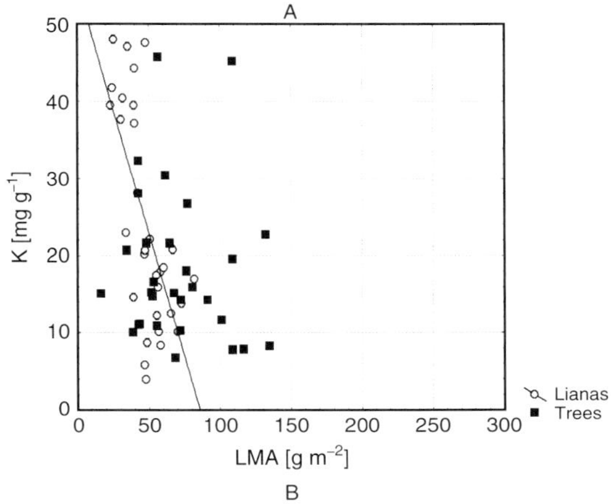

Fig. 22.3 Relationship between leaf mass per leaf area (LMA) and leaf potassium concentration in liana and tree leaves from: (A) Martin Garcia, Argentina (lianas: $y = 30.3 - 0.17*x$, $n = 40$, $p < 0.001$, trees $n = 25$, n.s.) and (B) El Rodeo, Costa Rica (lianas: $y = 55.1 - 0.64*x$, $n = 31$ $p < 0.001$, trees $n = 30$, n.s.). The highest LMA values above 180 g m^{-2} at Martin Garcia, Argentina, belong to *Jodina rhombifolia*, Santalacea. (Source: Based on Table 22.1; data from Kazda & Mehltreter (2001), Heinen et al. unpublished. Reproduced with permission.)

LMA and potassium concentrations for liana leaves ($R^2 = 0.45$, $p < 0.001$, Fig. 22.3B), but not for trees. Thus, even a slight decrease in LMA is counterbalanced by a sharp increase of leaf potassium to compensate for the lower LMA of liana leaves. Obviously, in both ecosystems, potassium concentration scales with LMA in liana leaves, howeve, the tree solution to the water shortage periods is to shed their leaves (Schnitzer 2005). Liana leaves show a different slope in the LMA vs. K relationship between Martin Garcia, Argentina (Fig. 22.3A) and El Rodeo, Costa Rica (Fig. 22.3B), which is presumably due to an extended dry season from December to April in El Rodeo when the average monthly maximum temperatures are 28°–30 °C.

High photosynthetic performance and drought stress mitigation may promote liana growth, particularly during drought conditions. Simultaneously, high radiation in the uppermost canopy requires adaptation to changes in leaf water potential but also to drought stress. Especially in habitats away from tropical rain forests, growth conditions may require leaf structural parameters to flexibly adapt to higher evaporative demands. Van der Sande et al. (2013) postulated that fast growth in lianas is facilitated by low leaf dry matter content and LMA, which would increase water limitation during drought, implying lower drought tolerance in lianas. However, this speculation is inconsistent with data for adult lianas, which had similar (Asner & Martin 2012) or higher water-use efficiency (WUE) than trees during the wet (Zhu & Cao 2009) and the dry season (Cai et al. 2009), and is also inconsistent with the relative abundance of lianas in seasonal forests (DeWalt et al., Chapter 11 in this volume). Higher potassium levels in adult liana leaves compared to trees (Santiago & Wright 2007; Asner & Martin 2011, Fig. 22.2) increase RuBP carboxylase activity and photosynthetic rate (Marschner 1986) and improve stomatal control (Morgan 1984; Marschner, 1986), leading to higher WUE in lianas, especially at dry sites.

Better adjustment between liana leaves of low LMA and osmotic-active potassium (Figs. 22.2, 22.3) may have consequences under drought stress. A comprehensive study in the Neotropics (van der Heijden & Phillips 2009) revealed a much lower decline in species richness with increasing dry season length for liana species compared to tree species richness (see also DeWalt et al., Chapter 11 in this volume). Liana species richness showed lower correlation with mean annual precipitation and dry season length compared to trees. Since no clear explanation for this trend could be given (van der Heijden & Phillips 2009), the less pronounced impact of dry season length on liana species richness might be linked to finer osmotic regulation and hence drought mitigation because of higher concentrations and well-allocated potassium in liana leaves. This would be especially effective under extended periods of drought. Indeed, in Ghana, the relative number of liana species (in terms of the vascular plant community) increased from 30% at 2000 mm annual rainfall to 43% at 1000 mm annual rainfall (Swaine & Grace 2007), consistent with the increase in liana density with decreasing rainfall (DeWalt et al., Chapter 11 in this volume).

LEAF ALUMINUM

Acid soils of tropical rainforests are known to be low in available nutrients and high in exchangeable aluminum, and thus plants have to cope with the toxic effects of Al^{3+} ions (Lüttge 1997; Ma et al. 1997). Our investigations in the Andean rain forest (Salzer et al. 2006) indicate that liana leaves have higher concentrations of nitrogen, phosphorus, and potassium, but lower aluminum compared to trees and shrubs. Tree leaf Al concentration was generally substantially higher (mean, median, lower and upper quartile respectively: 1229.8, 96.4, 55.3, 268.3 mg kg^{-1}) than in climbers (mean, median, lower and upper quartile: 89.9, 64.4, 34.1, 97.3 mg kg^{-1}). Out of the 70 host trees examined, 19 had leaf Al concentration far exceeding 1000 mg kg^{-1}. Maximum values measured in these Al hyper-accumulators (Jansen et al. 2002) were up to 15700 mg kg^{-1}. Especially pronounced was the aluminum accumulation in 10 self-supporting specimens belonging to five different species of the Melastomataceae, which is also known for high Al concentrations in xylem sap (Haridasan 1982; Lüttge 1997). In contrast, two climbing specimens of the genus *Topobea* (Melastomataceae) sampled in two different sites of the Andean rain forest in Ecuador (Salzer et al. 2006) had Al concentrations below 150 mg kg^{-1}.

Osaki et al. (2003) examined mineral characteristics of leaves from a variety of plant taxa grown in various soils and concluded that leaf Al concentrations are not determined by soil type but by plant-specific characteristics for Al accumulation or tolerance. Plant-internal Al storage is linked to higher level of phenolic substances (González-Santana et al. 2012), but there are other suggested mechanisms for internal detoxification of aluminum such as the formation of aluminum chelates, segregating aluminum in the vacuole, aluminum-tolerant proteins, and enzyme activity (reviewed in Jansen et al. 2002). Especially important for Al-tolerance are organic acids and other complex-forming compounds, such as phenolic substances, that bind Al^{3+} ions in the cytosol (Brunner & Sperisen 2013).

It is possible that climber leaves of low LMA are physiologically not well adapted for Al-storage. This view is supported by results from investigations of liana abundance at seven study sites in Ecuador (Homeier et al. 2010), where liana density was negatively correlated with exchangeable soil aluminum. Limited possibilities for leaf-internal Al storage may affect liana abundance on sites high in Al^{3+} ions in the soil, unless they can exclude aluminum from root uptake. Excluding aluminum from root uptake can be facilitated by exudation by chelating ligands (Jansen et al. 2002; Brunner & Sperisen 2013), which converts free Al^{3+} ions to less toxic Al complexes. Aluminum toxicity also diminishes at increased calcium concentrations in the soil solution (Kazda & Zwacek 1989), and lianas appear to have higher levels of leaf Ca (Asner & Martin, Chapter 21 in this volume).

Another mechanism for Al detoxification is the alkalinization of the rhizosphere through preferential NO_3 uptake (Marschner 1986). Increasing pH will shift Al speciation in the soil solution towards much less toxic Al hydroxides (Jansen et al. 2002). Preferential nitrate uptake may play an important role in this context because nitrate levels in the soil increase substantially after natural disturbance events (Hasselquist et al. 2009), especially after logging operations (Ivo et al. 1996). Under such circumstances, high nitrate levels after disturbance may enable lianas to perform better even on Al-rich soils.

DISTURBANCE AND LEAF NUTRIENT LEVELS

Malizia et al. (2010) found that total liana density increased in recent treefall gaps, and other studies confirm that many liana species respond positively to canopy disturbance (Putz 1984; Laurance et al. 2001; Schnitzer & Carson 2001; Dalling et al. 2012; Schnitzer et al., Chapter 7 in this volume). Disturbed sites, natural gaps, and newly created forest edges are characterized by high light conditions, variability in soil moisture and microclimate, and increased nutrient availability due to higher litter fall on the forest floor (van der Heijden & Phillips 2009; Schnitzer & Bongers 2011; Asner & Martin 2012; and others). Higher nutrient concentration in climber leaves compared to their host trees indicates that liana growth requires high soil nutrient availability (Nurfazliza et al. 2012), even though other investigations (Dalling et al. 2012) did not find a clear nutrient-related pattern in liana distribution and postulated their dependency on local disturbance history (see also Schnitzer et al., Chapter 7 in this volume).

Heinen (unpublished) compared leaf parameters of lianas and their host trees on neighboring intact and disturbed sites in two forests, Tirimbina and El Rodeo in Costa Rica (cf. Table 22.1). The mean LMA values of the lianas were significantly lower than trees on intact, undisturbed forest (liana 48.9, trees 64.1 $g\,m^{-2}$) as well as in disturbed areas (liana 38.7, trees 62.1 $g\,m^{-2}$). Tree LMA did not change in response to disturbance (only a 3% decrease), there was a clear trend towards lower LMA in lianas on disturbed sites (a 21% decrease), but this change was not significant ($p = 0.15$). Even though the species assembly was not identical on intact and disturbed sites, the trend for lower LMA on disturbed sites contrasts with the general increase in LMA under high light conditions after disturbance. Low LMA in lianas even after disturbance indicates the ability of lianas to build up their leaf area at low construction costs (Gallagher & Leishman 2012; Asner & Martin, Chapter 21 in this volume) and agrees with low per-volume investment for growing space acquisition (Kazda et al. 2009).

Leaf potassium concentrations of both lianas and trees were significantly higher in disturbed sites (lianas 26.6, trees 20.7 $mg\,g^{-1}$) than on intact sites (lianas 18.9, trees 13.4 $mg\,g^{-1}$). In this case, however, potassium decreased similarly in lianas and trees from intact to disturbed forest (29% and 35%, respectively). The change in potassium underlines the important role it plays in mitigating the impact of water stress after disturbance, similar to the compensatory role of potassium in dry sites. In contrast, leaf nitrogen concentration was not affected by disturbance (disturbed: lianas 28.8, trees 24.2 $mg\,g^{-1}$, intact: lianas 28.2, trees 22.7 $mg\,g^{-1}$), even though total soil nitrogen declined on disturbed sites from 5.1 to 3.1 $mg\,g^{-1}$ and from 6.7 to 5.1 $mg\,g^{-1}$ respectively. After disturbance, mineralization temporarily increases nitrogen availability (mainly as nitrate) but over the long run, the total nitrogen content in the soil decreases. Such dynamics of soil N as a consequence of organic matter mineralixation was also found by Mo et al. (2003) for tropical forests in southern China.

Combined evidence suggests that under high light availability and nutrient mineralisation following disturbance, lianas can grow quickly and capture resources effectively (Zhu & Cao 2010). Low LMA, high leaf area ratio (LAR) (Kazda et al. 2009), possible adjustment of LMA, leaf nitrogen (Salzer et al. 2006), photosynthetic apparatus (Avalos & Mulkey 1999; Avalos et al. 2007), and LAR (Kazda et al. 2009) enable lianas to grow well under the varied conditions after disturbance. Furthermore, large vessel elements in lianas keep the costs of water transport lower than in trees to per unit carbon

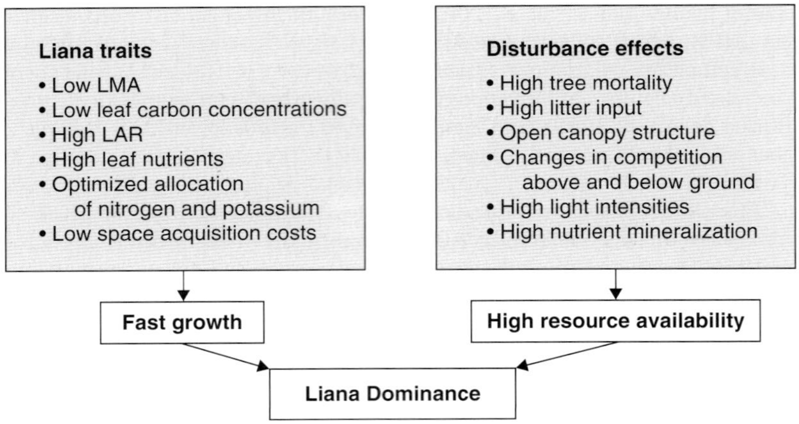

Fig. 22.4 Hypothesized schematic relationship between liana traits and disturbance.

investment (Gallagher & Leishman 2012). The resulting competitive advantage of lianas over trees in disturbed environments (Schnitzer 2005) is summarized in Fig. 22.4. This suggests a conceptual view of the high rate of climbers in space acquisition in disturbed sites (Dalling et al. 2012, Schnitzer et al., Chapter 7 in this volume).

Liana abundance, biomass, and species diversity increase after disturbance and close to forest edges (Hegarty & Caballé 1991; Laurance et al. 2001; Schnitzer, Chapter 30 in this volume). High rates of forest fragmentation over the past three decades have exaggerated this process, creating extended edge effects (Laurance et al. 2007). As a result, regeneration of self-supporting vegetation is suppressed by overgrowing lianas, creating long-term and large impacts on forest regeneration (Putz 1995; Laurance et al. 2001; Schnitzer et al. 2000; Schnitzer & Bongers 2002). Following disturbance, the remaining young trees and standing dead trees may offer structural support to lianas. The combination of available trellises and high light in forest gaps contributes to liana proliferation (Putz 1984). In growth experiments with three species of *Passiflora*, external support was correlated with a more efficient nitrogen allocation. Stable isotopes signatures indicated that supported plants were more strongly influenced by leaf nitrogen concentration than unsupported ones (Werth et al. 2013). Hättenschwiler (2002) also found strong growth response after nitrogen addition, and thus nitrogen appears to be one of the many factors, along with supporting structures, that influence liana proliferation.

IMPLICATIONS FOR NUTRIENT CYCLING

High nutrient levels are present in the foliage of various species of lianas. Lianas are thus a route for both storage and redistribution of nutrients in tropical forests (Santiago 2010; Powers, Chapter 14 in this volume), especially when liana canopies are distant from their rooting zone and the acquired soil nutrients are deposited via leaf litter far from their roots (Putz 1984). Such redistribution and homogenization of soil nutrients in tropical forests were postulated by Powers et al. (2004) and Schnitzer and Bongers (2011). As liana leaves are often higher in N, P and K than host trees, positive effects especially of litter P and K on decomposition can be expected (Santiago 2010). In a fertilization experiment at Barro Colorado Island, Kaspari et al. (2008) also found that P and K addition – via liana litter deposition – substantially increased decomposition rates.

In well-established, undisturbed tropical forests, most nutrients are locked in the standing biomass and litter layer. The soil litter layer is relatively poor in available nitrogen because plant tissues are rich in C compared to N. Constant litter fall keeps this layer N poor and it immobilizes bioavailable N (Hedin et al. 2009). The litter layer is even relatively poorer in phosphorus compared to nitrogen due to the high N:P ratio in leaf litter in tropical forests (Hedin et al. 2009). Therefore, under shaded conditions and limited nutrient availability, it may be difficult for lianas to obtain sufficient nutrients for fast growth and proliferation to the upper canopy.

This results in a tight spiral of decreased nutrient acquisition followed by decreased nutrient deposition per volume of leaf litter (Kazda et al. 2009).

Natural disturbances such as hurricanes and the collapse of large trees within a forest can result in a pulse of elevated nutrient litter fall because there was no opportunity for their re-translocation during leaf senescence (Hasselquist et al. 2009). Even more severe is the impact on nutrient cycling in tropical forests due to logging, where skidding trails increase soil erosion and nutrient mineralization and losses (Asner et al. 2004). Nitrate levels in soils increase following disturbance events (Hasselquist et al. 2009), especially after logging operations (Ivo et al. 1996). This may also be exacerbated by global nitrogen emissions (Hietz et al. 2011). Tropical rain forests are nitrogen-saturated (Brookshire et al. 2012) but this nitrogen is mostly locked in plant biomass and soil organic matter (Hedin et al. 2009). Even minor disturbances such as canopy gaps significantly enhance the mineralization of the soil litter and increase nutrient availability. High nitrate exports from tropical watersheds (Brookshire et al. 2012) are then accompanied by leaching of cations such as Mg and K.

Lianas are thus highly effective in capturing and sequestering nutrients after disturbance because of their rapid production of leaf area (Mooney & Gartner 1991; Hegarty 1991). Furthermore, fast growth and build-up of leaf cover by lianas may also prevent rain-related erosion of bare soil exposed by disturbance. Temporal storage in liana foliage and later re-distribution via litter fall may prevent nutrients from being leached from tropical soils. In relation to the low cation exchange capacity (CEC) of many tropical forest soils, the capturing and storage of potassium is crucial due to its high soil mobility. The high potassium concentrations in liana leaves may be of vital importance for ecosystem productivity. Thus, lianas are both structural parasites *sensu* Stevens (1987), and also play an important role in maintaining stable ecosystem productivity (Tang et al. 2012) via nutrient storage and re-distribution in forests (Schnitzer & Bongers 2011).

ACKNOWLEDGMENTS

I would like to thank the many students and co-workers from Ulm University, who have contributed to the work on liana–tree interactions in the field and laboratory over many years. For field work I am especially grateful to the organizers of the Operation Canopée for the opportunity to participate in expeditions to Gabon and Madagascar. I am especially indebted to Francis Hallé for his generous support and help in the systematic determination of plant specimens on these expeditions. I also thank Frans Bongers for his patience while finishing this work as well as to two anonymous reviewers for valuable and improving comments on earlier versions of this manuscript.

REFERENCES

Ackerly, D.D. (1992) Light, leaf age, and leaf nitrogen concentration in a tropical vine. *Oecologia*, **89**, 596–500.

Aerts, R. & Chapin, F.S. III (2000) The mineral nutrition of wild plants revisited: a re-evaluation of processes and patterns. *Advances in Ecological Research*, **30**, 1–67.

Asner, G.P., Keller, M. & Silva, J.N.M. (2004) Spatial and temporal dynamics of forest canopy gaps following selective logging in the Eastern Amazon. *Global Change Biology*, **10**, 765–783.

Asner, G.P. & Martin, R.E. (2011) Canopy phylogenetic, chemical and spectral assembly in a lowland Amazonian forest. *New Phytologist*, **189**, 999–1012.

Asner, G.P. & Martin, R.E. (2012) Contrasting leaf chemical traits in tropical lianas and trees: implications for future forest composition. *Ecology Letters*, **15**, 1001–1007.

Avalos, G. & Mulkey, S.S. (1999) Photosynthetic acclimation of the liana *Stigmaphyllon lindenianum* to light changes in a tropical dry forest canopy. *Oecologia*, **120**, 475–484.

Avalos, G., Mulkey, S.S. & Kitajima, K. (1999) Leaf optical properties of trees and lianas in the outer canopy of a tropical dry forest. *Biotropica*, **31**, 517–520.

Avalos, G., Mulkey, S.S., Kitajima, K. & Wright, S.J. (2007) Colonization strategies of two liana species in a tropical dry forest canopy. *Biotropica*, **39**, 393–399.

Brookshire, E.N.J., Gerber, S., Menge, D.N.L. & Hedin, O.L. (2012) Large losses of inorganic nitrogen from tropical rainforests suggest a lack of nitrogen limitation. *Ecology Letters*, **15**, 9–16.

Brunner, I. & Sperisen, C. (2013) Aluminum exclusion and aluminum tolerance in woody plants. *Frontiers in Plant Science*, **4**, 172. doi: 10.3389/fpls.2013.00172

Cai, Z. & Bongers, F. (2007) Contrasting nitrogen and phosphorus resorption efficiencies in trees and lianas from a tropical montane rain forest in Xishuangbanna, South-West China. *Journal of Tropical Ecology*, **23**, 115–118.

Cai, Z., Schnitzer, S. & Bongers, F. (2009) Seasonal differences in leaf-level physiology give lianas a competitive advantage over trees in a tropical seasonal forest. *Oecologia*, **161**, 25–33.

Castellanos, A., Mooney, H.A., Bullock, S.H. et al. (1989) Leaf, stem and metamer characteristics of vines in a tropical deciduous forest in Jalisco, Mexico. *Biotropica*, **21**, 41–49.

Cornelissen, J.H.C., Castro-Diez, P. & Hunt, R. (1996) Seedling growth, allocation, and leaf attributes in a wide range of woody plant species and types. *Journal of Ecology*, **84**, 755–765.

Dalling, J.W., Schnitzer, S.A., Baldeck, C. et al. (2012) Resource-based habitat associations in a neotropical liana community. *Journal of Ecology*, **100**, 1174–1182.

Ellsworth, D.S. & Reich, P.B. (1993) Canopy structure and vertical patterns of photosynthesis and related leaf traits in a deciduous forest. *Oecologia*, **96**, 169–178.

Gallagher, R.V. & Leishman, M.R. (2012) Contrasting patterns of trait-based community assembly in lianas and trees from temperate Australia. *Oikos*, **121**, 2026–2035.

Gartner, B.L. (1991) Relative growth rates of vines and shrubs of western Poison oak, *Toxicodendron diversilobum* (Anacardiaceae). *American Journal of Botany*, **78**, 1345–1353.

González-Santana, I., Hayde, J., Márquez-Guzmán, S. et al. (2012) *Conostegia xalapensis* (Melastomataceae): an aluminum accumulator plant. *Physiologia Plantarum*, **144**, 134–145.

Han, L., Xie, L.-J., Dai, K.-J. et al. (2011) Contrasting leaf characteristics of trees and lianas in secondary and mature forests in Southwestern China. *Photosynthetica*, **48**, 559–566.

Haridasan, M. (1982) Aluminium accumulation of some cerrado native species of central Brazil. *Plant and Soil*, **65**, 265–273.

Hasselquist, N.J., Santiago, L.S. & Allen, M.F. (2009) Belowground nitrogen dynamics in relation to hurricane damage along a tropical dry forest chronosequence. *Biogeochemistry*, **98**, 89–100.

Hättenschwiler, S. (2002) Liana seedling growth in response to fertilisation in a neotropical forest understorey. *Basic and Applied Ecology*, **143**, 135–143.

Hedin, L.O., Brookshire, E.N.J., Menge, D.N.L. & Barron, A.R. (2009) The nitrogen paradox in tropical forest ecosystems. *Annual Review of Ecology, Evolution, and Systematics*, **40**, 613–635.

Hegarty, E.E. (1991) Vine-host interactions, in *The Biology of Vines* (eds F.E. Putz & H.A. Mooney), Cambridge University Press, Cambridge, pp. 357–376.

Hegarty, E.E. & Caballé, G. (1991) Distribution, and abundance of vines in forest communities, in *The Biology of Vines* (eds F.E. Putz & H.A. Mooney), Cambridge University Press, Cambridge, pp. 313–335.

Hietz, P., Turner, B.L., Wanek, W. et al. (2011) Long-term change in the nitrogen cycle of tropical forests. *Science*, **334**, 664–666.

Homeier, J., Englert, F., Leuschner, C. et al. (2010) Factors controlling the abundance of lianas along an altitudinal transect of tropical forests in Ecuador. *Forest Ecology and Management*, **259**, 1399–1405.

Ichihashi, R. & Tateno, M. (2011) Strategies to balance between light acquisition and the risk of falls of four temperate liana species: to overtop host canopies or not? *Journal of Ecology*, **99**, 1071–1080.

Ivo, W.M., Ferreira, S., Biot, Y. & Ross, S. (1996) Nutrients in soil solution following selective logging of a humid tropical terra firme forest North of Manaus, Brazil. *Environmental Geochemistry and Health*, **8**, 69–75.

Jansen, S., Broadley, M.R., Robbrecht, R. & Smets, E. (2002) Aluminum hyperaccumulation in Angiosperms: a review of its phylogenetic significance. *The Botanical Review*, **68**, 235–269.

Kaspari, M., Garcia, M.N., Harms, K.E. et al. (2008) Multiple nutrients limit litterfall and decomposition in a tropical forest. *Ecology Letters*, **11**, 35–43.

Kazda, M. & Mehltreter, K. (2001) Contrasting leaf nutrition, and leaf mass per unit area in lianas, and trees from the subtropical Island Martin Garcia, Argentina. *Dissertationes Botaniceae*, **346**, 131–139.

Kazda, M., Miladera, J.C. & Salzer, J. (2009) Optimisation of spatial allocation patterns in lianas compared to trees used for support. *Trees*, **23**, 295–304.

Kazda, M. & Salzer, J. (2000) Leaves of lianas, and self-supporting plants differ in mass per unit area, and in nitrogen content. *Plant Biology*, **2**, 268–271.

Kazda, M. & Zwacek, L. (1989) Aluminium and manganese and their relation to calcium in soil solution and needles in three Norway spruce (*Picea abies*, L. Karst.) stands of Upper Austria. *Plant and Soil*, **267**, 257–267.

Kurzel, B.P., Schnitzer, S.A. & Carson, W.P. (2006) Predicting liana crown location from stem diameter in three Panamanian forests. *Biotropica*, **38**, 262–266.

Kusumoto, B. & Eonoki, T. (2008) Contribution of a liana species, *Mucuna macrocarpa* Wall., to litterfall production and nitrogen input in a subtropical evergreen broad-leaved forest. *Journal of Forest Research*, **13**, 35–42.

Laurance, W.F., Nascimento, H.E.M., Laurance, S.G. et al. (2007) Habitat fragmentation, variable edge effects, and the landscape-divergence hypothesis. *PloS One*, **2**, e1017. doi: 10.1371/journal.pone.0001017

Laurance, W.F., Perez-Salicrup, D., Delamonica, P. et al. (2001) Rain forest fragmentation and the structure of Amazonian liana communities. *Ecology*, **82**, 105–116.

Lüttge, U. (1997) *Physiological Ecology of Tropical Plants*, Springer Verlag, Berlin.

Ma, J.F., Hiradate, S., Nomoto, K. et al. (1997) Internal detoxification mechanism of Al in *Hydrangea*. *Plant Physiology*, **113**, 1033–1039.

Malizia, A., Grau, H.R. & Lichstein, J.W. (2010) Soil phosphorus and disturbance influence liana communities in a subtropical montane forest. *Journal of Vegetation Science*, **21**, 551–560.

Marschner, H. (1986) *Mineral Nutrition of Higher Plants*, Academic Press, London.

Mo, J., Brown, S., Peng, S. & Kong, G. (2003) Nitrogen availability in disturbed, rehabilitated and mature forests of tropical China. *Forest Ecology and Management*, **175**, 573–583.

Mooney, H.A. & Gartner, B.L. (1991) Reserve economy of vines, in *The Biology of Vines* (eds F.E. Putz & H.A. Mooney), Cambridge University Press, Cambridge, pp. 161–179.

Morgan, J.M. (1984) Osmoregulation and water-stress in higher plants. *Annual Review of Plant Physiology and Plant Molecular Biology*, **35**, 299–319.

Niinemets, Ü. (1999) Components of leaf dry mass per area; thickness and density alter leaf photosynthetic capacity in reverse directions in woody plants. *New Phytologist*, **144**, 35–47.

Niklas, K.J. (1994) Comparisons among biomass allocation and spatial distribution patterns of some vine, Pteridophyte, and Gymnosperm shoots. *American Journal of Botany*, **11**, 1416–1421.

Nurfazliza, K., Nizam, M.S. & Nur Supardi, M.N. (2012) Association of liana communities with their soil properties in a lowland forest of Negeri Sembilan, peninsular Malaysia. *Sains Malaysiana*, **41**, 679–690.

Osaki, M., Yamada, S., Ishizawa, T. et al. (2003) Mineral characteristics of the leaves of 166 plant species with different phylogeny in the temperate region. *Plant Foods for Human Nutrition*, **58**, 139–152.

Poorter, H. & Remkes, C. (1990) Leaf area ratio, and net assimilation rate of 24 wild species differing in relative growth rate. *Oecologia*, **83**, 553–559.

Powers, J.S., Kalicin, M.H. & Newman, M.E. (2004) Tree species do not influence local soil chemistry in a species-rich Costa Rica rain forest. *Journal of Tropical Ecology*, **20**, 587–590.

Putz, F.E. (1980) Lianas vs. trees. *Biotropica*, **12**, 224–225.

Putz, F.E. (1984) The natural history of lianas on Barro Colorado Island, Panama. *Biotropica*, **22**, 103–104.

Putz, F. (1990) Liana stem diameter growth and mortality rates on Barro Colorado Island, Panama. *Biotropica*, **22**, 103–105.

Putz, F.E. (1995) Vines in treetops: consequences of mechanical dependence, in *Forest Canopies* (eds M.D. Lohmann & N.M. Nadkarni), Academic Press, San Diego, pp. 311–323.

Reich, P.B., Kloeppel, B.D., Ellsworth, D.S. & Walters, M.B. (1995) Different photosynthesis-nitrogen relations in deciduous hardwood, and evergreen coniferous tree species. *Oecologia*, **104**, 24–30.

Salzer, J., Matezki, S. & Kazda, M. (2006) Nutritional differences, and leaf acclimation of climbing plants and the associated vegetation in different types of an Andean montane rainforest. *Oecologia*, **147**, 417–425.

Santiago, L.S. (2010) Can growth form classification predict litter nutrient dynamics and decomposition rates in lowland wet forest? *Biotropica*, **42**, 72–79.

Santiago, L.S. & Wright, S.J. (2007) Leaf functional traits of tropical forest plants in relation to growth form. *Functional Ecology*, **21**, 19–27.

Schieving, F. & Poorter, H. (1999) Carbon gain in a multispecies canopy: the role of specific leaf area and photosynthetic nitrogen-use efficiency in the tragedy of the commons. *New Phytologist*, **143**, 201–211.

Schnitzer, S.A. (2005) A mechanistic explanation for global patterns of liana abundance and distribution. *American Naturalist*, **166**, 262–276.

Schnitzer, S.A. & Bongers, F. (2002) The ecology of lianas and their role in forests. *Trends in Ecology & Evolution*, **17**, 223–230.

Schnitzer, S.A. & Bongers, F. (2011) Increasing liana abundance and biomass in tropical forests: emerging patterns and putative mechanisms. *Ecology Letters*, **14**, 397–406.

Schnitzer, S.A., Bongers, F. & Wright, S.J. (2011) Community and ecosystem ramifications of increasing lianas in neotropical forests. *Plant Signaling & Behavior*, **6**, 598–600.

Schnitzer, S.A. & Carson, W.P. (2001) Treefall gaps and the maintenance of species diversity in a tropical forest. *Ecology*, **82**, 913–919.

Schnitzer, S.A., Dalling, J.W. & Carson, W.P. (2000) The impact of lianas on tree regeneration in tropical forest canopy gaps: evidence for an alternative pathway of gap-phase regeneration. *Journal of Ecology*, **88**, 655–666.

Schnitzer, S.A., Kuzee, M.E. & Bongers, F. (2005) Disentangling above- and below-ground competition between lianas and trees in a tropical forest. *Journal of Ecology*, **93**, 1115–1125.

Stevens, G. (1987) Lianas as structural parasites: the *Bursera simarouba* example. *Ecology*, **68**, 77–81.

Suzuki, W. (1987) Comparative ecology of *Rubus* species (Rosaceae). I. Ecological distribution and life history characteristics of three species, *R. palmatus* var. *coptophyllus*, *R. microphyllus* and *R. crataegifolius*. *Plant Species Biology*, **2**, 85–100.

Swaine, M. & Grace, J. (2007) Lianas may be favoured by low rainfall; evidence from Ghana. *Plant Ecology*, **192**, 271–276.

Tang, Y., Kitching, R.L. & Cao, M. (2012) Lianas as structural parasites: a re-evaluation. *Chinese Science Bulletin*, **57**, 307–312.

Van der Heijden, G.M.F. & Phillips, O.L. (2008) What controls liana success in neotropical forests? *Global Ecology and Biogeography*, **17**, 372–383.

Van der Heijden, G.M.F. & Phillips, O.L. (2009) Environmental effects on neotropical liana species richness. *Journal of Biogeography*, **36**, 1561–1572.

Van der Sande, M.T., Poorter, L., Schnitzer, S. & Markesteijn, L. (2013) Are lianas more drought-tolerant than trees? A test for the role of hydraulic architecture and other stem and leaf traits. *Oecologia*, **172**, 961–972.

Werth, M., Spiegel, A.-K. & Kazda, M. (2013) Leaf allocation patterns and ^{13}C and ^{15}N natural abundances of tropical lianas (*Passiflora* sp.) as dependent on external climbing support. *Isotopes in Environmental and Health Studies*, **49**, 98–108.

Wright, I.J., Reich, P.B. & Westoby, M. (2001) Strategy shifts in leaf physiology, structure and nutrient content between species of high- and low-rainfall and high- and low-nutrient habitats. *Functional Ecology*, **15**, 423–434.

Wright, S.J., Caleron, O., Hernandez, A. & Paton, S. (2004) Are lianas increasing in importance in tropical forests? A 17-year record from Panama. *Ecology*, **85**, 484–489.

Zhu, S.D. & Cao, K.F. (2009) Hydraulic properties and photosynthetic rates in co-occurring lianas and trees in a seasonal tropical rainforest in southwestern China. *Plant Ecology*, **204**, 295–304.

Zhu, S.D. & Cao, K.F. (2010) Contrasting cost-benefit strategy between lianas and trees in a tropical seasonal rain forest in southwestern China. *Oecologia*, **163**, 591–599.

Chapter 23

STEM BIOMECHANICS, STRENGTH OF ATTACHMENT, AND DEVELOPMENTAL PLASTICITY OF VINES AND LIANAS

Nick P. Rowe[1] *and Thomas Speck*[2]

[1]Botanique et bioinformatique de l'architecture des plantes, UMR – AMAP, Montpellier, France
[2]University of Freiburg, Freiburg, Germany; Freiburg Materials Research Center (FMF) and Freiburg Center for Interactive Materials and Bio-Inspired Technologies (FIT), Freiburg, Germany

Ecology of Lianas, First Edition. Edited by Stefan A. Schnitzer, Frans Bongers, Robyn J. Burnham, and Francis E. Putz.

Companion Website: www.wiley.com/go/schnitzer/lianas

OVERVIEW

Forces of attachment and the mechanical properties of the stems and branches of climbing plants (vines and lianas) play essential roles in their life histories and ecology. In this chapter we compare some of the growth strategies associated with different attachment and stem mechanics, and we investigate some of the functional and developmentally variable traits that contribute to their ecological success. We then explore how different climbing plants attach mechanically to their supports and discuss the forces generated by leaf friction, hooks, branches, tendrils, roots, and twining stems. We provide quantitative comparisons of the kinds of forces theses structures can withstand, and the reliability of their connection with the host.

We further demonstrate how different species modify the mechanical properties of the stem from juvenile and "searcher" stages as they mature into older flexible stems. Most climbing plants develop large reductions in stem stiffness (Young's modulus) during development. The degree of initial stiffness as well as the rate and magnitude of its decrease are linked to the kind of attachment and species life history. Some climbers attach via gradual accumulations of weak attachments or by connections that are prone to disengagement, such as branch angles and open hooks. Such species tend to retain stiff stem properties for a large part of their early development, and often grow as self-supporting shrubs or treelets early in their lives. Climbers attaching rapidly and firmly via stem twining or tendrils tend to develop flexible stems rapidly and are less likely to grow as self-supporting stems early in the life cycle (Rowe et al. 2006).

The ability of climbing plants to undergo profound shifts in stem properties during development represents a high potential for ontogenetic and phenotypic plasticity. Many vines and lianas develop either self-supporting, climbing, scrambling or creeping axes during early growth according to different environmental cues. This likely represents an opportunistic advantage over trees and shrubs which are confined to self-supporting growth throughout the entire life-history. We suggest that this advantage may play a crucial role during periods of perturbation and environmental change.

INTRODUCTION: THE BIOMECHANICS OF LIANAS

Lianas and vines are well known for their specialized climbing behavior and opportunistic growth, and have been studied since the pioneering experiments and observations of Charles Darwin and his contemporaries (Darwin 1867; Schenck 1893; Haberlandt 1914). The flexible stems of these climbing plants that hang from the forest canopy emphasize the contrast in biomechanics with the trees that support them, and lianas in particular are emblematic of the large-scale functional diversity within tropical forests (Fig. 23.1).

Different liana species employ a variety of different structures and strategies that help them attach themselves to the surrounding vegetation and ascend to the forest canopy. These attachment structures and how they operate have fascinated botanists for centuries. The kind of attachment a climbing plant makes to its host – strong or weak, irreversible or detachable – influences a range of life history processes. Variations in attachment are therefore likely to be associated with the size and mass of climbing plants as well as the mechanical development of the stem.

The highly flexible, basal stems of mature lianas are readily visible in forest understory and often are counted and measured as indicators of their growth, proliferation, and contribution to tropical forest dynamics (e.g., Gerwing et al. 2006; Phillips et al. 2002; Schnitzer et al. 2006). However, the presence and absence of mature stems and measures of their increase in girth do not explain how climbing plants attach to their hosts and then spread and compete with their host trees. Determining how climbing plants attach to their host and ascend to the forest canopy is thus a key area in understanding the ecology and competitive ability of climbing plants.

Many liana species develop initially as mechanically stiff, self-supporting individuals that have not yet developed climbing organs. These species, when young, appear to be physiognomically identical to young trees or understory shrubs (Putz 1984; Putz & Chai 1987; Caballé 1998; Rowe & Speck 2005). At some point in their ontogeny, most climbers also develop stiff leaders known as "searchers," which they use to find a host plant and also to find additional branches to climb after they are already established on their host (Speck

Fig. 23.1 Growth forms of lianas. (A) *Bauhinia guianensis* (Fabaceae), ribbon-shaped flexible stem surviving tree fall damage. (B) Self-supporting treelet of *Strychnos* sp. (Loganiaceae), the plant is still mechanically stable and has not yet started to produce hooks. (C) Understory self-supporting shoots of *Strychnos*, stems are at the point of instability and producing paired branches with sensitive hooks. (D) Three meter-long searchers of an unidentified liana extending from a twining stem in the understory; such narrow stems have high Young's moduli in order to retain their orientation and to span gaps between potential supports. (E) Multiple sprouting of robust self-supporting shoots of *Strychnos* sp. arising from fallen stem. (F) Sensitive hooks of *Strychnos* form closed spring-like hooks without contact with host branches; these thicken and toughen on contact with host. (G) Twining stem of *Smilax* sp., the species produces sharp spines on twining axes that undoubtedly assist in anchoring the twining stem to the host. (H) Wide-angle branches of canopy-climbing *Manihot quinquepartita* (Euphorbiaceae); these branches form irreversible connections with canopy branches. (I) Wide-angled petioles of *Croton nuntians* interlocking with host branches. (J) Self-supporting searcher stem of *Croton nuntians* in forest understory. *See plate section for color representation of this figure.*

1991, 1994; Putz & Holbrook 1991; Putz 1995; Caballé 1998; Kennard 1998). These searcher stems extend and attach to nearby host plants, and, once attached, allow further colonization across numerous plant hosts (Putz 1984; Putz & Chai 1987). The mechanical development that enables leading shoots to first rigidly extend, attach, and then become highly flexible is a key sequence of adaptive traits and is played out hundreds of times across a single vine or liana canopy. Understanding these changes in development across diverse clades and grades as well as the mechanical constraints associated with attachment is essential for understanding liana biology on a broad ecological scale.

Lianas also have relatively flexible stems when mature, and these stems are believed to be highly adapted to life as climbers for a number of reasons: (1) climbers do not develop tree-sized diameters and therefore do not expend the same magnitude of biomass allocation to reach light at the canopy (e.g., Kurzel et al. 2006); (2) high flexibility combined with toughness results in stems that are less likely to break fatally when exposed to high loads in bending, tension, compression, and shear. High stresses occur often as a result of treefalls and branch failure, and flexible but tough stems are likely to survive large angles of curvature when individuals become dislodged and fall from their supports (Fisher & Ewers 1991; Putz & Holbrook1991); (3) resprouting and branching following trauma are also a feature of climbers that retain large proportions of soft living tissue within the stem (Dobbins & Fisher 1986; Putz 1984; Schnitzer et al, Chapter 7 in this volume) (Fig. 23.1E).

Putative cause and effect mechanisms that underlie the ecological success of lianas (in terms of growth, survival, abundance, and diversity) cannot therefore be deduced from simple traits. In this chapter we discuss, in unashamed detail, the biomechanics of attachment in some representative vines and lianas from both temperate and tropical environments. We then discuss how different attachment organs are linked to different mechanical strategies developed by the stems and branches that bear them. Next, we discuss how attachment and adaptive variations of mechanical properties contribute to profound developmental plasticity in vines and lianas. Finally, we discuss the need for detailed studies on the functional traits and life histories of climbing plants, which can be integrated with larger-scale studies inferring the cause and effect mechanisms that underlie the increase in abundance of lianas at large ecological scales (Schnitzer, Chapter 30 in this volume).

METHODS AND APPROACHES

Attachment

Forces of attachment are generated by highly diverse structures on variable host substrates. Attachment forces vary in strength from very light to extremely strong. They include, in order of increasing strength: leaf roughness, reflexed leaflets, leaf interlocking, petiole base angles, adventitious roots, roots, tendrillar organs, sensitive branch angles, sensitive twining branches and twining main stems.

Attachment mechanisms are often geometrically complex, thus measurements can be difficult to replicate under typical natural conditions. Furthermore, mechanical properties often change during ontogeny and so measurements depend on the maturity and length of time the attachment has been active. Therefore, the mechanical forces measured during the period necessary for the plant to effect an irreversible connection are part of a crucial and strategic life history trait for climbers. Unlike measurements of elastic bending and torsion of the stem, which can be carried out relatively easily using wholly mechanical devices (Rowe et al. 2006), measurements of attachment forces should ideally use electronic devices where forces and deformations are recorded continuously (Steinbrecher et al. 2010; Melzer et al. 2012). A variety of such tests have been applied recently to attachment processes in climbers using pull tests demonstrating the forces that hooks, roots and tendrils can withstand before failure (Figs. 23.2B, 23.2C) (Isnard & Rowe 2008b; Steinbrecher et al. 2010; Melzer et al. 2012).

Stem properties

Mechanical properties of plant stems can be measured using many parameters, including elastic resistance to bending, torsion, shear and tension, or a combination of these factors (Vincent 1990a, 1990b; Vogel, 1992; Ennos 1993; Gallenmüller et al. 2001). Characterization of the stem can also focus on time-dependent measurements of visco-elasticity and the forces and energies occurring at failure and fracture (Vincent 1990b; Niklas 1992). Methods for measuring stiffness in bending have been detailed elsewhere and have the advantage of being possible with either simple field apparatus or sophisticated devices with sensitive force transducers and computer interfaces (Gallenmüller et al. 2001; Lahaye et al. 2005; Rowe et al. 2006).

In general, elastic properties are characteristic of how the stem retains its stability during normal growth, while properties beyond the elastic limit are of interest for determining how the stem might respond and survive high levels of mechanical stress such as treefalls. Bending properties that characterize stems of climbing plants have been measured for a wide range of species with different modes of attachment

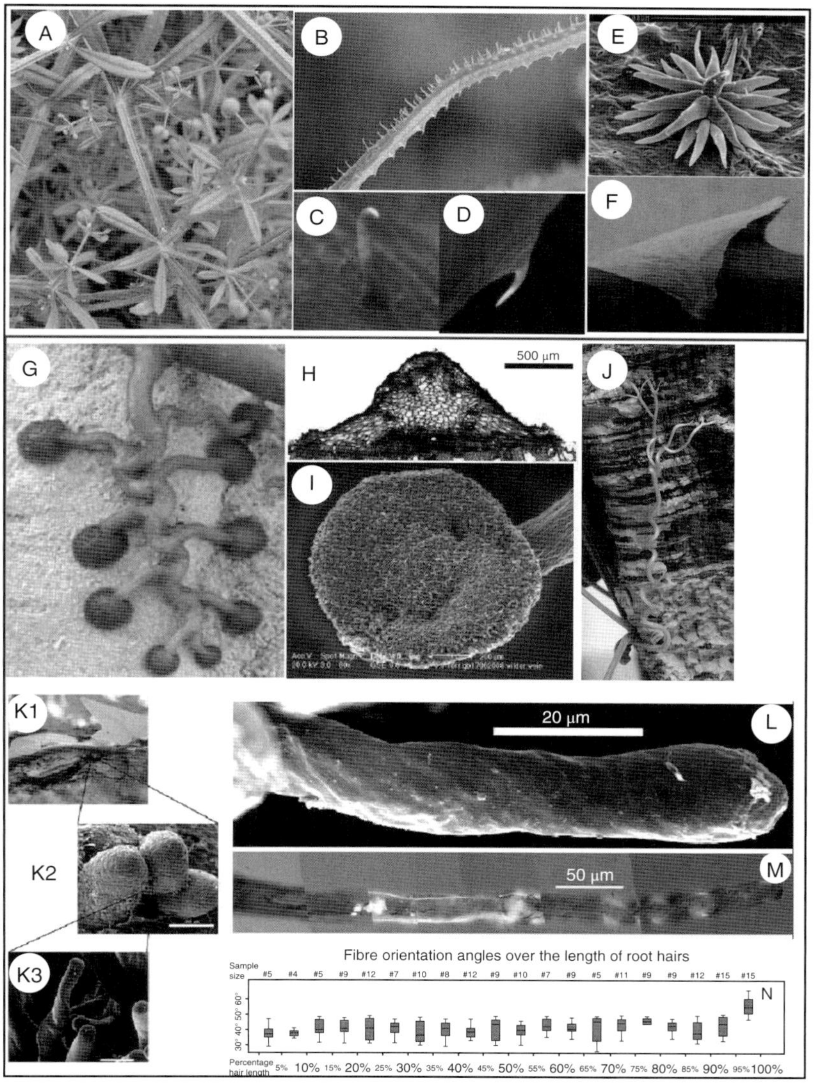

Fig. 23.2 Attachment structures in lianas. (A–D) *Galium aparine*; (A) tangle of shoots. (B) Lateral view of a leaf; hooks on the upper (adaxial) surface are curved towards the leaf tip (right), hooks on the underside of the leaf (abaxial surface) are curved towards the leaf base (left). (C) Adaxial hook. (D) Abaxial hook (C, D fluorescence micrographs, hooks stained with acridine orange). (E) Complex trichome on the underside of a leaf of *Croton pullei* (SEM); these structures increase friction with supporting structures. (F) Thorn on a shoot of *Asparagus setaceus* curved towards the shoot base (right). (G–I) *Parthenocissus tricuspidata*, (G) tendril with nine attachment pads on plaster. (H) Longitudinal section through an attachment pad on wood, at the interface between the wood and the underside of the pad, a thin layer of glue is visible. (I) Underside of an attachment pad (SEM). (J) Tendril of *Amphilophium crucigerum* with glueless attachment pads anchored in tree bark. (K–N) *Hedera helix*, (K) different hierarchical morphological levels of the attachment system: (K1) root cluster at the underside of an internode; (K2) three roots (SEM); (K3) root hairs (SEM). (L) Individual root hair demonstrating the orientation of cellulose microfibrils (SEM). (M) Stained single root hair in differential interference contrast, the darker lines on the root hair indicate cellulose microfibril orientation. (N) Bar plot diagram of mean orientation of cellulose microfibrils versus the relative position on the longitudinal axis of the root hair. (Sources: A, C, D from Bauer et al. 2011; H from Steinbrecher et al. 2011; K, L, M, N from Melzer et al. 2010. Reproduced with permission.) *See plate section for color representation of this figure.*

and at varying stages of development (Rowe et al. 2006). Values of elastic modulus (Young's modulus) in bending are generally represented by units of force over area (MN/m^2) – also depicted as Mega-Pascals (MPa). Changes in elastic properties from young to older phases of growth characterize how different vines and lianas maintain stability up to attachment and then change further, following attachment and establishment on the host. Values of elastic modulus are not directly linked with strength or toughness, two properties that are also important for vine and liana survival. Liana stems with large proportions of compliant tissues and segments of woody tissue generally bend or twist easily and are also resistant to catastrophic fracturing. In contrast, trees are comprised of homogeneous, stiff tissues that are more resistant to bending and twisting, but more susceptible to catastrophic and sudden buckling or fracturing if the forces exceed the linear elastic limit of the stem.

Anatomical, ultrastructural, and chemical analyses are ultimately required to more fully understand the processes that underlie changes in mechanical properties during growth and attachment (Hoffman et al. 2002). Cross-sections of each stem tissue at different growth stages can be readily quantified in terms of cross-sectional area and second moment of area of the stem, and then correlated with stem stiffness. Structural changes at the tissue level are of primary importance for identifying exactly what developmental traits underlie changes in mechanical properties. Such measurements are however time-consuming for large samples at large ecological scales. Finer-scale analyses such as tissue densities (wall: lumen ratios), microfibril angles (Köhler et al. 2000) and lignin compositions (Hoffman et al. 2002) can also be used to characterize mechanical properties. Such measurements, however, are time-consuming.

MECHANICS OF ATTACHMENT

Leaf friction and branch angles

Attachment by leaves is arguably among the weakest force produced by climbing plants. Nevertheless, leaf roughness provides sufficient friction to initially trap leaf surfaces between stem branches and petioles (Fig. 23.2I). Leaf friction has been observed in climbing species of *Croton* (Euphorbiaceae) (Gallenmüller et al. 2004) from French Guiana, where an initial clinging attachment by rough-surfaced leaves (Fig. 23.1E) is followed during later growth by more secure attachment with wide-angled branches. The latter then become the main anchoring structure for the climbing plant (Fig. 23.1H). Branch angles are highly effective in eventually anchoring climbers to the surrounding vegetation. Their efficiency depends on the angle, stiffness, and strength of the recurved branches as observed in species of *Manihot* (Ménard et al. 2009) and *Croton* (Gallenmüller et al. 2004). Branch angles can also be especially effective when produced in groups on the stem. *Manihot quinquepartita* develops curved branches in groups of three, which function like grappling hooks (Fig. 23.1H).

Climbing strategies using branch angles involve long phases of self-supporting growth as tall shrubs and treelets in the forest understory or forest margin. Young stems develop high values of Young's modulus (*Croton nuntians*, 8318 ± 2401 MPa; *Croton pullei*, 6741 ± 2206 MPa; *Manihot quinquepartita*, 4799 ± 2015 MPa). After attachment, older stems of *Manihot* and *Croton* develop flexible stems, more typical of mature lianas (*Croton nuntians* 849 ± 211 MPa; *Croton pullei*, 490 ± 203 MPa; *Manihot quinquepartita*, 3062 ± 1425) (Gallenmüller et al. 2001; Gallenmüller et al. 2004; Ménard et al. 2009).

Leaf friction and leaf ratcheting

Some climbers develop surprisingly derived forms of leaf friction. In *Galium aparine* (Rubiaceae), hooked trichomes on the upper leaf surface point (distally) towards the leaf tip, whereas those on the lower surface point (proximally) towards the leaf base (Figs. 23.2A, 23.2B) (Bauer et al. 2011; Gallenmüller et al. 2009). Vertical instability as well as wind, rain and animal movement cause movements and swaying of *Galium* plants and this initiates a mechanism that not only attaches the climber to its support (Fig. 23.2D), but also ensures that its leaves are not shaded by those of the host.

When leaves of *Galium* are positioned above a host leaf, proximal-pointing hooks on the underside act as a ratchet mechanism drawing the *Galium* leaf ever tighter to the host with each movement. When *Galium* leaves are positioned below a host leaf, the distal-pointing hooks on the upper leaf surface (Fig. 23.2C) tend to skid across the underside of the host leaves rather than ratcheting on to them. This mechanism (one surface ratcheting, the other surface slipping) creates a self-tightening attachment with the surrounding vegetation and at the same time ensures that *Galium* leaves tend to be positioned above host leaves rather than below them. This strategy allows *Galium* to effectively attach to the host, as well as position themselves advantageously for light capture.

The hook attachment in *Galium* is highly effective in terms of the mechanical strength and friction forces generated. Friction forces generated by the upper leaf surface in the direction of the hook curvature range from 9 to 22 mN (equivalent to forces of approximately 1–2 g weight). Forces generated by the lower surface are significantly higher, ranging from 33–71 mN, depending on surface characteristics (Table 23.1) (Bauer et al. 2011). Maximum contact separation force (the maximum force necessary to pull a single hook from a looped thread), as well as the angle (relative to the hook axis) at which the maximum force occurs, also differ between upper and lower surfaces. Forces are higher for hooks on the lower surface (ca. 20 mN at 45° to the hook axis) than for hooks on the upper surface (ca. 5 mN, at 180°). These results indicate that the ratcheting hooks are not only stronger than hooks on the upper surface but are also more efficient at the angles at which ratcheting hooks would be operating.

Previous reports of *Galium* stem biomechanics indicate that the basal-most part of the stem is highly flexible, with Young's moduli in the order of 235 ± 116 MPa (Goodman 2005). Basal segments also show remarkably high breaking strains extending up to 28% in length before failing. Overall this is consistent with many climbers that produce low Young's moduli following a firm attachment with host supports (Goodman 2005). Typically, *Galium aparine* individuals develop a large number of contact points along their length and the species demonstrates nicely that high flexibility and protection against damage under high forces are not confined to large woody lianas but can also be present in small-bodied climbers.

Hook climbers

Lianas that attach via open hooks require a continuous tensile force to remain engaged. In climbing palms, size, strength, geometry, and sharpness of hooks vary with different climbing habits (Isnard & Rowe 2008b; Rowe & Isnard 2009) (Table 23.1). Hook strength is finely tuned to increase in strength from the tip towards the base of cirri and flagella whip-like organs bearing them. This is consistent with a ratchet mode of attachment whereby climbing palms tighten their attachment to surrounding vegetation via swaying and movement and form stepwise attachments via stronger and stronger, more basal hooks (Putz 1990). Mature hooks and spines can have great strength for their size; acanthophyll hooks (modified leaflets) of *D. orthacanthos*, a large-bodied climbing palm of French Guiana, can support forces of up to 102 ± 64.5 N (equivalent to over 10 kg). In this case, the deployment of just one hook can support a substantial length of palm stem (Table 23.1). Hook size and strength also scale with overall body size: small-bodied species bear smaller hooks and large-bodied palms bear larger hooks. Size-specific hooks are probably linked to the fact that larger-bodied palms need to deploy larger hooks on larger diameter supports (Isnard & Rowe 2008b). However, hook functioning can change with changing body size. Small-bodied understory palms, such as *Desmoncus polycanthos* of French Guiana, bear small, sharply pointed, recurved spines (Table 23.1). These spines can attach to large diameter trunks like cat-claws when unstable palms lean or sweep against trunks and branches, allowing this understory palm to attach to trunks and branches far larger than the angle of curvature of its recurved hooks. Finally, mechanical tests indicate the presence of "failsafe" properties in climbing palms where, if tensile forces become too great, hooks will fail and break before the rachis bearing them does. Field observations indicate numerous cases where series of spines have been stripped from the rachis following disturbances and falls (Isnard & Rowe 2008a; Rowe & Isnard 2009).

Other climbers with open hooks include the semi-self-supporting species of *Rosa arvensis* (Rosaceae) and species of *Asparagus* (Asparagaceae). These species have shrub-like habits but mostly lean into vegetation and secure their support by stem-borne spines (*Rosa*) or leaf thorns (*Asparagus*). Spines of *R. arvensis* fail at mean forces of 3.5 N (long shoots), when loaded against the spine curvature and in the direction of the stem apex. Spines of the cultivar *R. arvensis* "Splendens" fail only after a mean force of 12 N (long shoots) or 21.5 N (water sprouts) (Table 23.1). Thorns of *Asparagus falcatus* and *Aspargus setaceus* fail at 7 and 8 N, respectively. Interestingly, the form of *Rosa*-spines and *Asparagus*-thorns is optimized so that under loading they show a nearly constant stress over their surface, and therefore local stress peaks are minimized. This

Table 23.1 Attachment mechanisms in lianas showing approximate ranges of forces.

Leaf friction	*Galium aparine* (adaxial surface)	8.8 ± 1.8 mN to 21.9 ± 13.4 mN[1]		(Bauer et al., 2011)
	Galium aparine (abaxial surface)	33.3 ± 15.1 mN to 71.4 ± 24.6 mN[1]		(Bauer et al., 2011)
Open hooks (spines, thorns, prickles)	*Galium aparine* (adaxial surface)	5 mN	(vertical pull)[2]	(Bauer et al., 2011)
	Galium aparine (abaxial surface)	20 mN	(pulled in 45° against curvature)[2]	(Bauer et al., 2011)
	Rosa arvensis (wild type)	3.4 ± 1.7 N	(spine on long shoot)[3]	[unpublished data]
	Rosa arvensis 'Splendens'	11.7 ± 4.1 N	(spine on long shoot)[3]	[unpublished data]
	Rosa arvensis 'Splendens'	21.6 ± 8.4 N	(spine on water sprout)[3]	[unpublished data]
	Asparagus falcatus	7.1 ± 4.1 N	(thron)[3]	[unpublished data]
	Asparagus seatceus	7.9 ± 4.4 N	(thron)[3]	[unpublished data]
	Desmoncus polyacanthos	8.0 ± 4.1 N	(acanthophylls on cirrus)[3]	(Isnard and Rowe, 2008a)
	Desmoncus polyacanthos	5.2 ± 3.8 N	(recurved hooks on cirrus)[3]	(Isnard and Rowe, 2008a)
	Desmoncus orthacanthos	102.0 ± 64.5 N	(acanthophylls on cirrus)[3]	(Isnard and Rowe, 2008a)
	Daemonorops jenkinsiana	89.6 ± 29.1 N	(recurved hooks on leaf)[3]	(Isnard and Rowe, 2008a)
	Daemonorops jenkinsiana	40.6 ± 20.4 N	(recurved hooks on cirrus)[3]	(Isnard and Rowe, 2008a)
	Calamus tetradactylus	14.5 ± 10.0 N	(recurved hooks on flagellum)[3]	(Isnard and Rowe, 2008a)

Tendrils with adhesive pads	*Parthenocissus tricuspidata*	5.9 ± 4.8 N	(old lignified pad on plaster)[4]	(Steinbrecher et al., 2010)
	Parthenocissus tricuspidata	1.6 ± 0.8 N	(young non lignified pad on wood)[4]	(Steinbrecher et al., 2011)
	Parthenocissus tricuspidata	5.3 ± 2.4 N	(tendril with 5-9 pads on plaster)[5]	(Steinbrecher et al., 2010)
Root climbers	*Hedera helix*	3.8 ± 2.4 N	(internodal root cluster on tree bark)[6]	(Steinbrecher et al., 2010)
	Hedera helix	0.2 to 0.4 N	(detachment/failure of single roots)[7]	(Melzer et al., 2010)
	Campsis radicans	18.3 ± 6.0 N	(nodal root cluster on wood)[8]	(Steinbrecher et al., 2010)

Notes:
[1] Mean friction force generated on different artificial surfaces.
[2] Median of contact separation forces for leaf hooks pulled in different directions.
[3] Mean force of hook failure if loaded in direction to the stem or leaf apex, i.e. against the hook curvature.
[4] Mean force of adhesive pad failure if loaded under uniaxial tension applied in normal direction to the climbing substrate.
[5] Mean force of failure of an entire tendril with five to nine adhesive pads if loaded under tensions applied in parallel direction to the main tendril axis.
[6,8] Mean force failure of root clusters if loaded under tension applied in normal direction to the climbing substrate.
[7] Median force of detachment resp. failure of single roots if loaded in different directions.

distribution of stress is potentially important for minimizing or avoiding failure via cracks that form at local geometric irregularities, which would increase local stress.

Tendril climbers

Plants with tendrils or sensitive hooks have typically undergone "reorganization" of all or part of the original structure that has adapted to become an attachment organ. Plants with twining tendrils, such as species of Cucurbitaceae and Fabaceae, have now been widely studied in terms of the physiology and biochemistry of the attachment process (Jaffe and Galston 1968; Jaffe 1970; Weiler 1993; Engelberth 2003; Meloche et al. 2007; Bowling & Vaughn 2009). From a biomechanics point of view, it is likely that attachment is secured in a similar way as described for twining stem climbers. In the following section, we present recent results on plants that develop pad-like structures, which form an attachment to the substrate and are borne at the tips of branched tendrils.

Tendril pads

Functional morphology and biomechanics of attachment in *Parthenocissus tricuspidata* (Vitaceae) and *Amphilophium crucigerum* (Bignoniaceae) have been analysed in detail over recent years (Figs. 23.2G–J). Boston ivy (*P. tricuspidata*) develops tendrils bearing apical attachment pads; it achieves permanent attachment to a wide variety of organic and inorganic surfaces, such as tree bark, rocks, and house plaster (Figs. 23.2G–I). Tendrils consist of a main axis with five to ten branches. During early growth, club-shaped tips of branches develop into flattened attachment pads after contact with the host (Steinbrecher et al., 2011). Adhesive fluid, mainly composed of an acidic mucopolysaccharide (Endress & Thomson 1976; Bowling & Vaughn 2008) is secreted by peripheral cells of the pad, which together effect a perfect seal with the substrate (Fig. 23.2H). Increased lignification of the pads takes place later, increasing mechanical strength.

Single, fully differentiated pads of *P. tricuspidata* grown on a substrate of house plaster fail under uniaxial tension at a mean force of 5.9 N (force applied in a normal direction to the climbing substrate) (Table 23.1). Maximum force at failure for young not fully differentiated or lignified attachment pads is significantly less at 1.6 N. Depending on the climbing substrate, mean forces for fully differentiated and lignified attachment pads vary between 1.7 and 5.9 N. The force-displacement curves typically increase almost linearly up to final failure (Fig. 23.3A) (Steinbrecher et al. 2009).

Entire tendrils of *P. tricuspidata* bearing numerous pads were tested as a whole unit to see how the entire structure responded to applied loads and to what extent it was adapted to absorb energy due to deflections of the main climbing stem. To mimic loading typical of a vertically orientated stem, a tensile force was applied parallel to the main axis of the tendril. During loading, extension of the coiled tendril was followed by consecutive failures of individual attachment pads (Fig. 23.3B). Mean attachment strength of the complete tendril system attached to plaster was 5.3 N.

The extension of the entire attachment system of 10–20 mm at final failure was much higher than that of individual pads. Thus, numerous pads contribute to the attachment of the complex structure, providing a high degree of energy dissipation. Mechanical damping of the tendril axis and its numerous pads represents a fail-safe-mechanism that ensures attachment even if one or more of the attachment pads fail (Steinbrecher et al. 2010, 2011).

The Monkey's comb (*Amphilophium crucigerum*) (Bignoniaceae) climbs via leaflet tendrils that develop attachment pads at their tips. The system represents the upper part of a modified compound leaf (Fig. 23.2J). A striking difference in comparison to *P. tricuspidata* is the absence of adhesive fluid. Attachment can occur by coiling of leaflet tendrils into a coiled spring-like structure as well as by attachment pads at the apices of the tendrils. Pads secure attachment physically by growing into fissures and gaps of the substrate, developing perfect form closure and strong interlocking connections between the pad and the substrate (Fig. 23.2J). Pads also develop vessel element-like tissue in areas of contact, which improves strength and toughness and increases friction with the substrate (Seidelmann et al. 2012).

Root climbers

Root climbers typically attach very tightly to host surfaces by attachment roots and root hairs (Figs. 23.2K1–K3). Attachment in English Ivy (*Hedera helix*) involves a four-phase process: (1) initial physical contact; (2) creation of a form closure with the substrate; (3) chemical adhesion; and (4) shape change of root hairs increasing the form closure with the substrate (Melzer et al., 2009, 2010). The different phases of attachment and their variations work together, to allow the attachment of English Ivy to various climbing substrates such as tree bark, rocks, and house plaster. The attachment of *Hedera helix* to the surface is often so firm that when stems are removed by force, pieces of the climbing substrate are torn off.

Initial contact is effected either by the longest root tip of a cluster or by root hairs covering them. It is not yet known which is more likely or what kind of signal initiates the contact with the substrate and stimulates the next phase. In phase two, the contact area between

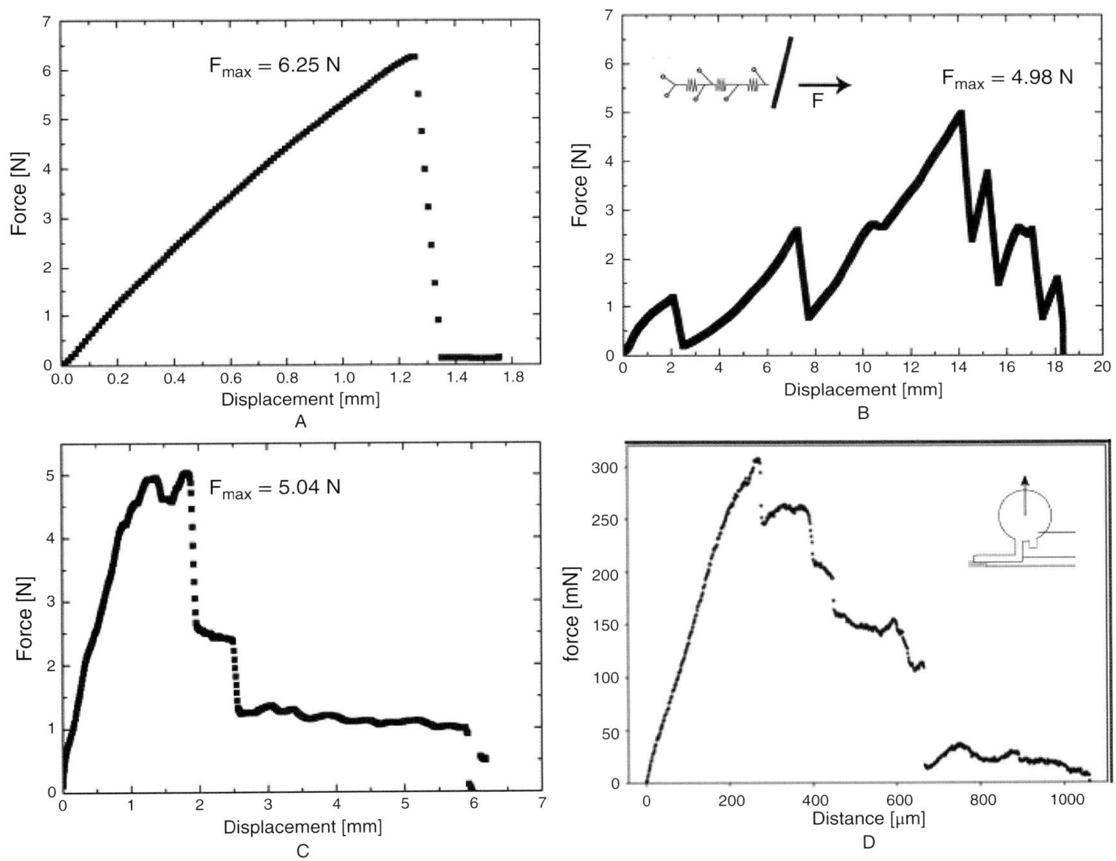

Fig. 23.3 Force displacements curves characterizing the mechanical behavior of selected attachment structures of plants. (A, B) *Parthenocissus tricuspidata*, (A) individual fully differentiated attachment pad loaded under uniaxial tension applied in normal direction to the climbing substrate (plaster). (B) Entire tendril with seven fully differentiated attachment pads attached on plaster loaded under tension applied in parallel direction to the main tendril axis, showing several pre-failure events due to the failure of individual pads or groups of pads. (C–E) *Hedera helix*, (C) root cluster at a 2-cm-long internode attached on tree bark loaded under tension applied in normal direction to the climbing substrate, the force dropped in different steps due to the failure of several individual root fields. (D, E) Single roots loaded in different directions as indicated by the inserts. (F) *Campsis radicans* nodal root cluster attached on wood loaded under tension applied in normal direction to the climbing substrate, the force dropped in different steps due to the failure of several individual root fields. (Sources: A, B, C, F from Steinbrecher et al. 2010. Reproduced with permission of Elsevier. D, E from Melzer et al. 2011. Reproduced with permission of Oxford University Press.)

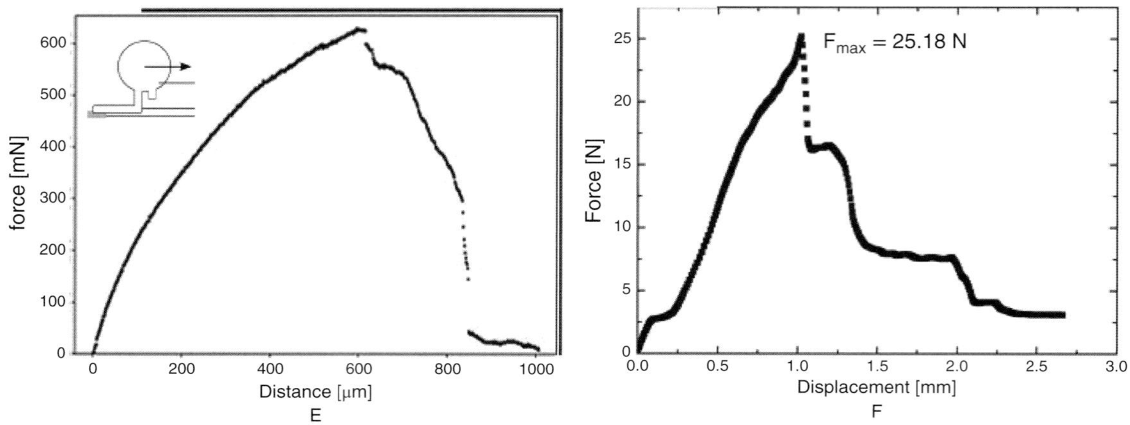

Fig. 23.3 (*continued*).

the attachment root and substrate is increased with additional form closure and deployment of additional root hairs into contact with the substrate. At this point, the root increases in diameter and begins to undergo lignification, improving the interlocking of the root with larger gaps in the climbing substrate. In the third phase, chemical adhesion takes place by glue excreted from round excrescences on the surface of the root hairs. It is likely that the excrescences are eventually obliterated by shear forces between the substrate and root hairs during the attachment process. In phase four, a passive shape change of the root hairs further improves the attachment strength. Shape changes are initiated by drying of the root hairs, which occurs by differential deformability of cellulose microfibrils and matrix in the cell wall of the root hairs (Fig. 23.2L). Shortening of the attached root hairs during the drying process draws the attachment root closer to the substrate and thereby further tightens the attachment (Melzer et al. 2009, 2010).

Attachment roots of English Ivy have a tensile strength of 3.4 MPa per individual root and a Young's modulus (measured in tension) of 109 MPa, which is relatively stiff for non-woody roots. The mean force under uniaxial tension to remove a 2-cm-long internode of *Hedera helix* with a typical root cluster from tree bark is 3.8 N (Table 23.1). Force-displacement curves of root clusters typically show several pre-failure events once the maximum force is reached. These pre-failure events are observed as a stepwise decrease in force and an increase in displacement and energy before final failure and detachment of the root cluster (Fig. 23.3C). Pre-failure events are caused by failures of individual root fields and represent another form of "fail-safe-mechanism," which increases the "reliability" of the attachment. At the finest level of attachment measured, values for detachment and failure of individual roots typically range from 200–400 mN (Figs. 23.3D, 23.3E) (Melzer et al. 2012; Melzer et al. 2009; Steinbrecher et al. 2010).

Similar attachment strategies are known, with some variations, in other root climbing species such as *Vanilla pompona*, *Vanilla planifolia*, and *Ficus pumilla* as well as for nodal root clusters of *Campsis radicans*. The force detachment curves of the latter also show several pre-failure events due to failing of single root fields of the nodal cluster after the maximum force is reached (Fig. 23.3F) (Steinbrecher et al. 2010).

Stem twiners

Twining is one of the most widespread attachment mechanisms in lianas, occurring in tendrils, branches, and stems. Stem climbing has fascinated botanists since Darwin and his contemporaries, and there has been much experimentation investigating the links between circumnutation, gravitropism, and the forces generated by twining (Isnard & Silk 2009).

An important mechanical aspect of twining is how the twining stem produces squeezing forces on the host stem. Recent studies have measured squeezing forces created by *Dioscorea bulbifera* on an artificial support equipped with load cells and deformable plates (Isnard et al. 2009). This research showed that squeezing forces became perceptible after only one half turn of the stem below the growing apex. Time-lapse photography indicated that the twining stem grew at just over 2 mm per hour taking approximately three days to produce a complete turn around the support of 33 mm in diameter. Squeezing forces reached a maximum of 100–300 mN after the twining stem had completed three turns.

The mechanism causing the forces generated by twining stems is not simple, and additional factors can contribute significantly to twining biomechanics. Two contributing factors are: (1) how the Young's modulus of the stem changes rapidly below the growing apex of the twining stem; and (2) how stem twiners co-opt additional structures to either increase the friction (Silk & Holbrook 2005) or increase tension of the twining stem. In *Dioscorea*, stipules develop on the inner surface of the twining stem and increase the tension and thus the grip on the support (Isnard et al. 2009). Recent studies have documented the role of friction in *Ipomoea purpurea*, in which phenotypes equipped with oriented trichomes significantly increase friction with the host support compared with phenotypes without trichomes (Silk & Holbrook 2005). We suspect that such mechanisms exist in a wide variety of forms among diverse species of twining climbers. Twining stems of hops (*Humulus lupulus*) are equipped with pointed emergences that appear to be spring-loaded, flipping open to engage the host surface when the stem contacts the host support (Isnard & Silk 2009).

Given the phylogenetic diversity of twining (Gentry 1991; Gianoli 2004), we suspect that the biomechanics behind initial fixation might be diverse in morphological and micro-anatomical mechanisms. While mechanisms of twining in young stages of growth have been uncovered over recent years, rather less is known about the effects of twining after significant secondary growth and how this later twining affects host plants. Observations suggest that tight twining and secondary growth might have a strangling effect and influence tree mortality (Lutz 1943; Putz 1991); however, the mechanical "competition" between cambial growth of trees and cambial growth of twining stems is lacking in published research.

Stem twining appears to take many shapes and forms, particularly in tropical rainforest environments, and might not always be functionally dependent on creating high friction, high tension, and the development of squeezing forces. Some kinds of twining might be better referred to as stem and branch winding, where axes form non-tightened coil-like structures around host plants. Depending on the geometry of host and climber, such attachment devices might be extremely effective after thickening and stiffening of the coiled segment. Further observations are needed to look into this aspect of stem twining and to investigate the ecological significance of different twining phenomena. The combination of twining or winding with other means of attachment such as spines, hooks, and tendrils in the same species or individual (e.g., Fig. 23.1G) is another aspect that has relevance to the diversity of host and habitat preferences among climbers.

MECHANICAL PROPERTIES OF THE STEM AND ATTACHMENT

A survey of over 40 species of climbing plants revealed how mechanical properties of stems varied among phylogenetically diverse groups, including woody dicotyledons, monocots, gymnosperms, and a lycopsid (Rowe et al. 2006). Woody dicotyledons that climb via twining (Lahaye et al. 2005; Rowe & Speck 1996; Speck 1994), tendril-like organs (Isnard et al. 2003), closed hooks (Rowe & Speck 1998), and wide angle branches (Gallenmüller et al. 2000; Gallenmüller et al. 2001; Ménard et al. 2009) generally show marked decreases in stiffness (Young's modulus) from young to old stages of growth (Fig. 23.4). Woody dicots that climb with hooks and spines that remain open, such as *Bougainvillea* (Nyctaginaceae) (Fig. 23.4) and *Rosa canina* (Rosaceae) retain relatively high levels of stiffness throughout their life span. This pattern was also found among woody species such as *Clematis recta* (Ranunculaceae) that simply lean against the surrounding vegetation (Isnard et al. 2003). The comparison among woody angiosperms demonstrated that the kind of mechanical architecture varied according to the kind of attachment: firm and irreversible attachment being linked to the development of high flexibility, whereas less secure attachment was linked to retention of stiffness (Young's modulus).

Climbing gymnosperm species of the genus *Gnetum* (Gnetales) are stem twiners that showed relatively low stiffness as young stems, but nevertheless developed even lower stiffness as mature attached stems

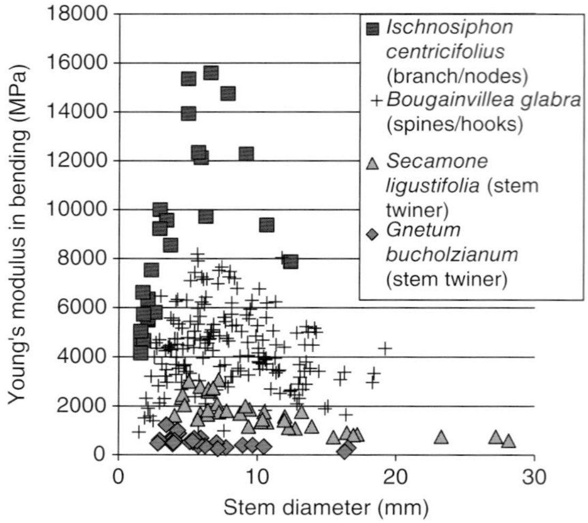

Fig. 23.4 Mechanical architectures of phylogenetically diverse lianas with different modes of attachment and habitat preferences. *Ischnosiphon centricifolius* (Marantaceae) is a monocotyledonous understory climber that develops extremely stiff cane-like stems that form angles via branched nodes that become flexible during ontogeny. *Bougainvillea glabra* (Nyctaginaceae) is a spiny, canopy hook climber that also develops high Young's moduli and retains relatively high values even after attachment. *Secamone ligustifiolia* (Apocynaceae), representative of a stem twiner of the forest canopy develops moderately stiff searcher stages of growth, which then show a large drop to highly flexible properties after attachment. *Gnetum bucholzianum* (Gnetales), representative of a genus of predominantly twining lianas (rare among gymnosperms) shows flexible stem properties as searchers and even higher flexibility after attachment.

(Fig. 23.4). Most climbing monocotyledons with open hooks, such as *Smilax* (Smilacaceae), rattans such as *Plectocomia* (Calamoideae), and branch-angle climbing *Ischnosiphon* (Marantaceae) (Fig. 23.4) developed high Young's moduli as young stems and generally retained high stiffness as mature climbing stems. At least one exception includes the arguably more "specialized" calamoid rattans of the genus *Calamus*, which develop highly flexible stems in old stages, resulting from loss of the stiff outer leaf sheath (Isnard & Rowe 2008a), and in which the inner cane-like stem is highly flexible (Dransfield et al. 2008).

A surprising finding from this comparison (Rowe et al. 2006) was the low values of Young's modulus in old basal growth stages of *Lycopodiella cernua* in comparison with the extremely stiff distal segments of stem (Rowe & Speck 1997). This pattern of development follows the overall trend seen in other clades with complex primary and secondary growth. This representative of the Lycopodiales develops one of the most ancient kinds of stem organization known among vascular plants – no secondary growth, a simple plectostele, a sheath of cortical mechanical tissue, and diminutive leaf-like microphylls. Nevertheless, the species develops flexible stem properties in individuals that can scramble to heights of 2 m in scrub vegetation using their microphylls and lateral branches as rudimentary attachment devices (Rowe et al. 2004). This example, along with the mechanical architectures found in *Gnetum* and different monocots, indicates that high flexibility in old stages of growth has developed among extremely wide-ranging lineages of vascular plants, and is not confined to woody angiosperms.

Both attachment type and phylogeny strongly influence climbing behaviour. The range of mechanical architectures also underlines a wide ecological diversity from canopy-climbing, treefall surviving, large-bodied woody liana niches to other habitat preferences, including shaded understory climbers and smaller bodied climbers at forest margins and in disturbed areas.

CONCLUSION AND FUTURE DIRECTIONS

Growth form plasticity and ecological amplitude

Some climbing species such as *Manihot* can grow to sexual maturity as shrubs or treelets in open conditions or as truly large-bodied climbing lianas in forest or covered habitats (Ménard et al. 2009). Some liana species such as *Croton* sp. can grow to quite considerable heights as self-supporting treelets (up to 7 m) in open conditions (Gallenmüller et al. 2004). Such high degrees of growth form plasticity and potential for opportunistic growth are linked to a facility for developing profoundly variable mechanical tissues in response to different environmental cues. Not all liana species show this degree of plasticity, and some large-bodied, woody liana species show little or no development of a truly self-supporting phase as shrub-like woody individuals (Rowe & Speck 1996). Such differences in plasticity could underlie important ecological processes, including ranges of habitat preference, successional phenomena, and resistance or survival to extreme perturbation, fire and forest clearance. Species showing high degrees of growth form plasticity might serve as interesting models for how lianas can survive and expand in disturbed and fragmented ecosystems.

Importance of non-elastic properties in surviving perturbation

Despite progress in understanding stem elastic properties and attachment modes in lianas, there have been relatively few studies that investigate aspects of stem toughness and just how liana stems survive extreme mechanical stresses (Dobbins & Fisher 1986; Fisher & Ewers 1989, 1991; Putz & Holbrook 1991). We suspect that some species, and developmental configurations of mature liana stems, are more resistant to breaking under extreme loading than others. An exploration of these differences has relevance to which species survive small and large-scale perturbation events underlying forest turnover, and which liana species might benefit from high levels of perturbation and clearance (Schnitzer et al., Chapter 7 in this volume).

Biomechanics of stem failure, toughness, and resistance to fracture and self-repair are complex features (Busch et al. 2010a, 2010b). They are difficult to measure critically, particularly in ecological and comparative contexts that require large sample observations. Furthermore, such measurements are problematic because of the technical difficulties of adequate fixing, applying appropriate forces, and interpreting results from geometrically complex and irregular structures. Mechanical failure of the plant stem in the event of mechanical, physical or physiological stress is a complex issue to measure meaningfully (Wainwright et al., 1976). From young to mature lianas, mechanical stem failure can mean a range of different things including loss of height, loss of stability, loss of rigidity, loss of hydraulic conductance, and loss of photosynthetic surface. Furthermore, we suspect that some liana species are adapted to fail either partially or completely in order for at least part of the stem to escape ruinous stresses and even aid propagation by vegetative reproduction (Gartner 1989). If lianas are indeed better at surviving widespread perturbation at ecological scales compared with other growth forms, part of the underlying mechanisms surely concern both elastic and non-elastic properties.

Mechanical effects on host vegetation

The wide diversity of mechanical architectures and attachment mechanisms is likely to have a variable influence on host plant development both on the individual and community scale. Long searchers, combined with twining, tendrillar or hook climbing mechanisms, can invade all orders of branching and extend to neighbors. These kinds of lianas are expected to have more effect on host plants compared with root-climbing species that are largely confined to the main trunk and branches (Schnitzler & Heuzé 2006). Attachment and growth of lianas affect host plants in terms of overall loading and thigmomorphogenetic effects, as well as the "strangling effects" on stem, leaf, fruit, and flower development. Investigation of how different liana species have variable effects on host plant growth and how much this depends on the kind of attachment and growth strategies of the liana is needed.

Functional traits, ecological scales and the increasing dominance of lianas

Three things stand out from our observations over recent years, which we believe have an important bearing on interpreting and predicting liana ecology. First, mechanical architectures of lianas are extremely diverse across different phylogenetic groups. Lianas (and vines) might generally all show slender, non-self-supporting stems at maturity, but the developmental and functional stages leading to mature climber establishment are incredibly variable.

Second, the changes in mechanical properties in both attachment and stem organization result from highly complex hierarchies of mechanical organization (Speck et al. 1996; Speck & Rowe 1999; Hoffman et al. 2002). These incorporate overall stem size and shape, tissue level organisation, and ultrastructural and chemical composition of the cell wall. While many lianas develop first stiffness, then flexibility, these changes are implemented by an incredible range and depth of developmental traits.

Third, from a broader perspective addressing woody plants in general, we have underlined the highly adaptive role of the bifacial vascular cambium, and how relatively simple changes in developmental timing can underpin profound changes in size, form, and shape of the plant body (Rowe & Speck 2005). We presented the idea that the presence of a vascular cambium within a lineage offers a potentially high degree of phenotypic plasticity, particularly in terms of growth form. This provides the potential for some lineages to survive periods of environmental and ecological instability by modulating the amount and nature of cambial products (wood and secondary phloem) within the plant stem.

Trees and shrubs are generally constrained to self-supporting architectures. They need to provision sufficient resources to support their own static load to stay upright, as well as resisting dynamic forces during their life history. Climbing plants, on the other hand, have largely escaped the need to sustain a self-supporting upright growth form, which is a key life history innovation that enables them to parasitize mechanical support from host plants.

As we have emphasized in this chapter, many groups of lianas express *both* mechanically stiff (self-supporting) *and* flexible (climbing) stem traits during their ontogeny and life history. It is therefore likely that in many climbing species, the genetic and developmental repertoire underlying the production of stiff *or* flexible stems can be modified in response to different environmental cues such as general perturbation and disturbance and climatic change. Changes in developmental timing among vines and lianas – between juvenile self-supporting and mature non-self-supporting traits – are undoubtedly linked to other functional traits and potential trade-offs involving water provisioning, leaf surface area, photosynthesis, and growth rate. We argue that high levels of phenotypic plasticity in climbers therefore provide greater potential for modulating potentially adaptive functional traits and exploiting environmental changes and perturbation compared to trees and shrubs that are constrained to self-supporting life histories. Of course, this advantage likely translates into a more direct competitive advantage in forest communities where climbers grow on trees and shrubs.

High phenotypic plasticity is therefore a potentially important factor underlying the current increase of lianas (Rowe & Speck 2005). However, there exist considerable challenges for unraveling cause–effect interactions between climbers and trees and inferring the impact of such interactions at the broader community and ecological scales. As we have shown in this chapter, part of the difficulty lies in the sheer diversity of climbers across many clades, and the mechanical and functional diversity these present. Nonetheless, broadly circumscribed functional traits are useful for detecting and identifying broad patterns of liana ecology. In order to understand the cause-effect mechanisms that undoubtedly exist and underlie the increasing dominance of vines and lianas, analyses at broad ecological scales need to be integrated with detailed studies based on functional biology, life histories, and developmental plasticity.

ACKNOWLEDGMENTS

We are grateful to numerous colleagues and institutions that have facilitated our research on lianas over recent years. These include, in France, the ANR funding (Project "Woodiversity") and the Franco-German PROCOPE funding; in Germany, the Ministry of Science, Research and the Arts Baden-Württemberg and the Baden-Württemberg-Stiftung. We are particularly grateful to the many colleagues, project students, thesis students and post-docs who have contributed to our projects over the years, in particular, in France: (P. Brion (*Bougainvillea* data), G. Caballé (*Gnetum* material), M. Guéroult, S. Isnard, R. LaHaye, L. Ménard, C. Paul-Victor; and in Germany: A. Barth (*Rosa* data), G. Bauer, K. Fiedler (*Asparagus* data), S. Rubach (*Ficus* data), V. Lauther (*Vanilla* data), H. Bohn, F. Gallenmüller, B. Melzer, and K. Seidelmann.

REFERENCES

Bauer, G., Klein, M.C., Gorn, S. et al. (2011) Always on the bright side: the climbing mechanism of *Galium aparine*. *Proceedings of the Royal Society London B*, **287**, 2233–2238.

Bowling, A.J. & Vaughn, K.C. (2008) Structural and immunocytochemical characterization of the adhesive tendril of Virginia creeper (*Parthenocissus quinquefolia* (L.) Planch.). *Protoplasma*, **232**, 153–163.

Bowling, A.J. & Vaughn, K.C. (2009) Gelatinous fibers are widespread in coiling tendrils and twining vines. *American Journal of Botany*, **96**, 719–727.

Busch, S., Schmidt, K., Erhardt, C. & Speck, T. (2010) Analysis of self-repair mechanisms of *Phasaeolus vulgaris* var. saxa using near-infrared surface enhanced Raman spectroscopy (SERS). *Journal of Raman Spectroscopy*, **41**, 490–497.

Busch, S., Seidel, R., Speck, T. & Speck, O. (2010) Morphological aspects of self-repair of lesions caused by internal growth stresses in stems of *Aristolochia macrophylla* and *Aristolochia ringens*. *Proceedings of the Royal Society London B*, **277**, 2113–2120.

Caballé, G. (1998) Le port autoportant des lianes tropicales: une synthèse des stratégies de croissance. *Canadian Journal of Botany*, **76**, 1703–1716.

Darwin, C. (1867) On the movements and habits of climbing plants. *Journal of the Linnean Society (Botany)*, **9**, 1–118.

Dobbins, D.R. & Fisher, J.B. (1986) Wound responses in girdled stems of lianas. *Botanical Gazette*, **147**, 278–289.

Dransfield, J., Uhl, N.W., Asmussen, C.B., et al. (2008) *Genera Palmarum*. Royal Botanic Gardens Kew, Kew.

Endress, A.G. & Thomson, W.W. (1976) Ultrastructural and cytochemical studies on the developing adhesive disc of Boston ivy tendrils. *Protoplasma*, **88**, 315–331.

Engelberth, J. (2003) Mechanosensing and signal transduction in tendrils. *Advances in Space Research*, **32**, 1611–1619.

Ennos, A.R. (1993) The mechanics of the flower stem of the sedge *Carex acutiformis*. *Annals of Botany*, **72**, 123–127.

Fisher, J.B. & Ewers, F.W. (1989) Wound healing in stems of lianas after twisting and girdling. *Botanical Gazette*, **150**, 251–265.

Fisher, B. & Ewers, F.W. (1991) Structural responses to stem injury in vines, in *The Biology of Vines* (eds F.E. Putz & H.A. Mooney), Cambridge University Press, Cambridge, pp. 99–124.

Gallenmüller, F., Bauer, G., Kubinski, K.-R. et al. (2009) Plant leaves as attachment devices: an experimental approach, in *Proceedings of the 6th Plant Biomechanics Conference, French Guiana, France* (ed B. Thibaut), Ecofog, Kourou, Cayenne, pp. 94–104.

Gallenmüller, F., Müller, U., Rowe, N.P. & Speck, T. (2000) Variability of habit and mechanical properties in the tropical liana *Croton nuntians*, in *Plant Biomechanics 2000– Proceedings of the 3rd International Plant Biomechanics Conference* (eds H.-C. Spatz & T. Speck), Thieme-Verlag, Stuttgart, pp. 63–70.

Gallenmüller, F., Müller, U., Rowe, N.P. & Speck, T. (2001) The growth form of *Croton pullei* (Euphorbiaceae): functional morphology and biomechanics of a neotropical liana. *Plant Biology*, **3**, 50–61.

Gallenmüller, F., Rowe, N.P. & Speck, T. (2004) Development and growth form of the neotropical liana *Croton nuntians*: the effect of light and mode of attachment on the biomechanics of the stem. *Journal of Plant Growth Regulation*, **23**, 83–97.

Gartner, B.L. (1989) Breakage and regrowth of *Piper* species in rain forest understory. *Biotropica*, **21**, 303–307.

Gentry, A.G. (1991) The distribution and evolution of climbing plants, in *The Biology of Vines* (eds F.E. Putz & H.A. Mooney), Cambridge University Press, Cambridge, pp. 73–97.

Gerwing, J.J., Schnitzer, S.A., Burnham, R.J. et al. (2006) A standard protocol for liana censuses. *Biotropica*, **38**, 256–261.

Gianoli, E. (2004) Evolution of a climbing habit promotes diversification in flowering plants. *Proceedings of the Royal Society of London, Series B, Biological Sciences Series B*, **271**, 2011–2015.

Goodman, A.M. (2005) Mechanical adaptation of cleavers (*Galium aparine*). *Annals of Botany*, **95**, 475–480.

Haberlandt, G. (1914) *Physiological Plant Anatomy*, Macmillan, London.

Hoffman, B., Chabbert, B., Monties, B. & Speck, T. (2002) Mechanical, chemical and X-ray analysis of wood in the two tropical lianas *Bauhinia guianensis* and *Condylocarpon guianense*: variations during ontogeny. *Planta*, **217**, 32–40.

Isnard, S., Cobb, A.R., Holbrook, N.M. et al. (2009) Tensioning the helix: a mechanism for force generation in twining plants. *Proceedings of the Royal Society series B*, **276**, 2643–2650.

Isnard, S. & Rowe, N.P. (2008a) The mechanical role of the leaf sheath in rattans. *New Phytologist*, **177**, 643–652.

Isnard, S. & Rowe, N.P. (2008b) The climbing habit in palms: biomechanics of the cirrus and flagellum. *American Journal of Botany*, **95**, 1538–1547.

Isnard, S. & Silk, W.K. (2009) Moving with climbing plants from Charles Darwin's time into the 21st century. *American Journal of Botany*, **96**, 1205–1221.

Isnard, S., Speck, T. & Rowe, N.P. (2003) Mechanical architecture and development in *Clematis*: implications for canalised evolution of growth forms. *New Phytologist*, **158**, 543–559.

Jaffe, M.J. (1970) Physiological studies on pea tendrils VI. *Plant Physiology*, **45**, 756–760.

Jaffe, M.J. & Galston, A.W. (1968) Physiological studies on pea tendrils V: Membrane changes and water movement associated with contact coiling. *Plant Physiology*, **43**, 537–542.

Kennard, D.K. (1998) Biomechanical properties of tree saplings and free-standing lianas as indicators of susceptibility to logging damage. *Forest Ecology and Management*, **102**, 179–191.

Köhler, L., Speck, T. & Spatz, H.-C. (2000) Micromechanics and anatomical changes during early ontogeny of two lianescent *Aristolochia* species. *Planta*, **210**, 691–700.

Kurzel, B.P., Schnitzer, S.A. & Carson, W.P. (2006) Predicting liana crown location from stem diameter in three Panamanian forests. *Biotropica*, **38**, 262–266.

Lahaye, R., Civeyrel, L., Speck, T. & Rowe, N.P. (2005) Evolution of shrub-like growth forms in the lianoid sub family Secamonoideae (Apocynaceae s.l.) of Madagascar: phylogeny, biomechanics and development. *American Journal of Botany*, **92**, 1381–1396.

Lutz, H.J. (1943) Injuries to trees caused by *Celastrus* and *Vitis*. *Bulletin of the Torrey Botanical Club*, **70**, 436–439.

Meloche, C., Knox, J. & Vaughn, K.C. (2007) A cortical band of gelatinous fibers causes the coiling of red vine tendrils: a model based upon cytochemical and immunocytochemical studies. *Planta*, **225**, 485–498.

Melzer, B., Seidel, R., Steinbrecher, T. & Speck, T. (2012) Biomechanics of the attachment system of English Ivy (*Hedera helix*): properties and structure of attachment roots in relation to the climbing substrate. *Journal of Experimental Botany*, **63**, 191–201.

Melzer, B., Steinbrecher, T., Seidel, R. et al. (2009) Mechanics and structure of the attachment system of English ivy

(Hedera helix L.), in *Proceedings of the 6th Plant Biomechanics Conference, French Guiana, France* (ed B. Thibaut), Ecofog, Kourou, Cayenne, pp. 205–210.

Melzer, B., Steinbrecher, T., Seidel, R. et al. (2010) The attachment strategy of English ivy: a complex mechanism acting on several hierarchical levels. *Journal of Royal Society Interface*, **7**, 1383–1389.

Ménard, L., McKey, D. & Rowe, N.P. (2009) Developmental plasticity and biomechanics of treelets and lianas in *Manihot* aff. *quinquepartita* (Euphorbiaceae): a branch-angle climber of French Guiana. *Annals of Botany*, **103**, 1249–1259.

Niklas, K.J. (1992) *Plant Biomechanics: An Engineering Approach to Plant Form and Function*, University of Chicago Press, Chicago.

Phillips, O.L., Martínez, R.V., Arroyo, L., et al. (2002) Increasing dominance of large lianas in amazonian forests. *Nature*, **418**, 770–774.

Putz, F.E. (1984) The natural history of lianas on Barro Colorado Island, Panama. *Ecology*, **65**, 1713–1724.

Putz, F.E. (1990) Growth habits and trellis requirements of climbing palms (*Calamus* spp.) in north-eastern Queensland. *Australian Journal of Botany*, **38**, 603–608.

Putz, F.E. (1991) Silvicultural effects of lianas, in *The Biology of Vines* (eds F.E. Putz & H.A. Mooney), Cambridge University Press, Cambridge, pp. 493–501.

Putz, F.E. (1995) Vines in treetops: consequences of mechanical dependance, in *Forest Canopies* (eds M.D. Lowman & N.M. Nadkarni), Academic Press, San Diego, pp. 311–323.

Putz, F.E. & Chai, P. (1987) Ecological studies of lianas in Lambir National Park, Sarawak, Malaysia. *Journal of Ecology*, **775**, 523–531.

Putz, F.E. & Holbrook, N.M. (1991) Biomechanical studies of vines, in *The Biology of Vines* (eds F.E. Putz & H.A. Mooney), Cambridge University Press, Cambridge, pp. 73–97.

Rowe, N.P. & Isnard, S. (2009) Biomechanics of climbing palms and how they climb. *Plant Signaling and Behavior*, **4**, 28–32.

Rowe, N.P., Isnard, S., Gallenmüller, F. & Speck, T. (2006) Diversity of mechanical architectures in climbing plants: an ecological perspective, in *Ecology and Biomechanics: A Mechanical Approach to the Ecology of Animals and Plants* (eds A. Herrel, T. Speck & N. Rowe), Taylor & Francis, Boca Raton, FL, pp. 35–59.

Rowe, N.P., Isnard, S. & Speck, T. (2004) Diversity of mechanical architectures in climbing plants: an evolutionary perspective. *Journal of Plant Growth Regulation*, **23**, 108–128.

Rowe, N.P. & Speck, T. (1996) Biomechanical characteristics of the ontogeny and growth habit of the tropical liana *Condylocarpon guianense* (Apocynaceae). *International Journal of Plant Sciences*, **157**, 406–417.

Rowe, N.P. & Speck, T. (1997) Biomechanics of *Lycopodiella cernua* and *Huperzia squarrosa*: implications for inferring growth habits of fossil small-bodied lycopsids. *Mededelingen Nederlands Instituut voor Toegepaste Geowetenschappen TNO*, **58**, 293–302.

Rowe, N.P. & Speck, T. (1998) Biomechanics of plant growth forms: the trouble with fossil plants. *Review of Palaeobotany and Palynology*, **102**, 43–62.

Rowe, N.P. & Speck, T. (2005) Plant growth forms: an ecological and evolutionary perspective. *New Phytologist*, **166**, 61–72.

Schenck, H. (1893) Beiträge zur Biologie und Anatomie der Lianen, im Besonderen der Brasilien einheimischen Arten. 1. Beiträge zur Biologie der Lianen/, in *Botanische Mitteilungen aus den Tropen* (ed A.F.W. Schimper), G. Fischer, Jena, pp. 1–271.

Schnitzer, S.A., DeWalt, S.J. & Chave, J. (2006) Censusing and measuring lianas: a quantitative comparison of the common methods. *Biotropica*, **38**, 581–591.

Schnitzler, A. & Heuzé, P. (2006) Ivy (*Hedera helix* L.) dynamics in riverine forests: effects of river regulation and forest disturbance. *Forest Ecology and Management*, **236**, 12–17.

Seidelmann, K., Melzer, B. & Speck, T. (2012) The complex leaves of the Monkey's comb (*Pithecoctenium crucigerum* (L.) A.H. Gentry): a climbing strategy without glue. *American Journal of Botany*, **99**, 1737–1744.

Silk, W.K. & Holbrook, N.M. (2005) The importance of frictional interactions in maintaining the stability of the twining habit. *American Journal of Botany*, **92**, 1820–1826.

Speck, T. (1991) Changes of the bending mechanics of lianas and self-supporting taxa during ontogeny. In *Natural Structures. Principles, Strategies and Models in Architecture and Nature, Proceedings of the II International Symposium Sonderforschungsbereich 230, Part I. Mitteilungen des SFB 230* Heft 6: 89–95.

Speck, T. (1994) Bending stability of plant stems: ontogenetical, ecological, and phylogenetical aspects. *Biomimetics*, **2**, 109–128.

Speck, T. & Rowe, N.P. (1999) A quantitative approach for analytically defining size, growth form and habit in living and fossil plants, in *The Evolution of Plant Architecture* (eds M.H. Kurmann & A.R. Hemsley), Royal Botanic Gardens, Kew, pp. 447–479.

Speck, T., Rowe, N.P., Brüchert, F., et al. (1996) How plants adjust the "material properties" of their stems according to differing mechanical constraints during growth: an example of smart design in nature. In *Bioengineering. Proceedings of the 1996 Engineering Systems and Design and Analysis Conference, PD-Vol. 77*, vol. 5, *ASME 1996* (ed. A.E. Engin), pp. 233–241. The American Society of Mechanical Engineers, New York.

Steinbrecher, T., Beuchle, G., Melzer, B. et al. (2011) Structural development and morphology of the attachment system of *Parthenocissus tricuspidata*. *International Journal of Plant Sciences*, **172**, 1120–1129.

Steinbrecher, T., Danninger, E., Harder, D. et al. (2010) Quantifying the attachment strength of climbing plants: a new approach. *Acta Biomaterialia*, **6**, 1497–1504.

Steinbrecher, T., Kraft, O., Speck, T. et al. (2009) Ontogenetic variations in morphology and attachment strength of permanent attachment pads of species of *Parthenocissus*, in *Proceedings of the 6th Plant Biomechanics Conference, French Guiana, France* (ed B. Thibaut), Ecofog, Kourou, Cayenne, pp. 444–449.

Vincent, J.F.V. (1990a) Plants, in *Biomechanics: Materials, a Practical Approach* (ed J.F.V. Vincent), IRL Press at Oxford University Press, Oxford, pp. 165–191.

Vincent, J.F.V. (1990b) *Structural Biomaterials*, Princeton University Press, Princeton, NJ.

Vogel, S. (1992) Twist-to-bend ratios and cross-sectional shapes of petioles and stems. *Journal of Experimental Botany*, **43**, 1527–1532.

Wainwright, S.A., Biggs, W.D., Currey, J.D. & Gosline, J.M. (1976) *Mechanical Design in Organisms*, John Wiley & Sons, Inc., New York.

Weiler, E.W. (1993) Evidence for the involvement of jasmonates and their octadecanoid precursors in the tendril coiling response of Bryonia dioica. *Phytochemistry*, **32**, 591–600.

Part VI

Liana–Animal Interactions

Chapter 24

EFFECTS OF LIANAS ON CANOPY ARTHROPOD COMMUNITY STRUCTURE

Stephen P. Yanoviak

University of Louisville, Louisville, KY, USA

Ecology of Lianas, First Edition. Edited by Stefan A. Schnitzer, Frans Bongers, Robyn J. Burnham, and Francis E. Putz.

Companion Website: www.wiley.com/go/schnitzer/lianas

OVERVIEW

This chapter reviews the known ecological associations between lianas and arboreal arthropods in tropical forests, and summarizes ongoing work focusing specifically on the ecological connections between lianas and arboreal ants. Lianas are abundant, conspicuous structural components of tropical forests, yet we know relatively little about their effects on local arthropod communities. Inventories of herbivorous taxa and their host specificity in tropical forest canopies show that lianas are significant contributors to local arthropod diversity via bottom-up effects. Ant associations with lianas generally are opportunistic; highly coevolved ant–liana relationships are rare, perhaps due to the greater longevity of arboreal ant colonies relative to liana life spans. Nonetheless, lianas provide important resources for canopy ants, especially physical connectivity and nutrients such as extrafloral nectar. From the perspective of arthropods, lianas are ecologically redundant with trees and epiphytes in many ways; however, only lianas provide persistent physical connections between tree crowns, which function as efficient pathways to distant resources for ants and other cursorial organisms. Preliminary results from a large-scale experiment in Panama show that liana removal decreases local canopy ant diversity, increases *Azteca* nest density, and alters ant species composition. Given that liana abundance is increasing in tropical forests, understanding their potential role as determinants of local species richness is essential for predicting future patterns of biodiversity more generally in tropical forest canopies.

INTRODUCTION: LIANAS AS A KEY STRUCTURAL ELEMENT OF TROPICAL FORESTS

The relationship between habitat structure and diversity depends on how organisms perceive and use their physical surroundings (Dean & Connell 1987; Bell et al. 1991; Tews et al. 2004). Terrestrial habitat structure largely is defined by vegetation physiognomy (Kenoyer 1929), and variation in vegetation characteristics within and among forests affects the diversity and distribution of both vertebrates and invertebrates (e.g., James & Wamer 1982; Fonseca 1989, Halaj et al. 1998, 2000). Although structure-diversity patterns are supported by observational and experimental data, the effects of individual forest components specifically on arthropod communities are poorly known. For example, the importance of plant architectural diversity (e.g., Lawton 1983), and foliage density and diversity on arthropod community parameters, remain largely undocumented. Such information is ecologically relevant and increasingly important given that disturbance and changing climatic conditions are unlikely to affect all forest components equally (e.g., Dale et al. 2001; Körner 2009).

By first-order approximation, tropical forests consist of four conspicuous living structural elements – trees, shrubs, lianas, and epiphytes – each with very different autecology and consumer faunas (Ødegaard 2000a). Lianas in particular are a key component of lowland forest ecosystems (Gentry & Dodson 1987; Gentry 1991; DeWalt et al., Chapter 11 in this volume). They provide diverse resources for numerous animal taxa (Gentry 1991; Yanoviak & Schnitzer 2013; Schnitzer et al., Chapter 1 this volume) and likely influence the evolution of arboreal animal morphology and behavior (Benson et al. 1975; Emmons & Gentry 1983). Moreover, liana abundance is increasing in neotropical forests due to climatic change, disturbance, and related factors (Phillips et al. 2002; Wright et al. 2004; Ingwell et al. 2010; Schnitzer & Bongers 2011; Schnitzer et al. 2011, 2012; Schnitzer, Chapter 30 in this volume). Thus, understanding the relationships between liana ecology and consumer diversity is essential for predicting patterns of biodiversity in tropical forests over the next century.

ARTHROPOD DIVERSITY IN THE TROPICAL FOREST CANOPY

Tropical rain forest canopies around the world are known for their high arthropod species richness (e.g., Erwin 1982; Stork et al. 1997; Basset et al. 2003; Ozanne et al. 2003; Basset et al. 2012). Many mechanisms have been proposed to explain patterns of arthropod diversity within and among tree crowns (Erwin 1982; Basset 1992b), but few have been adequately tested, in part because experimental manipulations in the canopy are logistically very difficult at ecologically relevant scales. Moreover, accurately estimating arboreal arthropod diversity in the canopy continues to be very challenging despite the development and application of diverse sampling methods (Basset et al. 2012; Lowman et al. 2012).

Although patterns of canopy arthropod diversity are difficult to quantify, the collective results of descriptive studies reveal some generalities. For example, arboreal arthropod diversity tends to be vertically stratified, such that more species occur in upper canopy vegetation than in lower forest levels (e.g., Elton 1973; Sutton et al. 1983; Basset et al. 1992, 2001; Su & Woods 2001). There are some notable exceptions to this pattern: tropical ant diversity is lower in the canopy than the litter (e.g., Yanoviak & Kaspari 2000), the abundance and diversity of cloud forest spiders are lower in the canopy than the understory (Yanoviak et al. 2003a), and the species richness of arthropods inhabiting water-filled tree holes tends to decline from the understory to the canopy (Yanoviak 1999). It is also generally evident that host tree specificity is low among canopy arthropods (Stork 1987; Basset 1992a; Basset et al. 1996; reviewed in Ødegaard 2000a), and that patchy microhabitats within forest canopies harbor a large fraction of local arthropod diversity (e.g., Yanoviak 2001; Ellwood et al. 2002; Yanoviak et al. 2003b, 2004, 2006). Finally, canopy arthropod diversity tends to be higher in primary forest than in secondary forest (e.g., Floren & Linsenmair 1999; Schonberg et al. 2004; Floren & Deeleman-Reinhold 2005; Klimes et al. 2012), which may seem intuitive, but is not true for all animal taxa (e.g., Fonseca 1989; Yanoviak et al. 2003a).

Whereas these and other patterns are relatively well established in the literature, many fundamental questions regarding arboreal arthropod ecology remain unanswered. For example, what are the relative contributions of local and regional processes in structuring canopy communities? How important is habitat structure (i.e., physical complexity) to the maintenance of diversity? And, is there correspondence between the distribution of plant traits (nutritional, phenological, etc.) and trait distributions within their arthropod communities (e.g., Basset 1991a, 1992b; Ødegaard 2000a; Yanoviak et al. 2004; Tanaka et al. 2010)? We can begin to address such questions by comparing the fauna of trees with and without specific components, such as lianas (this chapter), epiphytes (Yanoviak et al. 2011a), or young foliage (Basset 1991b). However, controlled experiments involving appropriate focal taxa and distinct structural elements are required to unambiguously separate the effects of physical habitat from nutritional and other contributions of individual vegetation components. Tropical arboreal arthropods and lianas collectively provide an excellent model for such investigations.

ECOLOGICAL ASSOCIATIONS BETWEEN ARTHROPODS AND LIANAS

Given that arthropods and the liana growth form have coexisted for millions of years (Stein et al. 2012), it is likely that some close ecological associations occur between them. If so, it is also likely that trees with lianas support more arthropod species than trees lacking lianas. Despite the large number of faunal inventories that have been conducted in tropical forest canopies, none explicitly addresses this assumption, though some studies do indicate that lianas broadly influence arthropod community parameters (i.e., species richness, abundance, and composition). For example, Stork (1987) showed that the local abundance of lianas (and epiphytes) was more important than tree identity in determining arthropod faunal similarity among trees in Borneo. In an Australian subtropical montane forest, Walter et al. (1994) recorded mite densities $>10{,}600\,m^2$ on liana foliage, which greatly exceeded densities observed on other canopy plants at the site. Likewise, arthropod abundance increased with liana foliage abundance in a dipterocarp forest (Dial et al. 2006), though the type of foliage (tree or liana) ultimately was less important than leaf vs. wood as a determinant of arthropod abundance (Dial et al. 2006). Faunal inventories such as these require enormous amounts of time and effort, and their results, though often correlative or anecdotal, provide a solid foundation for subsequent comparative and mechanistic studies. For practical and ecological reasons, most such studies have focused on herbivores and ants.

Herbivory is ubiquitous in tropical forests (Coley & Barone 1996), and specific contributions of lianas to arthropod community structure in the canopy should be most apparent in herbivorous species. Indeed, the best-known liana–arthropod associations involve highly specialized plant feeders. Commonly observed examples include *Heliconius* butterflies on *Passiflora* (Benson et al. 1975; Smiley 1987), Solanaceae-feeding Ithomiinae butterflies (e.g., Trigo & Motta 1990), and

Hylaeogena beetles on Bignoniaceae. In a broad study of liana–herbivore associations, Ødegaard (2000b) found that 24% of ca. 700 species of arboreal phytophagous beetles were associated with both lianas and trees in Panama. Although cumulative species richness was similar between trees and lianas, the number of beetle species specializing on green vs. woody tissues was higher on lianas, whereas the opposite was true on trees. Consequently, only 11% of species in the green tissue feeding guild were shared between trees and lianas (Ødegaard 2000b). Collectively, these results demonstrate that the presence of lianas significantly increases local arthropod diversity in the forest canopy via bottom-up effects. Wolda's (1979) observations for herbivorous, canopy-dwelling insects in Panama also support this conclusion. Specifically, the abundance and diversity of hemipterans in light trap samples increased with the number of vines in individual *Luehea* trees (Wolda 1979).

Elevated arthropod diversity in the presence of lianas apparently is a consequence of the relatively high production of new foliage (and tendrils) in lianas compared to trees (Wolda 1979; Hegarty & Caballé 1991; Ødegaard 2000b). Given that tropical plants often trade strong herbivore defense for rapid growth (Coley & Barone 1996; Stamp 2003), and that this trade-off is particularly strong for lianas (Asner & Martin, Chapter 21 in this volume), lianas should support more generalist herbivore species than trees based on their greater relative production of green tissue. Ødegaard's (2000b) results show the opposite pattern – more green-tissue specialists occurred on lianas than on trees in Panama – suggesting that lianas have stronger herbivore defenses (consequently excluding many generalist feeders) than do trees (Hegarty et al. 1991). The distribution of chemical defenses in a broad range of extant plant taxa (Hegarty et al. 1991), and specific comparisons between trees and lianas (Asner & Martin 2012; Asner & Martin, Chapter 21 in this volume), do not support this hypothesis. Nonetheless, the pattern may have been more prominent in ancient forests: Carboniferous fossils show that lianescent ferns were especially well defended against herbivorous arthropods (Krings et al. 2002). The conflicting results of these studies illustrate the need for more research on the ecology and evolution of herbivory specifically on lianas. To date, most studies addressing liana–arthropod associations and related questions have focused on ants, as discussed below.

ARBOREAL ANTS IN TROPICAL FORESTS

Ants are an excellent focal taxon for investigations of plant–animal relationships in tropical forests. They are both tractable for field research and relevant to understanding the functioning of tropical rain forests at the ecosystem level. They span approximately four orders of magnitude in body size (Kaspari & Weiser 1999) and play important roles in a wide range of ecological processes (e.g., seed dispersal, soil turnover, and nutrient cycling; Hölldobler & Wilson 1990; Lach et al. 2010). Ants are relatively easy to observe and collect (Bestelmeyer et al. 2000), their taxonomic composition and diversity can be determined without destroying entire colonies (Kaspari 2000), and species-level identifications are possible within many genera (e.g., Longino 2003; Wilson 2003). Moreover, molecularly based ant phylogenies are available to facilitate the interpretation of ecological results in an evolutionary context (Moreau & Bell 2013 and references therein). Finally, ants are particularly relevant to understanding forest canopy ecology because they represent a large fraction of arthropod abundance and consumer biomass specifically in tropical tree crowns (reviewed by Rico-Gray & Oliveira 2007).

Tropical canopy ants and litter ants form distinct communities (Yanoviak & Kaspari 2000), though arboreal ants (i.e., species nesting in aboveground vegetation) will occasionally forage in the understory (e.g., Hahn & Wheeler 2002), and some ground-nesting taxa forage in the canopy (e.g., *Atta* spp., *Paraponera clavata*, *Gigantiops destructor*). Many canopy ant species function as secondary herbivores, feeding on the honeydew provided by plant-sucking Hemiptera (Davidson et al. 2003), and thus are protein-limited relative to leaf litter ants (Davidson 1997; Yanoviak & Kaspari 2000; Davidson et al. 2007). Physical aspects of the canopy also uniquely shape the behavior and ecology of arboreal ants. Specifically, canopy ants live in a relatively sunny environment, forage along relatively permanent but highly exposed linear networks, and must cope with the hazard of falling (e.g., Yanoviak et al. 2005, 2011b). Whereas arboreal ants have traits that match these circumstances, not all species are found in all trees, and their distributions are likely determined by local processes such as niche determinism and competition.

It is widely accepted that ants compete for patchy resources (e.g., Hölldobler & Wilson 1990; Davidson

1998; Adler et al. 2007; Lach et al. 2010). Aggressive interspecific interactions among ants are readily observed at baits, and resource limitation likely promotes parabiosis (nest sharing; e.g., Longino 2003; Menzel et al. 2008) and trail parasitism (the use of foraging trails, and consequently food resources, by unrelated workers; Adams 1990). Competition among canopy ants often results in mosaic-like species distributions in plantations and forests having relatively low structural heterogeneity (Leston 1978; Adams 1994; Dejean et al. 2007). Such patterns tend to be diffuse or nonexistent among trees in complex and highly connected primary forests (Floren & Linsenmair 2000; Stuntz et al. 2003; Sanders et al. 2007; Klimes et al. 2012; but see Davidson et al. 2007). Likewise, tests of relationships specifically between vegetation structure and ant diversity have given mixed results; arboreal ant diversity increased with tree diversity (a surrogate for structural heterogeneity) in Brazilian savanna forest (Ribas et al. 2003), but decreased with habitat complexity in Australian woodlands (Lassau & Hochuli 2004), and showed no association with epiphyte abundance in trees of Panama (Stuntz et al. 2003).

Despite growing knowledge about the ecology of tropical arboreal ants, a unified framework for predicting local ant species richness and composition (i.e., in individual trees) is lacking. However, progress on this front is emerging; recent observational studies have added to our knowledge of ant species diversity and coexistence in individual trees. For example, comprehensive surveys of felled trees in New Guinea showed that habitat structure (tree density and tree size) is a stronger determinant of local ant species richness than is tree diversity (Klimes et al. 2012). Similarly, resident ants of individual Bornean trees appear to be spatially partitioned by micro-environmental heterogeneity associated with structural complexity (Tanaka et al. 2010). In the Atlantic Forest of Brazil, higher ant diversity in larger individuals of the tree *Anadenanthera macrocarpa* was attributed to ontogenic changes in tree architecture, which presumably influenced microclimate and the availability of nest and food resources (Campos et al. 2006). Studies of arboreal ants in Brazilian savanna forest suggest that structural variables such as tree size, tree density, and availability of appropriate nest sites are important determinants of diversity (Ribas et al. 2003; Powell et al. 2011). Collectively, these studies indicate that niche determinism mediated by habitat structure affects local arboreal ant diversity in tropical forests. Thus, variation in forest components like lianas should have predictable effects on ant communities. Almost all studies of arboreal ant ecology in lowland wet forests note the importance of lianas, but specific contributions of lianas to ant diversity have never been quantified.

ECOLOGICAL ASSOCIATIONS BETWEEN ANTS AND LIANAS

The evolutionary history of ants is closely associated with the diversification of angiosperms (Moreau et al. 2006), and ecological links between ants and tropical trees and epiphytes are common (e.g., Janzen 1966; Beattie 1985; Huxley & Cutler 1991; Dejean et al. 1995; Ellwood et al. 2002; Stuntz et al. 2003; Djiéto-Lordon et al. 2004; Frederickson & Gordon 2007). In contrast, coevolved (e.g., mutualistic) associations specifically between lianas and ants are rare (Moog et al. 2003; Djiéto-Lordon et al. 2005; Rico-Gray & Oliveira 2007). Climbing plants were abundant in ancient forests during the diversification of ants (Stein et al. 2012; Burnham, Chapter 5 in this volume), and use of liana extrafloral nectar is widespread among canopy ants (Bentley 1981; Blüthgen et al. 2000). However, the only examples of tight myrmecophytic associations are the defense of *Spatholobus* and other lianas by *Cladomyrma* spp. in Asia (Maschwitz et al. 1989; Moog et al. 1997, 2003; Moog 2009) and *Tetraponera tessmanni* inhabiting *Vitex thyrsiflora* in Africa (Djiéto-Lordon et al. 2005). In both cases, cavities in the plants created or maintained by the dominant ants also are occasionally occupied by other arboreal ant species. Similarly close associations between ants and lianas are unknown in the neotropics.

The low frequency of conspicuous coevolved associations between lianas and ants may be due to the high turnover rates of lianas relative to other forest components (e.g., Phillips et al. 2005; Ingwell et al. 2010). Few long-term life-history data exist specifically for tropical ants, but colonies of some ant species persist for decades (Keller 1998), and repeated surveys of individual mature trees in Panama over the past 15 years show that most arboreal ant species have long colony longevity (Yanoviak, pers. obs.). A long lifespan in combination with nest site limitation (Carroll 1979; Philpott & Foster 2005) places considerable selection pressure on arboreal ants that use plant cavities for nest sites (i.e., those that do not construct nests of leaves or carton). Specifically, selection should favor colonization of the most persistent substrates (e.g., trees and shrubs), or the subdivision of nests (i.e., polydomy) among unitary substrates that are short-lived but locally abundant (e.g., dead hollow twigs, live domatia, and epiphytes). Lianas do not neatly fit either scenario due to their patchy distribution and relatively high turnover rates. Also, domatia-like structures appear to be uncommon in the live stems of lianas. Thus, opportunistic use of lianas by ants is likely to be favored over obligate occupancy. This appears to be true within the predominantly arboreal genus *Pseudomyrmex*, which includes many myrmecophytic species, but none that is associated with lianas apart from opportunistic nesting in dead stems (Ward 1989, 1991, 1999).

In contrast to the rarity of mutualisms between lianas and ants, antagonistic interactions between ants and lianas may be quite common. Many neotropical plant-ants (e.g., *Pseudomyrmex* and *Myrmelachista*) maintain their hosts free of vines and other potential competitors by pruning or killing intruders (Janzen 1966; Frederickson & Gordon 2007). Although the best examples of this behavior involve understory trees and shrubs, Tanaka and Itioka (2011) showed experimentally that *Crematogaster difformis* ants exclude lianas from certain emergent trees in Borneo. Presumably the *Pseudomyrmex* ants inhabiting mature *Tachigali* and *Triplaris* spp. trees in the Amazon (Ward 1999) also actively prevent colonization of their hosts by lianas, though this has not been studied. Arboreal ants will wage prolonged attacks on any foreign objects secured to branches within their foraging territories, suggesting that even non-mutualistic canopy ants influence colonization of their home trees by lianas.

Regardless of the nature of liana–ant associations, lianas provide essential resources that are likely to enhance the local species richness of arboreal ants under most circumstances, especially physical connectivity, nutrients, and low-resistance pathways. Physical connectivity is a key feature of habitat structure at multiple scales, influencing the outcome of local interspecific interactions (Huffaker 1958; Davidson et al. 1988) and ecosystem processes across fragmented landscapes (e.g., Rosenberg et al. 1997, Fischer & Lindenmayer 2007). Physical connections in tropical forest canopies likely play a similarly important mechanistic role in ant species coexistence, though this has never been examined experimentally on a large scale (but see e.g., Davidson et al. 1988, Powell et al. 2011). Despite the high vegetation density of lowland tropical forests, the leaves and branches of neighboring canopy trees almost never have persistent physical contact with each other (commonly referred to as *crown shyness*; Ng 1977; Putz et al. 1984; Richards 1996). Consequently, in the absence of lianas or similar connectors (e.g., Snaddon et al. 2012), ants nesting in a given tree crown cannot access resources in nearby trees without first descending to the understory and traversing the leaf litter. This exposes workers to predator pressure on the ground between trees, and reduces foraging efficiency due to the increased distance to resources (Weber 1957; Catling 1997; Yanoviak et al. 2011b). Physical isolation of tree crowns also may enhance the local competitive effects of aggressive resident ants such as *Azteca* (Catling 1997). Lianas solve the crown shyness problem for ants and other arboreal animals by forming bridges between neighboring trees (e.g., Koenig et al. 2007; Tanaka et al. 2010; Camargo & Oliveira 2012), thus expanding the accessible resource base and producing a relatively continuous canopy surface (Emmons & Gentry 1983; Putz 1984).

Lianas also provide a variety of spatially and temporally predictable food sources that are consumed (or potentially consumed) by ants, including extrafloral nectaries (EFNs), insect exudates (i.e., hemipteran "trophobionts"), floral nectar, fruits, seeds, and pollen (Bentley 1981; Baroni Urbani & de Andrade 1997; Blüthgen et al. 2000; Pizo & Oliveira 2000; Davidson et al. 2003; Rico-Gray & Oliveira 2007). Many lianas flower and fruit out of phase with trees in seasonal forests (e.g., in Panama; Opler et al. 1991), thereby reducing temporal gaps in the availability of these nutritional resources. More importantly, EFNs and trophobionts tend to be more abundant on lianas than on trees (Blüthgen et al. 2000; Blüthgen & Fiedler 2002), and these resources fuel the aggressive activities of *Azteca* and other behaviorally dominant arboreal ants (Davidson 1997; Yanoviak & Kaspari 2000; Rico-Gray & Oliveira 2007). Thus, carbohydrates derived from lianas may determine local ant community structure in the tropical forest canopy by direct bottom-up effects or via tritrophic interactions (Blüthgen et al. 2004; Rico-Gray & Oliveira 2007).

The relatively narrow, linear structure of lianas also appears to be an important resource for arboreal ants (Clay et al. 2010; Yanoviak et al. 2012). Unlike flying central-place foragers (e.g., bees), for which foraging imposes significant energetic costs, *time* is the major cost of foraging in cursorial organisms like ants (Fewell 1988). Time is related to foraging distance, and a trade-off should exist between the benefits of rapid discovery or dominance of high quality (or distant) resources (Hölldobler & Wilson 1990; Davidson 1998; Adler et al. 2007), and the cost of time spent outside the nest. A worker's risk of becoming lost, and its exposure to competitors, predators and pathogens, increases with time spent away from home. Thus, ants should select the most efficient route when foraging. Whereas ground-dwelling ants tend to choose pathways that maximize their two-dimensional velocity (Fewell 1988; Dussutour et al. 2006; Farji-Brener et al. 2007), similar patterns remain relatively unexplored in arboreal ants. The forest canopy offers a particularly interesting evolutionary scenario for such studies because access to patchy resources is constrained by the reticulate structure of the habitat.

Many arboreal ant species preferentially use lianas as foraging trails even in the absence of EFNs or trophobionts (Bentley 1981; Yanoviak et al. 2012), suggesting that structural characteristics of liana stems enhance ant foraging efficiency. Results of our field studies in Costa Rica support this hypothesis; upon discovering a high-quality food item on a tree trunk, ants altered the location of their foraging trails to incorporate nearby climbing plant stems and hemiepiphyte roots, and behaviorally dominant ants defended such pathways (Clay et al. 2010). We also explored the effects of fine-scale liana stem characteristics on ant foraging by placing artificially high quality resources (tuna baits) on stems classified quantitatively by size (diameter), and qualitatively by roughness and type (herbaceous vs. woody). Liana substrate properties, especially stem diameter and roughness, influenced the distribution of foraging ants in the forest based on body size. Specifically (and not surprisingly), the largest ants were uncommon on the smallest stems except when enticed by baits (Yanoviak et al. 2012).

Although comparative data are not yet available, lianas generally appear to have relatively smooth surfaces that offer less costly foraging routes for ants than do tree branches and trunks. Our pilot studies show that ant running speed is significantly reduced when the height of the obstacles (i.e., the rugosity of a plant surface) is approximately 33% of the worker body's length (Silveri & Yanoviak, unpublished data). Similar patterns occur in cockroaches (Sponberg & Full 2007), and it is reasonable to expect the scaling of dynamic instability to be consistent among arthropods with a tripod gait. This relationship deserves further exploration specifically in light of the size-grain hypothesis, the surface properties of plant stems, and leg allometry in ants (Kaspari & Weiser 1999; Farji-Brener et al. 2004; Sarty et al. 2006).

ANT ASSOCIATIONS WITH OTHER CLIMBING PLANTS

Apart from lianas, various herbaceous climbing plants and hemiepiphytes also provide important resources for ants and other arthropods in tropical forests. Whereas lianas are a reliable and widespread source of horizontal connectivity within the forest canopy, ants commonly use the aerial roots of hemiepiphytes as vertical foraging pathways and canopy – understory corridors (Clay et al. 2010; Yanoviak, pers. obs.). Likewise, the slender ascending stems of climbing palms (rattans) in Asia provide connectivity and nest sites for ants and their associated Hemiptera (Moog et al. 2003; Edwards et al. 2010; Chan et al. 2012). One other example of ant use of tropical herbaceous climbers is particularly noteworthy for its specialization. The climbing insectivorous pitcher plant *Nepenthes bicalcarata* has a remarkable association with the ant *Camponotus schmitzi* in Borneo. As in other ant–plant mutualisms, the plant provides nectar and shelter for the ants while the ants provide protection for the plant against herbivory (Merbach et al. 2007). However, *C. schmitzi* workers also prevent trapped insect prey from escaping the pitchers, and forage on the victims by swimming in the plant's capture fluid (Clarke & Kitching 1995). Although this superficially appears detrimental to the plant, closer examination showed that the ants subsequently return unconsumed insect fragments to the pitcher (Clarke & Kitching 1995; Bonhomme et al. 2011). As noted above, similar mutualistic associations between ants and climbing vegetation are unknown in the neotropics.

In summary, lianas and other climbing plants play important ecological and evolutionary roles for arthropods in tropical forests. In particular, cursorial organisms like ants depend on the structural contributions of lianas to facilitate their foraging activities and territoriality, and the relatively abundant green tissues of lianas are a key resource for herbivores. However, most information regarding associations between arthropods and lianas remains observational or anecdotal. Given that liana abundance is increasing in tropical forests (e.g., Schnitzer & Bongers 2011; Schnitzer, Chapter 30 in this volume), there is a growing need for manipulative studies designed to isolate the specific contributions of lianas to arthropod diversity in the canopy.

EFFECTS OF LIANA EXTERMINATION ON ARBOREAL ANT COMMUNITIES

If lianas provide important resources for ants and other arboreal animals, we expect changes in liana abundance to have important bottom-up effects on local species richness in the forest canopy. My students and I are testing this hypothesis by measuring the effects of a replicated, large-scale liana removal experiment on arboreal ant community structure in a Panamanian forest (described below). Our ongoing experiments focus on three predictions. First, given that lianas provide food and nest sites that are used by ants, we expect the local (i.e., within-tree) species richness of ants to decline following liana removal. Second, we predict that effects of liana removal on ant species richness will be counteracted by the addition of artificial nest sites and inter-tree connectivity. Finally, we expect the increased isolation of tree crowns following liana removal to cause local extinctions of behaviorally dominant ants, specifically *Azteca* spp.

Stefan Schnitzer and colleagues established a large-scale liana removal experiment in Panama in 2008. The project consists of 16 forest plots 80 × 80 m on the Gigante Peninsula of the Barro Colorado Natural Monument (09.123°N, 79.857°W). Within each plot, all trees, lianas and shrubs >1 cm dbh (>25,000 individuals) were identified, tagged and measured following standard plot census protocols for lianas (Gerwing et al. 2006; Schnitzer et al. 2008) and trees (Condit 1998). In April 2011, all lianas in 8 of the 16 plots were exterminated by cutting their stems in the understory. The effects of the manipulation were conspicuous in the canopy within a few weeks (Fig. 24.1), and resulted in increased light in the understory over a period of months (Fig. 24.2). Approximately 95% of liana biomass had fallen from the Gigante canopy by June 2012 (Yanoviak, pers. obs.). We collected arboreal ants from the trunk and crown of five or six trees in each plot before liana cutting, as well as 2, 8, and 14 months after liana cutting. We climbed each tree using the single-rope method (Perry 1978) and collected ants by hand and with tuna baits for up to 2 hours per tree in fair weather. Variables recorded included air temperature, relative humidity, tree size

Fig. 24.1 Dead lianas were conspicuous in the canopy of liana removal plots within weeks of cutting their stems in the understory. (Source: S.P. Yanoviak. Reproduced with permission.) *See plate section for color representation of this figure.*

Fig. 24.2 Liana cutting increased light penetration to the understory in the months following the manipulation. (Source: S.P. Yanoviak. Reproduced with permission.) *See plate section for color representation of this figure.*

(dbh), tree species, total time spent in the tree, and a qualitative assessment of weather conditions.

We also quantified the abundance of *Azteca trigona/chartifex* nests (Fig. 24.3) by conducting systematic visual scans of the midstory and canopy (i.e., point counts) within each plot. *Azteca trigona/chartifex* are polydomous, and well-established colonies often have one or two very large nests and several smaller nests scattered within a 10–20 m radius (Fig. 24.3). Thus, to avoid artificial inflation of nest counts and to provide an accurate approximation of actual colony density, we counted multiple nests in a single tree and multiple nests located within a few meters of each other as a single nest.

Fig. 24.3 We conducted point counts to quantify the density of *Azteca trigona/chartifex* nests in plots. The nests often are quite large and accompanied by smaller nests (e.g., the small patch of carton to the left of the main nest in this image). The large nest shown here is ca. 1 m in length. (Source: S.P. Yanoviak. Reproduced with permission.)

Fig. 24.4 A field assistant secures used climbing ropes to a tree crown in Panama to mimic the connectivity provided by lianas. (Source: S.P. Yanoviak. Reproduced with permission.)

To test the hypothesis that tree-to-tree connectivity provided by lianas affects arboreal ant community structure, we added used climbing ropes (i.e., artificial liana stems) to several focal trees in each plot. The ropes were loosely tied around the main branches of neighboring trees (Fig. 24.4), and were used by ants to move from tree to tree almost immediately. We similarly tested for possible nest-site limitation in the canopy by adding 10 artificial twig nests to the crowns of up to three trees per plot. These nests consisted of aluminum or brass tubing 10 cm in length and closed at one end. Tubes were taped together in pairs or triplets and secured with nylon ties to the undersides or other protected portions of branches (Fig. 24.5). Finally, in 2013 we installed potted lianas in the crowns of selected trees. Over the coming years, these potted lianas will be pruned to confine their growth to a single tree crown, thus enabling us to test the hypothesis that lianas influence ant diversity via resources beyond connectivity and nest sites.

LIANA REMOVAL REDUCES LOCAL ANT SPECIES RICHNESS

To date, we have collected ca. 100 ant species or morphospecies from the Gigante plots. The number of ant species per tree ranged from 4–20, which is consistent with other studies of resident ants in tropical trees (Tanaka et al. 2010; Klimes et al. 2012). The number of ant species in our collections will likely increase as we continue with species identification and confirmation.

Average ant species richness in focal trees of liana removal plots declined by 22% relative to controls approximately one year after the lianas were cut. Two taxa, *Pachycondyla* spp. and *Cephalotes basalis*, showed the most consistent reduction in tree occupancy following liana removal. We attribute the loss of

Fig. 24.5 Approximately 500 artificial twig nests were constructed from brass and aluminum hobby tubing closed at one end and secured to tree branches in the canopy. Each tube is 10 cm long and tube diameters ranged 2–5 mm. (Source: S.P. Yanoviak. Reproduced with permission.)

Pachycondyla mainly to the reduction of physical connectivity between trees. *Pachycondyla* spp. are roaming, often specialized predators (e.g., Davidson et al. 2003), and their conspicuous use of artificial connectivity in liana removal plots (see below) suggests that their local distributions are dependent on access to broad foraging areas provided by inter-tree connections. Local declines in *Cephalotes basalis* were most likely due to loss of nest sites. Although they commonly inhabit cavities in living and dead tree branches, the *C. basalis* colonies we found on Gigante were almost always in large liana stems, and thus the natural history of this species and its association with lianas deserve further study. Despite the overall trend for reduced ant diversity following liana cutting, species richness increased in some trees in the liana removal plots. Almost all such cases were the result of *Camponotus* spp. (carpenter ant) colonization of dead lianas that had not yet fallen from the canopy. These effects are expected to disappear as the few remaining lianas decompose.

An exploratory Principal Components Analysis (PCA) showed a positive relationship between tree size, time spent in a tree, and ant species richness, which was driven by the tendency to spend less time in trees having fewer species, rather than by sampling error. The PCA also showed no positive associations between tree species identity or local environmental conditions on ant species richness. These results concur with other studies suggesting that arboreal arthropod communities are more strongly influenced by differences in structural properties (e.g., size) among trees rather than tree species identity (Stork 1987; Tanaka et al. 2010; Klimes et al. 2012).

Liana cutting had conspicuous short-term effects on ant behavior and the abundance of other taxa. We found significantly fewer ants foraging in trees two months after liana cutting; ant recruitment to baits was slower, and more time was required to thoroughly census each tree in the treatment plots. Also, the abundance of beetles (especially wood borers and fungus feeders), woodpeckers, and web-building spiders was conspicuously higher in liana removal plots shortly after liana cutting. Presumably, the pulse of dead wood and decaying foliage attracted the beetles, which subsequently attracted the woodpeckers. The conspicuous increase in spider density was likely a result of increased light availability (e.g., Lubin 1978), though data are needed to verify this observation and the other patterns described above.

The addition of artificial connectivity (old ropes) had positive effects on ant species richness in trees of both control and liana removal plots. Addition of connectivity to trees in control plots increased ant species richness by ca. 25% from June 2011 to June 2012, whereas trees in liana removal plots without added ropes lost ca. 30% of their ant species over the same time interval. Although connectivity did not completely compensate for the effects of liana removal, these trends suggest that the presence of connectivity is an important determinant of ant species richness in tropical trees. Species of all common arboreal ant genera readily used the added connectivity in both control and liana cut plots. *Azteca* spp. and *Pachycondyla* spp. consistently used the ropes, and *Azteca trigona/chartifex* constructed carton around the rope–tree connection point in some cases. We also commonly observed multiple genera (e.g., *Azteca*, *Camponotus*, *Cephalotes*, *Pachycondyla*, and *Pseudomyrmex*) simultaneously using a single rope. Rather than partition the rope into separate foraging pathways, unrelated ants simply avoided contact with each other and continued their forward progress.

In contrast to the connectivity addition experiment, the addition of artificial twig nests was unsuccessful. Approximately 10% of the ca. 500 artificial twig nests we installed in the canopy had fallen or were missing after one year. Less than 1% of the remaining artificial nests were colonized by ants, and ca. 10% were occupied by other taxa, especially spiders. Low occupancy of these artificial nests may be a consequence of using metal tubes, which we chose for their durability. However, other studies using wood or bamboo artificial nests also observed relatively low (<30%) occupancy by ants (Philpott & Foster 2005; Philpott 2010; Sagata et al. 2010). By contrast, naturally occurring cavities in living and dead twigs tend to have high ant occupancy, suggesting that ants are nest-limited, but also very selective with respect to nest characteristics.

Liana removal had strong effects on *Azteca* nest densities. The average number of *Azteca* nests increased at approximately twice the rate in liana removal plots as in control plots over the two years after cutting. Because *Azteca trigona/chartifex* are polydomous, we could not clearly distinguish between expansion of existing colonies and the founding of new colonies based on our field surveys. Regardless, it is clear that *Azteca* abundance increased on liana removal plots. We attribute this increase to three factors, all of which are purely speculative at this point. First, greater light availability and associated warmer temperatures may have stimulated colony activity and expansion. Second, the pulse of relatively labile dead wood in liana cut plots, which is used to construct carton nests, could have promoted the growth of existing nests. Third, the absence of lianas may have prevented anteaters (*Tamandua mexicana*) from accessing *Azteca* nests. This latter explanation (perhaps in combination with other mechanisms) seems most likely; our preliminary observations of foraging anteaters showed that they use lianas almost exclusively when accessing the canopy. Given that *Azteca* are among the most aggressive arboreal ants in neotropical forests, their increased local abundance and the increased isolation of their resident trees may result in competitive exclusion of subordinate species over the short term (Catling 1997; Vandermeer et al. 2010). We further expect that this process will occur more slowly in trees with added artificial connectivity, as explained above.

Although liana removals resulted in increased *Azteca* density over the two years after cutting, we suspect that the role of *Azteca* will become diminished in liana removal plots over the long term. This prediction is based on two assumptions that remain to be tested. First, liana-based EFNs and trophobionts (honeydew-producing herbivores) are important sources of the carbohydrates that fuel high-tempo, aggressive taxa like *Azteca* (Davidson 1997; Blüthgen & Fiedler 2002; Blüthgen et al. 2004). Assuming *Azteca*, their trophobionts, and lianas are involved in specialized tritrophic interactions, liana removal should cause localized *Azteca* extinctions, facilitating the coexistence of subordinate ant species via competitive release (though such effects are difficult to demonstrate, even in much simpler systems; e.g., Kareiva 1982; Riechert & Cady 1983). Second, we assume that physical connectivity in the canopy is essential for the maintenance of large *Azteca* colonies. If so, the physical isolation of tree crowns should dramatically reduce *Azteca* foraging efficiency, and eliminate their access to colony subunits and distant resources. The conspicuous use of artificial connectivity by *Azteca* foragers in liana removal plots (see above) suggests that the absence of such pathways will make large colonies unsustainable, though this remains to be determined.

Collectively, the preliminary results of this large-scale experiment suggest that lianas are important determinants of local species richness in arboreal ant communities. Although strong effects of liana removals and added connectivity appeared just one year after the manipulation, uncovering the mechanisms behind these effects will require many more years of data collection. Future censuses will enable us to determine if the observed changes in species richness are driven specifically by the absence of lianas, or by more generalized effects of large-scale disturbance. Liana removal clearly resulted in greater light penetration due to the significant reduction in canopy cover (Figs. 24.1, 24.2). This change resembles natural phenomena that also cause significant canopy defoliation while leaving standing trees, such as hurricanes (e.g., Tanner et al. 1991), seasonal drought stress (Dietrich et al. 1982; Boinski & Fowler 1989), large lightning strikes (Magnusson et al. 1996; Richards 1996; Yanoviak 2013), and pathogen outbreaks (e.g., Wong et al. 1990). Such disturbances should promote population growth or persistence of species that are tolerant of high light conditions and associated thermal and desiccation stress, as appears to be occurring for *Azteca* in the liana removal plots. Regardless, if the observed changes in ant communities are an artifact of collateral abiotic effects of liana removal, such effects will dissipate as tree crowns grow into the spaces left by liana foliage, which tends to occur relatively rapidly (Kira & Ogawa 1971). Ultimately, the great advantage of this large-scale experiment is that the specific contributions of lianas to arboreal ant community structure will become increasingly clear over time.

CONCLUSION AND FUTURE DIRECTIONS

Lianas are conspicuous components of tropical forests, yet we are just beginning to understand their effects on the ecology of arthropods and other consumers. Resolving this information deficit is important given that increasing liana abundance may be a key driver of changes in tropical forest structure over the next century. One fundamental question provides a solid basis for future work on this problem: Are lianas functionally redundant with trees, shrubs and epiphytes, or does the liana growth form provide a unique ecological template within tropical forests? To date, the only focused attempts at answering this question from the perspective of arthropods come from painstaking inventories of herbivorous insects and our experimental studies of arboreal ants. In particular, Ødegaard's (2000b) beetle surveys suggest that the large ratio of foliage to total plant biomass and the relatively high production of new growth make lianas important bottom-up contributors to arthropod diversity in the canopy, and our ongoing research in Panama indicates that a unique structural contribution of lianas – physical connectivity among tree crowns – is an important determinant of local arboreal ant diversity.

These studies are just the beginning of exploration into liana–arthropod interactions, and answers to even relatively basic questions would greatly enhance our understanding of the ecological contributions of lianas to arthropod diversity. For example, how do arthropod assemblages associated with lianas change temporally; i.e., does the composition of herbivores and other associates vary seasonally or over the lifetime of a liana? Is liana species richness a good predictor of arthropod species richness at regional scales? If so, what liana traits determine the composition and diversity of their arthropod associates? Finally, is the contribution of lianas to canopy arthropod diversity additive or synergistic with other forest components? The study of liana-arthropod relationships may provide a fruitful avenue for exploring specific relationships between arthropods and their specific plant hosts, as well as more general research questions on plant-animal and tritrophic interactions.

ACKNOWLEDGMENTS

Dana Frederick, Petrus Moreira, Samantha Nicholls, and Theo Sumnicht assisted in the field. Jack Longino and Phil Ward confirmed species identifications for selected ant taxa. The Little Rock Climbing Center donated essential equipment. Oris Acevedo, Belkys Jiménez, and the staff at the Smithsonian Tropical Research Institute provided logistical support. The Panamanian Autoridad Nacional del Ambiente (ANAM) provided permits. Comments from Petr Klimes, Stefan Schnitzer, Robin Verble, and anonymous reviewers improved the manuscript. This research was supported by a Franklin Award from the American Philosophical Society; NSF grants DEB-0845071, DEB-1019436, DEB-1252614, IOS-0843120; and the University of Wisconsin–Milwaukee Research Growth Initiative program.

REFERENCES

Adams, E.S. (1990) Interaction between the ants *Zacryptocerus maculatus* and *Azteca trigona*: interspecific parasitization of information. *Biotropica*, **22**, 200–206.

Adams, E.S. (1994) Territory defense by the ant *Azteca trigona*: maintenance of an arboreal ant mosaic. *Oecologia*, **97**, 202–208.

Adler, F.R., LeBrun, E.G. & Feener, D.H. Jr. (2007) Maintaining diversity in an ant community: modeling, extending, and testing the dominance-discovery trade-off. *The American Naturalist*, **169**, 323–333.

Asner, G.P. & Martin, R.E. (2012) Contrasting leaf chemical traits in tropical lianas and trees: implications for future forest composition. *Ecology Letters*, **15**, 1001–1007.

Baroni Urbani, C. & de Andrade, M.L. (1997) Pollen eating, storing, and spitting by ants. *Naturwissenschaften*., **84**, 256–258.

Basset, Y. (1991a) Influence of leaf traits on the spatial distribution of insect herbivores associated with an overstorey rainforest tree. *Oecologia*, **87**, 388–393.

Basset, Y. (1991b) Leaf production of an overstorey rainforest tree and its effects on the temporal distribution of associated insect herbivores. *Oecologia*, **88**, 211–219.

Basset, Y. (1992a) Host specificity of arboreal and free-living insect herbivores in rain forests. *Biological Journal of the Linnean Society*, **47**, 115–133.

Basset, Y. (1992b) Influence of leaf traits on the spatial distribution of arboreal arthropods within an overstorey rainforest tree. *Ecological Entomology*, **17**, 8–16.

Basset, Y., Aberlenc, H.-P., Barrios, H. et al. (2001) Stratification and diel activity of arthropods in a lowland rainforest in Gabon. *Biological Journal of the Linnean Society*, **72**, 585–607.

Basset, Y., Aberlenc, H.-P. & Delvare, G. (1992) Abundance and stratification of foliage arthropods in a lowland rain forest of Cameroon. *Ecological Entomology*, **17**, 310–318.

Basset, Y., Cizek, L., Cuénoud, P. et al. (2012) Arthropod diversity in a tropical forest. *Science*, **338**, 1481–1484.

Basset, Y., Novotny, V., Miller, S.E. & Kitching, R.L. (eds) (2003) *Arthropods of Tropical Forests: Spatio-temporal Dynamics and Resource Use in the Canopy*, Cambridge University Press, Cambridge.

Basset, Y., Samuelson, G.A., Allison, A. & Miller, S.E. (1996) How many species of host-specific insects feed on a species of tropical tree? *Biological Journal of the Linnean Society*., **59**, 201–216.

Beattie, A.J. (1985) *The Evolutionary Ecology of Ant-Plant Mutualisms*, Cambridge Univ Press, Cambridge.

Bell, S.S., McCoy, E.D. & Mushinsky, H.R. (eds) (1991) *Habitat Structure: The Physical Arrangement of Objects in Space*, Chapman & Hall, London.

Benson, W.W., Brown, Jr., K.S., & Gilbert, L.E. (1975) Coevolution of plants and herbivores: passion flower butterflies. *Evolution*, **29**, 659–680.

Bentley, B.L. (1981) Ants, extrafloral nectaries, and the vine life-form: an interaction. *Tropical Ecology*, **22**, 127–133.

Bestelmeyer, B.T., Agosti, D., Alonso, L.E. et al. (2000) Field techniques for the study of ground-dwelling ants: an overview, description, and evaluation, in *Ants: Standard Methods for Measuring and Monitoring Biodiversity* (eds D. Agosti, J.D. Majer, L.E. Alonso & T.R. Shultz), Smithsonian Institution Press, Washington, DC, pp. 122–145.

Blüthgen, N. & Fiedler, K. (2002) Interactions between weaver ants *Oecophylla smaragdina*, homopterans, trees and lianas in an Australian rain forest canopy. *Journal of Animal Ecology*, **71**, 793–801.

Blüthgen, N., Stork, N.E. & Fiedler, K. (2004) Bottom-up control and co-occurence in complex communities: honeydew and nectar determine a rainforest ant mosaic. *Oikos*, **106**, 344–358.

Blüthgen, N., Verhaagh, M., Goitía, W., et al. (2000) How plants shape the ant community in the Amazonian rainforest canopy: the key role of extrafloral nectaries and homopteran honeydew. *Oecologia*, **125**, 229–240.

Boinski, S. & Fowler, N.L. (1989) Seasonal patterns in a tropical lowland forest. *Biotropica*, **21**, 223–233.

Bonhomme, V., Gounand, I., Alaux, C. et al. (2011) The plant-ant *Camponotus schmitzi* helps its carnivorous host-plant *Nepenthes bicalcarata* to catch its prey. *Journal of Tropical Ecology*, **27**, 15–24.

Camargo, R.X. & Oliveira, P.S. (2012) Natural history of the Neotropical arboreal ant, *Odontomachus hastatus*: nest sites, foraging schedule, and diet. *Journal of Insect Science*, **12**, 48.

Campos, R.I., Vasconcelos, H.L., Ribeiro, S.P. et al. (2006) Relationship between tree size and insect assemblages associated with *Anadenanthera macrocarpa*. *Ecography*, **29**, 442–450.

Carroll, C.R. (1979) A comparative study of two ant faunas: the stem-nesting ant communities of Liberia, West Africa and Costa Rica, Central America. *The American Naturalist*, **113**, 551–561.

Catling, P.M. (1997) Influence of aerial *Azteca* nests on the epiphyte community of some Belizean orange orchards. *Biotropica*, **29**, 237–242.

Chan, C.S.-Y., Cheng, J., Loh, J.Y.Q., et al. (2012) Observations of ants and aphids in the rattan species *Korthalsia echinometra* Becc. and *Korthalsia rostrata* Blume. *The Raffles Bulletin of Zoology*, Supp. **25**, 133–139.

Clarke, C.M. & Kitching, R.L. (1995) Swimming ants and pitcher plants: a unique ant-plant interaction from Borneo. *Journal of Tropical Ecology*, **11**, 589–602.

Clay, N.A., Bauer, M., Solis, M. & Yanoviak, S.P. (2010) Arboreal substrates influence foraging in tropical ants. *Ecological Entomology*, **35**, 417–423.

Coley, P.D. & Barone, J.A. (1996) Herbivory and plant defenses in tropical forests. *Annual Review of Ecology and Systematics*, **27**, 305–335.

Condit, R. (1998) *Tropical Forest Census Plots: Methods and Results from Barro Colorado Island, Panama and a Comparison with Other Plots*, Springer Verlag, Berlin.

Dale, V.H., Joyce, L.A., McNulty, S. et al. (2001) Climate change and forest disturbances. *Bioscience*, **51**, 723–734.

Davidson, D.W. (1997) The role of resource imbalances in the evolutionary ecology of tropical arboreal ants. *Biological Journal of the Linnean Society*, **61**, 153–181.

Davidson, D.W. (1998) Resource discovery versus resource domination in ants: a functional mechanism for breaking the trade-off. *Ecological Entomology*, **23**, 484–490.

Davidson, D.W., Cook, S.C., Snelling, R.R. & Chua, T.H. (2003) Explaining the abundance of ants in lowland tropical rainforest canopies. *Science*, **300**, 969–972.

Davidson, D.W., Lessard, J.-P., Bernau, C.R. & Cook, S.C. (2007) The tropical ant mosaic in a primary Bornean rain forest. *Biotropica*, **39**, 468–475.

Davidson, D.W., Longino, J.T. & Snelling, R.R. (1988) Pruning of host plant neighbors by ants: an experimental approach. *Ecology*, **69**, 801–808.

Dean, R.L. & Connell, J.H. (1987) Marine invertebrates in an algal succession. II. Tests of hypotheses to explain changes in diversity with succession. *Journal of Marine Biology and Ecology*, **109**, 217–247.

Dejean, A., Corbara, B., Orivel, J. & Leponce, M. (2007) Rainforest canopy ants: the implications of territoriality and predatory behavior. *Functional Ecosystems and Communities*, **1**, 105–120.

Dejean, A., Olmsted, I. & Snelling, R.R. (1995) Tree-epiphyte-ant relationships in the low inundated forest of Sian Ka'an Biosphere Reserve, Quintana Roo, Mexico. *Biotropica*, **27**, 57–70.

Dial, R.J., Ellwood, M.D.F., Turner, E.C. & Foster, W.A. (2006) Arthropod abundance, canopy structure, and microclimate in a Bornean lowland tropical rain forest. *Biotropica*, **38**, 643–652.

Dietrich, W.E., Windsor, D.M. & Dunne, T. (1982) Geology, climate, and hydrology of Barro Colorado Island, in *The Ecology of a Tropical Forest* (eds E.G. Leigh, A.S. Jr. Rand & D.M. Windsor), Smithsonian Institution Press, Washington, DC, pp. 21–46.

Djiéto-Lordon, C., Dejean, A., Gibernau, M. et al. (2004) Symbiotic mutualism with a community of opportunistic ants: protection, competition, and ant occupancy of the myrmecophyte *Barteria nigritana* (Passifloraceae). *Acta Oecologica*, **26**, 109–116.

Djiéto-Lordon, C., Dejean, A., Ring, R.A. et al. (2005) Ecology of an improbable association: the pseudomyrmecine plant-ant *Tetraponera tessmanni* and the myrmecophytic liana *Vitex thyrsiflora* (Lamiaceae) in Cameroon. *Biotropica*, **37**, 421–430.

Dussutour, A., Nicolis, S.C., Deneubourg, J.-L. & Fourcassié, V. (2006) Collective decisions in ants when foraging under crowded conditions. *Behavioral Ecology and Sociobiology*, **61**, 17–30.

Edwards, D.P., Ansell, F.A., Woodcock, P. et al. (2010) Can the failure to punish promote cheating in mutualism? *Oikos*, **119**, 45–52.

Ellwood, M.D.F., Jones, D.T. & Foster, W.A. (2002) Canopy ferns in lowland dipterocarp forest support a prolific abundance of ants, termites, and other invertebrates. *Biotropica*, **34**, 575–583.

Elton, C.S. (1973) The structure of invertebrate populations inside neotropical rain forest. *Journal of Animal Ecology*, **42**, 55–104.

Emmons, L.H. & Gentry, A.H. (1983) Tropical forest structure and the distribution of gliding and prehensile-tailed vertebrates. *The American Naturalist*, **121**, 513–524.

Erwin, T.L. (1982) Tropical forests: their richness in Coleoptera and other arthropod species. *The Coleopterists Bulletin*, **36**, 74–75.

Farji-Brener, A.G., Barrantes, G., Laverde, O. et al. (2007) Fallen branches as part of leaf cutting ant trails: their role in resource discovery and leaf transport rates in *Atta cephalotes*. *Biotropica*, **39**, 211–215.

Farji-Brener, A.G., Barrantes, G. & Ruggiero, A. (2004) Environmental rugosity, body size and access to food: a test of the size-grain hypothesis in tropical litter ants. *Oikos*, **104**, 165–171.

Fewell, J.H. (1988) Energetic and time costs of foraging in harvester ants, *Pogonomyrmex occidentalis*. *Behavioral Ecology and Sociobiology*, **22**, 401–408.

Fischer, J. & Lindenmayer, D.B. (2007) Landscape modification and habitat fragmentation: a synthesis. *Global Ecology and Biogeography*, **16**, 265–280.

Floren, A. & Deeleman-Reinhold, C. (2005) Diversity of arboreal spiders in primary and disturbed tropical forests. *Journal of Arachnology*, **33**, 323–333.

Floren, A. & Linsenmair, K.E. (1999) Changes in arboreal arthropod communities along a disturbance gradient. *Selbyana*, **20**, 284–289.

Floren, A. & Linsenmair, K.E. (2000) Do ant mosaics exist in pristine lowland rain forests? *Oecologia*, **123**, 129–137.

Fonseca, G.A.B. (1989) Small mammal species diversity in Brazilian tropical primary and secondary forests of different sizes. *Revista brasileira de Zoologia*, **6**, 381–422.

Frederickson, M.E. & Gordon, D.M. (2007) The devil to pay: a cost of mutualism with *Myrmelachista schumanni* ants in 'devil's gardens' is increased herbivory on *Duroia hirsuta* trees. *Proceedings of Royal Society of London B*, **274**, 1117–1123.

Gentry, A.H. (1991) The distribution and evolution of climbing plants, in *The Biology of Vines* (eds F.E. Putz & H.A. Mooney), Cambridge University Press, Cambridge, pp. 3–49.

Gentry, A.H. & Dodson, C. (1987) Contribution of nontrees to species richness of a tropical rain forest. *Biotropica*, **19**, 149–156.

Gerwing, J.J., Schnitzer, S.A., Burnham, R.J. et al. (2006) Censusing lianas. *Biotropica*, **38**, 256–261.

Hahn, D.A. & Wheeler, D.E. (2002) Seasonal foraging activity and bait preferences of ants on Barro Colorado Island, Panama. *Biotropica*, **34**, 348–356.

Halaj, J., Ross, D.W. & Moldenke, A.R. (1998) Habitat structure and prey availability as predictors of the abundance and community organization of spiders in western Oregon forest canopies. *Journal of Arachnology*, **26**, 203–220.

Halaj, J., Ross, D.W. & Moldenke, A.R. (2000) Importance of habitat structure to the arthropod food-web in Douglas-fir canopies. *Oikos*, **90**, 139–152.

Hegarty, E.E. & Caballé, G. (1991) Distribution and abundance of vines in forest communities, in *The Biology of Vines* (eds F.E. Putz & H.A. Mooney), Cambridge University Press, Cambridge, pp. 313–335.

Hegarty, M.P., Hegarty, E.E., & Gentry, A.H. (1991) Secondary compounds in vines with an emphasis on those with defensive functions. In *The Biology of Vines* (eds. F.E. Putz & H.A. Mooney), pp. 287–310. Cambridge University Press, Cambridge.

Hölldobler, B. & Wilson, E.O. (1990) *The Ants*, Harvard University Press, Cambridge, MA.

Huffaker, C.B. (1958) Experimental studies on predation: dispersion factors and predator-prey oscillations. *Hilgardia*, **27**, 343–383.

Huxley, C.R. & Cutler, D.F. (eds) (1991) *Ant-Plant Interactions*, Oxford University Press, Oxford.

Ingwell, L.L., Wright, S.J., Becklund, K.K. et al. (2010) The impact of lianas on 10 years of tree growth and mortality on Barro Colorado Island, Panama. *Journal of Ecology*, **98**, 879–887.

James, F.C. & Wamer, N.O. (1982) Relationships between temperate forest bird communities and vegetation structure. *Ecology*, **63**, 159–171.

Janzen, D.H. (1966) Coevolution of mutualism between ants and acacias in Central America. *Evolution*, **20**, 249–275.

Kareiva, P. (1982) Exclusion experiments and the competitive release of insects feeding on collards. *Ecology*, **63**, 696–704.

Kaspari, M. (2000) A primer on ant ecology. In *Ants: Standard Methods for Measuring and Monitoring Biodiversity* (eds. D. Agosti, J.D. Majer, L.E. Alonso, & T.R. Shultz), pp. 9–24. Smithsonian Institution Press, Washington, DC.

Kaspari, M. & Weiser, M.D. (1999) The size-grain hypothesis and interspecific scaling in ants. *Functional Ecology*, **13**, 530–538.

Keller, L. (1998) Queen lifespan and colony characteristics in ants and termites. *Insectes Sociaux*, **45**, 235–246.

Kenoyer, L.A. (1929) Plant physiognomy. *Ecology*, **10**, 409–414.

Kira, T. & Ogawa, H. (1971) Assessment of primary production in tropical and equatorial forests, in *Productivity of Forest Ecosystems* (ed P. Duvigneaud), UNESCO, Paris, pp. 309–321.

Klimes, P., Idigel, C., Rimandai, M., et al. (2012) Why are there more arboreal ant species in primary than in secondary tropical forests? *Journal of Animal Ecology*, **81**: 1103–1112.

Koenig, S.E., Wunderle, J.M. Jr. & Enkerlin-Hoeflich, E.C. (2007) Vines and canopy contact: a route for snake predation on parrot nests. *Bird Conservation International*, **17**, 79–91.

Körner, C. (2009) Responses of humid tropical trees to rising CO_2. *Annual Review of Ecology, Evolution, and Systematics*, **40**, 61–79.

Krings, M., Taylor, T.N. & Kellogg, D.W. (2002) Touch-sensitive glandular trichomes: a mode of defence against herbivorous arthropods in the Carboniferous. *Evolutionary Ecology Research*, **4**, 779–786.

Lach, L., Parr, C.L. & Abbott, K. (eds) (2010) *Ant Ecology*, Oxford University Press, Oxford.

Lassau, S.A. & Hochuli, D.F. (2004) Effects of habitat complexity on ant assemblages. *Ecography*, **27**, 157–164.

Lawton, J.H. (1983) Plant architecture and the diversity of phytophagous insects. *Annual Review of Entomology*, **28**, 23–39.

Leston, D. (1978) A neotropical ant mosaic. *Annals of the Entomological Society of America*, **71**, 649–653.

Longino, J.T. (2003) The *Crematogaster* (Hymenoptera, Formicidae, Myrmicinae) of Costa Rica. *Zootaxa*, **151**, 1–150.

Lowman, M., Schowalter, T. & Franklin, J. (eds) (2012) *Forest Canopy Methods*, University of California Press, Berkeley, CA.

Lubin, Y.D. (1978) Seasonal abundance and diversity of web-building spiders in relation to habitat structure on Barro Colorado Island, Panama. *Journal of Arachnology*, **6**, 31–51.

Magnusson, W.E., Lima, A.P. & de Lima, O. (1996) Group lightning mortality of trees in a Neotropical forest. *Journal of Tropical Ecology*, **12**, 899–903.

Maschwitz, U., Fiala, B., Lee, Y.F. et al. (1989) New and little-known myrmecophytic associations from Bornean rain forests. *Malayan Nature Journal*, **43**, 106–115.

Menzel, F., Linsenmair, K.E. & Blüthgen, N. (2008) Selective interspecific tolerance in tropical *Crematogaster-Camponotus* associations. *Animal Behaviour*, **75**, 837–846.

Merbach, M.A., Zizka, G., Fiala, B. et al. (2007) Why a carnivorous plant cooperates with an ant: selective defense against pitcher-destroying weevils in the myrmecophytic pitcher plant *Nepenthes bicalcarata* Hook. f. *Ecotropica*, **13**, 45–56.

Moog, J. (2009) *The associations of the plant-ant* Cladomyrma *with plants in Southeast Asia. Doctoral dissertation*. Goethe University Frankfurt, Frankfurt.

Moog, J., Drude, T., Maschwitz, U. & Agosti, D. (1997) Flood control by ants: water-bailing behaviour in the Southeast Asian plant-ant genus *Cladomyrma* Wheeler (Formicidae, Formicinae). *Naturwissenschaften*, **84**, 242–245.

Moog, J., Fiala, B., Werner, M. et al. (2003) Ant-plant diversity in peninsular Malaysia, with special reference to the Pasoh Forest Reserve, in *Pasoh: Ecology of a Lowland Rrainforest in Southeast Asia* (eds T. Okuda, N. Manokaran, Y. Matsumoto et al.), Springer Press, Tokyo, pp. 459–494.

Moreau, C.S. & Bell, C.D. (2013) Testing the museum versus cradle tropical biological diversity hypothesis: phylogeny, diversification, and ancestral biogeographic range evolution of the ants. *Evolution*, **67-68**, 2240–2257.

Moreau, C.S., Bell, C.D., Vila, R. et al. (2006) Phylogeny of the ants: diversification in the age of angiosperms. *Science*, **312**, 101–104.

Ng, F.S.P. (1977) Shyness in trees. *Nature Malaysiana*, **2**, 35–37.

Ødegaard, F. (2000a) How many species of arthropods? Erwin's estimate revised. *Biological Journal of the Linnean Society*, **71**, 583–597.

Ødegaard, F. (2000b) The relative importance of trees versus lianas as hosts for phytophagous beetles (Coleoptera) in tropical forests. *Journal of Biogeography*, **27**, 283–296.

Opler, P.A., Baker, H.G. & Frankie, G.W. (1991) Seasonality of climbers: a review and example from Costa Rican dry forest, in *The Biology of Vines* (eds F.E. Putz & H.A. Mooney), Cambridge University Press, Cambridge, pp. 377–392.

Ozanne, C.M.P., Anhuf, D., Boulter, S.L. et al. (2003) Biodiversity meets the atmosphere: a global view of forest canopies. *Science*, **301**, 183–186.

Perry, D.R. (1978) A method of access into the crowns of emergent and canopy trees. *Biotropica*, **10**, 155–157.

Phillips, O.L., Vásquez Martínez, R., Arroyo, L., et al. (2002) Increasing dominance of large lianas in Amazonian forests. *Nature*, **418**, 770–774.

Phillips, O.L., Vásquez Martínez, R., Monteagudo Mendoza, A. et al. (2005) Large lianas as hyperdynamic elements of the tropical forest canopy. *Ecology*, **86**, 1250–1258.

Philpott, S.M. (2010) A canopy dominant ant affects twig-nesting ant assembly in coffee agroecosystems. *Oikos*, **119**, 1954–1960.

Philpott, S.M. & Foster, P.F. (2005) Nest-site limitation in coffee agroecosystems: artificial nests maintain diversity of arboreal ants. *Ecological Applications*, **15**, 1478–1485.

Pizo, M.A. & Oliveira, P.S. (2000) The use of fruits and seeds by ants in the Atlantic forest of southeast Brazil. *Biotropica*, **32**, 851–861.

Powell, S., Costa, A.N., Lopes, C.T. & Vasconcelos, H.L. (2011) Canopy connectivity and the availability of diverse nesting resources affect species coexistence in arboreal ants. *Journal of Animal Ecology*, **80**, 352–360.

Putz, F.E. (1984) The natural history of lianas on Barro Colorado Island, Panama. *Ecology*, **65**, 1713–1724.

Putz, F.E., Parker, G.G. & Archibald, R.M. (1984) Mechanical abrasion and intercrown spacing. *American Midland Naturalist*, **112**, 24–28.

Ribas, C.R., Schoereder, J.H., Pic, M. & Soares, S.M. (2003) Tree heterogeneity, resource availability, and larger scale processes regulating arboreal ant species richness. *Australian Ecolog*, **28**, 305–314.

Richards, P.W. (1996) *The Tropical Rain Forest*, second edn, Cambridge University Press, Cambridge.

Rico-Gray, V. & Oliveira, P.S. (2007) *The Ecology and Evolution of Ant-Plant Interactions*, University of Chicago Press, Chicago.

Riechert, S.E. & Cady, A.B. (1983) Patterns of resource use and tests for competitive release in a spider community. *Ecology*, **64**, 899–913.

Rosenberg, D.K., Noon, B.R. & Meslow, E.C. (1997) Biological corridors: form, function, and efficacy. *Bioscience*, **47**, 677–687.

Sagata, K., Mack, A.L., Wright, D.D. & Lester, P.J. (2010) The influence of nest availability on the abundance and diversity of twig-dwelling ants in a Papua New Guinea forest. *Insectes Sociaux*, **57**, 333–341.

Sanders, N.J., Crutsinger, G.M., Dunn, R.R. et al. (2007) An ant mosaic revisited: dominant ant species disassemble arboreal ant communities but co-occur randomly. *Biotropica*, **39**, 422–427.

Sarty, M., Abbott, K.L. & Lester, P.J. (2006) Habitat complexity facilitates coexistence in a tropical ant community. *Oecologia*, **149**, 465–473.

Schnitzer, S.A. & Bongers, F. (2011) Increasing liana abundance and biomass in tropical forests: emerging patterns and putative mechanisms. *Ecology Letters*, **14**, 397–406.

Schnitzer, S.A., Bongers, F. & Wright, S.J. (2011) Community and ecosystem ramifications of increasing lianas in neotropical forests. *Plant Signaling and Behavior*, **6**, 598–600.

Schnitzer, S.A., Mangan, S.A., Dalling, J.W. et al. (2012) Liana abundance, diversity, and distribution on Barro Colorado Island, Panama. *PLoS ONE*, **7**, e52114.

Schnitzer, S.A., Rutishauser, S. & Aguilar, S. (2008) Supplemental protocol for liana censuses. *Forest Ecology and Management*, **255**, 1044–1049.

Schonberg, L.A., Longino, J.T., Nadkarni, N.M. et al. (2004) Arboreal ant species richness in primary forest, secondary forest, and pasture habitats of a tropical montane landscape. *Biotropica*, **36**, 402–409.

Smiley, J. (1987) Plant chemistry and the evolution of host specificity: new evidence from *Heliconius* and *Passiflora*. *Science*, **201**, 745–747.

Snaddon, J.L., Turner, E.C., Fayle, T.M. et al. (2012) Biodiversity hanging by a thread: the importance of fungal litter-trapping systems in tropical rainforests. *Biology Letters*, **8**, 397–400.

Sponberg, S. & Full, R.J. (2007) Neuromechanical response of musculo-skeletal structures in cockroaches during rapid running on rough terrain. *Journal of Experimental Biology*, **211**, 433–446.

Stamp, N. (2003) Out of the quagmire of plant defense hypotheses. *Quarterly Review of Biology*, **78**, 23–55.

Stein, W.E., Berry, C.M., Hernick, L.V. & Mannolini, F. (2012) Surprisingly complex community discovered in the mid-Devonian fossil forest at Gilboa. *Nature*, **483**, 78–81.

Stork, N.E. (1987) Arthropod faunal similarity of Bornean rain forest trees. *Ecological Entomology*, **12**, 219–226.

Stork, N.E., Adis, J. & Didham, R.K. (eds) (1997) *Canopy Arthropods*, Chapman & Hall, London.

Stuntz, S., Linder, C., Linsenmair, K.E. et al. (2003) Do non-myrmocophilic epiphytes influence community structure of arboreal ants? *Basic and Applied Ecology*, **4**, 363–373.

Su, J.C. & Woods, S.A. (2001) Importance of sampling along a vertical gradient to compare the insect fauna in managed forests. *Environmental Entomology*, **30**, 400–408.

Sutton, S.L., Ash, C.P.J., & Grundy, A. (1983) The vertical distribution of flying insects in lowland rain-forests of Panama, Papua New Guinea and Brunei. *Zoological Journal of the Linnean Society*, **78**, 287–297.

Tanaka, H.O. & Itioka, T. (2011) Ants inhabiting myrmecophytic ferns regulate the distribution of lianas on emergent trees in a Bornean tropical rainforest. *Biology Letters*, **7**, 706–709.

Tanaka, H.O., Yamane, S. & Itioka, T. (2010) Within-tree distribution of nest sites and foraging areas of ants on canopy trees in a tropical rainforest in Borneo. *Population Ecology*, **52**, 147–157.

Tanner, E.V.J., Kapos, V. & Healey, J.R. (1991) Hurricane effects on forest ecosystems in the Caribbean. *Biotropica*, **23**, 513–521.

Tews, J., Brose, U., Grimm, V. et al. (2004) Animal species diversity driven by habitat heterogeneity/diversity: the importance of keystone structures. *Journal of Biogeography*, **31**, 79–92.

Trigo, J.R. & Motta, P.C. (1990) Evolutionary implications of pyrrolizidine alkaloid assimilation by danaine and ithomiine larvae (Lepidoptera: Nymphalidae). *Experientia*, **46**, 332–334.

Vandermeer, J., Perfecto, I. & Philpott, S. (2010) Ecological complexity and pest control in organic coffee production: uncovering an autonomous ecosystem service. *Bioscience*, **60**, 527–537.

Walter, D.E., O'Dowd, D. & Barnes, V. (1994) The forgotten arthropods: foliar mites in the forest canopy. *Memoirs of the Queensland Museum*, **36**, 221–226.

Ward, P.S. (1989) Systematic studies on pseudomyrmecine ants: revision of the *Pseudomyrmex oculatus* and *P. subtilissimus* species groups, with taxonomic comments on other species. *Quaestiones entomologicae*, **25**, 393–468.

Ward, P.S. (1991) Phylogenetic analysis of pseudomyrmecine ants associated with domatia-bearing plants, in *Ant-Plant Interactions* (eds C.R. Huxley & D.F. Cutler), Oxford University Press, Oxford, pp. 335–352.

Ward, P.S. (1999) Systematics, biogeography and host plant associations of the *Pseudomyrmex viduus* group (Hymenoptera: Formicidae), *Triplaris*- and *Tachigali*-inhabiting ants. *Zoological Journal of the Linnean Society*, **126**, 451–540.

Weber, N.A. (1957) The nest of an anomalous colony of the arboreal ant *Cephalotes atratus*. *Psyche*, **64**, 60–69.

Wilson, E.O. (2003) Pheidole *in the New World: A Dominant, Hyperdiverse Ant Genus*, Harvard University Press, Cambridge, MA.

Wolda, H. (1979) Abundance and diversity of Homoptera in the canopy of a tropical forest. *Ecological Entomolog*, **4**, 181–190.

Wong, M., Wright, S.J., Hubbell, S.P. & Foster, R.B. (1990) The spatial pattern and reproductive consequences of outbreak defoliation in *Quararibea asterolepis*, a tropical tree. *Journal of Ecology*, **78**, 579–588.

Wright, S.J., Calderón, O., Hernandéz, A. & Paton, S. (2004) Are lianas increasing in importance in tropical forests? A 17-year record from Panama. *Ecology*, **85**, 484–489.

Yanoviak, S.P. (1999) Community structure in water-filled tree holes of Panama: effects of hole height and size. *Selbyana*, **20**, 106–115.

Yanoviak, S.P. (2001) The macrofauna of water-filled tree holes on Barro Colorado Island, Panama. *Biotropica*, **33**, 110–120.

Yanoviak, S.P. (2013) Shock value: are lianas natural lightning rods?, in *Treetops at Risk: Challenges of Global Forest Canopies* (eds M. Lowman, S. Devy & T. Ganesh), Springer, New York, pp. 147–154.

Yanoviak, S.P., Berghoff, S.M., Linsenmair, K.E. & Zotz, G. (2011) Effects of an epiphytic orchid on arboreal ant community structure in Panama. *Biotropica*, **43**, 731–737.

Yanoviak, S.P., Dudley, R. & Kaspari, M. (2005) Directed aerial descent in canopy ants. *Nature*, **433**, 624–626.

Yanoviak, S.P. & Kaspari, M. (2000) Community structure and the habitat templet: ants in the tropical forest canopy and litter. *Oikos*, **89**, 259–266.

Yanoviak, S.P., Kragh, G. & Nadkarni, N.M. (2003) Spider assemblages in Costa Rican cloud forests: effects of forest level and forest age. *Studies on Neotropical Fauna and Environment*, **38**, 145–154.

Yanoviak, S.P., Munk, Y. & Dudley, R. (2011) Evolution and ecology of directed aerial descent in arboreal ants. *Integrative and Comparative Biology*, **51**, 944–956.

Yanoviak, S.P., Nadkarni, N.M., & Gering, J.C. (2003) Arthropods in epiphytes: a diversity component that is not effectively sampled by canopy fogging. *Biodiversity and Conservation*, **12**, 731–741.

Yanoviak, S.P., Nadkarni, N.M., & Solano R. (2006) Arthropod assemblages in epiphyte mats of Costa Rican cloud forests. *Biotropica*, **36**, 202–210.

Yanoviak, S.P. & Schnitzer, S.A. (2013) Functional roles of lianas for forest canopy animals, in *Treetops at Risk* (eds M. Lowman, S. Devy & T. Ganesh), Springer Verlag, New York, pp. 209–21.

Yanoviak, S.P., Silveri, C., Hamm, C.A. & Solis, M. (2012) Stem characteristics and ant body size in a Costa Rican rain forest. *Journal of Tropical Ecology*, **28**, 199–204.

Yanoviak, S.P., Walker, H., & Nadkarni, N.M. (2004) Arthropod assemblages in vegetative vs. humic portions of epiphyte mats in a neotropical cloud forest. *Pedobiologia*, **48**, 51–58.

Chapter 25

LIANA–BIRD RELATIONSHIPS: A REVIEW

Nicole L. Michel,[1] *W. Douglas Robinson,*[2] *and Thomas W. Sherry*[1]

[1]Tulane University, New Orleans, LA, USA
[2]Oregon State University, Corvallis, OR, USA

Ecology of Lianas, First Edition. Edited by Stefan A. Schnitzer, Frans Bongers, Robyn J. Burnham, and Francis E. Putz.

Companion Website: www.wiley.com/go/schnitzer/lianas

OVERVIEW

The abundance and biomass of lianas and birds are each undergoing rapid changes due to global change drivers, including forest fragmentation and climate change. Yet lianas and birds share a wide variety of important beneficial, neutral, or harmful relationships. Liana or bird species that are reliant on their symbiotic partner will be susceptible to changes in the abundance or biomass of the other. Here we review liana–bird relationships, and ask how reliant lianas are on birds for functional services, e.g., seed dispersal, and how reliant birds are on lianas for resources, e.g., food or protection from predators.

We found relationships between 457 bird species from 15 orders and 70 families, and 246 liana species in 27 orders and 48 families. Birds receive a variety of direct and indirect food sources (fruit, nectar, leaves, arthropods); nest material; substrates for display, nesting, and travel; and protection from – or increased exposure to – predators from lianas. Lianas are unique in providing both dense foliage and fruit in forest canopies, and dense tangles of stems and leaves in forest understory that capture falling leaf litter, support arthropod communities, and provide shelter to birds. Conversely, birds disperse seeds, pollinate flowers, and protect lianas by consuming herbivorous insects, but also rob nectar and predate liana seeds and flowers. Widespread increases in liana density and biomass are likely to be a net benefit for many bird species, while pervasive population declines of nectarivorous and insectivorous birds are likely to negatively impact many lianas, particularly those reliant upon birds for seed dispersal and pollination. Bird–liana relationships are likely to become of greater importance as global climate change and fragmentation pressures magnify, and birds and lianas increasingly rely upon one another.

INTRODUCTION: LIANAS AND BIRDS

Lianas and birds have a diverse array of obligate and facultative relationships that span most types of ecological relationships – mutualistic, commensalistic, parasitic, and amensalistic (Fig. 25.1). Liana–bird relationships occur in both temperate and tropical habitats, and may involve native and/or non-native species. Many liana–bird relationships have important implications for the conservation of lianas, birds, and forest communities.

While liana abundance and biomass have increased in forests throughout the Neotropics (Schnitzer & Bongers 2011; Schnitzer, Chapter 30 in this volume), they may have decreased in other forests, particularly in the Afro-tropical forests (Caballé & Martin 2001; Michel 2012, Bongers & Ewango, Chapter 3 in this volume, Thomas et al., Chapter 2 in this volume). Birds, however, are experiencing global declines in both abundance and species richness (Şekercioğlu et al. 2004). Consequently, important ecosystem functions provided by birds, including pollination and seed dispersal (Fig. 25.1), will likely be affected, as up to 39% of birds are predicted to be extinct or functionally extinct by 2100 (Şekercioğlu et al. 2004). To predict the effects of bird declines and changes in liana abundance, we first need to understand the nature of, and the degree of plasticity in, liana–bird relationships.

In this chapter, we address several questions regarding liana–bird relationships, and their roles within the context of global change. First, how dependent are birds on lianas for food, shelter, and other resources? Do lianas provide unique resources that are unmatched by other growth forms? Conversely, how dependent are lianas on birds for functional services such as pollination, seed dispersal, and control of arthropod herbivores? Finally, what are the conservation implications of these obligate and/or facultative relationships in a time of changes in abundance of both birds and lianas? Are birds contributing to liana increases (or declines), or lianas to bird declines? Conversely, can liana–bird relationships help mitigate effects of global change? We first review the benefits and costs to birds from using lianas, then the inverse, and finally, the conservation implications of these relationships.

BENEFITS AND COSTS OF LIANAS TO BIRDS

We found liana–bird relationships described in the literature for 457 bird species in 15 orders and 70 families (Figs. 25.2, 25.3, Appendix Table 25.1). Below we synthesize the published literature to describe the benefits and costs to birds of bird–liana relationships, including food, nesting and roosting, protection from or exposure to predators, movement, song or display perches, and general habitat associations.

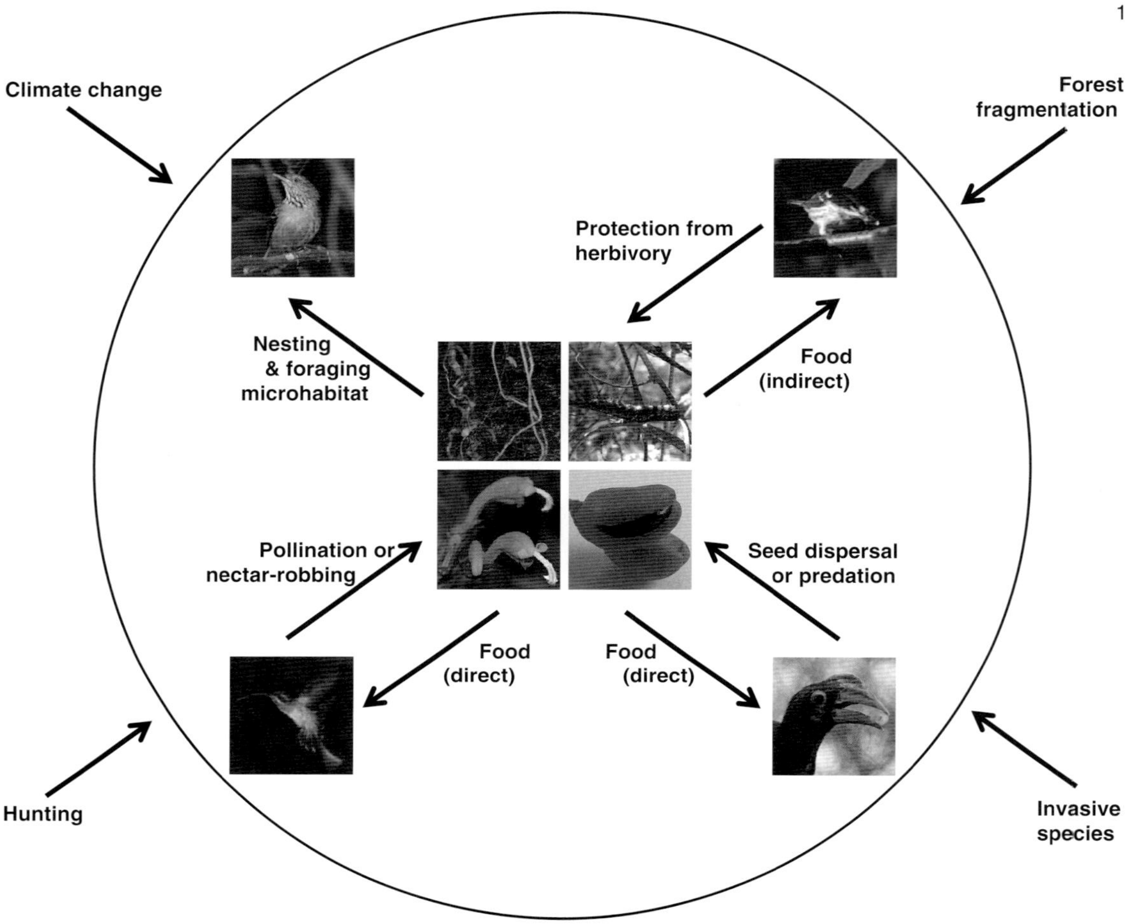

Fig. 25.1 Lianas and birds share many relationships that provide benefits and/or costs to each partner (clockwise, from top left). Lianas provide nesting and foraging microhabitat to birds, at no benefit or cost to themselves; insectivorous birds consume arthropods on lianas, while potentially protecting lianas from herbivores; frugivorous birds feed on liana fruits, while either dispersing or predating liana seeds; and nectarivorous birds feed on liana nectar, while either pollinating or nectar-robbing the liana. Yet these liana–bird relationships exist within a context of global change drivers such as climate change, forest fragmentation, invasive species, and hunting, which are increasing or reducing bird and liana abundance and diversity. Changes in bird and liana abundance will cascade to affect liana–bird relationships in unpredictable ways. Conversely, liana–bird relationships may help mitigate effects of global change drivers on their symbiotic partners. (Source: Photo attribution: birds (clockwise from top left): © Dominic Sherony/CC-BY-SA-2.0; © John Benson/CC-BY-SA-2.0; William Warby, CC; © Brian Gratwicke/CC-BY-SA-2.0; Lianas (clockwise from top left): Nicole Michel; © Patti Haskins/CC-BY-SA-2.0; © Alex Popovkin/CC-BY-SA-2.0; © Sam Beebe/CC-BY-SA-2.0, Ecotrust, CC By 3.0. This work is licensed under the Creative Commons Attribution 3.0 Unported License. To view a copy of this license, visit http://creativecommons.org/licenses/by/3.0/. Reproduced with permission.)

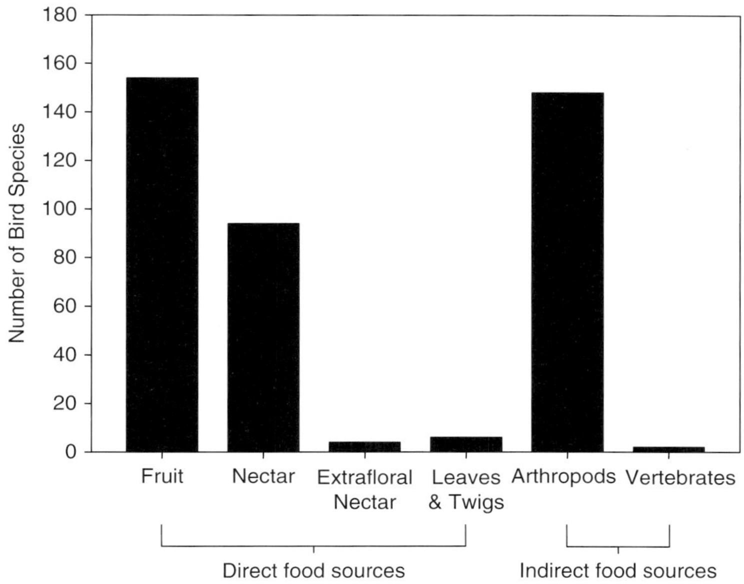

Fig. 25.2 Number of bird species known to obtain direct (fruit, nectar, extrafloral nectar, vegetative material) or indirect (arthropods, vertebrates) food resources from lianas.

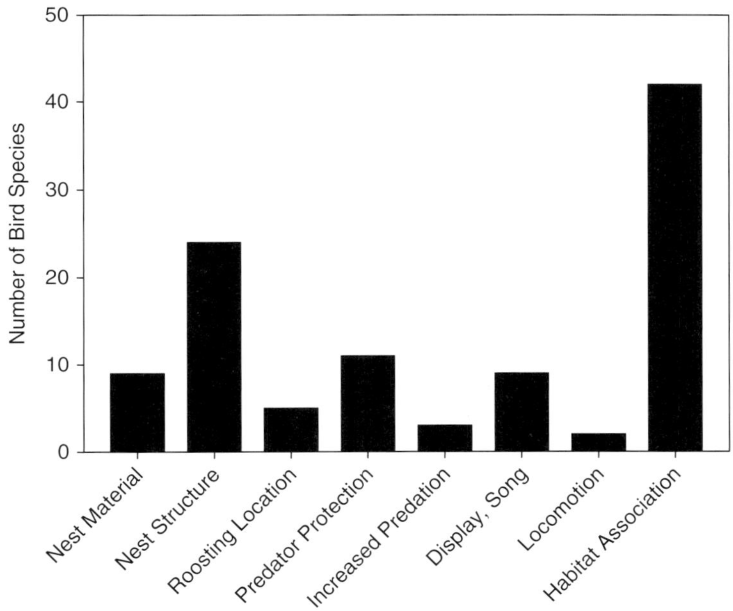

Fig. 25.3 Number of bird species known to obtain non-food resources and other benefits from lianas, including nest material, nest structural support, roosting, protection from predators, display or song perches, movement pathways, and general habitat associations.

Food

Lianas provide a wide variety of food resources to birds, some obtained directly (e.g., fruit or nectar) and some indirectly (e.g., arthropods feeding, sheltering, or reproducing in lianas). We found reports of 387 bird species from 58 families in 12 orders that obtained food resources from lianas either directly or indirectly (Fig. 25.2, Appendix Table 25.1).

Food obtained directly from lianas

Fruit: Many lianas produce nutrient-rich, fleshy fruits that are attractive to birds. The majority of the liana species in the Asian tropics provide fruit used by birds (LaFrankie 2005), ranging from 61% (Thailand; Sankamethawee et al. 2011) to 84% (Japan; Kominami et al. 2003). In contrast, most liana species in the Neotropics are wind-dispersed, and thus not as useful to birds in terms of food. For example, only 25% of lianas at Barro Colorado Island, Panama (BCI) produce fleshy fruits, while 60% are wind-dispersed (Muller-Landau & Hardesty 2005).

Fruit is one of the two most frequent food resources birds receive from lianas, second only to arthropods. At least 154 bird species from 42 families consume fruits from >160 liana species (Fig. 25.2, Appendix Table 25.1); and there are probably more bird species than previously reported due to the difficulty of observing birds feeding in forest canopies. In Thailand, 37% of frugivorous birds consumed liana fruits (Sankamethawee et al. 2011). Subcanopy and understory frugivores consume liana fruits (Putz et al. 2001), but liana fruit eating was also common among large canopy frugivores such as toucans (Skutch 1971) and hornbills (Whitney et al. 1998; Datta & Rawat 2003; Kitamura 2011; Sankamethawee et al. 2011). Fruiting lianas also draw ground- or understory-foraging frugivores and frugivore-insectivores to the canopy (Naka 2004).

Moreover, lianas are an important component of the diet of many avian frugivores. Fruits of climbing plants, including hemiepiphytic figs, supply sugars and calcium, and can be rich in lipids (Stiles 1993; Kitamura 2011). Though many frugivorous birds include liana fruits in their diets, no birds appear to strictly specialize on liana fruits. Averaged across the annual cycle, lianas account for only a small proportion of avian frugivore diets: at most, 22% of the total diet of 29 hornbill species consisted of liana fruits (Kitamura 2011).

However, frugivorous birds commonly switch their diet in response to fruiting phenology, thus, the importance of liana fruits to avian frugivores vary seasonally (van Schaik et al. 1993). For example, while only 20% of the fruits consumed annually by Ruwenzori Turacos (*Ruwenzorornis johnstoni*) came from lianas, liana fruits account for nearly 60% of their diet in June (early dry season; Sun & Moermond 1997). Moreover, lianas in many regions produce fruit less seasonally than trees (e.g., *Araucaria* forests of southern Brazil: Marques et al. 2004), or reach peak fruiting at times of lower tree fruit availability (e.g., Barro Colorado Island, Panama: Putz & Windsor 1987; semi-deciduous forests of southern Brazil: Morellato & Leitão-Filho 1996; Tanzania: Moscovice et al. 2007). Indeed, some liana fruits (e.g., *Celtis iguanea* [Ulmaceae], *Souroubea guianensis* [Marcgraviaceae]) are a particularly important resource for birds and other animals because they are regularly produced in abundance during periods of otherwise low food availability (van Schaik et al. 1993; Peres 2000). Consequently, changes in liana abundance could influence availability of seasonally critical food supplies for many frugivorous birds.

Floral nectar: Liana nectar is consumed by at least 94 bird species from 22 families (Fig. 25.2, Appendix Table 25.1). Hummingbirds are the most common liana nectar-feeders, accounting for 35% (29 species) of nectarivores, but parrots, honeyeaters, warblers, blackbirds, orioles, and cardinals also frequently feed on liana floral nectar. Sicklebill hummingbirds (*Eutoxeres aquila* and *E. condamini*) have highly specialized, coevolved relationships with tubular-flowered lianas such as *Centropogon* spp. (Stein 1992). Overall, vines and lianas account for 13% of the plants visited by hummingbirds (Fleming et al. 2005). Some lianas (e.g., *Passiflora longiracemosa;* Passifloraceae) provide such rich nectar that various hummingbird species, such as the Long-billed Hermit (*Phaethornis superciliosus*), forage at it for extended periods (7-second average visit; Snow 1973). The nectar of *Combretum fruticosum* (Combretaceae) is considered a particularly valuable resource for numerous passerines because it is produced reliably and in abundance at a time of otherwise low food availability (Peres 2000). In sum, coevolved nectarivorous birds and lianas – a mutualism found solely in the New World – would be greatly affected by changes in the abundance of their mutualistic partners.

Extrafloral nectar: While uncommon, some lianas produce nectar in extrafloral nectaries. Four bird species in three families – including Black Jacobins (*Florisuga fusca*), Bananaquits (*Coereba flaveola*), Green-headed Tanagers (*Tangara seledon*), and Brazilian Tanagers (*Ramphocelus bresilius*) – were observed foraging on extrafloral nectaries of the liana *Schwartzia brasiliensis* (Marcgraviaceae; Sazima et al. 1993). Use of extrafloral nectaries by vertebrates is little studied and may be underestimated, but appears to be rare enough that birds are unlikely to be affected by changes in liana abundance.

Vegetative material: Food and floral nectar account for the vast majority of direct food resources obtained from lianas by birds. However, six bird species from four families, including the Hoatzin (*Opisthocomus hoazin*), specialize on foliage, including that of lianas (Domínguez-Bello et al. 1994). Other birds eat both fruit and foliage of lianas (e.g., Ruffed Grouse, *Bonasa umbellus*; Stoll et al. 1980, in Senchina 2008). Ruwenzori Turacos (*Ruwenzorornis johnstoni*) are primarily frugivorous, but their diet is 6% leaves, over 60% of which come from lianas (Sun & Moermond 1997). Folivory is uncommon, and birds consume relatively small quantities of liana leaf material, so neither folivorous birds nor the lianas they consume would be greatly affected by changes in the abundance of the other.

Food obtained indirectly from lianas

Arthropods: Arthropods are the most frequent food resource birds obtained from lianas. We found reports of 148 bird species from 22 families that foraged for arthropods associated with lianas (Fig. 25.2, Appendix Table 25.1). In some cases, birds consumed arthropods that were themselves attracted to lianas for food resources (see Yanoviak, Chapter 24 in this volume). Many arthropods feed on floral and extrafloral nectar, particularly Hymenopterans, Dipterans, and Hemipterans, which in turn may attract insectivorous birds (e.g., Gryj et al. 1990). Over 1200 phytophagous beetle species were found on lianas in Panama (Ødegaard 2000). Plain-brown Woodcreepers (*Dendrocincla fuliginosa*) foraging on vines and lianas consumed more beetles than those foraging on other substrates (Chapman & Rosenberg 1991).

In many lowland tropical forests, lianas produce as much as 40% of the foliage (Putz 1984). Liana foliage forms dense mats on the top of the forest canopy, which serves to concentrate arthropods into relatively small areas and reduces travel and foraging times for insectivorous birds. Arthropod abundance increases with liana foliage cover in Borneo (Dial et al. 2006). In Madagascar, the Coquerel's Coua (*Coua coquereli*) preferentially forage in areas with high liana density, where arthropod abundance was greater than areas with lower stem density. Moreover, this species specialized on honeydew secreted by the larvae of *Phromnia rosea* (Hemiptera: Flatidae), which in turn was found only on the liana *Elachyptera minimiflora* (Celastraceae; Chouteau 2004).

Arthropods and their avian predators may also take advantage of the extensive bark surface area provided by lianas. Bark-gleaning insectivores (e.g., woodcreepers, treecreepers, woodpeckers, and spinetails) commonly forage on live and dead liana stems (Willis 1972a; Bierregaard 1990; Chapman & Rosenberg 1991; Pacheco et al. 1996). Several species of tanagers, including the Puerto Rican Tanager (*Nesospingus speculiferus*), the Paradise Tanager (*Tangara chilensis*), and the Black-capped Tanager (*Tangara heinei*) also forage for arthropods on liana stems (Isler & Isler 1987; Cruz 1988).

Over 100 bird species forage in dead-leaf clusters from 25% to ≥75% of the time (Remsen & Parker 1984; Rosenberg 1997). Dead-leaf clusters, and the arthropods that use them, accumulate in vine and liana tangles that are able to catch and retain more leaves than other less-dense understory vegetation (Gradwohl & Greenberg 1982); dead-leaf clusters may also accumulate on liana stem or live-leaf surfaces. Checker-throated Antwrens (*Epinecrophylla fulviventris*) forage near-exclusively (>75%) on dead-leaf clusters (Gradwohl & Greenberg 1982). The arthropods in dead leaf clusters provide such a distinctive resource that in the Cocos Finch (*Pinaroloxias inornata*), the only non-Galapagos Darwin's finch, some individual birds specialize on this abundant resource (particularly in *Ipomoea* spp. vines) relative to other finches with overlapping home ranges (Werner & Sherry 1987, unpublished data).

Other passerines and near-passerines forage preferentially in liana tangles on a combination of living and dead, leafy and woody substrates (Marra & Remsen 1997). In Ecuador, so-called vine-tangle mixed flocks forage nearly exclusively on vine and liana tangles

(English 1998). The Northern Bentbill (*Oncostoma cinereigularis*) in Costa Rica feeds abundantly in dense vine tangles with upward-strikes on relatively cryptic, immobile prey (Sherry 1984), often taken from woody substrates (T.W. Sherry, unpublished data), suggesting that its peculiar downwardly bent beak tip might be adapted to snatch arthropods from vines and lianas. Because so many insectivorous birds that forage on lianas are highly specialized, ecologically if not evolutionarily, changes in liana abundance are likely linked to changes in bird abundance (see Section 25.5 Conclusion).

Vertebrates: Tiny Hawks (*Accipiter superciliosus*) are specialized hummingbird hunters that use liana tangles as camouflage before "surprise attacking" hummingbirds that are foraging (traplining) or displaying nearby (Stiles 1977). Collared Forest-falcons (*Micrastur semitorquatus*) also hide in lianas and attack toucans and other frugivorous birds foraging in lianas (W.D. Robinson, unpublished data).

Nesting and roosting

Lianas and liana tangles provide nesting and/or roosting sites for a wide variety of birds. At least 24 bird species in 11 families build nests on lianas or in liana tangles (Fig. 25.3, Appendix Table 25.1). For example, Rufous-fronted Thornbirds (*Phacellodomus rufifrons*) build tall nests on upright lianas or vines that allow nests to be expanded vertically in subsequent years (Skutch 1969). Other birds build nests using liana material (Fig. 25.3, Appendix Table 25.1), including Golden-faced Tyrannulets (*Zimmerius chrysops*), which line nests with seed-arils from lianas in the genus *Odontadenia* (Apocynaceae; Goulding & Martin 2010). Scaled Antpittas (*Grallaria guatimalensis*; Grallariidae) both build nests on lianas and often include liana material in their nests (Rivera-Ortíz et al. 2010). At least five bird species roost in liana tangles (Fig. 25.3, Appendix Table 25.1), including Jungle Hawk-owls (*Ninox theomacha*; Mack & Wright 1996) and Spotted Antbirds (*Hylophylax naevioides*; Willis 1972b). Although some birds specialize on lianas in terms of nesting and roosting, most bird species appear to use lianas for nesting and roosting in a facultative manner, and thus ongoing changes in liana abundance may have minimal effects on the general nesting habits of birds.

Protection from predators

Dense liana tangles afford protection from predators to many birds by providing camouflage as well as reducing the amount of space needed by predators, particularly raptors, to navigate (Boinski et al. 2003; see also Lambert & Halsey, Chapter 26 in this volume for a similar phenomenon that appears to benefit small mammals). At least 11 bird species use liana tangles to avoid predators (Fig. 25.3, Appendix Table 25.1), including Lunulated Antbirds (*Gymnopithys lunulatus*; Willis 1968) and Spotted Antbirds (*H. naevioides*; Furnariidae; Willis 1972b). Ecuadorian vine-tangle mixed flocks lack a sentinel species, which watch for predators and alert flock members in many other mixed-flocks (English 1998), presumably because of the relative safety afforded by the vine-tangle foraging microhabitat.

Conversely, lianas may facilitate predator access to birds and, especially, bird nests. Boas have been observed using lianas as movement pathways and/or camouflage to attack nestling Black-billed Parrots (*Amazona agilis*) in Jamaica (Koenig et al. 2007), and adult Social Flycatchers (*Myiozetes similis*) and Red-crowned Woodpeckers (*Melanerpes rubricapillus*) in Panama (W.D. Robinson, unpublished data). Willis (1972a) observed a Neotropical bird-eating snake (*Pseustes poecilonotus*) traveling on a vine near a Plain-brown Woodcreeper (*D. fuliginosa*) cavity nest on BCI. As mentioned above, Tiny Hawks use liana tangles as camouflage during sit-and-wait hunting of hummingbirds (Stiles 1977). The protection provided by lianas is clearly of mixed benefit to birds. The effects of changes in liana abundance on birds will likely be species-specific, with both positive and negative effects.

Song or display perches

At least nine bird species sing or display while perched on lianas, or in a liana tangle (Fig. 25.3, Appendix Table 25.1). Manakins, bellbirds, and hermit hummingbirds frequently display while perched on or under liana tangles (Snow 1973; Stiles & Wolf 1979; Rosselli et al. 2002; Durães 2009). These birds frequently – but not exclusively – use lianas, so the impacts of changed liana abundance would likely be minimal.

Movement pathways

Lianas are used by many forest animals to traverse from tree to tree in the canopy without the need to descend to the forest floor (Yanoviak, Chaper 24 in this volume). For birds, the use of lianas as pathways is less common, but it does exist. For example, Hoatzin are poor fliers, and instead climb and scramble through vegetation, often using vines and lianas as pathways (Billerman 2012). Scale-backed Antbirds (*Hylophylax poecilinoata*) also travel on vines above the forest floor (Willis 1982b). The use of lianas as pathways by birds is generally facultative, and thus the impacts of changing liana abundance would be minimal.

Habitat association of unknown mechanism

We found reports of 42 bird species that show habitat associations with lianas but for which the underlying mechanism (e.g., food, protection) was not reported (Fig. 25.3, Appendix Table 25.1). Many bird species preferentially select habitats with high vine density, including Swainson's Warblers (*Limnothlypis swainsonii*; Peters et al. 2005) and Western Slaty-antshrikes (*Thamnophilus atrinucha*; Robinson et al. 2000). Other studies reported correlations between avian abundance or species richness and liana abundance, though the direction of these correlations (positive or negative) often varied among sites and foraging guilds. For example, the abundance and species richness of all birds and obligate frugivores increased with liana prevalence in logged but recovering rain forests in Borneo, while frugivore-insectivore bird abundance and richness, as well as sallying forager bird abundance, all decreased (Ansell et al. 2011). Liana density was positively correlated with the abundance of 8 bird species, negatively with 24 species, and showed no correlations with abundance of 30 bird species in western Africa (Manu et al. 2007). Total bird species richness was greater in liana-dense natural treefall gaps than anthropogenic gaps with vine cutting in Bolivia, with effects attributed to the loss of liana substrate (Felton et al. 2008). Conversely, Mason (1996) found that selectively logged forests with vine cutting had higher bird diversity and evenness than primary forest or logged forest without vine cutting. Therefore, bird species that select for or against habitats with dense lianas are susceptible to ongoing changes in liana abundance.

BENEFITS AND DISADVANTAGES THAT BIRDS PROVIDE TO LIANAS

Lianas benefit from an array of services from birds. Some liana–bird relationships are beneficial for lianas, e.g., pollination and seed dispersal mutualisms, while other interactions are negative, e.g., seed predation (Fig. 25.1). Here we review several types of liana–bird relationships, specifically: pollination, seed dispersal, seed predation, and protection from arthropod herbivores. We found liana–bird relationships reported for 245 liana species in 48 families across 2 classes, 12 clades, and 27 orders (Fig. 25.4, Appendix Table 25.2).

Bird pollination of lianas

We found reports of 56 species of bird-pollinated lianas (Fig. 25.4, Appendix Table 25.2). Hummingbirds are the most common pollinators, but passerines such as warblers, tanagers, and Bananaquits (*Coereba flaveola*) also pollinate lianas. Some lianas have highly coevolved relationships with birds, such as *Centropogon* spp. lianas (Campanulaceae) with sicklebill hummingbirds (*Eutoxeres* spp.; Stein 1992). *Passiflora mixta* (Passifloraceae) has apparently coevolved with the Sword-billed Hummingbird (*Ensifera ensifera*), upon which it is completely reliant for pollination (Stiles 1981). Other lianas are pollinated by a variety of bird species that may differ among sites. For example, *Combretum fruticosum* (Combretaceae) is visited by 31 species of hummingbirds, warblers, and orioles in Mexico, many of which pollinate it (Gryj et al. 1990); but is pollinated by only two species of hummingbirds and three passerines in Argentina (Bernardello et al. 1994). Some lianas are pollinated only by birds, while others are also pollinated by insects (e.g., *Passiflora mollissima*; Passifloraceae; LaRosa 1982) or bats (e.g., *Marcgravia trinitatus*; Marcgraviaceae; Zusi and Hamas 2001). The Cocos Finch (*P. inornata*) may have coevolved as the primary or sole pollinator of *Marcgravia waferi* (Marcgraviaceae; T.K. Sherry and T.W. Sherry, unpublished data). In the unusual case of *Schwartzia brasiliensis* (Marcgraviaceae), hummingbirds and passerines visiting extrafloral nectaries at pedicel bases gather and transfer pollen between plants (Sazima et al. 1993). Given the range of facultative and obligatory liana–pollinator relationships, the effects of changes in

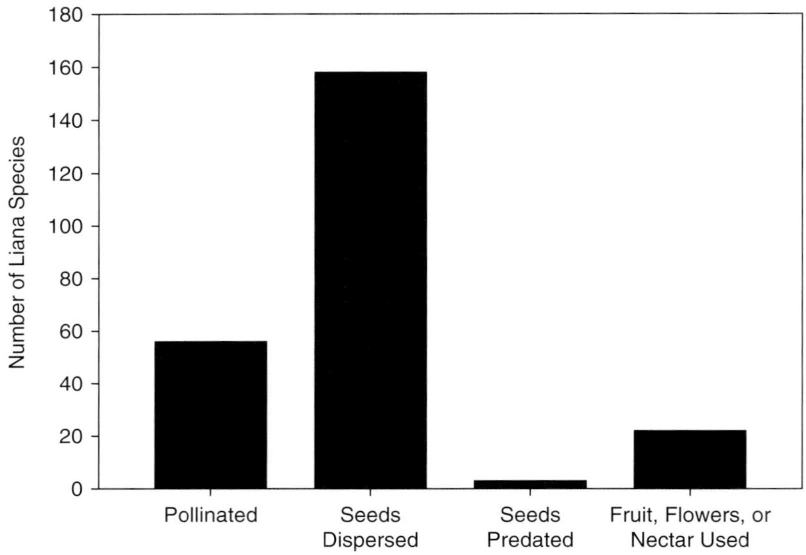

Fig. 25.4 Number of liana species known to have relationships with birds, including pollination, seed dispersal, seed predation, and other use of fruit, flowers, or nectar.

bird abundance on lianas would be species-specific and mixed, but predictable.

Bird seed dispersal and predation

Birds consume and disperse the seeds of many lianas. Many birds are capable of dispersing seeds long distances; Trumpeter Hornbills (*Bycanistes bucinator*) may disperse seeds up to 14.5 km (Lenz et al. 2011). Birds also frequently increase seed germination rates by removing pulp and scarifying seed coats (Muller-Landau & Hardesty 2005). Together, this means that birds greatly benefit some lianas by dispersing their seeds far away from parent trees and from the predators and parasites present there (Janzen 1970; Connell 1971) and by facilitating the colonization of new environments (e.g., recent treefall gaps; Levey 1988; Londré & Schnitzer 2006).

We found reports of 172 bird-dispersed (ornithochorous) liana species (Fig. 25.4, Appendix Table 25.2). Avian seed dispersal is most common in Asia and Australia, where most lianas produce fleshy fruit (Carsten et al. 2002; Kominami et al. 2003; Sankamethawee et al. 2011), unlike the predominantly wind-dispersed Neotropical lianas (Muller-Landau & Hardesty 2005). Annonaceae, Menispermaceae, Moraceae, and Vitaceae have the greatest number of bird-dispersed liana species (Kitamura 2011; Sankamethawee et al. 2011). Some lianas are dispersed by numerous bird species; for example, 57 bird species from 18 families consume poison oak and ivy (*Toxicodendron* spp., Anacardiaceae) fruits and disperse seeds (Senchina 2008).

Conversely, some birds predate liana seeds by crushing them during the digestive process. For example, birds eat the seeds of *Griffonia physocarpa* (Caesalpiniaceae) and *Urera cameroonensis* (Uritcaceae) in Gabon (Gautier-Hion et al. 1985). Avian predation of liana seeds is likely underestimated, but is nevertheless far more common among mammals than birds (Gautier-Hion et al. 1985; Muller-Landau & Hardesty 2005).

Because of their reliance on avian seed dispersal, lianas in Asia and Australia may be particularly hard-hit by bird declines. In contrast, since Neotropical lianas are more dependent on wind than birds for dispersal, the loss of bird species may affect them less. Furthermore, given that many lianas are dispersed by multiple bird, mammal, and/or arthropod species, the decline or extirpation of any one bird species may have minimal effects for the liana community in general

(Muller-Landau & Hardesty 2005); however, it may be potentially devastating to some liana species. Lianas may also have their seeds dispersed by introduced birds, or as a result of density compensation or dietary shifts in extant frugivores (Loiselle & Blake 2002). Therefore, the effects of frugivorous bird declines on bird-dispersed lianas will likely be highly variable, and species-, site-, and region-specific.

Protection from arthropod herbivores

Insectivorous birds also reduce herbivory on lianas by consuming folivorous arthropods. A meta-analysis of 113 studies showed that vertebrate insectivores (birds, bats, and lizards) reduced leaf damage on plants (mostly trees and shrubs) by 40% (Mooney et al. 2010). Insectivorous birds reduced herbivory on 14 understory plant species (including one liana species: *Paullinia ingafolia*; Sapindaceae) by 17% at La Selva (Michel et al. 2014). Checker-throated Antwrens (*E. fulviventris*) reduced abundance of arthropod prey by 44% from dead leaf clusters in liana tangles (Gradwohl & Greenberg 1982). Lianas also invest less energy into foliar chemical defenses and more towards growth and photosynthesis compared to trees (Asner & Martin this volume, Kazda this volume). Thus, the large number of insectivorous birds known to forage on lianas, and the known ability of insectivorous birds to reduce leaf damage on plants, and the lower amount of defensive chemicals in liana foliage, all suggest that lianas would be negatively affected by insectivorous bird declines – much more so than trees.

Other bird uses of liana fruit, flowers, nectar, and/or sap

Birds also often consume liana fruit, flowers, and/or nectar using means that may be either beneficial or harmful to the plant. Birds robbed nectar and/or predated flowers from 10 species of lianas including *Ipomoea* spp. (Convolvulaceae; Kodric-Brown et al. 1984; Lara & Ornelas 2001), activities that are clearly harmful to the plant. Birds consumed fruits and/or flowers of 14 liana species for which the frugivore's role as a seed disperser or predator was not reported (Fig. 25.4, Appendix Table 25.2). As described earlier, the Coquerel's Coua (*C. coquereli*) eats honeydew secreted by a flatid bug that feeds solely on the liana *Elachyptera minimiflora*, a likely commensalistic relationship (Chouteau 2004). The limited understanding of the nature of these relationships makes it difficult to predict the consequences of changes in bird abundance for lianas.

CONSERVATION IMPLICATIONS OF LIANA-BIRD RELATIONSHIPS

Numerous global change drivers, including forest fragmentation, climate change, and biotic homogenization via introduced (non-indigenous) species, are affecting both lianas and birds. Lianas are increasing in abundance and biomass in many tropical forests (Schnitzer & Bongers 2011; Schnitzer, Chapter 30 in this volume), but possibly declining in others (Bongers & Ewango, Chapter 3 in this volume; Michel & Sherry 2012). At the same time, birds are experiencing global declines. Currently, 21% of bird species are considered extinction-prone, and 13–39% of bird species are predicted to be extinct or functionally extinct by 2100 (Şekercioğlu et al. 2004). Avian frugivores and nectarivores have more extinction-prone species than other guilds, and, given their key roles as seed dispersers and pollinators, these declines are certain to affect lianas, as well as other key plant groups. Conversely, some generalist bird species are expanding their ranges and/or increasing in abundance, including species that pollinate and/or disperse seeds of lianas (Biamonte et al. 2011). Given the ubiquity of liana–bird relationships described herein, the changes in each group are almost certain to affect the other.

Increasing liana abundance and biomass

Lianas are increasing in abundance and biomass in many Neotropical and subtropical forests, due to mechanisms that include forest fragmentation, hunting (which benefits wind-dispersed lianas), and climate change (Schnitzer & Bongers 2011; Schnitzer, Chapter 30 in this volume; Fig. 25.1). For example, liana infestations and litter production on

Barro Colorado Island, Panama (BCI) have increased by 50–140% in recent years (Wright et al. 2004; Ingwell et al. 2010; Schnitzer et al. 2012). At the same time, BCI's liana-specialized insectivorous birds (e.g., Checker-throated Antwrens [*E. fulviventris*] and Dot-winged Antwrens [*Microrhopias quixensis*]) increased by 167–300% (Robinson 2001). This case of lianas and a liana-specialized bird increasing simultaneously on BCI may be an example of a liana–bird relationship mitigating the effects of forest fragmentation on insectivorous birds, which have declined or been extirpated in many other tropical forest reserves (e.g., La Selva; Sigel et al. 2006; Michel 2012). In contrast, other birds avoid lianas, e.g., sallying insectivores that prefer open understory where they can locate and pursue their prey (Marra & Remsen 1997; Ansell et al. 2011). Therefore, avian responses to increasing liana abundance and biomass are likely to be both species- and site-specific, with both positive and negative effects on bird abundance and species richness.

Forest-interior birds, particularly in the tropics, typically decline following disturbances such as logging that open up the canopy, in part, due to their adaptation to relatively dark, microclimatically stable, and structurally complex environments (Janzen 1967; Karr & Freemark 1983; Endler & Théry 1996; Canaday 1996; Marra & Remsen 1997; Putz et al. 2001; Walther 2002; Stratford & Robinson 2005). Increasing liana abundance and biomass in disturbed forests may help restore the light environment, microclimatic conditions, and vegetation structure that forest-interior species prefer more quickly than in forests with fewer lianas, thus mitigating the effects of forest fragmentation on forest-interior birds. Though it is unclear whether the shading and structure provided by lianas will be sufficient to create the conditions preferred by forest-interior birds, the possibility is tantalizing enough to warrant further investigation.

Conversely, liana abundance and biomass may be decreasing in some sites. Liana density recently declined by 20–33% in two African forests (Caballé & Martin 2001; Bongers & Ewango, Chapter 3 in this volume); however, the decline noted by Bongers and Ewango was driven largely by the collapse of one dominant liana species. At La Selva, the abundant collared peccaries (*Pecari tajacu*) reduced liana density, cover, and tangle frequency by 24–57% relative to peccary exclosures (Michel 2012). Though lianas appear to be increasing in abundance in some parts of La Selva (Yorke et al. 2013), the reported rate of increase appears to be lower than that of some other sites, such as BCI, and may be due to clonal colonization by stems that escape peccary disturbance (Michel 2012; Yorke et al. 2013). La Selva's peccary-induced liana tangle rarity is implicated in the declines of several understory insectivorous birds: Checker-throated and Dot-winged Antwrens recently declined to near extirpation (Sigel et al. 2006; Michel 2012). Changing liana abundance is an important but little-studied mechanism of bird population changes in tropical and temperate forests.

Birds may be contributing to the observed increases in liana abundance in neotropical and temperate forests. Many seed-dispersing frugivorous birds forage on lianas in treefall gaps, then rest on forest edges, where they deposit the liana seeds. Lianas are then dispersed to their preferred forest-edge habitat, and as fragmentation increases forest edge, bird dispersal of lianas to edges may be contributing to increasing liana abundance (Londré & Schnitzer 2006).

Gap dynamics and liana-bird relationships

Lianas often originate and reach their greatest density and species richness in treefall gaps (Putz 1984; Schnitzer & Carson 2001; Schnitzer et al. 2012; Dalling et al. 2012; Ledo & Schnitzer 2014; Schnitzer et al., Chapter 7 in this volume), where they often suppress shade-tolerant tree regeneration (Schnitzer et al. 2000; Schnitzer & Carson 2010). Many species of birds also specialize on treefall gaps, particularly tropical frugivores that disperse seeds of lianas and trees (Schemske & Brokaw 1981; Levey 1988). As reviewed herein, many other bird species specialize on liana tangles, which often originate in treefall gaps (Schnitzer et al. 2000). Therefore, treefall gaps may maintain the diversity of avian as well as liana and tree communities. Meanwhile, frugivorous birds' associations with treefall gaps may aid in the dispersal of lianas and pioneer trees to treefall gaps in the first place, thus continuing the cycle.

Dispersal of invasive exotic lianas by birds

Many non-native, invasive lianas are dispersed by native and non-native birds. This problem has been best studied in North America (Leicht-Young and Pavlovic, Chapter 28 in this volume), where invasive lianas, such as English ivy (*Hedera helix*) and Oriental bittersweet (*Celastrus orbiculatus*), are dispersed by native and exotic birds. Non-native European Starlings (*Sturnus vulgaris*) and invasive oriental bittersweet have developed a mutualistic relationship in the northeastern United States. Bittersweet fruits sustain starlings through the winter food-limited months, and in return starlings disperse bittersweet seeds by a mean 2.1 km, facilitating the spread of this liana throughout the region (e.g., Merow et al. 2011). Elsewhere, English ivy has become so dominant, thanks in part to bird dispersal, that it forms "ivy deserts" in which ivy has choked out nearly all ground cover and understory, as well as many canopy trees (Okerman 2000). Ultimately, by spreading highly invasive lianas, birds are contributing to biotic homogenization of liana communities.

CONCLUSION

There is a great deal of available natural history information regarding liana–bird relationships, but little in the way of rigorous experimental or observational studies on the benefits of birds to lianas, and lianas to birds. Though many birds preferentially select lianas for foraging and/or nesting, the degree of avian dependence upon lianas is unclear. Experimental liana removal experiments, such as the experiment currently underway on Gigante Peninsula, Panama (S.A. Schnitzer, pers. comm.) provide an ideal opportunity for investigating the degree to which birds are dependent upon lianas. Frugivorous bird declines will negatively affect ornithochorous lianas, particularly in Australia and Asia, unless other animals are able to compensate by replacing the positive effects of birds on lianas. Observational studies or avian exclosure experiments are urgently needed in order to assess the degree of liana dependence on seed-dispersing birds. Investigations into the role of seed-dispersing birds in increasing abundance of native and non-native lianas are also called for.

Lianas have commonly been viewed as detrimental to forests because they decrease the growth and increase the mortality rates of trees. These effects notwithstanding, the benefits birds provide lianas cascade in ways that affect other members of the biotic community (Wright et al. 2004; Schnitzer & Bongers 2011; Michel 2012; Schnitzer et al., Chapter 1 in this volume; Yanoviak, Chapter 24 in this volume; Lambert & Halsey, Chapter 26 in this volume; Arroyo-Rodríguez et al., Chapter 27 in this volume). Liana–bird relationships are likely to become increasingly important as both lianas and birds are affected by diverse global change phenomena.

APPENDIX

Appendix Table 25.1 List of bird species associated with lianas. Bird species are listed in phylogenetic order as per Gill and Donsker (2012).

Bird species	Food	Nest	Other	Source
ORDER GALLIFORMES				
Family Cracidae				
Mitu salvini	F			Santamaría & Franco 2000
M. tuberosum		M		Udoye & Schulenberg 2012
Crax alector			H	Vath 2008
C. blumenbachii		M		Bernardo 2011
Family Odontophoridae				
Callipepla californica	F			Senchina 2008
Colinus virginianus	F			Senchina 2008
Family Phasianidae				
Meleagris gallopavo	F			Senchina 2008
Bonasa umbellus	F, L			Senchina 2008
Tympanuchus phasianellus	F			Senchina 2008
T. cupido	F			Senchina 2008
Syrmaticus soemmerringii	F			Kominami et al. 2003
Phasianus colchicus	F			Senchina 2008
Argusianus argus		D		Winarni et al. 2009
ORDER ACCIPITRIFORMES				
Family Accipitridae				
Accipiter superciliosus	V			Stiles 1977
ORDER FALCONIFORMES				
Family Falconidae				
Micrastur semitorquatus	V			W.D. Robinson, *unpubl. data*
ORDER GRUIFORMES				
Family Rallidae				
Aramides wolfi		S		Karubian et al. 2011
ORDER COLUMBIFORMES				
Family Columbidae				
Patagioenas leucocephala		S		Rivera-Milán 1996
P. squamosa		S		Rivera-Milán 1996
P. inornata		S		Rivera-Milán 1996
Macropygia amboinensis	N			Terborgh & Diamond 1970
M. nigrirostris	N			Terborgh & Diamond 1970
Zenaida macroura		S		Rivera-Milán 1996
Z. aurita		S		Rivera-Milán 1996
Z. asiatica		S		Rivera-Milán 1996
Columbina passerina	N	S		Gryj et al. 1990; Rivera-Milán 1996
Geotrygon chrysia		S		Rivera-Milán 1996
G. mystacea		S		Rivera-Milán 1996
G. montana		S		Rivera-Milán 1996
Treron sieboldii	F			Kominami et al. 2003
Ducula badia	F			Kitamura et al. 2002

(continued overleaf)

Appendix Table 25.1 *(continued).*

Bird species	Food	Nest	Other	Source
ORDER PSITTACIFORMES				
Family Psittacidae				
Loriculus galgulus	N			Sheldon et al. 2001
Trichoglossus haematodus	N			Terborgh & Diamond 1970
Psitteuteles goldiei	N			Mack & Wright 1996
Charmosyna placentis	N			Terborgh & Diamond 1970
Amazona agilis			Pi	Koenig et al. 2007
Cyclopsitta gulielmitertii	N			Terborgh & Diamond 1970
Psittaculirostris desmarestii	N			Terborgh & Diamond 1970
ORDER OPISTHOCOMIFORMES				
Family Opisthocomidae				
Opisthocomus hoazin	L		T	Billerman 2012; Domínguez-Bello et al. 1994
ORDER MUSOPHAGIFORMES				
Family Musophagidae				
Corythaeola cristata	F, L			Sun & Moermond 1997
Tauraco schuetti	F, L			Sun & Moermond 1997
Ruwenzorornis johnstoni	F, L			Sun & Moermond 1997
ORDER CUCULIFORMES				
Family Cuculidae				
Crotophaga major			H	Willis & Eisenmann 1979
Neomorphus geoffreyi			H	Willis & Eisenmann 1979
Coua coquereli	A			Chouteau 2004
Piaya cayana	A			Remsen & Parker 1984; Willis & Eisenmann 1979
Eudynamys scolopaceus	N			Terborgh & Diamond 1970
ORDER STRIGIFORMES				
Family Strigidae				
Strix virgata			R	Willis & Eisenmann 1979
Lophostrix cristata			R	Willis & Eisenmann 1979
Ninox theomacha			R	Mack & Wright 1996
ORDER APODIFORMES				
Family Trochilidae				
Eutoxeres aquila	N			Stein 1992
E. condamini	N			Stein 1992
Glaucis hirsutus	N			Snow 1973
Phaethornis spp. (6 spp.)	N	D	Pi	Snow 1973; Stiles 1977
P. superciliosus	N			Snow 1973
P. pretrei	N			Lopes et al. 2006
P. ruber	A, N			Lopes et al. 2006; Remsen & Parker 1984
Phaeochroa cuvierii		S		Skutch 1964
Eupetomena macroura	N			Lopes et al. 2006
Florisuga fusca	E			Sazima et al. 1993
Colibri thalassinus	N			Kodric-Brown et al. 1984; Lara & Ornelas 2001
Anthracothorax dominicus	N			Kodric-Brown et al. 1984
A. viridis	N			Kodric-Brown et al. 1984
Eulampis jugularis	N			Zusi & Hamas 2001
Orthorhyncus cristatus	N			Zusi & Hamas 2001

Appendix Table 25.1 *(continued)*.

Bird species	Food	Nest	Other	Source
Chlorostilbon canivetii	N			Gryj et al. 1990; Schemske 1975a
C. lucidus	N			Bernardello et al. 1994
C. maugaeus	N			Kodric-Brown et al. 1984
Cynanthus latirostris	N			Gryj et al. 1990
Cyanophaia bicolor	N			Zusi & Hamas 2001
Hylocharis chrysura	N			Bernardello et al. 1994
Amazilia rutila	N			Gryj et al. 1990; Schemske 1975a,1975b
A. fimbriata	N			Lopes et al. 2006
A. saucerrottei	N			Schemske 1975a
Lampornis amethystinus	N			Kodric-Brown et al. 1984; Lara & Ornelas 2001
Heliodoxa jacula	N			Zusi & Hamas 2001
Eugenes fulgens	N			Kodric-Brown et al. 1984; Lara & Ornelas 2001
Ensifera ensifera	N			Stiles 1981
Sephanoides sephanoides	N			Magrach et al. 2012; Urcelay et al. 2006
Heliothryx barroti		S		Willis & Eisenmann 1979
Heliomaster constantii	N			Gryj et al. 1990; Schemske 1975a
Archilochus colubris	N			Gryj et al. 1990
ORDER CORACIIFORMES				
Family Coraciidae				
Eurystomus glaucurus			H	Manu et al. 2007
Family Alcedinidae				
Halcyon senegalensis			H	Manu et al. 2007
ORDER BUCEROTIFORMES				
Family Bucerotidae				
Anorrhinus austeni	F			Kitamura et al. 2002; Sankamethawee et al. 2011
A. galeritus	F			Kitamura et al. 2011
Anthracoceros albirostris	F			Kitamura et al. 2002; Datta & Rawat 2003; Sankamethawee et al. 2011
Buceros bicornis	F			Kitamura et al. 2002, 2004, 2011; Datta & Rawat 2003;
B. rhinoceros	F			Kitamura et al. 2011
Aceros cassidix	F			Kinnaird 1998
Rhyticeros undulatus	F			Kitamura et al. 2002, 2011; Datta & Rawat 2003;
Bycanistes fistulator	F			Whitney et al. 1998
B. cylindricus	F			Whitney et al. 1998
Ceratogymna atrata	F			Whitney et al. 1998
ORDER PICIFORMES				
Family Capitonidae				
Capito aurovirens	A			Remsen & Parker 1984
C. niger	A			Remsen & Parker 1984; Rosenberg 1997
Eubucco richardsoni	A			Remsen & Parker 1984; Rosenberg 1997
E. bourcierii	A			Remsen & Parker 1984
E. tucinkae	A			Remsen & Parker 1984
E. versicolor	A			Remsen & Parker 1984
Family Bucconidae				
Nonnula frontalis			H	Willis & Eisenmann 1979

(continued overleaf)

Appendix Table 25.1 *(continued)*.

Bird species	Food	Nest	Other	Source
Family Ramphastidae				
Ramphastos sulfuratus	F			Skutch 1971
Family Megalaimidae				
Megalaima incognita	F			Sankamethawee et al. 2011
Family Lybiidae				
Pogoniulus subsulphureus			H	Manu et al. 2007
P. bilineatus			H	Manu et al. 2007
Family Picidae				
Picumnus rufiventris	A		Pr	English 1998
Melanerpes lewis	F			Senchina 2008
M. chrysogenys	N			Gryj et al. 1990
M. carolinus	F			Senchina 2008
Sphyrapicus varius	F			Senchina 2008
S. ruber	F			Senchina 2008
Dendrocopus kizuki	F			Kominami et al. 2003
Picoides nuttallii	F			Senchina 2008
P. pubescens	F			Senchina 2008
P. villosus	F			Senchina 2008
P. borealis	F			Senchina 2008
Colaptes auratus	F			Senchina 2008
Dryocopus pileatus	F			Senchina 2008
Blythipicus rubiginosus	A			Styring & Zakaria 2004
Picus awokera	F			Kominami et al. 2003
Meiglyptes tristis	A			Styring & Ickes 2001
ORDER PASSERIFORMES				
Family Eurylaimidae				
Philepitta castanea		D		Prum & Razafindratsita 1997
Family Pittidae				
Pitta sordida			R	Sheldon et al. 2001
Family Furnariidae				
Synallaxis azarae	A			Remsen & Parker 1984
S. brachyura	A			Remsen & Parker 1984
Cranioleuca albiceps	A			Remsen & Parker 1984
C. pallida	A			Pacheco et al. 1996; Remsen & Parker 1984
C. erythrops	A			Remsen & Parker 1984
C. gutturata	A			Remsen & Parker 1984; Rosenberg 1997
Thripophaga macroura	A			Remsen & Parker 1984
T. fusciceps	A			Remsen & Parker 1984
T. berlepschi	A			Remsen & Parker 1984
Phacellodomus rufifrons	A	S		Skutch 1969
Margarornis rubiginosus	A			Remsen & Parker 1984
M. squamiger	A			Remsen & Parker 1984
Pseudocolaptes boissonneautii	A			Remsen & Parker 1984
Anabacerthia striaticollis	A			Remsen & Parker 1984
A. amaurotis	A			Bencke 1996
Syndactyla subalaris	A			Remsen & Parker 1984
Ancistrops strigilatus	A			Remsen & Parker 1984

Appendix Table 25.1 *(continued)*.

Bird species	Food	Nest	Other	Source
Hyloctistes subulatus	A			Remsen & Parker 1984; Rosenberg 1997
Philydor ruficaudatum	A			Remsen & Parker 1984; Rosenberg 1997
P. erythrocercum	A			Remsen & Parker 1984; Rosenberg 1997
P. lichtensteini	A			Remsen & Parker 1984
P. atricapillus	A			Parrini et al. 2010
P. rufum	A			Remsen & Parker 1984
P. pyrrhodes	A			Remsen & Parker 1984
Anabazenops dorsalis	A			Remsen & Parker 1984
A. fuscus	A			Remsen & Parker 1984
Thripadectes holostictus	A			Remsen & Parker 1984
Automolus ochrolaemus	A	M		Remsen & Parker 1984; Rosenberg 1997; Skutch 1952
A. infuscatus	A			Remsen & Parker 1984; Rosenberg 1997
A. melanopezus	A			Remsen & Parker 1984; Rosenberg 1997
A. rubiginosus			H	Thiollay 2003
A. rufipileatus	A			Remsen & Parker 1984; Rosenberg 1997
Xenops minutus	A		Pr	Bierregaard 1990; English 1998; Remsen & Parker 1984
Dendrocincla fuliginosa	A		Pi	Chapman & Rosenberg 1991; Willis 1972b, 1979
Sittasomus griseicapillus	A			Remsen & Parker 1984
Dendrocolaptes certhia	A			Remsen & Parker 1984
Dendrocolaptes hoffmannsi	A			Willis 1982a
D. picumnus	A			Willis 1982a
D. platyrostris	A			Willis 1982a
Xiphorhynchus guttatus	A			Remsen & Parker 1984; Rosenberg 1997; Willis 1972
X. flavigaster	A			Remsen & Parker 1984
X. lachrymosus	A			Willis 1972a
Family Thamnophilidae				
Cymbilaimus lineatus	A		H, Pr	English 1998; Terborgh et al. 1990; Thiollay 2003; Willis & Eisenmann 1979
C. sanctaemariae	A			Remsen & Parker 1984
Hypoedaleus guttatus			H	Lopes et al. 2006
Mackenziaena severa			H	Lopes et al. 2006
Taraba major	A			Remsen & Parker 1984
Thamnophilus cryptoleucus	A			Remsen & Parker 1984
T. aethiops	A		Pr	English 1998
T. schistaceus	A		H	Remsen & Parker 1984; Terborgh et al. 1990
T. aroyae	A			Remsen & Parker 1984
T. atrinucha	A		H	Robinson et al. 2000; Tarwater & Kelley 2010
T. punctatus	A			Remsen & Parker 1984
T. amazonicus	A		H	Thiollay 2003; Zimmer & Hilty 1997
T. caerulescens			H	Lopes et al. 2006
Xenornis setifrons	A			Whitney & Rosenberg 1993
Dysithamnus stictothorax	A			Remsen & Parker 1984
D. mentalis	A		H	Lopes et al. 2006; Remsen & Parker 1984
Pygiptila stellaris	A			Remsen & Parker 1984; Rosenberg 1997
Epinecrophylla fulviventris	A		H	Gradwohl & Greenberg 1980, 1982; Michel 2012; Remsen & Parker 1984; Robinson et al. 2000
E. leucophthalma	A			Remsen & Parker 1984; Rosenberg 1997

(continued overleaf)

Appendix Table 25.1 *(continued)*.

Bird species	Food	Nest	Other	Source
E. haematonota	A			Remsen & Parker 1984; Rosenberg 1997; Willis 1972a
E. ornata	A		Pr	English 1998; Remsen & Parker 1984; Rosenberg 1997
E. erythrura	A			Remsen & Parker 1984; Willis 1972a
Myrmotherula brachyura	A			Willis & Eisenmann 1979
M. cherriei	A			Zimmer & Hilty 1997
M. longicauda	A			Remsen & Parker 1984
M. axillaris	A			Remsen & Parker 1984
M. schisticolor	A			Remsen & Parker 1984
M. iheringi	A			Remsen & Parker 1984
Herpsilochmus dorsimaculatus	A			Zimmer & Hilty 1997
H. rufimarginatus			H	Lopes et al. 2006
Microrhopias quixensis	A		H, Pr	English 1998; Gradwohl & Greenberg 1980; Michel 2012; Robinson et al. 2000; Thiollay 2003; Willis 1972a; Willis & Eisenmann 1979
Formicivora iheringi	A		H	Willis 1985
Drymophila squamata	A			Willis 1985
Terenura humeralis	A			Remsen & Parker 1984
Cercomacra cinerascens	A		H	Terborgh et al. 1990; Thiollay 2003; Willis 1985
C. tyrannina	A		H	Robinson et al. 2000; Willis 1985
C. nigrescens	A			Willis 1985
C. serva	A			Willis 1985
C. nigricans	A			Willis 1985
Pyriglena leucoptera	A			Hansbauer et al. 2008, 2010; Del Hoyo et al. 2003
Hypocnemis cantator	A		H	Marra & Remsen 1997; Remsen & Parker 1984; Willis 1985
Schistocichla schistacea	A			Remsen & Parker 1984
S. leucostigma	A			Willis 1985
S. rufifacies	A			Willis 1985
S. caurensis	A			Zimmer 1999
Myrmeciza disjuncta	A			Zimmer 1999
M. exsul	A			Remsen & Parker 1984; Willis 1972a; Willis & Oniki 1972
M. hemimelaena	A			Remsen & Parker 1984
Gymnopithys salvini	A			Willis 1968
G. lunulatus	A		Pr	Willis 1968
Hylophylax naevioides	A		Pr, R	Willis 1972a
H. poecilinoata			T	Willis 1982b
Family Grallariidae				
Grallaria guatimalensis		M, S		Rivera-Ortíz et al. 2010
Hylopezus perspicillatus		S		Willis & Eisenmann 1979
Family Rhinocryptidae				
Liosceles thoracicus			H	Terborgh et al. 1990
Family Tyrannidae				
Elaenia albiceps	F			Magrach et al. 2012
Zimmerius chrysops		M		Goulding & Martin 2010
Mionectes macconnelli	F			Naka 2004

Appendix Table 25.1 (*continued*).

Bird species	Food	Nest	Other	Source
Myiophobus flavicans		M		Goulding & Martin 2010
M. phoenicomitra		M		Goulding & Martin 2010
M. roraimae		M		Goulding & Martin 2010
Myiornis auricularis			H	Willis 2003
M. atricapillus		S		Willis & Eisenmann 1979
Oncostoma olivaceum	A		H	Robinson et al. 2000; Willis 1972a
Lophotriccus vitiosus	A		Pr	English 1998
Poecilotriccus capitalis	A		Pr	English 1998
Todirostrum poliocephalum			H	Willis 2003
Rhynchocyclus olivaceus			H	Willis 1972a; Willis & Eisenmann 1979
Sayornis phoebe	F			Senchina 2008
Silvicultrix diadema		S		Miller & Greeney 2008
S. jelskii		S		Miller & Greeney 2008
Tyrannus melancholicus	F			Skutch 1980
Attila spadiceus			H	Willis & Eisenmann 1979
Family Cotingidae				
Carpornis cucullata	F			Pizo et al. 2002
Carpornis melanocephala	F			Pizo et al. 2002
Rupicola rupicola	F			Enard et al. 1989
Procnias albus		D		Snow 1961, 1973
P. nudicollis	F			Pizo et al. 2002
Lipaugus vociferans	F			Naka 2004
Family Pipridae				
Tyranneutes virescens	F			Naka 2004
Corapipo gutturalis	F			Naka 2004
C. altera		D		Rosselli et al. 2002
Lepidothrix coronata		D		Durães 2009
Manacus aurantiacus	F			Skutch 1980
Heterocercus flavivertex		D		Zimmer & Hilty 1997
Pipra pipra	F			Naka 2004
P. erythrocephala	F			Naka 2004; Snow & Snow 1971
Dixiphia chloromeros		D		Tello 2001
Family Tityridae				
Oxyruncus cristatus	F			Pizo et al. 2002
Onychorhynchus mexicanus		S		Peck 1910
Terenotriccus erythrurus	F		H	Michel 2012
Laniocera rufescens		D		Willis & Eisenmann 1979
Family Maluridae				
Sipodotus wallacii	A			Mack & Wright 1996
Family Meliphagidae				
Xanthotis polygrammus	F, N			Terborgh & Diamond 1970
X. flaviventer	F, N			Terborgh & Diamond 1970
Lichenostomus obscurus	N			Terborgh & Diamond 1970
Meliphaga spp. (6 spp.)	F			Terborgh & Diamond 1970
Pycnopygius ixoides	F, N			Terborgh & Diamond 1970
Glycichaera fallax	N			Terborgh & Diamond 1970

(*continued overleaf*)

Appendix Table 25.1 *(continued)*.

Bird species	Food	Nest	Other	Source
Family Acanthizidae				
Pachycare flavogriseum	F			Terborgh & Diamond 1970
Family Melanocharitidae				
Melanocharis striativentris	F			Terborgh & Diamond 1970
Oedistoma pygmaeum	N			Terborgh & Diamond 1970
Family Pachycephalidae				
Pachycephala rufiventris	F			Terborgh & Diamond 1970
Family Vireonidae				
Vireo griseus	F		H	Marra & Remsen 1997; Senchina 2008
V. hypochryseus	N			Gryj et al. 1990
V. gilvus	F			Senchina 2008
V. philadelphicus	F			Skutch 1980
V. olivaceus	F			Skutch 1980
V. flavoviridis	F			Skutch 1980
Hylophilus decurtatus	A			Remsen & Parker 1984
Family Corvidae				
Cyanocorax sanblasianus	N			Gryj et al. 1990
C. cristatellus		M		Amaral & Macedo 2003, Baran 2010
Psilorhinus morio	A, F			Remsen & Parker 1984; Skutch 1980
Cyanocitta stelleri	F			Swearingen & Dietrich 2006
Garrulus gl&arius	F			Kominami et al. 2003
Pica hudsonia	F			Senchina 2008
P. nuttalli	F			Senchina 2008
Corvus brachyrhynchos	F			Senchina 2008
C. hawaiiensis	N			LaRosa 1982
Family Monarchidae				
Hypothymis azurea		S		Sheldon et al. 2001
Neolalage banksiana	A			Kratter et al. 2006
Family Paradisaeidae				
Drepanornis bruijnii	F			Beehler & Beehler 1986
Diphyllodes magnificus	N			Terborgh & Diamond 1970
Cicinnurus regius	A, F			Bell 1983
Paradisaea apoda	N			Terborgh & Diamond 1970
Family Bombycillidae				
Bombycilla cedrorum	F			Senchina 2008; Swearingen & Dietrich 2006
Family Paridae				
Poecile carolinensis	A, F			Remsen & Parker 1984; Senchina 2008
P. atricapillus	F			Senchina 2008
P. gambeli	F			Senchina 2008
P. rufescens	F			Senchina 2008
P. varius	F			Kominami et al. 2003
Parus major	F			Kominami et al. 2003
Baeolophus inornatus	F			Senchina 2008
B. bicolor	F			Senchina 2008

Appendix Table 25.1 *(continued).*

Bird species	Food	Nest	Other	Source
Family Pycnonotidae				
Pycnonotus atriceps	F			Kitamura et al. 2002
P. melanicterus	F			Kitamura et al. 2002
P. flaviventris	F			Sankamethawee et al. 2011
P. jocosus	F			Kitamura et al. 2002; Mandon-Dalger et al. 2004
P. finlaysoni	F			Kitamura et al. 2002; Sankamethawee et al. 2011
Alophoixus pallidus	F			Sankamethawee et al. 2011
Iole propinqua	F			Kitamura et al. 2002; Sankamethawee et al. 2011
Hemixos flavala	F			Kitamura et al. 2002; Sankamethawee et al. 2011
Hypsipetes amaurotis	F			Kominami et al. 2003
Family Cettidae				
Horornis diphone	F			Kominami et al. 2003
Family Aegithalidae				
Psaltriparus minimus	F			Senchina 2008
Family Cisticolidae				
Camaroptera superciliaris			H	Manu et al. 2007
Family Timaliidae				
Pomatorhinus schisticeps	F			Sankamethawee et al. 2011
Family Pellorneidae				
Illadopsis rufipennis			H	Manu et al. 2007
Malacopteron affine	A			Sheldon et al. 2001
Family Leiothrichidae				
Phyllanthus atripennis			H	Manu et al. 2007
Family Sylviidae				
Chamaea fasciata	F			Senchina 2008
Family Zosteropidae				
Zosterops japonicus	F, N			Kominami et al. 2003, LaRosa 1982
Z. everetti	F			Sankamethawee et al. 2011
Z. minor	N			Terborgh & Diamond 1970
Family Regulidae				
Regulus calendula	F			Senchina 2008
Family Troglodytidae				
Campylorhynchus rufinucha	N			Schemske 1975a
C. brunneicapillus	F			Senchina 2008
C. turdinus			H	Terborgh et al. 1990
Pheugopedius fasciatoventris	A			Logue 2007
P. genibarbis	A			Remsen & Parker 1984; Rosenberg 1997
P. coraya	A		H	Remsen & Parker 1984; Thiollay 2003
P. rutilus	A			Remsen & Parker 1984
Thryothorus ludovicianus	F		H	Marra & Remsen 1997; Senchina 2008
Cantorchilus leucotis		S		Gill 2011
C. longirostris	A			Remsen & Parker 1984
Microcerculus marginatus			H	Terborgh et al. 1990

(*continued overleaf*)

Appendix Table 25.1 *(continued).*

Bird species	Food	Nest	Other	Source
Family Polioptilidae				
Ramphocaenus melanurus	A			Thiollay 2003
Family Mimidae				
Dumetella carolinensis	F			Senchina 2008; Skutch 1980
Mimus polyglottos	F			Senchina 2008; Swearingen & Dietrich 2006
Toxostoma rufum	F			Senchina 2008
T. redivivum	F			Senchina 2008
Family Sturnidae				
Sturnus vulgaris	F			Merow et al. 2011; Okerman 2000; Senchina 2008; Swearingen & Dietrich 2006
Family Turdidae				
Geokichla citrina	F			Sankamethawee et al. 2011
G. guttata		S	Pr	Ndang'ang'a et al. 2005
Ixoreus naevius	F			Senchina 2008
Sialia sialis	F			Senchina 2008
Myadestes obscurus	N			LaRosa 1982
M. townsendi	F			Senchina 2008
Catharus aurantiirostris	F			Skutch 1980
C. ustulatus	F			Senchina 2008; Skutch 1980
C. guttatus	F			Senchina 2008
Turdus pallidus	F			Kominami et al. 2003
T. rufiventris	N			Bernardello et al. 1994
T. falckl&ii	F			Magrach et al. 2012
T. assimilis	N			Gryj et al. 1990
T. albicollis	F			Naka 2004
T. rufopalliatus	N			Gryj et al. 1990
T. migratorius	F			Senchina 2008; Swearingen & Dietrich 2006
Family Muscicapidae				
Tarsiger cyanurus	F			Kominami et al. 2003
Ficedula narcissina	F			Kominami et al. 2003
Family Chloropseidae				
Chloropsis cyanopogon	F			Sheldon et al. 2001
C. cochinchinensis	F			Sankamethawee et al. 2011
Family Dicaeidae				
Dicaeum chrysorrheum	F			Sankamethawee et al. 2011
D. geelvinkianum	F			Terborgh & Diamond 1970
Family Nectariniidae				
Anthreptes simplex	F			Sheldon et al. 2001
Hedydipna collaris			H	Manu et al. 2007
Hypogramma hypogrammicum	F			Sheldon et al. 2001
Family Passeridae				
Passer domesticus	F			Swearingen & Dietrich 2006

Appendix Table 25.1 *(continued).*

Bird species	Food	Nest	Other	Source
Family Fringillidae				
Euphonia imitans	A			Remsen & Parker 1984
Carpodacus purpureus	F			Senchina 2008
C. mexicanus	F			Senchina 2008
Eophona personata	F			Kominami et al. 2003
Hemignathus virens	N			LaRosa 1982
Loxops coccineus	N			LaRosa 1982
Vestiaria coccinea	N			LaRosa 1982
Family Parulidae				
Helmitheros vermivorum	A			Remsen & Parker 1984
Vermivora chrysoptera	A			Ch&ler & King 2011; Remsen & Parker 1984
V. cyanoptera	A			Remsen & Parker 1984
Mniotilta varia	N			Gryj et al. 1990
Protonotaria citrea	A			Remsen & Parker 1984
Limnothlypis swainsonii		S	H	Anich et al. 2010; Benson et al. 2009; Peters et al. 2005
Leiothlypis peregrina	N			Schemske 1975a
L. ruficapilla	N			Gryj et al. 1990
Geothlypis tolmiei	N			Gryj et al. 1990
G. formosa	A		H	Willis 1972a; Marra & Remsen 1997
Setophaga tigrina	F			Senchina 2008
S. pitiayumi	N			Gryj et al. 1990
S. petechia	N			Schemske 1975a
S. coronata	F, N			Gryj et al. 1990; Senchina 2008
S. auduboni	F			Senchina 2008
S. nigrescens	N			Gryj et al. 1990
Myiothlypis luteoviridis	A			Remsen & Parker 1984
Cardellina pusilla	N			Gryj et al. 1990
Icteria virens	N			Gryj et al. 1990
Family Icteridae				
Cacicus uropygialis	A			Remsen & Parker 1984
C. melanicterus	N			Gryj et al. 1990
Amblycercus holosericeus	A			Remsen & Parker 1984
Icterus spurius	N			Gryj et al. 1990
I. cucullatus	N			Gryj et al. 1990
I. wagleri	N			Gryj et al. 1990
I. galbula	A, N			Gryj et al. 1990; Remsen & Parker 1984; Schemske 1975a
I. pustulatus	N			Gryj et al. 1990; Schemske 1975a
Agelaioides badius	N			Bernardello et al. 1994
Agelaius xanthomus	A			Post 2011
Euphagus carolinus	F			Senchina 2008
Family Coerebidae				
Coereba flaveola	E, N			Cruz 1980; Gross 1958; Kodric-Brown et al. 1984; Sazima et al. 1993; Zusi & Hamas 2001
Family Emberizidae				
Passerella iliaca	F			Senchina 2008
Zonotrichia capensis	N			Bernardello et al. 1994

(continued overleaf)

Appendix Table 25.1 *(continued).*

Bird species	Food	Nest	Other	Source
Z. leucophrys	F			Senchina 2008
Z. albicollis	F			Senchina 2008
Z. atricapilla	F			Senchina 2008
Junco hyemalis	F			Senchina 2008
Pipilo maculatus	F			Senchina 2008
Atlapetes rufinucha	A			Remsen & Parker 1984
Family Thraupidae				
Mitrospingus cassinii	A			Remsen & Parker 1984
Hemispingus calophrys	A			Remsen & Parker 1984
Thlypopsis sordida	A			Willis 2003
T. ruficeps	A			Remsen & Parker 1984
Tachyphonus surinamus	F			Naka 2004
T. luctuosus	A			Remsen & Parker 1984; Willis 1972a
Ramphocelus bresilius	E			Sazima et al. 1993
R. passerinii	F			Skutch 1980
Buthraupis montana	A			Remsen & Parker 1984
Tangara chilensis	A			Isler & Isler 1987
T. seledon	E			Sazima et al. 1993
T. schrankii	A			Remsen & Parker 1984
T. icterocephala	F			Skutch 1980
T. chrysotis	A			Remsen & Parker 1984
T. guttata	F			Skutch 1980
T. gyrola	F			Skutch 1980
T. heinei	A			Isler & Isler 1987
Dacnis albiventris	N			Terborgh & Diamond 1970
D. cayana	F			Skutch 1980
Cyanerpes lucidus	A, F			Remsen & Parker 1984; Skutch 1980
C. cyaneus	F			Skutch 1980
Chlorophanes spiza	F			Skutch 1980
Piezorina cinerea	A			Remsen & Parker 1984
Volatinia jacarina	N			Gryj et al. 1990
Loxigilla noctis	L, N			Zusi & Hamas 2001
Pinaroloxias inornata	A, N			Sherry, T.K., & Sherry, T.W., unpubl. data
Nesospingus speculiferus	A			Cruz 1988
Family Cardinalidae				
Piranga rubra	N			Gryj et al. 1990
P. ludoviciana	N			Gryj et al. 1990; Schemske 1975a
Habia rubica	A			Remsen & Parker 1984
H. fuscicauda	A			Chiver & Morton 2010
H. gutturalis	A			Remsen & Parker 1984
H. cristata	A			Remsen & Parker 1984
Cardinalis cardinalis	A			Marra & Remsen 1997
Caryothraustes poliogaster	A			Remsen & Parker 1984
Saltator maximus	F			Skutch 1980
Cyanocompsa parellina	N			Gryj et al. 1990
Passerina leclancherii	N			Gryj et al. 1990
P. versicolor	N			Gryj et al. 1990

Notes: Food: A = arthropods; E = extrafloral nectar; F = fruit; L = leaves/twigs; N = nectar; V = vertebrates
Nest: M = liana material used in nest; S = structural (nests built on/in lianas)
Other: D = display, song; H = Habitat association; Pi = increased predation; Pr = reduced predation (protection); R = roost/rest on lianas; T = travel

Appendix Table 25.2 List of liana species with bird associations. Plant families are listed in phylogenetic order as per APG III (Angiosperm Phylogeny Group 2009).

Liana species	Association	Source
CLASS GNETOPSIDA		
ORDER GNETALES		
Family Gnetaceae		
Gnetum cuspidatum	Sd	Kitamura et al. 2011
G. latifolium	Sd	Kinnaird 1998
G. macrostachyum	Sd	Sankamethawee et al. 2011
G. microcarpum	Sd	Kitamura et al. 2011
G. montanum	Sd	Kitamura et al. 2002
CLASS EQUISITOPSIDA		
CLADE ANGIOSPERMS		
ORDER AUSTROBAILEYALES		
Family Schisandraceae		
Kadsura japonica	Sd	Kominami et al. 2003
CLADE MAGNOLIIDS		
ORDER PIPERALES		
Family Piperaceae		
Piper kadzura	Sd	Kominami et al. 2003
P. novae-hollandiae	Sd	Carsten et al. 2002
P. retrofractum	Sd	Sankamethawee et al. 2011
P. ribesioides	Sd	Sankamethawee et al. 2011
ORDER MAGNOLIALES		
Family Annonaceae		
Cyathostemma micranthum	Sd	Kitamura et al. 2002; Sankamethawee et al. 2011
Desmos chinensis	Sd	Kitamura et al. 2002, 2011; Kitamura 2011
D. cochinchinensis	Sd	Kitamura 2011; Kitamura et al. 2011
D. dumosus	Sd	Sankamethawee et al. 2011
Melodorum leichhardtii	Sd	Carsten et al. 2002
Uvaria cordata	Sd	Kitamura et al. 2002
U. lurida	Sd	Kitamura et al. 2002, 2011; Kitamura 2011
CLADE MONOCOTS		
ORDER ALISMATALES		
Family Araceae		
Pothos chinensis	Sd	Sankamethawee et al. 2011
ORDER LILIALES		
Family Philesiaceae		
Luzuriaga polyphylla	Sd	Magrach et al. 2012
L. radicans		Magrach et al. 2012
Family Smilacaceae		
Smilax elastica	Sd	Pizo et al. 2002
S. glabra	F	Kitamura et al. 2002
S. tamnoides	Sd	Londré & Schnitzer 2006

(*continued overleaf*)

Appendix Table 25.2 *(continued).*

Liana species	Association	Source
CLADE COMMELINIDS		
ORDER ARECALES		
Family Arecaceae		
Calamus deerratus	Sd	Clark et al. 2001
Eremospatha macrocarpa	Sd	Whitney et al. 1998
Laccosperma secundiflorum	Sd	Clark et al. 2001
ORDER POALES		
Family Flagellariaceae		
Flagellaria indica	Sd	Carsten et al. 2002
CLADE EUDICOTS		
ORDER RANUNCULALES		
Family Lardizabalaceae		
Akebia trifoliata	Sd	Kominami et al. 2003
Stauntonia hexaphylla	Sd	Kominami et al. 2003
Family Menispermaceae		
Abuta selloana	Sd	Pizo et al. 2002
Anamirta cocculs	Sd	Kitamura et al. 2002
Diploclisia glaucescens	Sd	Kitamura et al. 2002; Sankamethawee et al. 2011
Hypserpa nitida	Sd	Kitamura et al. 2002; Sankamethawee et al. 2011
Menispermum canadensis	Sd	Londré & Schnitzer 2006
M. dauricum	Sd	Kominami et al. 2003
Stephania japonica	Sd	Sankamethawee et al. 2011
Tinomiscium petiolare	Sd	Kitamura et al. 2011
[ORDER UNPLACED]		
Family Sabiaceae		
Sabia limoniacea	Sd	Sankamethawee et al. 2011
CLADE CORE EUDICOTS		
[ORDER UNPLACED]		
Family Dilleniaceae		
Davilla kunthii	Sd	Skutch 1980
Doliocarpus dentatus	Sd	Skutch 1980
Tetracera indica	Sd	Sankamethawee et al. 2011
T. loureiri	Sd	Sankamethawee et al. 2011
ORDER LORANTHACEAE		
Family Loranthaceae		
Phthirusa stellis	F	Naka 2004
CLADE ROSIDS		
ORDER VITALES		
Family Vitaceae		
Ampelocissus martinii	Sd	Kitamura et al. 2002
Ampelopsis cantoniensis	Sd	Kitamura et al. 2002; Sankamethawee et al. 2011
Cayratia sp. 1	F	Kitamura et al. 2002
Cissus convolvulacea	Sd	Kitamura et al. 2002

Appendix Table 25.2 *(continued)*.

Liana species	Association	Source
C. hypoglauca	Sd	Carsten et al. 2002
C. sterculiifolia	Sd	Carsten et al. 2002
C. dinklagei	Sd	Gautier-Hion et al. 1985
Cyathostemma sp. 1	Sd	Kitamura et al. 2011
Parthenocissus quinquefolia	Sd	Ladwig & Meiners 2010; Londré & Schnitzer 2006; Robinson & Handel 1993
P. tricuspidata	Sd	Kominami et al. 2003
Tetrastigma spp.	Sd	Sankamethawee et al. 2011
T. sp. 1	Sd	Kitamura et al. 2002
T. cruciatum	Sd	Kitamura et al. 2002
T. lanceolarium	Sd	Sankamethawee et al. 2011
Vitis spp.	Sd	Ladwig & Meiners 2010; Londré & Schnitzer 2006; Robinson & Handel 1993
V. flexuosa	Sd	Kominami et al. 2003
CLADE FABIDS (EUROSIDS I)		
ORDER CELASTRALES		
Family Celastraceae		
Celastrus orbiculatus	Sd	Kominami et al. 2003; Ladwig & Meiners 2010; Merow et al. 2011; Silveri 2001
Elachyptera minimiflora	O	Chouteau 2004
Salacia erecta	Sd	Sun & Moermond 1997
ORDER OXALIDALES		
Family Connaraceae		
Agelaea trinervis	Sd	Kitamura 2011; Kitamura et al. 2011
Cnestidium rufescens	Sd	Skutch 1971
Rourea minor	Sd	Kitamura et al. 2004, 2011
Roureopsis obliquifoliolata	Sd	Clark et al. 2001
ORDER MALPIGHIALES		
Family Clusiaceae		
Clusia spp.	F	Naka 2004
Havetiopsis flavida	F	Naka 2004
Family Linaceae		
Hugonia platysepala	Sd	Gautier-Hion et al. 1985
H. spicata	Sd	Gautier-Hion et al. 1985
Family Passifloraceae		
Adenia sp. 1	Sd	Kitamura 2011; Kitamura et al. 2011
Passiflora spp.	Sd	Skutch 1980
P. mixta	P	Stiles 1981
P. mollissima	N, P, Sd	LaRosa 1982
ORDER CUCURBITALES		
Family Cucurbitaceae		
Trichosanthes tricuspidata	Sd	Kitamura et al. 2002

(continued overleaf)

Appendix Table 25.2 (*continued*).

Liana species	Association	Source
ORDER FABALES		
Family Fabaceae		
Abrus precatorius	Sd	Meisenburg & Fox 2002
Dalbergia lactea	Sd	Sun & Moermond 1997
Griffonia physocarpa	Sp	Gautier-Hion et al. 1985
Neorudolphia volubilis	P	Kodric-Brown et al. 1984
ORDER ROSALES		
Family Elaeagnaceae		
Elaeagnus glabra	Sd	Kominami et al. 2003
E. latifolia	Sd	Kitamura et al. 2002
Family Moraceae		
Broussonetia kaempferi	Sd	Kominami et al. 2003
Ficus spp.	F, Sd	Naka 2004; Whitney et al. 1998
F. aurantiacea	Sd	Kitamura et al. 2011
F. deltoidea	Sd	Kitamura et al. 2011
F. sagittata	Sd	Sankamethawee et al. 2011
F. sarmentosa	Sd	Kominami et al. 2003
F. villosa	Sd	Sankamethawee et al. 2011
F. sp. 1	Sd	Clark et al. 2001
F. sp. II no. 460	Sd	Kitamura et al. 2002
Ficus sp. PH281	Sd	Whitney et al. 1998
Maclura cochinchinensis	Sd	Kitamura et al. 2002; Kominami et al. 2003
Trophis sp. 1	Sd	Kitamura et al. 2002
Family Rhamnaceae		
Berchemia racemosa	Sd	Kominami et al. 2003
Family Rosaceae		
Rubus alceifolius	Sd	Kitamura et al. 2002; Mandon-Dalger et al. 2004
R. kerrii	F	Kitamura et al. 2002
Family Ulmaceae		
Celtis iguanea	F	Peres 2000
Family Urticaceae		
Coussapoa spp.	F	Naka 2004
Poikilospermum spp.	Sd	Sheldon et al. 2001
P. sauveolens	Sd	Sheldon et al. 2001
Urera cameroonensis	Sp (Sd?)	Gautier-Hion et al. 1985
U. hypselodendron	Sd	Senbeta et al. 2005
U. trinervis	Sd	Senbeta et al. 2005
CLADE MALVIDS (EUROSIDS II)		
ORDER MYRTALES		
Family Combretaceae		
Combretum farinosum	P	Schemske 1975a,1975b
C. fruticosum	F, N	Bernardello et al. 1994; Domínguez-Bello et al. 1994; Gryj et al. 1990; Peres 2000

Appendix Table 25.2 *(continued).*

Liana species	Association	Source
Family Onagraceae		
Fuchsia regia	Sd	Pizo et al. 2002
ORDER MALVALES		
Family Malvaceae		
Grewia laevigata	Sd	Sankamethawee et al. 2011
ORDER SAPINDALES		
Family Anacardiaceae		
Toxicodendron spp.	Sd, Sp	Londré & Schnitzer 2006; Senchina 2008
T. radicans	Sd	Ladwig & Meiners 2010; Robinson & Handel 1993
Rhus ambigua	Sd	Kominami et al. 2003
Family Sapindaceae		
Allophylus spp.	Sd	Sun & Moermond 1997
Matayba elaeagnoides	Sd	Pizo et al. 2002
Paullinia spp.	Sd	Pizo et al. 2002
CLADE ASTERIDS		
ORDER ERICALES		
Family Actinidiaceae		
Actinidia arguta	Sd	Kominami et al. 2003
A. polygama	Sd	Kominami et al. 2003
Family Marcgraviaceae		
Marcgravia lineolata	P	Zusi & Hamas 2001
M. rectiflora	P	Zusi & Hamas 2001
M. schippii	N	Zusi & Hamas 2001
M. trinitatus	N, P	Zusi & Hamas 2001
M. umbellata	N, P	Zusi & Hamas 2001
M. waferi	P	Sherry, T.K., & Sherry, T.W., unpubl. data
Norantea guianensis	F	Naka 2004
Schwartzia brasiliensis	E, P	Sazima et al. 1993
Souroubea guianensis	Sd	Skutch 1980; Peres 2000
Family Myrsinaceae		
Embelia schimperi	Sd	Sun & Moermond 1997
E. sessiliflora	Sd	Sankamethawee et al. 2011
CLADE LAMIIDS (EUASTERIDS I)		
ORDER GENTIANALES		
Family Apocynaceae		
Melodinus australis	Sd	Carsten et al. 2002
Odontadenia gr&iflora	N	Gross 1958
Family Loganiaceae		
Gardneria nutans	Sd	Kominami et al. 2003
Family Rubiaceae		
Canthium spp.	Sd	Sun & Moermond 1997
Coelospermum paniculatum	Sd	Carsten et al. 2002
C. truncatum	Sd	Sankamethawee et al. 2011

(continued overleaf)

Appendix Table 25.2 *(continued).*

Liana species	Association	Source
Morinda jasminoides	Sd	Carsten et al. 2002
M. umbellata	Sd	Sankamethawee et al. 2011
Morinda sp. SK042	Sd	Kitamura et al. 2002
Mussaenda sanderiana	F	Kitamura et al. 2002
Oxyceros longiflorus	Sd	Sankamethawee et al. 2011
ORDER LAMIALES		
Family Bignoniaceae		
Campsidium valdivianum	P	Urcelay et al. 2006
Lundia coronata	N, P	Lopes et al. 2002
Family Gesneriaceae		
Drymonia pilifera	P, Sd?	Levey 1988
Mitraria coccinea	P	Magrach et al. 2012
Family Lamiaceae		
Callicarpa glandulosa	Sd	Kitamura et al. 2002
Premna sp. 1	F	Kitamura et al. 2002
Family Oleaceae		
Jasminum elongatum	Sd	Kitamura et al. 2002
Family Verbenaceae		
Premna corymbosa	Sd	Sankamethawee et al. 2011
P. flavescens	Sd	Sankamethawee et al. 2011
ORDER SOLANALES		
Family Convolvulaceae		
Ipomoea repanda	N, P	Kodric-Brown et al. 1984
I. arenaria	N, P	Kodric-Brown et al. 1984
I. hederifolia	N, P	Lara & Ornelas 2001
CLADE CAMPANULIDS (EUASTERIDS II)		
ORDER ASTERALES		
Family Campanulaceae		
Centropogon spp. (36 spp.)	P	Stein 1992
C. baezanus	P	Stein 1992
C. granulosus	P	Stein 1992
C. loretensis	P	Stein 1992
ORDER DIPSACALES		
Family Caprifoliaceae		
Lonicera hypoglauca	Sd	Kominami et al. 2003
L. japonica	Sd	Kominami et al. 2003; Ladwig & Meiners 2010; Robinson & Handel 1993
ORDER APIALES		
Family Araliaceae		
Hedera helix	Sd	Okerman 2000; Swearingen & Dietrich 2006
Schefflera sp.	Sd	Sankamethawee et al. 2011
S. goetzenii	Sd	Sun & Moermond 1997

Notes: E = extrafloral nectar used by birds; F = fruits and/or flowers used by birds; N = nectar robbed and/or flower predated; O = other; P = pollinated; Sd = seed dispersed; Sp = seed predated.

REFERENCES

Angiosperm Phylogeny Group (2009) An update of the Angiosperm Phylogeny Group classification for the orders and families of flowering plants: APG III. *Botanical Journal of the Linnean Society*, **161**, 105–121.

Amaral, M. & Macedo, R.H.F. (2003) Breeding patterns and habitat use in the endemic Curl-crested Jay of central Brazil. *Association of Field Ornithologists*, **74**, 331–340.

Anich, N.M., Benson, T.J. & Bednarz, J.C. (2010) Factors influencing home-range size of Swainson's Warblers in eastern Arkansas. *Condor*, **112**, 149–158.

Ansell, F.A., Edwards, D.P. & Hamer, K.C. (2011) Rehabilitation of logged rain forests: avifaunal composition, habitat structure, and implications for biodiversity-friendly REDD+. *Biotropica*, **43**, 504–511.

Baran, M.A. (2010) Curl-crested Jay (*Cyanocorax cristatellus*). In *Neotropical Birds Online*. (ed. T.S. Schulenberg). [online] URL: http://neotropical.birds.cornell.edu/portal/species/overview?p_p_spp=516716.

Beehler, B.M. & Beehler, C.H. (1986) Observations on the ecology and behavior of the Pale-billed Sicklebill. *Wilson Bulletin*, **98**, 505–515.

Bell, H.L. (1983) A bird community of lowland rainforest in New Guinea. 6 Foraging ecology and community structure of the avifauna. *Emu*, **84**, 142–158.

Bencke, G.A. (1996) New records and habits of the White-browed Foliage-gleaner (*Anabacerthia amaurotis*) (Passeriformes: Furnariidae) in Rio Grande do Sul, southern Brazil. *Ararajuba*, **4**, 24–28.

Benson, T.J., Anich, N.M., Brown, J.D. & Bednarz, J.C. (2009) Swainson's Warbler nest-site selection in eastern Arkansas. *Condor*, **111**, 694–705.

Bernardello, L., Galetto, L. & Rodriguez, I.G. (1994) Reproductive biology, variability of nectar features and pollination of *Combreturn fruticosum* (Combretaceae) in Argentina. *Botanical Journal of the Linnean Society*, **114**, 293–308.

Bernardo, C.S.S. (2011) Red-billed Curassow (*Crax blumenbachii*). In *Neotropical Birds Online* (ed. T.S. Schulenberg). [online] URL: http://neotropical.birds.cornell.edu/portal/species/overview?p_p_spp=82631.

Biamonte, E., Sandoval, L., Chacón, E. & Barrantes, G. (2011) Effect of urbanization on the avifauna in a tropical metropolitan area. *Landscape Ecology*, **26**, 183–194.

Bierregaard, R.O. Jr. (1990) Species composition and trophic organization of the understory bird community in a central Amazonian terra fiorg forest, in *Four Neotropical Rainforests* (ed A.H. Gentry), Yale University Press, New Haven, CT, pp. 161–181.

Billerman, S. (2012) Hoatzin (*Opisthocomus hoazin*). In *Neotropical Birds Online* (ed. T.S. Schulenberg), [online] URL: http://neotropical.birds.cornell.edu/portal/species/overview?p_p_spp=201176.

Boinski, S., Kauffman, L., Westoll, A. et al. (2003) Are vigilance, risk from avian predators and group size consequences of habitat structure? A comparison of three species of squirrel monkey (*Saimiri oerstedii*, *S. boliviensis*, and *S. sciureus*). *Behaviour*, **140**, 1421–1467.

Caballé, G. & Martin, A. (2001) Thirteen years of change in trees and lianas in a Gabonese rainforest. *Plant Ecology*, **152**, 167–173.

Canaday, C. (1996) Loss of insectivorous birds along a gradient of human impact in Amazonia. *Biological Conservation*, **77**, 63–77.

Carsten, L.D., Juola, F.A., Male, T.D. & Cherry, S. (2002) Host associations of lianas in a South-east Queensland Rain Forest. *Journal of Tropical Ecology*, **18**, 107–120.

Chandler, R.B. & King, D.I. (2011) Habitat quality and habitat selection of Golden-winged Warblers in Costa Rica: an application of hierarchical models for open populations. *Journal of Applied Ecology*, **48**, 1038–1047.

Chapman, A. & Rosenberg, K.V. (1991) Diets of four sympatric Amazonian woodcreepers (Dendrocolaptidae). *Condor*, **93**, 904–915.

Chiver, I. & Morton, E.S. (2010) Red-throated Ant-Tanager (*Habia fuscicauda*). In *Neotropical Birds Online* (ed. T.S. Schulenberg). [online] URL: http://neotropical.birds.cornell.edu/portal/species/overview?p_p_spp=%0A%09%09%09%09595596.

Chouteau, P. (2004) The impacts of logging on the microhabitats used by two species of couas in the western forest of Madagascar. *Comptes Rendus Biologies*, **327**, 1157–1170.

Clark, C.J., Poulsen, J.R. & Parker, V.T. (2001) The role of arboreal seed dispersal groups on the seed rain of a lowland tropical forest. *Biotropica*, **33**, 606–620.

Connell, J.H. (1971) On the roles of natural enemies in preventing competitive exclusion in some marine animals and in rain forest trees, in *Dynamics of Populations* (eds P.J. den Boer & G.R. Gradwell), Centre for Agricultural Publishing and Documentation, Wageningen, the Netherlands, pp. 298–312.

Cruz, A. (1988) Avian resource use in a Caribbean pine plantation. *Journal of Wildlife Management*, **52**, 274–279.

Dalling, J.W., Schnitzer, S.A., Baldeck, C. et al. (2012) Resource-based habitat associations in a Neotropical liana community. *Journal of Ecology*, **100**, 1174–1182.

Datta, A. & Rawat, G.S. (2003) Foraging patterns of sympatric hornbills during the nonbreeding season in Arunachal Pradesh, northeast India. *Biotropica*, **35**, 208–218.

Del Hoyo, J., Elliott, A. & Christie, D.A. (2003) *Handbook of the Birds of the World: Broadbills to Tapaculos*, vol. **8**, Lynx Editions, Barcelona.

Dial, R.J., Ellwood, M.D.F., Turner, E.C. & Foster, W.A. (2006) Arthropod abundance, canopy structure, and microclimate in a Bornean lowland tropical rain forest. *Biotropica*, **38**, 643–652.

Domínguez-Bello, M.G., Michelangeli, F., Ruiz, M.C. et al. (1994) Ecology of the folivorous Hoatzin (*Opisthocomus hoazin*) on the Venezuelan plains. *Auk*, **111**, 643–651.

Durães, R. (2009) Lek structure and male display repertoire of Blue-Crowned Manakins in Eastern Ecuador. *Condor*, **111**, 453–461.

Enard, C., Thery, M. & Sabatier, D. (1989) The diet of the Guianan Cock-of-the-rock *Rupicola rupicola*: the relationships between frugivory and zoochory. *Revue d'Ecologie la Terra et la Vie*, **44**, 47–74.

Endler, J.A. & Théry, M. (1996) Interacting effects of lek placement, display behavior, ambient light, and color patterns in three Neotropical forest-dwelling birds. *American Naturalist*, **148**, 421–452.

English, P.H. (1998) Ecology of mixed-species understory flocks in Amazonian Ecuador. Doctoral dissertation, University of Texas, Austin.

Felton, A., Wood, J.T., Felton, A.M. et al. (2008) A comparison of bird communities in the anthropogenic and natural-tree fall gaps of a reduced-impact logged subtropical forest in Bolivia. *Bird Conservation International*, **18**, 129–143.

Fleming, T.H., Muchhala, N. & Ornelas, J.F. (2005) New World nectar-feeding vertebrates: community patterns and processes, in *Contribuciones mastozoológicas en homenaje a Bernardo Villa* (eds V. Sanchez-Cordero & R.A. Medellín), Institutos de Biología y Ecología, UNAM, and CONABIO, Mexico, pp. 163–186.

Gautier-Hion, A., Duplantier, J.-M., Quris, R. et al. (1985) Fruit characters as a basis of fruit choice and seed dispersal in a tropical forest vertebrate community. *Oecologia*, **65**, 324–337.

Gill, S. (2011) Buff-breasted Wren (*Cantorchilus leucotis*). In *Neotropical Birds Online* (ed. T.S. Schulenberg). [online] URL: http://neotropical.birds.cornell.edu/portal/species/overview?p_p_spp=535596.

Gill, F. & Donsker, D. (eds.) (2012) *IOC World Bird Names*, vol. **3.1**. [online] URL: http://www.worldbirdnames.org.

Goulding, W. & Martin, T.E. (2010) Breeding biology of the Golden-faced Tyrannulet (*Zimmerius chrysops*) in Venezuela. *Wilson Journal of Ornithology*, **122**, 689–698.

Gradwohl, J. & Greenberg, R. (1982) The effect of a single species of avian predator on the arthropods of aerial leaf litter. *Ecology*, **63**, 581–583.

Gross, A.O. (1958) Life history of the Bananaquit of Tobago Island. *Wilson Bulletin*, **70**, 257–279.

Gryj, E., Martinez del Rio, C. & Baker, I. (1990) Avian pollination and nectar use in *Combretum fruticosum* (Loefl.). *Biotropica*, **22**, 266–271.

Hansbauer, M.M., Storch, I., Leu, S. et al (2008) Movements of Neotropical understory passerines affected by anthropogenic forest edges in the Brazilian Atlantic rainforest. *Biological Conservation*, **141**, 782–791.

Hansbauer, M.M., Végvári, Z., Storch, I. et al. (2010) Microhabitat selection of three forest understory birds in the Brazilian Atlantic rainforest. *Biotropica*, **42**, 355–362.

Ingwell, L.L., Wright, S.J., Becklund, K.K. et al. (2010) The impact of lianas on 10 years of tree growth and mortality on Barro Colorado Island, Panama. *Journal of Ecology*, **98**, 879–887.

Isler, M.L. & Isler, P.R. (1987) *The Tanagers: Natural History, Distribution, and Identification*, Smithsonian Institution Press, Washington, DC.

Janzen, D.H. (1967) Why mountain passes are higher in the tropics. *American Naturalist*, **101**, 233–249.

Janzen, D.H. (1970) Herbivores and the number of tree species in tropical forests. *American Naturalist*, **104**, 501–528.

Karr, J.R. & Freemark, K.E. (1983) Habitat selection and environmental gradients: Dynamics in the "stable" tropics. *Ecology*, **64**, 1481–1494.

Karubian, J., Carrasco, L., Mena, P. et al. (2011) Nesting biology, home range, and habitat use of the Brown Wood Rail (*Aramides wolfi*) in northwest Ecuador. *Wilson Journal of Ornithology*, **123**, 137–141.

Kinnaird, M.F. (1998) Evidence for effective seed dispersal by the Sulawesi Red-knobbed Hornbill, *Aceros cassidix*. *Biotropica*, **30**, 50–55.

Kitamura, S. (2011) Frugivory and seed dispersal by hornbills (Bucerotidae) in tropical forests. *Acta Oecologia*, **37**, 531–541.

Kitamura, S., Thong-Aree, S., Madsri, S. & Poonswad, P. (2011) Characteristics of hornbill-dispersed fruits in lowland Dipterocarp forests of southern Thailand. *Raffles Bulletin of Zoology*, **24**, 137–147.

Kitamura, S., Yumoto, T., Poonswad, P. et al. (2002) Interactions between fleshy fruits and frugivores in a tropical seasonal forest in Thailand. *Oecologia*, **133**, 559–572.

Kitamura, S., Yumoto, T., Poonsawat, P. et al. (2004) Pattern and impact of hornbill seed dispersal at nest trees in a moist evergreen forest in Thailand. *Journal of Tropical Ecology*, **20**, 545–53.

Kodric-Brown, A., Brown, J.H., Byers, G.S. & Gori, D.F. (1984) Organization of a tropical lowland community of hummingbirds and flowers. *Ecology*, **65**, 1358–1368.

Koenig, S.E., Wunderle, J.M. Jr. & Enkerlin-Hoeflich, E.C. (2007) Vines and canopy contact: a route for snake predation on parrot nests. *Bird Conservation International*, **17**, 79–91.

Kominami, Y., Sato, T., Takeshita, K. et al. (2003) Classification of bird-dispersed plants by fruiting phenology, fruit size, and growth form in a primary lucidophyllous forest: an analysis, with implications for the conservation of fruit–bird interactions. *Ornithological Society of Japan*, **2**, 3–23.

Ladwig, L.M. & Meiners, S.J. (2010) Liana host preference and implications for deciduous forest regeneration. *The Journal of the Torrey Botanical Society*, **137**, 103–112.

LaFrankie, J.V. (2005) Lowland tropical rain forests of Asia and America: parallels, convergence, and divergence, in *Pollination Ecology and the Rain Forest: Sarawak Studies* (eds D.W.

Roubik, S. Sakai & A.A. Hamid Karim), Springer, New York, pp. 178–190.

Lara, C. & Ornelas, J.F. (2001) Preferential nectar robbing of flowers with long corollas: experimental studies of two hummingbird species visiting three plant species. *Oecologi*, **128**, 263–273.

LaRosa, A.M. (1982) *The biology and ecology of* Passiflora mollissima *in Hawaii*. Technical Report 50, Cooperative National Park Resources Studies Unit, University of Hawaii, Manoa.

Ledo, A. & Schnitzer, S.A. (2014) Disturbance, not negative density dependence or habitat specialization maintains liana diversity in a tropical forest. *Ecology*, in press.

Lenz, J., Fiedler, W., Caprano, T. et al. (2011) Seed-dispersal distributions by trumpeter hornbills in fragmented landscapes. *Proceedings of the Royal Society of London, Series B*, **278**, 2257–2264.

Levey, D.J. (1988) Tropical wet forest treefall gaps and distributions of understory birds and plants. *Ecology*, **69**, 1076–1089.

Logue, D.M. (2007) Duetting in space: a radio-telemetry study of the Black-bellied Wren. *Proceedings of the Royal Society of London, Series B*, **274**, 3005–3010.

Loiselle, B.A. & Blake, J.G. (2002) Potential consequences of extinction of frugivorous birds for shrubs of a tropical wet forest, in *Seed Dispersal and Frugivory: Ecology, Evolution and Conservation* (eds D.J. Levey, W.R. Silva & M. Galetti), CABI, Oxford, pp. 397–406.

Londré, R.A. & Schnitzer, S.A. (2006) The distribution of lianas and their change in abundance in temperate forests over the past 45 years. *Ecology*, **87**, 2973–2978.

Lopes, A.V., Vogel, S. & Machado, I.C. (2002) Secretory trichomes, a substitutive floral nectar source in *Lundia* A. DC. (Bignoniaceae), a genus lacking a functional disc. *Annals of Botany*, **90**, 169–174.

Lopes, E.V., Volpato, G.H., Mendonça, L.B. et al. (2006) Abundância, microhabitat, e repartição ecológica de papa-formigas (Passeriformes, Thamnophilidae) na bacia hidrográfica do rio Tibagi, Paraná, Brasil. *Revista Brasiliera de Zoologia*, **23**, 395–403.

Mack, A.L. & Wright, D.D. (1996) Notes on occurrence and feeding of birds at Crater Mountain Biological Research Station, Papua New Guinea. *Emu*, **96**, 89–101.

Magrach, A., Larrinaga, A.R. & Santamaría, L. (2012) Effects of matrix characteristics and interpatch distance on functional connectivity in fragmented temperate rainforests. *Conservation Biology*, **26**, 238–247.

Mandon-Dalger, I., Clergeau, P., Tassin, J. et al. (2004) Relationships between alien plants and an alien bird species on Reunion Island. *Journal of Tropical Ecology*, **20**, 635–642.

Manu, S., Peach, W. & Cresswell, W. (2007) The effects of edge, fragment size and degree of isolation on avian species richness in highly fragmented forest in West Africa. *Ibis*, **149**, 287–297.

Marques, M.C.M., Roper, J.J. & Salvalaggio, A.P.B. (2004) Phenological patterns among plant life-forms in a subtropical forest in southern Brazil. *Plant Ecology*, **173**, 203–213.

Marra, P.P. & Remsen, J.V. Jr. (1997) Insights into the maintenance of high species diversity in the Neotropics: habitat selection and foraging behavior in understory birds of tropical and temperate forests. *Ornithological Monographs*, **48**, 445–483.

Mason, D. (1996) Responses of Venezuelan understory birds to selective logging, enrichment strips, and vine cutting. *Biotropica*, **28**, 296–309.

Meisenburg, M.J. & Fox, A.M. (2002) What role do birds play in dispersal of invasive plants? *Wildland Weeds*, **5**(3), 8–14.

Merow, C., LaFleur, N., Silander, J.A. Jr. et al. (2011) Developing dynamic mechanistic species distribution models: Predicting bird-mediated spread of invasive plants across Northeastern North America. *American Naturalist*, **178**, 30–43.

Michel, N.L. (2012) Mechanisms and consequences of avian understory insectivore population decline in fragmented Neotropical rainforest. Doctoral dissertation, Tulane University.

Michel, N.L. & Sherry, T.W. (2012) Human-altered mesoherbivore densities and cascading effects on plant and animal communities in fragmented tropical forests, in *Tropical Forests* (eds P. Sudarshana, M. Nageswara-Rao & J.R. Sonegi), InTech Open Access Publisher, pp. 177–202.

Michel, N.L., Sherry, T.W. & Carson, W.P. (2014). The omnivorous collared peccary negates an insectivore-generated trophic cascade in Costa Rican wet tropical forest understorey. *Journal of Tropical Ecology*, **30**, 1–11.

Miller, E.T. & Greeney, H.F. (2008) Clarifying the nest architecture of the *Silvicultrix* clade of *Ochthoeca* chat-tyrants (Tyrannidae). *Ornitología Neotropical*, **19**, 361–370.

Mooney, K.A., Gruner, D.S., Barber, N.A. et al. (2010) Interactions among predators and the cascading effects of vertebrate insectivores on arthropod communities and plants. *Proceedings of the National Academy of the Sciences of the USA*, **107**, 7335–7340.

Morellato, P.C. & Leitão-Filho, H.F. (1996) Reproductive phenology of climbers in a southeastern Brazilian forest. *Biotropica*, **28**, 180–191.

Moscovice, L.R., Issa, M.H., Petrzelkova, K.J. et al. (2007) Fruit availability, chimpanzee diet, and grouping patterns on Rubondo Island, Tanzania. *American Journal of Primatology*, **69**, 487–502.

Muller-Landau, H.C. & Hardesty, B.D. (2005) Seed dispersal of woody plants in tropical forests: concepts, examples, and future directions, in *Biotic Interactions in the Tropics* (eds D.F.R.P. Burslem, M.A. Pinard & S. Hartley), Cambridge University Press, Cambridge, pp. 267–309.

Ndang'ang'a, P.K., Sande, E., Evans, S.W., et al. (eds.) (2005) *International Species Action Plan for the Spotted Ground Thrush* Zoothera guttata. BirdLife International, Nairobi, Kenya and Royal Society for the Protection of Birds, Sandy, Bedfordshire, UK.

Naka, L.N. (2004) Structure and organization of canopy bird assemblages in Central Amazonia. *Auk*, **121**, 88–102.

Ødegaard, F. (2000) The relative importance of trees versus lianas as hosts for phytophagous beetles (Coleoptera) in tropical forests. *Journal of Biogeography*, **27**, 283–296.

Okerman, A. (2000) Combating the "ivy desert": the invasion of *Hedera helix* (English Ivy) in the Pacific Northwest United States. *Restoration and Reclamation Review*, **6**, 1–10.

Pacheco, J.F., Whitney, B.M. & Gonzaga, L.P. (1996) A new genus and species of furnariid (Aves: Furnariidae) from the cocoa-growing region of southeastern Bahia, Brazil. *Wilson Bulletin*, **108**, 397–433.

Parrini, R., Pacheco, J.F. & Mallet-Rodrigues, F. (2010) Comportamento de forrageamento de *Philydor atricapillus* (Passeriformes: Furnariidae) na Floresta Atlântica do Estado do Rio de Janeiro, região Sudeste do Brasil. *Atualidades Ornitológicas*, **153**, 55–61.

Peck, M.E. (1910) The effect of natural enemies on the nesting habits of some British Honduras birds. *Condor*, **12**, 53–60.

Peres, C.A. (2000) Identifying keystone plant resources in tropical forests: the case of gums from *Parkia* pods. *Journal of Tropical Ecology*, **16**, 287–317.

Peters, K.A., Lancia, R.A. & Gerwin, J.A. (2005) Swainson's Warbler habitat selection in a managed bottomland hardwood forest. *Journal of Wildlife Management*, **69**, 409–417.

Pizo, M.A., Silva, W.R., Galetti, M. & Laps, R. (2002) Frugivory in cotingas of the Atlantic Forest of southeast Brazil. *Ararajuba*, **10**, 175–185.

Post, W. 2011. Yellow-shouldered Blackbird (*Agelaius xanthomus*). In *Neotropical Birds Online* (ed. T.S. Schulenberg). [online] URL: http://neotropical.birds.cornell.edu/portal/species/overview?p_p_spp=666636.

Prum, R.O. & Razafindratsita, V.R. (1997) Lek behavior and natural history of the Velvet Asity (*Philepitta castanea*: Eurylaimidae). *Wilson Bulletin*, **109**, 371–392.

Putz, F.E. (1984) The natural history of lianas on Barro Colorado Island, Panama. *Ecology*, **65**, 1713–1724.

Putz, F.E., Sirot, L.K. & Pinard, M.A. (2001) Tropical forest management and wildlife: silvicultural effects on forest structure, fruit production, and locomotion of non-volant arboreal animals, in *The Cutting Edge: Conserving Wildlife in Managed Tropical Forests* (eds R. Fimbel, A. Grajal & J. Robinson), Columbia University Press, New York, pp. 11–34.

Putz, F.E. & Windsor, D.M. (1987) Liana phenology on Barro Colorado Island, Panama. *Biotropica*, **19**, 334–341.

Remsen, J.V. Jr. & Parker, T.A. III (1984) Arboreal dead-leaf searching birds of the Neotropics. *Condor*, **86**, 36–41.

Rivera-Milán, F.F. (1996) Nest density and success of Columbids in Puerto Rico. *Condor*, **98**, 100–113.

Rivera-Ortíz, F.A., Rodríguez-Flores, C., Soberanes-González, C., & Arizmendi, M.C. (2010) Scaled Antpitta (*Grallaria guatimalensis*). In *Neotropical Birds Online* (ed. T.S. Schulenberg). [online] URL: http://neotropical.birds.cornell.edu/portal/species/ overview?p_p_spp=402441.

Robinson, W.D. (2001) Changes in abundances of birds in a Neotropical forest fragment over 25 years. *Animal Biodiversity and Conservation*, **24**, 51–65.

Robinson, W.D., Brawn, J.D. & Robinson, S.K. (2000) Forest bird community structure in central Panama: influence of spatial scale and biogeography. *Ecological Monographs*, **70**, 209–235.

Rosenberg, K.V. (1997) Ecology of dead-leaf foraging specialists and their contribution to Amazonian bird diversity. *Ornithological Monographs*, **48**, 673–700.

Rosselli, L., Vasquez, P. & Ayub, I. (2002) The courtship displays and social system of the White-ruffed Manakin in Costa Rica. *Wilson Bulletin*, **114**, 165–178.

Sankamethawee, W., Pierce, A.J., Gale, G.A. & Hardesty, B.D. (2011) Plant-frugivore interactions in an intact tropical forest in north-east Thailand. *Integrative Zoology*, **6**, 195–212.

Sazima, I., Buzato, S. & Sazima, M. (1993) The bizarre inflorescence of *Norantea brasiliensis* (Marcgraviaceae): visits of hovering and perching birds. *Botanica Acta*, **106**, 507–513.

Schemske, D.W. (1975a) Territoriality in a nectar feeding Northern Oriole in Costa Rica. *Auk*, **92**, 594–595.

Schemske, D.W. (1975b) Time budget and foraging site preference of the Cinnamon Hummingbird in Costa Rica. *Condor*, **77**, 216–217.

Schemske, D.W. & Brokaw, N. (1981) Treefalls and the distribution of understory birds in a tropical forest. *Ecology*, **62**, 938–945.

Schnitzer, S.A. & Bongers, F. (2011) Increasing liana abundance and biomass in tropical forests: Emerging patterns and putative mechanisms. *Ecology Letters*, **14**, 397–406.

Schnitzer, S.A. & Carson, W.P. (2001) Treefall gaps and the maintenance of diversity in a tropical forest. *Ecology*, **82**, 913–919.

Schnitzer, S.A. & Carson, W.P. (2010) Lianas suppress tree regeneration and diversity in treefall gaps. *Ecology Letters*, **13**, 849–857.

Schnitzer, S.A., Dalling, J.W. & Carson, W.P. (2000) The impact of lianas on tree regeneration in tropical forest canopy gaps: evidence for an alternative pathway of gap-phase regeneration. *Journal of Ecology*, **8**, 655–666.

Schnitzer, S.A., Mangan, S.A., Dalling, J.W. et al. (2012) Liana abundance, diversity, and distribution on Barro Colorado Island, Panama. *PLoS One*, **7**, e52114.

Şekercioğlu, Ç.H., Daily, G.C. & Ehrlich, P.R. (2004) Ecosystem consequences of bird declines. *Proceedings of the National Academy of Sciences of the United States of America*, **101**, 18042–18047.

Senbeta, F., Schmitt, C., Denich, M. et al. (2005) The diversity and distribution of lianas in the Afromontane rain forests of Ethiopia. *Diversity and Distributions*, **11**, 443–452.

Senchina, D.S. (2008) Fungal and animal associates of *Toxicodendron* spp. (Anacardiaceae) in North America. *Perspectives in Plant Ecology, Evolution and Systematics*, **10**, 197–216.

Sheldon, F.H., Moyle, R.G. & Kennard, J. (2001) Ornithology of Sabah: history, gazetteer, annotated checklist, and bibliography. *Ornithological Monographs*, **52**, 1–285.

Sherry, T.W. (1984) Comparative dietary ecology of sympatric, insectivorous Neotropical flycatchers (Tyrannidae). *Ecological Monographs*, **54**, 313–338.

Sigel, B.J., Sherry, T.W. & Young, B.E. (2006) Avian community response to lowland tropical rainforest isolation: 40 years of change at La Selva Biological Station, Costa Rica. *Biotropica*, **20**, 111–121.

Silveri, A., Dunwiddie, P.W. & Michaels, H.J. (2001a) Logging and edaphic factors in the invasion of an Asian woody vine in a mesic North American forest. *Biological Invasions*, **3**, 379–389.

Skutch, A.F. (1969) A study of the Rufous-fronted Thornbird and associated birds: Part 1. Life history of the Rufous-fronted Thornbird. *Wilson Bulletin*, **81**, 5–43.

Skutch, A.F. (1971) Life history of the Keel-billed Toucan. *Auk*, **88**, 381–424.

Silveri, A., Dunwiddie, P.W. & Michaels, H.J. (2001b) Logging and edaphic factors in the invasion of an Asian woody vine in a mesic North American forest. *Biological Invasions*, **3**, 379–389.

Snow, B.K. (1973) The behavior and ecology of hermit hummingbirds in the Kanaku Mountains, Guyana. *Wilson Bulletin*, **85**, 163–177.

Snow, B.K. & Snow, D.W. (1971) The feeding ecology of tanagers and honeycreepers in Trinidad. *Auk*, **88**, 291–322.

Stein, B.A. (1992) Sicklebill hummingbirds, ants, and flowers. *BioScience*, **42**, 27–33.

Stiles, E.W. (1993) The influence of pulp lipids on fruit preference by birds. *Vegetatio*, **107/108**, 227–235.

Stiles, F.G. (1977) Possible specialization for hummingbird-hunting in the Tiny Hawk. *Auk*, **95**, 550–553.

Stiles, F.G. (1981) Geographical aspects of bird-flower coevolution, with particular reference to Central America. *Annals of the Missouri Botanical Garden*, **68**, 323–351.

Stiles, F.G. & Wolf, L.L. (1979) Ecology and evolution of lek mating behavior in the Long-tailed Hummingbird. *Ornithological Monographs*, **27**, 1–78.

Stoll, R.J., McClain, M.W., Nixon, C.M. & Worley, D.M. (1980) Foods of ruffed grouse in Ohio. *Ohio Fish and Wildlife Report*, **7**, 1–17.

Stratford, J.A. & Robinson, W.D. (2005) Gulliver travels to the fragmented tropics: geographic variation in mechanisms of avian extinction. *Frontiers in Ecology and the Environment*, **3**, 85–92.

Sun, C. & Moermond, T.C. (1997) Foraging ecology of three sympatric turacos in a montane forest in Rwanda. *Auk*, **114**, 396–404.

Swearingen, J.M. & Dietrich, S. (2006) Fact sheet: English Ivy. *Plant Conservation Alliance's Alien Plant Working Group* [online]. URL: http://www.nps.gov/plants/alien/fact/pdf/hehe1.pdf.

Tarwater, Corey E., & Patrick Kelley, J. (2010) Western Slaty-Antshrike (*Thamnophilus atrinucha*). In *Neotropical Birds Online* (ed. T.S. Schulenberg). [online] URL: http://neotropical.birds.cornell.edu/portal/species/overview?p_p_spp=369186.

Tello, J.G. (2001) Lekking behavior of the Round-tailed Manakin. *Condor*, **103**, 298–321.

Terborgh, J. & Diamond, J.M. (1970) Niche overlap in feeding assemblages of New Guinea birds. *Wilson Bulletin*, **82**, 29–52.

Terborgh, J., Robinson, S.K., Parker, T.A. et al. (1990) Structure and organization of an Amazonian forest bird community. *Ecological Monographs*, **60**, 213–238.

Thiollay, J.-M. (2003) Comparative foraging behavior between solitary and flocking insectivores in a Neotropical forest: does vulnerability matter? *Ornitologia Neotropical*, **14**, 47–65.

Udoye, K.C. & Schulenberg, T.S. (2012) Razor-billed Curassow (*Mitu tuberosum*). In *Neotropical Birds Online* (ed. T.S. Schulenberg). [online] URL: http://neotropical.birds.cornell.edu/portal/species/overview?p_p_spp=81191.

Urcelay, C., Morales, C.L. & Chalcoff, V.R. (2006) Relationship between corolla length and floral larceny in the South American hummingbird-pollinated *Campsidium valdivianum* (Bignoniaceae). *Annales Botanici Fennici*, **43**, 205–211.

Van Schaik, C.P., Terborgh, J.W. & Wright, S.J. (1993) The phenology of tropical forests: adaptive significance and consequences for primary consumers. *Annual Review of Ecology and Systematics*, **24**, 353–377.

Walther, B.A. (2002) Vertical stratification and use of vegetation and light habitats by Neotropical forest birds. *Journal of Ornithology*, **143**, 64–81.

Werner, T.K. & Sherry, T.W. (1987) Behavioral feeding specialization in Pinaroloxias inornata, the 'Darwin's Finch' of Cocos Island, Costa Rica. *Proceedings of the National Academy of Sciences of the United States of America*, **84**, 5506–5510.

Whitney, K.D., Fogiel, M.K., Lamperti, A.M. et al. (1998) Seed dispersal by *Ceratogymna* hornbills in the Dja Reserve, Cameroon. *Journal of Tropical Ecology*, **14**, 351–371.

Whitney, B.M. & Rosenberg, G.H. (1993) Behavior, vocalizations, and possible relationships of *Xenornis setifrons* (Formicariidae), a little-known Chocó endemic. *Condor*, **95**, 227–231.

Willis, E.O. (1968) Studies of the behavior of Lunulated and Salvin's Antbirds. *Condor*, **70**, 128–148.

Willis, E.O. (1972a) The behavior of Plain-brown Woodcreepers, *Dendrocincla fuliginosa*. *Wilson Bulletin*, **84**, 377–420.

Willis, E.O. (1972b) The behavior of Spotted Antbirds. *Ornithological Monographs*, **10**, 1–162.

Willis, E.O. (1979) Behavior and ecology of two forms of White-chinned Woodcreepers (*Dendrocincla merula*, Dendrocolaptidae) in Amazonia. *Papeis Avulsos Zoologia São Paulo*, **33**, 27–66.

Willis, E.O. (1982a) The behavior of Black-banded Woodcreepers (*Dendrocolaptes picumnus*). *Condor*, **84**, 272–285.

Willis, E.O. (1982b) The behavior of Scale-backed Antbirds. *Wilson Bulletin*, **94**, 447–462.

Willis, E.O. (1985) *Cercomacra* and related antbirds (Aves, Formicariidae) as army ant followers. *Revista Brasiliera de Zoologia*, **2**, 427–432.

Willis, E.O. (2003) Birds of a Neotropical woodlot after fire. *Ornitología Neotropical*, **14**, 233–246.

Willis, E.O. & Eisenmann, E. (1979) A revised list of birds of Barro Colorado Island, Panama. *Smithsonian Contributions to Zoology*, **291**, 1–31.

Willis, O.W. & Oniki, Y.O. (1972) Ecology and nesting behavior of the Chestnut-backed Antbird (*Myrmeciza exsul*). *Condor*, **74**, 87–98.

Winarni, N.L., O'Brien, T.G., Carroll, J.P. & Kinnaird, M.F. (2009) Movements, distribution, and abundance of Great Argus Pheasants (*Argusianus argus*) in a Sumatran rainforest. *Auk*, **126**, 341–350.

Wright, S.J., Calderón, O., Hernandéz, A. & Paton, S. (2004) Are lianas increasing in importance in tropical forests? A 17-year record from Panama. *Ecology*, **85**, 484–489.

Yorke, S.R., Schnitzer, S.A., Mascaro, J. et al. (2013) Increasing liana abundance and basal area in a tropical forest: the contribution of long-distance clonal colonization. *Biotropica*, **45**, 317–324.

Zimmer, K.J. (1999) Behavior and vocalizations of the Caura and the Yapacana antbirds. *Wilson Bulletin*, **111**, 195–209.

Zimmer, K.J. & Hilty, S.L. (1997) Avifauna of a locality in the upper Orinoco drainage of Amazonas, Venezuela. *Ornithological Monographs*, **48**, 865–885.

Zusi, R.L. & Hamas, M.J. (2001) Bats and birds as potential pollinators of three species of *Marcgravia* lianas on Dominica. *Caribbean Journal of Science*, **37**, 274–278.

Chapter 26

RELATIONSHIP BETWEEN LIANAS AND ARBOREAL MAMMALS: EXAMINING THE EMMONS–GENTRY HYPOTHESIS

Thomas D. Lambert and Michaela K. Halsey

Frostburg State University, Frostburg, MD, USA

Ecology of Lianas, First Edition. Edited by Stefan A. Schnitzer, Frans Bongers, Robyn J. Burnham, and Francis E. Putz.

Companion Website: www.wiley.com/go/schnitzer/lianas

OVERVIEW

In a seminal publication, Emmons and Gentry (1983) were the first to suggest that differences in liana structure and abundance between tropical Africa, Southeast Asia, and the neotropics created selection pressure differences that led to the evolution of gliding forms in Southeast Asia, and prehensile-tailed species in the neotropics, as means of predator avoidance. In this chapter, we briefly review the Emmons–Gentry hypothesis, and address the following questions: (1) Does the available biogeographical and evolutionary evidence support the notion that the pattern observed by Emmons and Gentry is anything other than an evolutionary accident? (2) Does research on the contemporary ecological relationships between liana abundance, canopy structure, and mammal communities suggest lianas played an important role in mammal evolution? (3) Is Emmons and Gentry's (1983) contention that escape from predation was the primary driving mechanism supported and does it provide for a plausible evolutionary pathway to arrive at gliding and prehensile tails? (4) Are any of the proposed alternatives to Emmons and Gentry's predation escape hypothesis sufficiently supported by empirical evidence, and thus are they more likely evolutionary drivers towards gliding and prehensile tails? We conclude by suggesting that the adaptations that would eventually lead to gliding and prehensile tails first benefitted non-volant arboreal species by reducing injury risk associated with accidental falling. Our mechanism better allows for the development of intermediate forms before true gliding or fully prehensile tails were obtained, than either the Emmons-Gentry hypothesis or the proposed alternatives which are based largely on the contemporary benefits of gliding and prehensile tails.

INTRODUCTION: THE EMMONS–GENTRY HYPOTHESIS

Since it was first published, the Emmons–Gentry hypothesis has been extensively reviewed and critiqued (Archer 1984; Kemper & Bell 1985; Dudley & DeVries 1990; Jackson 1999; Dial et al. 2004; Wells et al. 2004, Schiebe et al. 2006; Byrnes et al. 2011), thus we will provide only a general overview focusing on components central to the development of our hypothesis. Emmons and Gentry (1983) suggested that differences in liana abundance and mechanical strength among different tropical zones may have led to the evolution of many species of mammals that glide in Southeast Asian forests, and many species of prehensile-tailed mammals in the neotropics. Specifically, they suggested that the low relative abundance of lianas in Southeast Asian forests allowed for clear gliding pathways, which favored this means of predator escape. In contrast, in African and neotropical forests where lianas are abundant, gliding is not as feasible as clear pathways through the canopy are scarce. They further suggested that the presence of mega-herbivores in Africa, such as elephants, selected for structurally stronger, more firmly attached vegetation, especially lianas, than those found in the neotropics. The more fragile, less firmly attached lianas in the neotropics favored prehensile tails.

Emmons and Gentry based their hypothesis on two key contentions. First, both gliding membranes and prehensile tails are complex structures requiring numerous muscular and nervous system adaptations; thus, strong selective pressures must be responsible for their evolution. Second, tropical forests across regions differ in both liana abundance and the structural strength of lianas. Indeed, gliding membranes are controlled by elaborate muscular systems (Johnson-Murray 1977) that allow for aerial maneuverability and deceleration prior to landing (Jackson 1999), and prehensile tails are heavily muscled, equaling the weight of hind limbs in some species (Grand 1983). Additionally, while liana abundance varies considerably, even across small spatial scales within tropical regions, in general, Southeast Asian forests are typically taller, more open in the upper canopy, and support fewer lianas than forests in the neotropics and Africa (Dudley & DeVries 1990; Jackson 1999; Dial et al. 2004; DeWalt et al. Chapter 11 in this volume). The notion that neotropical vegetation is more mechanically fragile than African vegetation is more subjective, but has been supported by numerous authors (Cristoffer 1987; Prins & Reitsma 1989; Cristoffer & Peres 2003). Thus, the key components of Emmons and Gentry's (1983) argument appear to be supported.

While the key contentions of the Emmons–Gentry (1983) hypothesis seem well supported, their proposed selective mechanism for gliding and prehensile tail has received a fair amount of criticism (see discussion below). They proposed that escape from predators was the driving evolutionary force behind the development of gliding membranes and prehensile tails.

They speculated that in the open forests of Southeast Asia, gliding evolved as a means of predator avoidance from individuals deliberately dropping off branches as a means of escape. Prehensile tails evolved in the neotropics as a locomotive aid to avoid predation by allowing utilizing thin, fragile vegetation as escape routes. In Africa, in contrast, the dense canopies connected with strong, firmly attached lianas allowed for ease of movement, and therefore neither of these specialized locomotory adaptions were favored.

DOES THE AVAILABLE BIOGEOGRAPHICAL AND EVOLUTIONARY EVIDENCE SUPPORT THE NOTION THAT THE PATTERN OBSERVED BY EMMONS AND GENTRY IS ANYTHING OTHER THAN AN EVOLUTIONARY ACCIDENT?

Key to examining the Emmons–Gentry hypothesis is to rule out the possibility that the observed pattern is simply an accident of evolution, similar to the general lack of native eutherian mammals in Australia. Indeed, a strong biogeographical-evolutionary pattern supports the Emmons–Gentry hypothesis. Gliding mammals evolved twice in Southeast Asia and thrice in Australia and can be found in three Orders and six Families, whereas there is only one gliding family in Africa (Anomaluridae) and no gliding mammals in the lowland neotropics (Table 26.1). Similarly, prehensile tails evolved independently seven times in the neotropics and are found in six Orders and seven Families, whereas Africa and Southeast Asia each have only two families have prehensile tails, with one of these families shared (Table 26.2). Additionally, large forest herbivores are generally lacking in the neotropics (Bakker & Kelt 2000; Cristoffer & Peres 2003), supporting the contention that such species may have been selected for differences in strength of vegetation between the neotropics and Africa. More fragile vegetation in the neotropics is further supported by the generally smaller size of neotropical mammals when compared to African mammals, potentially indicating that vegetation cannot support large arboreal mammals (Cristoffer 1987; Cristoffer & Peres 2003).

In addition to the broad trends, closer examination strengthens the notion that the observed pattern is the result of differing selection pressures among the tropical regions. In their discussion, Emmons and Gentry (1983) differentiate between prehensile tails used in locomotion from those found in arboreal myrmecophagous species that use their tails primarily as support when feeding on insect nests. Only one family in Africa (Viverridae) and one in Southeast Asia (Muridae) have prehensile tails used in locomotion,

Table 26.1 Orders and families of tropical gliding mammals (genera in italics and numbers of species in parentheses) from Wilson and Reeder (2005).

South America		Africa		Southeast Asia		Australia	
none		**Anomaluridae**		**Dermoptera**		**Acrobatidae**	
		Anomalurus	(4)	*Cynocephalus*	(1)	*Acrobates*	(1)
		Idiurus	(2)	*Galeopterus*	(1)	*Distoechurus*	(1)
		*Zenkerella**	(1)				
				Petauristidae		**Pseudocheiridae**	
				Petaurista	(8)	*Hemibelidius**	(1)
						Petauroides	(1)
						Petropseudes	(1)
						Pseudocheirus	(1)
						Pseudochirops	(5)
						Pseudochirulus	(8)
						Petauridae	
						Dactylopsida	(4)
						*Gymnobelidus**	(1)
						Petaurus	(6)
Total Species	(0)		(6)		(10)		(28)

Note: *Family has gliding and non-gliding species.

Table 26.2 Orders and families of tropical prehensile-tailed mammals (genera in italics and species in parentheses) from Wilson and Reeder (2005).

South America		Africa		Southeast Asia		Australia	
Didelphimorphia		**Manidae**		**Manidae**		**Diprodontia**	
Didelphis	(6)	*Manis*	(4)	*Manis*	(4)	Phalangeridae	
Caluromys	(3)					*Ailurops*	(2)
Caluromysiops	(1)					*Phalanger*	(13)
Monodelphis	(20)					*Spilocuscus*	(5)
Philander	(7)					*Strigocuscus*	(2)
Marmosa	(19)					*Trichosurus*	(5)
Metachirus	(1)					*Wyulda*	(1)
Glironia	(1)						
Cebidae		**Muridae**		**Viverridae**		**Potoroidae**	
*Cebus**	(4)	*Dendromus**	(12)	*Arctictis*	(1)	*Aepyprymnus*	(1)
						Bettongia	(4)
						Potorous	(3)
Atelidae							
Alouatta	(15)						
Ateles	(7)						
Lagothrix	(4)						
						Macropodidae	
Myrmecophagidae						*Dendrolagus*	(14)
*Tamandua**	(2)						
*Cyclopes**	(1)						
						Burramyidae	
Erethizontidae						*Burramys*	(1)
Coendou	(6)					*Cercatetus*	(4)
Procyonidae						**Acrobatidae**	
Potos	(1)					*Acrobates*	(1)
*Nasua**	(2)					*Distoechurus*	(1)
Microbiotheria						**Pseudocheiridae**	
Dromiciops	(1)					*Hemibelidius*	(1)
						Petauroides	(1)
						Petropseudes	(1)
						Pseudocheirus	(1)
						Pseudochirops	(5)
						Pseudochirulus	(8)
						Petauridae	
						Dactylopsida	(4)
						Gymnobelidus	(1)
						Petaurus	(6)
						Tarsipedidae	
						Tarsipes	(1)
Total species	(101)		(16)		(5)		(86)

Note: *Indicates weakly prehensile tail.

thus strengthening the trend. Additionally, within New World porcupines (Erethizontidae) prehensile tails are found only in the neotropical species and not in the North American porcupine (*Erethizon dorsatum*), suggesting the locomotive advantage of a prehensile tail is not present in temperate forests. Among the many squirrels (Sciuridae) that invaded Central and South America from North America (Marshall 1988), the sole glider, the Southern Flying Squirrel (*Glaucomys volans*) is restricted to high elevation oak and oak-pine forests

(Reid 2009), areas structurally different than the generally liana-rich lowland neotropical forests which are the focus of this discussion. This suggests gliding is not favored in forests with high liana abundance.

DOES RESEARCH ON THE CONTEMPORARY ECOLOGICAL RELATIONSHIPS BETWEEN LIANA ABUNDANCE, CANOPY STRUCTURE AND MAMMAL COMMUNITIES SUGGEST LIANAS COULD HAVE PLAYED AN IMPORTANT ROLE IN MAMMAL EVOLUTION?

If the pattern observed by Emmons and Gentry (1983) cannot be explained as a historical or evolutionary accident, a causal mechanism must exist. Emmons and Gentry suggested that differences in liana structural strength and abundance are the primary difference among the tropical regions and this was the primary selective factor. If this is the case, then we would expect strong associations between mammals and lianas, indicating that liana structure and abundance could strongly influence mammal evolution. Fortunately, this relationship has been heavily studied as anthropogenic disturbance often leads to predictable changes in habitat structure, including an increase in edge habitat and an associated increase in herbaceous vegetation, denser understories, a lower and more broken canopy, and increased liana abundance (Schnitzer, Chapter 30 in this volume). These structural changes can have profound effects on animal communities (Lovejoy et al. 1986; Lambert et al. 2005a), which can in turn disrupt normal ecological processes and ultimately affect overall biodiversity. Especially important may be increases in the abundance of lianas (Schnitzer & Bongers 2011; Schnitzer, Chapter 30 in this volume) and the impact of this increase on mammal communities.

Numerous studies have documented strong associations between mammal species and habitat features indicative of edge or disturbed habitats, including liana abundance (Lambert & Adler 2000; Malcolm & Ray 2000; Lambert et al. 2003, 2006; Malcolm 2004). Furthermore, many studies suggest that these same mammals may play crucial roles in forest regeneration through their actions as seed dispersers and predators (Adler 1995; Asquith et al. 1997; Hoch & Adler 1997; Struhsaker 1997; Adler & Kestell 1998; Terborgh et al. 2001; Lambert et al. 2005b). Several studies in both the Old and New World tropics have documented suppression of forest regeneration after disturbance, an effect at least partially attributed to increased mammal abundance (Struhsaker 1997; Terborgh et al. 2001). If lianas increase in abundance, not just as the result of forest disturbance and edge effects (Lovejoy et al. 1986), but also in response to anthropogenic climate change (Phillips et al. 2002; Wright et al. 2004; Schnitzer 2005; Schnitzer & Bongers 2011, Schnitzer, Chapter 30 in this volume), then the ecological importance of mammals may be increasing throughout the tropics.

By connecting adjacent trees, lianas provide pathways for non-volant arboreal animals (Yanoviak, Chapter 24 in this volume). Numerous mammal species rarely descend even into the subcanopy in intact forest, as evidenced by their lack of captures in this strata (Lambert et al. 2005b; Lambert et al. 2006). In contrast, in areas with more open canopies where dense tangles of lianas that commonly reach the ground as the euphotic zone moves downward, captures of many arboreal small species increase (Lambert et al. 2005b; Lambert et al. 2006). Indeed, the impact on trapping results of canopy structure and the vertical complexity created by lianas can even influence which methods should be applied in small mammal surveys. In areas with tall unbroken canopies, some form of canopy trapping (e.g., Lambert et al. 2005b) is likely necessary to gain an adequate picture of the small mammal fauna, whereas in lower-statured, more broken, or younger forest, understory traps are likely sufficient (Lambert et al. 2005b). Failure to deploy canopy trapping in mature undisturbed forests may lead to overestimates of the importance of gaps and disturbed areas to mammal diversity, as numerous highly arboreal species are more likely to venture nearer the ground in gaps where liana tangles are abundant (Lambert et al. 2005b; Lambert et al. 2006).

While the importance of lianas to arboreal species is intuitive, terrestrial species also show strong positive correlations with liana abundance. Many liana species produce fruits or seeds that are utilized by a wide variety of forest animals (e.g., Michel et al., Chapter 25 in this volume; Yanoviak, Chapter 24 in this volume), including mammals (Adler, 1995). Additionally, dense tangles of lianas can provide areas of cover and protection for terrestrial species when these tangles occur near the ground in gaps and along edges.

Given these factors, it is not surprising that recent work has shown associations not just between lianas and mammal abundance, but also with mammal fruit and seed exploitation (Kilgore et al. 2010). Liana abundance is strongly correlated with both seed removal rates and seed cache locations in a forest in Central Panama (Kilgore et al. 2010). Given these strong associations it is reasonable to suggest that liana structure and abundance could have played an important role mammal evolution.

IS EMMONS AND GENTRY'S (1983) CONTENTION THAT ESCAPE FROM PREDATION WAS THE PRIMARY DRIVING MECHANISM SUPPORTED, AND DOES IT PROVIDE FOR A PLAUSIBLE EVOLUTIONARY PATHWAY TO ARRIVE AT GLIDING AND PREHENSILE TAILS?

The available evidence strongly suggests that the pattern observed by Emmons and Gentry (1983) is in fact the result of differing selection pressures among the tropical regions, and that liana structure and abundance could greatly influence mammal fitness and evolution. Thus, we must now turn our attention to Emmons and Gentry's proposed mechanism of predator avoidance, and whether it could explain the observed pattern. This aspect of the Emmons–Gentry hypothesis has received a great deal of criticism, with this criticism focusing nearly exclusively on the evolution of gliding (e.g., Jackson 1999; Dudley et al. 2007).

Despite the strong criticism, some lines of evidence do support Emmons and Gentry's contention that gliding evolved from purposely dropping from branches to escape predators. Indeed, some non-gliding species are known to drop to avoid predators, and some non-gliding species have been observed to flatten their bodies to control or slow their descent during falls (Jackson 1999; Dudley et al. 2007). Additionally, gliding is quieter than moving on the forest floor and leaves less of a scent trail (Jackson 1999). Thus, once the ability to glide has been achieved, gliders could face lower predation pressure.

Critics point out that, for species such as flying squirrels (*Glaucomys* spp), that are arboreal and cavity nesters, predation pressure should already be relatively low (Stapp 1994), and therefore the gain from gliding in terms of predator avoidance would be slight. Additionally, the notion that gliding evolved from dropping, would apply mostly to arboreal mammalian predators, while numerous other groups, including raptors, owls, snakes, lizards and bats all feed on mammals. Each of these predators hunts by different means, using different cues, thus the predator escape benefits of dropping (and later gliding) would largely be determined by the predator community present when gliding evolved. Unfortunately, gliding forms are not easily recognized in the fossil record, thus the early evolution of gliding remains obscure and reconstruction of the predator communities is not yet possible (Jackson 1999; Thorington Jr. et al. 2005).

While the gliding membranes in various gliding taxa differ in terms of their attachment points and musculature, their similarities suggest that gliding evolved from a flattening of the body and increased skin surface area as a means to slow or control descent while falling (Johnson-Murray 1977; Jackson 1999). This interpretation is consistent with the mechanism proposed by Emmons and Gentry (1983). However, as gliding evolved in areas with tall-statured, less vertically complex forests, the notion that purposeful dropping began this evolution seems less likely. Prior to the ability to glide, falls in these types of forest would be much longer than falls in denser, more liana-packed forests, as such, the risk of injury would be greater. Thus, the anti-predator benefits of gliding would be realized only after true gliding had been obtained. Additionally, the contention that predator avoidance also led to the evolution of prehensile tails is more difficult to support or explain.

ARE ANY OF THE PROPOSED ALTERNATIVES TO EMMONS AND GENTRY'S PREDATION CONTENTION BETTER SUPPORTED OR DO THEY PROVIDE FOR MORE LIKELY EVOLUTIONARY PATHWAYS TO REACH BOTH GLIDING AND PREHENSILE TAILS?

Criticisms of the Emmons and Gentry (1983) hypothesis have focused almost exclusively on the benefit of gliding and can largely be divided into two related alternatives: (1) energy efficiency; and (2) food resources. Authors have suggested that gliding is an energy-efficient way to move through the forest (Jackson 1999; Scheibe et al. 2006). While some dismiss this notion, suggesting that once the costs of climbing are figured in with the glide, there are no energy-saving benefits (Byrnes et al. 2011), these arguments ignore that gliding certainly evolved from an arboreal lifestyle, thus the costs of climbing have already been paid. Therefore, if one considers gliding as an alternative to an organism climbing down from one tree and then into the canopy of an adjacent tree, gliding can be considered an energy-saving locomotive specialization. In forests with few inter-crown liana connections, such descent and ascension movements between trees would often be necessary for non-volant arboreal species. Thus, energy efficiency could be an explanation for large number of gliders seen in generally liana-free forests.

As a continuation of the energy-efficiency hypothesis, some have argued that the energy savings benefits of gliding allow exploitation of low-quality and widely scattered food resources, which would otherwise not sufficiently provide the species needs (Goldingay 2000). Indeed, some gliding species such as colugos (Dermoptera) and Northern Flying Squirrels (*Glaucomys sabrinus*) do utilize these types of resources, perhaps suggesting that food resources could play a role in the selective advantage of gliding.

While energy efficiency and the potential to exploit widely scattered, low-quality resources likely represent some of the contemporary benefits of gliding, they are unlikely explanations for how gliding evolved. These benefits are only gained after the species is able to glide, intermediate pre-gliding forms would still drop to the forest floor and still need to ascend the adjacent tree, and thus these intermediate forms would enjoy none of the energy-saving benefits nor would they be able to exploit these lower-quality scattered resources. Additionally, explanations based on energy savings and food resources would only explain the evolution of gliding and thus ignore the pattern seen with prehensile tails. A complete explanation must account for both the evolution of gliding and prehensile tails, therefore these proposed alternative mechanisms are inadequate.

Risk of falling

Life in the canopy provides numerous benefits, including access to food resources and exposure to a reduced suite of predators. However, arboreal lifestyles have inherent costs that are not present for terrestrial species, including the inherent risk of injury from accidental falls. Falls resulting from high winds, vegetation breaking, or conflicts with other individuals, can result in serious injury and even death. Indeed, studies of museum specimens reveal that breaks in long bones are common in arboreal primate species, and that these breaks may be fatal if they occur in adults (Bulstrode 2000). Thus, natural selection should favor individuals that can minimize the risk of injury occurring from accidental falls. Differences in open space and landing substrates among tropical regions could select for differing means to minimize this cost. In the forests of Southeast Asia where lianas are less abundant, fall distances are likely to be great; therefore being able to slow or redirect a fall should be favored. Indeed, ants with no specialized adaptions have been observed to direct their fall back to the trunk of the tree from which they fell (Yanoviak et al. 2005; Dudley et al. 2007).

Intermediate pre-gliding forms could have started by flattening their bodies, and contracting lateral muscles to slow and redirect their fall back towards the tree. Later, the size of the gliding membrane could increase and become equipped with more muscle control, allowing for slower descents and better maneuverability. Slower, more controlled falls, would clearly reduce the severity of injury resulting from accidental falls. Once a certain level of membrane area, muscular control, and maneuverability was achieved, then active dropping or leaping as a means of predator escape could have further pushed toward the evolution of true gliding. In the liana-rich yet structurally fragile forests of the neotropics, fall distances are likely to be shorter than in the tall forests of Southeast Asia, and falling animals are likely to land on vegetation that gives way upon impact. Under these conditions, animals benefit by maximizing contact with the landing substrate.

During the initial phases of evolution of prehensile tails, tail musculature might have increased so as to allow the tail to be stiffened and directed towards vegetation. Increasing contact area would slow and potentially stop the animal's descent during an accidental fall, thereby reducing the risk of severe and potentially fatal injuries. Over time, greater strength and tail control would allow the animals to grab vegetation and stop the fall. In Africa, where lianas are common and firmly attached, fall distances are likely to be short and the landing substrate is likely to catch and stop the falling animal, thus less benefit can be gained by developing the intermediates to gliding membranes or prehensile tails.

CONCLUSION

The pattern in the distribution of gliding and prehensile-tailed mammals recognized by Emmons and Gentry (1983) cannot be explained away as an evolutionary accident. Strong associations between mammals and lianas suggest that lianas could have played a major role in mammalian evolution. However, the selective mechanisms put forth by Emmons and Gentry (1983) and their critics (Jackson 1999; Goldingay 2000; Dudley et al. 2007) are inadequate to explain how intermediate forms of gliding would confer benefits on the evolutionary pathway to these adaptations, and they generally fail to explain the evolution of prehensile tails. Our proposed selective mechanism, which is based on reducing injury from accidental falls, provides for both a selective advantage to intermediate forms of gliding, and also explains the pattern observed with prehensile tails. Our proposed mechanism remains untested and, as with any attempt to decipher the evolutionary past, tests are difficult. However, we suggest that detailed examination of the use of vertical strata by mammals with different characteristics could yield valuable insights. Especially fruitful might be studies of the three-dimensional space use by species such as the large American opossums (in the genus *Didelphis*), which utilize all levels of the forests and have a long evolutionary history in tropical America. If the risk of injury from falling has been an important factor in the evolution of arboreal species, then these species should avoid using the canopy in areas or at times when the risk of falls is high, or fall lengths are greater.

ACKNOWLEDGMENTS

We would like the thanks the editors for their invitation and support throughout the development of this manuscript. We would like to thank Frostburg State University and the Smithsonian Tropical Research Institution and their staffs for their logistical and material support. Finally, we would like to thank Scott Mangan and F.E. Putz for comments on earlier versions of this manuscript.

REFERENCES

Adler, G.H. (1995) Fruit and seed exploitation by Central American spiny rats, *Proechimys semispinosus*. *Studies on Neotropical Fauna and Environment*, **30**, 237–244.

Adler, G.H. & Kestell, D.W. (1998) Fates of Neotropical tree seeds influenced by spiny rats (*Proechimys semispinosus*). *Biotropica*, **30**, 677–681.

Archer, M. (1984) Evolution of arid Australia and its consequences for vertebrates, in *Vertebrate Zoogeography and Evolution in Australia* (eds M. Archer & G. Clayton), Hesperian Press, Perth, pp. 97–108.

Asquith, N.M., Wright, S.J. & Clauss, M.J. (1997) Does mammal community composition control recruitment in Neotropical forests? Evidence from Panamá. *Ecology*, **78**, 941–946.

Bakker, V.J. & Kelt, D.A. (2000) Scale-dependent patterns in body size distributions of Neotropical mammals. *Ecology*, **81**, 3530–3547.

Bulstrode, C. (2000) What happens to wild animals with broken bones. *The Iowa Orthopaedic Journal*, **10**, 19–23.

Byrnes, G., Lim, N.T.-L., Yeong, C., & Spence, A.J. (2011) Sex differences in the locomotor ecology of a gliding mammal, the Malayan colugo (*Galeopterus variegatus*). *Journal of Mammalogy*, **92**, 444–451.

Cristoffer, C. (1987) Body size differences between New World and Old World, arboreal, tropical vertebrates: cause and consequences. *Journal of Biogeography*, **14**, 165–172.

Cristoffer, C. & Peres, C.A. (2003) Elephants versus butterflies: the ecological role of large herbivores in the evolutionary history of two tropical worlds. *Journal of Biogeography*, **30**, 1357–1380.

Dial, R., Bloodworth, B., Lee, A. et al. (2004) The distribution of free space and its relation to canopy composition at six forest sites. *Forest Science*, **50**, 312–325.

Dudley, R., Byrnes, G., Yanoviak, S.P. et al. (2007) Gliding and the functional origins of flight: biomechanical novelty or necessity? *Annual Review of Ecology, Evolution, and Systematics*, **38**, 179–201.

Dudley, R. & DeVries, P. (1990) Tropical rain forest structure and the geographical distribution of gliding vertebrates. *Biotropica*, **22**, 432–434.

Emmons, L.H. & Gentry, A.H. (1983) Tropical forest structure and the distribution of gliding and prehensile-tailed vertebrates. *The American Naturalist*, **121**, 513–524.

Goldingay, R.L. (2000) Gliding mammals of the world: diversity and ecological requirements, in *Biology of Gliding Mammals* (eds R.L. Goldingay & J.S. Scheibe), Filander Verlag, Fürth.

Grand, T.I. (1983) Motion economy within the canopy: four strategies for mobility, in *Adaptations for Foraging in Non-Human Primates* (eds P.S. Rodman & J.G.H. Cant), Columbia University Press, New York.

Hoch, G.A. & Adler, G.H. (1997) Removal of black palm (*Astrocaryum standleyanum*) seeds by spiny rats (*Proechimys semispinosus*). *Journal of Tropical Ecology*, **13**, 51–58.

Jackson, S.M. (1999) Glide angle in the genus Petaurus and a review of gliding in mammals. *Mammal Review*, **30**, 9–30.

Johnson-Murray, J.L. (1977) Myology of the gliding membranes of some Petauristine rodents (Genera: *Glaucomys*,

Pteromys, Petinomys, and *Petaurista*). *Journal of Mammalogy*, **58**, 374–384.

Kemper, C. & Bell, D.T. (1985) Small mammals and habitat structure in lowland rain forest of peninsular Malaysia. *Journal of Tropical Ecology*, **1**, 5–22.

Kilgore, A., Lambert, T.D. & Adler, G.H. (2010) Lianas influence fruit and seed use by rodents in a tropical forest. *Tropical Ecology*, **51**, 265–271.

Lambert, T.D. & Adler, G.H. (2000) Microhabitat use by a tropical forest rodent, Proechimys semispinosus, in Central Panama. *Journal of Mammalogy*, **81**, 70–76.

Lambert, T.D., Adler, G.H., Riveros, C.M. et al. (2003) Rodents on tropical land-bridge islands. *Journal of Zoology*, **260**, 179–187.

Lambert, T.D., Malcolm, J.R. & Zimmerman, B.L. (2005a) Effects of mahogany (*Swietenia macrophylla)* logging on small mammal communities, habitat structure, and seed predation in the southeastern Amazon Basin. *Forest Ecology and Management*, **206**, 381–398.

Lambert, T.D., Malcolm, J.R. & Zimmerman, B.L. (2005b) Variation in small mammal richness by trap height and trap type in a southeastern Amazonian site, with notes on a new method of canopy trapping. *Journal of Mammalogy*, **86**, 982–990.

Lambert, T.D., Malcolm, J.R. & Zimmerman, B.L. (2006) Amazonian small mammal abundances in relation to habitat structure and research abundance. *Journal of Mammalogy*, **87**, 766–776.

Lovejoy, T.E., Bierregaard, R.O., Rylands, A.B. et al. (1986) Edge and other effects of isolation on Amazon forest Fragments, in *Conservation Biology: The Science of Scarcity and Diversity* (ed M.E. Soule), Sinauer Associates, Sunderland, MA, pp. 257–285.

Malcolm, J.R. (2004) Ecology and conservation of canopy mammals. In *Forest Canopies*. 2nd edition, (eds. M.D. Lowman & H.B. Rinker), pp. 297–331. Elsevier Academic Press, New York.

Malcolm, J.R. & Ray, J.C. (2000) Influence of timber extraction routes on Central African small mammal communities, forest structure and tree diversity. *Conservation Biology*, **14**, 1623–1638.

Marshall, L.G. (1988) Land mammals and the Great American Interchange. *American Scientist*, **76**, 380–388.

Phillips, O.L., Martinez, R.V., Arroyo, L. et al. (2002) Increasing dominance of large lianas in Amazonian forests. *Nature*, **418**, 770–774.

Prins, H.H.T. & Reitsma, J.M. (1989) Mammalian biomass in an African equatorial rain forest. *Journal of Animal Ecology*, **58**, 851–861.

Reid, F.A. (2009) *A Field Guide to the Mammals of Central America & Southeast Mexico*, 2nd edn, Oxford University Press, New York.

Scheibe, J.S., Smith, W.P., Bassham, J. & Magness, D. (2006) Locomotor performance and cost of transport in the northern flying squirrel *Glaucomys sabrinus*. *Acta Theriologica*, **51**, 169–178.

Schnitzer, S.A. (2005) A mechanistic explanation for global patterns of liana abundance and distribution. *American Naturalist*, **166**, 262–276.

Schnitzer, S.A. & Bongers, F. (2011) Increasing liana abundance and biomass in tropical forests: emerging patterns and putative mechanisms. *Ecology Letters*, **14**, 397–406.

Stapp, P. (1994) Can predation explain life-history strategies in mammalian gliders? *Journal of Mammalogy*, **77**, 227–228.

Struhsaker, T.T. (1997) *Ecology of an African Rain Forest: Logging in Kabale and the Conflict Between Conservation and Exploitation*, University of Florida Press, Gainesville, FL.

Terborgh, J., Lopez, L., Nunez, P. et al. (2001) Ecological meltdown in predator-free forest fragments. *Science*, **294**, 1923–1926.

Thorington, R.W. Jr., Schennum, C.E. et al. (2005) The difficulties of identifying flying squirrels (Sciuridae: Pteromyini) in the fossil record. *Journal of Vertebrate Paleontology*, **25**, 950–961.

Wells, K., Pfeiffer, M., Lakim, M.B. & Linsenmair, K.E. (2004) Use of arboreal and terrestrial space by a small mammal community in a tropical rain forest in Borneo, Malaysia. *Journal of Biogeography*, **31**, 641–652.

Wilson, D.E. & Reeder, D.M. (eds) (2005) *Mammal Species of the World: A Taxonomic and Geographic Reference*, 3rd edn, Johns Hopkins University Press, Baltimore, MD.

Wright, S.J., Calderon, O., Hernandez, A. & Paton, S. (2004) Are lianas increasing in importance in tropical forests? A 17-year record from Barro Colorado Island, Panamá. *Ecology*, **85**, 484–489.

Yanoviak, S.P., Dudley, R. & Kaspari, M. (2005) Directed aerial descent in arboreal ants. *Nature*, **433**, 624–626.

Chapter 27

USE OF LIANAS BY PRIMATES: MORE THAN A FOOD SOURCE

Víctor Arroyo-Rodríguez,[1] *Norberto Asensio,*[2] *Jacob C. Dunn,*[3,4] *Jurgi Cristóbal-Azkarate,*[3] *and Arturo Gonzalez-Zamora*[5,6]

[1]Centro de Investigaciones en Ecosistemas, Universidad Nacional Autónoma de México, Morelia, Michoacán, Mexico
[2]Faculty of Environment and Resource Studies, Mahidol University, Nakorn Pathom, Thailand
[3]Primate Immunogenetics and Molecular Ecology (PrIME) Research Group, University of Cambridge, Cambridge, UK
[4]Centro de Investigaciones Tropicales, Universidad Veracruzana, Xalapa, Veracruz, Mexico
[5]Instituto de Ecología A.C., Xalapa, Veracruz, Mexico
[6]Instituto de Investigaciones Biológicas, Universidad Veracruzana, Xalapa, Veracruz, Mexico

Ecology of Lianas, First Edition. Edited by Stefan A. Schnitzer, Frans Bongers, Robyn J. Burnham, and Francis E. Putz.

Companion Website: www.wiley.com/go/schnitzer/lianas

OVERVIEW

Lianas are becoming an increasingly prominent component of tropical forests. Therefore, understanding the impact that lianas may have on other organisms has significant ecological and conservation implications. Here we evaluated the use of lianas by primates – an abundant, species-rich, and ecologically important order of mammals. Our literature review demonstrates that lianas represent an important food resource for primates. Sixteen primate species in 10 countries used a total of 150 liana species from 47 families. Lianas can account for a high percentage of the total diet, reaching more than 20% of total feeding time in several cases. Lianas were used mainly as a source of fruits and leaves, but primates also consumed flowers, nectar, piths, rhizomes, stalks, bark, and buds. Furthermore, lianas can act as fallback foods for primates; i.e., abundant foods of relatively low quality that are used at sites where and/or during times when food availability is scarce. However, the use of lianas by primates goes beyond their importance as food sources. We found evidence suggesting that lianas may affect primate niche partitioning, sleeping tree selection, and home range use. Several studies demonstrate that lianas can also affect several aspects of primate behavior, including foraging, reproduction, locomotion, positional behavior, and grouping patterns. Overall, this review supports the idea that lianas play an important role in the ecology and conservation of primates, and hence, further studies are necessary to fully determine how the predicted increases in liana abundance and biomass will affect primate populations.

INTRODUCTION

The order Primates represents one of the most conspicuous and ecologically important groups of mammals in the tropics (e.g., Stoner et al. 2007, Russo & Chapman 2011). The great majority of primate species are highly arboreal, and they depend on the different strata above the forest floor for feeding, traveling, resting, breeding, and protection from predators. However, primatological research has traditionally focused on trees and has paid little attention to the ways in which primates use other life forms such as lianas. Nevertheless, lianas produce a large quantity of leaves (up to 40% of the total productivity in tropical forests, Hegarty & Caballé 1991), flowers and fruits, which are used as valuable food resources by a wide range of animal taxa (e.g., invertebrates: Aide & Zimmerman 1990; Yanoviak, Chapter 24 in this volume; birds: Snow 1981; bats: Hodgkison et al. 2003). Moreover, lianas facilitate the movement of animals through the canopy, and are thought to have contributed to the independent evolution of prehensile tails and suspensory postures in several mammalian lineages (Emmons & Gentry 1983; Lambert & Halsey, Chapter 26 in this volume). Finally, the presence of lianas can also increase the complexity of forests, creating microhabitats suitable for different animal species (Kilgore et al. 2010; Lambert & Halsey, Chapter 26 in this volume; Michel et al., Chapter 25 in this volume; Yanoviak, Chapter 24 in this volume).

Here, our main objective is to better understand the ecological importance of lianas to primates and to uncover general patterns that could be put to use in ecology and conservation. Such knowledge is particularly relevant as liana abundance and biomass appear to be increasing in tropical forests (Phillips et al. 2002; Schnitzer 2005; Schnitzer & Bongers 2011, Schnitzer, Chapter 30 in this volume), particularly in human-dominated landscapes (Putz 1984; Hegarty & Caballé 1991; Laurance et al. 1998; Schnitzer & Bongers 2002) in which an increasing number of primate populations are forced to live (e.g., Arroyo-Rodríguez & Mandujano 2009; Estrada et al. 2012).

We reviewed the available literature within the SCOPUS (©Elsevier, The Netherlands) database on 15 May 2012 for original research papers (without filtering for particular journals) containing the words "primate" and "lianas" in the title, the abstract and/or the keywords. This search revealed a total of 50 papers. However, only 35 of these actually gave information on the use of lianas by primates. We also reviewed eight additional studies cited within this primary literature, totaling 43 reviewed papers. Our literature review suggests that primates use lianas in several different ways (Table 27.1). Thus, we first assessed the role that lianas play in the diet of primates. Then, we evaluated the impact that lianas have on habitat use, and, finally, we assessed how lianas affect primate behavior.

Table 27.1 Studies showing the importance of lianas for primates[a].

Topic	Primate species (family)	Site (country)	Ref.
Positional behavior and movement	*Saguinus mystax* (Callitrichinae)	Padre Isla field site & Rio Blanco (Peru)	1
Positional behavior (foraging postures)	*Cebuella pygmaea* (Callitrichinae)	Yasuni National Park (Ecuador)	2
Habitat use (sleeping site selection)	*Saguinus midas midas* (Callitrichinae)	Field station of the MNHN, Saint Eugène (French Guiana)	3
Habitat use (population distribution)	*Eulemur fulvus* (Lemuridae)	Parc National de Ranomafana and Midongy-Sud (Madagascar)	4
Diet	*Procolobus badius*, *P. verus* and *Colobus polykomos* (Cercopithecidae)	Tiwai Island (Sierra Leone)	5
Habitat use and niche partitioning	*Chiropotes satanas* (Pitheciidae); *Alouatta belzebuth* (Atelidae); *Cebus apella* (Cebidae); *Saguinus midas* (Callitrichinae)	Xingu-Tocantus interfluvium, Eastern Amazonia (Brazil)	6
Behavior (locomotion)	*Lagothrix lagotricha* and *Ateles belzebuth* (Atelidae)	Yasuni National Park (Ecuador)	7
Habitat use (sleeping site selection)	*Microcebus murinus* and *M. revelobensis* (Cheirogaleidae)	Jardim Botanique Ampijoroa (Madagascar)	8
Diet, habitat use, and niche partitioning	*Macaca fuscata* (Cercopithecidae)	Kinkazan island (Japan)	9
Habitat use (population distribution)	*Cebuella pygmaea* (Callitrichinae)	Nourages Station (French Guiana); Yasuni (Ecuador)	10
Habitat use (population distribution)	*Cheirogaleus major* (Cheirogaleidae)	Vohibola III Classified Forest (Madagascar)	11
Habitat use (population distribution)	*Saguinus bicolor* (Callitrichinae)	Ducke Reserve, Manaus (Brazil)	12
Diet and seed dispersal	*Alouatta guariba* and *Brachyteles arachnoides* (Atelidae)	Fazenda Barreiro Rico (Brazil)	13
Food availability and behavior (grouping patterns)	*Pan troglodytes schweinfurthii* (Hominidae)	Mahale Mountains National Park (Tanzania)	14
Population density and habitat use	*Pongo pygmaeus morio* (Hominidae)	Sangkulirang Peninsula, east Kalimantan (Indonesia)	15
Diet	*Pan troglodytes* (Hominidae)	Rubondo Island, Lake Victoria (Tanzania)	16
Diet and habitat use	*Microcebus berthae* and *M. murinus* (Cheirogaleidae)	Forêt de Kirindy/CFPF (Madagascar)	17
Diet and habitat use (ranging behavior)	*Hylobates albibarbis* (Hylobatidae); *Pongo pygmaeus wurmbii* (Hominidae)	Tuanan, Kalimantan Tengah (Indonesia)	18
Diet	*Hylobates lar* (Hylobatidae)	Khao Yai National Park (Thailand)	19
Diet	*Callimico goeldii* (Callitrichinae)	Field camp Callimico, Pando (Bolivia)	20
Diet	*Cercocebus galeritus* (Cercopithecidae)	Tana River (Kenya)	21

Diet	*Alouatta guariba* and *Brachyteles arachnoides* (Atelidae)	Fazenda Barreiro Rico (Brazil)	22
Population density and habitat use	*Nycticebus bengalensis* (Lorisidae)	Khao Ang Rue Nai Wildlife Sanctuary (Thailand)	23
Habitat use	*Saguinus geoffroyi* (Callitrichinae)	Barro Colorado Island (Panama)	24
Habitat use and ranging patterns	*Pan troglodytes schweinfurthii* (Hominidae)	Rubondo Island, Lake Victoria (Tanzania)	25
Habitat use (sleeping site selection)	*Hylobates pileatus* (Hylobatidae)	Khao Ang Rue Nai Wildlife Sanctuary (Thailand)	26
Diet	*Rungwecebus kipunji* (Cercopithecidae)	Kitulo National Park (Tanzania)	27
Diet	*Papio anubis* (Cercopithecidae)	Comoé National Park (Ivory Coast)	28
Diet and ranging pattern	*Hylobates lar* (Hylobatidae)	Khao Yai National Park (Thailand)	29
Diet	*Hylobates muelleri* × *agilis/albibarbis* (Hylobatidae)	Baritu Ulu (Borneo)	30
Diet	*Trachypithecus francoisi* (Cercopithecidae)	Mayanghe Nature Reserve (China)	31
Diet	*Callicebus coimbrai* (Pitheciidae)	Fazenda Trapsa, Itaporanga d'Ajuda (Brazil)	32
Diet and habitat use	*Nomascus nasutus* (Hylobatidae)	Guangxi Province (China) - Cao Bang Province (Vietnam)	33
Diet and behavior	*Pongo abelli* (Hominidae)	Ketambe, Sumatra (Indonesia)	34
Habitat use (population distribution)	*Aotus azarai* (Cebidae)	Estancia Guaycolec, Chaco (Argentina)	35
Diet	*Rhinopithecus brelichi* (Cercopithecidae)	Fanjingshan National Nature Reserve, Guizhou (China)	36
Diet	*Presbytis rubicunda* (Cercopithecidae)	Danum Valley (Borneo)	37
Diet	*Alouatta palliata* (Atelidae)	Los Tuxtlas (Mexico)	38
Diet	*Hylobates lar* (Hylobatidae)	Khao Yai National Park (Thailand)	39
Diet	*Trachypithecus johnii* (Cercopithecidae)	Waynad (Kelgara, India)	40
Diet	*Alouatta palliata* (Atelidae)	Los Tuxtlas (Mexico)	41
Habitat use	*Pongo abelii* (Hominidae)	Gunung Leuser National Park (Sumatra, Indonesia)	42
Habitat use	*Hylobates klossi* (Hylobatidae)	Siberut Island, Indonesia	43

[a] We indicate the topic, primate species, and sites evaluated within each study.

References: 1. Garber & Pruetz 1995; 2. Youlatos 1999; 3. Day & Elwood 1999; 4. Johnson & Overdorff 1999; 5. Davies et al. 1999; 6. Bobadilla & Ferrari 2000; 7. Cant et al. 2001; 8. Radespiel et al. 2003; 9. Tsuji & Takatsuki 2004; 10. Youlatos 2004; 11. Lehman et al. 2006; 12. Vidal & Cintra 2006; 13. Martins 2006; 14. Itoh & Nishida 2007; 15. Marshall et al. 2007; 16. Moscovice et al. 2007; 17. Dammhahn & Kappeler 2008; 18. Vogel et al. 2009; 19. Kanwatanakid-Savini et al. 2009; 20. Porter et al. 2009; 21. Wieczkowski 2009; 22. Martins 2009; 23. Pliosungnoen et al. 2010; 24. Madden et al. 2010; 25. Moscovice et al. 2010; 26. Phoonjampa et al. 2010; 27. Davenport et al. 2010; 28. Kunz & Linsenmair 2010; 29. Asensio et al. 2011; 30. McConkey 2009; 31. Hu 2011; 32. Souza-Alves et al. 2011; 33. Fan et al. 2011; 34. Hardus et al. 2012; 35. van der Heide et al. 2012; 36. Xiang et al. 2012; 37. Hanya & Bernard 2012; 38. Dunn et al. 2012; 39. Brockelman 2011; 40. Sujana et al. 2012; 41. Asensio et al. 2007; 42. Thorpe et al. 2007; 43.Tenaza & Tilson 1985.

THE ROLE OF LIANAS IN THE DIET OF PRIMATES

Overall, we found that lianas represent an important food resource for primates, as: (1) many different primate species use a large number of liana species; (2) lianas can account for a high proportion of total diet; (3) lianas provide different food items, including fruits, leaves and flowers; and (4) lianas serve as fallback foods for primates, i.e., abundant foods of relatively low quality that are used during periods of low overall food availability (Marshall et al. 2009). We divide our discussion of lianas as a food source for primates below into these four different areas.

Lianas used as food sources by primates

Nineteen papers reported that 16 primate species in 10 different countries used a total of 150 liana species from 47 families as food sources (Table 27.2). The liana families with the highest number of consumed species were Leguminosae (18 species), Vitaceae (13), Bignoniaceae (10), Apocynaceae (10), and Celastraceae (9), together representing 40% of the consumed liana species. Primates used a mean (±SD) of 9.8 ± 8.1 liana species per study (Table 27.2), but some primates used a larger number of liana species. For example, kipunjis (*Rungwecebus kipunji*) in Tanzania (Davenport et al. 2010) and mantled howler monkeys (*Alouatta palliata*) in Mexico (Asensio et al. 2007; Dunn et al. 2012) used 16 and 28 liana species, respectively. Some liana species were used by more than one primate species (Table 27.2). For example, fruits of the liana *Saba comorensis* (Apocynaceae) were eaten by Tana River mangabeys (*Cercocebus galeritus*) in Kenya (Wieczkowski, 2009) and chimpanzees (*Pan troglodytes*) in Tanzania (Itoh & Nishida 2007; Moscovice et al. 2007). Similarly, leaves and fruits of *Diclidanthera* sp. (Polygalaceae) and *Pereskia aculeata* (Bignoniaceae) were frequently consumed both by brown howler monkeys (*Alouatta guariba*) and muriquis (*Brachyteles arachnoides*) in Brazil (Martins 2006, 2009). Another example is the liana *Tinospora caffra* (Menispermaceae), with fruits eaten by chimpanzees and leaves eaten by kipunjis in Tanzania (Moscovice et al. 2007; Davenport et al. 2010).

Representation of lianas in the diet of primates

Six of the 19 papers (32%) we reviewed in Table 27.2 reported the proportional representation of lianas in the diet of primates. Proportional representation was described in terms of percentage of total feeding time (TFT) or total feeding frequency (TFF), depending whether the studies recorded feeding time or number of feeding episodes. In general, particular liana species represented a very small proportion of primate diets (usually <1% TFT); however, there were notable exceptions. For example, in the Mahale Mountains National Park, Tanzania, fruits of the liana *Saba comorensis* (Apocynaceae) accounted for 20% of the TFT of chimpanzees (Itoh & Nishida 2007), and in Los Tuxtlas, Mexico, mantled howler monkeys spent 9.3% of TFT on fruits of the liana *Vitis tiliifolia* (Vitaceae) (Asensio et al. 2007). The Cao vit gibbon (*Nomascus nasutus*) in China spent a large percentage of feeding records consuming fruits and leaves of *Tetrastigma pubinerve* (Vitaceae) (6.3% TFF) and *Trichosanthes kirilowii* (Cucurbitaceae) (3.1% TFF) (Fan et al. 2011). Similarly, white-handed gibbons (*Hylobates lar*) in the Khao Yai National Park, Thailand, consumed fruits of *Desmos chinensis* (Annonaceae) and *Eleagnus latifolia* (Eleagnaceae) for 4.6% and 2.4% of TFT, respectively (Kanwatanakid-Savini et al. 2009).

The reason for the observed differences in the consumption of particular liana species is not clear, as studies generally do not report the abundance or availability of lianas within the home ranges of primates. Nevertheless, the high consumption of *Vitis tiliifolia* by howler monkeys in Los Tuxtlas may be linked to its high abundance in the region (Arroyo-Rodríguez et al. 2009). In other cases, the consumption of lianas may be driven by the dietary preferences of primates, rather than availability. For example, Fan et al. (2011) observed that the consumption of *Tetrastigma pubinerve* and *Trichosanthes kirilowii* by Cao vit gibbons was higher than predicted by its availability. Similarly, in Mahale National Park, Tanzania, the liana *Saba comorensis* produced fruit at the same time as many large tree species, but Mahale chimpanzees preferred to consume these liana fruits over tree fruits (Itoh & Nishida 2007).

Table 27.2 Liana taxa used as food resources by 16 primate species in 10 countries throughout the world[a].

Family	Liana species	America			Africa					Asia								Ref.
		Brazil		Mexico	Ivory Coast	Kenya	Madagascar	Tanzania		Borneo		China			India	Thailand		
		Agu	Bar	Apa	Pan	Cga	Cma	Ptr	Rki	Hmu	Pru	Rbr	Tfr	Nna	Tjo	Hla	Nbe	
Actinidiaceae	*Actinidia arguta* (Siebold & Zucc.) Planch. ex Miq											X						36
	Actinidia callosa Lindl.											X						36
	Actinidia latifolia (Gardner & Champ.) Merr.													X				33
Annonaceae	*Artabotrys lanuginosus* Boerl.									X								30
	Desmos chinensis Lour.															X		19,29
	Monanthotaxis barteri (Baill.) Verdc.				X													28
	Uvaria angolensis Welw. ex Oliv.							X										14,16
	Uvaria cordata (Dunal) Alston															X		29
	Uvaria sp.							X										16
	Uvaria welwitschii (Hiern) Engl. & Diels							X										16
Apocynaceae	*Ancylobothrys amoena* Hua.				X													28
	Ceropegia sp.								X									27
	Cynanchum rungweense Bullock								X									27
	Forsteronia acouci (Aubl.) A.DC.			X														38,41
	Landolphia buchananii (Hallier f.) Stapf								X									27
	Landolphia owariensis P.Beauv.							X										14
	Melodinus suaveolens (Hance) Champ. ex Benth													X				33
	Saba comorensis (Bojer ex A.DC.) Pichon					X		X										14,16,21
	Saba senegalensis (A.DC.) Pichon				X													28
	Willughbeia sp.									X								30
Araliaceae	*Schefflera stolzii* Harms								X									27
Arecaceae	*Calamus* sp.									X								30
Aristolochiaceae	*Aristolochia ovalifolia* Duch.			X														38,41
Bignoniaceae	*Adenocalymma* sp.	X																22
	Amphilophium crucigerum (L.) L.G.Lohmann			X														38,41
	Callichlamys latifolia (Rich.) K. Schum.			X														38,41

(continued overleaf)

Table 27.2 *(continued).*

Family	Liana species	America			Africa					Asia								Ref.
		Brazil		Mexico	Ivory Coast	Kenya	Madagascar	Tanzania		Borneo		China			India	Thailand		
		Agu	Bar	Apa	Pan	Cga	Cma	Ptr	Rki	Hmu	Pru	Rbr	Tfr	Nna	Tjo	Hla	Nbe	
	Dolichandra unguis-cati (L.) L.G.Lohmann		X															22
	Fridericia samydoides (Cham.) L.G.Lohmann	X																22
	Fridericia schumanniana (Loes.) L.G.Lohmann			X														38,41
	Lundia obliqua Sond.		X															22
	Stizophyllum riparium (Kunth) Sandwith		X															22
	Tanaecium pyramidatum (Rich.) L.G.Lohmann			X														38,41
	Tanaecium selloi (Spreng.) L.G.Lohmann	X																22
Boraginaceae	*Tournefortia hirsutissima* L.			X														41
Cactaceae	*Pereskia aculeata* Mill.	X	X															13,22
Campanulaceae	*Canarina eminii* Asch. & Schweinf.								X									27
Capparaceae	*Capparis erythrocarpos* Isert							X										16
	Capparis tomentosa Lam.							X										16
	Ritchiea capparoides (Andrews) Britten				X													28
Caprifoliaceae	*Lonicera acuminata* Wall.											X						36
	Lonicera japonica Thunb.											X						36
Celastraceae	*Celastrus glaucophyllus* Rehder & E.H.Wilson											X						36
	Celastrus hypoleucus (Oliv.) Warb. ex Loes.											X						36
	Celastrus orbiculatus Thunb.													X				33
	Celastrus stylosus Wall.											X						36
	Salacia chinensis L.															X		19
	Salacia erecta (G.Don) Walp							X							X			16,40
	Salacia leptoclada Tul.							X										16
	Salacia oblonga Wall. ex Wight & Arn.														X			40
	Sarcostigma kleinii Wight & Arn. Odalvalli														X			40
Combretaceae	*Combretum* sp.				X													28
Composite	*Mikania aromatica* Oerst.			X														38,41
Connaraceae	*Rourea minor* (Gaertn.) Alston									X								30

Convolvulaceae	*Erycibe elliptilimba* Merr. & Chun										X		19
	Erycibe maingayi C.B. Clarke				X								30
	Ipomoea batatas (L.) Poir.	X											38,41
	Ipomoea batatoides Choisy	X											38,41
	Ipomoea philomega (Vell.) House	X											38,41
Elaeagnaceae	*Elaeagnus conferta* Roxb.										X		29
Eleagnaceae	*Elaeagnus latifolia* L.										X		19
Euphorbiaceae	*Mallotus* sp.								X				33
	Briedelia stipularis (L.) Blume Nendravalli									X			40
	Croton caudatus Geiseler Panjipoo									X			40
Gnetaceae	*Gnetum montanum* Markgr.										X		19,29
	Gnetum sp.				X								30
Icacinaceae	*Iodes vitiginea* (Hance) Hemsl.								X				33
	Pyrenacantha sylvestris S.Moore		X										16
Lardizabalaceae	*Akebia trifoliata* (Thunb.) Koidz.						X						33
	Holboellia grandiflora Réaub.						X						33
Leguminosae	*Acacia caesia* (L.) Willd. Eangavalli									X			40
	Acacia pennata (L.) Willd.										X		29
	Acacia sinuata (Lour.) Merr. Cheevakka									X			40
	Bauhinia rubro-villosa K.Larsen & S.S.Larsen								X				33
	Bauhinia sp.											X	23
	Bauhinia phoenicea Wight & Arn.									X			40
	Dalbergia lactea Vatke			X									27
	Derris brevipes (Benth.) Baker in Hook. f.									X			40
	Caesalpinia bonduc (L.) Roxb. Kazhanjikuru									X			40
	Machaerium falciforme Rudd	X											38
	Entada redeye Spreng. Kakkumvalli									X			40
	Millettia nitida Benth.								X				33
	Millettia reticulata Benth.								X				33
	Millettia speciosa Champ.							X					31
	Rhynchosia minima (L.) DC.	X											38,41
	Spatholobus macropterus Miq.					X							37

(*continued overleaf*)

Table 27.2 *(continued)*.

Family	Liana species	America			Africa					Asia								Ref.
		Brazil		Mexico	Ivory Coast	Kenya	Madagascar	Tanzania		Borneo		China			India	Thailand		
		Agu	Bar	Apa	Pan	Cga	Cma	Ptr	Rki	Hmu	Pru	Rbr	Tfr	Nna	Tjo	Hla	Nbe	
	Spatholobus purpureus Benth.														X			40
	Strongylodon craveniae R.Baron & Baker						X											11
Loganiaceae	morphospecies 1				X													28
	Strychnos colubrina Blume									X								30
	Strychnos lucens Baker							X										16
Loranthaceae	*Dendropthoe* sp.															X		29
Malpighiaceae	*Hiraea fagifolia* (DC.) A.Juss.			X														38,41
	Mascagnia vacciniifolia Nied.			X														41
Malvaceae	*Grewia flavescens* Juss.							X										16
	Grewia hirsuta Vahl															X		29
Menispermaceae	*Abuta panamensis* (Standl.) Krukoff & Barneby			X														38,41
	Cissampelos pareira L.			X														38,41
	Diploclisia glaucescens (Blume) Diels													X				33
	Tiliacora funifera (Miers) Oliv.								X									27
	Tinospora caffra (Miers) Troupin							X	X									16,27
	Triclisia sp.								X									27
Moraceae	*Maclura cochinchinensis* (Lour.) Corner													X				33
	Maclura fruticosa (Roxb.) Corner															X		29
Myrsinaceae	*Embelia ribes* Burm. f.Manivalli														X			40
	Embelia tsjeriam-cottam (Roem. & Schult.)														X			40
Nyctaginaceae	*Pisonia aculeata* L.			X														38,41
Oleaceae	*Jasminum abyssinicum* Hochst. ex DC.								X									27
Olacaceae	*Erythropalum scandens* Blume														X			40
Opiliaceae	*Opilia amentacea* Roxb.				X													28
Passifloraceae	*Adenia stolzii* Harms								X									27
	Passiflora edulis Sims								X									27
Polygalaceae	*Diclidanthera* sp.	X	X															13,22
Primulaceae	*Embelia schimperi* Vatke								X									27
	Embelia subcoriacea (C.B.Clarke) Mez													X				33

Rahmnaceae	*Gouania microcarpa* DC. Pattamparathy									X		40
	Gouania lupuloides (L.) Urb.		X									38,41
	Sageretia rugosa Hance							X				31
	Ventilago maderaspatana Gaertn.									X		40
	Ziziphus suluensis Merr.*						X					30
	Ziziphus horsfieldii Miq.*						X					30
	Ziziphus oenopolia (L.) Mill. Mullanpazham									X		40
	Ziziphus rugosa Lam. Thodalipazham									X		40
Ranunculaceae	*Clematis uncinata* Champ. ex Benth.								X			33
Rubiaceae	*Keetia venosa* (Oliv.) Bridson				X							16
	Thunbergia mysorensis (Wight) T. And.									X		40
	Rutidea orientalis Bridson					X						27
Rutaceae	*Toddalia asiatica* (L.) Lam.					X						27,40
Sapindaceae	*Deinbollia fulvotomentella* Baker f.				X							16
	Paullinia clavigera Schltdl.		X									38,41
	Paullinia costata Schltdl. & Cham.		X									38,41
	Paullinia sp.		X									38,41
	Serjania goniocarpa Radlk.		X									38,41
	Serjania mexicana (L.) Willd.		X									38,41
	Serjania sp.	X										22
Schisandraceae	*Kadsura heteroclita* (Roxb.) Craib								X			33
Solanaceae	*Lycianthes purpusii* (Brandegee) Bitter		X									38,41
Urticaceae	*Urera hypselodendron* Wedd*					X						27
Vitaceae	*Cissus gossypiifolia* Standl.		X									38,41
	Cissus populnea Guill. & Perr			X								28
	Cissus quadrangularis L.				X							16
	Cissus sp.			X								28
	Cissus verticillata (L.) Nicolson & C.E.Jarvis		X									38,41
	Tetrastigma delavayi Gagnep.								X			33
	Tetrastigma kwangsiense C.L. Li								X			33
	Tetrastigma laoticum Gagnep.*										X	29

(continued overleaf)

Table 27.2 (*continued*).

Family	Liana species	America			Africa					Asia								Ref.
		Brazil		Mexico	Ivory Coast	Kenya	Madagascar	Tanzania		Borneo		China			India	Thailand		
		Agu	Bar	Apa	Pan	Cga	Cma	Ptr	Rki	Hmu	Pru	Rbr	Tfr	Nna	Tjo	Hla	Nbe	
	Tetrastigma planicaule (Hook. f.) Gagnep.													X				33
	Tetrastigma pubinerve Merr. & Chun													X				33
	Tetrastigma trifoliolatum Merr.*									X								30
	Vitis tiliifolia Humb. & Bonpl. ex Schult.			X														38,41
	Vitis wilsoniae H.J. Veitch											X						36
	Total number of lianas species per primate	6	5	28	9	1	1	16	16	8	1	12	2	17	20	13	1	

Note: [a]Plant nomenclature was corrected according to The Plant List (2010) update database (http://www.theplantlist.org). Species with asterisk (*) represent unresolved names.

Primates species = Agu, *Alouatta guariba*; Apa, *Alouatta palliata*; Bar, *Brachyteles arachnoides*; Cga, *Cercocebus galeritus*; Cma, *Cheirogaleus major*; Hla, *Hylobates lar*; Hmu, *Hylobates muelleri* × *agilis/albibarbis*; Nna, *Nomascus nasutus*; Nbe, *Nycticebus bengalensis*; Ptr, *Pan troglodytes*; Pan, *Papio anubis*; Pru, *Presbytis rubicunda*; Rbr, *Rhinopithecus brelichi*; Rki, *Rungwecebus kipunji*; Tfr, *Trachypithecus francoisi*.; Tjo, *Trachypithecus johnii*.

Reference numbers (Refs.) correspond to those in Table 27.1.

Although individual liana species generally represented a small percentage of the TFT of primates, the contribution of lianas was often important when considered collectively. For example, mantled howler monkeys used 26 liana species in Los Tuxtlas, Mexico, which represented 29% of TFT (Asensio et al. 2007). At the border between China and Vietnam, the eastern black crested gibbon (*Nasutus nomascus*) consumed fruits, leaves and buds from 25 liana species, which represented 19% of TFF (Fan et al. 2011). Finally, in Thailand, the consumption of five species of lianas by white-handed gibbons accounted for 14% of TFT (Kanwatanakid-Savini et al. 2009).

Plant items eaten from lianas

Thirteen of the 43 studies (30%) we reviewed provided information about the liana items consumed by primates. In these studies, lianas were most commonly used as a source of fruit (79 out of 105 species, 75%), followed by leaves (37%), and other plant items, such as flowers, nectar, piths, rhizomes, stalks, bark, and buds (22%, Fig. 27.1). The majority of liana species (72 out of 105 species, 69%) represented a single item, 21 species (20%) represented a source of two items, and 12 species (11%) contributed to three or more items.

Information regarding the percentage of time using each plant item was even scarcer. Dunn et al. (2012) showed that in Los Tuxtlas, Mexico, mantled howler monkeys' liana consumption was focused mainly on leaves (i.e., between 70% and 97% of total liana feeding time in four different monkeys' groups). The high proportion is consistent with other studies that highlight the importance of liana leaves in the diet of folivorous primates. For example, in the Danum Valley, Borneo, the young leaves of the liana *Spatholobus macropterus* (Leguminosae) were by far the most important food resource used by red leaf monkeys (*Presbytis rubicunda*), representing 27.9% of TFT (Hanya & Bernard 2012). However, liana fruit can also represent an important component of primate TFT. For example, all of the liana species (n = 5) used by white-handed gibbons in Thailand, were sources of fruits, and, together, accounted for 14% of TFT (Kanwatanakid-Savini et al. 2009). Furthermore, the majority of fruit consumed by brown howler monkeys and muriquis in Brazil comes from just two liana species (*Pereskia aculeata* and *Diclidanthera* sp.) (Martins 2009). Similarly, fruit consumption by chimpanzees in Tanzania (Itoh & Nishida 2007; Moscovice et al. 2007) and langurs in China (Hu 2011) and India (Sujana et al. 2012) is based largely on lianas.

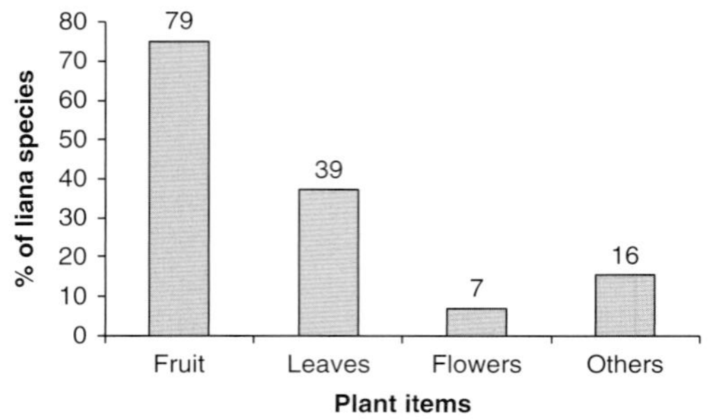

Fig. 27.1 Distribution of liana items used as food sources by 12 primate species in 9 neotropical and paleotropical countries. The numbers above the columns indicate the total number of species per item (n = 105 species). The "Other" category includes bark, buds, nectar, pith, rhizome, and stalk. (Sources: Asensio et al. 2011, Davenport et al. 2010, Fan et al. 2011, Hanya & Bernard 2012, Itoh & Nishida 2007, Kanwatanakid-Savini et al. 2009, Kunz & Linsenmair 2010, Lehman et al. 2006, Martins 2009, McConkey 2009, Moscovice et al. 2007, Wieczkowski 2009, Xiang et al. 2012. Reproduced with permission.)

All primates need to procure appropriate quantities of certain macro-nutrients and micro-nutrients from their habitat, while avoiding/regulating plant secondary metabolites. These requirements vary among different primate species, depending on factors such as body size, metabolic requirements, lifestyle, and digestive system (Parra 1978; Milton 1993). Thus, different primates may prioritize different nutritional parameters when choosing the types and quantities of foods they consume (reviewed by Felton et al. 2009a). It is widely recognized that fruits are the most important sources of carbohydrates (and energy) for primates (Felton et al. 2009a, 2009c). In general, leaves have a much lower energetic content than fruits, and a higher concentration of secondary compounds (Milton 1980; Felton et al. 2009a). However, recent research has shown that a stable daily protein intake may be more important to some primates than energy maximization (e.g., Felton et al. 2009b). Interestingly, evidence indicates that liana leaves are poorer in structural and chemical defenses than tree leaves, making protein more readily available, as well as being richer in many minerals (i.e., K, Ca, Mg, Zn, Mn, B, and Fe) (Asner & Martin 2012; Asner & Martin, Chapter 21 in this volume). Other items, such as flowers (e.g., Castellanos & Chanin 1996; Silver et al. 2000) and bark (e.g., Chaves et al. 2011), may also provide important minerals to primates.

Lianas as fallback foods

Lianas often reproduce during periods that are unfavorable to trees, and have a much higher growth rate during the dry season than trees (Schnitzer 2005). Through such competitive advantages, lianas can provide important food resources for primates where and when food availability is scarce (Onderdonk & Chapman 2000; Preece 2006; Moscovice et al. 2007; Marshall et al. 2009; Bartlett 2009; Vogel et al. 2009). In the Danum Valley, Borneo, during the period of fruit scarcity, red leaf monkeys used a large number of plant species as fallback foods, most of which were lianas (Hanya & Bernard 2012). The consumption of young leaves of *Spatholobus macropterus* (Leguminosae) was negatively related to fruit availability, indicating that this liana species is an important fallback food for this primate species. Similarly, in Los Tuxtlas, Mexico, mantled howler monkeys consumed liana leaves more frequently during months of low fruit consumption, which also corresponded with months of low fruit availability in the region (Dunn et al. 2012). Equally, on Rubondo Island, Tanzania, chimpanzees increased their reliance on the fruits of the liana *Saba comorensis* during the long dry season (Moscovice et al. 2007, 2010). Also, in Tuanan, Indonesia, gibbons (*Hylobates albibarbis*) increased the amount of liana food items in their diet as fruit abundance decreased (Vogel et al. 2009), indicating that lianas are also important fallback foods for this primate species (see also Marshall et al. 2009). Finally, the Goeldi's marmoset (*Callimico goeldii*) in Bolivia, and the Coimbra Filho's titi (*Callicebus coimbrai*) in Brazil consumed different liana items as fallback food during periods of resource scarcity (Porter et al. 2009; Souza-Alves et al. 2011).

Unlike large trees, the diversity and abundance of lianas are higher in fragmented and degraded tropical forests (Putz 1984; Laurance et al. 1998; Schnitzer & Bongers 2002; Schnitzer & Carson 2010; Letcher, Chapter 10 in this volume; Schnitzer et al., Chapter 7 in this volume). Accordingly, lianas can represent an important food resource for primates inhabiting transformed habitats. For example, in Mexico, groups of howler monkeys living in smaller fragments spent more time feeding from a greater number of liana species than their counterparts in bigger forest fragments (Dunn et al. 2012).

THE IMPACT OF LIANAS ON PRIMATE BEHAVIOR

Studies of the impact of lianas on primate behavior indicate that lianas can affect modes of locomotion, foraging postures, ranging behavior, and grouping patterns (Table 27.1). Lianas are important structural elements during locomotion, as the systems of bridges that lianas form among trees facilitate the movement of arboreal primates (see also Lambert & Halsey, Chapter 26 in this volume; Michel et al., Chapter 25 in this volume; Yanoviak, Chapter 24 in this volume). For example, orangutans (*Pongo abelii*) in the Gunung Leuser National Park, Sumatra, Indonesia, frequently use lianas for support during locomotion, which reduces the energetic cost of gap crossing (Thorpe et al. 2007). One example given by the authors shows how lianas can act like a catapult, recoiling and adding power to a swinging motion, allowing individuals to enter the canopy of an adjacent tree at a higher level than their current position. Similarly, in Mexico, where we have studied howler monkeys and spider monkeys for several years, it is very common to see these primate species using lianas as supporting structures during locomotion, resting, and feeding (Figs. 27.2A–F). Interestingly, in the Lacandona rain forest, southeast Mexico, wood from decaying trees (*Licania platypus*) represents an important source of sodium and/or calcium for spider monkeys, particularly for females (Chaves et al. 2011), and we observed that in most cases the consumption of this food source was aided by the presence of lianas near the trunk, which were used as support during feeding bouts (Fig. 27.2F).

There is ample evidence that callitrichids (family Callitrichidae) use lianas both for locomotion and to gain access to food sources. For example, tamarins use lianas for foraging in holes, consuming plant gums, searching for prey in rotting bark, palm fronds and bromeliad whorls, gleaning food from leaves, selecting fruits, and snatching insects from the ground (Rylands et al. 1989; Garber 1993). Likewise, some of the foraging postures used by pygmy marmosets (*Cebuella pygmaea*) in Yasuni National Park, Ecuador, occur while using horizontal lianas to gain access to food resources (Youlatos 1999). In fact, Youlatos (1999)

Fig. 27.2 Black howler monkeys (*Alouatta pigra*, left side) and Geoffroyi's spider monkeys (*Ateles geoffroyi*, right side) using lianas as supporting structures during locomotion (A, B), resting (C, D) and feeding (E, F) in the Lacandona rain forest, Mexico. (Source: A. González-Zamora. Reproduced with permission.) *See plate section for color representation of this figure.*

suggests that the locomotor anatomy of pygmy marmosets may be the result of feeding and traveling in liana forests, which favor claw clinging and climbing instead of quadrupedal postures or locomotion. Similarly, also in Yasuni National Park, woolly monkeys (*Lagothrix lagotricha*) and spider monkeys (*Ateles belzebuth*) both used lianas more frequently while moving between trees than while moving within trees, which was associated with "bridging and hoisting" and "drop and leap" postural modes (Cant et al. 2001).

With regard to ranging behavior, evidence from forest fragments in Los Tuxtlas, Mexico, indicates that howler monkeys are forced to visit a significantly greater number of food patches when feeding from lianas than when feeding from trees, which results in increased travel time (Dunn et al. 2012). This phenomenon is thought to be related to: (1) lianas being used mainly as a source of leaves (which contain higher concentrations of secondary compounds than fruits; Freeland & Janzen 1974), and (2) lianas representing smaller food patches than trees. Thus, though lianas may represent important fallback foods for howler monkeys, liana consumption may also be related to increased foraging effort, which may increase stress and have important consequences for their fitness (Dunn et al. 2013).

Finally, lianas have also been reported to affect primate grouping patterns. For example, Asensio et al. (2011) showed that the spatiotemporal distribution of a few preferred food species (>10% of feeding time), that included the lianas *Acacia pennata* (Leguminosae), *Eleagnus conferta* (Eleagnaceae), *Tetrastigma laotican* (Vitaceae), and *Toddalia asiatica* (Rutaceae), determined lar gibbons' (*Hylobates lar*) ranging patterns. Similarly, Itoh and Nishida (2007) reported that the size of chimpanzee subgroups in the Mahale Mountains National Park, Tanzania, depended on the spatial distribution of the liana *Saba comorensis*.

LIANAS AS DETERMINANTS OF HABITAT USE BY PRIMATES

Overall, we found 18 studies showing the importance of lianas as potential determinants of habitat use by primates (Table 27.1). These studies suggest that lianas predominantly affect niche partitioning, sleeping tree selection, and home range use. For example, sympatric Japanese macaques (*Macaca fuscata*) and sika deer (*Cervus nippondeer*) compete for fibrous plants foods on Kinkazan Island, Japan. However, the deer avoid forested areas where the liana *Berchemia racemosa* (Rhamnaceae) is abundant, as these areas are also abundant in *Illicium anisatur* (Schisandraceae), a tree which is toxic to deer. The home range of macaques in summer corresponds to the location of this liana resource, which is an important staple food during this season, reducing direct food competition with deer through niche segregation (Tsuji & Takatsuki 2004).

In Madagascar, liana density was the only variable positively correlated with brown lemur (*Eulemur fulvus*) abundance (Johnson & Overdorff 1999), probably because in this region high densities of lianas are associated with high plant species richness (Zurring et al. 1993). The distribution of greater dwarf lemurs (*Cheirogaleus major*) can also be influenced by edge-related variations in the abundance and spatial arrangement of the liana *Strongylodon craveniae* (Leguminosae), which provides an important source of nectar in the diet of this primate (Lehman et al. 2006). In eastern Thailand, Pliosungnoen et al. (2010) evaluated the density and microhabitat use of Bengal slow lorises (*Nycticebus bengalensis*) in primary forest and non-native forest plantations and found that loris densities in older plantations were nearly identical to primary forest. These authors argue that this is related (among other factors) to older plantations containing higher densities of *Bauhinia* spp. lianas, a food source commonly eaten by lorises.

In other cases, lianas may affect the distribution of primates through their positive effect on the abundance of preferred food sources. For example, in Argentina, fig tree (*Ficus* spp.) establishment and growth are favored in liana-infested forests, and this has been related to the presence of owl monkeys (*Aotus azarai*) (van der Heide et al. 2012). In most cases, however, lianas have a significant negative effect on tree growth and reproduction (Stevens 1987; Kainer et al. 2006; Wright et al. 2005; van der Heijden et al., Chapter 13 in this volume).

In fragmented forests, small primates such as tamarins (*Saguinus*) often forage in canopies interconnected by lateral branches and from lianas that drape over trees and extend into forest gaps (Garber 1993; Peres 1996; Vidal & Cintra 2006). The presence of lianas seems to be important for callitrichids because they often used areas with clumps of lianas as roosting sites (Day & Elwood 1999, Vidal & Cintra 2006), presumably for safety (Day & Elwood 1999). Accordingly, *Saguinus* in French Guiana (Youlatos 2004) and Panama (Madden et al. 2010), and *Cebuella* in Ecuadorian Amazonia (Youlatos 2004) are frequently associated with liana forests.

The presence and abundance of lianas can also affect the selection of sleeping trees or roosting sites for other primates. For example, in the Khao Ang Rue Nai Wildlife Sanctuary, Thailand, pileated gibbons (*Hylobates pileatus*) preferentially selected tall (mean = 38.5 m), emergent trees without lianas as sleeping trees (Phoonjampa et al. 2010). In Siberut Island, Indonesia, Kloss's gibbons (*H. klossi*) also preferred sleeping trees without lianas (Tenaza & Tilson 1985). In both cases, the preference of sleeping trees without lianas can represent a defense strategy, as predators and human hunters can gain access to the trees via lianas (see Michel et al., Chapter 25 in this volume).

On the other hand, Radespiel et al. (2003) reported niche segregation between golden-brown mouse lemurs (*Microcebus ravelobensis*) and gray mouse lemurs (*M. murinus*) in Madagascar, as *M. ravelobensis* used lianas as sleeping sites more frequently than *M. murinus*, avoiding direct competition for this resource. In eastern Amazonia, Bobadilla and Ferrari (2000) observed that compared to bearded saki monkeys (*Chiropotes satanas utahicki*), brown capuchins (*Sapajus apella*) showed a significant preference for liana forest and flooded forests over primary or secondary terra firme forests, suggesting that lianas can contribute to niche partitioning in this primate assemblage. However, these authors did not discuss the possible causes of the observed difference in habitat preference.

CONCLUSION: CONSERVATION IMPLICATIONS FOR PRIMATES

The data and examples described here probably represent an underestimation of the full role that lianas may play in primate biology, ecology, behavior, and conservation, as there may be studies which our Scopus search failed to detect. Overall we found that lianas are highly valuable for primates, and that primates use lianas for foraging, shelter, and locomotion. Different primate species in both neotropical and paleotropical forests use fruits, leaves and other liana items from a large number of liana species as food sources.

Lianas are often available during times of overall food scarcity, acting as fallback foods (e.g., Dunn et al. 2012), which can influence primate carrying capacity within forest habitats (Marshall et al. 2009). Another notable aspect of lianas is that they increase canopy-connectivity, acting as important pathways for primate locomotion, feeding, and resting, as well as increasing the complexity of forests, creating microhabitats suitable for different animal species (Kilgore et al. 2010; Lambert & Halsey, Chapter 26 in this volume; Yanoviak, Chapter 24 in this volume). Therefore, lianas may play an important role in several different aspects of primates' lives.

Evidence suggests that lianas are proliferating in many tropical forests around the world, presumably due to climate change, increasing atmospheric CO_2 concentrations, and anthropogenic disturbances such as deforestation and forest fragmentation (Schnitzer & Bongers 2011; Schnitzer, Chapter 30 in this volume). Such changes could have significant consequences on the survival, distribution, habitat use, and behavior of primates. The response of primates to changing liana abundance and biomass is a potentially important avenue for future research.

In general, we found that lianas have a positive effect on primates, and hence, the proliferation of lianas within the forest could ultimately be advantageous for primates and contribute to their survival and conservation. For example, on Rubondo Island (Lake Victoria, Tanzania), the high density of fleshy fruit-producing lianas was probably the most important factor in the long-term success of captive chimpanzees that were released on the island and adapted to the novel habitat (Moscovice et al. 2010). Similarly, several studies of fragmented landscapes suggest that lianas may provide key resources to primates, allowing them to tolerate the reduction in food availability associated with habitat perturbation (e.g., Souza-Alves et al. 2011; Dunn et al. 2012).

In this sense, it has been suggested that the high light-tolerance of lianas, along with their fast growth rate, make them ideal species for primates in perturbed or human-managed forested habitats (Pliosungnoen et al. 2010), as well as for reforestation projects directed towards increasing food availability and connectivity in perturbed forests (Cristóbal-Azkarate & Arroyo-Rodríguez 2007; Asensio et al. 2007). Nevertheless, evidence indicates that lianas can enhance tree mortality and suppress tree growth, and reduce forest biomass accumulation (Laurance et al. 2001; Ingwell et al. 2010; Schnitzer & Carson 2010; Schnitzer et al. 2014; van der Heijden et al., Chapter 13 in this volume), and, hence, the proliferation of lianas could ultimately reduce the availability of trees as food sources. Likewise, small food patches are likely to be depleted more quickly, forcing primates to move more frequently among patches, increasing foraging effort, with potential negative effects on primates (e.g., Suarez 2006; Dunn et al. 2012). Therefore, further studies are needed to provide a more comprehensive understanding of the impact that lianas may have on primate fitness and long-term survival.

ACKNOWLEDGMENTS

The Centro de Investigaciones en Ecosistemas (Universidad Nacional Autónoma de México) provided logistical support to VAR. The Instituto de Investigaciones Biológicas (Universidad Veracruzana) and Conacyt provided support to AGZ, and the Centro de Investigaciones Tropicales (Universidad Veracruzana), the University of Barcelona, the University of Cambridge, and the Fundaccion BBVA provided support to JD. We thank Stefan Schnitzer, the members of his lab, and an anonymous reviewer for valuable criticism and suggestions that improved this manuscript.

REFERENCES

Aide, T.M. & Zimmerman, J.K. (1990) Patterns of insect herbivory, growth, and survivorship in juveniles of a neotropical liana. *Ecology*, **8**, 1412–1421.

Arroyo-Rodríguez, V., Dunn, J., Mandujano, S. & Benítez-Malvido, J. (2009) Angiosperms, Los Tuxtlas Biosphere Reserve, Veracruz, Mexico. *Check List*, **5**, 787–799.

Arroyo-Rodríguez, V. & Mandujano, S. (2009) Conceptualization and measurement of rainforest fragmentation from the primates' perspective. *International Journal of Primatology*, **30**, 497–514.

Asensio, N., Brockelman, W.Y., Malaivijitnond, S. & Reichard, U. (2011) Gibbon travel paths are goal oriented. *Animal Cognition*, **14**, 395–405.

Asensio, N., Cristóbal-Azkarate, J., Dias, P.A.D. et al. (2007) Foraging habits of *Alouatta palliata mexicana* in three forest fragments. *Folia Primatologica*, **78**, 141–153.

Asner, G.P. & Martin, R.E. (2012) Contrasting leaf chemical traits in tropical lianas and trees: implications for future forest composition. *Ecology Letters*, **15**, 1001–1007.

Bartlett, T.Q. (2009) *The Gibbons of Khao Yai: Seasonal Variation in Behavior and Ecology*. Pearson Prentice Hall, Upper Saddle River, NJ.

Bobadilla, U.L. & Ferrari, S.F. (2000) Habitat use by *Chiropotes satanas utahicki* and syntopic platyrrhines in eastern Amazonia. *American Journal of Primatology*, **50**, 215–224.

Brockelman, W.Y. (2011) Rainfall patterns and unpredictable fruit production in seasonally dry evergreen forest, and their effects on gibbons, in *The Unique Ecology and Conservation of Tropical Dry Forests in Asia* (eds W.J. McShea, S. Davies & N. Phumpakphan), Smithsonian Institution Press, Washington, DC, pp. 195–216.

Cant, J.G.H., Youlatos, D. & Rose, M.D. (2001) Locomotor behavior of *Lagothrix lagotricha* and *Ateles belzebuth* in Yasuni National Park, Ecuador: general patterns and non suspensory modes. *Journal of Human Evolution*, **41**, 141–166.

Castellanos, H.G. & Chanin, P. (1996) Seasonal differences in food choice and patch preference of long-haired spider monkeys (*Ateles belzebuth*), in *Adaptive Radiations of Neotropical Primates* (eds M.A. Norconk, A.L. Rosenberger & P.A. Garber), Plenum Press, New York, pp. 451–466.

Chaves, O.M., Stoner, K.E., Ángeles-Campos, S. & Arroyo-Rodríguez, V. (2011) Wood consumption by Geoffroyi's spider monkeys and its role in mineral supplementation. *PLoS ONE*, **6**, e25070.

Cristóbal-Azkarate, J. & Arroyo-Rodríguez, V. (2007) Diet and activity patterns of howler monkeys (*Alouatta palliata*) in Los Tuxtlas, Mexico: effects of habitat fragmentation and implications for conservation. *American Journal of Primatology*, **69**, 1–17.

Dammhahn, M. & Kappeler, P.M. (2008) Small-scale coexistence of two mouse lemur species (*Microcebus berthae* and *M. murinus*) within a homogeneous competitive environment. *Oecologia*, **157**, 473–483.

Davenport, T.R.B., de Luca, D.W., Bracebridge, C.E. et al. (2010) Diet and feeding patterns in the kipunji (*Rungwecebus kipunji*) in Tanzania's Southern Highlands: A first analysis. *Primates*, **51**, 213–220.

Davies, A.G., Oates, J.F. & Dasilva, G.L. (1999) Patterns of frugivory in three West African colobine monkeys. *International Journal of Primatology*, **20**, 327–357.

Day, R.T. & Elwood, R.W. (1999) Sleeping site selection by the golden-handed tamarin *Saguinus midas midas*: the role of predation risk, proximity to feeding sites, and territorial defence. *Ethology*, **105**, 1035–1051.

Dunn, J.C., Asensio, N., Arroyo-Rodríguez, V. et al. (2012) The ranging costs of a fallback food: liana consumption supplements diet but increases foraging effort in howler monkeys. *Biotropica*, **44**, 704–714.

Dunn, J.C., Cristóbal-Azkarate, J., Chavira, R. & Veà, J. (2013) Travel time predicts fecal glucocorticoid levels in free-ranging howler monkeys. *International Journal of Primatology*, **34**, 246–259.

Emmons, L.H. & Gentry, A.H. (1983) Tropical forest structure and the distribution of gliding and prehensile-tailed vertebrates. *The American Naturalist*, **121**, 513–524.

Estrada, A., Raboy, B.E. & Oliveira, L.C. (2012) Agroecosystems and primate conservation in the tropics: a review. *American Journal of Primatology*, **74**, 696–711.

Fan, P.F., Fei, H.L., Scott, M.B. et al. (2011) Habitat and food choice of the critically endangered Cao vit gibbon (*Nomascus nasutus*) in China: Implications for conservation. *Biological Conservation*, **144**, 2247–2254.

Felton, A.M., Felton, M., Lindenmayer, D.B. & Foley, W.J. (2009a) Nutritional goals of wild primates. *Functional Ecology*, **23**, 70–78.

Felton, A.M., Felton, A., Raubenheimer, D. et al. (2009b) Protein content of diets dictates the daily energy intake of a free-ranging primate. *Behavioural Ecology*, **20**, 685–690.

Felton, A.M., Felton, A., Wood, J.T. et al. (2009c) Nutritional ecology of *Ateles chamek* in lowland Bolivia: how macronutrient balancing influences food choices. *International Journal of Primatology*, **30**, 675–696.

Freeland, W.J. & Janzen, D.H. (1974) Strategies of herbivory in mammals: the role of plant secondary compounds. *The American Naturalist*, **108**, 269–189.

Garber, P.A. (1993) Feeding, ecology, and behaviour of the genus *Saguinus*, in *Marmosets and Tamarins: Systematics, Behavior, and Ecology* (ed A.B. Rylands), Oxford University Press, Oxford, pp. 273–295.

Garber, P.A. & Pruetz, J.D. (1995) Positional behavior in moustached tamarin monkeys: effects of habitat on locomotor variability and locomotor stability. *Journal of Human Evolution*, **28**, 411–426.

Gentry, A.H. (1991) The distribution and evolution of climbing plants, in *The Biology of Vines* (eds F. Putz & H.A. Mooney), Cambridge University Press, Cambridge, pp. 3–52.

Hanya, G. & Bernard, H. (2012) Fallback foods of red leaf monkeys (*Presbytis rubicunda*) in Danum Valley, Borneo. *International Journal of Primatology*, **33**, 322–337.

Hardus, M.E., Lameira, A.R., Menken, S.B.J. & Wich, S.A. (2012) Effects of logging on orangutan behavior. *Biological Conservation*, **146**, 177–187.

Hegarty, E.E. & Caballé, G. (1991) Distribution and abundance of vines in forest communities, in *The Biology of Vines* (eds F.E. Putz & H.A. Mooney), Cambridge University Press, Cambridge, pp. 313–335.

Hodgkison, R., Balding, S.T., Zubaid, A. & Kunz, T.H. (2003) Fruit bats (Chiroptera: Pteropodidae) as seed dispersers and pollinators in a lowland Malaysian rain forest. *Biotropica*, **35**, 491–502.

Hu, G. (2011) Dietary breadth and resource use of François' langur in a seasonal and disturbed habitat. *American Journal of Primatology*, **73**, 1176–1187.

Ingwell, L.L., Wright, S.J., Becklund, K.K. et al. (2010) The impact of lianas on 10 years of tree growth and mortality on Barro Colorado Island, Panama. *Journal of Ecology*, **98**, 879–887.

Itoh, N. & Nishida, T. (2007) Chimpanzee grouping patterns and food availability in Mahale Mountains National Park, Tanzania. *Primates*, **48**, 87–96.

Johnson, S.E. & Overdorff, D.J. (1999) Census of brown lemurs (*Eulemur fulvus*) in Southeastern Madagascar: methods-testing and conservation implications. *American Journal of Primatology*, **47**, 51–60.

Kainer, K.A., Wadt, L.H.O., Gomes-Silva, D.A.P. & Capanu, M. (2006) Brazil nut-liana relations in the Western Brazilian Amazon. *Journal of Tropical Ecology*, **22**, 147–154.

Kanwatanakid-Savini, C., Poonswad, P. & Savini, T. (2009) An assessment of food overlap between gibbons and hornbills. *Raffle's Bulletin of Zoology*, **57**, 189–198.

Kilgore, A., Lambert, T.D. & Adler, G.H. (2010) Lianas influence fruit and seed use by rodents in a tropical forest. *Tropical Ecology*, **51**, 265–271.

Kunz, B.K. & Linsenmair, K.E. (2010) Fruit traits in baboon diet: a comparison with plant species characteristics in West Africa. *Biotropica*, **42**, 363–371.

Laurance, W.F., Ferreira, L.V., Rankin-de Merona, J.M. & Laurance, S.M. (1998) Rain forest fragmentation and the dynamics of Amazonian tree communities. *Ecology*, **79**, 2032–2040.

Laurance, W.F., Pérez-Salicrup, D., Delamónica, P. et al. (2001) Rain forest fragmentation and the structure of Amazonian liana communities. *Ecology*, **82**, 105–116.

Lehman, S.M., Rajaonson, A. & Day, S. (2006) Edge effects on the density of *Cheirogaleus major*. *International Journal of Primatology*, **27**, 1569–1588.

Madden, D., Garber, P.A., Madden, S.L. & Snyder, C.A. (2010) Rain-forest canopy-connectivity and habitat selection by a small Neotropical primate, Geoffroy's tamarin (*Saguinus geoffroyi*). *Journal of Tropical Ecology*, **26**, 637–644.

Marshall, A.J., Boyko, C.J., Feilen, K.L. et al. (2009) Defining fallback foods and assessing their importance in primate ecology and evolution. *American Journal of Physical Anthropology*, **140**, 603–614.

Marshall, A.J., Salas, L.A., Stephens, S. et al. (2007) Use of limestone karst forests by Bornean orangutans (*Pongo pygmaeus morio*) in the Sangkulirang Peninsula, East Kalimantan, Indonesia. *American Journal of Primatology*, **69**, 212–219.

Martins, M.M. (2006) Comparative seed dispersal effectiveness of sympatric *Alouatta guariba* and *Brachyteles arachnoides* in southeastern Brazil. *Biotropica*, **38**, 57–63.

Martins, M.M. (2009) Lianas as a food resource for brown howlers (*Alouatta guariba*) and southern muriquis (*Brachyteles arachnoides*) in a forest fragment. *Animal Biodiversity and Conservation*, **32**, 51–58.

McConkey, K.R. (2009) The seed dispersal niche of gibbons in Bornean dipterocarp forests, in *The Gibbons: New Perspectives on Small Ape Socioecology and Population Biology* (eds D.J. Whittaker, S.M. Lappan & T. Geissmann), Springer, New York, pp. 189–210.

Milton, K. (1980) *The Foraging Strategy of Howler Monkeys: A Study in Primate Economics*, Columbia University Press, New York.

Milton, K. (1993) Diet and primate evolution. *Scientific American*, **Aug**, 86–93.

Moscovice, L.R., Issa, M.H., Petrzelkova, K.J. et al. (2007) Fruit availability, chimpanzee diet, and grouping patterns on Rubondo Island, Tanzania. *American Journal of Primatology*, **69**, 487–502.

Moscovice, L.R., Mbago, F., Snowdon, C.T. & Huffman, M.A. (2010) Ecological features and ranging patterns at a chimpanzee release site on Rubondo Island, Tanzania. *Biological Conservation*, **143**, 2711–2721.

Onderdonk, D.A. & Chapman, C.A. (2000) Coping with forest fragmentation: the primates of Kibale National Park, Uganda. *International Journal of Primatology*, **21**, 587–611.

Parra, R. (1978) Comparison of foregut and hindgut fermentation in herbivores, in *The Ecology of Arboreal Folivores* (ed G.G. Montgomery), Smithsonian Institution Press, Washington, DC, pp. 205–230.

Peres, C.A. (1996) Food patch structure and plant resource partitioning in interspecific associations of Amazonian tamarins. *International Journal of Primatology*, **17**, 695–723.

Phillips, O.L., Vásquez Martínez, R., Arroyo, L. et al. (2002) Increasing dominance of large lianas in Amazonian forests. *Nature*, **418**, 770–774.

Phoonjampa, R., Koenig, A., Borries, C. et al. (2010) Selection of sleeping trees in Pileated gibbons (*Hylobates pileatus*). *American Journal of Primatology*, **72**, 617–625.

Pliosungnoen, M., Gale, G. & Savini, T. (2010) Density and microhabitat use of Bengal slow Loris in primary forest and non-native plantation forest. *American Journal of Primatology*, **72**, 1108–1117.

Porter, L.M., Garber, P.A. & Nacimento, E. (2009) Exudates as a fallback food for *Callimico goeldii*. *American Journal of Primatology*, **71**, 120–129.

Preece, G.A. (2006) Factors influencing variation in the population densities of *Colobus guereza* within selectively logged forest at the Budongo Forest Reserve: the importance of lianas during a subsistence diet, in *Primates of Western Uganda* (eds N.E. Newton-Fisher, H. Notman, J.D. Paterson & V. Reynolds), Springer, New York, pp. 23–43.

Putz, F.E. (1984) How trees avoid and shed lianas. *Biotropica*, **16**, 19–23.

Radespiel, U., Ehresmann, P. & Zimmermann, E. (2003) Species-specific usage of sleeping sites in two sympatric mouse lemur species (*Microcebus murinus* and *M. ravelobensis*) in northwestern Madagascar. *American Journal of Primatology*, **59**, 139–151.

Russo, S.E. & Chapman, C.A. (2011) Primate seed dispersal: linking behavioral ecology with forest community structure, in *Primates in Perspective*, 2nd edn (eds C.J. Campbell, A.F. Fuentes, K.C. MacKinnon et al.), Oxford University Press, Oxford, pp. 523–534.

Rylands, A.B., Dacruz, O.M. & Ferrari, S.F. (1989) An association between marmosets and army ants in Brazil. *Journal of Tropical Ecology*, **5**, 113–116.

Schnitzer, S.A. (2005) A mechanistic explanation for global patterns of liana abundance and distribution. *The American Naturalist*, **166**, 262–276.

Schnitzer, S.A. & Bongers, F. (2002) The ecology of lianas and their role in forests. *Trends in Ecology and Evolution*, **17**, 223–230.

Schnitzer, S.A. & Bongers, F. (2011) Increasing liana abundance and biomass in tropical forests: emerging patterns and putative mechanisms. *Ecology Letters*, **14**, 397–406.

Schnitzer, S.A. & Carson, W.P. (2010) Lianas suppress tree regeneration and diversity in treefall gaps. *Ecology Letters*, **13**, 849–857.

Schnitzer, S.A., van der Heijden, G.M.F., Mascaro, J., & Carson, W.P. (2014) Lianas in gaps reduce carbon accumulation in a tropical forest. *Ecology* (in press, **http://dx.doi.org/10.1890/13-1718.1**).

Silver, S.C., Ostro, L.E.T., Yeager, C.P. & Dierenfeld, E.S. (2000) Phytochemical and mineral components of foods consumed by black howler monkeys (*Alouatta pigra*) at two sites in Belize. *Zoo Biology*, **19**, 95–109.

Snow, D.W. (1981) Tropical frugivorous birds and their food plants: a world survey. *Biotropica*, **13**, 1–4.

Souza-Alves, J.P., Fontes, I.P., Chagas, R.R. & Ferrari, S.F. (2011) Seasonal versatility in the feeding ecology of a group of titis (*Callicebus coimbrai*) in the northern Brazilian Atlantic Forest. *American Journal of Primatology*, **73**, 1199–1209.

Stevens, G. (1987) Lianas as structural parasites: the *Bursera simaruba* example. *Ecology*, **68**, 77–81.

Stoner, K.E., Vulinec, K., Wright, S.J. & Peres, C.A. (2007) Hunting and plant community dynamics in tropical forests: a synthesis and future directions. *Biotropica*, **39**, 385–392.

Suarez, S.A. (2006) Diet and travel costs for spider monkeys in a nonseasonal, hyperdiverse environment. *International Journal of Primatology*, **27**, 411–436.

Sujana, K.A., Sivaperuman, C. & Anil-Kumar, N. (2012) Lianas as a food resource for nilgiri langur (*Trachypithecus johnii* Fischer 1829) in forest of Wayanad, Western Ghats, India. *International Journal of Plant, Animal and Environmental Sciences*, **2**, 239–243.

Tenaza, R. & Tilson, R.L. (1985) Human predation and Kloss's gibbon (*Hylobates klossii*) sleeping trees in Siberut Island, Indonesia. *American Journal of Primatology*, **8**, 299–308.

Thorpe, S.K.S., Crompton, R.H. & Alexander, R.M. (2007) Orangutans use compliant branches to lower the energetic cost of locomotion. *Biology Letters*, **3**, 253–256.

Tsuji, Y. & Takatsuki, S. (2004) Food habits and home range use of Japanese macaques on an island inhabited by deer. *Ecological Research*, **19**, 381–388.

Van der Heide, G., Fernandez-Duque, E., Iriart, D. & Juarez, C.P. (2012) Do forest composition and fruit availability predict demographic differences among groups of territorial Owl monkeys (*Aotus azarai*)? *International Journal of Primatology*, **33**, 184–207.

Vidal, M.D. & Cintra, R. (2006) Effects of forest structure components on the occurrence, group size and density of groups of bare-face tamarin (*Saguinus bicolor*-Primates: Callitrichinae) in Central Amazonia. *Acta Amazonica*, **36**, 237–248.

Vogel, E.R., Haag, L., Mitra-Setia, T. et al. (2009) Foraging and ranging behavior during a fallback episode: *Hylobates albibarbis* and *Pongo pygmaeus wurmbii* compared. *American Journal of Physical Anthropology*, **140**, 716–726.

Wieczkowski, J. (2009) Tree mortality due to an El Niño flood along the lower Tana River, Kenya. *African Journal of Ecology*, **47**, 56–62.

Wright, S., Jaramillo, M., Pavon, J. et al. (2005) Reproductive size thresholds in tropical trees: variation among individuals, species and forests. *Journal of Tropical Ecology*, **21**, 307–315.

Xiang, Z.F., Liang, W.B., Nie, S.G. & Li, M. (2012) Diet and feeding behavior of *Rhinopithecus brelichi* at Yangaoping, Guizhou. *American Journal of Primatology*, **74**, 551–560.

Youlatos, D. (1999) Positional behavior of *Cebuella pygmaea* in Yasuni National Park, Ecuador. *Primates*, **40**, 543–550.

Youlatos, D. (2004) Multivariate analysis of organismal and habitat parameters in two Neotropical primate communities. *American Journal of Physical Anthropology*, **123**, 181–194.

Zurring, H., Clinebell, R.R. & Phillips, O.L. (1993) Prediction of Neotropical trees and liana species richness from soil and climatic data. *Biodiversity and Conservation*, **4**, 56–62.

Part VII

Lianas and Forest Management and Conservation

Chapter 28

LIANAS AS INVASIVE SPECIES IN NORTH AMERICA

Stacey A. Leicht-Young[1] *and Noel B. Pavlovic*[2]

[1]Arnold Arboretum of Harvard University, Roslindale, MA, USA
[2]United States Geological Survey, Great Lakes Science Center, Lake Michigan Ecological Research Station, Porter, IN, USA

Ecology of Lianas, First Edition. Edited by Stefan A. Schnitzer, Frans Bongers, Robyn J. Burnham, and Francis E. Putz.

Companion Website: www.wiley.com/go/schnitzer/lianas

OVERVIEW

Liana diversity is typically low in the temperate zones; however, the influx of non-native invasive liana species in North America has increased local diversity at the expense of native habitats and species. Like many other invasive species, invasive liana species grow rapidly, but their plastic aboveground growth strategy allows them to exploit three dimensions and survive damage in ways that seriously impede species with other growth habits. In addition, the low diversity of native lianas in North America compared with other regions suggests that there may be empty niches for these species to exploit. For example, invasive lianas in North America have successfully colonized a variety of habitats, including old-growth forests, which are often considered resistant to invasion. In closed canopy forests, most invasive lianas persist in the understory until a gap opens to give them canopy access. Studies using both empirical and modeled data suggest that invasive lianas have not yet reached their full invasion potential; increased landscape fragmentation will increase opportunities for additional infiltration.

INTRODUCTION: TRAITS AND DISTRIBUTIONS OF TEMPERATE LIANAS

Lianas display some of the most visually dramatic growth habits in the plant kingdom by ascending their hosts and often overtopping them (Fig. 28.1). Their ability to climb allows them to grow in a diversity of habitats, most commonly those that are disturbed (e.g., treefall gaps) or edges (Schnitzer et al. 2000; Schnitzer et al. 2004, 2012; Schnitzer et al, Chapter 7 in this volume; Letcher & Chazdon 2009; Letcher, Chapter 10 in this volume), but also in forest understories (Leicht-Young et al. 2007; Valladares et al. 2011). While their dependence on external support can be a disadvantage in some circumstances, it is largely advantageous because it frees lianas from the spatial constraints of plants with "sedentary" habits (i.e., trees, shrubs) and allows for vertical growth in high densities due to their ability to escape some of the consequences of competition (Leicht-Young et al. 2011).

Lianas are structural parasites (Stevens 1987) that have negative impacts on host plant growth through canopy competition (Ladwig & Meiners 2009; Ladwig & Meiners, Chapter 15 in this volume; Schnitzer et al., Chapter 7 in this volume; Toledo-Aceves, Chapter 12 in this volume; van der Heijden et al., Chapter 13 in this volume), trunk girdling (Fig. 28.1A; Lutz 1943), and, for temperate species, increasing susceptibility to ice and storm damage (Siccama et al. 1976; Allen et al. 1997). Unseen belowground competition also decreases the fitness of their hosts (Dillenburg et al. 1993b; Schnitzer et al. 2005; Chen et al. 2008; Toledo-Aceves, Chapter 12 in this volume), as does their release of allelopathic chemicals (Pisula & Meiners 2010; Ladwig et al. 2012). Lianas are often targeted for removal in forests and forestry plantations because of their negative effects on tree growth (Dillenburg et al. 1993b; Gerwing 2001; Ladwig & Meiners 2009; van der Heijden & Phillips 2009; Paul & Yavitt 2011; Schnitzer et al., Chapter 1 in this volume; Toledo-Aceves, Chapter 12 in this volume), tree diversity (Schnitzer & Carson 2010), and forest succession (Schnitzer et al. 2000; Letcher, Chapter 10 in this volume). While both native and invasive lianas can negatively affect desirable host plant species, native liana species may be at a disadvantage because they evolved with the plant communities and are consequently prone to native enemies (Blossey & Notzold 1995).

Lianas have a combination of traits that make them some of the most formidable of invaders (Harris et al. 2007). They possess an inherent plasticity by virtue of their climbing growth habit (Putz 1995b). They can survive treefall damage (Putz 1984; Putz & Holbrook 1991), support large total leaf areas (Selaya et al. 2007; Isnard & Feild, Chapter 17 in this volume), and generally display substantial resprouting potential (Gerwing 2001; Schnitzer et al. 2012; Schnitzer et al., Chapter 7 in this volume). And like many other invasive plant species, they are capable of rapid growth (Forseth Jr. & Innis 2004; Leicht-Young et al. 2011), have high tolerance to herbivory (Schierenbeck et al. 1994; Ashton & Lerdau 2008), and high reproductive output (Bois et al. 2012b.). Invasive lianas cause considerable damage to native ecosystems around the world (Harris et al. 2007; Jiménez-Castillo & Lusk 2009; Wang et al. 2011), but they may be particularly detrimental in temperate North America, where native liana diversity is low (Gentry 1991; Schnitzer 2005). The paucity of liana species in North America is at least partially a result of cold winter temperatures leading to freezing-induced embolism in their wide vessel elements (Sperry et al. 1987; Ewers et al. 1991;

Fig. 28.1 Impacts of the liana, *Celastrus orbiculatus*, on trees and habitats. (A) *C. orbiculatus* girdling a tree sapling. (B) *C. orbiculatus* on trees it has killed. (C) Aggressive *C. orbiculatus* resprouts after fire. (D) Large stem of *C. orbiculatus* (ca. 18 cm diameter) in hands of first author. (Sources: Photographs A–C taken at the Indiana Dunes National Lakeshore by Leicht-Young and Pavlovic, D at the University of Connecticut. Reproduced with permission.)

Schnitzer et al. 2005; Ladwig & Meiners, Chapter 15 in this volume).

Non-native invasive lianas have been reported from nearly every region in North America, with perhaps the greatest concentration and diversity in the eastern United States (Table 28.1). The invasive liana species in the eastern USA were most often introduced for horticultural purposes (Mehrhoff et al. 2003), with a few others brought in for erosion control and wildlife feed (Forseth Jr. & Innis 2004). A tally of lianas on state or federal invasive species lists for temperate North America (i.e., not including subtropical Florida or Hawaii) revealed 18 species, of which ten were from eastern Asia, four from meso and South America, and four from Europe or western Asia (Table 28.1; Center for Invasive Species and Ecosystem Health 2012). In all of temperate North America there are only approximately 75 native species of lianas (Table 28.2). In contrast, liana diversity in temperate habitats in Asia is substantially higher, which may result in greater competition among species (Qian and Ricklefs 1999; Kodani 2006; Cai et al. 2009; Bois et al. 2012b). For example, the genus *Celastrus*, which only has one native species in North America, is represented by at least 25 species in China. Similarly, there are only two native species of *Ampelopsis* in the USA but 13 species in China (eFloras 2008). Other genera, such as *Clematis*, have over 90 species in China compared with 8 in the USA, and *Vitis*, which is represented by 14 species in the USA, has about 30 species in China and East Asia (eFloras 2008). The higher diversity of temperate lianas in Asia apparently resulted from its different evolutionary and geological histories (Qian & Ricklefs 1999), along with its less severe climate such that tropical and subtropical species can grow fairly far north (D. Boufford, pers. comm.). The lack of a long evolutionary history of a high diversity and abundance of lianas in North American forests may render them susceptible to introduced lianas.

INVASIVE TEMPERATE LIANAS IN NORTH AMERICA: BIOLOGICAL ATTRIBUTES

Some of the most illustrative studies of invasive lianas in temperate North America compared the biological traits of invasive lianas with native congeners or ecological analogs. The majority of these studies focused on two species, *Celastrus orbiculatus* (oriental bittersweet) and *Lonicera japonica* (Japanese honeysuckle). While comparisons with native congeners are preferable from a phylogenetic standpoint (Schierenbeck & Marshall 1993; Schierenbeck et al. 1994; Larson 2000; Daehler 2003; Leicht & Silander 2006; Leicht-Young et al. 2007), comparisons with native species that are ecological correlates are also informative from a community perspective (Carter & Teramura 1988b;

Table 28.1 List of non-native, invasive lianas of temperate North America generated from the www.invasive.org website (see literature for citation). For inclusion in the list, species had to be non-native to North America, be a liana (woody vine) and be on a state or federal invasive species list.

Scientific name	Common name	Family	Invaded region	Origin
Akebia quinata	chocolate vine	Lardizabalaceae	NE, MA, MW	East Asia
Ampelopsis brevipedunculata	Amur peppervine	Vitaceae	NE, MA, MW	East Asia
*Bougainvillea glabra**	paperflower	Nyctaginaceae	FL, TX	SA
Cardiospermum halicacabum	balloonvine	Sapindaceae	South	SA
Celastrus orbiculatus	oriental bittersweet	Celastraceae	NE, MA, MW, SE	East Asia
Clematis orientalis	oriental virginsbower	Ranunculaceae	Mountain west	West Asia
Clematis terniflora	sweet autumn virginsbower	Ranunculaceae	MA, SE, South	East Asia
Clematis vitalba	evergreen clematis	Ranunculaceae	NW	Europe, SW Asia
Euonymus fortunei	winter creeper	Celastraceae	NE, MA, MW, SE	East Asia
Hedera helix	English ivy	Araliaceae	NE, MA, MW, SE, NW	Europe, W Asia
Hedera hibernica	Atlantic Ivy	Araliaceae	NW	W Europe
Lonicera japonica	Japanese honeysuckle	Caprifoliaceae	NE, MA, MW, SE, South	East Asia
Macfadyena unguis-cati	catclaw-vine	Bignoniaceae	SE, South, FL	SA
*Rosa multiflora***	multiflora rose	Rosaceae	NE, MA, MW, SE, South, NW	East Asia
*Passiflora edulis**	passionflower	Passifloraceae	AL, FL, HI	SA
Pueraria montana var. *lobata*	kudzu	Fabaceae	MA, SE, South	East Asia
Wisteria floribunda	Japanese wisteria	Fabaceae	NE, MA, SE, South	East Asia
Wisteria sinensis	Chinese wisteria	Fabaceae	NE, MA, SE, South, NW	East Asia

Notes: * = species which are bordering on only subtropical species.
** = climbs passively by means of thorns getting caught up in nearby vegetation; may not be considered a true liana.
For invaded region: NE = northeastern North America, MA = mid-Atlantic USA, MW = Midwestern USA, SE = southeastern NA, South = southern gulf states, NW = Pacific northwestern North America, FL = Florida only, TX = Texas only, AL = Alabama. For origin: SA = South-/meso-America.

Tibbetts & Ewers 2000; Bois et al. 2012b). For invasive lianas to be successful, they presumably would have biological or ecological traits that make them superior competitors.

Biophysical traits

Lianas have some of the widest and longest vessels in the plant kingdom (Ewers et al. 1991; Angyalossy et al., Chapter 19 in this volume; Isnard & Feild, Chapter 17 in this volume). These vessels present a problem for temperate lianas when winter-induced freezing embolisms cause dieback (Schnitzer 2005). Some species native to the region (e.g., *Vitis* spp.) use positive root pressure to repair embolisms that form from freezing temperatures (Sperry et al. 1987). In contrast, the sympatric invasive, *Celastrus orbicualtus*, does not use positive root pressures but instead produces new xylem vessels yearly to replace those that were damaged by cavitation (Tibbetts & Ewers 2000); a trait that may explain why this species reaches large

Table 28.2 Native genera of temperate lianas in the United States. List was developed using the USDA PLANTS database advanced search on native vines, then eliminating non-woody species. Species only found in Hawaii, subtropical Florida and extreme south Texas were excluded from the tally.

Family	Genus	Number of species
Anacardiaceae	*Toxicodendron*	3
Apocynaceae	*Trachelospermum*	1
Aristolochiaceae	*Aristolochia*	2
Asteraceae	*Ampelaster**	1
Bignonaceae	*Bignonia*	1
	Campsis	1
Caprifoliaceae	*Lonicera*	11
Celastraceae	*Celastrus*	1
Ericaceae	*Pieris**	1
Fabaceae	*Dioclea*	1
	Wisteria	1
Hydrangeaceae	*Decumaria*	1
Loganiaceae	*Gelsemium*	2
Menispermaceae	*Calycocarpum*	1
	Cocculus	2
	Menispermum	1
Polygonaceae	*Brunnichia*	1
Ranunculaceae	*Clematis*	9
Rhamnaceae	*Berchemia*	1
	*Sageretia**	1
Rosaceae	*Rosa**	1
Schisandraceae	*Schisandra*	1
Smilacaceae	*Smilax***	10
Vitaceae	*Ampelopsis*	2
	Cissus	1
	Parthenocissus	3
	Vitis	14

Notes: *These species climb passively by getting thorns or long stems caught up in host vegetation or by leaning. **Although *Smilax* is a monocot and technically not woody, they possess the characteristics of lianas and are commonly included in liana censuses in both temperate and tropical systems (e.g., Schnitzer et al., Chapter 1 in this volume).

diameters relatively quickly (e.g., 18 cm in approximately 30 years; Fig. 28.1D). Other liana species such as invasive *L. japonica* and native *L. sempervirens* (trumpet honeysuckle) actually have fairly narrow vessels, which may be an adaptation to cold winter temperatures (Chiu & Ewers 1992).

Another important trait of lianas is their ability to resprout after damage. Lianas can sprout from fallen stems in treefall gaps (Appanah & Putz 1984; Putz & Holbrook 1991; Schnitzer et al. 2012; Schnitzer et al, Chapter 7 in this volume) as well as from cut stumps, roots, and even fragments of cut stems (Gerwing 2001; Alvira et al. 2004). Resprouting allows persistence after disturbances and can promote rapid regeneration (Mandle et al. 2011; Schnitzer et al. 2012; Schnitzer et al., Chapter 7 in this volume). Similar to other plant species, resprouting in lianas is aided by root reserves. In the invasive *C. orbiculatus*, approximately 50% of the total nonstructural carbohydrates (TNC) are depleted in the spring at the initiation of growth and are then replenished by photosynthesis during the growing season. Cutting *C. orbiculatus* during the growing season reduced TNC reserves by an additional 25%, an effect that lasted until the next growing season (N. Pavlovic, unpublished data). Thus, the ability to resprout does not come without a cost for lianas, and may indicate a method of liana control.

Photosynthesis

Lianas have evolved to take advantage of canopy light with minimal investment in woody biomass for support (Rowe & Speck 2005; Rowe & Speck, Chapter 23 in this volume). In contrast, lianas in the "waiting phase" below the canopy must be able to photosynthesize at low light levels. Furthermore, even individual leaves on lianas in the canopy may be exposed to a highly variable light regime; from 1–100% full sun (Teramura et al. 1991). An example of an adaptation to this unequal exposure of leaves to light is seen in *Hedera helix* (English ivy). Leaves of adult and juvenile *H. helix* photosynthesize at different rates, with the juvenile leaves being more adapted to low light, as these leaves are typically exposed to more shaded forest understory conditions while searching for a host, compared with adult leaves, which exhibit a high photosynthetic rate in the sun when the plant has reached the canopy. (Hoflacher & Bauer 1982; Okerman 2000).

Temperate zone lianas generally have higher photosynthetic rates than other early successional species and their host trees (Teramura et al. 1991). However, invasive lianas do not appear to display higher photosynthetic rates than native lianas. In studies that compared photosynthetic capacity of suites of native and invasive lianas (Carter & Teramura 1988b; Carter et al. 1989; Schierenbeck & Marshall 1993; Leicht 2005), only one invasive liana had a higher photosynthetic rate compared with the native (*Celastrus* spp.; Leicht 2005). The photosynthetic rates of the two *Lonicera* species studied by Schierenbeck and Marshall (1993) did not differ; however, the introduced species had greater annual carbon gain due to its greater leaf area, retention of leaves over the winter, and higher photosynthetic rates of new leaves. In addition, the native North American species are deciduous, whereas *L. japonica* is semi-evergreen and winter photosynthesis contributes to its total annual carbon balance (Carter & Teramura 1988a).

Growth rates

The rapid growth of invasive lianas can effectively smother underlying vegetation and stall succession (Hardt 1986; Fike & Niering 1999; Schnitzer et al. 2000). For example, in open, sunny habitats, such as roadsides, old-fields, forest edges, and highly disturbed forest, *Pueraria montana* var. *lobata* (kudzu) can grow 3–19 cm per day, resulting in up to 30 m of growth in a single growing season (Forseth Jr. & Innis 2004). However, this liana is less successful in forest understories because of its inability to tolerate shade (Carter & Teramura 1988b; Carter et al. 1989). In forested habitats, where light levels are low, growth rate is critical for a liana to climb into the canopy and to increase total leaf area, which will allow it to increase its resource acquisition and ultimately determine its survival. Several studies have shown that invasive lianas, such as *C. orbiculatus* can grow both in full sun to very deep shade (Ellsworth et al. 2004; Leicht-Young et al. 2007; Leicht-Young et al. 2011; Bois et al. 2012b), and across a fairly wide moisture gradient (Leicht-Young et al. 2007; Bois et al. 2012b). In fact, for *C. orbiculatus*, there was no significant difference in growth rate between full sun and low light levels (ca. 25%; Ellsworth et al. 2004; Leicht-Young et al. 2011). In comparison, the native *C. scandens* (American bittersweet) has much lower growth and survival under low light conditions (Leicht-Young et al. 2007; Leicht-Young et al. 2011).

Searching behavior and ascension to the forest canopy

The ability of these invasive lianas to sit and wait, as well as actively search for a support and forest, is a key to their success in undisturbed temperate forests. Leicht and Silander (2006) simulated understory light conditions with respect to the ratio of red to far-red light and, when the ratio of red:far red light was reduced, thus simulating an intact canopy, the native *C. scandens* did not show growth or "searching" behavior, while the invasive *C. orbiculatus* still exhibited positive growth, thus increasing the probability of finding a support. In an examination of native and invasive *Lonicera*, Larson (2000) reported that the shoots of the invasive *L. japonica* remained prostrate for longer than the native *L. sempervirens*, resulting in greater lateral spread through the formation of roots where the vine contacted the ground. The ability to spread through the forest understory likely increases the probability of finding a canopy gap or a small diameter tree that will serve as a trellis for the liana to ascend to the canopy.

Invasive lianas may grow faster than natives once they find a trellis. For example, *Lonicera japonica* showed more rapid growth when it reached its supports compared with native *L. sempervirens* (Schweitzer & Larson 1999). Stem twining lianas, such as *L. japonica* and *C. orbiculatus* may be at a disadvantage in more mature forests that are dominated by large trees, since they are unable to attach to large host trees due to biomechanical constraints (Putz & Holbrook 1991; Rowe & Speck, Chapter 23 in this volume). However, if a canopy gap forms, stem twining lianas will be able to use small trees or other lianas to use as ladders (Putz 1995a; Leicht-Young et al. 2010). Reaching the canopy, these species can overtop and intercept light that would be used by the host species. *Celastrus orbiculatus* in its native Japan was able to intercept 40–80% of the PPFD relative to the top of the host canopy (Ichihashi & Tateno 2011). That percentage of light interception is likely higher in the USA where *C. orbiculatus* exhibits more rampant growth (Bois et al. 2012a). On forest edges, species like kudzu (*P. montana* var. *lobata*) can smother their host trees, effectively intercepting nearly all of the light (Forseth Jr. & Innis 2004).

Reproduction/population dynamics

Invasive lianas and their native counterparts often have high reproductive output in the form of fleshy fruits that can be dispersed long distances via birds and mammals (Aronson et al. 2007; LaFleur et al.

2007; LaFleur et al. 2009; Merow et al. 2011; Michel et al., Chapter 25 in this volume). However, the reproductive potential of invasive lianas may be much higher than native analogs. For example, in a common garden study, 46% of the invasive *C. orbiculatus* individuals were flowering compared to just 5% of the native *C. scandens* plants after three growing seasons (Leicht-Young et al. 2011). In habitats where *C. orbiculatus* and *C. scandens* co-occur, there were approximately 56 times more pistillate flowers per *C. orbiculatus* females compared with *C. scandens* females. This difference in reproductive output translates into approximately 3580 seeds per invasive plant vs. 44 for the native species (D. Zaya 2013). In this same system, pollen of *C. orbiculatus* swamps that of *C. scandens*, resulting in approximately 40% hybrid seeds (D. Zaya 2013). In addition, *C. orbiculatus* females, in almost all cases were unreceptive to *C. scandens* pollen (D. Zaya & S. Leicht-Young, unpublished data). The pollen swamping of *C. scandens* coupled with high reproductive output of *C. orbiculatus* are likely major contributors to the decline of the former species where both overlap in range. In contrast, hybrid vigor resulting from hybridization between two invasive species of *Wisteria* (*sinensis* and *floribunda*; Chinese and Japanese wisteria) is a likely contributor to their success on the landscape, especially in the southern USA (Trusty et al. 2007). Germination rates of *C. orbiculatus* are significantly greater than *C. scandens* (Dreyer et al. 1987; Leicht 2005), as well as a native ecological correlate *Vitis labrusca* (fox grape; Bois et al. 2012b), thus further augmenting the advantage of the invasive species.

Vegetative spread is another trait that gives invasive species an advantage. Invasive lianas such as *P. montana* var. *lobata* often exhibit rampant vegetative spread and can cover as much as 50,000 ha yr^{-1} of new infestations (Forseth Jr. & Innis 2004). Lianas can also spread vegetatively in forest understories via root sprouts, even after the rest of the plant has grown into the canopy (Ichihashi & Tateno 2011; Schnitzer et al. 2012; Schnitzer et al., Chapter 7 in this volume) or if the liana has been damaged (Pavlovic et al. unpublished data). Extensive vegetative spread *H. helix* creates “ivy deserts” where little to no understory forest species can grow (Okerman 2000). Similarly, the stems of *L. japonica* root more successfully than its native conger (*L sempervirens*), contributing to *L. japonica*'s more rapid spread (Larson 2000).

Herbivory

Evolution of increased competitive ability (EICA) is a common hypothesis for the success of invasive plants in their new ranges through escaping their natural enemies (Blossey & Notzold 1995). This hypothesis largely holds for some invasive lianas. When grown in its native Japan, *C. orbiculatus* experienced significantly greater herbivore loads than in the USA, thus supporting the EICA hypothesis (Bois, unpublished data). However, herbivory did not differ between *C. orbiculatus* and co-occurring *V. labrusca* in the USA (Bois et al. 2012b), suggesting that herbivory alone may not explain the relative increase in *C. orbiculatus*. Invasive lianas that sustained simulated herbivory had no change in relative growth rate but increased leaf number, while native and naturalized species had decreased growth rates (Ashton & Lerdau 2008). *Pueraria montana* var. *lobata* decreased above- and belowground growth in response to 75% simulated defoliation within its introduced range (Frye et al. 2012). In the absence of herbivores, the growth rate of *L. japonica* was equal to or even less than the native *L. sempervirens*, but with herbivory, *L. japonica* increased its allocation to secondary leaves and stems, and had greater overall total biomass allocation (Schierenbeck et al. 1994). Thus, compensatory responses to damage and resprouting ability may give invasive lianas an advantage over their native counterparts.

ECOLOGICAL INTERACTIONS OF INVASIVE TEMPERATE LIANAS IN NORTH AMERICA

Competition

The most striking images of lianas, in both tropical and temperate forests, are the dense liana tangles forming “topiaries” of the vegetation below (Fig. 28.1B; see also Fig. 7.2 on p. 83). In this context, there are two types of competition taking place – competition among the lianas and competition between the liana and their host plant. There is very little information on liana-to-liana competition from either the temperate or tropical zones. Liana competition has been assumed to be comparable to other plant species: in high densities, lianas would have lower growth rates and increased mortality. Interestingly, in a common garden competition experiment between invasive *C. orbiculatus* and native *C. scandens*, growing in high densities did not have an adverse effect on the relative height growth of either species (Leicht-Young et al. 2011). Most striking was that both species had significantly higher survival in high-density treatments compared to low-density treatments, revealing a positive density effect. However, the native *C. scandens* experienced much lower survival compared with *C. orbiculatus* in the same treatments (43% vs. 81%), pointing to its lower competitive ability. Since lianas can escape neighbors by climbing, they may not be bound to the self-thinning law in the same way as non-climbing species (Leicht-Young et al. 2011).

Lianas reduce tree growth via both above- and belowground competition (Toledo-Aceves, Chapter 12 in this volume). In open-grown trees and lianas, trees experiencing competition belowground and above- and belowground had lower growth rates than trees with no competition or only aboveground competition (Dillenburg et al. 1993a, 1993b, 1995). However, the invasive *L. japonica* decreased tree growth more than the native *Parthenocissus quinquefolia* (Virginia creeper; Dillenburg et al. 1993b), suggesting that invasive species compete more intensely. In Japan, trees infested by *C. orbiculatus* showed decreased growth compared with control trees (Ichihashi & Tateno 2011). Decreased growth in trees where lianas (native and invasive) dominate the canopy was also shown in US forests (Ladwig & Meiners 2009; Ladwig & Meiners, Chapter 15 in this volume). This reduction in tree growth may lead to reduced forest productivity and reduced tree health (Ladwig & Meiners 2009; Ladwig & Meiners, Chapter 15 in this volume; van der Heijden et al., Chapter 13 in this volume; Schnitzer et al. in press).

Recent studies have highlighted the allelopathic potential of invasive lianas. The invasive *C. orbiculatus* inhibited the germination of test species more than another invasive (*L. japonica*), and had mid-range allelopathic potential when ranked with invasive species of other life forms (Pisula & Meiners 2010; Ladwig & Meiners, Chapter 15 in this volume). In sunny conditions, *C. orbiculatus* and *L. japonica* had the greatest allelopathic potential compared with other native lianas, which may aid their successful invasion of new habitats from edges (Ladwig et al. 2012). In the shade, lianas that persist in the understory later in succession (e.g., *P. quinquefolia*) more negatively impacted the germination of test species, though the allelopathic potential of *C. orbiculatus* remained the same regardless of light conditions (Ladwig et al. 2012). Extracts from kudzu (*P. montana* var. *lobata*) also suppressed the germination of other plant species (Rashid et al. 2010a, 2010b). Positive feedback effects from soil conditioning may also aid lianas. For example, *Euonymus fortunei* (wintercreeper) had increased growth when grown in soil conditioned by conspecifics compared to growing in soil conditioned by the herb *Asarum canadense* (wild ginger). Conversely, *A. canadense* had no change in growth in either soil (Smith & Reynolds 2012).

Representation and distribution in forest communities

Invasive liana habitat preferences have direct impacts on species composition and successional trajectories in native forests. Lianas are more prominent and abundant in early successional forests and along edges (Robertson et al. 1994; Londré & Schnitzer 2006; Ladwig & Meiners 2009, 2010a; Ladwig & Meiners, Chapter 15 in this volume; Letcher, Chapter 10 in this volume; Allen et al. 2013); however, they are also present in old-growth forests, often waiting in the understory, or as members of the canopy (Ladwig & Meiners 2010b; Leicht-Young et al. 2010; Pavlovic & Leicht-Young 2011). In mature forests in the Great Lakes region, the most common invasive liana was *C. orbiculatus*, and in the majority of cases, it was not actively climbing trees, but instead was present in the forest understory. Native *C. scandens*, however, was more restricted to open dunes or forest edges (Pavlovic & Leicht-Young 2011). *Lonicera* species showed a similar pattern, with the invasive *L. japonica* growing in a wider variety of habitats than the native *L. sempervirens* (Schweitzer & Larson 1999).

In a study that followed old-field succession for 50 years, liana cover peaked during mid-succession when the herbaceous to woody plant transition occurred (Ladwig & Meiners 2010b; Ladwig & Meiners, Chapter 15 in this volume). Interestingly, invasive lianas, such as *L. japonica*, were not any more prominent or invasive than their native counterparts. Liana climbing methods (e.g., stem twining, tendrils) did not predict their successional placement. *Celastrus orbiculatus*, a stem twiner, that presumably needs small diameter supports for climbing, occurred later in the time series (Ladwig & Meiners 2010b) again indicating that invasive lianas like *C. orbiculatus* can be successful in late successional stages. The abundance of these invasive lianas, such as *C. orbiculatus*, in the understory may point to a significant change in future forest composition when new gaps open or fragmentation occurs (Ladwig & Meiners 2010b; Pavlovic & Leicht-Young 2011; Allen et al. 2013).

Responses to disturbance

The idea of increasing fragmentation resulting in increasing liana abundance is supported by a retrospective study in Wisconsin, where there was no increase in liana abundance in the last 45 years (Londré & Schnitzer 2006). However, lianas were more abundant near the edges of the forest, due to increased light availability (Robertson et al. 1994; Londré & Schnitzer 2006). Thus, increasing fragmentation of forests may be a driver of increases in liana abundance (Londré & Schnitzer 2006; Allen et al. 2013; Schnitzer, Chapter 30 in this volume). *Celastrus orbiculatus* was more abundant in landscapes that experienced recent land use change from forest fragmentation and residential development vs. forest landscapes that had developed after agricultural abandonment in the 1930s (Mosher et al. 2009; Allen et al. 2013). Models of *C. orbiculatus* suggest that invasions began in agricultural and developed settings and then infiltrated forested habitats via bird dispersal (Merow et al. 2011). In the model, rapid expansion of this species into forests does not occur until the 1980s, and thus the full invasion potential of these liana species in forested habitats, has yet to be realized (Merow et al. 2011).

Natural (e.g., hurricanes) and human-caused fragmentation can result in higher numbers of gaps in forests. One threat is that of positive feedback resulting from invasive twining lianas bringing down trees after major disturbances (Allen et al. 1997), opening up gaps for the invasion and resprouting of additional lianas (Gordon 1998; Forseth Jr. & Innis 2004). Surveys in secondary and old-growth forests in the south-eastern United States indicated an increase in native liana abundance (Allen et al. 2007). In old-growth forest, hurricane damage initially decreased liana densities, but ultimately resulted in an overall increase (Allen et al. 2005; Allen et al. 2007). In tropical systems, lianas can arrest succession in gaps for 13 years or more (Schnitzer et al. 2000), and have been documented to stall succession in old fields in temperate climates (Fike & Niering 1999). Therefore, invasive lianas will likely have an important role to play in succession after disturbance in temperate forests.

How lianas respond to other types of disturbances, such as fire, is of theoretical and practical importance in determining the reciprocal interactions between vegetation and fire, and whether this type of disturbance can control invasive lianas (Mandle et al. 2011). Resprouting in other woody species after fire has been well studied (Bellingham & Sparrow 2000; Del Tredici 2001; Bond & Midgley 2003; Vesk & Westoby 2004; Poorter et al. 2010), but work on temperate lianas is limited (Proença et al. 2010). Fire has been shown to both increase (e.g., *Toxicodendron radicans* (poison ivy); Faulkner et al. 1989) and decrease (e.g., *P. quinquefolia*; Nuzzo et al. 1996) native liana abundance. For invasive lianas, fire reduces the cover of *L. japonica*, but does not eliminate it (Mandle et al. 2011). In contrast, *C. orbiculatus* stem density increased more than twofold after burning compared to control and cut plots, mostly due to copious resprouting and root suckering (Pavlovic et al. unpublished data; Fig. 28.1C). The proliferation of roots by 5 m or more in *C. orbiculatus* (Ichihashi & Tateno 2011), and its ability to rootsucker, allows fire to promote lianas' local spread (Pavlovic unpublished data). Invasive lianas have the potential to act as ladder fuels (Dibble et al. 2008), though this has not been extensively studied. Anderson (1972), however, found that *L. japonica* vines can act as ladder fuels up to 10–15 feet above the ground. Whether crown fires initiated by invasive lianas will become more common due to changes in forest structure by their presence depends on the flammability of dominant tree species (e.g. *Quercus* vs. *Pinus*) and needs further investigation.

CONCLUSION AND FUTURE DIRECTIONS

Invasive lianas are having an increasing impact on the dynamics and trajectories of North American plant communities. They often exhibit superior growth and survival compared to their native counterparts, and in some cases, invasive lianas may directly contribute to the decline of their native correlates (e.g., the detrimental effect of *C. orbiculatus* on *C. scandens*). Invasive lianas have the characteristics of invasive plant species, as well as a more plastic and resilient liana growth form, which gives them the ability to grow in ways trees, shrubs, and herbs cannot. In high densities, lianas retain high growth rates and escape competition, and exhibit high survival (Leicht-Young et al. 2011). Perhaps the most important characteristic of invasive lianas is the ability to sit and wait in the forest understory until a canopy gap forms. Gap formation provides the opportunity for many liana species to begin a cycle of establishment and dominance, bringing down trees and continuing the cycle of liana infestation (Gordon 1998; Forseth Jr. & Innis 2004). Thus, positive density dependence in liana tangles, coupled with attaining the canopy when a gap opens up, makes these species formidable competitors and contributes greatly to their success. In the limited number of studies that have surveyed temperate lianas (e.g., Londré & Schnitzer 2006; Ladwig & Meiners 2010b; Ladwig & Meiners, Chapter 25 in this volume; Leicht-Young et al. 2010), there has not been a marked spike in invasive lianas. However, historical landscape studies (e.g., Mosher et al. 2009; Allen et al. 2013) have indicated that more recent disturbances often pave the way for the invasion of these species, thus their full potential may yet be realized (Merow et al. 2011). Although the lack of native competitors should theoretically allow for invasive lianas to take advantage of this empty niche, there are few studies that show this quantitatively. It could be that with intense propagule pressure and increased fragmentation, North American temperate forests will look very different in the future (Londré & Schnitzer 2006).

Further opportunities for research on temperate invasive lianas are vast. Studies are needed on individual species whose biological and ecological attributes are not fully understood (e.g., *Ampelopsis brevipedunculata*; porcelainberry). It is important that these studies compare the invasive species with their native congeners or ecological correlates, as well as their relative impact on these in order to have a control for comparison. Another urgent, but often overlooked area of research is the successful management and control of these damaging species in natural settings. Because of their plasticity and resprouting ability, invasive lianas are some of the most challenging plant invaders to control and eradicate. Further investigation into the relative competitive abilities of invasive and native lianas in natural settings is critical to understanding the advantage of invasive lianas over native species. How lianas can impede succession in gaps is important for forest health and regeneration, and while studies on gap dynamics have been conducted in tropical systems (e.g., Schnitzer et al. 2000; Schnitzer & Carson 2010), few experimental studies have been done in temperate forests (Ladwig & Meiners, Chapter 15 in this volume). Liana surveys, such as those taken in the southern USA before and after hurricanes (e.g., Allen et al. 1997; Allen et al. 2005), are important not only for future demographic studies following disturbance, but also to study the potential effects of future climate change (Wright et al. 2004). There also focus on the phenology of invasive or native lianas and how the timing of life-cycle events may contribute to their success. Collaboration with researchers in Asia, where most of these species originate, is critical to understanding their behavior in their native range compared with that of the USA (e.g., Bois et al. 2012a). Ultimately, research on invasive temperate lianas needs to be directed so that it can assist land managers in prevention as well as to curb further spread and damage to North American habitats.

ACKNOWLEDGMENTS

We wish to thank Sarah Treanor Bois for her review of a previous draft of this chapter, and for sharing her insights and data on temperate lianas. Jack Putz also made helpful edits on this chapter. Dave Boufford of the Arnold Arboretum for his guidance on temperate lianas in Asia. We are also grateful to David Zaya and Jenica Allen who shared data for this chapter and gave us some valuable insights, John Silander for useful discussions on liana ecology, and to all the summer research assistants who made research on the *Celastrus* species possible through the years. This chapter is Contribution 1871 of the U.S. Geological Survey Great Lakes Science Center.

REFERENCES

Allen, B.P., Pauley, E.F. & Sharitz, R.R. (1997) Hurricane impacts on liana populations in an old-growth southeastern bottomland forest. *Journal of the Torrey Botanical Society*, **124**, 34–42.

Allen, B.P., Sharitz, R.R. & Goebel, P.C. (2005) Twelve years post-hurricane liana dynamics in an old-growth southeastern floodplain forest. *Forest Ecology and Management*, **218**, 259–269.

Allen, B.P., Sharitz, R.R. & Goebel, P.C. (2007) Are lianas increasing in importance in temperate floodplain forests in the southeastern United States? *Forest Ecology and Management*, **242**, 17–23.

Allen, J.M., Leininger, T.J., Hurd, J. Jr. et al. (2013) Socioeconomics drive woody invasive plants in New England through forest fragmentation. *Landscape Ecology*, **28**, 1671–1686.

Alvira, D., Putz, F.E. & Fredericksen, T.S. (2004) Liana loads and post-logging liana densities after liana cutting in a lowland forest in Bolivia. *Forest Ecology and Management*, **190**, 73–86.

Anderson, R.C. (1972) Prairie history, management and restoration in southern Illinois, in *Proceedings of the Second Midwest Prairie Conference*, Madison, WI, pp. 15–21.

Appanah, S. & Putz, F.E. (1984) Climber abundance in virgin dipterocarp forest and the effect of pre-felling climber cutting on logging damage. *Malaysian Forester*, **47**, 335–342.

Aronson, M., Handel, S. & Clemants, S. (2007) Fruit type, life form and origin determine the success of woody plant invaders in an urban landscape. *Biological Invasions*, **9**, 465–475.

Ashton, I.W. & Lerdau, M.T. (2008) Tolerance to herbivory, and not resistance, may explain differential success of invasive, naturalized, and native North American temperate vines. *Diversity and Distributions*, **14**, 169–178.

Bellingham, P.J. & Sparrow, A.D. (2000) Resprouting as a life history stategy in woody plant communities. *Oikos*, **89**, 409–416.

Blossey, B. & Notzold, R. (1995) Evolution of increased competitive ability in invasive nonindigenous plants: a hypothesis. *Journal of Ecology*, **83**, 887–889.

Bois, S.T., Holsinger, K., Kaproth, M.A., et al. (2012) Phenotypic variation and local site adaptation in native and invasive ranges: responses of two contrasting woody ornamentals. Unpublished manuscript.

Bois, S.T., Ibáñez, I., & Silander Jr., J.A. (2012) Colonization potential of three invasive species and their ecological analogs in New England; capitalizing on environmental stochasticity to understand seedling establishment and spread. Unpublished manuscript.

Bond, W.J. & Midgley, J.J. (2003) The evolutionary ecology of sprouting in woody plants. *International Journal of Plant Sciences*, **164**, S103–S114.

Cai, Z.Q., Schnitzer, S.A., Wen, B. et al. (2009) Liana communities in three tropical forest types in Xishuangbanna, South-West China. *Journal of Tropical Forest Science*, **21**, 252–264.

Carter, G.A. & Teramura, A.H. (1988a) Nonsummer stomatal conductance for the invasive vines kudzu and Japanese honeysuckle. *Canadian Journal of Botany*, **66**, 2392–2395.

Carter, G.A. & Teramura, A.H. (1988b) Vine photosynthesis and relationships to climbing mechanics in a forest understory. *American Journal of Botany*, **75**, 1011–1018.

Carter, G.A., Teramura, A.H. & Forseth, I.N. (1989) Photosynthesis in an open field for exotic versus native vines of the southeastern United States. *Canadian Journal of Botany/Revue Canadienne de Botanique*, **67**, 443–446.

Center for Invasive Species and Ecosystem Health (2012) http://www.invasive.org/. University of Georgia. Tifton, GA. August 15.

Chen, Y.J., Bongers, F., Cao, K.F. & Cai, Z.Q. (2008) Above- and below-ground competition in high and low irradiance: tree seedling responses to a competing liana Byttneria grandifolia. *Journal of Tropical Ecology*, **24**, 517–524.

Chiu, S.-T. & Ewers, F.W. (1992) Xylem structure and water transport in a twiner, a scrambler, and a shrub of *Lonicera* (Caprifoliaceae). *Trees: Structure and Function*, **6**, 216–224.

Daehler, C.C. (2003) Performance comparisons of co-occurring native and alien invasive plants. *Annual Review of Ecology, Evolution, and Systematics*, **34**, 183–211.

Del Tredici, P. (2001) Sprouting in temperate trees: a morphological and ecological review. *Botanical Review*, **67**, 121–140.

Dibble, A.C., Zouhar, K., & Smith, J.K. (2008) Fire and nonnative invasive plants of the northeast bioregion. In *Wildland Fire in Ecosystems: Fire and Nonnative Invasive Plants* (eds. K.C. Zouhar, J.K. Smith, S. Sutherland, & M.L. Brooks). U.S. Department of Agriculture, Forest Service, Rocky Mountain Research Station, Ogden, UT.

Dillenburg, L.R., Teramura, A.H., Forseth, I.N. & Whigham, D.F. (1995) Photosynthetic and biomass allocation responses of *Liquidambar Styraciflua* (Hamamelidaceae) to vine competition. *American Journal of Botany*, **82**, 454–461.

Dillenburg, L.R., Whigham, D.F., Teramura, A.H. & Forseth, I.N. (1993a) Effects of vine competition on availability of light, water and nitrogen to a tree host (*Liquidambar styraciflua*). *American Journal of Botany*, **80**, 244–252.

Dillenburg, L.R., Whigham, D.F., Teramura, A.H. & Forseth, I.N. (1993b) Effects of below- and aboveground competition from the vines *Lonicera japonica* and *Parthenocissus quinquefolia* on the growth of the tree host *Liquidambar styraciflua*. *Oecologia*, **93**, 48–54.

Dreyer, G.D., Baird, L.M. & Fickler, C. (1987) *Celastrus scandens* and *Celastrus orbiculatus*: comparisons of reproductive potential between a native and an introduced woody vine. *Bulletin of the Torrey Botanical Club*, **114**, 260–264.

eFloras. (2008) http://www.efloras.org (accessed 9 September 2012). *Missouri Botanical Garden*, St. Louis, MO & Harvard University Herbaria, Cambridge, MA.

Ellsworth, J.W., Harrington, R.A. & Fownes, J.H. (2004) Survival, growth and gas exchange of *Celastrus orbiculatus* seedlings in sun and shade. *American Midland Naturalist*, **151**, 233–240.

Ewers, F.W., Fisher, J.B. & Fichtner, K. (1991) Water flux and xylem structure in vines, in *The Biology of Vines* (eds F.E. Putz & H.A. Mooney), Cambridge University Press, Cambridge, pp. 127–160.

Faulkner, J.L., Clebsch, E.E.C. & Sanders, W.L. (1989) Use of prescribed burning for managing natural and historic resources in Chickamauga and Chattanooga National Military Park, USA. *Environmental Management*, **13**, 603–612.

Fike, J. & Niering, W.A. (1999) Four decades of old field vegetation development and the role of *Celastrus orbiculatus* in

the northeastern United States. *Journal of Vegetation Science*, **10**, 483–492.
Forseth, I.N. Jr. & Innis, A.F. (2004) Kudzu (*Pueraria montana*): history, physiology, and ecology combine to make a major ecosystem threat. *Critical Reviews in Plant Sciences*, **23**, 401–413.
Frye, M.J., Hough-Goldstein, J. & Kidd, K.A. (2012) Response of kudzu (*Pueraria montana* var. *lobata*) to different types and levels of simulated insect herbivore damage. *Biological Control*, **61**, 71–77.
Gentry, A.H. (1991) The distribution and evolution of climbing plants, in *The Biology of Vines* (eds F.E. Putz & H.A. Mooney), Cambridge University Press, Cambridge, pp. 3–52.
Gerwing, J.J. (2001) Testing liana cutting and controlled burning as silvicultural treatments for a logged forest in the eastern Amazon. *Journal of Applied Ecology*, **38**, 1264–1276.
Gordon, D.R. (1998) Effects of invasive, non-indigenous plant species on ecosystem processes: lessons from Florida. *Ecological Applications*, **8**, 975–989.
Hardt, R.A. (1986) Japanese honeysuckle: from "one of the best" to ruthless pest. *Arnoldia*, **25**, 27–34.
Harris, C.J., Murray, B.R., Hose, G.C. & Hamilton, M.A. (2007) Introduction history and invasion success in exotic vines introduced to Australia. *Diversity and Distributions*, **13**, 467–475.
Hoflacher, H. & Bauer, H. (1982) Light acclimation in leaves of the juvenile and adult life phases of ivy (*Hedera helix*). *Physiologia Plantarum*, **56**, 177–182.
Ichihashi, R. & Tateno, M. (2011) Strategies to balance between light acquisition and the risk of falls of four temperate liana species: to overtop host canopies or not? *Journal of Ecology*, **99**, 1071–1080.
Jiménez-Castillo, M. & Lusk, C.H. (2009) Host infestation patterns of the massive liana *Hydrangea serratifolia* (Hydrangeaceae) in a Chilean temperate rainforest. *Australian Ecology*, **34**, 829–834.
Kodani, J. (2006) Species diversity of broad-leaved trees in *Cryptomeria japonica* plantations in relation to the distance from adjacent broad-leaved forests. *Journal of Forest Research*, **11**, 267–274.
Ladwig, L.M. & Meiners, S.J. (2009) Impacts of temperate lianas on tree growth in young deciduous forests. *Forest Ecology and Management*, **259**, 195–200.
Ladwig, L.M. & Meiners, S.J. (2010a) Liana host preference and implications for deciduous forest regeneration. *The Journal of the Torrey Botanical Society*, **137**, 103–112.
Ladwig, L.M. & Meiners, S.J. (2010b) Spatiotemporal dynamics of lianas during 50 years of succession to temperate forest. *Ecology*, **91**, 671–680.
Ladwig, L., Meiners, S., Pisula, N. & Lang, K. (2012) Conditional allelopathic potential of temperate lianas. *Plant Ecology*, 1–9. doi: 10.1007/s11258-012-0087-4
LaFleur, N.E., Rubega, M.A. & Elphick, C.S. (2007) Invasive fruits, novel foods, and choice: an investigation of European starling and American Robin frugivory. *Wilson Journal of Ornithology*, **119**, 429–438.
LaFleur, N., Rubega, M. & Parent, J. (2009) Does frugivory by European starlings (*Sturnus vulgaris*) facilitate germination in invasive plants? *Journal of the Torrey Botanical Society*, **136**, 332–341.
Larson, K.C. (2000) Circumnutation behavior of an exotic honeysuckle vine and its native congener: influence on clonal mobility. *American Journal of Botany*, **87**, 533–538.
Leicht, S.A. (2005) The comparative ecology of an invasive bittersweet species (*Celastrus orbiculatus*) and its native congener (*C. scandens*). PhD dissertation. University of Connecticut.
Leicht, S.A. & Silander, J.A. (2006) Differential responses of invasive *Celastrus orbiculatus* (Celastraceae) and native *C. scandens* to changes in light quality. *American Journal of Botany*, **93**, 972–977.
Leicht-Young, S.A., Latimer, A.M. & Silander, J.A. (2011) Lianas escape self-thinning: Experimental evidence of positive density dependence in temperate lianas *Celastrus orbiculatus* & *C. scandens*. *Perspectives in Plant Ecology Evolution and Systematics*, **13**, 163–172.
Leicht-Young, S.A., Pavlovic, N.B., Frohnapple, K.J. & Grundel, R. (2010) Liana habitat and host preferences in northern temperate forests. *Forest Ecology and Management*, **260**, 1467–1477.
Leicht-Young, S.A., Silander, J.A. & Latimer, A.M. (2007) Comparative performance of invasive and native *Celastrus* species across environmental gradients. *Oecologia*, **154**, 273–282.
Letcher, S.G. & Chazdon, R.L. (2009) Lianas and self-supporting plants during tropical forest succession. *Forest Ecology and Management*, **257**, 2150–2156.
Londré, R.A. & Schnitzer, S.A. (2006) The distribution of lianas and their change in abundance in temperate forests over the past 45 years. *Ecology*, **87**, 2973–2978.
Lutz, H.J. (1943) Injuries to trees caused by *Celastrus* & *Vitis*. *Bulletin of the Torrey Botanical Club*, **70**, 436–439.
Mandle, L., Bufford, J.L., Schmidt, I.B. and Daehler, C.C. (2011) Woody exotic plant invasions and fire: reciprocal impacts and consequences for native ecosystems. *Biological Invasions*, **13**, 1815–1827.
Mehrhoff, L.J., Silander, J.A., Leicht, S.A., et al. (2003) *IPANE: Invasive Plant Atlas of New England* (http://www.ipane.org). Department of Ecology and Evolutionary Biology. University of Connecticut, Storrs, CT.
Merow, C., LaFleur, N., Silander, J.A. et al. (2011) Developing dynamic mechanistic species distribution models: predicting bird-mediated spread of invasive plants across northeastern North America. *American Naturalist*, **178**, 30–43.
Mosher, E.S., Silander, J.A. & Latimer, A.M. (2009) The role of land-use history in major invasions by woody plant species in the northeastern North American landscape. *Biological Invasions*, **11**, 2317–2328.

Nuzzo, V.A., McClain, W. & Strole, T. (1996) Fire impact on groundlayer flora in a sand forest 1990–1994. *The American Midland Naturalist*, **136**, 207–221.

Okerman, A. (2000) Combating the "ivy desert": the invasion of *Hedera helix* (English ivy) in the Pacific Northwest United States [Online]. Restoration and Reclamation Review. University of Minnesota–Twin Cities, Department of Horticultural Science, St. Paul, Minnesota http://purl.umn.edu/59738 (accessed 11 September 2012).

Paul, G.S. & Yavitt, J.B. (2011) Tropical vine growth and the effects on forest succession: a review of the ecology and management of tropical climbing plants. *Botanical Review*, **77**, 11–30.

Pavlovic, N.B. & Leicht-Young, S.A. (2011) Are temperate mature forests buffered from invasive lianas? *Journal of the Torrey Botanical Society*, **138**, 85–92.

Pisula, N.L. & Meiners, S.J. (2010) Relative allelopathic potential of invasive plant species in a young disturbed woodland. *The Journal of the Torrey Botanical Society*, **137**, 81–87.

Poorter, L., Kitajima, K., Mercado, P. et al. (2010) Resprouting as a persistence strategy of tropical forest trees: Relations with carbohydrate storage and shade tolerance. *Ecology*, **91**, 2613–2627.

Proença, V., Pereira, H.M. & Vicente, L. (2010) Resistance to wildfire and early regeneration in natural broadleaved forest and pine plantation. *Acta Oecologica*, **36**, 626–633.

Putz, F.E. (1984) The natural history of lianas on Barro Colorado Island, Panama. *Ecology*, **65**, 1713–1724.

Putz, F.E. (1995a) Relay ascension of big trees by vines in Rock Creek Park, District of Columbia. *Castanea*, **60**, 167–169.

Putz, F.E. (1995b) Vines in treetops: consequences of mechanical dependence. In *Forest Canopies* (eds. M.D. Lowman & N.M. Nadkarni), pp. 311–323. Academic Press, Inc., San Diego, CA.

Putz, F.E. & Holbrook, N.M. (1991) Biomechanical studies of vines, in *The Biology of Vines* (eds F.E. Putz & H.A. Mooney), Cambridge University Press, Cambridge, pp. 73–98.

Qian, H. & Ricklefs, R.E. (1999) A comparison of the taxonomic richness of vascular plants in China and the United States. *The American Naturalist*, **154**, 160–181.

Rashid, M.H., Asaeda, T. and Uddin, M.N. (2010a) The allelopathic potential of kudzu (*Pueraria montana*). *Weed Science*, **58**, 47–55.

Rashid, M.H., Asaeda, T., & Uddin, M.N. (2010b) Litter-mediated allelopathic effects of kudzu (*Pueraria montana*) on *Bidens pilosa* & *Lolium perenne* and its persistence in soil. *Weed Biology and Management*, **10**, 48–56.

Robertson, D.J., Robertson, M.C. and Tague, T. (1994) Colonization dynamics of four exotic plants in a northern Piedmont natural area. *Bulletin of the Torrey Botanical Club*, **121**, 107–118.

Rowe, N. & Speck, T. (2005) Plant growth forms: an ecological and evolutionary perspective. *New Phytologist*, **166**, 61–72.

Schierenbeck, K.A., Mack, R.N. & Sharitz, R.R. (1994) Effects of herbivory on growth and biomass allocation in native and introduced species of *Lonicera*. *Ecology*, **75**, 1661–1672.

Schierenbeck, K.A. & Marshall, J.D. (1993) Seasonal and diurnal patterns of photosynthetic gas exchange for *Lonicera sempervirens* & *L. japonica* (Caprifoliaceae). *American Journal of Botany*, **80**, 1292–1299.

Schnitzer, S.A. (2005) A mechanistic explanation for global patterns of liana abundance and distribution. *The American Naturalist*, **166**, 262–276.

Schnitzer, S.A. & Carson, W.P. (2010) Lianas suppress tree regeneration and diversity in treefall gaps. *Ecology Letters*, **13**, 849–857.

Schnitzer, S.A., Dalling, J.W. & Carson, W.P. (2000) The impact of lianas on tree regeneration in tropical forest canopy gaps: evidence for an alternative pathway of gap-phase regeneration. *Journal of Ecology*, **88**, 655–666.

Schnitzer, S.A., Kuzee, M.E. & Bongers, F. (2005) Disentangling above- and below-ground competition between lianas and trees in a tropical forest. *Journal of Ecology*, **93**, 1115–1125.

Schnitzer, S.A., Mangan, S.A., Dalling, J.W. et al. (2012) Liana abundance, diversity, and distribution on Barro Colorado Island, Panama. *PLoS ONE*, **7**, e52114.

Schnitzer, S.A., Parren, M.P.E. & Bongers, F. (2004) Recruitment of lianas into logging gaps and the effects of pre-harvest climber cutting in a lowland forest in Cameroon. *Forest Ecology and Management*, **190**, 87–98.

Schnitzer, S.A., van der Heijden, G.M.F., Mascaro, J., & Carson, W.P. (2014) Lianas reduce biomass accumulation in a tropical forest. *Ecology*. In press.

Schweitzer, J.A. & Larson, K.C. (1999) Greater morphological plasticity of exotic honeysuckle species may make them better invaders than native species. *Journal of the Torrey Botanical Society*, **126**, 15–23.

Selaya, N.G., Anten, N.P.R., Oomen, R.J. et al. (2007) Above-ground biomass investments and light interception of tropical forest trees and lianas early in succession. *Annals of Botany*, **99**, 141–151.

Siccama, T.G., Weir, G. & Wallace, K. (1976) Ice damage in a mixed hardwood forest in Connecticut in relation to *Vitis* infestation. *Bulletin of the Torrey Botanical Club*, **103**, 180–183.

Smith, L.M. & Reynolds, H.L. (2012) Positive plant-soil feedback may drive dominance of a woodland invader, *Euonymus fortunei*. *Plant Ecology*, **213**, 853–860.

Sperry, J.S., Hollbrook, N.M., Zimmerman, M.H. & Tyree, M.T. (1987) Spring filling of xylem vessels in wild grapevine. *Plant Physiology*, **83**, 414–417.

Stevens, G.C. (1987) Lianas as structural parasites: the *Bursera simaruba* example. *Ecology*, **68**, 77–81.

Teramura, A.H., Gold, W.G. & Forseth, I.N. (1991) Physiological ecology of mesic, temperate woody vines, in *The Biology of Vines* (eds F.E. Putz & H.A. Mooney), Cambridge University Press, Cambridge, pp. 245–285.

Tibbetts, T.J. & Ewers, F.W. (2000) Root pressure and specific conductivity in temperate lianas: Exotic *Celastrus orbiculatus* (Celastraceae) vs. native *Vitis riparia* (Vitaceae). *American Journal of Botany*, **87**, 1272–1278.

Trusty, J.L., Lockaby, B.G., Zipperer, W.C. & Goertzen, L.R. (2007) Identity of naturalised exotic *Wisteria* (Fabaceae) in the south-eastern United States. *Weed Research*, **47**, 479–487.

Valladares, F., Gianoli, E. & Saldana, A. (2011) Climbing plants in a temperate rainforest understorey: searching for high light or coping with deep shade? *Annals of Botany*, **108**, 231–239.

Van der Heijden, G.M.F. & Phillips, O.L. (2009) Liana infestation impacts tree growth in a lowland tropical moist forest. *Biogeosciences*, **6**, 2217–2226.

Vesk, P.A. & Westoby, M. (2004) Sprouting ability across diverse disturbances and vegetation types worldwide. *Journal of Ecology*, **92**, 310–320.

Wang, R.L., Zeng, R.S., Peng, S.L. et al. (2011) Elevated temperature may accelerate invasive expansion of the liana plant *Ipomoea cairica*. *Weed Research*, **51**, 574–580.

Wright, S.J., Calderón, O., Hernandéz, A. & Paton, S. (2004) Are lianas increasing in importance in tropical forests? A 17-year record from Panama. *Ecology*, **85**, 484–489.

Zaya, D.N. (2013) Genetic characterization of invasion and hybridization: a bittersweet (*Celastrus* spp.) story, PhD dissertation, University of Illinois at Chicago.

Chapter 29

ECOLOGICAL EFFECTS OF LIANAS IN FRAGMENTED FORESTS

Mason Campbell, William F. Laurance, and Ainhoa Magrach

James Cook University, Cairns, Queensland, Australia

Ecology of Lianas, First Edition. Edited by Stefan A. Schnitzer, Frans Bongers, Robyn J. Burnham, and Francis E. Putz.

Companion Website: www.wiley.com/go/schnitzer/lianas

OVERVIEW

Understanding how biodiversity persists in the small fragments of forest that remain in many tropical regions is a vital priority. If lianas flourish in fragmented forests, as is expected, then they might have a wide array of ecological effects, including those on biodiversity. In this chapter, we review available studies on liana communities and liana–tree interactions in fragmented tropical forests. Although much remains unknown, it is apparent that lianas often increase dramatically in abundance in fragmented forests, especially those with large amounts of forest edge or recurring canopy disturbance. Where lianas are particularly abundant, they reduce tree survival, growth, fecundity, and regeneration. Abundant lianas also alter tree-community composition and reduce forest carbon storage, though the magnitude of these effects is variable and not fully understood. Finally, liana proliferation at the expense of trees affects rain forest fauna that are dependent on resources provided by trees such as fruits, nectar, foliage, and tree cavities, as well as fauna that capitalize on liana resources. If lianas benefit markedly in the future from rising atmospheric CO_2 levels or other global change phenomena, then they will become even more dominant in fragmented forests.

INTRODUCTION: TROPICAL FOREST FRAGMENTATION

Whether by happenstance or design, deforestation rarely removes all pre-existing vegetation in a given area (Laurance & Bierregaard 1997), but leaves isolated fragments of the original vegetation surrounded by new habitat types (Wilcove et al. 1986). This process of habitat fragmentation leads to the modification of a variety of biological and physical processes within the fragmented forests that can be deleterious to the constituent species and their ecological interactions (e.g., see reviews by Fahrig 2003; Fischer & Lindenmayer 2007; Laurance et al. 2011). However, remnant forest fragments now represent a large proportion of the remaining tropical forested area (Achard et al. 2002; Broadbent et al. 2008) and despite their degradation, they provide an important biodiversity repository for many landscapes including the preservation of many rare and endangered species and threatened ecosystems (e.g., Guindon 1996; Tabanez & Viana 2000; Arroyo-Rodríguez & Mandujano 2006; Muthuramkumar et al. 2006; Arroyo-Rodríguez et al. 2009). The importance of remnant tropical forest fragments for biodiversity conservation increases with continued worldwide tropical forest loss (Achard et al. 2002; Broadbent et al. 2008).

Maximization of the conservation values of forest fragments requires that they are not only retained, but are managed effectively, which necessitates an understanding of their ecology. One potentially important, yet minimally examined component of fragmented tropical forests is the liana community: how lianas respond to forest fragmentation and the antagonistic interaction that they have with their tree hosts (Fig. 29.1, Jacobs 1976; Schnitzer & Bongers 2002; Toledo-Aceves, Chapter 12 in this volume).

Fig. 29.1 Lianas proliferating along an abrupt forest edge in Gabon, central Africa. (Source; Photo by William Laurance. Reproduced with permission.)

FRAGMENTATION EFFECTS ON LIANA DIVERSITY

Lianas generally comprise 20–25% of the woody species diversity in undisturbed tropical forest (Putz 1984b; Gentry 1991; Appanah et al. 1992) but can be as high as 35% of the species in some forests, such as the one on Barro Colorado Island, Panama (Schnitzer et al., Chapter 7 in this volume). During the loss of forest concurrent with the initial fragmentation process, local extirpation of many sparsely distributed species of trees and lianas can occur, resulting in decreased landscape-scale species diversity (Laurance et al. 1999; Zhu et al. 2004). After the initial forest loss, however, liana diversity is usually proportionally greater in forest fragments (relative to trees) than in comparable undisturbed forest, with this enhanced diversity linked to increased forest edge area and elevated disturbance levels (Laurance 1997; Oliveira et al. 1997; Laurance et al. 2001; Schnitzer & Bongers 2002; Zhu et al. 2004).

Although an increase in local liana diversity compared to trees within forest fragments is the more common trend, two major factors can potentially depress liana diversity in fragments. First, fragmentation results in a number of potential impacts on populations and communities, such as genetic drift, isolation of breeding populations, propagule-dispersal limitation, and pollination limitation through local extirpation of obligatory dispersers or pollinators, which may decrease the diversity of both trees and lianas within forest fragments (e.g., Aizen & Feinsinger 1994; Young et al. 1996; Benitez-Malvido & Martínez-Ramos 2003). Second, a collapse in the availability of structural hosts (available trees) within heavily disturbed forest fragments can also reduce liana diversity (Muthuramkumar et al. 2006; Arroyo-Rodríguez & Toledo-Aceves 2009; Addo-Fordjour et al. 2012). This host loss can occur through both continued anthropogenic disturbances within fragments (e.g., repeated logging; Muthuramkumar et al. 2006; Arroyo-Rodríguez & Toledo-Aceves 2009) or through cascades of impacts initiated by forest loss and fragmentation (e.g., enhanced tree mortality (Laurance et al. 1998a; Laurance et al. 2000; Laurance et al. 2006a), such as edge effects, that alter forest microclimate and increase wind damage (Kapos 1989; Williams-Linera 1990; Laurance & Curran 2008).

FOREST FRAGMENTATION EFFECTS ON LIANA ABUNDANCE

Fragmentation of once-continuous primary forests results in a considerable increase in landscape-wide liana abundance and rates of tree infestation (Laurance 1997; Oliveira et al. 1997; Viana et al. 1997; Laurance et al. 2001; Benitez-Malvido & Martínez-Ramos 2003). There are three main reasons for the increase in liana abundance. First, the area of forest edge greatly increases within fragmented forest landscapes (Laurance & Yensen 1991) and forest edges are preferential liana habitat (Figs. 29.2, 29.3, Laurance 1997; Laurance et al. 2001; Oliveira et al. 1997; Londré & Schnitzer 2006). Second, elevated rates of large tree mortality, turnover and treefall-gap creation (Laurance et al. 1997; Laurance et al. 1998a; Laurance et al. 2000; Hill & Curran 2003; Laurance et al. 2006a) occur in many forest fragments, which again enhance the amount of available disturbed and well-lit habitat preferred by lianas (Schnitzer & Bongers 2002, 2011). Finally, initial forest loss and fragmentation may lead to a greater area of forest regeneration ("younger" forest),

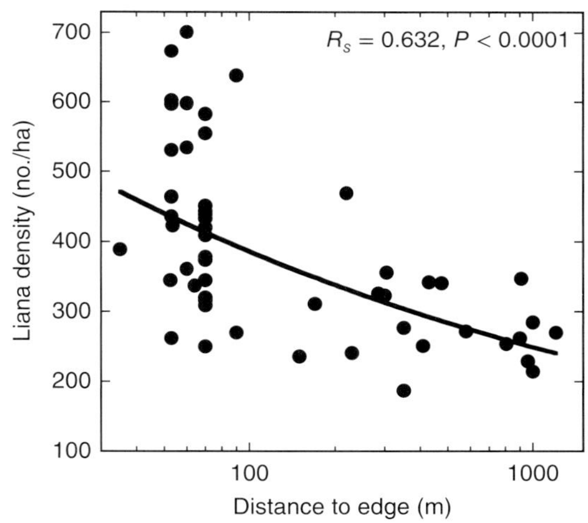

Fig. 29.2 Lianas tend to increase near the abrupt, artificial boundaries of forest fragments. Shown is the number of liana stems (≥2 cm diameter) within 1-hectare plots as a function of distance of plots from the nearest forest edge, in the rain forests of the central Amazon (results from a Spearman rank correlation).

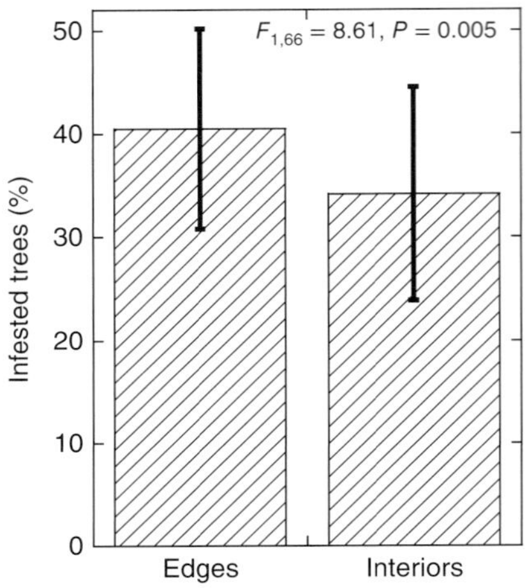

Fig. 29.3 In Amazonian forests, liana infestation rates (the proportion of trees with at least one liana) are higher near forest edges than in forest interiors (results from a one-way ANOVA). (Source: Adapted from Laurance et al. 2001. Reproduced with permission of Ecological Society of America.)

which again harbors an increased liana abundance and diversity when compared to equivalent unfragmented ("older") forest (Dewalt et al. 2000; Letcher & Chazdon 2009; Letcher, Chapter 10 in this volume).

LIANA IMPACT UPON FRAGMENTED VEGETATION COMMUNITIES

Any increase in liana abundance may be highly detrimental to the tree community of a fragmented forest due to the enhanced structural stress and increased resource competition experienced by trees infested with lianas (Fig. 29.1, Putz 1984b; Stevens 1987; Schnitzer & Bongers 2002; Schnitzer et al. 2005; Toledo-Aceves, Chapter 12 in this volume). Liana infestation may even contribute to the death of individual trees (Putz 1984b; Clark & Clark 1990; Schnitzer & Bongers 2002; Ingwell et al. 2010; van der Heijden et al., Chapter 13 in this volume). Consequently, a fragment-wide decline or extirpation of vulnerable tree species may occur, changing the composition and diversity of the tree community (Laurance et al. 1997; Laurance et al. 2001; Laurance et al. 2006b; van der Heijden et al., Chapter 13 in this volume). This problem is particularly pertinent for "vulnerable" tree species whose morphology leads to a high probability of liana infestation (Putz 1980, 1984a; Hegarty 1991; Talley et al. 1996; Schnitzer & Bongers 2002) and for tree species that are isolated from other meta-populations or otherwise restricted in their potential for recruitment (Young et al. 1996).

Lianas may also hasten the decline of tree species diversity within forest fragments via their differential impact on different successional guilds of trees. Traits that serve as liana "defence," such as fast growth, large leaves, and few branches, occur more often in pioneer or secondary succession species than in large mature-phase (shade-tolerant) tree species (Putz 1980, 1984a; Clark & Clark 1990; Schnitzer & Bongers 2002). Consequently, mature-phase tree species host lianas more frequently (Clark & Clark 1990; Laurance et al. 2001; Schnitzer & Bongers 2002; Schnitzer & Carson 2010), and increased liana abundance within forest fragments may contribute to their decline or loss (Phillips & Gentry 1994; Laurance et al. 2000; Laurance et al. 2001; Laurance et al. 2006a). The loss of mature-phase tree species may be exacerbated through a synergism with the enhanced wind shear and altered microclimates that forest fragments experience (Kapos 1989; Williams-Linera 1990; Laurance & Curran 2008), accelerating the alteration of the tree community composition (Laurance et al. 1998a; Laurance et al. 2006a).

A third mechanism by which lianas may alter the tree composition of a forest fragment is via their impact upon the succession process itself. Lianas can alter the succession pathway and eventual vegetation type, or arrest succession within treefall gaps (Schnitzer et al. 2000; Schnitzer & Carson 2001, 2010; Schnitzer & Bongers 2002, 2005; Toledo-Aceves & Swaine 2008; Letcher, Chapter 10 in this volume; Toledo-Aceves, Chapter 12 in this volume). Additionally, lianas may promote treefall-gap formation by elevating tree mortality (Putz 1984b; Clark & Clark 1990; Schnitzer & Bongers 2002; Schnitzer, Chapter 30 in this volume) and collateral damage during a treefall event (Appanah & Putz 1984; Putz 1984b). Accordingly, an increased liana abundance may alter the succession of the tree community and the formation of a canopy gap, promoting the loss of mature-phase tree species in fragments (Oliveira et al. 1997; Viana et al. 1997).

FOREST BIOMASS

In addition to altering the composition of the tree community, lianas can also suppress tree biomass in fragments (Laurance et al. 1997; Nascimento & Laurance 2004; van der Heijden et al., Chapter 13 in this volume). When lianas are abundant, they can kill or reduce growth in trees and thereby reduce the ability of fragmented forests to sequester and store carbon (Fig. 29.4; Laurance et al. 1997; Laurance et al. 1998b; Laurance et al. 2001; Phillips et al. 2002; Nascimento & Laurance 2004). Any increase in liana biomass is relatively negligible and does not offset the loss of tree biomass, because lianas generally comprise less than one-tenth of the aboveground biomass even in disturbed forests (Hegarty & Caballé 1991; DeWalt et al. 2000; Gerwing & Farias 2000). Given that tropical forests store ~44% (or 228 billion tons of carbon, Baccini et al. 2012) of the globe's terrestrial vegetation-derived carbon (Dixon et al. 1994; Phillips et al. 1998; Malhi & Grace 2000), liana effects on fragmented forest biomass could have nontrivial impacts on the global carbon cycle (van der Heijden et al., Chapter 13 in this volume).

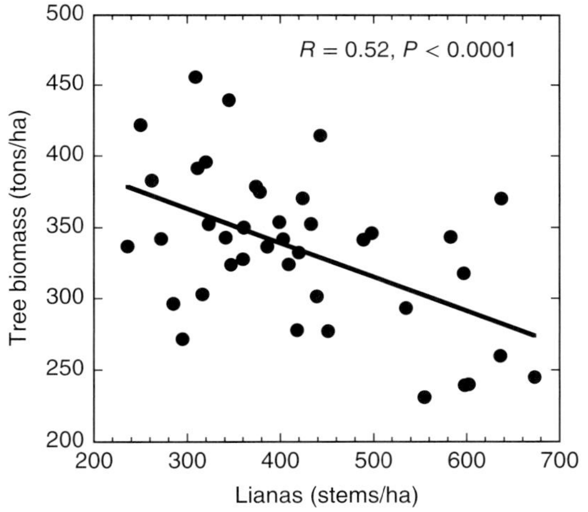

Fig. 29.4 Lianas are negatively correlated with the aboveground biomass of live trees in Amazonian forests (results from a Pearson correlation; adapted from Laurance et al. 2001). This negative relationship can arise both because lianas reduce tree survivorship and growth and because external disturbances, such as windstorms, can fell or damage trees and thereby create disturbed conditions favored by lianas. (Source: Adapted from Laurance et al. 2001. Reproduced with permission of Ecological Society of America.)

FUTURE LIANA INCREASE WITHIN FRAGMENTED FORESTS

Lianas seem likely to increase in abundance in fragmented forests for three reasons. First, future climatic predictions suggest tropical storms will become more frequent and increase in intensity (Emanuel 2005; Elsner et al. 2008) and trees in fragmented forests are known to display an increased vulnerability to wind damage (Laurance & Curran 2008). Therefore, in the future, forest fragments are likely to experience elevated tree turnover rates, greater vegetation disturbance, and substantial changes in forest microclimates (Webb 1958; Turton & Siegenthaler 2004; Laurance & Curran 2008), all of which could favor lianas (Putz 1984b; Schnitzer & Bongers 2002).

Second, lianas achieve their peak abundance in tropical forests with a pronounced dry season (Gentry 1991; Schnitzer 2005; DeWalt et al. 2010; DeWalt et al, Chapter 11 in this volume) and rainfall in many tropical regions is projected to increase in seasonality (Malhi & Wright 2004). A transition from wetter to more seasonal forest types would favor an increase in liana abundance (see also Schnitzer & Bongers 2011; Schnitzer, Chapter 30 in this volume). This process might be magnified in fragmented forests because forest edges are often prone to desiccation (Kapos 1989; Williams-Linera 1990).

Finally, liana growth rates may increase proportionately more than tree growth rates in response to rising atmospheric CO_2 levels (Granados & Korner 2002; Phillips et al. 2002; Schnitzer & Bongers 2011). If so, there could be a shift in the competitive interactions between trees and lianas in fragmented forests even more in favor of lianas (Tabanez & Viana 2000; Dalling et al. 2012). However, the extent of any potential increase in liana growth might be somewhat ameliorated if increased air temperatures limit the duration of transpiration periods (Betts et al. 1997).

CONCLUSION

Fragmented forests are ubiquitous in tropical landscapes and are occasionally the last surviving remnants of rare habitats and species for which conservation is an urgent priority. Given the potent role that lianas can play in some fragmented forests, and are likely to play in the future, understanding how lianas affect forest ecology and ecological interactions is a key priority. In some circumstances, control and management of lianas might be necessary if lianas are having major deleterious impacts on rare ecosystems or species.

ACKNOWLEDGMENTS

This research was supported by an ARC Discovery Grant awarded to William Laurance. We thank Will Edwards and Francis Putz for kindly reviewing earlier drafts of the manuscript.

REFERENCES

Achard, F., Eva, H.D., Stibig, H.-J. et al. (2002) Determination of deforestation rates of the world's humid tropical forests. *Science*, **297**, 999–1002.

Addo-Fordjour, P., Rahmad, Z.B., Amui, J. et al. (2012) *Patterns of liana community diversity and structure in a tropical rain forest reserve, Ghana: effects of human disturbance*, African Journal of Ecology.

Aizen, M.A. & Feinsinger, P. (1994) Forest fragmentation, pollination, and plant reproduction in a chaco dry forest, Argentina. *Ecology*, **75**, 330–351.

Appanah, S., Gentry, A.H. & LaFrankie, J.V. (1992) Liana diversity and species richness of Malaysian rain forests. *Journal of Tropical Forest Science*, **6**, 116–123.

Appanah, S. & Putz, F.E. (1984) Climber abundance in virgin dipterocarp forest and the effect of pre-felling climber cutting on logging damage. *Malaysian Forester*, **47**, 335–342.

Arroyo-Rodríguez, V. & Mandujano, S. (2006) The importance of tropical rain forest fragments to the conservation of plant species diversity in Los Tuxtlas, Mexico. *Biodiversity and Conservation*, **15**, 4159–4179.

Arroyo-Rodríguez, V., Pineda, E., Escobar, F. & Benitez-Malvido, J. (2009) Value of small patches in the conservation of plant-species diversity in highly fragmented rain forest. *Conservation Biology*, **23**, 729–739.

Arroyo-Rodríguez, V. & Toledo-Aceves, T. (2009) Impact of landscape spatial pattern on liana communities in tropical rain forests at Los Tuxtlas, Mexico. *Applied Vegetation Science*, **12**, 340–349.

Baccini, A., Goetz, S.J., Walker, W.S. et al. (2012) Estimated carbon dioxide emissions from tropical deforestation improved by carbon-density maps. *Nature Climate Change*, **2**, 182–185.

Benitez-Malvido, J. & Martinez-Ramos, M. (2003) Impact of forest fragmentation on understory plant species richness in Amazonia. *Conservation Biology*, **17**, 389–400.

Betts, R.A., Cox, P.M., Lee, S.E. & Woodward, F.I. (1997) Contrasting physiological and structural vegetation feedbacks in climate change simulations. *Nature*, **387**, 796–799.

Broadbent, E.N., Asner, G.P., Keller, M. et al. (2008) Forest fragmentation and edge effects from deforestation and selective logging in the Brazilian Amazon. *Biological Conservation*, **141**, 1745–1757.

Clark, D.B. & Clark, D.A. (1990) Distribution and effects on tree growth of lianas and woody hemiepiphytes in a Costa Rican tropicl wet forest. *Journal of Tropical Ecology*, **6**, 321–331.

Dalling, J.W., Schnitzer, S.A., Baldeck, C. et al. (2012) Resource-based habitat associations in a neotropical liana community. *Journal of Ecology*, **100**, 1174–1182.

DeWalt, S.J., Schnitzer, S.A., Chave, J. et al. (2010) Annual rainfall and seasonality predict pan-tropical patterns of liana density and basal area. *Biotropica*, **42**, 309–317.

DeWalt, S.J., Schnitzer, S. and Denslow, J.S. (2000) Density and diversity of lianas along a chronosequence in a central Panamanian lowland forest. *Journal of Tropical Ecology*, **16**, 1–19.

Dixon, R.K., Brown, S., Houghton, R.A. et al. (1994) Carbon pools and flux of global forest ecosystems. *Science*, **263**, 185–190.

Elsner, J.B., Kossin, J.P. & Jagger, T.H. (2008) The increasing intensity of the strongest tropical cyclones. *Nature*, **455**, 92–95.

Emanuel, K. (2005) Increasing destructiveness of tropical cyclones over the past 30 years. *Nature*, **436**, 686–688.

Fahrig, L. (2003) Effects of habitat fragmentation on biodiversity. *Annual Review of Ecology, Evolution, and Systematics*, **34**, 487–515.

Fischer, J. & Lindenmayer, D.B. (2007) Landscape modification and habitat fragmentation: a synthesis. *Global Ecology and Biogeography*, **16**, 265–280.

Gentry, A.H. (1991) The distribution and evolution of climbing plants, in *The Biology of Vines* (eds F.E. Putz & H.A. Mooney), Cambridge University Press, Cambridge, pp. 3–49.

Gerwing, J.J. & Farias, D.L. (2000) Integrating liana abundance and forest stature into an estimate of total aboveground biomass for an eastern Amazonian forest. *Journal of Tropical Ecology*, **16**, 327–335.

Granados, J. & Korner, C. (2002) In deep shade, elevated CO_2 increases the vigor of tropical climbing plants. *Global Change Biology*, **8**, 1109–1117.

Guindon, C.F. (1996) The importance of forest fragments to the maintenance of regional biodiversity in Costa Rica, in *Forest Patches in Tropical Landscapes* (eds J. Schelhas & R. Greenberg), Island Press, Washington, DC, pp. 168–186.

Hegarty, E.E. (1991) Vine-host interactions, in *The Biology of Vines* (eds F.E. Putz & H.A. Mooney), Cambridge University Press, Cambridge, pp. 357–376.

Hegarty, E.E. & Caballé, G. (1991) Distribution and abundance of vines in forest communities, in *The Biology of Vines* (eds F.E. Putz & H.A. Mooney), Cambridge University Press, Cambridge, pp. 313–335.

Hill, J.L. and Curran, P.J. (2003) Area, shape and isolation of tropical forest fragments: effects on tree species diversity and implications for conservation. *Journal of Biogeography*, **30**, 1391–1403.

Ingwell, L.L., Wright, S.J., Becklund, K.K. et al. (2010) The impact of lianas on 10 years of tree growth and mortality on Barro Colorado Island, Panama. *Journal of Ecology*, **98**, 879–887.

Jacobs, M. (1976) The study of lianas. *Flora Malesiana Bulletin*, **29**, 2610–2618.

Kapos, V. (1989) Effects of isolation on the water status of forest patches in the Brazilian Amazon. *Journal of Tropical Ecology*, **5**, 173–185.

Laurance, W.F. (1997) Hyper-disturbed parks: edge effects and the ecology of isolated rain forest reserves in tropical Australia, in *Tropical Forest Remnants: Ecology, Management, and Conservation of Fragmented Communities* (eds W.F. Laurance & R.O. Bierregaard Jr.), University of Chicago Press, Chicago, pp. 71–83.

Laurance, W.F. and Bierregaard, R.O. Jr. (eds) (1997) *Tropical Forest Remnants: Ecology, Management, and Conservation of Fragmented Communities*, University of Chicago Press, Chicago.

Laurance, W.F., Camargo, J.L.C., Luizão, R.C.C. et al. (2011) The fate of Amazonian forest fragments: a 32-year investigation. *Biological Conservation*, **144**, 56–67.

Laurance, W.F. & Curran, T.J. (2008) Impacts of wind disturbance on fragmented tropical forests: a review and synthesis. *Australian Ecology*, **33**, 399–408.

Laurance, W.F., Delamonica, P., Laurance, S.G. et al. (2000) Conservation: rain forest fragmentation kills big trees. *Nature*, **404**, 836–836.

Laurance, W.F., Ferreira, L.V., Rankin-de Merona, J.M. & Laurance, S.G. (1998) Rain forest fragmentation and the dynamics of Amazonian tree communities. *Ecology*, **79**, 2032–2040.

Laurance, W.F., Gascon, C. & Rankin-de Merona, J.M. (1999) Predicting effects of habitat destruction on plant communities: a test of a model using Amazonian trees. *Ecological Applications*, **9**, 548–554.

Laurance, W.F., Laurance, S.G. & Delamonica, P. (1998) Tropical forest fragmentation and greenhouse gas emissions. *Forest Ecology and Management*, **110**, 173–180.

Laurance, W.F., Laurance, S.G., Ferreira, L.V. et al. (1997) Biomass collapse in Amazonian forest fragments. *Science*, **278**, 1117–1118.

Laurance, W.F., Nascimento, H.E.M., Laurance, S.G. et al. (2006a) Rapid decay of tree-community composition in Amazonian forest fragments. *Proceedings of the National Academy of Sciences of the United States of America*, **103**, 19010–19014.

Laurance, W.F., Nascimento, H.E.M., Laurance, S.G. et al. (2006b) Rain forest fragmentation and the proliferation of successional trees. *Ecology*, **87**, 469–482.

Laurance, W.F., Perez-Salicrup, D., Delamonica, P. et al. (2001) Rain forest fragmentation and the structure of Amazonian liana communities. *Ecology*, **82**, 105–116.

Laurance, W.F. & Yensen, E. (1991) Predicting the impacts of edge effects in fragmented habitats. *Biological Conservation*, **55**, 77–92.

Letcher, S.G. & Chazdon, R.L. (2009) Lianas and self-supporting plants during tropical forest succession. *Forest Ecology and Management*, **257**, 2150–2156.

Londré, R.A. & Schnitzer, S.A. (2006) The distribution of lianas and their change in abundance in temperate forests over the past 45 years. *Ecology*, **87**, 2973–2978.

Malhi, Y. & Grace, J. (2000) Tropical forests and atmospheric carbon dioxide. *Trends in Ecology and Evolution*, **15**, 332–337.

Malhi, Y. & Wright, J. (2004) Spatial patterns and recent trends in the climate of tropical rain forest regions. *Philosophical Transactions of the Royal Society B: Biological Sciences*, **359**, 311–329.

Muthuramkumar, S., Ayyappan, N., Parthasarathy, N. et al. (2006) Plant community structure in tropical rain forest fragments of the western Ghats, India. *Biotropica*, **38**, 143–160.

Nascimento, H.E.M. & Laurance, W.F. (2004) Biomass dynamics in Amazonian forest fragments. *Ecological Applications*, **14**, S127–S138.

Oliveira, A.T., de Mello, J.M. & Scolforo, J.R.S. (1997) Effects of past disturbance and edges on tree community structure and dynamics within a fragment of tropical semideciduous forest in south-eastern Brazil over a five-year period (1987–1992). *Plant Ecology*, **131**, 45–66.

Phillips, O.L. & Gentry, A.H. (1994) Increasing turnover through time in tropical forests. *Science*, **263**, 954–954.

Phillips, O.L., Malhi, Y., Higuchi, N. et al. (1998) Changes in the carbon balance of tropical forests: evidence from long-term plots. *Science*, **282**, 439–442.

Phillips, O.L., Martinez, R.V., Arroyo, L. et al. (2002) Increasing dominance of large lianas in Amazonian forests. *Nature*, **418**, 770–774.

Putz, F.E. (1980) Lianas vs trees. *Biotropica*, **12**, 224–225.

Putz, F.E. (1984a) How trees shed and avoid lianas. *Biotropica*, **16**, 19–23.

Putz, F.E. (1984b) The natural history of lianas on Barro-Colorado island, Panama. *Ecology*, **65**, 1713–1724.

Schnitzer, S.A. (2005) A mechanistic explanation for global patterns of liana abundance and distribution. *American Naturalist*, **166**, 262–276.

Schnitzer, S.A. & Bongers, F. (2002) The ecology of lianas and their role in forests. *Trends in Ecology and Evolution*, **17**, 223–230.

Schnitzer, S.A. & Bongers, F. (2005) Lianas and gap phase regeneration: implications for forest dynamics and species diversity, in *Forest Climbing Plants of West Africa: Diversity, Ecology and Management* (eds F. Bongers, M.P.E. Parren & D. Traore), CABI Publishing, Wallingford, pp. 59–72.

Schnitzer, S.A. & Bongers, F. (2011) Increasing liana abundance and biomass in tropical forests: emerging patterns and putative mechanisms. *Ecology Letters*, **14**, 397–406.

Schnitzer, S.A. & Carson, W.P. (2001) Treefall gaps and the maintenance of species diversity in a tropical forest. *Ecology*, **82**, 913–919.

Schnitzer, S.A. & Carson, W.P. (2010) Lianas suppress tree regeneration and diversity in treefall gaps. *Ecology Letters*, **13**, 849–857.

Schnitzer, S.A., Dalling, J.W. & Carson, W.P. (2000) The impact of lianas on tree regeneration in tropical forest canopy gaps: evidence for an alternative pathway of gap-phase regeneration. *Journal of Ecology*, **88**, 655–666.

Schnitzer, S.A., Kuzee, M.E. & Bongers, F. (2005) Disentangling above- and below-ground competition between lianas and trees in a tropical forest. *Journal of Ecology*, **93**, 1115–1125.

Stevens, G.C. (1987) Lianas as structural parasites: the *Bursera simaruba* example. *Ecology*, **68**, 77–81.

Tabanez, A.A.J. & Viana, V.M. (2000) Patch structure within Brazilian Atlantic forest fragments and implications for conservation. *Biotropica*, **32**, 925–933.

Talley, S.M., Setzer, W.N. & Jackes, B.R. (1996) Host associations of two adventitious-root-climbing vines in a north Queensland tropical rain forest. *Biotropica*, **28**, 356–366.

Toledo-Aceves, T. & Swaine, M.D. (2008) Effect of lianas on tree regeneration in gaps and forest understorey in a tropical forest in Ghana. *Journal of Vegetation Science*, **19**, 717–U770.

Turton, S.M. & Siegenthaler, D.T. (2004) Immediate impacts of a severe tropical cyclone on the microclimate of a rain forest canopy in north-east Australia. *Journal of Tropical Ecology*, **20**, 583–586.

Viana, V.M., Tabanez, A.A.J. & Batista, J.L. (1997) Dynamics and restoration of forest fragments in the Brazilian Atlantic Moist Forest, in Tropical Forest *Remnants* (eds W.F. Laurance & R.O. Bierregaard Jr.), University of Chicago Press, Chicago, pp. 351–365.

Webb, L.J. (1958) Cyclones as an ecological factor in tropical lowland rain forest, North Queensland. *Australian Journal of Botany*, **6**, 220–230.

Wilcove, D.S., McLellan, C.H., & Dobson, A.P. (1986) Habitat fragmentation in the temperate zone. In *Conservation Biology: The Science of Scarcity and Diversity* (ed. M.E. Soule), pp. 237–256. Sinauer Associates, Sunderland, MA.

Williams-Linera, G. (1990) Vegetation structure and environmental conditions of forest edges in Panama. *Journal of Ecology*, **78**, 356–373.

Young, A., Boyle, T. & Brown, T. (1996) The population genetic consequences of habitat fragmentation for plants. *Trends in Ecology & Evolution*, **11**, 413–418.

Zhu, H., Xu, Z.F., Wang, H. & Li, B.G. (2004) Tropical rain forest fragmentation and its ecological and species diversity changes in southern Yunnan. *Biodiversity and Conservation*, **13**, 1355–1372.

Chapter 30

INCREASING LIANA ABUNDANCE IN NEOTROPICAL FORESTS: CAUSES AND CONSEQUENCES

Stefan A. Schnitzer

University of Wisconsin-Milwaukee, WI, USA; Smithsonian Tropical Research Institute, Balboa, Panama

Ecology of Lianas, First Edition. Edited by Stefan A. Schnitzer, Frans Bongers, Robyn J. Burnham, and Francis E. Putz.

Companion Website: www.wiley.com/go/schnitzer/lianas

OVERVIEW

Lianas are increasing in density, biomass, and productivity in neotropical forests and there are 12 studies to date documenting this pattern. The pattern of increase seems to be consistent among a variety of mature tropical forests and has been reported for wet, moist, and dry forests. Lianas compete intensely with trees, which may have wide-ranging effects on tropical forest plant and animal communities, as well as on such ecosystem-level factors as carbon, nutrient, and water fluxes, as has been described in many of the chapters in this book. Therefore, the increase in liana abundances has potentially profound ramifications for tropical forest composition, diversity, and functioning.

Several hypotheses have been proposed to explain the phenomenon of increasing liana abundance in neotropical forests, including increases in forest disturbance and turnover (e.g., changes in land use and fragmentation), intensity of seasonal drought and lower annual rainfall, elevated atmospheric CO_2, and nutrient deposition. Each of these factors may contribute separately and synergistically to the observed patterns of increasing liana density and biomass. To date, however, there is compelling empirical evidence for a mechanistic response of lianas to only two of the proposed factors: increasing forest disturbance and increasing seasonal drought intensity. Lianas respond strongly to disturbance, which may provide an important regeneration niche for liana establishment. Lianas are also able to grow more rapidly than trees during seasonal drought, and thus lianas might increase proportionally more than trees from an increase in the duration and intensity of seasonal droughts. In addition, both drought and disturbance may interact synergistically to benefit lianas. Drought increases canopy tree mortality, forest turnover, and disturbance, all of which benefit lianas both directly and indirectly. Elevated atmospheric CO_2 and nutrient deposition may also contribute to liana increases by enabling lianas to grow more efficiently both during seasonal drought and in the hot, dry conditions that follow disturbance.

Lianas may be acting as niche constructors, creating the very environment that provides their particular regeneration niche. As lianas increase in abundance and biomass, they compete more intensely with trees, covering tree crowns and competing belowground for water and nutrients. Increasing competition from lianas increases tree mortality, elevating disturbance rates and thus promoting liana proliferation. Lianas appear to beget ever more lianas, and an initial increase in lianas following disturbance may set in motion a positive feedback of increasing liana-induced tree death and disturbance followed by greater liana proliferation.

To gain insight into the causes and rates of increasing liana abundance in tropical forests, ecologists urgently need to establish a widespread network of large-scale, long-term liana and tree monitoring plots in multiple forest types (e.g., tropical dry, moist, and wet forest). These plots, arrayed across forest types, can help us determine: (1) in which forest types lianas are increasing relative to trees; (2) the rate of liana increases in different forest types; (3) whether all liana species are increasing or, more likely, which subset of liana species is increasing; and (4) the functional traits associated with (or responsible for) the increase in certain liana species. To understand the causes of liana increases, we need direct experimental tests of the hypothesized mechanisms: (1) disturbance; (2) drought severity; (3) atmospheric CO_2; and (4) nutrients. Liana removal experiments will help us determine the effects of liana increases on tropical forests. The combination of large-scale monitoring plots and controlled experimental approaches will allow us to determine the rates, causes, and consequences of liana increases in tropical forests. Because lianas play an important role in nearly all aspects of tropical forest dynamics and functioning, a liana-monitoring network should be a priority for tropical biology.

INTRODUCTION

The increase in liana abundance may be one of the most striking structural changes now occurring in neotropical forests. Lianas are naturally abundant and diverse in lowland tropical forests, where they commonly constitute approximately 25% of the woody stems and species (Schnitzer & Bongers 2002; Paul & Yavitt 2010; Schnitzer et al. 2012; Schnitzer et al., Chapter 1 in this volume). Lianas compete with trees, decreasing tree growth and survival (Grauel & Putz 2004; Ingwell et al. 2010; Schnitzer & Carson 2010; Toledo-Aceves, Chapter 12 in this volume). They also dramatically reduce forest-level carbon uptake (van der Heijden & Phillips 2009; Schnitzer et al. 2014; van der Heijden et al, Chapter 13 in this volume), potentially alter soil fluxes (Powers, Chapter 14 in this volume), and influence movement and community structure of animals

(Arroyo-Rodríguez et al., Chapter 27 in this volume; Lambert & Halsey, Chapter 26 in this volume; Michel et al., Chapter 25 in this volume; Yanoviak, Chapter 24 in this volume). Therefore, changes in liana density and biomass will likely have important repercussions on forest plant and animal diversity and community composition, as well as on important ecosystem functions such as carbon uptake and storage, and water and nutrient fluxes. Because tropical forests contain most of the Earth's terrestrial species, store more than one-third of global terrestrial carbon, and are responsible for approximately one-third of global terrestrial net primary productivity (Dixon et al. 1994; Field et al. 1998; Pan et al. 2011), the ramifications of increasing liana abundance extend far beyond the tropics.

Over the past decade (since 2002), 12 studies have been published documenting liana increases in neotropical forests. Eight of these publications were highlighted in a comprehensive review and analysis of increasing lianas in tropical forests (Schnitzer & Bongers 2011; Schnitzer et al. 2011). Four additional publications have provided supporting data for the emerging pattern of increasing lianas in forests of Central and South America (Enquist & Enquist 2011; Schnitzer et al. 2012; Yorke et al. 2013; Laurance et al. 2014). These studies have been conducted in different forest types, different geographic areas, and have used different metrics of liana increase, and yet they all support the pattern of increasing lianas in neotropical forests.

This chapter synthesizes the available supporting evidence that lianas have increased in abundance and biomass over the past three decades in neotropical forests. It reviews the hypotheses for the most likely drivers of increasing liana abundance in tropical forests and provides the evidence (or lack thereof) for each of them. It concludes with recommendations for future studies of the patterns, mechanisms, and potential large-scale ramifications of liana increases in neotropical forests.

INCREASING LIANA ABUNDANCE IN NEOTROPICAL FORESTS

Nearly all of the studies documenting increases in liana abundance have focused on the longitudinal change in the liana community at a single site in the neotropics. Specifically, studies have been conducted in the Bolivian Amazon, the Brazilian Amazon, Costa Rica (in both dry and wet forest), French Guiana, Panama (four separate studies on Barro Colorado Island), and in the subtropical southeastern USA (Table 30.1). Below is a summary of the main findings of each of the studies.

At Nouragues Biological Station in French Guiana, Chave et al. (2008) reported that from 1992 to 2002 the abundance of large (≥10 cm diameter) lianas increased 1.8%, while large tree abundance decreased 4.6%. Both liana and tree biomass increased at Nouragues during this period, but liana biomass increased 60% faster (4.8% increase for lianas versus 3% increase for trees). In the highly seasonal forest of Noel Kempff Mercado National Park, in the Bolivian Amazon, Foster et al. (2008) used high-resolution, color, infrared videography and hyper-spectral imagery to quantify the regeneration of low-canopy, liana-suppressed areas from 1986 and 2000. The authors reported that the area of low-canopy, liana-suppressed areas increased nearly 60% across the landscape, from 1.64 ha to 2.61 ha.

Two studies from a seasonal, non-fragmented, old-growth forest in the central Amazon reported strong increases in liana abundance and biomass. For example, Benítez-Malvido and Martínez-Ramos (2003) reported a 500% increase in liana seedling density from 1993–1999, while tree and herbaceous plant seedling recruitment decreased over this period. In this same forest, Laurance et al. (2014) examined the change in liana density (stems ≥2 cm diameter) from 1999 until 2012 in 36 1-ha plots covering approximately 600 km^2 and reported that liana density had increased approximately 1% per year during the 13-year study period. Liana density increased predominantly in the smallest size class (2–3 cm), consistent with the findings for seedlings of Benítez-Malvido and Martínez-Ramos (2003).

Lianas are also increasing in Central American forests. There are now four separately published studies documenting the increase in lianas in the seasonally moist, semi-deciduous forest on Barro Colorado Island, Panama (BCI; Schnitzer et al., Chapter 7 in this volume). From 1986 until 2002, the proportion of liana productivity (measured as leaf litter production) increased by 57% (Wright et al. 2004), and liana flower production increased more than 125% faster than that of trees (Wright & Calderon 2006). Schnitzer et al. (2012) compared liana density on the BCI 50-ha plot (censused in 2007) with previous studies that had surveyed lianas in the same area nearly 30 years earlier and reported that liana density had increased

Table 30.1 Published studies documenting the increase in lianas in neotropical forests. Columns represent the original reference, the geographic location of the study, the period examined, the metric of liana response, and a brief synopsis of the study's main finding. The studies are presented chronologically by publication date.

Study	Location	Period	Forest type	Liana response	Main finding
Phillips et al. 2002	Central & South America	1979–2002	Various	Change in large (≥10 cm) and smaller (≥2.5 cm) liana stem density in 1-ha plots and 0.1-ha transects, respectively	Both large and small lianas increased in density and basal area over time (both absolute values and relative to trees)
Benítez-Malvido & Martínez-Ramos 2003	Amazonia; 70 km north of Manaus, Brazil	1993–1999	Seasonal tropical moist forest	Change in liana seedlings <1 m tall in 40 1 m^2 plots (2 blocks of 20 1 m^2 plots)	Liana seedling density increased 500%, from 2 to 12 seedlings m^{-2}. Tree seedling recruitment decreased in the same plots.
Wright et al. 2004	Barro Colorado Island, Panama (BCI)	1986–2002	Seasonal tropical moist forest	Liana seedling density and leaf and seed productivity (relative to trees). Leaf litter was collected in 59 0.25 m^2 litter traps located in 4 2.25-ha areas. Seed production was collected in 200 0.5 m^2 litter traps on the BCI 50-ha plot. Seedling density in 600 1 m^2 plots (3 per litter trap)	Proportion of liana leaf litter production increased 57% (10.9–17.1%). The change in liana seed production did not differ from that of trees. Liana seedling density did not increase relative to trees.
Wright & Calderon 2006	BCI, Panama	1987–2005	Seasonal tropical moist forest	Liana and tree flower and seed production in 200, 0.5 m^2 litter traps on the BCI 50 ha plot. A total of 33 liana species and 48 tree species were examined	Liana flower production increased 125% faster than trees; seed production did not vary among growth forms
Allen et al. 2007	Savannah River (SRS) & Congaree National Park (CNP), South Carolina, USA	1979–2001 (SRS) and 1990–2002 (CNP)	Subtropical floodplain forest. In 1989, CNP experienced significant hurricane damage.	Liana stem density and basal area (≥4.5 cm) in 5, 1-ha forest plots (SRS) and for stems (≥2.5 cm) in 6, 1-ha plots (CNP)	Liana stem density increased in both sites. Relative liana density and BA increased at SRS.
Chave et al. 2008	Nouragues Biological Station, French Guiana	1992–2002	Wet tropical forest	Change in density of large (≥10 cm diameter) lianas and trees in 22 ha (two large plots)	Liana density increased 1.8% while tree density decreased 4.6%; liana biomass increased 4.8% while tree biomass increased 3%

Foster et al. 2008	Noel Kempff Mercado National Park, Bolivia	1986–2000	Evergreen liana forest	Change in low-canopy, liana-suppressed forest area using remote sensing	Liana-suppressed areas increased from 1.64 ha to 2.61 ha across the landscape
Ingwell et al. 2010	BCI, Panama	1979–2007 and 1996–2007	Seasonal tropical moist forest	Change in the proportion of trees infested by lianas and the severity of liana infestation	Proportion of trees with liana infestation increased from 32% to 75% (from 1979–2007). Severity of tree infestation (trees with >75 of their canopy covered by lianas) increased 65% from (1996–2007)
Enquist & Enquist 2011	San Emilio Forest, Santa Rosa, Costa Rica	1976–1996	Tropical dry forest	Change in liana and tree density and basal area in a 16-ha plot.	Relative liana density (as a proportion to all stems) increased 45%, from 7.0% to 10.2%
Schnitzer et al. 2012	BCI, Panama	1979–2007	Seasonal tropical moist forest	Change in rooted liana density (stems ≥1 cm diameter) scaled to 1 ha	Liana density (stems ≥1 cm) increased 75%; liana density (≥5 cm) increased 140%
Yorke et al. 2013	La Selva Biological Station, Costa Rica	1999–2007	Tropical wet forest	Change in rooted liana density (stems ≥1 cm diameter) in 9,864 m^2 plots (6 old-growth, 3 selectively logged)	Liana density and basal area increased 14% and 20%, respectively, in the old-growth plots, but neither liana metric changed in the selectively logged plots
Laurance et al. 2014	Amazonia; 70 km north of Manaus, Brazil	1999–2012	Seasonal tropical moist forest	Change in rooted liana density (stems ≥2 cm diameter) in 36 1-ha plots	Liana density increased ~1% annually

75% for stems ≥1 cm diameter and 140% for stems >5 cm diameter (see also Schnitzer et al., Chapter 7 in this volume). In this same forest, Ingwell et al. (2010) found that the percentage of trees with liana infestation increased from 32% in 1967–1968 to 47% in 1979, and then to nearly 75% in 2007. In addition, the percentage of trees with severe liana infestation (i.e., >75% of their canopy covered by lianas) increased 65% from 1996–2007. Collectively, these studies provide substantial evidence that lianas are increasing on BCI.

In Costa Rica, liana increases have been reported in both dry and wet forests. In a seasonally dry tropical forest (San Emilio, Guanacaste Provence) in Costa Rica, Enquist and Enquist (2011) reported that over a 20-year period (1976–1996) the percentage of lianas in terms of total woody stem density (lianas plus trees) increased 45%, from 7.0% to 10.2%. The increase in the proportion of lianas in this forest was due to the combination of a 6.5% increase in liana stem density and a concomitant significant decrease in tree density. In a wet forest at La Selva Biological Station in Costa Rica, Yorke et al. (2013) found a 2.9% annual increase in mean liana basal area and a 2.0% annual increase in mean liana density (stems ≥1 cm diameter) in six old-growth plots over an 8-year period from 1999 to 2007. Thus, lianas in Costa Rican forests appear to be increasing in density and basal area both in absolute terms and relative to trees. Additional longitudinal data will be necessary to determine whether the liana increase trend is maintained.

In two subtropical bottomland hardwood forests in South Carolina, USA, Allen et al. (2007) reported that both absolute and relative liana basal area (compared to trees), as well as relative liana stem density increased significantly from the initial census. At the Congaree National Park, liana stem density and basal area increased faster than those of trees from 1990–2002, a period following a particularly large disturbance in 1989 by Hurricane Hugo, which killed ~20% of the trees ≥10 cm diameter (Allen et al. 2007).

One study documented the increase in liana density (relative to trees) in forests in northwest South America, Amazonia, and Central America. Specifically, Phillips et al. (2002) examined the change in lianas ≥2.5 cm diameter in 58 0.1-ha plots and lianas ≥10 cm diameter in 47 1-ha forest plots from 1979–2002 and found that both small and large lianas had increased significantly in stem density and basal area, both absolute values and relative to trees. The authors also found that the most rapid increases had occurred in the last decade of the study, which led the authors to predict that liana increases would likely continue into the future.

In contrast to neotropical forests, several studies in Africa have documented decreases in liana abundance, suggesting that lianas may be changing in an entirely different direction in Africa (Caballé & Martin 2001; Bongers & Ewango, Chapter 3 in this volume; Thomas et al., Chapter 2 in this volume). If so, the reason for changes in liana abundance may be linked to large-scale differences between the continents in terms of the abiotic and biotic changes that they are now experiencing. To date, however, there are still insufficient data from Africa to determine whether these initial findings constitute a consistent continental pattern (see Bongers & Ewango, Chapter 3 in this volume), and thus additional data are urgently needed to verify this interesting potential deviation from neotropical forests. Nonetheless, the pattern of increasing liana abundance in the neotropics is now compelling, and the remainder of the chapter focuses on neotropical forests.

LIANA INCREASES AND TROPICAL FOREST SUCCESSION

Forest age and successional status may determine the pattern of liana increase. Lianas recruit rapidly into tropical forests, and liana densities increase rapidly with forest age. For example, forests as young as five years old in Panama can have liana densities equal to or higher than adjacent old-growth forests (Barry et al. in review). Lianas continue to increase in density and biomass until forests are between 30–70 years old, after which liana density begins to decrease as the forest continues to mature (DeWalt et al. 2000; Letcher & Chazdon 2009; Barry et al. in review; Letcher, Chapter 10 in this volume). Long-term censuses in a young or recently disturbed forest may result in a pattern of increasing lianas that is driven by natural successional changes rather than climate forcing. By contrast, long-term censuses of forests that are just past peak liana density (i.e., forests between 30 and 70 years old) or forests that are recovering from large-scale disturbance may show a natural decrease in liana abundance, even when lianas are increasing in nearby old-growth forests. For example, from 1999 to 2008, Yorke et al. (2013) followed the fate of lianas at La Selva Biological Station in six old-growth forest plots and three adjacent forest plots that had been selectively logged decades early and found that liana density had increased in five of the six old-growth plots, but that liana density decreased slightly in all three selectively logged forest plots. Yorke and colleagues concluded that the natural decrease in lianas 30–40 years following disturbance (selective logging, in this case) was stronger than the forces causing increases in nearby old-growth forests. The increase in liana seedling recruitment in central Amazonia (~80 km north of Manaus, Brazil) was also found to decrease with the level of disturbance (habitat fragmentation), supporting the hypothesis that liana increases are most prevalent in undisturbed, old-growth forest (Benítez-Malvido & Martínez-Ramos 2003). From 1993–1999, liana seedlings increased by 500% in continuous primary forest, but liana increases were not as substantial in nearby fragmented forests: 150% increase in 100-ha fragments, 50% increase in 10-ha fragments, and 7% in 1-ha fragments (Benítez-Malvido & Martínez-Ramos 2003). Therefore, the successional status of a forest is an important consideration in predicting whether liana density will increase or decrease.

Currently, most studies documenting increasing neotropical liana density have been conducted in forests that are thought to be old-growth and largely undisturbed. The decrease in liana density in the Ituri forest in Democratic Republic of Congo between 1994 and 2007 may be an example of the effect of successional status on changes in liana abundance. Bongers and Ewango (Chapter 3 in this volume) found that from 1994 until 2007, liana density in two 10-ha plots in the Ituri forest decreased from 750 to 499 individuals ≥2 cm diameter per ha. The decrease of lianas was driven by the population collapse of the most abundant liana species (*Manniophyton fulvum*, Müll. Arg., Euphorbiaceae), which lost a stunning 97% of the individuals over the 13-year period. This species is present in closed-canopy forest, but it has also been reported to be an opportunistic species that can rapidly capitalize on disturbance and open areas adjacent to forests (e.g., roadsides, farmers' fields, and heavily disturbed forests). Although it is unclear why *M. fulvum* became the most dominant species at Ituri or why its population subsequently collapsed, one hypothesis is that the population increased dramatically after a previous disturbance and its decrease was the result of a predictable successional maturation of the forest (Bongers & Ewango, Chapter 3 in this volume). The *Manniophyton fulvum* example illustrates the need to determine the disturbance history and successional status of a forest in order to accurately interpret changes in forest composition and liana abundance.

PUTATIVE REASONS FOR INCREASING LIANA ABUNDANCE, BIOMASS, AND PRODUCTIVITY

Four hypotheses have been proposed to explain increasing liana density and biomass: (1) increasing levels of disturbance; (2) increasing evapotranspirative demand from increasingly severe seasonal drought; (3) increasing levels of atmospheric CO_2; and (4) increasing nutrient deposition. These proposed explanations are not mutually exclusive and are likely operating synergistically (Schnitzer & Bongers 2011). Below, I briefly describe each of the hypothesized mechanisms and provide available empirical support.

Increasing disturbance

Both natural and anthropogenic disturbances appear to promote liana regeneration and proliferation, and increasing disturbance could explain increasing liana abundance. Lianas recruit in high numbers and diversity and grow rapidly in the resource-rich conditions that exist following disturbance (Putz 1984; Schnitzer et al. 2000; Schnitzer & Carson 2001, 2010; Schnitzer 2005), and therefore disturbed areas may provide a key regeneration niche for lianas (Ledo & Schnitzer 2014). Indeed, liana distribution within a forest appears to be closely related to disturbance, with high liana densities in the most disturbed areas (Dalling et al. 2012; Schnitzer et al. 2012; Ledo & Schnitzer 2014). Many liana species are wind dispersed and are able to colonize disturbed areas (Puerto-Piñero et al. 2013); however, the additional capacity of many liana species for copious clonal stem production following disturbance increases with their ability to colonize disturbed areas (Schnitzer et al. 2012; Ledo & Schnitzer 2014).

Natural disturbance from canopy tree mortality may be increasing in many tropical forests because of increases in temperature, intensity of seasonal drought, nutrient deposition, and atmospheric CO_2 concentrations (Phillips & Gentry 1994; Wright 2010; Schnitzer & Bongers 2011; Schnitzer et al. 2011). Stronger and more intense droughts, including El Niño-caused droughts (Phillips et al. 2009; Lee & McPhaden 2010; Fu et al. 2013) result in greater tree mortality, leading to more canopy gaps in which lianas would proliferate. Increased nutrient deposition and atmospheric CO_2 concentrations can increase forest productivity, which could increase canopy tree growth and mortality (Phillips & Gentry 1994; Körner 2006). Increased temperature has also been linked to increased tree mortality. Although the tropics have been largely buffered from high temperature increases, even small temperature changes can increase night-time respiration rates in trees, which can lower tree growth and survival (Clark 2004).

Higher levels of canopy disturbance may also increase long-distance liana clonal colonization (Yorke et al. 2013), which can increase liana abundance. Large lianas can extend through the canopy for hundreds of meters and fall from the canopy long distances from their rooting point (Putz 1984). When lianas fall, they can produce numerous independently rooted clonal stems that can climb back to the canopy (Peñalosa 1984; Schnitzer et al. 2004, 2012), thus increasing the number of lianas in the forest. Clonal stem production is a common phenomenon in treefall gaps in which liana stems survive the fall of their host tree, re-sprout copiously, re-root, and eventually climb back to the forest canopy (Putz 1984; Schnitzer et al. 2000, 2012; Schnitzer et al., Chapter 7 in this volume). Increasing disturbance due to greater intensity of seasonal droughts, higher atmospheric CO_2, and land-use changes could increase the number of lianas that recruit as clonal stems, possibly explaining the pattern of increasing liana abundance in neotropical forests. Indeed, Yorke et al. (2013) reported that long-distance clonal dispersal accounted for 19% of the increase in liana density and 60% of the increase in liana basal area in an old-growth forest in Costa Rica over an 8-year period (1999–2007). Long-distance clonal colonization is one of the specialized ways that lianas may be responding to disturbance through prolific vegetative reproduction (e.g., Schnitzer et al. 2012).

Lianas themselves may be contributing to their own increase by acting as niche constructors (*sensu* Odling-Smee et al. 2003), killing their host trees and increasing forest turnover and gap formation – creating the very environment in which they proliferate. If so, then the increase in lianas, regardless of the initial cause, may set in motion a positive feedback of increasing liana-induced tree death and disturbance followed by greater liana regeneration. Indeed, lianas quickly colonize and dominate small natural disturbances such as treefall gaps, where they proliferate in density. Once established, lianas can dominate forest patches, smothering and eventually killing surrounding trees and remaining at high density for decades (Schnitzer et al. 2000; Foster et al. 2008).

Deforestation and logging are also common throughout the tropics (e.g., Wright 2005; Laurance et al. 2009), which may benefit lianas in terms of increased abundance and biomass. Lianas aggressively colonize young, regenerating successional forests following logging, pasture, and agriculture; liana densities can remain high for many decades thereafter. For example, Barry et al. (in review) found that forests in central Panama as young as 4 years old had liana densities on a par with nearby old-growth forests, though mean liana diameter was much smaller in the young forest. Even low-level logging, which is now evident in large portions of forests that were previously considered to be unaffected by humans (Asner et al. 2006), may promote liana regeneration but leave little trace of disturbance other than a tangle of lianas. Forest fragmentation resulting from road construction, clear-cutting, establishment of power lines, and other activities also increases liana density because lianas aggressively colonize the dry, high-light forest edge, where they provide a propagule source for liana colonization into the forest interior (Londré & Schnitzer 2006; Laurance et al. 2009; Arroyo-Rodríguez & Toledo-Aceves 2009).

Increasing severity of seasonal drought

The ability of lianas to grow rapidly in the dry, hot conditions following disturbance may also give an insight into their distribution within and among forests. Lianas peak in abundance in forests with low rainfall and high seasonality (Schnitzer 2005; DeWalt et al. 2010; DeWalt et al., Chapter 11 in this volume). One possible explanation for this distribution is that lianas grow more during the dry season (and in treefall gaps) when light and temperature are relatively high (high evapotranspirative demand, *sensu* Schnitzer & Bongers 2011), which results in higher liana abundance in drier forests (Schnitzer 2005). Thus, lianas may have a dry season growth advantage because they can grow during seasonal droughts when evapotranspirative demand is high and competing trees are dormant (Schnitzer 2005).

Empirical evidence for the hypothesis of dry season growth advantage is rapidly accumulating from precise growth and physiology measurements, phenological studies, and experimental approaches. For example, Schnitzer (2005) found that lianas in a Panamanian forest grew twice as fast as trees during the wet season and seven times as fast during the dry season, suggesting that lianas capitalized on dry season resources and suffered substantially less water stress than did trees. In a follow-up study measuring seasonal diameter growth of lianas and trees over a two-year period, Schnitzer and van der Heijden (unpublished data) found that lianas grew more than trees during the dry season, but that liana growth and tree growth were similar during the wet season. Furthermore, liana mortality was greater in the wet season than the dry season – a trend that was opposite that of trees. Using liana removal experiments, both Alvarez-Cansino et al. (in review) and Tobin et al. (2012) reported that lianas reduced tree sap velocity far more during the dry season than the wet season, presumably due to higher competition for soil water. Cai et al. (2009) found that in a seasonal tropical forest in southwestern China, lianas fixed carbon and used water more efficiently than trees and that this effect was particularly strong during the dry season. Also in southwestern China, Zhu & Cao (2009) found that lianas had higher hydraulic conductivity, maximum carbon assimilation rate, and predawn leaf water potential than those of trees during the dry season. Phenological studies document that, compared to trees, lianas have much lower levels of deciduousness, lianas maintain leaves longer into the dry season in very dry forests, and produce new leaves later in the dry season, presumably when water is most scarce (Putz & Windsor 1987; Opler et al. 1991; Condit et al. 2000; Kalácska et al. 2005). Collectively, these studies, using very different approaches, all support the hypothesis of dry season growth advantage.

Many tropical forests are getting hotter and drier, which is a viable explanation for the increase in liana abundance and biomass relative to competing trees. Air temperature and the severity of seasonal drought have increased, whereas rainfall has decreased in many parts of the tropics (Malhi & Wright 2005). Rainfall and dry season intensity appear to be particularly variable among and within tropical regions (Malhi & Wright 2005; Asner & Alencar 2010; Fu et al. 2013). For example, many Amazonian forests are experiencing decreasing precipitation both annually and during seasonal droughts (Phillips et al. 2009; Fu et al. 2013), which promotes liana growth relative to trees and may explain their documented increases. In southern Amazon, dry season length has increased significantly between 1979 and 2011 (Fu et al. 2013). In Panama, where four separate studies document profound increases in liana density, levels of tree infestation, and leaf and flower productivity (Wright et al. 2004; Wright & Calderon 2006; Ingwell et al. 2010; Schnitzer et al. 2012), annual precipitation decreased nearly 20% during 60 years (10-year running averages from 1930–1990; cited in Schnitzer 2005). In the dry forest at San Emilio forest in Costa Rica, where relative liana density increased 45% from 1976–1996, rainfall decreased significantly from 1920 until 1997 and, between 1976 and 1996, the number of dry years increased significantly (Enquist & Enquist 2011). Consequently, the concomitant long-term decrease in rainfall and increase in temperature and dry season length have resulted in higher evapotranspirative demand for longer periods, providing a viable hypothesis to explain recently documented increases in liana density and biomass.

Elevated atmospheric CO_2

The hypothesis that increasing CO_2 fertilization is causing the increase in liana abundance is attractive theoretically because it would explain the increase of lianas across forest type and broad geographic area. Also, lianas appear to have the functional traits and characteristics to respond rapidly to fertilization (Schnitzer et al. 2011; Asner & Martin, Chapter 21 in this volume). There are two hypothesized mechanisms by which CO_2 fertilization would increase liana abundance relative to trees: (1) greater liana leaf area ratio (LAR); and (2) greater liana water use efficiency.

The LAR hypothesis states that lianas should grow proportionately faster with elevated CO_2 because lianas have a relatively high ratio of leaf area to total plant mass compared to trees, and lianas therefore fix more carbon per total plant biomass. There is a positive feedback component to the LAR hypothesis because the rate of liana growth from CO_2 fertilization accelerates as lianas add more photosynthetic tissue per unit biomass gain compared to trees, thus accelerating the rate of whole-plant photosynthesis faster for lianas than for trees (Schnitzer & Bongers 2011). Indeed, lianas appear to have greater leaf area ratio than trees (Wyka et al. 2013), and there is strong evidence that lianas have much greater leaf area per unit sap wood area (Huber Value (HV); Isnard & Feild, Chapter 17 in this volume). However, it is currently unknown whether the lianas that have greater LAR and HV are the ones that are also increasing in abundance and biomass – a pattern that must be established before the LAR hypothesis can be considered viable.

The Water Use Efficiency–CO_2 (WUE–CO_2) hypothesis states that elevated CO_2 will increase water-use efficiency because plants will be able to uptake a higher concentration of carbon and lose less water per unit time that their stomata are open (reviewed by Eamus 1991). Increased water use efficiency is particularly important during the dry season when both lianas and trees become water-stressed (Schnitzer 2005). Because lianas are better adapted to grow under drought conditions than trees (Schnitzer 2005; Domingues et al. 2007; Cai et al. 2009; Zhu & Cao 2009), lianas may be pre-adapted to capitalize on the additional carbon gain per unit water loss during the dry season under elevated CO_2 conditions (Schnitzer & Bongers 2011).

Empirical support for elevated atmospheric CO_2 as the mechanism for liana increases remains weak. There is some evidence that lianas grow rapidly under elevated CO_2 concentrations (Granados & Körner 2002; Körner 2006), which has been used to justify increasing atmospheric CO_2 as an explanation for increases in liana abundance in tropical forests (e.g., Phillips et al. 2002; Laurance et al. 2014). While there is evidence that some temperate lianas may respond faster to elevated CO_2 concentrations than trees (Belote et al. 2003; Hättenschwiler & Körner 2003; Mohan et al. 2006; Zotz et al. 2006; but see Schnitzer et al. 2008a), there is little evidence that tropical lianas and trees are responding differently. In a recent study, Marvin et al. (in review) compared the growth response of seedlings of 11 liana and 10 tree species and reported no statistical differences between the two growth forms in their response to elevated CO_2. Consequently, both the LAR and WUE-CO_2 hypotheses are theoretically possible; however, there is currently little empirical evidence directly linking either mechanism to observed increases in liana abundance or biomass. Until the CO_2 fertilization hypotheses are tested explicitly, their contribution to increasing liana abundance in tropical forests will remain speculative.

Increased nutrient deposition

Another fertilization hypothesis to explain increases in liana abundance in tropical forests is the nutrient deposition hypothesis. This hypothesis posits that increases in liana abundance may be explained by greater nutrient deposition (Schnitzer et al. 2011, Asner & Martin, Chapter 21 in this volume), particularly nitrogen, which has been increasing rapidly in the tropics (Galloway et al. 2004; Hietz et al. 2011). The leaves of lianas tend to be smaller and thinner and have higher levels of nitrogen, phosphorus, calcium, and potassium per unit area and mass than do the leaves of their host trees (Salzer et al. 2006; Wyka et al. 2013; Asner & Martin, Chapter 21in this volume; Kazda, Chapter 22 in this volume). Lianas may be able to respond strongly to nutrient fertilization by increasing leaf nutrient levels, resulting in greater growth and photosynthetic capacity (Schnitzer et al. 2011), which could explain their relative increases in recent decades. Indeed, Asner and Martin (Chapter 21 in this volume) argue that greater nutrient concentration in liana leaves compared to trees may allow lianas to respond more rapidly to increasing CO_2. If so, there may be a single fertilization hypothesis that states that increased levels of both nutrients and CO_2 favor lianas more than trees and is thus responsible for increasing liana abundance.

Currently, there is little evidence for soil nutrients as a cause of increasing liana abundance and biomass. Dalling et al. (2012) tested the hypothesis that the distribution of liana species on the Barro Colorado Island (BCI) 50-ha forest dynamics plot varies with natural availability in soil nutrients but found little evidence that lianas responded strongly to any pattern of soil chemistry. Likewise, Ledo and Schnitzer (2014) found no relationship between liana species preference and habitat type on the BCI 50-ha plot, consistent with the findings of Dalling and colleagues. Using a 15-year N-P-K nutrient addition experiment on Barro Colorado Nature Monument, Panama (Kaspari et al. 2008; Santiago et al. 2012), liana density and basal area did not differ among treatment or control plots after 15 years of aggressive nutrient addition (Schnitzer, Wright et al. in prep). These data, however, are all from central Panama, and thus additional tests on the responsive of lianas and trees to nutrient addition in other regions representing different climate and soil conditions are necessary. While much additional research remains to be done on the fertilization hypothesis, especially testing growth form-specific responses to experimental nutrient fertilization across climate and soil fertility gradients (Asner & Martin, Chapter 21 in this volume); to date, there is little evidence to support the fertilization hypothesis (in terms of either increasing nutrients or CO_2) as the cause for increasing liana abundance in tropical forests.

CONCLUSION AND FUTURE DIRECTIONS

Evidence for the increase in liana density and basal area in neotropical forests, first proposed in 2002 by Phillips et al., is now compelling. Lianas have increased in abundance in many parts of the neotropics, including dry, moist, and wet lowland tropical forests. However, current datasets are limited because most provide only a single longitudinal datapoint on the change in liana density, biomass, or productivity. The next step is to collect longer-term data to determine the *rate* of liana change: whether the rate of liana change is accelerating, remaining the same, or decreasing, and how the rate of liana change varies with forest type. Determining the forest types in which lianas are increasing most (and least) may lead to insights into the mechanisms responsible for these changes, as well as providing valuable information about the forests that are most vulnerable to future liana infestation.

To further expose the pattern of liana change throughout the tropics, we need a widespread, systematic network of large-scale, long-term liana and tree monitoring plots using established sampling protocols (e.g., Gerwing et al. 2006; Schnitzer et al. 2008b). Data at the continental scale (Africa, America, Asia, and Australia) are particularly important to reveal whether the pattern emerging from the neotropics is also present in the rest of the tropics (Schnitzer & Bongers 2011; Bongers & Ewango, Chapter 3 in this volume). Furthermore, large-plot censuses across a range of size classes that allow for the collection of detailed data on many individuals of many liana species will permit us to determine which species are increasing and whether phylogenetic and functional trait differences among liana species are responsible for these increases. Because of the importance of lianas in tropical forest diversity, dynamics, and functioning, illustrated by the chapters in this volume, a worldwide liana monitoring network should be a priority for tropical biologists.

Determining the mechanistic causes of liana increases is also an important priority because it will provide the ability to predict where, when, and why lianas are increasing. Correlations between the changing rates of liana increase and disturbance, the seasonality of rainfall, temperature, nutrient deposition, and atmospheric CO_2 can help formulate hypotheses to explain increasing liana abundance. However, correlations do not provide a causal explanation and, in some cases, may be misleading. A plethora of factors are changing globally, yet their effects on liana density and basal area may be minimal. Therefore, direct experimental approaches to compare physiological and growth responses of liana and tree species to direct manipulations of the putative abiotic drivers (disturbance, seasonality of rainfall, temperature, atmospheric CO_2, and nutrients), ideally in conjunction with each other, are critical to determine the causes of liana increases.

Experimental studies should account for several different conditions, including liana size, age, and species identity. Detailed experimental manipulations that use only juvenile lianas may prove misleading if liana increases are driven by adults. Likewise, it may be naïve to suspect that all lianas are increasing, regardless of species identity and life-history strategy. Thus, we need large-scale, long-term, repeated liana censuses to determine which liana populations are increasing, decreasing, and remaining the same. Ultimately, the ability to determine where, when, and why lianas will increase is predicated upon our fundamental understanding of the processes that control liana distributions and maintain their diversity within and among various types of forest.

ACKNOWLEDGMENTS

Thanks to K. Barry, S. DeWalt and A. Ercoli for constructive comments on this manuscript. This research was supported financially by the US National Science Foundation (DEB-0613666, DEB-0845071, and DEB-1019436).

REFERENCES

Allen, B.P., Sharitz, R.R. & Goebel, P.C. (2007) Are lianas increasing in importance in temperate floodplain forests in the southeastern United States? *Forest Ecology and Management*, **242**, 17–23.

Arroyo-Rodríguez, V. & Toledo-Aceves, T. (2009) Impact of landscape spatial patterns on liana communities in tropical rainforests at Los Tuxtlas, Mexico. *Applied Vegetation Science*, **12**, 340–348.

Asner, G.P. & Alencar, A. (2010) Drought impacts on the Amazon forest: the remote sensing perspective. *New Phytologist*, **187**, 569–578.

Asner, G.P., Broadbent, E.N., Oliveira, P.J.C. et al. (2006) Condition and fate of logged forests in the Brazilian Amazon. *Proceedings of the National Academy of Sciences, USA*, **103**, 12947–12950.

Barry, K.E., Schnitzer, S.A., van Bruegal, M., & Hall, J.S. (in review). Rapid liana colonization and community development along a secondary forest chronosequence. *Biotropica*.

Belote, R.T., Weltzen, J.F. & Norby, R.J. (2003) Response of an understory plant community to elevated CO_2 depends on differential responses of dominant invasive species and is mediated by soil water availability. *New Phytologist*, **161**, 827–835.

Benítez-Malvido, J. & Martínez-Ramos, M. (2003) Impact of forest fragmentation on understory plant species richness in Amazonia. *Conservation Biology*, **17**, 389–400.

Caballé, G. & Martin, A. (2001) Thirteen years of change in trees and lianas in a Gabonese rainforest. *Plant Ecology*, **152**, 167–173.

Cai, Z.-Q., Schnitzer, S.A. & Bongers, F. (2009) Seasonal differences in leaf-level physiology give lianas a competitive advantage over trees in a tropical seasonal forest. *Oecologia*, **161**, 25–33.

Chave, J., Olivier, J., Bongers, F. et al. (2008) Aboveground biomass and productivity in a rain forest of eastern South America. *Journal of Tropical Ecology*, **24**, 355–366.

Clark, D.A. (2004) Tropical forests and global warming: slowing it down or speeding it up? *Frontiers in Ecology & the Environment*, **2**(2), 73–80.

Condit, R., Watts, K., Bohlman, S.A. et al. (2000) Quantifying the deciduousness of tropical canopies under varying climates. *Journal of Vegetation Science*, **11**, 649–658.

Dalling, J.W., Schnitzer, S.A., Baldeck, C. et al. (2012) Resource-based habitat associations in a neotropical liana community. *Journal of Ecology*, **100**, 1174–1182.

DeWalt, S.J., Schnitzer, S.A., Chave, J. et al. (2010) Annual rainfall and seasonality predict pan-tropical patterns of liana density and basal area. *Biotropica*, **42**, 309–317.

DeWalt, S.J., Schnitzer, S.A. & Denslow, J.S. (2000) Density and diversity of lianas along a chronosequence in a central Panamanian tropical forest. *Journal of Tropical Ecology*, **16**, 1–19.

Dixon, R.K., Solomon, A.M., Brown, S. et al. (1994) Carbon pools and flux of global forest ecosystems. *Science*, **263**, 185–190.

Domingues, T., Martinelli, L. & Ehleringer, J. (2007) Ecophysiological traits of plant functional groups in forest and pasture ecosystems from eastern Amazonia, Brazil. *Plant Ecology*, **193**, 101–112.

Durán, S.M. & Gianoli, E. (2013) Carbon stocks in tropical forests decrease with liana density. *Biology Letters*, **9**, 20130301.

Eamus, D. (1991) The interaction of rising CO_2 and temperatures with water use efficiency. *Plant, Cell, & Environment*, **14**, 843–852.

Enquist, B.J. & Enquist, C.A.F. (2011) Long-term change within a neotropical forest: assessing differential functional and floristic responses to disturbance and drought. *Global Change Biology*, **17**, 1408–1424.

Field, C.B., Behrenfeld, M.J., Randerson, J.T. & Falkowski, P. (1998) Primary production of the biosphere: integrating terrestrial and oceanic components. *Science*, **281**, 237–240.

Foster, J.R., Townsend, P.A. & Zganjar, C.E. (2008) Spatial and temporal patterns of gap dominance by low-canopy lianas detected using EO-1 Hyperion and Landsat Thematic Mapper. *Remote Sensing of Environment*, **112**, 2104–2117.

Fu, R. et al. (2013) Increased dry-season length over southern Amazonia in recent decades and its implications for future climate projection. *Proceedings of National Academy of Sciences of the USA*, **110**(45), 18110–18115.

Galloway, J.N., Dentener, F.J., Capone, D.G. et al. (2004) Nitrogen cycles: past, present, and future. *Biogeochemistry*, **70**, 153–226.

Gerwing, J.J., Schnitzer, S.A., Burnham, R.J. et al. (2006) A standard protocol for liana censuses. *Biotropica*, **38**, 256–261.

Granados, J. & Körner, C. (2002) In deep shade, elevated CO_2 increases the vigor of tropical climbing plants. *Global Change Biology*, **8**, 1–9.

Grauel, W.T. & Putz, F.E. (2004) Effects of lianas on growth and regeneration of *Prioria copaifera* in Darien, Panama. *Forest Ecology and Management*, **190**, 99–108.

Hättenschwiler, S. & Körner, C. (2003) Does elevated CO_2 facilitate naturalization of the non-indigenous *Prunus laurocerasus* in Swiss temperate forests? *Functional Ecology*, **17**, 778–785.

Hietz, P., Turner, B.L., Wanek, W., et al. (2011) Long-term change in the nitrogen cycle of tropical forests. *Science*, **334**, 664–666.

Ingwell, L.L., Wright, S.J., Becklund, K.K., et al. (2010) The impact of lianas on 10 years of tree growth and mortality on Barro Colorado Island, Panama. *Journal of Ecology*, **98**, 879–887.

Kalácska, M., Calvo-Alvarado, J.C. & Sánchez-Azofeifa, G.A. (2005) Calibration and assessment of seasonal changes in leaf area index of a tropical dry forest in different stages of succession. *Tree Physiology*, **25**, 733–744.

Kaspari, M., Garcia, M.N., Harms, K.E. et al. (2008) Multiple nutrients limit litterfall and decomposition in a tropical forest. *Ecology Letters*, **11**, 35–43.

Körner, C. (2006) Forests, biodivsity and CO_2: surprises are certain. *Biologist*, **53**, 82–90.

Laurance, W.F., Andrade, A.S., Magrach, A., et al. (2014) Long-term changes in liana abundance and forest dynamics in undisturbed Amazonian forests. *Ecology*, in press.

Laurance, W.F., Goosem, M. & Laurance, S.G.W. (2009) Impacts of roads and linear clearings on tropical forests. *Trends in Ecology & Evolution*, **24**, 659–669.

Ledo, A. & Schnitzer, S.A. (2014) Disturbance, not negative density dependence or habitat specialization maintains liana diversity in a tropical forest. *Ecology*, in press.

Lee, T., & McPhaden, M.J. (2010) Increasing intensity of El Niño in the central-equatorial Pacific. *Geophysical Research Letters*, **37**, L14603, doi:10.1029/2010GL044007

Letcher, S.G. & Chazdon, R.L. (2009) Lianas and self-supporting plants during tropical forest succession. *Forest Ecology and Management*, **257**, 2150–2156.

Londré, R.A. & Schnitzer, S.A. (2006) The distribution of lianas and their change in abundance in temperate forests over the past 45 years. *Ecology*, **87**, 2973–2978.

Malhi, Y. & Wright, J. (2005) Late twentieth-century patterns and trends in the climate of tropical forest regions, in *Tropical Forests and Global Amospheric Change* (eds Y. Malhi & O. Phillips), Oxford University Press, Oxford.

Mohan, J., Ziska, L., Schlesinger, W. et al. (2006) Biomass and toxicity responses of poison ivy (*Toxicodendron radicans*) to elevated atmospheric CO_2. *Proceedings of the National Academy of Sciences of the USA*, **103**, 9086–9089.

Odling-Smee, F.J., Laland, K.N. & Feldman, M.W. (2003) Niche construction, the neglected process in evolution, in *Monographs in Population Biology*, vol. **37**, Princeton University Press, Princeton, NJ.

Opler, P.A., Baker, H.G. & Frankie, G.W. (1991) Seasonality of climbers: a review and example from Costa Rican dry forest, in *The Biology of Vines* (eds F.E. Putz & H.A. Mooney), Cambridge University Press, Cambridge, pp. 377–391.

Paul, G.S. & Yavitt, J.B. (2010) Tropical vine growth and the effects on forest succession: a review of the ecology and management of tropical climbing plants. *The Botanical Review*. doi: 10.1007/s12229-010-9059-3

Peñalosa, J. (1984) Basal branching and vegetative spread in two tropical rain forest lianas. *Biotropica*, **16**, 1–9.

Phillips, O.L., Aragao, L.E.O.C., Lewis, S.L. et al. (2009) Drought sensitivity of the Amazon rainforest. *Science*, **323**, 1344–1347.

Phillips, O.L. & Gentry, A.H. (1994) Increasing turnover through time in tropical forests. *Science*, **263**, 954–957.

Phillips, O.L., Vásquez, M.R., Arroyo, L. et al. (2002) Increasing dominance of large lianas in Amazonian forests. *Nature*, **418**, 770–774.

Puerto-Piñero, C., Mueller-Landau, H.C., Calderon, O. & Wright, S.J. (2013) Seed arrival in tropical forest tree fall gaps. *Ecology*, **94**, 1552–1562.

Putz, F.E. (1984) The natural history of lianas on Barro Colorado Island, Panama. *Ecology*, **65**, 1713–1724.

Putz, F.E. & Windsor, D.M. (1987) Liana phenology on Barro Colorado Island, Panama. *Biotropica*, **19**, 334–341.

Salzer, J., Matezki, S. & Kazda, M. (2006) Nutritional differences and leaf acclimation of climbing plants and the associated vegetation in different types of an Andean montane rainforest. *Oecologia*, **147**, 417–425.

Santiago, L.S., Wright, S.J., Harms, K.E. et al. (2012) Tropical tree seedling growth resonses to nitrogen, phosphorus and potassium addition. *Journal of Ecology*, **100**, 309–316.

Schnitzer, S.A. (2005) A mechanistic explanation for global patterns of liana abundance and distribution. *American Naturalist*, **166**, 262–276.

Schnitzer, S.A. & Bongers, F. (2002) The ecology of lianas and their role in forests. *Trends in Ecology and Evolution*, **17**, 223–230.

Schnitzer, S.A. & Bongers, F. (2011) Increasing liana abundance and biomass in tropical forests: emerging patterns and putative mechanisms. *Ecology Letters*, **14**, 397–406.

Schnitzer, S.A., Bongers, F. & Wright, S.J. (2011) Community and ecosystem ramifications of increasing lianas in neotropical forests. *Plant Signaling & Behavior*, **6**, 598–600.

Schnitzer, S.A. & Carson, W.P. (2001) Treefall gaps and the maintenance of species diversity in a tropical forest. *Ecology*, **82**, 913–919.

Schnitzer, S.A. & Carson, W.P. (2010) Lianas suppress tree regeneration and diversity in treefall gaps. *Ecology Letters*, **12**, 849–857.

Schnitzer, S.A., Dalling, J.W. & Carson, W.P. (2000) The impact of lianas on tree regeneration in tropical forest canopy gaps: evidence for an alternative pathway of gap-phase regeneration. *Journal of Ecology*, **88**, 655–666.

Schnitzer, S.A., Londré, R.A., Klironomos, J. & Reich, P.B. (2008) Biomass and toxicity responses of poison ivy (*Toxicodendron radicans*) to elevated atmospheric CO_2: Comment. *Ecology*, **89**, 581–585.

Schnitzer, S.A., Mangan, S.A., Dalling, J.W. et al. (2012) Liana abundance, diversity, and distribution on Barro Colorado Island, Panama. *PLoS One*, **7**, e52114.

Schnitzer, S.A., Parren, M.P.E. & Bongers, F. (2004) Recruitment of lianas into logging gaps and the effects of pre-harvest climber cutting in a lowland forest in Cameroon. *Forest Ecology and Management*, **190**, 87–98.

Schnitzer, S.A., Rutishauser, S. & Aguilar, S. (2008) Supplemental protocol for censusing lianas. *Forest Ecology and Management*, **255**, 1044–1049.

Schnitzer, S.A., van der Heijden, G.M.F., Mascaro, J., & Carson, W.P. (2014) Lianas reduce biomass accumulation in a tropical forest. *Ecology*. In press.

Tobin, M.F., Wright, A.J., Mangan, S.A., & Schnitzer, S.A. (2012) Lianas have a greater competitive effect than trees of similar biomass on tropical canopy trees. *Ecosphere*, **3**, Article 20: 1–11. http://dx.doi.org/10.1890/ES11-00322.1

Van der Heijden, G.M.F. & Phillips, O.L. (2009) Liana infestation impacts tree growth in a lowland tropical moist forest. *Biogeosciences*, **6**, 2217–2226.

Van der Heijden, G.M.F., Schnitzer, S.A., Powers, J.S. & Phillips, O.L. (2013) Liana impacts on carbon cycling, storage and sequestration in tropical forests. *Biotropica*, **45**, 682–692.

Wright, S.J. (2005) Tropical forests in a changing environment. *Trends in Ecology and Evolution*, **20**, 553–560.

Wright, S.J. (2010) The future of tropical forests. *Annals of the New York Academy of Sciences*, **1195**, 1–27.

Wright, S.J. & Calderon, O. (2006) Seasonal, El Niño and longer-term changes in flower and seed production in a moist tropical forest. *Ecology Letters*, **9**, 35–44.

Wright, S.J., Calderon, O., Hernandéz, A. & Paton, S. (2004) Are lianas increasing in importance in tropical forests? A 17-year record from Panamá. *Ecology*, **85**, 484–489.

Wright, S.J., Jaramillo, M.A., Pavon, J. et al. (2005) Reproductive size thresholds in tropical trees: variation among individuals, species and forests. *Journal of Tropical Ecology*, **21**, 307–315.

Wyka, T.P., Oleksy, J., Karolewski, P. & Schnitzer, S.A. (2013) Phenotypic correlates of the lianescent growth form: a review. *Annals of Botany*. doi: 10.1093/aob/mct236

Yorke, S.R., Schnitzer, S.A., Mascaro, J. et al. (2013) Increasing liana abundance and biomass in a tropical forest: the contribution of long distance clonal colonization. *Biotropica*, **45**, 317–324.

Zhu, S.-D. & Cao, K.-F. (2009) Hydraulic properties and photosynthetic rates in co-occuring lianas and trees in a seasonal tropical rainforest in southwestern China. *Plant Ecology*, **204**, 295–304.

Zotz, G., Cueni, N. & Körner, C. (2006) In situ growth stimulation of a temperate zone liana (*Hedera helix*) in elevated CO_2. *Functional Ecology*, **20**, 763–769.

INDEX

Page numbers in **bold** refer to tables.

Ecology of Lianas, First Edition. Edited by Stefan A. Schnitzer, Frans Bongers, Robyn J. Burnham, and Francis E. Putz.

Companion Website: www.wiley.com/go/schnitzer/lianas